ELEMENTS

OF

CHEMISTRY;

INCLUDING THE

APPLICATIONS OF THE SCIENCE IN THE ARTS.

BY

THOMAS GRAHAM, F.R.S.,

PROFESSOR OF CHEMISTRY IN UNIVERSITY COLLEGE, LONDON; CORRESPONDING MEMBER OF THE INSTITUTE OF FRANCE; VICE-PRESIDENT OF THE CHEMICAL SOCIETY, ETC. ETC.

SECOND AMERICAN,

FROM AN

ENTIRELY REVISED AND GREATLY ENLARGED ENGLISH EDITION.

WITH NUMEROUS WOOD ENGRAVINGS.

EDITED, WITH NOTES,

BY ROBERT BRIDGES, M.D.,

PROFESSOR OF CHEMISTRY IN THE PHILADELPHIA COLLEGE OF PHARMACY; LECTURER ON CHEMISTRY IN THE PHILADELPHIA ASSOCIATION FOR MEDICAL INSTRUCTION, ETC.

PHILADELPHIA:

BLANCHARD AND LEA.

1852.

PREFACE

TO THE SECOND EDITION.

(PART I.)

If the Inorganic department of chemistry has not recently been expanded in the same vast proportion as the Organic branch of the science, still the former has been far from stationary of late years. The advance observed is partly in the old direction of enlarging the list of elements, partly and more conspicuously in supplying deficient members to familiar series of compounds, and in thus enlarging these series,—as in the compounds of chlorine with oxygen, and of sulphur with oxygen. But the most important feature in the recent progress of Inorganic Chemistry has been the rigorous verification which numerical data of all kinds have received, whether relating to physical laws, such as the specific heat of substances, or to chemical properties and composition. The statement of properties and relations has thus acquired a fulness and precision for many substances, which contrasts strongly with the history that could be offered of the same substances even but a very few years ago. The correction and revision of every minute branch of the science was never, indeed, more general and rapid than at the present time. The enlarged means of practical instruction in chemistry, now everywhere provided for the student, and the consequent increase in the number of able investigators, have no doubt contributed much to this result.

Progress of this description cannot fail to affect the theoretical views of chemists, and to promote sound conclusions by affording an extended and safe foundation for reasoning, in a body of well-established facts. It must be admitted that the fundamental views respecting the constitution of salts are at present in a state of transition, but the great questions of chemical theory, if not yet solved, have at least been correctly enunciated, and a general assent obtained to the facts upon which they rest.

In preparing a new edition of his Elements of Chemistry, the Author has incorporated much new and accurate information with the old, while he has endeavoured to give to both the space and the measure of importance which their true value demanded. In such a work, judicious selection of matter is as necessary as careful condensation, while the grounds of the selection are changed with the shifting point of view from which, in a progressive science, the retrospect is taken.

The important bearings of the laws of Heat, particularly in reference to the physical condition of matter, have led to their consideration before the chemical properties of substances, in this as in most other elementary treatises on chemistry. Light is then shortly considered, chiefly in reference to its chemical relations. The principles of its Nomenclature, in which, compared with many sciences, chemistry has been highly fortunate, are then explained, together with the symbolical notation and chemical formulæ in use, by means of which the composition of highly compound bodies is expressed with the same palpable distinctness which, in arithmetic, attends the use of figures, in the place of words, for the expression of numerical sums.

A considerable section of the present volume is then devoted to the consideration of the fundamental doctrines of chemistry, under the heads of combining proportions, atomic theory, doctrine of volumes, isomorphism, isomerism, constitution of salts, chemical affinity and polarity, including the propagation of affinity through metallic and saline media, in the voltaic circle, with the new subject of the atomic volume of solids.

The materials of the inorganic world are then described under two great divisions of non-metallic elements and their compounds, and metallic elements and their compounds.

UNIVERSITY COLLEGE, LONDON,
September, 1850.

PREFACE

TO THE

SECOND AMERICAN EDITION.

THE "Elements of Chemistry," of which a second edition is now presented, attained, on its first appearance, an immediate and deserved reputation. The copious selection of facts from all reliable sources, and their judicious arrangement, render it a safe guide for the beginner, while the clear exposition of theoretical points, and frequent references to special treatises, make it a valuable assistant for the more advanced student.

From this high character the present edition will in no way detract. The great changes which the science of Chemistry has undergone during the interval, have rendered necessary a complete revision of the work, and this has been most thoroughly accomplished by the author. Many portions will therefore be found essentially altered, thereby increasing greatly the size of the work, while the series of illustrations has been entirely changed in style, and nearly doubled in number.

Under these circumstances but little has been left for the editor. Owing, however, to the appearance of the London edition in parts, some years have elapsed since the first portions were published, and he has therefore found occasion to introduce the more recent investigations and discoveries in some subjects, as well as to correct such inaccuracies or misprints as had escaped the author's attention, and to make a few additional references. Such matter as he has thus introduced has been enclosed in brackets, with his initials appended.

PHILADELPHIA, *March*, 1852.

In consequence of numerous inquiries for the new edition of GRAHAM'S CHEMISTRY, the publishers issue separately the first half, which may be regarded as complete in itself. The concluding portion may be expected for publication during the present year, when the whole will be presented in one volume.

CONTENTS.

CHAPTER I.

CHAPTER II.

CHAPTER III.

CHAPTER IV.

CHAPTER V.

CHAPTER VI.

ORDER I.

METALLIC BASES OF THE ALKALIES.

ORDER II.

METALLIC BASES OF THE ALKALINE EARTHS.

ORDER III.

METALLIC BASES OF THE EARTHS.

LIST OF WOOD CUTS.

ELEMENTS OF CHEMISTRY.

CHAPTER I.

HEAT.

THE objects of the material world are altered in their properties by heat in a very remarkable manner. The conversion of ice into water, and of water into vapour, by the application of heat, affords a familiar illustration of the effects of this agent in changing the condition of bodies. All other material substances are equally under its influence; and it gives rise to numerous and varied phenomena, demanding the attention of the chemical inquirer.

Heat is very readily communicated from one body to another; so that when hot and cold bodies are placed near each other, they speedily attain the same temperature. The obvious transference of heat in such circumstances impresses the idea that it possesses a substantial existence, and is not merely a quality of bodies, like colour or weight; and when thus considered as a material substance, it has received the name *caloric*. It would be injudicious, however, to enter at present into any speculation on the nature of heat; it is sufficient to remark that it differs from matter, as usually conceived, in several respects. Our knowledge of heat is limited to the different effects which it produces upon bodies, and the mode of its transmission; and these subjects may be considered without reference to any theory of the nature of this agent.

The subject of Heat will be treated of under the following heads: —

1. Expansion, the most general effect of heat, and the Thermometer.
2. Specific heat.
3. The communication of heat by Conduction and Radiation.
4. Liquefaction, as an effect of heat.
5. Vaporization, or the gaseous state, as an effect of heat.
6. Speculative notions respecting the nature of heat.

EXPANSION AND THE THERMOMETER.

All bodies in nature, solids, liquids, or gases, suffer a temporary increase of dimension when heated, and contract again into their original volume on cooling.

1. *Expansion of solids.* — The expansion of solid bodies, such as the metals, is by no means considerable, but may readily be made sensible. A bar of iron which fits easily when cold into a gauge, will be found, on heating it to redness, to have increased sensibly both in length and thickness. The expansion and contraction of metals, indeed, and the immense force with which these changes take place, are matters of familiar observation, and are often made available in the arts. The iron hoops of carriage wheels, for instance, are applied to the frame while they are red hot, and in a state of expansion, and being then suddenly cooled by dashing water upon them, they contract, and bind the wood-work of the wheel with great force. The expansion of solids, however, is very small, and requires nice measurement to ascertain its amount. The expansion in length only has generally been

determined, but it must always be remembered that the body expands also in its other dimensions in an equal proportion. The first general fact observable is, that the amount of dilatation by heat is different in different bodies. No two solids expand alike. The metals expand most, and their rates of expansion are best known. Rods of the undermentioned substances, on being heated from the freezing to the boiling point of water, elongate as follows:—

Zinc (cast)	1 on	323	Pure Gold	1 on	682
Zinc (sheet)	1 "	340	Iron Wire	1 "	812
Lead	1 "	351	Palladium	1 "	1000
Tin	1 "	516	Glass without lead	1 "	1142
Silver	1 "	524	Platinum	1 "	1167
Copper	1 "	581	Flint Glass	1 "	1248
Brass	1 "	584	Black Marble (Lucullite)	1 "	2833

This is the increase which these bodies sustain in length. Their increase in general bulk is about three times greater. Thus, if glass elongates 1 part in 1248 from the freezing to the boiling point of water, it will dilate in cubic capacity 3 parts in 1248, or 1 part in 416. The expanded bodies return to their original dimensions on cooling. Wood does not expand much in length; hence it is occasionally used as a pendulum rod. For the same reason a slip of marble, of the variety mentioned in the preceding table, was employed for that purpose, in constructing the clock of the Royal Society of Edinburgh. Glass without lead expands by the table $\frac{1}{1142}$ part, while the metal platinum expands very little less, $\frac{1}{1167}$. Hence the possibility of cementing glass and platinum together, as is done in many chemical instruments. Other metals pushed through the glass when it is red hot and soft, shrink afterwards so much more than glass on cooling, as to separate from it, and become loose. Zinc is the most expansible of the metals; it expands nearly four times more than platinum from the same heat. But ice, of which the contraction by cold has been observed for 30 or 40 degrees under the freezing point, proves to be more dilatable even than the metals, the rate of this solid being in the proportion of $\frac{1}{267}$th part, while that of zinc is $\frac{1}{323}$d part only. (Brunner (fils), Ann. de Chim. et de Phys., 3 sér. t. 14, p. 377.)

The most important discovery, in a theoretical point of view, that has been made on the subject of the dilatation of solids by heat, is the observation of Professor Mitscherlich, of Berlin, that the angles of some crystals are affected by changes of temperature. This proves that some solids in the crystalline form do not expand uniformly, but more in one direction than in another. Indeed, Mitscherlich has shown that while a crystal is expanding in length by heat, it may actually be contracting at the same time in another dimension. An angle of rhomboidal calcareous spar alters eight and a half minutes of a degree between the freezing and boiling points of water. But this unequal expansion does not occur in crystals of which all the sides and angles are alike, as the cube, the regular octohedron, the rhomboidal dodecahedron. In investigating the laws of expansion among solids, it is advisable, therefore, to make choice of crystallized bodies. For, in a substance not regularly crystallized, the expansion of different specimens may not be precisely the same, as the internal structure may be different. Hence the expansions of the same substance, as given by different experimenters, do not always exactly correspond. The same glass has been observed to dilate more when in the form of a solid rod, than in that of a tube; and the numerous experiments on uncrystallized bodies, which we possess, have afforded no ground for general deductions.

It has been further observed, that the same solid is more expansible at high than at low temperatures, although the increase in the rate of expansion is in general not considerable. Thus, if we mark the progress of the dilatation of a bar of iron under a graduated heat, we find that the increase in dimension is greater for one degree of heat near the boiling point of water, than for one degree near its freezing point. Solids are observed to expand at an accelerated rate, in particular, when heated up to near their fusing points. The cohesion or attraction which subsists between the particles of a solid is supposed to resist the expansive power of heat. But many solids become less tenacious, or soften before melting, which may account for their increasing expansibility. Platinum is the most uniform in its expansions of the metals.

Such changes in bulk, from variations in temperature, take place with irresistible force. This is well illustrated in an experiment, which was first made upon a gallery in the Museum of Arts and Manufactures in Paris, in order to preserve it, and has been successfully repeated in many other buildings. The opposite walls of the edifice referred to were bulging outwards, from the pressure of the floors and roof, which endangered its stability. By the directions of an ingenious mechanic, stout iron rods were laid across the building, with their extremities projecting through the opposite walls so as to bind them together. Half the number of the rods were then strongly heated by means of lamps, and, when in an expanded condition, a disc on either extremity of each rod was screwed firmly up against the external surface of the wall. On afterwards allowing the rods to cool, they contracted, and drew the walls to which they were attached somewhat nearer together. The process was several times repeated, till the walls were restored to a perpendicular position.

The force of expansion always requires to be attended to in the arts, when iron is combined in any structure with less expansible materials. The cope-stones of walls are sometimes held together with clamps, or bars of iron: such bars, if of cast iron, which is brittle, often break on the first frost, from a tendency to contract more than the stone will permit; if of malleable iron, they generally crush the stone, and loosen themselves in their sockets. When cast iron pipes are employed to conduct hot air or steam through a factory, they are never allowed to abut against a wall or an obstacle which they might in expanding overturn. Lead, from its extreme softness, is permanently expanded when repeatedly heated; a waste steam pipe of that metal being elongated several inches in a few weeks.

Fig. 1.

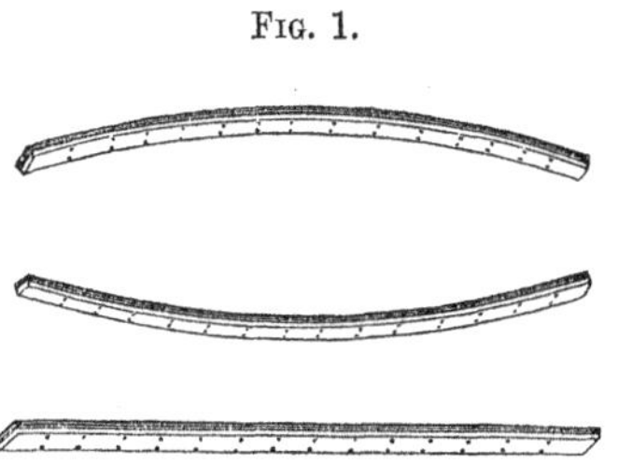

A compound bar, made by riveting or soldering together two thin plates of copper and platinum, affords a good illustration of unequal expansion by heat. The copper plate, being much more expansible than platinum, the bar is bent upon the application of heat to it; and in such a manner, that the copper is on the outside of the curve. The reverse is produced when the bar is cooled. It may easily be conceived, that by a proper attention to the expansions of the metals of which it is composed, a bar of this kind might be so constructed, that although it was heated and expanded, its extreme points should always remain at the same distance from each other, the lengthening being compensated for by the bending. The balance-wheels of chronometers are preserved invariable in their diameters, at all temperatures, by a contrivance of this kind. It has also been applied to the construction of a thermometer of solid materials — that of Breguet.

When hot water is suddenly poured upon a thick plate of glass, the upper surface is heated and expanded before the heat penetrates to the lower surface of the plate. There is here unequal expansion, as in the slip of copper and platinum. The glass tends to bend, with the hot and expanded surface on the outside of the curve, but is broken from its want of flexibility. The occurrence of such fractures is best avoided by applying heat to glass vessels in a gradual manner, so as to occasion no great inequality of expansion; or by using very thin vessels, through the substance of which heat is rapidly transmitted.

This effect of heat on glass may by a little address be turned to advantage. Watch-glasses are cut out of a thin globe of glass, by conducting a crack in a proper direction, by means of an iron rod, or piece of tobacco pipe, heated to redness. Glass vessels damaged in the laboratory may often be divided in the same manner, and still made available for useful purposes.

Both cast iron and glass are peculiarly liable to accidents from unequal expansion, when in the state of flat plates. Plate glass, indeed, can never be heated without risk of its breaking. The flat iron plates placed across chimneys as dampers, are also very apt to split when they become hot, and much inconvenience has often been experienced in manufactories from this cause. A slight curvature in their form has been found to protect them most effectually.

Expansion of liquids.—In liquids the expansive force of heat is little resisted by cohesive attraction, and is much more considerable than in solids. This fact is strikingly exhibited by filling the bulb and part of the stem of a common thermometer tube with a liquid, and applying heat to it. The liquid is seen immediately to mount in the tube.

The first law, in the case of liquids, is that some expand much more considerably by heat than others. Thus, on being heated to the same extent, namely, from the freezing to the boiling point of water—

Spirit of wine expands	$\frac{1}{9}$, that is,	9 measures become		10
Fixed oils	$\frac{1}{12}$,	" 12	"	13
Water	$\frac{1}{22 \cdot 76}$,	" 22·76	"	23·76
Mercury	$\frac{1}{55 \cdot 5}$,	" 55·5	"	56·5

Spirit of wine is, therefore, six times more expansible by heat than mercury is. The difference in the heat of the seasons affects sensibly the bulk of spirits. In the height of summer, spirits will measure 5 per cent. more than in the depth of winter.

The new liquids produced by the condensation of gases appear to be characterized by an extraordinary dilatability. M. Thilorier has observed, that fluid carbonic acid is more expansible by heat than air itself; heated from 32° to 86°, twenty volumes of this liquid increase to twenty-nine, which is a dilatation four times greater than is produced in air, by the same change of temperature. (Annales de Chimie et de Physique, t. 60, p. 427.) Mr. Kemp extended this observation to liquid sulphurous acid and cyanogen, which, although not possessing the excessive dilatability of liquid carbonic acid, are still greatly more expansible than ordinary liquids. Sir D. Brewster had several years before discovered certain fluids in the minute cavities of topaz and quartz, which seemed to bear no analogy to any other then known liquid in their extraordinary dilatability. They do not appear to have been entirely liquefied gases, but probably were so in part. (Edinburgh Phil. Journ. vol. ix. p. 94, 1824; vol. xvi. p. 11, 1845.)

A singular correspondence has been observed, by M. Gay-Lussac, (Ann. de Chimie, t. 2, p. 130,) between two particular liquids—alcohol and bisulphuret of carbon, in the amount of their expansion by heat: although each of these liquids has a peculiar temperatnre at which it boils—

Alcohol at	173°
Sulphuret of carbon at	116°

still the ratios of expansions from the addition, and of contraction from the loss of heat, are found to be uniformly the same in these two liquids, compared at the same distance from their respective boiling points. A similar relation has lately been observed by M. Isidore Pierre, between the bromide of ethyl and bromide of methyl, and between the iodide of ethyl and iodide of methyl, which does not appear to exist between a pair of isomeric bodies, which were also compared,—namely, the formiate of oxide of ethyl and the acetate of oxide of methyl. The observations made with this view on four different groups of liquids, including those mentioned, are thus exhibited, the degrees of temperature being of Fahrenheit's scale:[1]—

[1] M. Pierre has also examined the dilatations of water, oxide of ethyl (ether), and chloride of ethyl. The results he has already published are the most exact and valuable we possess on the subject of the dilatation of liquids; and he is proceeding with his experiments. Ann. de Chimie, &c., 3 série, t. 15, p. 325. 1845.

CONTRACTION OF LIQUIDS FROM THE BOILING POINT (PIERRE).

NAMES OF THE LIQUIDS.	BOILING POINT.	TEMPERATURES equidistant from the boiling point for each group.	INTERVAL between the two preceding temperatures.	VOLUME at boiling point.	VOLUMES at the equidistant temperatures.
I. GROUP.					
Sulphuret of carbon	118·22°	—22·72°	140·94°	1	0·913099
Alcohol	172·94°	32°	140·94°	1	0·914452
Wood-spirit	151·34°	—10·4°	140·94°	1	0·905819
II. GROUP.					
Bromide of ethyl	105·26°	32°	73·26°	1	0·944375
Bromide of methyl	55·4°	—17·86°	73·26°	1	0·944575
III. GROUP.					
Iodide of ethyl	158°	32°	126°	1	0·918704
Iodide of methyl	110·84°	—15·16°	126°	1	0·916643
IV. GROUP.					
Formiate of oxide of ethyl	127·22°	—20·12°	107·1°	1	0·910223
Acetate of oxide of methyl	139·10°	—15·8° 32°	107·1°	1	0·918750

I have only to add the following results obtained by M. Muncke, of St. Petersburgh:[1]—

EXPANSION OF LIQUIDS, VOLUME AT 32° FAHR. BEING 1.

Solution of ammonia (sp. gr. 0·9465) ...	1·0198310	at 113° (45° Centig.)
Hydrochloric acid (sp. gr. 1·1978).........	1·0253598	" "
Nitric acid (sp. gr. 1·4405)	1·0479512	" "
" "	1·1148853	at 212° (100° Centig.)
Sulphuric acid (sp. gr. 1·836)...............	1·0578495	at 212°
" "	1·1388577	at 446° (230° Centig.)
Rectified petroleum (sp. gr. 0·7813)	1·1060059	at 203° (95° Centig.)
Almond oil	1·0787005	at 212° (100° Centig.)

The second law is, that liquids are progressively more expansible at higher than at lower temperatures. This is less the case with mercury, perhaps, than with any other liquid. The expansions of that liquid are, indeed, so uniform, as to render it extremely proper for the construction of the thermometer, as will afterwards appear. The rate of expansion of mercury was determined with extraordinary care by Dulong and Petit.

From 0° to 100° Centigrade, mercury expands 1 measure on 55½
" 100° " 200° " " " 1 " 54¼
" 200° " 300° " " " 1 " 53

According to the same experimenters, the expansion of mercury, confined in glass tubes, is only 1 on 64·8. The dilatation of the glass causes the capacity of the instrument to be enlarged, so that the whole expansion of the mercury is not indicated. The only mode in which the error introduced by the expansion of the enclosing vessel can be

FIG. 2.

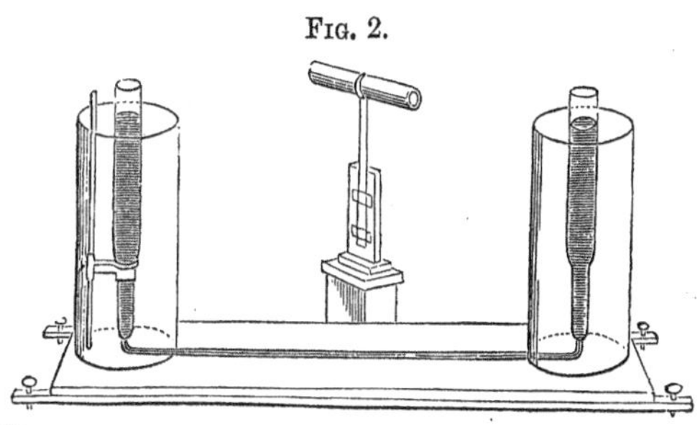

[1] See the Handwörterbuch der Chemie of Liebig, Poggendorff, and Wöhler, vol. i. p. 632; article Ausdehnung (Dilatation).

avoided, in ascertaining the expansion of liquids, is that practised by Dulong and Petit: namely, heating the liquid in one limb of a syphon (see fig. 2), and observing how high it rises above the level of the same liquid in the other limb, kept at a constant temperature. The columns of course balance each other, and the shorter column of dense fluid supports a longer column of dilated fluid. All other modes of obtaining the absolute expansions of liquids are fallacious.

No progress has yet been made in discovering the law by which expansions of liquids are regulated; for the complicated mathematical formulæ of Biot, Dr. Young, and others, are mere general expressions for these expansions, which proceed upon no ascertained physical principle. Some theory must be formed of the constitution of liquids, before we can hope to account for their expansions.

Count Rumford ascertained the contraction of water for every 22½ degrees, in cooling from 212° to 32°. The results are as follows:—

2000 measures of water contract—

In cooling 22½ degrees, or from	212°	to	189½°	18	measures.
" "	189½	"	167	16·2	"
" "	167	"	144½	13·8	"
" "	144½	"	122	11·5	"
" "	122	"	99½	9·3	"
" "	99½	"	77	7·1	"
" "	77	"	54½	3·9	"
" "	54½	"	32	0·2	"

The expansion of water by heat is subject to a remarkable peculiarity, which occasions it to be extremely irregular, and demands special notice. This liquid, in a certain range of temperature, becomes an exception to the very general law that bodies expand by heat. When heat is applied to ice-cold water, or water at the temperature of 32°, this liquid, instead of expanding, contracts by every addition of heat, till its temperature rises to 40°, at or very near which temperature water is as dense as it can be. And, conversely, when water of the temperature of 40° is exposed to cold, it actually expands with the progress of the refrigeration. Water may, with caution, be cooled 20 or 25 degrees below its freezing point, in the fluid form, and still continue to expand. It is curious that this liquid, in a glass bulb, expands as nearly as possible to the same amount on each side of 40°, when either heated or cooled the same number of degrees. Hence, when cooled to 36° it rises to the same point in the stem as when heated to 44°; at 32° it stands at the same point as at 48°; at 20°, at the same point as at 60°, temperatures (fig. 3). The expansion of water by cold, under 40°, is certainly not very great, being little more than 1 part in 10,000 at 32°; hence it was early suspected that it might be an illusion, from the contraction of the glass bulb (in which the experiment was always made) forcing up the water in the stem. But all grounds of objection on this score have been removed by the mode in which the experiment has subsequently been conducted, particularly in the researches of the late Dr. Hope, of Edinburgh, on this subject. (Phil. Trans. vol. v. p. 379.)

Fig. 3.

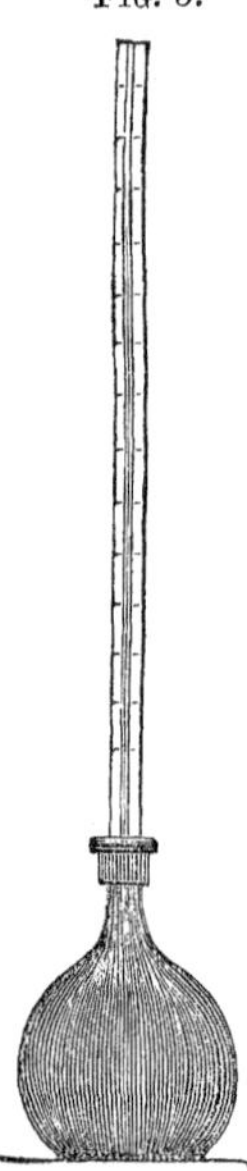

Dr. Hope carried a deep glass jar, filled with water of the temperature of 50°, into a very cold room; and having immersed two small thermometers in the water, one near the surface, and the other at the bottom of the jar, watched their indications as the cooling proceeded. The thermometer above indicated a temperature higher by several degrees than the thermometer below, till the temperature fell to 40°, that is, the chilled water fell as usual to the bottom of the jar, or became denser as it lost heat, as illustrated in fig. 4. At 40° the two thermometers were

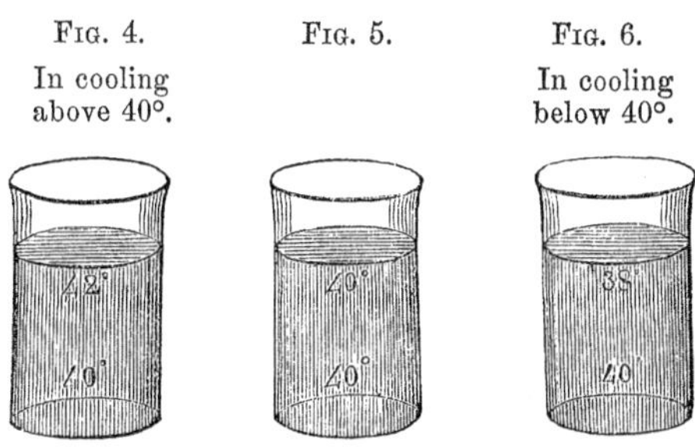

Fig. 4. In cooling above 40°.

Fig. 5.

Fig. 6. In cooling below 40°.

for some time steady (fig. 5), but as the cooling proceeded beyond that point, the instrument in the higher situation indicating the lower temperature, (fig. 6); or the water now as it became colder, became lighter, and rose to the top. A better demonstration of the fact in question could not be devised.

Great pains have been taken by several philosophers to determine the exact temperature of this turning point at which water possesses its maximum density. By the elaborate experiments of both Hällstrom and of Muncke and Stampfer, as calculated by Hällstrom, this point is 39°·38, or 4°·1 Centigrade. Rudberg has more recently obtained 4°·02 C·, and Despretz 4°·00 C., or 39°·2 Fahr., the number now generally taken. Sir C. Blagden and Mr. Gilpin had made it 39°. Dr. Hope had estimated it at 39½°.[1]

When salt is dissolved in water, the temperature of maximum density becomes lower and lower, in proportion to the quantity of salt in solution, and sinking below the freezing point of the liquid, the anomaly disappears. This is the reason why the property in question cannot be observed in sea water.

There is a solid body which presents the only other known parallel case of progressive contraction by heat; this is Rose's fusible metal, which is an alloy of—

2 parts by weight of Bismuth
1 part " " " Lead
1 " " " " Tin

A bar of this metal expands progressively, like other bodies, till it attains the temperature of 111°; it then rapidly contracts by the continued addition of heat, and at 156° attains its maximum density, occupying less space than it does at the freezing point of water. It afterwards progressively expands, melting at 201°. It may be remarked, however, of this body, that it is a chemical compound, of a kind in which a change of constitution is very likely to occur from a change in temperature; and that it cannot, therefore, be fairly compared with water.

The dilatation which water undergoes below 39° has been supposed to be connected with its sudden increase in volume in freezing, for ice is lighter than water, bulk for bulk, in the proportion of 92 to 100. The water, it is said, may begin to pass partially into the solid form at 39°, although the change is not complete till the temperature sinks to 32°. But such an assumption is altogether gratuitous, and improbable in the extreme.

The extraordinary irregularity in the dilatation of water by heat is not only curious in itself, but also of the utmost consequence in the economy of nature. When the cold sets in, the surface of our rivers and lakes is cooled by the contact of the cold air and other causes. The superficial water so cooled, sinks and gives place to warmer water from below, which, chilled in its turn, sinks in like manner. The progress of cooling in the lake goes on with considerable rapidity, so long as the cold water descends and exposes that not hitherto cooled. But this *circulation*, which accelerates the cooling of a mass of water in so extraordinary a degree, ceases entirely when the whole water has been cooled down to the temperature of 40°, which is still eight degrees above the freezing point. Thereafter the chilled surface water expands as it loses its heat, and remains at the top, from its lightness, while the cold is very imperfectly propagated downwards. The surface in the end freezes, and the ice may thicken, but at the depth of a few feet the temperature is not under

[1] For tables of the volume of water at different temperatures, see Appendix 1.

40°, which is high when compared with that frequently experienced, even in this climate, during winter.

If water continued to become heavier, until it arrived at the freezing temperature, the whole of it would be cooled to that point before ice began to be formed; and the consequence would be, that the whole body of water would rapidly be converted into ice, to the destruction of every being that inhabits it. Our warmest summers would make but little impression upon such masses of ice; and the cheerful climate, which we at present enjoy, would be less comfortable than the frozen regions of the pole. Upon such delicate and beautiful adjustments do the order and harmony of the universe depend.

Expansion of gases.—The expansion by heat in the different forms of matter is exceedingly various.

By being heated from 32° to 212°,

1000	cubic inches of	iron	become	1004
1000	"	water	"	1045
1000	"	air	"	1366

Gases are, therefore, more expansible by heat than matter in the other two conditions of liquid and solid. The reason is, that the particles of air or gas, far from being under the influence of cohesive attraction, like solids or liquids, are actuated by a powerful repulsion for each other. The addition of heat mightily enhances this repulsive tendency, and causes great dilatation.

The rate of the expansion of air and gases from increase of temperature, was long involved in considerable uncertainty. This arose from the neglect of the early experimenters to *dry* the air or gas upon which they operated. The presence of a little water by rising in the state of steam into the gas, on the application of heat, occasioned great and irregular expansions. But in 1801, the law of the dilatation of gases was discovered by M. Gay-Lussac, of Paris, and by our countryman, Dr. Dalton, independently of each other. It was discovered by these philosophers that *all gases* experience the same increase in volume by the application of the same degree of heat, and that the rate of expansion continues uniform at all temperatures.

Dr. Dalton confined a small portion of dry air over mercury in a graduated tube. He marked the quantity by the scale, and the temperature by the thermometer. He then placed the whole in circumstances where it was uniformly heated up to a certain temperature, and observed the expansion. Gay-Lussac's apparatus was more complicated, but calculated to give very precise results. He found that 1000 volumes of air, on being heated from 32° to 212°, become 1375, which agreed very closely with Dalton's result. The expansion was lately corrected by Rudberg, who found that 1000 volumes of air expand to 1365.

The still more recent and exact researches of Magnus and of Regnault give as the expansion of air from 32° to 212°, $\frac{366\cdot5}{1000}$, or $\frac{11}{30}$ of its volume at 32°. The dilatation for every degree of Fahrenheit is 0·002036 (Regnault); or $\frac{1}{491\cdot2}$ part.

It follows, consequently, that air at the freezing point expands $\frac{1}{491}$ part of its bulk for every added degree of heat on Fahrenheit's scale: that is—

491	cubic inches	of air	at 32°	become
492	"	"	33°	
493	"	"	34°, &c.	

increasing one cubic inch for every degree. A contraction of one cubic inch occurs for every degree below 32°.

491	cubic inches	of air	at 32°	become
490	"	"	31°	
489	"	"	30°	
488	"	"	29°, &c.	

We can easily deduce, from this law, the expansion which a certain volume of gas at a given temperature will undergo, by heating it up to any particular temperature;

or the contraction that will result from cooling.[1] Air of the temperature of freezing water, has its volume doubled when heated 491 degrees, and when heated 982 degrees, or twice as intensely, its volume is tripled, which is the effect of a low red heat.

A slight deviation from exact uniformity in the expansion of different gases was established by the rigorous experiments of both Magnus (Ann. de Chim. &c. 3 sér. t. 4, p. 330; et t. 6, p. 353) and Regnault (ibid. t. 4, p. 5; et t. 6, p. 370). The more easily liquefied gases, which exhibit a sensible departure from the law of Mariotte, are more expansible by heat than air, as will appear by the following table:—

Names of the Gases.	Expansion upon 1 volume from 32° to 212°. Regnault.	Magnus.
Atmospheric air	0·36650	0·366508
Hydrogen	0·36678	0·365659
Carbonic acid	0·36896	0·369087
Sulphurous acid	0·36696	9·385618
Nitrogen	0·36682	
Nitrous oxide	0·36763	
Carbonic oxide	0·36667	
Cyanogen	0·36821	
Hydrochloric acid	0·36812	

The expansion is also found to be sensibly greater when the gas is in a compressed than when in a rare state; and the results above strictly apply only to the gases under the atmospheric pressure.

THE THERMOMETER,

An instrument for indicating variations in the intensity of heat, or degrees of temperature, by their effect in expanding some body, was invented more than two centuries ago, and has received successive improvements.

The expansions of solids are too minute to be easily measured, and cannot, therefore, be conveniently applied to mark degrees of heat. Air and gases, on the other hand, are so much dilated by a slight increase of heat, that they are not calculated for ordinary purposes. The first thermometer constructed, however, that of Sanctorio, was an air one. A glass tube, open at one end, with a bulb blown upon the other (fig. 7), was slightly heated, so as to expel a portion of the air from it, and then the open end of the tube was dipped under the surface of a coloured fluid, which was allowed to rise into the tube, as the air cooled and contracted. When heat, the heat of the hand for instance, is applied to the bulb, the air in it is expanded, and depresses the column of coloured fluid in the tube. A useful modification of the air thermometer, for researches of great delicacy, was contrived by Sir John Leslie, under the name of the Differential Thermometer. In this instrument two close bulbs are connected by a syphon containing a coloured liquid (fig. 8). If both bulbs be equally heated, the air in each is equally expanded, and the liquid between them remains stationary. But if the upper bulb only be heated, then the air in that bulb is expanded, and the column of liquid depressed. It is, therefore, the difference of temperature between the two bulbs which is indicated.

Fig. 7. Fig. 8.

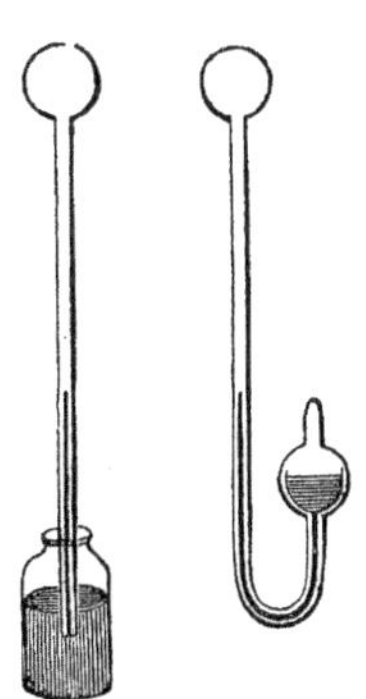

[1] As 491 cubic inches of air at 32° become 459 cubic inches at 0°, air may be stated to expand $\frac{1}{459}$th part of its volume at the zero of Fahrenheit for each degree. That is, 459 volumes of air at 0° become at 50°, 459 + 50 volumes, or 509 volumes; at 60°, 459 + 60 volumes, or 519 volumes. Hence the expansion of 100 volumes of air from 50° to 60° is obtained by the proportion —

Meas. at 50°.		Meas. at 60°.		Meas. at 50°.		Meas. at 60°.
509	:	519	::	100	:	101·96

But liquids fortunately are intermediate in their expansions between solids and gases, and when contained in a glass vessel of a proper form, the changes of bulk which they undergo can be indicated to any degree of precision.

A hollow glass stem or tube is selected, the calibre or bore of which may be of any convenient size, but must be uniform, or not wider at one place than another. Tubes of very narrow bore, and which are called *capillary*, the bore being like a hair in magnitude, are now alone employed. Such tubes are made by rapidly drawing out a hollow mass of glass while soft and ductile under the influence of heat. The central cavity still continues, becoming the bore of the tube, and would not cease to exist although the tube were drawn out into the finest thread. From the mode in which capillary tubes are made, their equality of bore, and suitableness for thermometers, cannot always be depended upon. The bore is frequently conical, or wider at one end than at the other. It is tested by drawing up into the tube a little mercury, as much as fills a few lines of the cavity. The little column is then moved progressively along the tube, and its length accurately measured, at every stage, by a pair of compasses. The column will measure the same in every part of the tube, provided the bore does not alter. Not more than one-sixth part of the tubes made are found to possess this requisite.

Satisfied with the regularity of the bore, the thermometer-maker softens one extremity of the tube, and blows a ball upon it. This is not done by the mouth, which would moisten the interior, by introducing watery vapour, but by means of an elastic bag of caoutchouc, which is fitted to the open end of the tube. He then marks off the length which the thermometer ought to have, and above that point expands the tube into a second bulb a little larger than the first. It has the form of fig. 9. After cooling, the open extremity of the tube is plunged into distilled and well-boiled mercury, and one of the bulbs heated so as to expel air from it. During the cooling, the mercury is drawn up and rises into the ball *a*. It is made to pass from thence into the ball *b*, by turning the instrument, so that *b* is undermost, and then expelling the air from that bulb by applying heat to it, after which the mercury descends, from the effect of cooling. The ball *b*, being entirely filled with mercury, and a portion left in *a*, the tube is supported by an iron wire, as represented in the figure, over a charcoal fire, where it is heated throughout its whole length, so as to boil the mercury, the vapour of which drives out all the air and humidity, and the balls contain at the end nothing but the metal and its vapour. The open end of the tube, which must not be too hot, is then touched with sealing-wax, which is drawn into the tube on melting, and solidifies there on protecting that end of the tube from the heat. That being done, the thermometer is immediately withdrawn from the fire, and being held with the end sealed with wax uppermost, during the cooling the ball *b*, and the portion of the tube below the ball *a*, are filled with mercury. After cooling, the instrument is inclined a little, and by warming the lower ball, a portion of mercury is expelled from it, so that the mercury may afterwards stand at a proper height in the tube when the instrument is cold. The tube is then melted with care by the blow-pipe flame below the ball *a*, and closed, or hermetically sealed, as in *c*. The thermometer is in this way properly filled with mercury, and contains no air.

Fig. 9.

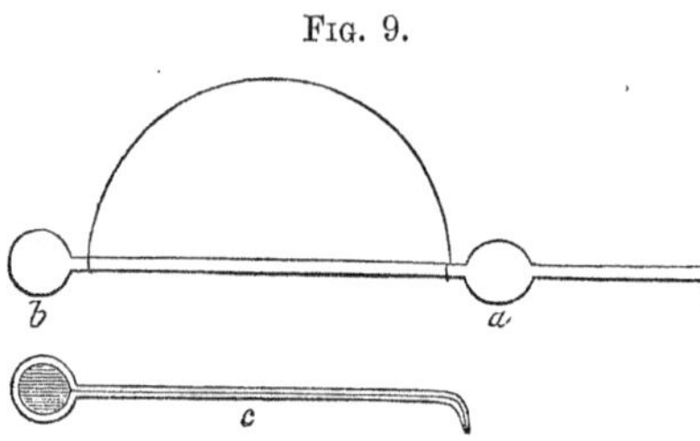

We have now an instrument in which we can nicely measure and compare any change in the bulk of the included fluid metal. Having previously made sure of the equality of the bore, it is evident that if the mercury swells up and rises two, three, four, or five inches in the tube, it has expanded twice, thrice, four, or five times more than if it had risen only one inch in the tube. By placing a graduated scale against the tube, we can therefore learn the quantity of expansion by simple inspection.

In order to have a fixed point on the scale, from which to begin counting the expansion of mercury by heat, we plunge the bulb of the thermometer into melting ice, and put a mark on the stem at the point to which the mercury falls. However frequently we do so with the same instrument, we shall find that the mercury always falls to the same point. This is, therefore, a fixed starting point. We obtain another fixed point by plunging the thermometer into boiling water. With certain precautions, this point will be found equally fixed on every repetition of the experiment. The most important of these precautions is, that the barometer be observed to stand at 30 inches,[1] when the boiling point is taken. It will afterwards be explained that the boiling point of water varies with the atmospheric pressure to which it is subject at the time.

Thermometers which are properly closed, and contain no air, can be inverted without injury, and the mercury falls into the tube, producing a sound as water does in the water-hammer. When the instrument contains air, the thread of mercury is apt to divide on inversion, or from other circumstances. When this accident occurs, it is best remedied by attaching a string to the upper end of the instrument, and whirling it round the head. The detached little column of mercury generally acquires in this way a centrifugal force, which enables it to pass the air, and rejoin the mercury in the bulb.

When the glass of the bulb is thin, it is proper to seal the tube as described, and to retain it for a few weeks before marking upon it the fixed points. Thermometers, however carefully graduated at first, are found in a short time to stand above the mark in melting ice, unless this precaution be attended to. Old instruments often err by as much as half a degree, or even a degree and a half, in this way.[2] The effect is supposed to arise from the pressure of the atmosphere upon the bulb, which, when not truly spherical, seems to yield slightly, and in a gradual manner. The chance of this defect may be avoided by giving the bulb a certain thickness. Mr. Crichton's thermometers, of which the freezing point has not altered in forty years, were all made unusually thick in the glass. But this thickness has the disadvantage of diminishing the sensibility of the instrument to the impression of heat.

We have in this way the expansion marked off on the tube, which takes place between the freezing and boiling points of water. On the thermometer which is used in this country, and called Fahrenheit's, this space is subdivided into 180 equal parts, which are called degrees. This division appears empirical, and different reasons are given why it was originally adopted. But as Fahrenheit, who was an instrument-maker in Amsterdam, kept his process for graduating thermometers a secret, we can only form conjectures as to what were the principles that guided him.

It is more convenient to divide the space between the freezing and boiling of water into 100 equal parts, which was done in the instrument of Celsius, a Swedish philosopher. This division was adopted at a later period in France, under the designation of the Centigrade scale, and is now generally used over the continent. The freezing point of water is called 0, or zero, and the boiling point 100. But in our scale, the point is arbitrarily called 32°, or the 32d degree; and consequently the boiling point is 32 added to 180, or the 212th degree.[3]

---

[1] More exactly 29·92 inches, that is, 760 millimètres; the latter number being universally assumed on the continent as the standard height of the barometer.

[2] Many thermometers cannot be heated 60 or 80 degrees, without a sensible displacement of the zero point, as remarked by Regnault (Ann. de Chimie, &c., 3 sér., t. 6, p. 378), and by Is. Pierre (Ib. 3 sér., t. 5, p. 427; et t. 15, p. 332), who indicate the extraordinary precautions requisite in the construction of thermometers for accurate research.

[3] A simple rule may be given for converting Centigrade degrees into degrees Fahrenheit. 100 degrees Centigrade being equal to 180 degrees Fahrenheit, 10 degrees C. = 18 degrees

The scale can easily be prolonged to any extent, above or below these points, by marking off equal lengths of the tube for 180 degrees, either above or below the space first marked. The degrees of contraction below zero, or 0°, are marked by the minus sign (—), and called negative degrees, in order to distinguish them from degrees of the same name above zero, or positive degrees. Thus, 47° means the 47th degree above zero, —47°, the 47th degree under zero.

The only other scale in use is that of Reaumur, in the north of Germany. The expansion between the freezing and boiling of water is divided into 80 parts in this thermometer. The relation between the three scales is illustrated in the following diagram.

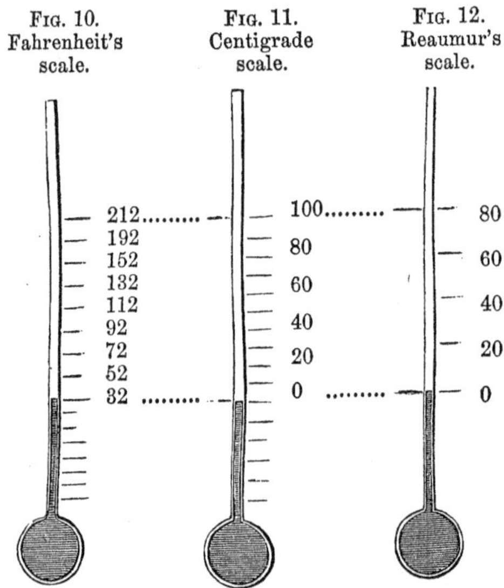

Fig. 10. Fahrenheit's scale.

Fig. 11. Centigrade scale.

Fig. 12. Reaumur's scale.

The zero of our scale is 32 degrees below the freezing point of water, and the expansions of mercury are available in the thermometer from —39° to 600°; but about the latter degree, mercury rises in the tube in the state of vapour, so as to derange the indications, and at about 660° it boils, and can no longer be retained in the glass vessel; while at the former low point it freezes or becomes solid. For degrees of cold below the freezing point of mercury, we must be guided by the contractions of alcohol or spirits of wine, a liquid which has not been frozen by any degree of cold we are capable of producing. There is no reason, however, for believing that we have ever descended more than 160 or 170 degrees below zero of Fahrenheit.

The zero of these scales has, therefore, no relation to the *real zero* of heat, or point at which bodies have lost all heat. Of this point we know nothing, and there is no reason to suppose that we have ever approached it. The scale of temperature may be compared to a chain, extended both upwards and downwards beyond our sight. We fix upon a particular link, and count upwards and downwards from that link, and not from the beginning of the chain.

The means of producing heat are much more at our command, but we have no measure of it, of easy application and admitted accuracy, above the boiling point of mercury. Recourse has been had to the expansion of solids at high temperatures, and various pyrometers, or "measures of fire," have been proposed. Professor

F., or 5 degrees C. = 9 degrees F.; multiply the Centigrade degrees by 9, and divide by 5, and add 32. Thus to find the degree F. corresponding with 50° C.

$$\begin{array}{r} 50 \\ 9 \\ \hline 5)450 \\ \hline 90 \\ \text{add } 32 \\ \hline \end{array}$$

Or the 50° C. corresponds with the 122° F.

For facility of reference a table of the corresponding degrees is given in Appendix II.

FIG. 13.

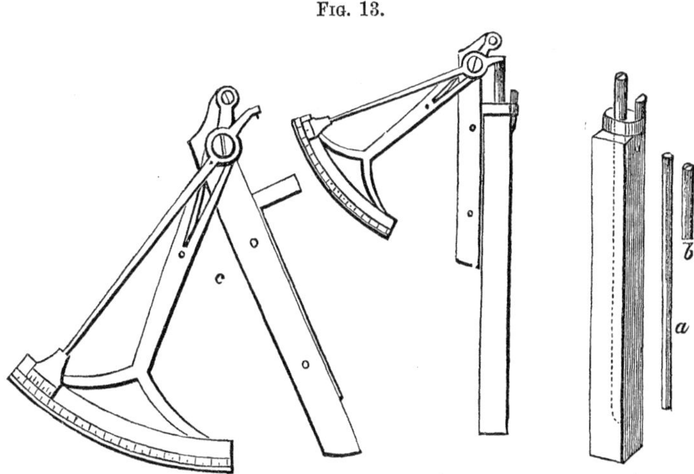

Daniell's pyrometer is a valuable instrument of this kind, of which the indications result from the difference in the expansion by heat of an iron or platinum bar, and a tube of well-baked black-lead ware, in which the bar is contained. The metallic bar *a* is shorter than the tube, and a short plug of earthenware *b* is placed in the mouth of the tube above the iron bar, and so secured by a strap of platinum foil and a little wedge, that it slides with difficulty in the tube. By the expansion of the metallic bar, the plug of earthenware is pushed outwards, and remains in its new position after the contraction of the metallic bar on cooling. The expansion of the iron bar thus obtained, is measured by adapting to the instrument an index, *c*, which traverses a circular scale, before and after the earthenware plug has been moved outwards by the expansion of the metallic bar. The degrees marked on the scale are in each instrument compared experimentally with those of the mercurial scale, and the ratio marked on the instrument, so that its degrees are convertible into those of Fahrenheit, (Philosophical Transactions, 1830–31). An air thermometer, of which the bulb and tube were of metal, has also been employed to explore high temperatures. In the old pyrometer of Wedgwood, the degree of heat was estimated by the permanent contraction which it produced upon a pellet of pipe-clay; but the indications of this instrument are fallacious, and it has long gone out of use.

The applicability of the mercurial thermometer to measure degrees of heat, depends upon two important circumstances, which involve the whole theory of the instrument:—

1st. The hollow glass ball, with its fine tube of uniform bore, is a nice fluid measure. The ball and part of the stem being filled with a fluid, the slightest change in the bulk of the fluid, which may arise from the application of heat or of cold to it, is conspicuously exhibited by the rise or fall of the fluid column in the stem. No more delicate measure of the bulk of an included fluid could be devised.

2d. It fortunately happens that the expansions of mercury, which can thus be measured so accurately, are proportional to the quantities of heat which produce them. But the mode in which this is proved requires a little attention. Suppose we had two reservoirs, one containing cold, and the other hot water. Plunge a thermometric bulb containing mercury first into the cold water, and mark at what point in the stem the mercury stands. Then plunge it into the hot water, and mark also the point to which the mercury now rises in the stem. We can obviously make a heat which will be half way exactly between the hot and cold water, by taking the same quantity of the hot and cold water, and mixing them together. Now, does this half heat produce a half expansion in mercury? On trial we find that it does. In the mixture of equal parts of the hot and cold water, the mercury stands exactly half way between the marks, supposing the experiment to be conducted with the proper precautions. This proves that the dilatations of mercury are proportional to the intensity of the heat which produces them. In the mercurial thermometer, therefore, quantities or degrees of expansion may be taken to indicate quantities or degrees of heat; and that is the principle of the instrument.

The same correspondence exists between the expansions of *air* and the quantities of heat which produce them. Indeed, in air, the correspondence is rigidly exact, while in mercury it is only a close approximation. Thus Dulong and Petit found that the boiling point of mercury was,

As measured by mercury in a syphon	680°
" " the air thermometer (true temp.)	662°
" " mercury in glass (Mr. Crichton)	660°

A short table exhibiting the increasing rate of the expansions of mercury has already been given, but glass expands in a ratio increasing quite as rapidly as this metal; so that the greater expansion of the mercury in the thermometer at high temperatures is fortunately corrected by the increasing capacity of the glass bulb.[1]

Fixed oils and spirit of wine do not deviate far from uniformity in their expansions, at least at low temperatures, and therefore are sometimes used as thermometric liquids. Spirit of wine thermometers, however, are often found to vary 6 or 8 degrees from each other at temperatures so low as —30° or —40°.

Thermometers have been devised which indicate the highest and lowest temperature which has occurred between two observations, or are self-registering. A thermometer, which was invented by Dr. Rutherford, is of this kind. This instrument consists, properly speaking, of two thermometers, one, *a*, of spirit of wine, and the other, *b*, of mercury, which are placed in the position represented in the figure, their stems being horizontal. The thermometer *b* is intended to indicate the maximum temperature. It contains, in advance of the mercury, a short piece of iron wire, which the mercury carries forward with it in dilating, and which remains in its advanced position, marking the highest temperature that has occurred, when the mercury withdraws. The minimum temperature is indicated by the spirit of wine thermometer *a*, which contains, immersed in the spirit, a small cylinder of ivory, or enamel, which, by a slight inclination of the instrument, falls to the surface of the liquid without being able to pass out of it. When the thermometer sinks, the ivory is carried back in the spirit; but when the temperature rises, the alcohol only advances, leaving the ivory where it was. Its extremity most distant from the bulb then indicates the lowest temperature to which the thermometer had been exposed. Before another observation is made, the ivory must be brought again to the surface of the alcohol by a slight percussion of the instrument.

Fig. 14.

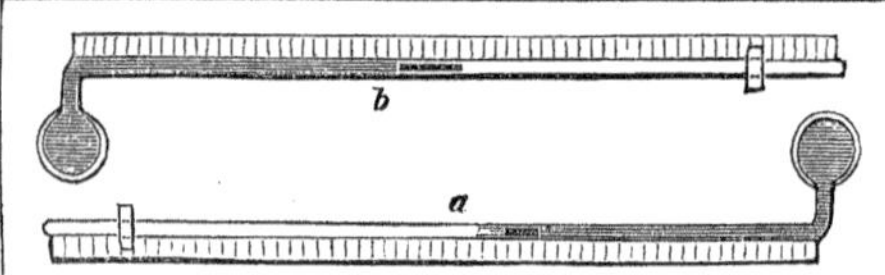

Another self-registering instrument, known in London as Six's, has the great advantage over the preceding instrument of being much less liable to go out of order. It consists of one thermometer only (fig. 15), filled with colourless spirit of wine, having a large cylindrical bulb. The stem is twice bent, and contains a column of mercury, *b*, in the lower bend, which is in contact with the alcohol, and

[1] In a note on the Comparison of the Air and Mercurial Thermometers; by M. Regnault. Annales de Chimie, &c. 3 sér. t. 6, p. 470.

advances or recedes with it. On either side of this mercury there is placed a little iron cylinder, or index, *c* and *d*, which has a fine hair projecting from it, so as to press against the sides of the tube, and cause the cylinder to move with a little difficulty. These iron cylinders, which have flattened ends covered with a vitreous matter, are brought into contact with the mercury by means of a magnet, and are pushed along by the column of mercury, when the latter is moved by the alcohol. The minimum temperature is indicated by *c*, and the maximum by *d*. The tube is expanded at *e*, and sealed after filling that space partly with alcohol, for no other purpose than to facilitate the movement of the index, *d*.

Fig. 15.

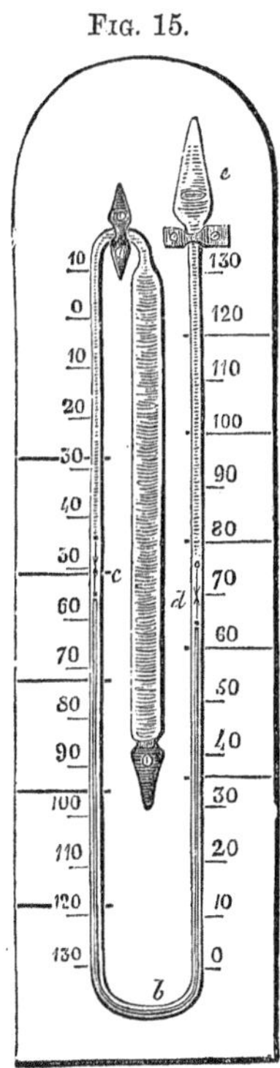

Our notions of the range of temperature acquire all their precision from the use of the thermometer. Cold, for instance, is allowed a substantial existence, as well as heat, in popular language. What is cold? it is the absence of heat, as darkness is the absence of light. The absence of heat, however, is never complete, but only partial. Water, after it is frozen into ice, cold as it is in relation to our bodies, has not lost all its heat, for it is easy to cool a thermometer far below the temperature of ice, and have it in such a condition as that it shall acquire heat, and be expanded by contact with ice; thus proving that the ice contains heat. Spirits of wine have not been frozen at the lowest temperature that has hitherto been attained; but even then this liquid possesses heat, and there is no doubt that if a sufficiently large portion of its heat were withdrawn, it would freeze like other bodies. The following are interesting circumstances in the range of temperature —

—220°	Fahr.	Greatest artificial cold measured. (Natterer.)
—166°	"	" " " " (Faraday.)
—150°	"	Liquid nitrous oxide freezes. "
—122°	"	Liquid sulphuretted hydrogen freezes. "
—105°	"	Liquid sulphurous acid freezes. "
— 71°	"	Liquid carbonic acid freezes. "
— 91°	"	Greatest artificial cold measured by Walker.
— 56°	"	Greatest natural cold observed by a "verified" thermometer. (Sabine.)
— 70°	"	Greatest natural cold observed at Fort Reliance by Back. Doubtful.
— 58°	"	Estimated temperature of planetary space. (Fourier.)
— 47°	"	Sulphuric ether freezes.
— 39°	"	Mercury freezes.
— 30°	"	Liquid cyanogen freezes. (Faraday.)
— 7°	"	A mixture of equal parts of alcohol and water freezes.
+ 7°	"	A mixture of one part of alcohol and three parts of water freezes.
20°	"	Strong wine freezes.
32°	"	Ice melts.
50°·7	"	Mean temperature of London.
81°·5	"	Mean temperature at the Equator.
98°	"	Heat of the human blood.
117°·3	"	Highest natural temperature observed — of a hot wind in Upper Egypt. (Burckhardt.)
151°·34	"	Wood-spirit boils. (Is. Pierre.)
172°·94	"	Alcohol boils. "
212°	"	Water boils.
442°	"	Tin melts.
594°	"	Lead melts.
662°	"	Mercury boils.
980°	"	Red heat. (Daniell.)

1141°	Fahr.	Heat of a common fire.	(Daniell.)
1869°	"	Brass melts.	"
2283°	"	Silver melts.	"
3479°	"	Cast iron melts.	"

SPECIFIC HEAT.

Equal bulks of different substances, such as water and mercury, require the addition of different quantities of heat to produce the same change in their temperature. This appears evident from a variety of circumstances. If two similar glass bulbs, like thermometers, one containing mercury and the other water, be immersed at the same time in a hot water-bath, it will be found that the mercury bulb is heated up to the temperature of the water-bath in half the time that the water bulb requires; and if the two bulbs, after having both attained the temperature of the water-bath, be removed from it and exposed to the air, the mercury bulb will cool twice as rapidly as the other. These effects must arise from the mercury absorbing only half the quantity of heat which the water does in being heated up to the same degree in the water-bath, and from having, consequently, only half the quantity of heat to lose in the subsequent cooling. Again, if we mix equal measures of water at 70° and 130°, the temperature of the whole will be 100°; or the hot measure of water, in losing 30°, elevates the temperature of the cold measure by an equal amount. But if we substitute for the hot water, in this experiment, an equal measure of mercury at 130°, on mixing it with the measure of water at 70°, the temperature of the whole will not be 100°, but more nearly 90°. Here the mercury is cooled from 130° to 90°, or loses 40° of heat, which have been transferred to the water, but which raise the temperature of the latter only 20°, or from 70° to 90°. To heat the measure of water at 70° to 100°, we must mix with it two, or a little more than two, equal measures of mercury at 130°, although one measure of water at 130° would answer the purpose. If, therefore, *two* measures of mercury, by losing 30° of temperature, heat only *one* measure of water 30°, it follows that hot mercury possesses only half the heat of equally hot water; or that water requires double the quantity of heat that is required by mercury, to raise it a certain number of degrees. This is expressed by saying that water has twice the *capacity for heat* that mercury possesses.

It is more convenient to express the capacities of different bodies for heat, with reference to equal weights than equal measures of the bodies. On accurate trial, it is found that a pound of water absorbs thirty times more heat than a pound of mercury, in being heated the same number of degrees: the capacity of water for heat is, therefore, thirty times greater than that of mercury. The capacities of these two bodies are in the relation of 1000 to 33; and it is convenient to express the capacities for heat of all bodies, in relation to that of water, as 1000. Such numbers are the *specific heats* of bodies.

There are two methods usually followed in determining capacity for heat. The first, which was that practised by MM. Dulong and Petit, consists in allowing different substances to cool the same number of degrees in circumstances which are exactly similar; to inclose them, for instance, in a polished silver vessel, containing the bulb of a thermometer in its centre, and to place this vessel under a bell-jar in which a vacuum is made. The time which the different substances take to cool, enables us to calculate the quantity of heat which they give out. The second, or method of mixture, consists in heating up the metal or other substance to 212°, and then throwing it into a vessel containing a considerable weight of cold water, to which a quantity of heat will be communicated, and a rise of temperature occasioned proportional to the capacity for heat of the substance. The following table contains results of M. Regnault, which closely coincide with the prior determinations of Dulong and Petit:—

Substances.	Specific heat of equal weights.
Water	1000
Ice[1]	513
Oil of turpentine, at 63·5° Fahr	426[2]
" " at 50° "	414
Wood charcoal	241[2]
Sulphur	203
Glass	198
Diamond	147[2]
Iron	113·79
Nickel	108·63
Cobalt	106·96
Zinc	95·55
Copper	95·15
Arsenic	81·40
Silver	57·01
Tin	56·23
Iodine	54·12
Antimony	50·77
Gold	32·44
Platinum	32·43
Mercury	33·32
Lead	31·40
Bismuth	30·84

The method of cooling gives results so exact, as to allow the detection of an increase of capacity with the temperature. The capacity of iron, when tried between 32° and 212°, as was the case with all the bodies in the table, was 110; but 115 between 32° and 392°, and 126 between 32° and 662°. It hence follows, that the capacity for heat, like dilatation, augments in proportion as the temperature is elevated. Dulong and Petit likewise established a relation between the capacity for heat of metallic bodies and the proportion by weight in which they combine with oxygen, or any other substance, which will again be adverted to.

Of all liquid or solid bodies, water has much the greatest capacity for heat. Hence the sea, which covers so large a proportion of the globe, is a great magazine of heat, and has a beneficial influence in equalizing atmospheric temperature. Mercury has a small specific heat, so that it is quickly heated or cooled, another property which recommends it as a liquid for the thermometer, imparting, as it does, great sensibility to the instrument.

The determination of the specific heat of gases is a problem involved in the greatest practical difficulties; so that notwithstanding its having occupied the attention of some of the ablest chemists, our knowledge on the subject is still of the most uncertain nature. It has been concluded by Delarive and Marcet (Annales de Ch. et de Ph. t. 35, p. 5; t. 41, p. 78; and t. 75, p. 113), and by Mr. Haycraft (Edinburgh Phil. Journ. vol. x. p. 351), that the specific heat of all gases is the same for equal volumes. But this opinion has been controverted by Dulong (Annales de Ch. et de Ph. t. 41, p. 113), by Dr. Apjohn (Transactions of the Royal Irish Academy, 1837), and by Suermann (Ann. de Ch. et de Ph. t. 63, p. 315), who have followed Delaroche and Berard in this inquiry (Annales de Chimie, t. 75; or Annals of Philosophy, vol. ii.) Their method was to transmit known quantities of the gases, heated to 212° in an uniform current, through a serpentine tube, surrounded by water, the temperature of which was observed, by a delicate thermometer at the beginning and end of the process. The results obtained by the different experimenters are contained in the following table: —

---

[1] Ed. Desains, Annales de Chimie et de Physique, 3me sér. t. 14, p. 306 (1845). By another method, the number 465 was obtained. The capacity of ice is, therefore, sensibly one-half that of water. This is a valuable paper, which will be referred to with advantage.

[2] Regnault, ibid. t. ix. pp. 339 and 324.

SPECIFIC HEAT OF GASES.

Name of the gas.	Capacity for equal volumes. Air = 1.	Capacity for equal weights. Air = 1.	Capacity for equal weights. Water = 1.	Authority.
Air	1·0000	1·0000	0·2669	Delaroche and Berard.
			0·3046	Suermann.
Oxygen	0·8080	0·7328	0·1956	Apjohn.
	0·9765	0·8848	0·2361	Delaroche and Berard.
	0·9954	0·9028	0·2750	Suermann.
	1·0000	0·9069	......	Delarive and Marcet, Haycraft, Dulong.
Hydrogen	0·9033	12·3401	3·2936	Delaroche and Berard.
	1·0000	14·4930	......	D. and M., Haycraft, Dulong.
	1·3979	20·3121	6·1892	Suermann.
	1·4590	21·2064	......	Apjohn.
Chlorine	1·0000	0·4074	......	Delarive and Marcet.
Nitrogen	1·0000	1·0318	0·2754	Delaroche and Berard.
	1·0005	1·0293	0·3138	Suermann.
	1·0480	1·0741	......	Apjohn.
Steam	1·9600	3·1360	0·8470	Delaroche and Berard.
Carbonic oxide	0·9925	1·0253	0·3123	Suermann.
	0·9960	1·0239	......	Apjohn.
	1·0000	1·0802	......	D. and M., Dulong.
	1·0340	1·0805	0·2884	Delaroche and Berard.
Carbonic acid	1·0000	0·6557	......	Haycraft.
	1·0655	0·6925	0·2124	Suermann.
	1·1750	......	......	Dulong.
	1·1950	0·7838	......	Apjohn.
	1·2220	......	......	Delarive and Marcet.
	1·2583	0·8280	0·2210	Delaroche and Berard.
Sulphurous acid	1·0000	0·4507	......	Delarive and Marcet.
Sulphuretted hydrog.	1·0000	0·8485	......	" "
Hydrochloric acid	1·0000	0·7925	......	" "
Nitrous oxide	1·0000	0·6557	......	" "
	1·0229	0·7354	0·2240	Suermann.
	1·1600	......	......	Dulong.
	1·1930	0·7827	......	Apjohn.
	1·3503	0·8878	0·2369	Delaroche and Berard.
Nitric oxide	1·7000	0·9616	......	Delarive and Marcet.
Ammonia	1·0000	1·6968	......	" "
Cyanogen	1·0000	0·5547	......	" "
Olefiant gas	1.0660	......	......	Haycraft.
	1·5310	......	......	Dulong.
	1·5530	1·5763	0·4207	Delaroche and Berard.
	1·5300	......	......	Delarive and Marcet.

It will be observed, that the capacity for heat of steam, as well as of ice, is less than that of an equal weight of water. Hence the specific heat of a body may change with its physical state. Delaroche and Berard likewise observed that the capacity of a gas is increased by its rarefaction. When the volume of a gas is doubled, by withdrawing half the pressure upon it, its specific heat is not quite so much as doubled. This is the reason why a gas becomes cold in expanding. In the expanded state it requires more heat to sustain it at its former temperature, from the augmentation which has occurred in its capacity. Air expanded into double its volume is cooled 40 or 50 degrees; and it has its temperature raised to that extent by compression into half its volume; suddenly condensed to one-fifth of its volume by a piston in a small cylinder, so much heat is evolved as to cause the ignition of a readily inflammable substance, such as tinder.

COMMUNICATION OF HEAT BY CONDUCTION AND RADIATION.

1. *Conduction.* — When one extremity of a bar of iron is plunged into a fire, the heat passes through the bar in a gradual manner, being communicated from particle to particle, and after passing through the whole length of the bar, may arrive at the other extremity. Heat, when conveyed in this way, is said to be conducted.

In solid substances, the phenomenon of the conduction of heat is so simple and familiar, that little need be said on the subject. Different solid substances vary exceedingly from each other in their power to conduct heat. Dense or heavy substances are generally good conductors, while light and porous bodies conduct heat imperfectly. Hence the universal use of substances of the latter class for the purposes of clothing. Count Rumford observed, that the finer the fabric of woollen cloth is, the more imperfectly does it conduct (Phil. Trans. 1792). The down of the eider-duck appears to be unrivalled in this respect. Bad conductors are also the most suitable for keeping bodies cool, protecting them from the access of heat. Hence to preserve ice in summer, we wrap it in flannel. Among good conductors of heat, the metals are the best. The relative conducting power of several bodies is expressed by the numbers in the following table, from the experiments of Despretz (Ann. de. Ch. et Ph. t. xxxvi. p. 422): —

Gold	1000	Tin	303·9
Silver	973	Lead	179·6
Copper	898	Marble	23·6
Iron	374·3	Porcelain	14·2
Zinc	363	Clay	11·4

Glass is an imperfect conductor, for we can fuse the point of a glass rod in a lamp, holding it within an inch of the extremity. On the contrary, we find it difficult to heat any part of a thick metallic wire to redness in a lamp, owing to the rapidity with which the heat is carried away by the contiguous parts.

The following table of the conducting power of various materials used in the construction of houses, as observed by Mr. Hutchinson, is of considerable utility for practical purposes. The substances are arranged in the order in which they resist most the passage of heat; the warmest substances, which are most valuable in construction, being placed first.[1]

Name of Substance.	Conducting power referred to that of slate = 100.	Name of Substance.	Conducting power referred to that of slate = 100.
Plaster and Sand	18·70	Bath Stone	61·08
Keene's Cement	19·01	Fire Brick	61·70
Plaster of Paris	20·26	Painswick Stone (H. P.)	71·36
Roman Cement	20·88	Malm Brick	72·92
Beech Wood	22·44	Portland Stone	75·10
Lath and Plaster	25·55	Lunelle Marble	75·41
Fir Wood	27·61	Bolsover Stone (H. P.)	76·35
Oak Wood	33·66	Norfal Stone (H. P.)	95·36
Asphalt	45·19	Slate	100·00
Chalk (soft)	56·38	Yorkshire Flag	110·94
Napoleon Marble	58·27	Lead	521·34
Stock Brick	60·14		

[1] New Experiments on Building Materials, by J. Hutchinson: Taylor and Walton. The three substances marked H. P. are the building stones employed in the construction of the New Houses of Parliament.

Certain vibrations were observed by Mr. Trevelyan to take place between metallic masses having different temperatures, occasioning particular sounds, which appear to be connected with the conducting power of the metal (Phil. Mag. 3d Series, vol. iii. 321). Thus, if a heated curved bar of brass *b*, be laid upon a cold support of lead *l*, of which the surface is flat, as represented in the figure, the brass bar, while communicating its heat to the lead, is thrown into a state of vibration, accompanied with a rocking motion and the production of a musical note, like that of the glass harmonicon. The rocking motion of the brass bar, accidentally commenced, appears to be continued from a repulsion which exists between heated surfaces, enhanced in this case by the low conducting power of the lead, which allows its surface to be strongly heated by the brass. Professor Forbes finds that the most intense vibrations are produced between the best conductors and the worst conductors of heat, the latter being the cold bodies (Edinburgh Phil. Trans. vol. xii.)

Fig. 16.

Our ordinary conceptions of the actual temperature of different bodies are much affected by their conducting power. If we apply the hand, at the same time, to a good and to a bad conductor, such as a metal and a piece of wood, which are exactly of the same temperature by the thermometer, the good conductor will feel colder or hotter than the other, from the greater rapidity with which it conducts away heat from, or communicates heat, to our body, according as the temperature of the metal and wood happens to be above or below that of the hand applied to them.

Fig. 17.

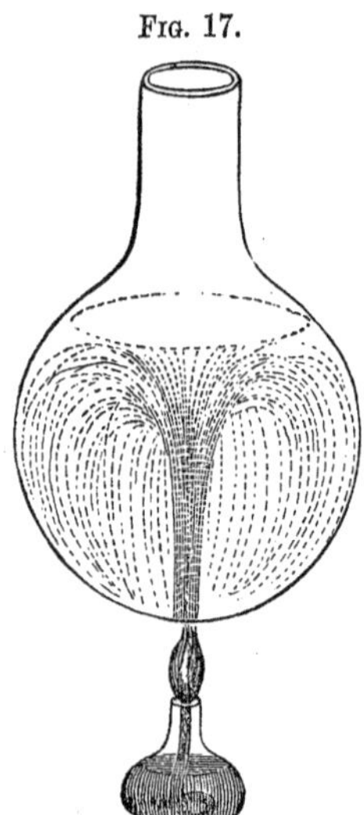

The diffusion of heat through liquids and gases is effected, in a great measure, by the motion of their particles among each other. When heat is applied to the lower part of a mass of liquid, the heated portions become lighter than the rest, and ascend rapidly, conveying or carrying the heat through the mass of the fluid. In a glass flask, for instance, containing water, with which a small quantity of any light insoluble powder has been mixed, a circulation of the fluid may be observed upon the application of the flame of a lamp to the bottom of the vessel, the heated liquid rising in the centre of the vessel, and afterwards descending near its sides, as represented in the annexed figure. But when heat is applied to the surface of a liquid, this circulation does not occur, and the heat is propagated very imperfectly downwards. It has even been doubted whether liquids conduct heat downwards at all, or, indeed, in any other way than by conveying it as above described. It can be proved, however, that heat passes downwards in fluid mercury, and hence it is probable that all liquids possess a slight conducting power similar to that of solids.

Fig. 18.

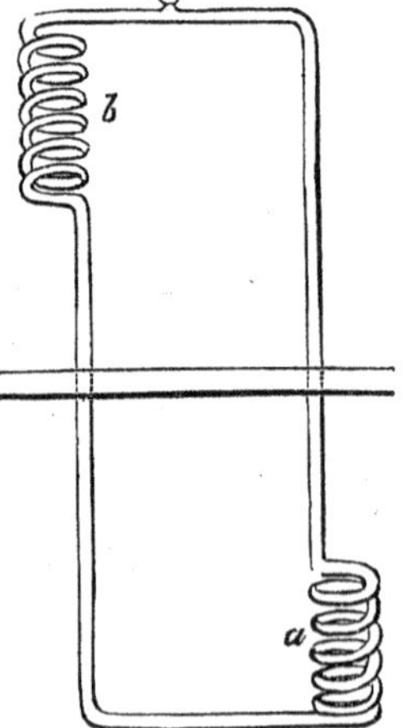

Let the endless tube represented in the accompanying figure be supposed to be entirely filled with water, and the heat of a fire be applied to the lower portion of it at *a*, which is twisted into a spiral form, the water will immediately be set in motion, and made to circulate through the tube, from the expansion and ascent of the portion in *a*, and the whole of the water in the tube will be brought in succession to the source of heat. The tube may be led into an apartment above *d*, and being twisted into another spiral at *b*, a quantity of the heat of the circulating water will be discharged in proportion to the extent of surface of tube exposed. Water of a temperature considerably above 212° is made to circulate in this manner through a very strong drawn-iron tube of about one inch in diameter, for the purpose of heating houses and public buildings. A slight waste of the water is found to occur, so that it is necessary to introduce a small quantity every few weeks by an opening and stopcock *c*, in the upper part of the tube. Tubes of larger calibre, with water circulating below the boiling point, are likewise much used for warming large buildings.

Air and gases are very imperfect conductors. Heat appears to be propagated through them almost entirely by conveyance, the heated portions of air becoming lighter, and diffusing the heat through the mass in their ascent, as in liquids. Hence, in heating an apartment by hot air, the hot air should always be introduced at the floor or lowest part. The advantage of double windows for warmth depends in a great measure on the sheet of air confined between them, through which heat is very slowly transmitted. In the fur of animals, and in clothing, a quantity of air is detained among the loose fibres, which materially enhances their non-conducting property. In dry air, the human body can resist a temperature of 250° without inconvenience, provided it is not brought into contact with good conductors at the same time.

Radiation of Heat.—Heat is also emitted from the surface of bodies in the form of rays, which pass through a vacuum, air, and certain other transparent media, with the velocity of light. It is not necessary that a body be heated to a visible redness to enable it to discharge heat in this manner. Rays of heat, unaccompanied by light, continue to issue from a hot body through the whole process of its cooling, till it sinks to the actual temperature of the air or surrounding medium. The circumstance that bodies suspended in a perfect vacuum cool rapidly and completely, without the intervention of conduction, places the fact of the dissipation of heat by radiation, at low temperatures, beyond a doubt.

The most valuable observations which we possess on this subject, were published by Sir John Leslie, in his Essay on Heat, in 1804. Leslie proved that the rate of cooling of a hot body is more influenced by the state of its *surface* than by the nature of its *substance*. He filled a bright tin globe with hot water, and observed its rate of cooling in a room of which the air was undisturbed. A thermometer placed in the water cooled half way to the temperature of the apartment in 156 minutes. The experiment was repeated, after covering the globe with a thin coating of lamp-black. The whole now cooled to the same extent as in the first experiment, in 81 minutes; the rapidity of cooling being nearly doubled merely by this change of surface.

An experiment of Count Rumford is even more singular. Water, of the same temperature, was allowed to cool in two similar brass cylinders, one of which was covered by a tight investiture of linen, and the other left naked. The covered vessel cooled 10° in 36½ minutes, while the naked vessel required 55 minutes; or the covering of linen, like the coating of lamp-black, greatly expedited the cooling, instead of retarding the escape of heat, as might be expected. The cooling was accelerated in the same manner, when the cylinder was coated with black or white paint, or smoked by a candle.

Fig. 19.

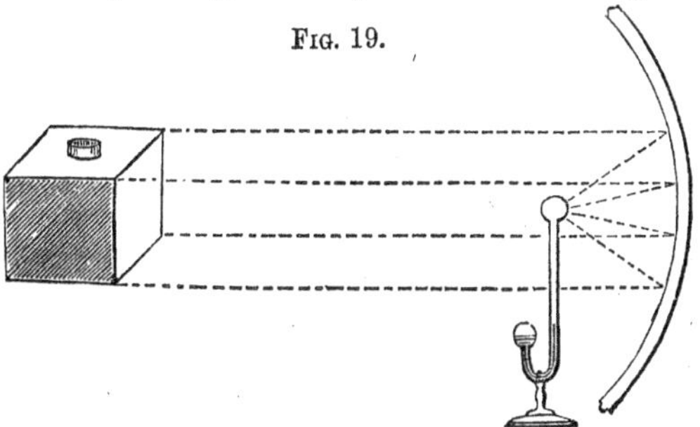

In determining the radiating power of different surfaces, Leslie generally made use of square tin canisters, of which the surfaces were variously coated, and which he filled with hot water. Instead of watching the rate of cooling, as in the experiments already mentioned, he presented the side of a canister, having its surface in any particular condition, to a concave metallic mirror, which concentrated the heat falling upon it into a focus, where the bulb of an air thermometer was placed to receive it, as represented in figure 19. The differential thermometer answered admirably for this purpose, as from its construction it is unaffected by the temperature of the room, while the slightest change in the temperature of the focal spot is immediately indicated by it.

Two metallic mirrors were occasionally used in conducting these experiments. The mirrors being arranged so as to face each other (fig. 20), with their principal axes in the same line; when a lighted lamp or hot canister is placed in the focus of one mirror, the incident rays are reflected by that mirror against the other, and collected in its focus.

FIG. 20.

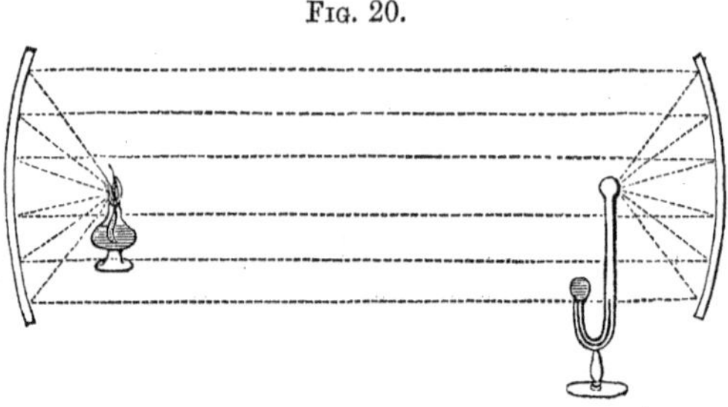

The following table exhibits the relative radiating power of various substances with which the surface of the canister was coated, as indicated by the effect upon the differential thermometer:—

Lamp-black	100	Plumbago	75
Water by estimate	100+	Tarnished lead	45
Writing-paper	98	Clean lead	19
Sealing-wax	95	Iron, polished	15
Crown glass	90	Tin plate, gold, silver, copper	12

It thus appears that lamp-black radiates five times more of the heat of boiling water than clean lead, and eight times more than bright tin. The metals have the lowest radiating power, which arises from their brightness and smoothness. If allowed to tarnish, their radiating power is greatly increased. Thus the radiating power of lead with its surface tarnished is 45, and with its surface bright, only 19; but glass and porcelain radiate most powerfully, although their surface is smooth. When the actual radiating surface is metallic, it is not affected in a sensible manner by the substance under it. Thus, glass covered with gold-leaf possesses the radiating power of a bright metal.

It is placed beyond doubt, by the recent experiments of Prof. A. D. Bache, that the radiating power of any surface is not affected by its *colour*, at least in an appreciable degree. Hence, no particular colour of clothes can be recommended for superior warmth in winter. But the absorbent powers of bodies for the heat of the sun depend entirely upon their colour. (Journ. Franklin Inst., May and November, 1835.)

The faculty which different surfaces possess of *absorbing* or of *reflecting* heat radiated against them, is connected with their own radiating power. Those surfaces which radiate heat freely, such as lamp-black, glass, &c., also absorb a large proportion of the heat falling upon them, and reflect little of it; while surfaces which have a feeble radiating and absorbing faculty, such as the bright metals, reflect a large proportion, as they absorb little, and form the most powerful reflectors. So that the good absorbents are found at the top, and the good reflectors at the bottom of the preceding table. The efficiency of a reflector depending upon its low absorbing power, reflectors of glass are totally useless in conducting experiments upon radiant heat. Metallic reflectors remain cold, although they collect much heat in their foci.

These laws of the radiation of heat admit of some practical applications. If we wish to retard, as much as possible, the cooling of a hot fluid or other substance, in what sort of vessel should we inclose it? In a metallic vessel, of which the surface is not dull and sooty, but clean and highly polished; for it has been observed, that hot water cools twice as fast in a tin globe of which the surface is covered with a thin coating of lamp-black, as in the same globe when the surface is bright and clean. Hence the advantage of bright metallic covers at table, and the superiority of metallic tea-pots over those of porcelain and stone-ware.

TRANSMISSION OF RADIANT HEAT THROUGH MEDIA, AND THE EFFECT OF SCREENS.

It has been shown by Dulong and Petit, that hot bodies radiate equally in all gases, or exactly as they radiate in a vacuum. Hot bodies certainly cool more rapidly in some gases than in others; but this is owing to the mobility and conducting powers of the gases being different.

Light of every colour, and from every source, is equally transmitted by all transparent bodies in the liquid or solid form; but this is not true of heat. The heat of the sun passes through any transparent body without loss; but of heat from terrestrial sources, a certain variable proportion only is allowed to pass, which increases as the temperature of the radiant body is elevated. Thus, it was observed by Delaroche that, from a body heated to 182°, only 1-40th of all the heat emitted passed through a glass screen: from a body at 346°, 1-16th of the whole; and from a body at 960°, so large a proportion as 1-4th appeared to pass through a glass screen. M. Melloni has, within the last few years, greatly extended our knowledge respecting the transmission of heat through media, in a series of the most profound researches.[1] In his experiments, he made use of the thermo-electric pile to detect changes of temperature; an instrument which, in his hands, exhibited a sensibility to the impressions of heat vastly greater than that of the most delicate mercurial or air thermometer.

His instrument, or the thermo-multiplier (fig. 21), consists of an arrangement of thirty pairs of bismuth and antimony bars contained in a brass cylinder, *t*, and

Fig. 21.

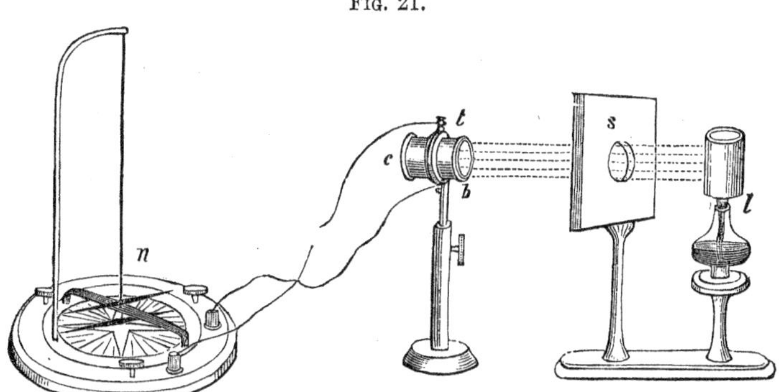

having the wires from its poles connected with an extremely delicate magnetic galvanometer, *n*. The extremities of the bars at *b* being exposed to any source of radiant heat, such as the copper cylinder *d*, heated by the lamp *l*, while the temperature of the other extremities of the bars at *c* is not changed, an electric current passes through the wires from the poles of the pile, and causes the magnetic needle of the galvanometer to deflect. The force of the electric current increases in proportion to the difference of the temperatures of the two ends, *b* and *c*, that is, in

[1] The complete series of Melloni's Memoirs is given in Taylor's Scientific Memoirs, Vols. I. and II.

proportion to the quantity of heat falling upon *b*; and the effect of this current upon the needle, or the deviation produced, is proportional to the force of the current, and consequently to the heat itself; at least, Melloni finds this correspondence to be exact through the whole arc, from zero to 20°, when the needle is truly astatic.

Melloni proved that heat, which has passed through one plate of glass, becomes less subject to absorption in passing through a second. Thus, of 1000 rays of heat from an oil flame, 451 rays being intercepted in passing through four plates of glass of equal thickness—

381	rays were	intercepted	by the first plate.
43	"	"	by the second.
18	"	"	by the third.
9	"	"	by the fourth.
451			

The rays appear to lose considerably when they enter the first layers of a transparent medium; but that portion of heat, which has forced its passage through the first layers, may penetrate to a great depth. Transparent liquids are found to be less penetrable to radiant heat than solids.

The capacity which bodies possess of transmitting heat does not depend upon their transparency; or bodies are not at all *transparent to heat* in the same proportion that they are transparent to light. Thus, plates of the following transparent minerals, having a common thickness of 0·1031 of an inch, allowed very different proportions of the heat from the flame of an argand oil-lamp to pass through them.

Of 100 incident rays there were transmitted:—

By Rock-salt	92 rays.
Mirror glass	62 "
Rock-crystal	62 "
Iceland spar	62 "
Rock-crystal, smoky and brown	57 "
Carbonate of lead	52 "
Sulphate of barytes	33 "
Emerald	29 "
Gypsum	20 "
Fluor spar	15 "
Citric acid	15 "
Rochelle salt	12 "
Alum	12 "
Sulphate of copper	0 "

A piece of smoky rock-crystal, so brown that the traces of letters on a printed page covered by it could not be seen, and which was fifty-eight times thicker than a transparent plate of alum, transmitted 19 rays, while the alum transmitted only 6. One substance, which is perfectly opaque, a kind of black glass used for the polarization of light by reflection, was found by Melloni to allow a considerable quantity of rays of heat to pass through it. He applied the term *diathermanous* to bodies which transmit heat, as *diaphanous* is applied to bodies which transmit light. Of all *diaphanous* or transparent bodies, water is in the least degree diathermanous. With the exception of the opaque glass referred to above, all diathermanous bodies belong also to the class of diaphanous bodies; for those kinds of metal, wood and marble, which totally obstruct the passage of light, obstruct that of heat also.

The proportion of heat from various sources which radiates through a plate of glass 1-50th of an inch in thickness, was observed by Melloni to be as follows:—

Of 100 rays	Transmitted.	Absorbed.
From the flame of an oil-lamp there were	54	46
" red hot platinum	37	63
" blackened copper, heated to 732° F.	12	88
" " " " 212°	0	100

But the power of transmission of rock-salt is the same for heat from all these sources, or for heat of all intensities; 92 per cent. of the incident heat being transmitted by that body, whether it be the heat radiated from the hand or from a bright argand lamp. Rock-salt stands alone in this respect among diathermanous bodies. This substance may be cut into lenses or prisms, and be used in concentrating heat of the very lowest intensity, or in decomposing it by double refraction, in the same manner as glass is employed with the light of the sun. Indeed, rock-salt has become quite invaluable in researches upon the transmission of heat.

It thus appears that a body at different temperatures emits different species of rays of heat, which may be sifted or separated from each other by passing them through certain transparent media. They are all emitted simultaneously, and in different proportions, by flame; but in heat from sources of lower intensity some of them are always absent. The calorific rays of the sun are chiefly of the kind which passes through glass; but Melloni shows that the other species are not altogether wanting. The rays of heat emitted by the sun and other luminous bodies are quite different rays from the rays of light with which they are accompanied.

Of the equilibrium of temperature.—When several bodies of various temperatures, some cold and some hot, are placed near each other, their temperatures gradually approximate, and, after a certain period has elapsed, they are found all to be of one and the same temperature. To account for the production and continuation of this equilibrium of temperature, it is necessary to assume that all bodies are at all times radiating heat in great abundance in all directions, although their temperature does not exceed or even falls below the temperature of the atmosphere. Hence, there is an incessant interchange of heat between neighbouring bodies; and a general equalization of temperature is produced when every object receives as much radiated heat as it emits.

This theory, which was first proposed by Prevost, of Geneva, enables us to account for the apparent radiation of cold. Cold, we know, is a negative quality, being merely the absence of heat, and cannot therefore be radiated. Yet, when a lump of ice is placed in the focus of a reflecting mirror, a thermometer in the focus of the opposite conjugate mirror is chilled. To account for this phenomenon we must remember that the temperature of the thermometer is stationary only so long as it receives as much heat as it radiates. It is in that state before the experiment is made with the ice; for the air or any object which may happen to be in the other focus is of the same temperature as the ball of the thermometer. But it is evident that the moment ice is introduced into one focus less heat will be sent from that to the other focus than was previously transmitted, and than is necessary to sustain the thermometer at a constant temperature. The thermometer ball, therefore, giving out as much heat as formerly, and receiving less in return, must fall in temperature. This is an experiment in which the thermometer ball is in fact the *hot body.*

The doctrine of the radiation of heat is happily applied to account for the deposition of *dew.* A considerable refrigeration of the surface of the ground below the temperature of the air resting upon it, amounting to 10 or 20 degrees, occurs every calm and clear night, and is caused by the radiation of heat from the earth (which is a good radiator) into empty space. Now, on becoming colder than the air above it, the ground will condense the moisture of the air in contact with it, and be covered with dew. For the air, however clear, is never destitute of watery vapour, and the quantity of vapour which air can retain depends upon its temperature; air at 52°, for instance, being capable of retaining 1-86th of its volume of vapour, while at 32° it can retain no more than 1-150th of its volume. The greatest difference between the temperature of the day and night takes place in spring and autumn, and these are the seasons in which the most abundant dews are deposited.

That the deposition of dew-drops depends entirely upon radiation is fully established by the following circumstances:—1. It is on clear and calm nights only that dew is observed to fall. When the sky is overcast with clouds, no dew is formed; for then the heat which radiates from the earth is returned by the clouds above, and prevented from escaping into space; so that the ground never becomes colder than the air. 2. The slightest screen, such as a thin cambric handkerchief, stretched between pins, at the height of several inches above the ground, is sufficient to protect the objects below it from this chilling effect of radiation, and to prevent the formation of dew or of hoar-frost upon them. This fact was well known to gardeners, and they had long availed themselves of it in protecting their tender plants from frost, before the laws of the radiation of heat came to be explained. 3. Dr. Wells proved by numerous experiments that the quantity of dew which condenses on different objects exposed in the same circumstances is proportional to the radiating power of those substances. Thus, when a polished plate of metal and a quantity of wool are exposed together in favourable circumstances, scarcely a trace of dew is to be observed on the metal, while a large quantity condenses in the wool, the latter substance being incomparably the best radiator, and therefore falling to a much lower temperature than the metal.

The same theory has been applied to explain a process for making ice followed by the Indian natives near Calcutta. In that climate the temperature of the air rarely falls below 40° in the coldest nights; but the sky is clear, and a powerful radiation takes place from the surface of the ground. Hence, water contained in shallow pans imbedded in straw is often sheeted over with ice by a night's exposure. The water is certainly cooled by radiation from its surface, and not by evaporation; for the process succeeds best when the pans are placed in shallow trenches dug in the ground, an arrangement which retards evaporation; and no ice forms in windy weather, when evaporation is greatest.

The morning frosts of autumn are first felt in sequestered situations, as in ravines closed on all sides, or along the low courses of rivers, where the cooling of the earth's surface by radiation is in the least degree checked by the movement of the air over it. These are also the very situations upon which the sun's rays produce the greatest effect in summer.

Reverting again to the subject of conduction of heat through solid bodies, it may now be stated, that there is every reason to believe that heat is propagated, even in that case, in a manner not unlike radiation. Heat, in its passage through a bar of iron, is probably radiated from particle to particle; for the material atoms, of which the bar consists, are not supposed to be in absolute contact, although held near each other by a strong attraction. Radiation, as observed in air or a vacuum, may thus pass into conduction in solids, without any breach of continuity in the natural law to which heat in motion is subject. Baron Fourier proceeds upon such an hypothesis in his mathematical investigation of the law of cooling by conduction in solid bodies.[1]

We are now in a condition to advert with advantage to the equilibrium of the temperature of the earth. There can be no doubt of the existence, in this globe of ours, of a *central heat.* At a depth under the surface of the earth, not in general exceeding twenty feet, the thermometer is perfectly stationary, not being affected by the change of the seasons; but at greater depths the temperature progressively rises. M. Cordier, to whom we are indebted for a most profound investigation of this interesting subject, considers the two following conclusions to be established by all the observations on temperature which have been made at considerable depths. 1st. That below the stratum where the annual variations of the solar heat cease to be sensible, a notable increase of temperature takes place as we descend into the interior of the earth. 2dly. That a certain irregularity must be admitted in the distribution of the subterraneous heat, which occasions the progressive increase of temperature to vary at different places. Fifteen yards has been provisionally assumed as the average depth which

---

[1] See a report by Professor Kelland, On the present state of our Theoretical and Experimental Knowledge of the Laws of the Conduction of Heat, in the Reports of the British Association for the Advancement of Science, for 1841, p. 1.

corresponds to an increase of one degree Fahrenheit. This is about 116 degrees for each mile. Admitting this rate of increase, we have at the depth of 30½ miles below the surface a temperature of 3500°, which would melt cast iron, and which is amply sufficient to melt the lavas, basalts, and other rocks, which have actually been erupted from below in a fluid state. But this central heat has long ceased to affect the surface of the earth. Fourier demonstrates, from the laws of conduction, that although the crust of the globe were of cast iron, heat would require myriads of years to be transmitted to the surface from a depth of 150 miles. But the crust of the globe is actually composed of materials greatly inferior to cast iron in conducting power. The temperature of the surface of the globe now depends upon the amount of heat which it receives from the sun, compared with the heat radiated away from its surface into free space. There is reason to believe that no material change has occurred in the quantity of heat received from the sun during the historical epoch. The radiation from the surface of the earth has its limit in the temperature of the planetary space in which it moves, which Fourier deduces, from calculation, to lie between —58° and —76°, and which Schwanberg, from a calculation on totally different principles, estimates at —58°.6; a close coincidence. This low temperature appears to be attained in the long absence of the sun during a polar winter, as Captain Parry found the thermometer to fall so low as —55° or —56° at Melville Island; and Captain Back has recorded a temperature observed on the North American continent so low as —70°.

FLUIDITY AS AN EFFECT OF HEAT.

One of the general effects of heat upon bodies has already been adverted to, namely its power of causing them to expand, which demanded our earliest attention, as it involves the principle of the thermometer. But heat, besides effecting changes in the bulk, is capable of effecting changes in the condition of bodies. Matter is presented to us in three very dissimilar conditions, or forms, namely, in the solid, liquid, and gaseous forms. It is believed that no body is restricted to any of these forms, but that the state of bodies depends entirely upon the temperature in which they are placed. In the lowest temperatures, they are all solid, in higher temperatures they are converted into liquids, and in the highest of all they become elastic gases. The particular temperatures at which bodies undergo these changes are exceedingly various, but they are always constant for the same body. The first effect, then, of heat on the state of bodies is the conversion of solids into liquids; or heat is the cause of fluidity.

Some substances, in liquefying, pass through an intermediate condition, in which it is difficult to say whether they are liquids or solids. Thus tallow, wax, and several other bodies, pass through every possible degree of softness before they attain complete fluidity. Such bodies, however, are in general mixtures of two or more substances, which crystallize imperfectly. But ice, and the great majority of bodies, pass immediately from the solid into the liquid state. The temperatures at which bodies undergo this change are exceedingly various.

	Melts at		Melts at
Lead	594°	Olive oil	36°
Bismuth	476	Ice	32
Tin	442	Milk	30
Sulphur	232	Wines	20
Wax	142	Oil of turpentine	14
Spermaceti	112	Mercury	—39
Phosphorus	108	Liquid ammonia	—46
Tallow	92	Ether	—47
Oil of anise	50		

If the bodies are in the fluid form, they freeze upon being cooled below the temperatures set against them.

It may be added, in reference to this table, first, that in certain circumstances liquids can be cooled down several degrees below their usual freezing point before they begin to congeal. Thus we may succeed, by taking certain precautions, in cooling a small quantity of water, in a glass tube, so low as the temperature 8°, or even as 5°, without its freezing; that is, 24 or 27 degrees under its proper freezing point 32°. The water must be cooled without the slightest agitation, and no sand or angular body be in contact with it; for the instant any solid body is dropped into water cooled below its freezing point, or a tremor is communicated to it, congelation commences, and the temperature of the liquid starts up to 32°. But, on the other hand, we cannot heat a solid the smallest fraction of a degree above its proper melting point, without occasioning liquefaction. Hence it is not the freezing of water, but the melting of ice, which takes place with rigorous constancy at 32° Fahrenheit.

All salts dissolved in water have the effect of lowering the freezing temperature of that liquid. Common culinary salt appears to depress this point lower than any other saline body; and the effect appears to be closely proportional to the quantity of salt in solution. A solution of 1 part of salt in 4 of water freezes at 4°; and sea-water, which contains 1-30th of its weight of salt, freezes at 28°.

But the principal fact to be adverted to in liquefaction is the disappearance of a large quantity of heat during the change. Heat pours into a body during its melting, without raising its temperature in the most minute degree. This heat, which enters the body and becomes insensible or latent, serves merely to melt the body. We are indebted to Dr. Black for this observation, which involves consequences of greater importance than any other announcement in the theory of heat.

Before Dr. Black's views were made known, fluidity was considered as produced by a very small addition to the quantity of heat which a body contains, when it is once heated up to its melting point. But if we attend to the manner in which ice and snow melt, when exposed to the air of a warm room, we can perceive that, however cold they may be at first, they are soon heated up to their melting point, and begin at their surface to be changed into water. Now, if the complete change of these bodies into water required only the farther addition of a very small quantity of heat, a mass of them, though of considerable size, ought all to be melted in a few minutes or seconds more, the heat continuing to be communicated from the air around. But masses of ice and snow melt with extreme slowness, especially if they be of a large size, as are those collections of ice and wreaths of snow that are formed in some places during winter. These, after they begin to melt, often require many weeks of warm weather, before they are totally dissolved into water. The slow manner in which ice melts in ice-houses is also familiarly known.

By examining what happens in these cases, it may easily be perceived that a very great quantity of heat must enter the melting ice, to form the water into which it is changed, and that the length of time necessary for the collection of so much heat from surrounding bodies is the reason of the slowness with which the ice is liquefied. When melting ice is suspended in warm air, the entrance of heat into it is made sensible by a stream of cold air descending constantly from the ice, which may be perceived by the hand. It is, therefore, evident that the melting ice receives heat very fast; but the only effect of this heat is to change it into water, which is not in the least sensibly warmer than the ice was before. A thermometer applied to the drops or small streams of water as they come immediately from the melting ice, will point to the same degree as when applied to the ice itself. A great quantity of the heat, therefore, which enters into the melting ice, has no other effect than that of giving it fluidity. The heat appears to be absorbed or concealed within the water, and cannot be detected by the thermometer.

When ice is melted by means of warm water, this absorption of heat is made exceedingly obvious. Thus, on mixing a pound of water at 172° with a pound of snow at 32°, the snow is all melted, and the mixture is two pounds of water of the temperature of 32°. In being cooled down from 172° to 32°, the hot water loses 140 degrees of heat, which convert the snow into water, indeed, but produce no rise of temperature in the mixture above the 32 degrees originally possessed by the snow.

Dr. Black proved that the heat which disappears in this manner is not extinguished or destroyed, but remains latent in the water so long as it is fluid, and is extricated again when it freezes.

In water that has been cooled below its usual freezing point, when the congelation is once determined, quantities of icy spiculæ are produced in proportion to the depression of temperature, whilst at the same instant the temperature of ice and water starts up to 32°. The heat which thus appears was previously latent in that portion of the water which is frozen. The same disengagement of latent heat may be conveniently illustrated by means of a supersaturated solution of sulphate of soda, formed by dissolving, at a high temperature, three pounds of the salt in two pounds of water. When this liquid is allowed to cool undisturbed, and with a stratum of oil on its surface, it remains fluid, although containing a much greater quantity of salt in solution than the water could dissolve at the temperature to which it has fallen. But the suspended congelation of the salt being determined by the introduction of any solid substance into the solution, the temperature then often rises 30 and even 40 degrees, while crystals of sulphate of soda shoot rapidly through the liquid.

Wax, tallow, sulphur, and all other solid bodies, are melted in the same manner as water, by the assumption of a certain dose of heat. The latent heat which the following substances possess in the fluid form was, with the exception of water, determined by Dr. Irvine.

	Latent heat.
Water	142 degrees.[1]
Sulphur	145 "
Lead	162 "
Bees'-wax	175 "
Zinc	493 "
Tin	500 "
Bismuth	550 "

Even in the solid form certain bodies admit of a variation in their structure and properties from the assumption or loss of latent heat. Dr. Black made it appear probable that metals owe their malleability and ductility to a quantity of latent heat combined with them. When hammered they become hot from the disengagement of this heat, and at the same time become brittle. Their malleability is restored by heating them again in a furnace. Sugar, it is well known, may exist as a transparent and colourless body, with the physical properties of glass, or as a white and opaque, because a granular or crystalline mass. The transition from the glassy to the granular state is attended by a very remarkable evolution of heat, which appears to have escaped the notice of scientific men. If melted sugar be allowed to cool to about 100°, and then, while it is still soft and viscid, be rapidly and frequently extended and doubled up, till at last it consists of threads, as in drawn sugar, the temperature of the mass quickly rises so as to become insupportable to the hand. After this liberation of heat, the sugar on again cooling is no longer a glass, but consists of minute crystalline grains, and has a pearly lustre. The same change may occur in a gradual manner, as when a clear stick of barley-sugar becomes white and opaque in the atmosphere; but then we have no means of observing the escape of the latent heat on which the change depends. It may be inferred that glass itself, like transparent barley-sugar, owes its peculiar constitution and properties to the permanent retention of a certain quantity of latent heat. Of this heat glass can be deprived by keeping it long in a soft state; it then becomes granular, and, passing into the condition of Reaumur's porcelain, loses all the characters of glass.

It is not unlikely that the *dimorphism* of a body, or its property to assume two different crystalline forms, may likewise depend upon the retention of a certain

[1] De la Provostaye and Regnault, Annales de Chimie, &c., 3 sér. t. 8, p. 1.

quantity of latent heat by the body in the one form, and not in the other. Thus, sulphur assumes two forms, one on cooling from a state of fusion by heat, another in crystallizing at a lower temperature, and probably with the retention of less latent heat, from a solution of sulphuret of carbon. In charcoal and plumbago, again, we have carbon which has assumed the solid form at a high temperature, and possibly with the fixation of a quantity of latent heat which does not exist in the diamond, another form of the same body.

When a solid body is melted by the intervention of some affinity, without heat being applied to it, cold is generally produced. Thus, most salts occasion a reduction of temperature, in the act of dissolving in water, which requires them to become fluid. Nitre, for instance, cools the water in which it is dissolved 15 or 18 degrees. A mixture of five parts of sal ammoniac and five of nitre, both finely powdered, dissolved in nineteen parts of water, may reduce its temperature from 50° to 10°, or considerably below the freezing point of pure water. These salts are necessitated, by their affinity for water, to dissolve when mixed with it, and to become fluid, a change which implies the assumption of latent heat. Most of our artificial processes for producing cold are founded upon this disappearance of heat during liquefaction. A very convenient process for freezing a little water, without the use of ice, is to drench finely powdered sulphate of soda with the undiluted hydrochloric acid of the shops. The salt dissolves to a greater extent in this acid than in water, and the temperature may sink from 50° to 0°. The vessel in which the mixture is made becomes covered with hoar frost, and water in a tube immersed in the mixture is speedily frozen.

The same affinity between salts and water may be taken advantage of to cause the liquefaction of ice. On mixing snow with a third of its weight of salt, the snow is instantly melted, and the temperature sinks nearly to 0°. It was in this way that Fahrenheit is supposed to have obtained the zero of his scale. Ices for the table are always made in summer by mixing roughly pounded ice and salt together, and immersing the cream, or other liquid to be frozen, contained in a thin metallic pan, in the cold brine which is produced by the melting of the ice.

The liquefaction of snow by means of the salt, chloride of calcium, occasions a still greater degree of cold. To prepare this salt, marble or chalk is dissolved in hydrochloric acid, and the solution evaporated by a temperature not exceeding 300°. It should be stirred, as it becomes dry at this temperature; and is obtained in a crystalline powder, being the combination of chloride of calcium with two atoms of water. When three parts of this salt are mixed with two of dry snow, the temperature is reduced from 32° to —50°. In attempting to freeze mercury by means of this mixture, it is advisable to make use of not less than three or four pounds of the materials. When the materials are divided, and the mercury is first cooled considerably by one portion, it rarely fails in being frozen when transferred into another portion of the mixture. For producing still more intense degrees of cold, the evaporation of highly volatile liquids, of liquid carbonic acid, for instance, affords the most efficient means.

VAPORIZATION.

We have now to consider the second general effect of heat—Vaporization, or the conversion of solids and liquids into vapour. Vapours, of which steam is the most familiar to us, are light, expansible, and generally invisible gases, resembling air completely in their mechanical properties, while they exist, but subject to be condensed into liquids or solids by cold. Water undergoes a great expansion when converted into steam, a cubic inch of water becoming, in ordinary circumstances, a cubic foot of steam; or, more strictly, one cubic inch of water, when converted into steam, expands into 1694 cubic inches.

This change, like fluidity, is produced by the addition of heat to the body which undergoes it. But a much larger quantity of heat enters into vapours than into liquids, into steam than into water. If, over a steady fire, a certain quantity of ice-cold water requires one hour to bring it to the boiling point, it will require a continuance of the same heat for five hours more to boil it off entirely. Yet liquids do not become hotter after they begin to boil, however long, or with whatever violence the boiling is continued: for if a thermometer be plunged into water, and the point marked where it stands at the beginning of the boiling, it will be found to rise no higher, although the boiling be continued for a long time.

This fact is of importance in domestic economy, particularly in cookery; and attention to it would save much fuel. Soups, &c., made to boil in a gentle way, by the application of a moderate heat, are just as hot as when they are made to boil on a strong fire with the greatest violence; when water in a copper is once brought to the boiling point, the fire may be reduced, as having no further effect in raising its temperature, and a moderate heat being sufficient to preserve it.

The steam from boiling water, when examined by the thermometer, is found to be no hotter than the water itself. What, then, becomes of all the heat which is communicated to the water, since it is neither indicated in the steam nor in the water? It enters into the water, and converts it into steam, without raising its temperature. As much heat disappears as is capable of raising the temperature of the portion of water converted into steam 1000 degrees, or, what is the same thing, as would raise the temperature of one thousand times as much water by one degree. This is now generally assumed to be the amount of the latent heat of steam. Dr. Black found it to be about 960 degrees, Mr. Watt 940 degrees, and Lavoisier rather more than 1000 degrees.

Several circumstances may be remarked during the occurrence of this change in water. On heating water gradually in a vessel, we first observe minute bubbles to form in the liquid, and rise through it, which consist of air. As the temperature increases, larger bubbles are formed at the bottom of the vessel, which rise a little way in the liquid, and then contract and disappear, producing a hissing or simmering sound. But, as the heating goes on, these bubbles, which are steam, rise higher and higher in the liquid, till at last they reach its surface and escape, producing a bubbling agitation, or the phenomenon of *ebullition*. The whole process of boiling is beautifully seen in a glass vessel. It will be remarked that steam itself is invisible; it only appears when condensed again into minute drops of water by mixing with the cold air.

It was first observed by Gay-Lussac, that liquids are converted more easily into vapour when in contact with angular and uneven surfaces, than when the surfaces which they touch are smooth and polished. He also remarked that water boils at a temperature two degrees higher in glass than in metal; so that if into water, in a glass flask, which has ceased to boil, a twisted piece of cold iron be dropped, the boiling is resumed: it is only in vessels of metal that the boiling point is regular, and should be taken in graduating thermometers. It has been remarked by Mr. Scrymgeour, of Glasgow, that if oil be present with water, the boiling point of the water is raised a few degrees, in any kind of vessel. A much greater elevation of the boiling point has been observed by M. Marcet, (Ann. de Chimie, &c., 3 sér. t. 5, p. 449), in a glass flask, having its inner surface coated with a thin film of shellac, in which the temperature often rises to 221°, or even higher, before a burst of vapour occurs; it then sinks a few degrees, after which it rises again.[1] The reason why water in these circumstances does not pass into vapour at its usual boiling point, is not distinctly understood. The water appears to be in a precarious state of equilibrium, as in the other analogous case, when cooled with caution in a smooth glass vessel considerably under its usual freezing point. The introduction of an angular body into the water is sufficient, in either instance, to induce the sus-

---

[1] The author has quoted incorrectly the results of Marcet's experiments, as referred to above. The statements made, are to the effect that in glass vessels deprived of all foreign matter on their surface, a marked elevation of the temperature of ebullition may be obtained, distilled water not boiling below 105° C. (221° F.); but in vessels coated with shellac or sulphur, this temperature is inferior by some tenths of a degree to that in metal vessels. — R. B.

pended change. The same irregular deviation of the boiling point in glass vessels takes place in other liquids as well as water, and in some of them to a much greater extent.

There is a curious circumstance in regard to boiling, which is a matter of common observation in some shape or other. When a little water (a few drops) is thrown into a metallic cup considerably above the boiling point of water, the liquid assumes a spheroidal form, and rolls about the cup like melted crystal, without visible ebullition, being only slowly dissipated. The cause of the phenomenon appears to be this. Water exhibits an attraction for the surface of almost all solids at low temperatures, and wets them. Fluid mercury exhibits the opposite property, or a repulsion for most surfaces. The attraction of water for surfaces brings it into the closest contact with them, and greatly promotes the communication of heat by a heated vessel to the water contained in it. But heat appears to develope a repulsive power in bodies, and it is probable that above a particular temperature the heated metal no longer possesses this attraction for water. The water, not being attracted to the surface of the hot metal, and induced to spread over it, is not rapidly heated, and therefore boils off slowly. A rude method of judging of the degree of heat is founded on the same principle, and is seen familiarly exemplified in the laundry. The heat of the smoothing iron is judged of by its effects upon a drop of saliva let fall upon it. If the drop do not boil, but run along the surface of the metal, the iron is considered sufficiently hot; but if the drop adheres and is rapidly dissipated, the temperature is considered low.

The spheroidal ebullition of liquids, which was first examined by Leidenfrost, in 1756, has lately received from M. Boutigny some striking experimental illustrations (Annales de Chimie, &c., 3 sér. t. ix. p. 350; et t. xi. p. 16). He has observed that water may pass into spheroidal ebullition at any temperature above 340°, and remain in that state till the temperature falls to 288°; then it moistens the metallic capsule in which the experiment is made, and evaporates rapidly. The corresponding temperatures at which alcohol and ether pass into the spheroidal form in a heated capsule were found to be proportional to their points of ebullition; the temperature for the first being 273°, and for the second 142°. The ball of a thermometer being plunged in liquids while in the spheroidal state, indicated the temperatures — in water, of 205·7°; in absolute alcohol, of 167·9°; in ether, 93·6°; in hydrochloric ether, 50·9°; in sulphurous acid, 13·1°; which are all several degrees below the ordinary temperatures of ebullition of these liquids. When distilled water is allowed to fall drop by drop into sulphurous acid in the spheroidal state, the water is immediately congealed into a spongy mass of ice, even when the containing capsule is visibly red-hot.

The temperature at which any liquid boils is not fixed (like the melting point of solids), but depends entirely upon a particular circumstance,—the degree of pressure to which the liquid is at the time subject. Liquids are in general subject to the pressure of the atmosphere; for although the air is an exceedingly light substance, being 815 times lighter than water, yet by reason of its great quantity and height, it comes to weigh with considerable force upon the earth. This is called the atmospheric pressure, and amounts to about fifteen pounds upon each square inch of surface. The force with which air presses upon a man of ordinary size has been estimated at fifty tons; yet, from all the cavities of the animal frame being filled with equally elastic air, we support this great pressure without being sensible of it; indeed, we should suffer the greatest inconvenience from its sudden removal. Now the pressure of the atmosphere is not always the same at the same place, but is found by the barometer to vary within the limits of one-tenth of the whole pressure. This difference affects the boiling point to the extent of 4½ degrees. Thus, when the height of the mercury in the barometer is expressed by the numbers in the first column, water boils at the temperatures placed against them in the second column.

Barometer in inches of mercury.	Water boils.
27·74	208°
28·29	209
28·84	210
29·41	211
29·92	212
30·6	213

On this account the pressure of the atmosphere must be attended to in fixing the boiling point of water on thermometers. Water boils at 212° only when the pressure of the atmosphere is equivalent to a column of 29·92 inches of mercury.

The pressure of the atmosphere will be greatest at the level of the sea, and will diminish as we ascend to any height above it, for then we have less of the atmosphere above and pressing upon us, part of it being below us. Hence, water boils on the tops of mountains at a considerably lower temperature than at their bases. On the top of Mont Blanc, which is the pinnacle of Europe, water was observed by Saussure to boil at 184°. In deep pits, on the other hand, water requires a higher temperature to boil it than at the surface of the earth. An instrument has been constructed for ascertaining the heights of mountains on this principle. It consists essentially of a thermometer, graduated with great care about the boiling point of water, by means of which the temperature at which water boils at different altitudes can be ascertained with minute accuracy. A difference of one degree of temperature is occasioned by an ascent of about 550 feet, and the depression of the boiling point is accurately proportional to the elevation above the earth's surface, according to the observations of Prof. Forbes (Edinburgh Phil. Trans. xv. 409).[1]

When the pressure on liquids is removed by artificial means, they boil at greatly reduced temperatures. This may be done by placing them under the receiver of an air-pump, and exhausting. When the whole air is withdrawn, liquids in general boil at about 145° under the temperature which they require to make them boil when subject to the atmospheric pressure. In a good vacuum water will boil at 67°. This fact is also illustrated by a simple experiment which any one may perform. A flask, containing boiling water, is closed with a cork, while the upper part is filled with steam. The boiling in the flask may be renewed by plunging it into cold water; and the colder the water the brisker will the ebullition become. But the boiling is instantly checked by removing the flask from the cold water and immersing it in very hot water. On corking the flask the ebullition ceased from the pressure exerted by the confined steam upon the surface of the water; but on plunging the flask into cold water, the steam was condensed, and the water began to boil under the reduced pressure. On removing the flask to the hot water, the steam above ceased to be condensed, and by its pressure stopped the boiling. On the other hand, in a Papin's digester, which is a tight and strong kettle with a safety valve, water may be raised to 3 or 400° without ebullition: but the instant that this great pressure is removed, the boiling commences with prodigious violence.

FIG. 22.

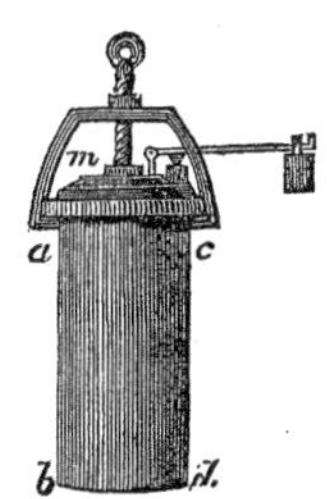

The facility with which liquids boil under reduced pressure is frequently taken advantage of in the arts, in concentrating liquors which would be injured in flavour or colour by the heat necessary to boil them under the pressure of the atmosphere. Mr. Howard applied this principle in concentrating the syrup of sugar, which is apt to be browned when made to boil under the usual pressure. He thus boiled syrup at 150°, applying heat to it in a pan covered by an air-tight lid, and pumping off the air and steam from the upper part

[1] For the most recent minute determinations of the boiling point of water, under variations of atmospheric pressure, see the memoir of M. Regnault; Ann. de Chimie, &c., 3 série, t. xiv. p. 196. A simple portable apparatus for the experiment is also described there.

of the pan by means of a steam-engine. This was the most essential part of his patent process, by which nearly the whole of the loaf sugar consumed in this country has been manufactured for many years.

In the same apparatus vegetable infusions may be inspissated, or reduced to the state of extracts, for medical purposes, with great advantage. When an extract is prepared in the ordinary way, by boiling down an infusion or expressed juice in an open vessel under atmospheric pressure, a considerable and variable proportion of the active principle is always destroyed by the high temperature and exposure to the air. But the extract is not injured when the infusion or juice is evaporated at a low temperature, and without access of air, and is generally found to be a more active medicine.

The temperatures at which different liquids are converted into vapour are exceedingly various; but other things remaining the same, the boiling temperature is constant for any particular liquid. The following table exhibits the boiling points of a few liquids, in which that point has been determined with precision: —

	Boiling point.
Hydrochloric ether	52°
Ether	96
Sulphuret of carbon	118
Ammonia (sp. gr. 0·945)	140
Alcohol	173
Water	212
Nitric acid (sp. g. 1·42)	248
Crystallized chloride of calcium	302
Oil of turpentine	314
Naphtha	320
Phosphorus	554
Sulphuric acid (sp. gr. 1·843)	620
Whale oil	630
Mercury	662

The boiling point of water is uniformly elevated by the solution of salts in the fluid; but much more so by some salts than others. Tables have been constructed of the boiling points of saline liquors, which are of useful application when it is wished to maintain a steady temperature somewhat above 212°. Thus, water saturated with common salt (100 water to 30 salt), boils at 224°; saturated with nitrate of potash (100 water to 74 salt), it boils at 238°; saturated cold with chloride of calcium, at 264°.

When steam from water is confined, it increases in temperature, and acquires great force; and the experiment can only be performed with safety in a boiler possessed of a safety-valve. This is a small lid in the upper part of the boiler, properly loaded, according to the force of the steam to be generated. The steam of boiling water occasions a severe scald, if allowed to condense upon the body. But when steam from water under pressure, or "high pressure" steam, which may be of a much higher temperature than boiling water, issues into the air, the hand may be directly exposed to it with impunity; and a thermometer placed in it shows that its temperature is greatly below that of boiling water. This singular property of high pressure steam is connected with the great expansion which it undergoes on escaping into the air from the vessel in which it was confined; elastic bodies having a tendency, when escaping from a state of compression, to fly asunder, not only to their original dimensions, but beyond them. The steam is greatly expanded, and at the same time mixed with air, which prevents it from afterwards collapsing. Now, after being incorporated with several times its bulk of air, steam is not easily condensed, but becomes low-pressure steam, and may have its condensing point reduced from above 212° to 120° or 130°. Hence the heat which it is capable of communicating, while condensing upon the hand held in it, is of much less intensity than that of ordinary steam, and inadequate to occasion scalding.

Steam, when heated by itself, apart from the liquid which produced it, does not possess a greater elasticity than an equal bulk of air confined and heated to the same degree, and may be heated to the temperature at which the containing vessel becomes red hot, without acquiring great elastic force. But if water be present, then more and more steam continues to rise, adding its elastic force to that of the vapour previously existing, so that the pressure becomes excessive.

Fig. 23.

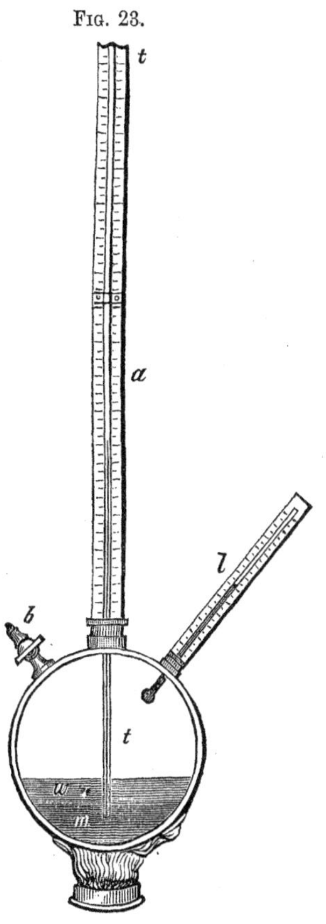

The elastic force of steam at temperatures above 212° is determined by heating water in a stout globular vessel containing mercury, *m*, (see fig. 23,) and water, *w*, and having a long glass tube, *t t*, screwed into it, open at both ends, and dipping into the mercury, with a scale, *a*, divided into inches, applied to it. The globular vessel has two other openings, into one of which a stopcock, *b*, is screwed, and into the other thermometer, *l*, having its bulb within the globe. The water is boiled in this vessel for some time, with the stopcock open so as to expel all the air. On shutting the stopcock, and continuing the heat, the temperature of the interior, as indicated by the thermometer, now rises above 212°, at which it was stationary while the steam generated was allowed to escape. The steam in the upper part of the globe becomes denser, more and more steam being produced, and forces the mercury to ascend in the gauge tube, *t*, to a height proportional to the elastic force of the steam. The height of the mercurial column is taken to express the elastic force or pressure of the steam produced at any particular temperature above 212°. The weight of the atmosphere itself is equivalent to a column of mercury of 30 inches, and this pressure has been overcome by the steam at 212°, before it began to act upon the mercurial gauge. For every thirty inches that the mercury is forced up in the gauge tube by the steam, it is said to have the pressure or elastic force of another atmosphere. Thus, when the mercury in the tube stands at thirty inches, the steam is said to be of two atmospheres; at 45 inches, of two and a half atmospheres; at 60 inches, of three atmospheres, and so on.

Experiments have been made on the elastic force of steam by Professor Robison, Mr. Southern, Mr. Watt, and others; but all preceding results have been superseded by those of a commission of the French Academy, consisting of MM. Dulong and Arago, appointed by the French government to investigate the subject, from its importance in connexion with the steam engine (Annales de Chimie, &c. 2 sér. t. xliii. p. 74). Their results, which are expressed in the following table, were obtained by experiment, up to a pressure of 25 atmospheres. The higher pressures were calculated by extending the progression observed at lower temperatures:—

Elasticity of Steam taking Atmospheric Pressure as Unity.	Temp. Fahr.
1	212.0
1½	233.96
2	250.52
2½	263.84
3	275.18
3½	285.08
4	293.72
4½	300.28
5	307.5
5½	314.24
6	320.36
6½	326.26
7	331.20
7½	336.50
8	341.78
9	350.78
10	358.28
11	366.85
12	374.00
13	386.66
14	386.94
15	392.46
16	398.48
17	403 82
18	408.92
19	413.78
20	418.46
21	422.96
22	427.28
23	431.42
24	435.56
25	439.34
30	457.16
35	472.73
40	486.59
45	499.14
50	510.60

Some curious experiments were made by M. Cagniard de la Tour on the vapour from various liquids at very high temperatures, and under great pressures. He filled a small glass tube in part with ether, alcohol, or water, and sealed it hermetically. The tube was then exposed to heat, till the liquid passed entirely into vapour. Ether became gaseous in a space scarcely double its volume at a temperature of 320°, and the vapour exerted a pressure of no more than 38 atmospheres. Alcohol became gaseous in a space about thrice its volume at the temperature of 404½°, with a pressure of about 139 atmospheres. Water acted chemically on the glass, and broke it; but adding a little carbonate of soda to it, the water became gaseous in a space four times its volume at the temperature at which zinc melts, or about 648°. These results are singular, in so far as the pressure or elastic force of the vapours proves to be much smaller than that which corresponds with their calculated density. It thus appears that highly compressed vapours lose a portion of their elasticity, or yield more to a certain pressure than air, by calculation, would do.

A measure is obtained of the quantity of latent heat in steam by observing the degree to which it heats up a mass of water when condensed in it. Cold water is easily made to boil by placing the open end of a pipe from a steam-boiler in it, and causing the steam to blow through it for a sufficient time. If a measured quantity of water at 32°, amounting to 11 cubic inches, is heated up to 212° in this manner, it is found that the volume is increased to 13 cubic inches by the condensed steam. Consequently, 11 cubic inches of water are heated up from 32° to 212°, or one hundred and eighty degrees, by 2 cubic inches of water in the form of steam. But if, for comparison, 2 cubic inches of boiling hot water be substituted for the steam, and added to 11 cubic inches of cold water, the temperature of the latter is raised no more than about twenty-eight degrees. In both experiments, however, the temperature of the steam, and of the boiling water added, was the same, or 212°; the difference of their heating effects depends entirely upon the latent heat which the former possesses, in addition to its sensible temperature, and abandons to the cold water on condensing.

In the condensing experiment 2 cubic inches of water in the form of steam raised the temperature of 11 cubic inches of water one hundred and eighty degrees, or 1 of steam raised the temperature of 5½ of water to that amount. As it follows that one part of steam would heat one part of water, 5½ times 180, or 990 degrees, it appears that steam possesses as much heat latent as might raise its own temperature to that amount on becoming sensible.

The latest, and probably most exact, determinations which we possess of the latent heat of the vapours of water, and other liquids, are those of M. Brix, of

Berlin, (Poggendorff's Annalen, lv.) He employed the apparatus represented in Fig. 24. The refrigeratory to contain the cold condensing water consists of a cylindrical vessel, A C, 3 inches in diameter and 3 inches deep. The steam from a small retort R does not pass directly into the water of the refrigeratory, but is conveyed by the spout M into an inner hollow cylinder E G, of a ring-formed basis, which has an opening into the atmosphere by the tube L, by which the air it contains finds vent on the arrival of the vapour. The condensing water is agitated by means of a thin disc of metal B, attached to a vertical rod, the upper end of which passes through the cover of the refrigeratory. A known quantity of cold water being introduced into this refrigeratory, its temperature is accurately observed by the including thermometer. In conducting the experiments it was arranged that the temperature of the condensing water should at first be a few degrees below that of the atmosphere, and vapour was thrown into the inner receiver by boiling a weighed portion of liquid in R, till the temperature of the condensing water rose as many degrees above that point. The weight of liquid distilled is then found by weighing the retort R with what remains in it, and ascertaining the loss; and the latent heat calculated by increasing the rise of temperature observed in the refrigeratory, in the same proportion as the weight of the condensing water in the refrigeratory exceeds that of the liquid distilled from the retort.

Fig. 24.

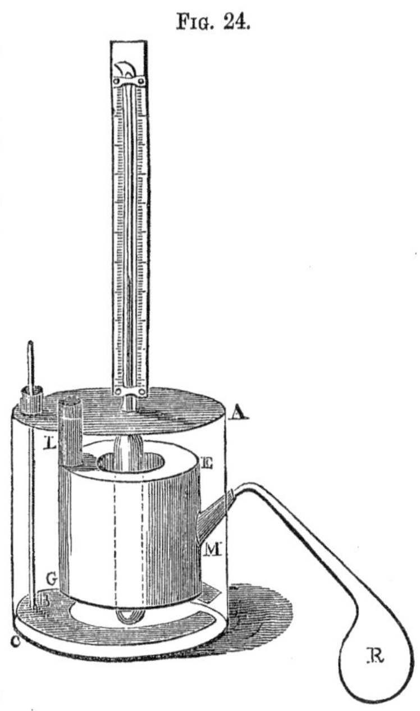

The following are the mean results which M. Brix obtained by this method, several experiments being made upon each liquid:—

Equal weights.	Latent heat of vapour.
Water	972 degrees.
Alcohol	385.2 "
Ether	162 "
Oil of turpentine	133.2 "
Oil of lemons	144 "

Despretz, who at an earlier period had also made very careful experiments on several of the same liquids, gave the following estimations of latent heat:—

Equal weights.	Latent heat of vapour.
Water	955.8 degrees.
Alcohol	374.4 "
Ether	174.6 "
Oil of turpentine	138.6 "

Dulong obtained for the latent heat of the vapour of water 977.4 degrees.

It is to be further remarked, that equal weights of these liquids yield very different volumes of vapour, owing to the different specific gravities of the latter; and the densest vapours appear to have generally the least latent heat. According to the table of M. Brix, the latent heat of the vapour of water is 972 degrees, while that of the vapour of alcohol is 385 degrees: or water-vapour has for equal weights about 2.5 times more latent heat than alcohol-vapour. The specific gravity of alcohol-vapour, on the other hand, is about 2.5 times greater than that of water-vapour, taking the former at 1589.4, and the latter at 622; consequently, equal volumes of these two vapours possess equal quantities of latent heat.

If the latent heat of different vapours be proportional to their volume, as these numbers seem to indicate, the same bulk of vapour will be produced from all liquids with the same expenditure of heat; and hence there can be no advantage in substituting any other liquid for water, as a source of vapour, in the steam-engine.

The latent heat of the vapour of water itself increases with its rarity at low temperatures, and diminishes with its increasing density at high temperatures. Water may easily be made to boil in a vacuum at the temperature of 100°, but the steam produced is much more expanded and rare than that produced at 212°, and has a greater latent heat. Hence there is no fuel saved by distilling in vacuo. It has been shown, by Mr. Sharpe, of Manchester, that whatever be the temperature of steam, from 212° upwards, if the same weight of it be condensed by water, the temperature of the water will always be raised the same number of degrees; or the latent and sensible heat of steam, added together, amount to a constant quantity. We may hence deduce a simple rule for ascertaining the latent heat of steam at any particular temperature. The sensible heat of steam at 212° may be assumed at 212 degrees, neglecting the heat which it has below zero Fahrenheit, and the latent heat of such steam is 972 degrees, of which the sum is 1184 degrees. To calculate the latent heat of steam at any particular temperature above 212°, subtract the sensible heat from this constant number 1184. Thus the latent heat of steam at 300° is 1184—300, or 884 degrees. The same relation between the latent and sensible heat of vapour appears to exist at temperatures below 212°, and the latent heat of vapour, below that temperature, may therefore be calculated by the same rule.

Temperature.	Latent heat of Equal Weights of Steam.
0°	1184 degrees.
32°	1152 "
100°	1084 "
150°	1034 "
212°	972 "
250°	934 "

The latent heat of other vapours, such as that of alcohol, ether, and oil of turpentine, has been found by Despretz to vary according to the same law.

From the large quantity of heat which steam possesses, and the facility with which it imparts it to bodies colder than itself, it is much used as a vehicle for the communication of heat. The temperature of bodies heated by it can never be raised above 212°; so that it is much preferable to an open fire for heating extracts and organic substances, all danger of empyreuma being avoided. When applied to the cooking of food, the steam is generally conveyed into a shallow tin box, in the upper surface of which are cut several round apertures, of such sizes as admit exactly the pans with the materials to be heated. The pans are thus surrounded by steam, which condenses upon them with great rapidity, till their temperature rises to within a degree or two of 212°. For some purposes, a pan containing the matters to be heated is placed within another and similar larger one, and steam admitted between the two vessels. Manufactured goods also are often dried by passing them once over a series of metallic cylinders, or of square boxes filled with steam. Factories are now very generally heated by steam, conveyed through them in cast-iron pipes. It has been found by practice that the boiler to produce steam for this purpose must have one cubic foot of capacity for every 2,000 cubic feet of space to be heated to a temperature of 70° or 80°; and that of the conducting steam pipe, one square foot of surface must be exposed for every 200 cubic feet of space to be heated.

The expansion of water into steam is used as a moving power in the steam engine. The application is made upon two different principles, both of which may

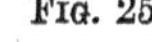

FIG. 25.

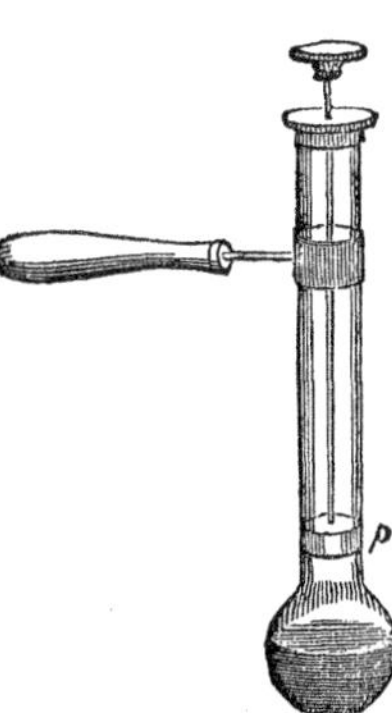

be illustrated by the little instrument depicted on the margin. It consists of a glass tube, about an inch in diameter, slightly expanded into a bulbous form at one extremity, and open at the other (fig. 25); a piston is made, by twisting tow about the end of a piece of straight wire, which must be fitted tightly in the tube by the use of grease. Upon heating a little water in the bulb below piston *p*, steam is generated, which raises the piston to the top of the cylinder. Here the simple elastic form of the steam is the moving power; and in this manner steam is employed in the high pressure engine. The greater the load upon the piston, and the more the steam is confined, the greater does its elastic force become. Again: the piston being at the top of the cylinder, if we condense the steam with which the cylinder is filled, by plunging the bulb in cold water, a vacuum is produced below the piston, which is now forced down to the bottom of the cylinder by the pressure of the atmosphere. In this second part of the experiment, the power is acquired by the condensation of the steam, or the production of a vacuum; and this is the principle of the common condensing engine. In the first efficient form of the condensing engine (that of Newcomen) the steam was condensed by injecting a little cold water below the piston, which then descended, from the pressure of the atmosphere upon its upper surface, exactly as in the instrument. But Mr. Watt introduced two capital improvements into the construction of the condensing engine; the first was, the admitting steam, instead of atmospheric air, to press down the piston through the vacuous cylinder, which steam itself could afterwards be condensed, and a vacuum produced above the piston, of which the same advantage might be taken as of the vacuum below the piston. The second was, the effecting the condensation of the steam, not in the cylinder itself, which was thereby greatly cooled, and occasioned the waste of much steam in being heated again at every stroke; but in a separate air-tight chamber, called the condenser, which kept cool and vacuous. Into this condenser the steam is allowed to escape from above and from below the piston alternately, and a vacuum is obtained without ever reducing the temperature of the cylinder below 212°.

A third improvement in the employment of steam as a moving power consists in using it *expansively;* a mode of application which will be best understood by being explained in a particular case. Let it be supposed that a piston, loaded with one ton, is raised four feet by filling the cylinder in which it moves with low-pressure steam, or steam of the tension of one atmosphere. An equivalent effect may be produced at the same expense of steam, by filling one-fourth of the cylinder with steam of the tension of four atmospheres, and loading the piston with four tons, which will be raised one foot. But the piston being raised one foot by steam of four atmospheres, and in the position represented in fig. 26, the supply of steam may be cut off, and the piston will continue to be elevated in the cylinder by the simple expansion of the steam below it, although with a diminishing force. When the piston has been raised another foot in the cylinder, or two feet from the bottom, the volume of the steam will be doubled, and its tension consequently reduced from four to $\frac{4}{2}$, or two atmospheres. At a height of three feet in the cylinder, the piston will have steam below it of the tension of $\frac{4}{3}$ or $1\frac{1}{3}$ atmosphere, and when the piston is elevated four feet, or reaches the top of the

FIG. 26.

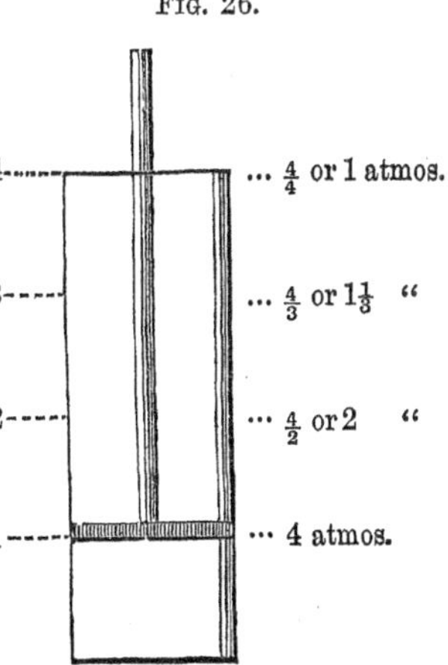

cylinder, the tension of the steam below it will still be $\frac{4}{4}$, or one atmosphere. The piston has, therefore, been raised to a height of three feet, with a force progressively diminishing from four atmospheres to one, or with an average force of two atmospheres, by means of a power acquired without any consumption of steam; but by the expansion merely of steam that had already produced its usual effect.[1]

Fig. 27.

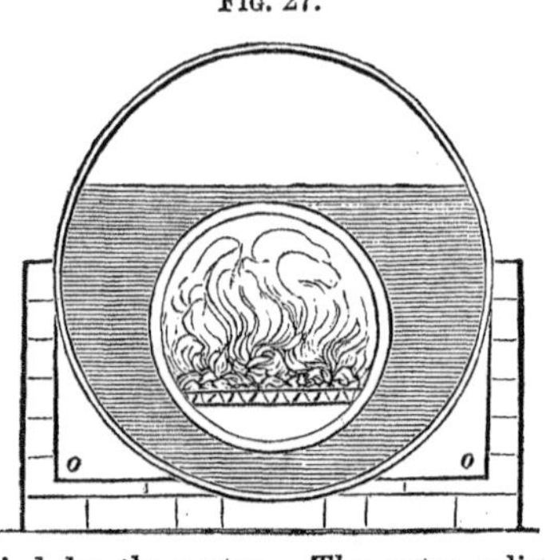

The boiler used to produce the steam is constructed of different forms. The cylinder boiler, of which a section is given in fig. 27, was found the most economical for the great steam-engines at the Cornish mines, and its use is extending in other quarters. It consists of two cylinders, one within the other, the smaller cylinder containing the fire, and the space between the two cylinders being occupied by the water. The outer cylinder may be six feet in diameter, and is often fifty or sixty feet in length. The heated air from the fire, after traversing the inner cylinder, is conducted under the boiler by the flues *o*, *o*, before it is conveyed to the chimney.

Fig. 28.

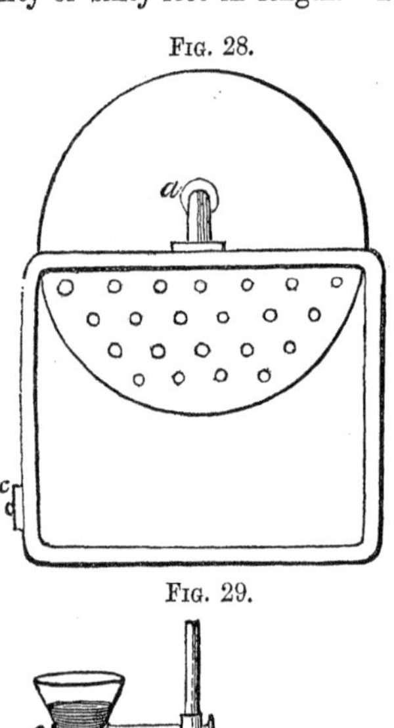

In the locomotive steam-engines, where the principal object is to generate steam in a small and compact apparatus with great rapidity, a different construction is adopted. Here the boiler consists of two parts, a square box with a double casing (of which a section or end view is given in figure 28), which contains the fire *f*, surrounded by a thin shell of water in the space *e e*, between the casings; and a cylinder *a*, through the lower part of which pass a number of copper tubes of small size, which communicate at one end with the fire-box, and at the other with the chimney, and form a passage for the heated air from the fire to the chimney. By means of these tubes, the object is accomplished of exposing to a source of heat the greatest possible quantity of surface in contact with the water. (See Dr. Lardner on the Steam-Engine: Cabinet Cyclopœdia.)

Fig. 29.

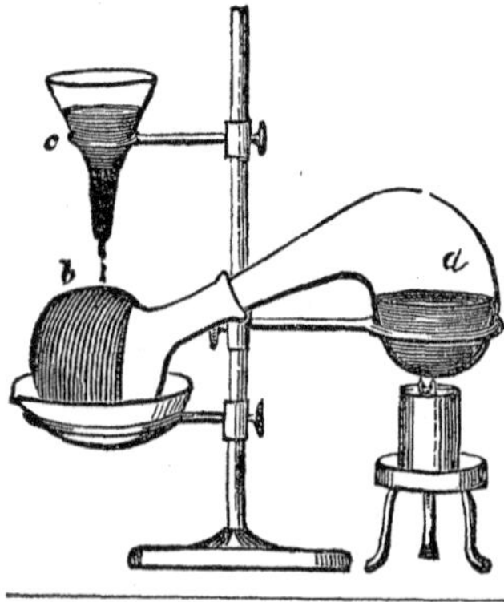

The subject of distillation is a natural sequel to vaporization; but it is unnecessary to enter into much detail. The principal point to be attended to is the most efficient mode of condensing the vapour. Figure 29 represents the ordinary arrangement in distilling a liquid from a retort *a*, and condensing the vapour in a glass flask *b*, which is kept cool by water dropping upon it from a funnel above, *c*. The condensing flask is covered by bibulous paper, so that the water falling upon it may be made to pass

[1] For the mathematical theory of the steam-engine, see a Memoir on the Motive Power of Heat, by E. Clapeyron, Taylor's Scientific Memoirs, vol. i. p. 347; a Memoir on the Heat and Elasticity of Gases and Vapours, by C. Holtzmann, ibid. vol. iv. p. 169; Experiments on the Expansive Force of Steam, by Prof. G. Magnus, ibid. p. 218; and on the Force requisite for the Production of Vapours, by the same, ibid. p. 235.

equally over its surface, and it is supported in a basin likewise containing cold water.

But a much superior instrument to the condensing flask is the condensing tube of Professor Liebig (fig. 30). This is a plain glass tube, *t t*, about thirty inches in

Fig. 30.

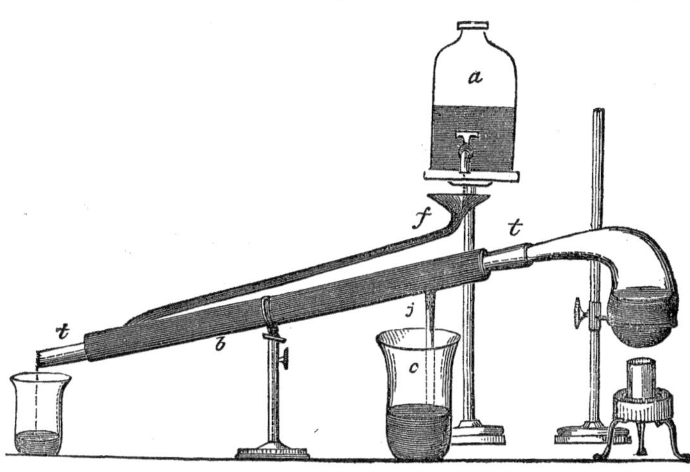

length, and one inch internal diameter, which is enclosed in a larger tube, *b*, of brass or tin-plate, about two feet long and two inches in diameter, the ends of which are closed by perforated corks, made fast by a mixture of white and red lead with a drying oil, a resinous cement being useless for such a junction. Or, the lower opening may be contracted by a collar of tin-plate, not much wider than the glass tube, and the two be united by a strong ring of sheet caoutchouc. A constant supply of cold condensing water from a vessel *a* is introduced into the space between the two tubes, being conveyed to the lower part of the instrument by the funnel and tube *f*, and flowing out from the upper part by the tube *j*. The condensed liquid drops quite cool from the lower extremity of the glass tube, where a vessel *c* is placed to receive it. The spiral copper tube or worm which is used for condensing in the common still is commonly made longer than is necessary, and, from its form, cannot be examined and cleared like a straight tube. Much vapour may be condensed by a small extent of surface, provided it is kept cold by an ample supply of condensing water.

Fig. 31.

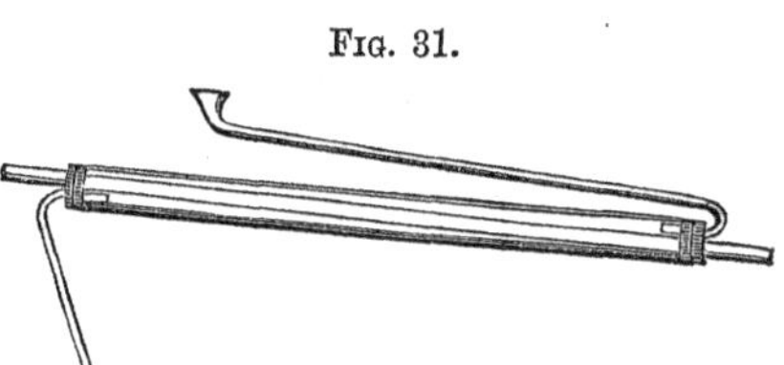

Both the outer and inner tube may be of glass in the condensing apparatus which has been described, and then the small tubes to bring and carry off the condensing water may be made to pass through openings in the corks, which they fit, as represented in figure 31.

EVAPORATION IN VACUO.

Water rises rapidly in vapour into a vacuous space, without the appearance of ebullition, at all temperatures, even at 32°, and greatly lower. Its elastic force increases as the temperature is elevated, till at 212° it is equal to that of the atmosphere, or capable of supporting a column of mercury thirty inches in height.

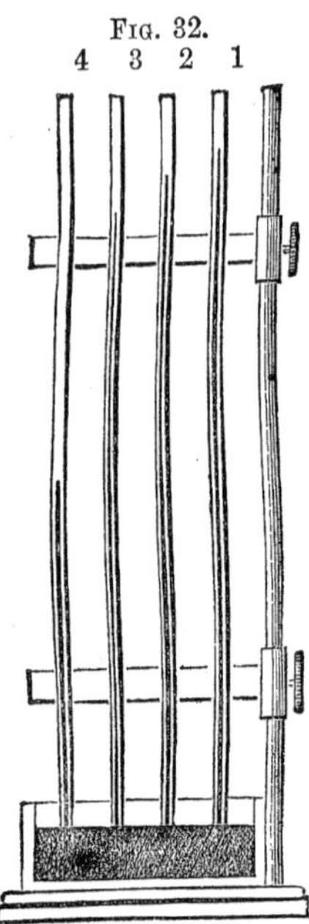

Fig. 32.

Various other solid and liquid substances emit vapour in similar circumstances; such as camphor, alcohol, ether, and oil of turpentine. Such bodies are said to be *volatile*, and other bodies, such as marble, the metals, &c. which do not emit a sensible vapour at the temperature of the air, are said to be *fixed*. All bodies which boil at low temperatures belong to the volatile class. An accurate estimate of the volatility of different bodies is obtained by determining the elastic force of the vapour which they emit in the vacuous space above the column of mercury in the barometer. If we pass up a bubble of air into the vacuum of the barometer, above the mercurial column, standing at the time at a height of 30 inches, the mercury is depressed, we may suppose, to the level of 29 inches, or by one inch. This would indicate that the air, by rising above the mercury, has been expanded into thirty times its former bulk, or that the elastic force of this rare air is equal to a column of one inch of mercury. The elastic force of vapour is estimated in the same manner. A few drops of the liquid operated upon are passed up into the vacuum above the mercurial column, which is depressed in proportion to the elastic force of the vapour. The depression produced by various liquids is very different, as illustrated in the annexed figure, representing four barometer tubes, in which the mercury is at its proper height in No. 1; is depressed by the vapour of water of the temperature 60° in No. 2; and by alcohol and ether at the same temperature in Nos. 3 and 4 respectively.

The depression of the mercurial column produced by water at every degree of temperature, between 32° and 212°, was first determined by Dr. Dalton, afterwards by M. Kaemtz, (Kaemtz, Meteorology, edited by C. Walker, p. 69), and again quite recently by M. Regnault, (Annales de Chimie, 3d sér. t. xi. p. 333; and t. xv. p. 139). The following selected observations prove that the elasticity increases at a very rapid rate with the temperature.

VAPOUR OF WATER IN VACUO (*Regnault*).

Temperature.		Tension in Millimeters and English inches of Mercury.	
Centig.	Fahr.	Millimeters.	English Inches.
—30°	—22°	0·365	0·0144
—25°	—13°	0·553	0·0218
—20°	—4°	0·841	0·0331
—15°	—5°	1·284	0·0506
—10°	14°	1·963	0·0818
—5°	23°	3·004	0·1233
0°	32°	4·600	0·1811
5°	41°	6·534	0·2573
10°	50°	9·165	0·3608
15°	59°	12·699	0·5000
20°	68°	17·391	0·6847
25°	77°	23·550	0·9272
30°	86°	31·548	1·2421
35°	95°	41·827	1·6468
60°	140°	148·791	5·8583
85°	185°	433·041	7·0488
100°	212°	760·000	29·9220

The vapours of other liquids increase in density and elastic force with the temperature, as well as the vapour of water; but each vapour appears to follow a rate of progression peculiar to itself.[1]

The assumption of latent heat by such vapours is evinced in some processes for producing cold. Water may be frozen by the evaporation of ether in the air-pump, and a cold produced of 55 degrees under the zero of Fahrenheit by the evaporation of that fluid. The ether vapour derives its store of latent heat from the remaining fluid and contiguous bodies, which being robbed of their heat, suffer a great refrigeration. To sustain the evaporation of this fluid, it is necessary to withdraw the vapour as it is produced by continual pumping. The volatile liquid, sulphuret of carbon, substituted for ether, produces even greater effects.

On the same principle is founded Leslie's elegant process for the freezing of water by its own evaporation, within the exhausted receiver of an air-pump, the evaporation being kept up by the absorbent power of sulphuric acid. (Supp. Encycloped. Britt., Art. Cold). A little water in a cup of porous stone-ware is supported over a shallow basin containing sulphuric acid (fig. 33). All that is necessary is to produce a good exhaustion at first: the processes of evaporation and absorption then go on spontaneously, in an uninterrupted manner. Various bodies, which have a powerful attraction for watery vapour, may be used as absorbents, such as parched oatmeal, the powder of mouldering whinstone, and even dry sole leather, by means of any one of which a small quantity of water may be frozen, during summer, in the exhausted receiver of an air-pump. No substance, however, is superior, in this respect, to concentrated sulphuric acid. When this liquid becomes too dilute to act powerfully as an absorbent, it may be rendered again fit for use, by boiling it and driving off the water. Ice might be procured in quantity, in a warm climate, by this process. The necessary vacuum would be most easily commanded, on the large scale, by allowing the receivers to communicate with a strong drum, filled with steam which could be condensed.

Fig. 33.

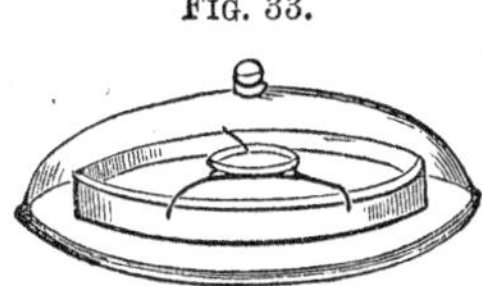

In the *Cryophorus* of Dr. Wollaston, water is also frozen by its own evaporation. This instrument consists of two glass bulbs, connected by a tube, and containing a portion of water, as represented in the figure. The air is first entirely expelled from the instrument by boiling the water, in both bulbs, at the same time, and allowing the steam to escape by a small opening at the extremity of the little projecting tube *e*. While the instrument is entirely filled with steam, the point of *e* is fused by the blow-pipe flame, and the opening hermetically closed. In experimenting with this instrument, the water is all poured into one bulb, and the other, or empty bulb, placed in a basin containing a mixture of ice and salt. The vapour in the cooled bulb is condensed, but its place is supplied by vapour from the water in the other bulb. A rapid evaporation takes place in the water bulb, and condensation in the empty bulb, till the water in the former bulb is cooled so low as to freeze. The instrument derives its name of the *cryophorus*, or frost-bearer, from this transference of the cold of the bulb in the freezing mixture to the bulb at a distance from it.

Fig. 34.

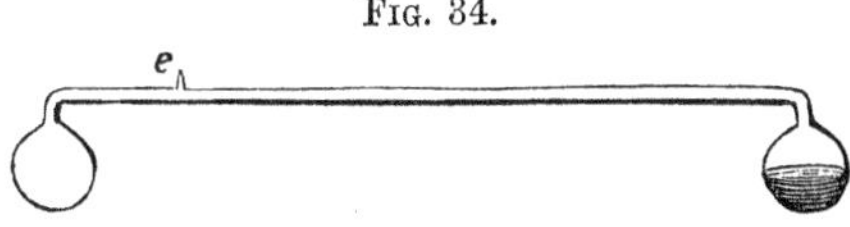

The question arises, do those bodies which evaporate at a moderate temperature continue to evaporate at all temperatures, however low. The opinion has prevailed,

[1] For the tension of the vapour of mercury at different temperatures, see a memoir of M. Avogadro, Annales de Chimie, &c., t. xlix. p. 369. For other vapours, the article *Dampf*, in the Handwörterbuch der Chemie, &c. of Liebig, Poggendorff, and Wöhler; and the memoir by Mr. Faraday, On the Liquefaction and Solidification of Bodies generally existing as Gases, (Philos. Trans. 1845, p. 155).

that bodies which are decidedly vaporous at high temperatures, such as sulphuric acid and mercury, never cease to evolve vapour, however far their temperature may be depressed, although the quantity emitted becomes less and less, till it ceases to be appreciable by our senses. Even fixed bodies, such as metals, rocks, &c., have been supposed to allow an escape of their substance into air at the ordinary temperature; and hence the atmosphere has been supposed to contain traces of the vapours of all the bodies with which it is in contact. Certain researches of Mr. Faraday, published in the Philosophical Transactions for 1826, on the existence of a limit to vaporization, establish the opposite conclusion. Mercury was found to yield a small quantity of vapour during summer, at a temperature varying from 60° to 80°, but in winter no trace of vapour could be detected. Mr. Faraday has proved that several chemical agents, which may be volatized by a heat between 300° and 400°, did not undergo the slightest evaporation when kept in a confined space with water during four years.

Bodies, therefore, cease all at once to emit vapour, at some particular temperature. In mercury, this temperature lies between 40° and 60° Fahrenheit. But a progressive and endless diminution of vaporizing power is certainly more natural than an abrupt cessation. What puts a stop to vaporization? it may be asked. Liquids, we know, have a certain attraction for their own particles, evinced in their disposition to collect into drops. The particles of solids are attracted more powerfully, and cohere strongly together. Mr. Faraday is of opinion, that when the vaporizing power becomes weak, at low temperatures, it may be overcome and negatived completely by this cohesive attraction, and no escape of particles in the vaporous form be permitted.

This supposition is conformable with the views of corpuscular philosophy which were entertained by Laplace. According to that profound philosopher, the form of aggregation which à body affects depends upon the mutual relation of three forces: 1. The attraction of each particle for the other particles which surround it, which induces them to approach as near as possible to each other. 2. The attraction of each particle for the heat which surrounds the other particles in its neighbourhood. 3. The repulsion between the heat which surrounds each particle, and that which surrounds the neighbouring particles—a force which tends to disunite the particles of bodies. When the first of these forces prevails, the body is solid; if the quantity of heat augments, the second force becomes dominant, the particles then move among each other with facility, and the body is liquid. While this is the case, the particles are still retained by the attraction for the neighbouring heat, within the limits of the space which the body formerly occupied, except at the surface, where the heat separates them, that is to say, occasions evaporation, till the influence of some pressure prevents the separation from being effected. When the heat increases to such a degree that the reciprocal repulsive force prevails over the attraction of the particles for one another, they disperse in all directions, as long as they meet no obstacle, and the body assumes the gaseous form. Berzelius adds the reflection, that if, in that gaseous state into which Cagnard de la Tour reduced some volatile liquids, the pressure does not correspond with the result of calculation, that difference may depend on this: that, as the particles have not an opportunity to recede much, the two first forces continue always to act, and oppose the tension of the gas, which does not establish itself in all its power unless when the particles are so distant from each other as to be out of the sphere of the influence of these forces. (Traité de Chimie, par J. J. Berzelius, t. i. p. 85).

GASES.

Permanent gases, such as atmospheric air, unquestionably owe their elastic state to the possession of latent heat. But the theory of the similar constitution of gases and vapours, although supported by strong analogies, was not generally adopted by chemists, till it was experimentally confirmed by Faraday, who first liquefied several of the gases. (Philosophical Transactions, 1823, pp. 160, 189; and 1845, p. 155). His method was to generate the gas in one end of a strong glass tube, bent in the middle, as represented (fig. 35); and hermetically sealed. The gas accumulating in a confined space, comes to exert a prodigious pressure; an effect of which is, that a portion of the gas itself condenses into a liquid in the end of the tube most remote from the materials, which is kept cool with that view. Considerable danger is to be apprehended by the operator in conducting such experiments, from the bursting of the glass tubes, and the face ought always to be protected by a wire-gauze mask from the effects of an explosion. The names of the gases which were liquefied in this manner, are sulphurous acid, cyanogen, chlorine, ammoniacal gas, sulphuretted hydrogen, carbonic acid, muriatic acid, and nitrous oxide; which required a degree of pressure varying, in the different gases, from two atmospheres, in the first mentioned, to fifty atmospheres, in the last mentioned gas, at the temperature of 45°. The liquefaction of several of these gases has since been effected by the application of cold alone, without compression.

FIG. 35.

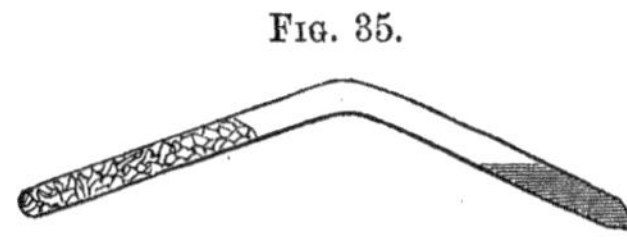

The principle of Faraday's condensing tube has been embodied in the machine of Thilorier for the liquefaction of carbonic acid gas. (Annales de Chimie, &c. 1835, lx. 427, 432). It consists (fig. 36) of two similar cylindrical vessels of wrought iron, made exceedingly strong, of the capacity of about three-fourths of a gallon, each of which is provided with a peculiarly constructed stopcock, being a spherical plug of lead on a spindle which can be screwed down, by turning the handle above, into a spherical cavity of brass-work, having at its base a tubular opening into the cylinder, which is thus closed. There is also a connecting tube of copper, the ends of which can be attached by screws to the discharging orifices of the stopcocks, so as to unite the two cylinders when necessary. The stopcock being

FIG. 36.

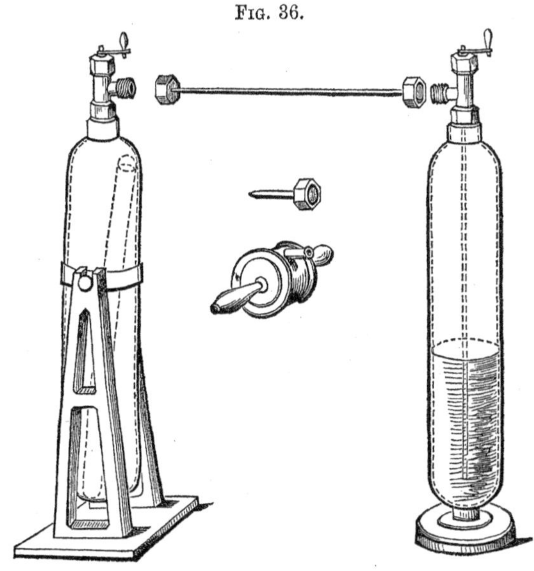

removed from one of the cylinders *a*, which is called the generator, a charge is introduced, consisting of two pounds of pulverulent bicarbonate of soda and three pounds of water at the temperature of 90°. After stirring these well together with a wooden rod, a quantity amounting to one pound three ounces of undiluted oil of vitriol is added, the latter being contained in a long cylindrical vessel of brass, sufficiently narrow to enter the generator, into which it is carefully let down by a hook without spilling. The stopcock being now applied to the mouth of the generator, and firmly screwed down upon it, with the intervention of a leaden washer, the generator is turned round upon its supporting pivots, so as completely to invert it; the brass measure within is thus canted over, and the acid which it contained mixed with the solution of soda. The carbonic acid of the salt, which amounts to half its weight, is thus disengaged, and accumulates with great elastic force in the vacant part of the generator. The charge of gas is then transferred to the other large cylinder, which is used as a receiver, by attaching it to the generator by the connecting tube, and after the lapse of five minutes, opening the stopcocks of both. It is advisable to have a woollen case or bag about the receiver, to hold fragments of ice for cooling it. The cylinders may again be separated, after shutting the stopcocks, and the same operations repeated. After two or three charges of gas are conveyed into the receiver, the pressure of the latter becomes sufficient to liquefy the gas; and after five or six charges the receiver may contain several pints of liquid carbonic acid. The receiver being finally detached is set aside, and the liquid it contains preserved for use.

When this highly volatile liquid is allowed to escape into air it evaporates so readily that one portion is instantly resolved into gas, and another portion is cooled so low by the heat thus abstracted as to freeze. From the stopcock of the receiver, a small tube, shown in the figure, descends to near the bottom and dips into the liquid; so that upon opening the former it is the liquid, and not gaseous carbonic acid, which escapes. A nozzle, being applied to the receiver, the stream of liquid is directed into a small cylindrical box of thin copper, with hollow wooden handles, which is soon filled with solid carbonic acid, in the form of a white substance like snow, or more closely resembling anhydrous phosphoric acid, from its opacity and entire want of crystallization.

Solid carbonic acid is a very bad conductor of heat, and may, therefore, be handled without injury, although its temperature is supposed to be so low as —100° C., or —148° Fahr.; and also preserved in the air for hours, if a considerable mass of it in a glass vessel be placed within another similar and larger glass vessel, with any non-conducting material between them. When applied to produce cold, in order to give it contact the solid carbonic acid is mixed with a little ether, with which it unites and forms a soft semifluid mass like half melted snow, capable of abstracting heat and evaporating rapidly, by means of which mercury can be frozen in large quantities, and an alcohol thermometer sunk in the open air so low as —135° (Thilorier). The apparatus of Thilorier forms thus an invaluable cold-producing machine.

Mr. Faraday has since produced a still lower degree of cold by placing a bath of Thilorier's mixture of solid carbonic acid and ether in the receiver of an air-pump, from which the air and gaseous carbonic acid were rapidly removed. The bath consisted of an earthenware dish of the capacity of four cubic inches or more, which was fitted into a similar dish somewhat larger, with three or four folds of dry flannel intervening; with the mixture in the inner dish such a bath lasted for twenty or thirty minutes, retaining solid carbonic acid the whole time. An alcohol thermometer placed in the bath, merely covered with paper, fell to —106°; and in the air-pump receiver, exhausted to within 1·2 inch mercury of a vacuum, the thermometer fell to —166°; or a cold of 60 degrees additional was produced by promoting the evaporation in this manner. At this low temperature the solid carbonic acid mixed with ether, was not more volatile than water at the temperature of 86°, or alcohol at ordinary temperatures.

By combining this extreme cooling power with the effect of mechanical pressure upon gases, several most interesting results were obtained. To produce the pressure, Mr. Faraday employed two condensing syringes, fixed to a table, the first having a piston of an inch in diameter, and the second a piston of only half an inch in diameter; and these were so associated by a connecting pipe, that the first pump forced the gas into and through the valves of the second, and then the second could be employed to throw forward this gas, already condensed to ten or twenty atmospheres, into its final recipient, the condensing tube, at a much higher pressure.

The condensing tubes were of green bottle-glass, being from $\frac{1}{6}$th to $\frac{1}{4}$th of an inch external diameter, and from $\frac{1}{42}$d to $\frac{1}{30}$th of an inch in thickness. They were of two kinds, about nine and eleven inches in length: one, in form of an inverted syphon (fig. 37), could have the bend cooled by immersion into a cold bath, and the other, horizontal (fig. 38), having a curve downward near one end to be cooled in the same manner. Into the longest leg of the syphon tube, and the straight part of the horizontal tube, minute pressure gauges were introduced when required. The caps, stopcocks, and connectors, were attached to the tubes by common cement,[1] and the screw joints made tight by leaden washers.

Fig. 37.

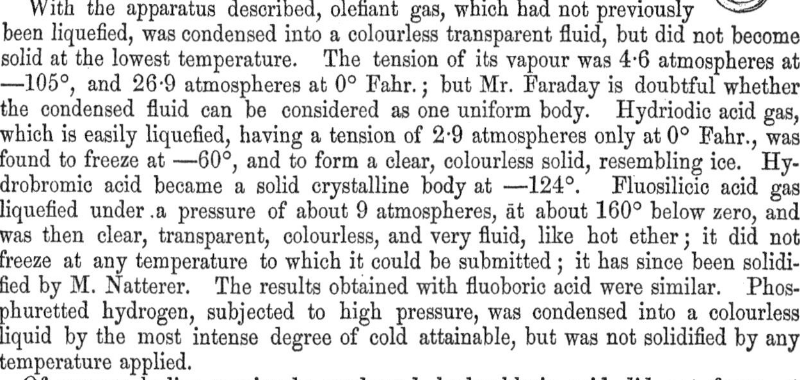

Fig. 38.

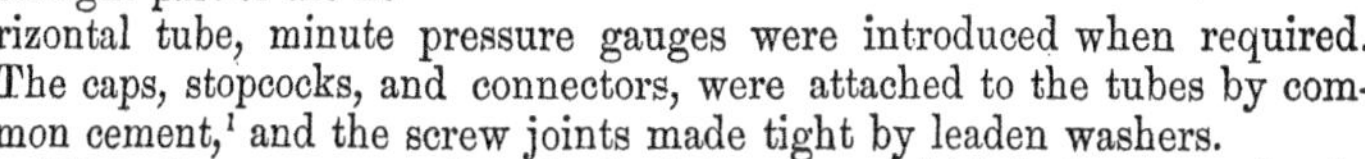

With the apparatus described, olefiant gas, which had not previously been liquefied, was condensed into a colourless transparent fluid, but did not become solid at the lowest temperature. The tension of its vapour was 4·6 atmospheres at —105°, and 26·9 atmospheres at 0° Fahr.; but Mr. Faraday is doubtful whether the condensed fluid can be considered as one uniform body. Hydriodic acid gas, which is easily liquefied, having a tension of 2·9 atmospheres only at 0° Fahr., was found to freeze at —60°, and to form a clear, colourless solid, resembling ice. Hydrobromic acid became a solid crystalline body at —124°. Fluosilicic acid gas liquefied under a pressure of about 9 atmospheres, at about 160° below zero, and was then clear, transparent, colourless, and very fluid, like hot ether; it did not freeze at any temperature to which it could be submitted; it has since been solidified by M. Natterer. The results obtained with fluoboric acid were similar. Phosphuretted hydrogen, subjected to high pressure, was condensed into a colourless liquid by the most intense degree of cold attainable, but was not solidified by any temperature applied.

Of gaseous bodies previously condensed, hydrochloric acid did not freeze at the lowest attainable temperature; the tension of its vapour was 1·8 atmospheres at —100°, 15·04 atmospheres at 0°, 26·20 atmospheres at 32°, and 30·67 atmospheres at 40°. Sulphurous acid became a crystalline, transparent, and colourless solid body at —105°; the pressure of the vapour of liquid sulphurous was 0·726 atmospheres at 0° Fahr., 1·53 atmospheres at 32°, 2 atmospheres at 46°·5, 3 atmospheres at 68°, 4 atmospheres at 85°, 5 atmospheres at 98°, and 6 atmospheres at 110°.

Sulphuretted hydrogen solidified at —122°, forming a white crystalline translucent substance, more like nitrate of ammonia solidified from the melted state, or camphor, than ice. The pressure of the vapour from the solid is not more, probably, than 0·8 of an atmosphere, so that the liquid allowed to evaporate in the air would not solidify as carbonic acid does. The tension of sulphuretted hydrogen vapour was 1·02 atmosphere at —100°, 2 atmospheres at —58°, 6·1 atmospheres at 0°, 9·94 atmospheres at 30°, and 14·6 atmospheres at 52°, which form a progression considerably different from that of water or carbonic acid.

---

[1] Five parts of resin, one part of yellow bees'-wax, and one part of red ochre, by weight, melted together.

Mr. Faraday observed, that when carbonic acid is melted and resolidified by a bath of low temperature, it appears as a clear transparent crystalline colourless body, like ice. It melts at —70° or —72°, and the solid carbonic acid is heavier than the liquid bathing it. The solid or liquid carbonic acid, at this temperature, has a pressure of 5·33 atmospheres. Hence the facility with which liquid carbonic acid, when allowed to escape into air, exerting only a pressure of one atmosphere, freezes a part of itself by the evaporation of another part. The following are the pressures of the vapours of carbonic acid which Mr. Faraday has obtained:—

CARBONIC ACID VAPOUR.

Temp. Fahr.	Tension in Atmospheres.	Temp. Fahr.	Tension in Atmospheres.
—111°	1·14	—15°	17·80
—107	1·36	— 4	21·48
— 95	2·28	0	22·84
— 83	3·60	5	24·75
— 75	4·60	10	26·82
— 56	6·97	15	29·09
— 34	12·50	23	33·15
— 23	15·45	32	38·50

Nitrous oxide was obtained solid, as a beautiful clear crystalline colourless body, by a temperature estimated at about —150°, when the pressure of its vapour was less than one atmosphere. Mr. Faraday believes that liquid nitrous oxide may be used instead of carbonic acid, to produce degrees of cold far below those which the latter body can supply. This idea was verified by M. Natterer, who has liquefied nitrous oxide, and several other gases, by mechanical compression. He found that liquid nitrous oxide may be mixed with sulphuret of carbon in all proportions, and on placing a mixture of these two liquids under the receiver of an air-pump, he saw an alcohol thermometer fall to —140° C., or —220° Fahr.; at this extremely low temperature neither chlorine nor the sulphuret of carbon lost its fluidity. He also succeeded in freezing liquid fluosilicic acid by the same means (Poggendorff's Annalen, t. xii. p. 132: and Liebig's Annalen, t. liv. p. 254). The tension of its vapour was observed by Faraday to be, atmosphere at —125°, 19·34 atmospheres at 0°, and 33·4 atmospheres at 35°.

Liquid cyanogen, when cooled, becomes a transparent crystalline solid, as Bussy and Bunsen had previously observed,[1] which liquefies at —30°. The tension of its vapour was 1·25 atmospheres at 0°, 2·37 atmospheres at 32°, and 6·9 atmospheres at 63°.

Ammonia formed a white, translucent, crystalline solid, melting at —103°. The density of the liquid was 0·731 at 60°; its tension 2·48 atmospheres at 0°, 4·44 atmospheres at 32°, and 6.9 atmospheres at 60°.

Arsenietted hydrogen, which was liquefied by Dumas and Soubeiran, did not solidify at —166°. The tension of its vapour was 0·94 atmospheres at —75°, 5·21 atmospheres at 0°, 8·95 atmospheres at 32°, and 13·19 atmospheres at 60°.

The following gases showed no signs of liquefaction when cooled by the carbonic acid bath in vacuo, at the pressure expressed:—

	Atmospheres.
Hydrogen at	27
Oxygen	58·5
Nitrogen	50
Nitric oxide	50
Carbonic oxide	40
Coal gas	32

Several gases were submitted by M. G. Aimé to still higher pressures, rising for nitrogen and hydrogen gases to 220 atmospheres, by immersion in the depths of the

[1] For Bunsen's results on the liquefaction of several of the gases, see Bibliothèque Universelle, 1839, t. xxxii. p. 185.

sea, where the results under pressure could not be observed (Annales de Chimie, &c. 1843, 3d ser. t. viii. p. 275). Most of them were diminished in bulk in a ratio greatly exceeding the pressure; but this has been shown to be often the case whilst the substance retains the gaseous form. No sufficient evidence of the liquefaction of any of the gases just enumerated has yet been produced. The same may be said of light carburetted hydrogen. At the lowest temperatures attainable, alcohol, ether, sulphuret of carbon, chloride of phosphorus, and chlorine, also retained the liquid form.

Sir H. Davy threw out the idea that the prodigious elastic force of the liquid gases might be used as a moving power. But supposing the application practicable, it may be doubted, from what we know of the constancy of the united sum of the latent and sensible heat of high pressure steam, whether any saving of heat would be effected by such an application of the vapours of these fluids.

All gases whatever are absorbed and condensed by water in a greater or less degree, in which case they certainly assume the liquid form. The quantity condensed is widely different in the different gases; and in the same gas the quantity condensed depends upon the pressure to which the gas is subject, and the temperature of the absorbing water. Dr. Henry proved that with carbonic acid gas the *volume* absorbed by water is the same, whatever be the pressure to which the gas is subject. Hence, we double the weight or quantity of gas absorbed, by subjecting it, in contact with water, to the pressure of two atmospheres; and this practice is adopted in impregnating water with carbonic acid, to make soda-water. The colder the water, the greater also the quantity of gas absorbed.

In the physical theory of gases, they are assumed to be expansible to an indefinite extent, in the proportion that pressure upon them is diminished, and to be contractible under increased pressure exactly in proportion to the compressing force — the well-known law of Mariotte. The bulk of atmospheric air has been found rigidly to correspond with this law, when it was expanded to 300 volumes, and also when compressed into 1-25th of its primary volume. But there is reason to doubt whether the law holds with absolute accuracy, in the case of a gas either in a state of extreme rarefaction, or of the greatest density. Thus atmospheric air does not appear to be indefinitely expansible, as the law of Mariotte would require; for there is certainly a limit to the earth's gaseous atmosphere, and it does not expand into all space. Dr. Wollaston supposed that the material particles of air are not indefinitely minute, but have a certain magnitude and weight. These particles are under the influence of a powerful mutual repulsion, as is always the case in gaseous bodies, and, therefore, tend to separate from each other; but as this repulsive force diminishes as the distance of the particles from each other increases, Dr. Wollaston imagined that the weight of the individual particles might come at last to balance it, and thus prevent their further divergence. On this view, which is probable on other grounds, the expansion of a gas, caused by the removal of pressure, will cease at a particular point of rarefaction, and the gas not expanding farther, will come to have an upper surface, like a liquid. The earth's atmosphere has probably an exact limit, and true surface.

The deviation from the law of Mariotte, in gases under a greater pressure than that of the atmosphere, has been distinctly observed in the more liquefiable gases. Thus, Professor Oersted, of Copenhagen, found that sulphurous acid gas diminishes, under increased pressure, more rapidly than common air. The volumes of atmospheric air and of the gas were equal at the following pressures:—

Pressure upon air in atmospheres.	Pressure upon sulphurous gas in atmospheres.
1	1
1.175	1.173
2.821	2.782
3.319	3.189

It will be observed that less pressure always suffices to reduce the sulphurous acid gas to the same bulk than is required by air. If the pressure upon the air and gas were made equal, then the gas would be compressed into less bulk than the air, and deviate from the law of Mariotte. Despretz observed an equally conspicuous deviation from this law under increasing pressures, in several other gases, particularly sulphuretted hydrogen, cyanogen, and ammonia, which are all easily liquefied. There is no reason, however, to suppose that any partial liquefaction of the gases occurs under the pressure applied to them in such experiments. They remain entirely gaseous, and their superior compressibility must be referred to a law of their constitution. It is the phenomenon beginning to show itself in a gas under moderate pressure, which was observed in all its excess by Cagnard de la Tour, in the vapours confined by him under great pressure (page 68).

Those gases which exhibit this deviation must occupy less bulk than they ought to do under the pressure of the atmosphere itself; which may be the reason why the liquefiable gases are generally found by experiment specifically heavier than they ought by theory to be.

M. Regnault accordingly finds, that at the temperature of 32°, and under more feeble pressures than that of the atmosphere, carbonic acid deviates from the law of Mariotte in a marked manner; while it appears to follow that law when heated to 212° under more feeble pressures than that of the atmosphere.

The density of carbonic acid at 33° (air = 1000) was :—

Under the pressure of 760 millimeters (30 inches).. 1529.10
374.13 .... 1523.66
224.17 .... 1521.45

The density of the gas at 212° (that of air at the same temperature being 1000) was :—

Under the pressure of 760 millimeters (30 inches).. 1524.18
338.39 .... 1524.10

The theoretical density of carbonic acid, calculated in a manner which shall be afterwards explained, and taking for the atomic weight of carbon the number 6, is 1520.24; to which the numbers for the density of the gas under greatly reduced pressures appear to be converging. M. Regnault verified at the same time the exactness of the law of Mariotte for atmospheric air. (Annales de Ch., xiv. 227, and 234).

Such are the most remarkable features which gases exhibit in relation to pressure and temperature. These properties are independent of the specific weights of the gases, which are very different in the various members of the class, and they are but little connected with the nature of the particular substance or material which exists in the gaseous form. But when gases differing in composition are presented to each other, a new property of the gaseous state is developed, namely, the forcible disposition of dissimilar gases to intermix, or to diffuse themselves through each other. This is a property which interferes in a great variety of phenomena, and is no less characteristic of the gaseous state than any we have considered. It appears in the spontaneous diffusion of gases through each other, and in the diffusion of vapours into gases, or the ascent of vapours from volatile bodies into air and other gases, of which the spontaneous evaporation of water into the air is an instance. Related closely to this subject, and preliminary to its consideration, is the passage of different gases into a vacuum, through a small aperture, which takes place with different degrees of facility; with their rates of transmission by capillary tubes. The whole may be briefly treated under the heads of, (1) Effusion of gases (their pouring out), by which I express their passage into a vacuum by a small aperture in a thin plate; (2) Transpiration of gases, or their passage through tubes of fine bore of greater or less length; (3) The diffusion of gases; and (4) Evaporation in air.

EFFUSION OF GASES.

The specific weights, or weights of an equal measure, of the different gases vary exceedingly. The numbers representing these weights are always referred to the weight of a gas, generally air, as 1 or 1000, instead of water, which is the standard comparison for liquids and solids. The operation of taking the specific gravity of a gas is simple in principle, but the accurate execution of it is attended with great practical difficulties. A light glass globe *g* (fig. 39) from 50 to 100 cubic inches in capacity, is weighed full of air, then exhausted by an air-pump and weighed empty, the loss being taken as the weight of its volume of air. It is then, in its exhausted state, united with a bell-jar *c*, containing the gas to be weighed and standing over a mercurial trough, by a union screw between the stopcocks *d* and *e* of the two vessels; and filled with the gas, which rushes from the jar to the vacuous globe on opening both stopcocks. A supply of gas is conveyed to the jar by the bent tube *b*, after being deprived of moisture by passing through a drying tube *a*, containing fragments of chloride of calcium. The globe is again weighed when full of gas of the atmospheric pressure and temperature, and the weight of a volume of the gas obtained by deducting the weight of the vacuous globe. The specific gravity is then calculated by the proportion, as the weight of air first found, to the weight of gas, so 1.000 (density of air), to a number which expresses the density of the gas required. MM. Dumas and Boussingault, in their late careful observations of the density of oxygen, nitrogen, and hydrogen, employed a capacious glass globe, of which the cubic contents were first ascertained by measuring in an accurate manner the volume of water required to fill it (Annales de Chimie, 3d sér. viii. 201). In the refined experiments of M. Regnault, lately published, a light glass balloon of about ten litres or 616 cubic inches in capacity, was employed as the weighing globe. It was counterpoised, when weighed, by a similar globe formed of the same glass; by which arrangement numerous and somewhat uncertain corrections for variations in the density, temperature, and hygrometric state of the air, during the continuance of an experiment, the film of moisture which adheres to glass, and the displacement of air by the solid materials of the balloon, were entirely avoided. (Ibid., 1845, 3d sér. t. iv. 211).

FIG. 39.

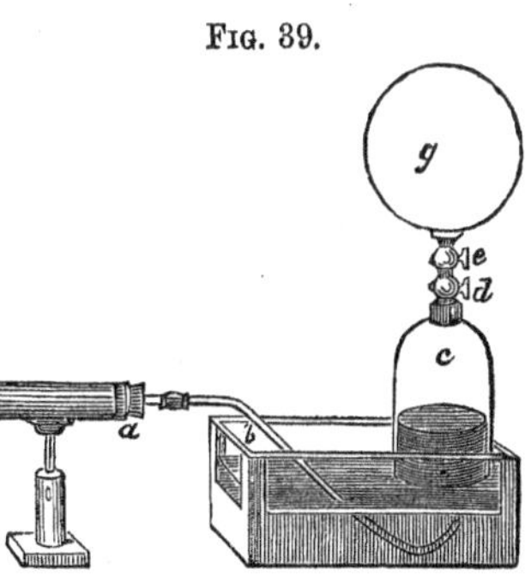

The following tables exhibit the specific gravity of those gases to which reference will most frequently be made, air being taken as the standard of comparison in the first table, and oxygen in the second. To each specific gravity is added, in a second column, the square root of the number, and in a third column 1 divided by the square root, or the reciprocal of the square root.

TABLE I. DENSITY OF GASES, AIR = 1.

	DENSITY.	SQUARE ROOT OF DENSITY.	1 / SQUARE ROOT.	AUTHORITY.
Nitrogen	0·97137	0·9856	1·0147	Regnault.
Oxygen	1·10563	1·0515	0·9510	"
Hydrogen	0·06926	0·2632	3·7994	"
Carbonic acid	1·52901	1·2365	0·8087	"
Carbonic oxide	0·9712	1·9855	1·0147	Calculated.
Light carburetted hydrogen	0·5549	0·7449	1·3424	"
Olefiant gas	0·9712	0·9855	1·0147	"
Nitrous oxide	1·5261	1·2353	0·8095	"
Nitric oxide	1·0405	1·0205	0·9799	"
Sulphuretted hydrogen	1·1793	1·0860	0·9208	"
Chlorine	2·4573	1·5676	0·6379	"

TABLE II. DENSITY OF GASES, OXYGEN = 1.

GASES.	DENSITY.	SQUARE ROOT OF DENSITY.	1 / SQUARE ROOT.	AUTHORITY.
Air	0·9038	0·9507	1·0518	Regnault.
Nitrogen	0·8785	0·9373	1·0669	"
Hydrogen	0·6626	0·2502	3·9968	"
Carbonic acid	1·3830	1·1760	0·8503	"
Carbonic oxide	0·8750	0·9354	1·0691	Calculated.
Light carburetted hydrogen CH^2	0·5000	0·7071	1.4142	"
Olefiant gas	0·8750	0·9354	1·0691	"
Nitrous oxide	1·3750	1·1705	0·8545	"
Nitric oxide	0·9375	0·9682	1·0328	"
Sulphuretted hydrogen	1·0625	1·0308	8·9701	"
Chlorine	2·2129	1·4876	0·6722	"

A jar on the plate of an air-pump is kept vacuous by continued exhaustion, and a measured quantity of air, or any other gas, allowed to find its way into the vacuous jar through a minute aperture in a thin metallic plate, such as platinum foil, made by a fine steel point, and not more than 1-300dth of an inch in diameter. With an imperfect exhaustion, it is found that the velocity with which the gas flows into the jar rapidly increases till the aspiration power or degree of exhaustion amounts to about one-third of an atmosphere. Higher degrees of exhaustion do not produce a corresponding increase of velocity, and the difference of an inch of the mercurial column of the gauge barometer scarcely affects the rate at which the gas enters, when the vacuum is nearly complete, and the pressure to which the gas is subject approaches that of a whole atmosphere. By a perforated plate such as described, 60 cubic inches of dry air entered the vacuous, or nearly vacuous air-pump receiver, in about 1000 seconds, and in successive experiments the time of passage did not vary more than one or two seconds.

The time of passage into a vacuum of a constant volume varied in the different gases, the lightest passing in the shortest time. The time corresponded very closely for each gas with the square root of its density. Thus the square root of the density of oxygen being 1·0515, and that of air 1, (Table I.), the time of passage of the constant volume of oxygen was observed to be 1·0519, 1·0519, 1·0506, 1·0502, in experiments made on different occasions, the time of passage of the same volume of air being 1. Compared with the time of the passage of a constant volume of oxygen taken as 1, the time of hydrogen was 0·2631, instead of 0·25 (Table II.); the time of nitrogen was 0·9365 and 0·9345, instead of 0·9373; the time of carbonic oxide, of which the theoretical density is the same as the last gas, was 0·9345, instead of 0.9354; of carburetted hydrogen 0·7023, instead of 0·7071; of carbonic acid 1·1675, instead of 1·1705. The time of nitrous oxide was always the same, as nearly as could be observed, as that of carbonic acid; while these two gases have the same specific gravity. For gases which do not differ greatly from air in specific gravity, the times correspond so closely with the law, that the densities of these gases, it appears, might be deduced as accurately from an effusion experiment as by actually weighing them. The sensible deviation from the law in the times of both the very light and very heavy gases can be shown to be occasioned by the tubularity of the aperture arising from the unavoidable thickness of the metallic plate.

The times of passage into a vacuum of equal volumes of different gases varying, then, as the square root of their densities, the velocities of passage will consequently be in the inverse proportion, or as 1 divided by the square root of the gas. This is the physical law of the passage of fluids generally under pressure, which has been long established for liquids of different densities by observation, but had not previously received an experimental verification in the case of gases.

Mixtures of nitrogen and oxygen in different proportions were found to have the mean rāte of their constituent gases. This is also true of mixtures of carbonic acid, nitrous oxide, and carbonic oxide, with each other or with the preceding gases. But hydrogen and carburetted hydrogen lose more or less of their peculiar rate, and pass slower, when mixed with other gases. Thus the time of passage of a mixture of equal volumes of oxygen and hydrogen is 0·7255; instead of 0·6315, the mean of the times, 1 and 0·2631, of those gases individually. Supposing the rate of the oxygen in the mixture to remain unchanged, and that the alteration takes place on the hydrogen exclusively, then the time of passage of the hydrogen has increased from 0·2631 to 0·4510, or been nearly doubled. But it is in mixtures where the proportion of hydrogen is large compared with that of the other gas, that the departure from the mean velocity is most conspicuous. Thus the addition of half a per cent. of air or oxygen has an effect in retarding the passage of hydrogen at least three times greater than what it should produce from its greater density by calculation. The time of the effusion of hydrogen thus becomes a delicate test of the purity of that gas. This want of mechanical equivalency in hydrogen mixtures is exceedingly remarkable, being a marked departure from the usual uniformity of gaseous properties.

TRANSPIRATION OF GASES.

The arrangement exhibited (fig. 40), was adopted in examining the rates of passage of different gases into a vacuum through a capillary tube. The gas is taken from a

Fig. 40.

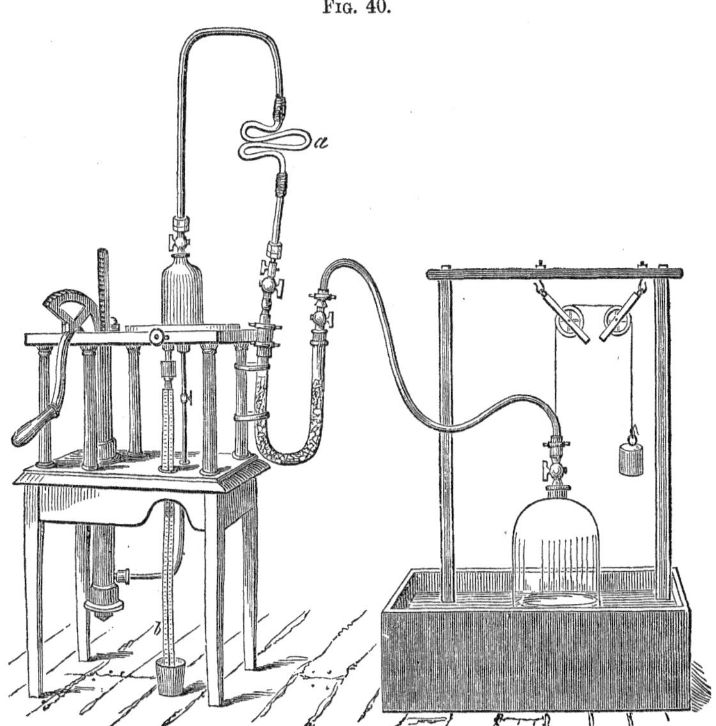

counterpoised bell-jar, standing over the water of a pneumatic trough, and passes first by a flexible tube to a U-shaped drying tube filled with fragments of chloride of calcium, in order to be deprived of aqueous vapour before entering the capillary glass tube *a*. The last is connected by means of a tube of block tin with a receiver on the plate of an air-pump, provided with a gauge barometer *b*, as represented. Gas is allowed to enter the exhausted receiver by the capillary tube, and the time observed which the gauge barometer requires to fall a certain number of inches from the admission of a constant volume.

It is found that for a tube of any given diameter, the times of passage of different gases approximate the more closely to their respective times of effusion, the more the tube is shortened and made to approximate to an aperture in a thin plate. While, as the tube is elongated, a deviation from those rates is observed, which is rapid with the first additions in length, but becomes gradually less; and, finally, with a certain length of tube, the gases attain rates of which the relation remains constant, or nearly so, for any farther increase of length. The same relation in velocity between the different gases is then found to extend also through a considerable range of pressure, as from one to one-tenth of an atmosphere.

The ultimate rates of transpiration differ considerably from the rates of effusion of the same gases, and have no uniform relation to their density. Of all the gases tried, oxygen passes with least velocity through a capillary tube. The time of passage into a vacuum, under the atmospheric pressure, of a volume of oxygen being 1, that of air was 0.9010, of nitrogen 0.8704, and carbonic oxide 0.8671. The transpiration times of these gases approach so closely to their specific gravities, as will be seen by Table II., as to lead to the inference that the transpiration times are directly as the density for these gases. Nitric oxide appears to coincide in transpiration time with nitrogen, although denser, the specific gravity of the former being the mean between the densities of the nitrogen and oxygen. The transpiration time of carbonic acid approached very closely to 0.75, or three-fourths of that of oxygen. Nitric oxide, which has the same specific gravity as carbonic acid, coincides perfectly with that gas also in time of transpiration. The densities of these two gases are to that of oxygen as 22 to 16, but their times of transpiration are to the time of transpiration of oxygen, as 12 to 16.

The transpiration time of hydrogen, by several capillary tubes, varied but very little from 0.44, the time of oxygen being 1. The number for hydrogen therefore approaches 0.4375, which is 7-16ths of the oxygen time. The time of light carburetted hydrogen was also remarkably constant at 0.550 to 0.555; which approach, although not very closely, to 0.5625, or 9-16ths of the oxygen time. Olefiant gas has probably sensibly the same specific gravity as nitrogen and carbonic oxide, but it is much more transpirable than these gases; the transpiration time of olefiant gas being found so low as 0.512. This result is not inconsistent with the true number for olefiant gas, being 0.5, or one-half the time of oxygen; for the gas operated upon was found always to contain either a trace of a heavy hydrocarbon, or a few per cent. of carbonic oxide, both of which increase the time of transpiration. Hydrogen with five per cent. of air was less rapidly transpired than olefiant gas, the time of that mixture being 0.5237.

The transpiration time of mixtures of the following gases was exactly the mean of the times of the mixed gases, namely, oxygen, nitrogen, hydrogen, carbonic oxide, nitrous oxide, and carbonic acid; but the transpiration time of hydrogen and carburetted hydrogen, particularly the former, is greatly increased when these gases are in a state of mixture with each other, or with gases of the former class. Thus the transpiration time of a mixture of equal volumes of oxygen and hydrogen was 0.9008, instead of 0.72, the mean time of the two gases. The transpiration time of hydrogen in such a mixture is as high as 0.8016; or, its transpiration is then less rapid than that of pure carbonic acid.

The effusion of a given measure of air into a vacuum takes place always in the same time, whatever may be its density, from one-fourth of an atmosphere up to two atmospheres. But the transpiration of air of different densities was observed to take place in times which are inversely as the densities; or, the denser air is, the more rapidly is a given volume of it transpired. Hence the transpiration of air and all gases is greatly affected by variations of the barometer; the higher the barometer the more quickly are the gases transpired. The difference in this respect separates completely the phenomena of effusion and transpiration. Nor can the phenomena of transpiration be an effect of friction, for the greater the density of air, the more should its passage be resisted by friction. The transpirability of a gas appears to be a constitutional property, like its density, or its combining volume; and the investigation is of peculiar interest from supplying a new class of constants for the gases, namely, their coefficients of transpiration. The rates of transpiration of different gases were further observed to be the same through a fine capillary tube of copper of eleven feet in length, and a mass of dry stucco, as through capillary tubes of glass.

DIFFUSION OF GASES.

When a light and heavy gas are once mixed together, they do not exhibit any tendency to separate again, on standing at rest; differing in this respect from mixed liquids, many of which speedily separate, and arrange themselves according to their densities, the lightest uppermost, and the heaviest undermost — as in the familiar example of oil and water, unless they have combined together. This peculiar property of gases has repeatedly been made the subject of careful experiment. Common air, for instance, is essentially a mixture of two gases, differing in weight in the proportion of 971 to 1105; but the air in a tall close tube of glass several feet in length, kept upright in a still place, has been found sensibly the same in composition at the top and bottom of the tube, after a lapse of months. Hence, there is no reason to imagine that the upper strata of the air differ in composition from the lower; or that a light gas, such as hydrogen, escaping into the atmosphere, will rise, and ultimately possess the higher regions; — suppositions which have been the groundwork of meteorological theories at different times.

The earliest observations we possess on this subject are those of Dr. Priestley, to whom pneumatic chemistry stands so much indebted. Having repeated occasion to transmit a gas through stoneware tubes surrounded by burning fuel, he perceived that the tubes were porous, and that the gas escaped outwards into the fire; while at the same time the gases of the fire penetrated into the tube, although the gas within the tube was in a compressed state.

Fig. 41.

h

c

Dr. Dalton, however, first perceived the important bearings of this property of aërial bodies, and made it the subject of experimental inquiry. He discovered that any two gases, allowed to communicate with each other, exhibit a positive tendency to mix or to penetrate through each other, even in opposition to the influence of their weight. Thus, a vessel *h*, containing a light gas (hydrogen), being placed above a vessel *c*, containing a heavy gas (carbonic acid), and the two gases allowed to communicate by a narrow tube, as represented (fig. 41), an interchange speedily took place of a portion of their contents, which it might be supposed that their relative position would have prevented. Contrary to the solicitation of gravity, the heavy gas continued spontaneously to ascend, and the light gas to descend, till in a few hours they became perfectly mixed, and the proportion of the two gases was the same in the upper and lower vessels. This disposition of different gases to intermix, appeared to Dr. Dalton so decided and strong, as to justify the inference that different gases afforded no resistance to each other; but that one gas spreads or expands into the space occupied by another gas, as it would rush into a vacuum. At least, that the resistance which the particles of one gas offer to those of another is of a very imperfect kind, to be compared to the resistance which stones in the channel of a stream oppose to the flow of running water. Such is Dalton's theory of the miscibility of the gases. (Manchester Memoirs, Vol. V.)

In entering upon this inquiry, I found, first, that gases diffuse into the atmosphere, and into each other, with different degrees of ease and rapidity. This was observed by allowing each gas to diffuse from a bottle into the air through a narrow tube, taking care, when the gas was lighter than air, that it was allowed to escape from the lower part of the vessel, and when heavier from the upper part, so that it had, on no occasion, any disposition to flow out, but was constrained to diffuse in opposition to the effect of gravity. The result was, that the same volume of different gases escapes in times which are exceedingly unequal, but have a relation to the specific gravity of the gas. The light gases diffuse or escape most rapidly: thus, hydrogen escapes five times quicker than carbonic acid, which is twenty-two times heavier than that gas. Secondly, in an intimate mixture of two gases, the most diffusive gas separates from the other, and leaves the receiver in the greatest proportion. Hence, by availing ourselves of the tendencies of mixed gases to diffuse with different degrees of rapidity, a sort of mechanical separation of gases may be effected. The mixture must be allowed to diffuse for a certain time into a confined gaseous or vaporous atmosphere, of such a kind as may afterwards be condensed or absorbed with facility. (Quarterly Journal of Science, New Series, Vol. V.)

But the nature of the process of diffusion is best illustrated when the gases communicate with each other through minute pores or apertures of insensible magnitude.

A singular observation belonging to this subject was made by Professor Dobereiner, of Jena, on the escape of hydrogen gas by a fissure or crack in glass receivers. Having occasion to collect large quantities of that light gas, he had accidentally made use of a jar which had a slight fissure in it. He was surprised to find that the water of the pneumatic trough rose into this jar one and a half inches in twelve hours; and that after twenty-four hours the height of the water was two inches two-thirds above the level of that in the trough. During the experiment, neither the height of the barometer nor the temperature of the place had sensibly altered. (Annales de Chimie et de Physique, 1825.) He ascribed the phenomenon to capillary action, and supposed that hydrogen only is attracted by the fissures, and escapes through them on account of the extreme smallness of its atoms. It is unnecessary to examine this explanation, as Dobereiner did not observe the whole phenomenon. On repeating the experiment, and varying the circumstances, it appeared to me that hydrogen never escapes outwards by the fissure without a certain portion of air penetrating at the same time inwards, amounting to between one-fourth and one-fifth of the volume of the hydrogen which leaves the receiver. It was found by an instrument which admits of much greater precision than the fissured jar, that when hydrogen gas communicates with air through such a chink, the air and hydrogen exhibit a powerful disposition to exchange places with each other; a particle of air, however, does not exchange with a particle of hydrogen of the same magnitude, but of 3.83 times its magnitude. We may adopt the word *diffusion-volume*, to express this diversity of disposition in gases to interchange particles, and say that the diffusion-volume of air being 1, that of hydrogen gas is 3.83. Now every gas has a diffusion-volume peculiar to itself, and depending upon its specific gravity. Of those gases which are lighter than air, the diffusion-volume is greater than 1, and of those which are heavier, the diffusion-volume is less than 1. The diffusion volumes are, indeed, inversely as the square root of the densities of the gases. Hence the times of the effusion and diffusion of gases follow the same law.

Exact results are obtained by means of a simple instrument, which may be called a diffusion tube, and which is constructed as follows. A glass tube, open at both ends, is selected, half an inch in diameter, and from six to fourteen inches in length. A cylinder of wood, somewhat less in diameter, is introduced into the tube, so as to occupy the whole of it, with the exception of about one-fifth of an inch at one extremity, which space is filled with a paste of Paris plaster, of the usual consistence for casts. In the course of a few minutes the plaster sets, and on withdrawing the wooden cylinder the tube forms a receiver, closed by an immoveable plate of stucco. In the wet state, the stucco is air-tight; it is therefore dried, either by exposure to the air for a day, or by placing it in a temperature of 200° for a few hours; and is thereafter found to be permeable by gases, even in the most humid atmosphere, if not positively wetted. When such a diffusion-tube, six inches in length, is filled with hydrogen over mercury, the diffusion, or exchange of air for hydrogen, instantly commences through the minute pores of the stucco, and proceeds with so much force and velocity, that within three minutes the mercury attains a height in the receiver of more than two inches above its level in the trough; within twenty minutes, the whole of the hydrogen has escaped. In conducting such experiments over water, it is necessary to avoid wetting the stucco. With this view, before filling the diffusion-tube with hydrogen, the air is withdrawn by placing the tube upon the short limb of an empty syphon (see figure 42), which does not reach, but comes within half an inch of the stucco, and then sinking the instrument in the water trough, so that the air escapes by the syphon, with the exception of a small quantity, which is noted. The diffusion tube is then filled up, either entirely or to a certain extent, with the gas to be diffused.

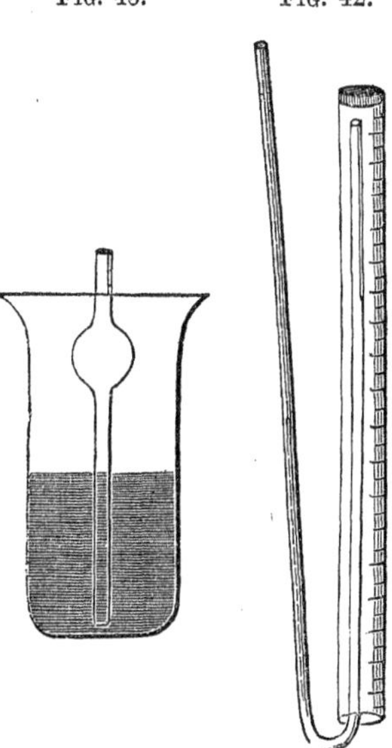
FIG. 43. FIG. 42.

The ascent of the water in the tube, when hydrogen is diffused, forms a striking experiment. But in experiments made with the purpose of determining the proportion between the gas diffused and the air which replaces it, it is necessary to guard against any inequality of pressure, by placing the diffusion tube in a jar of water as in figure 43, and filling the jar with water in proportion as it rises in the tube.

In this instrument we may substitute many other porous substances for the stucco; but few of them answer so well. Dry and sound cork is very suitable, but permits the diffusion to go on very slowly, not being sufficiently porous; so do thin slips of many granular foliated minerals, such as flexible magnesian limestone. Charcoal, woods, unglazed earthenware, dry bladder, may all be used for the same purpose.

It can be shown, on the principles of pneumatics, that gases should rush into a vacuum with velocities corresponding to the numbers which have been found to express their diffusion volumes; that is, with velocities inversely proportional to the square root of the densities of the gases. The law of the diffusion of gasses has on this account been viewed by my friend, Mr. T. S. Thomson, of Clitheroe, as a confirmation of Dr. Dalton's theory, that gases are inelastic towards each other (L. Ed. and D., Phil. Mag. 3d series, iv. 321). It must be admitted that the ultimate result in diffusion is in strict accordance with Dalton's law, but there are certain circumstances which make me hesitate in adopting it as a true representation of the phenomenon, although it affords a covenient mode of expressing it. 1. It is supposed, on that law, that when a cubic foot of hydrogen gas is allowed to communicate with a cubic foot of air, the hydrogen *expands* into the space occupied by the air, as it would do into a vacuum, and becomes two cubic feet of hydrogen of half density. The air, on the other hand, expands in the same manner into the space occupied by the hydrogen, so as to become two cubic feet of air of half density. Now if the gases actually expanded through each other in this manner, cold should be produced, and the temperature of the mixed gases should fall 40 or 45 degrees. But not the slightest change of temperature occurs in diffusion, however rapidly the process is conducted. 2. Although the ultimate result of diffusion is always in conformity with Dalton's law, yet the diffusive process takes place in different gases with very different degrees of rapidity. Thus, the external air penetrates into a diffusion tube with velocities denoted by the following numbers, 1277, 623, 302, according as the diffusion tube is filled with hydrogen, with carbonic acid, or with chlorine gas. Now, if the air were rushing into a vacuum in all these cases, why should it not always enter it with the same velocity? Something more, therefore, must be assumed than that gases are vacua to each other, in order to explain the whole phenomena observed in diffusion.

Passage of gases through membranes.—In connexion with diffusion, the passage of gases through humid membranes may be noticed. If a bladder, half filled with air, with its mouth tied, be passed up into a large jar filled with carbonic acid gas, standing over water, the bladder, in the course of twenty-four hours, becomes greatly distended, by the insinuation of the carbonic acid through its substance, and may even burst, while a very little air escapes outwards from the bladder. But this is not simple diffusion. The result depends upon two circumstances: first, upon carbonic acid being a gas easily liquefied by the water in the substance of the membrane,—the carbonic acid penetrates the membrane as a liquid; secondly, this liquid is in the highest degree volatile, and, therefore, evaporates very rapidly from the inner surface of the bladder into the air confined in it. The air in the bladder comes to be expanded in the same manner as if ether or any other volatile fluid was admitted into it. The phenomenon was observed by Dalton in its simplest form. Into a very narrow jar, half filled with carbonic acid gas over water, he admitted a little air. The air and gas were accidentally separated by a water-bubble, and thus prevented from intermixing. But the carbonic gas immediately began to be liquified by the film of water, and passing through it, evaporated into the air below. The air was in this way gradually expanded, and the water-bubble ascended in the tube. Here the particular phenomenon in question was observed to take place, but without the intervention of membrane. It is to be remembered that the thinnest film of water or any liquid is absolutely impermeable to a gas as such.

In the experiments of Drs. Mitchell and Faust, and others, in which gases passed through a sheet of caoutchouc, it is to be supposed that the gases were always liquefied in that substance, and penetrated through it in a fluid form. Indeed, few bodies are more remarkable than caoutchouc for the avidity with which they imbibe various liquids. The absorption of ether, of naphtha, of oil of turpentine, softening the substance of the caoutchouc, without dissolving it, may be referred to. It is likewise always those gases which are more easily liquified by cold or pressure that pass most readily through both caoutchouc and humid membranes. Dr. Mitchell found that the time required for the passage of equal volumes of different gases through the same membrane, was

1	minute, with	ammonia.
2½	minutes, with	sulphuretted hydrogen.
3¼	"	cyanogen.
5½	"	carbonic acid.
6½	"	nitrous oxide.
27½	"	arsenietted hydrogen.
28	"	olefiant gas.
37½	"	hydrogen.
113	"	oxygen.
160	"	carbonic oxide.

and a much greater time with nitrogen.

DIFFUSION OF VAPOURS INTO AIR, OR SPONTANEOUS EVAPORATION.

Volatile bodies, such as water, rise into air as well as into a vacuum, and obviously according to the law by which gases diffuse through each other. Thus, if a small quantity of the volatile liquid ether be conveyed into two tall jars standing over water, one half filled with air, and the other with hydrogen gas, the air and hydrogen immediately begin to expand, from the ascent of the ether-vapour into them, and the two gases in the end have their volume increased exactly in the same proportion. But the hydrogen gas undergoes this expansion in half the time that the air requires;

that is to say, ether-vapour follows the usual law of diffusion in penetrating more rapidly through the lighter gas.

We are indebted to Dr. Dalton for the discovery that the evaporation of water has the same limit in air as in a vacuum. Indeed, the quantity of vapour from a volatile body which can rise into a confined space, is the same, whether that space be a vacuum, or be already filled with air or gas, in any state of rarefaction or condensation. The vapour rises, and adds its own elastic force, such as it exhibits in a vacuum, to the elastic force of the other gases or vapours already occupying the same space. Hence, it is only necessary to know what quantity of any vapour rises into a vacuum at any particular temperature; — the same quantity rises into air. Thus the vapour from water, which rises into a vacuum at 80°, depresses the mercurial column one inch, or its tension is one-thirtieth of the usual tension of the air. Now, if water at 80° be admitted into dry air, it will increase the tension of that air by 1-30th, if the air be confined; or increase its bulk by 1-30th, if the air be allowed to expand. M. Regnault has, indeed, observed that the tension of the vapour of water in air, and in pure nitrogen gas, is always a little more feeble (2 or 3 per cent.) than in a vacuum for the same temperature, (Annales de Ch. et Ph., xv. 137); from which may be inferred the existence of some physical obstacle to the full diffusion of vapours, of which the nature is at present unknown. The density of the vapour of water in air saturated with it may also be taken as the same as it has been found in a vacuum, or 622 (air = 1000), M. Regnault having observed it to deviate not more than one-hundredth part from that density, at all temperatures between 32° and 72° Fahr. — (Ibid., p. 160).

The spontaneous evaporation of water into air is much affected by three circumstances:—1. The previous state of dryness of the air — for a certain fixed quantity only of vapour can rise into air, as much as into the same space if vacuous; and if a portion of that quantity be already present, so much the less will be taken up by the air; and no evaporation whatever takes place into air which contains this fixed quantity, and is already saturated with humidity. 2. By warmth — for the higher the temperature the more considerable is the quantity of vapour which rises into any accessible space. Thus water emits so much vapour at 40° as expands the air in contact with it 1-114th part, and at 60° as much as expands air 1-57th part, or double the quantity emitted at the lower temperature. Hence, humid hot air contains a much greater portion of moisture than humid cold air. 4. The evaporation of water is greatly quickened by the removal of the incumbent air in proportion as it becomes saturated; and hence a current of air is exceedingly favourable to evaporation.

When air saturated with humidity at a high temperature is cooled, it ceases to be able to sustain the large portion of vapour which it possesses, and the excess assumes the liquid form, and precipitates in drops. Many familiar appearances depend upon the condensation of the vapour in the atmosphere. When a glass of cold water, for instance, is brought into a warm room, it is often quickly covered with moisture. The air in contact with the glass is chilled, and its power to retain vapour so much reduced as to occasion it to deposit a portion upon the cold glass. It is from the same cause that water is often seen in the morning running down in streams upon the inside of the glass panes of bed-room windows. The glass has the low temperature of the external air, and by contact cools the warm and humid air of the apartment so as to occasion the precipitation of its moisture. Hence also, when a warm thaw follows after frost, thick stone walls which continue to retain their low temperature are covered by a profusion of moisture.

Hygrometers. — As water evaporates at all temperatures, however low, the atmosphere cannot be supposed to be ever entirely destitute of moisture. The proportion present varies with the temperature, the direction of the wind, and other circumstances, but is generally greater in summer than in winter. There are various means by which the moisture in the air may be indicated, and its quantity estimated, affording principles for the construction of different hygroscopes or hygrometers.

1. The chemical method consists in passing a known measure of air over a highly hygrometric substance, such as chloride of calcium, contained in a glass tube, which has been weighed; the increase of weight is that of the vapour absorbed. The experiment admits of being made with rigorous accuracy, but is seldom had recourse to, except to check other methods which are more expeditious, but less certain.[1]

2. Many solid substances swell on imbibing moisture, and contract again on drying: such as wood, parchment, hair, and most dry organic substances. The hygrometer of Deluc consisted of an extremely thin piece of whalebone, which in expanding and contracting moved an index. The principle of this instrument is illustrated in the transparent shavings of whalebone cut into figures, which bend and crumple up when laid upon the warm hand. Saussure made use of human hair boiled in a solution of carbonate of soda, as a hygrometric body, and it appears to answer better than any other substance of the class. Regnault does not make any essential change in the construction of Saussure, but prefers to deprive the hairs of unctuous matter by leaving them for twenty-four hours in a tube filled with ether. They preserve in this way all their tenacity, and acquire at the same time nearly as much sensibility as if they had been prepared by an alkali. He finds that each instrument must be graduated experimentally by placing it in a confined space with air kept in a known state of humidity by the presence of dilute sulphuric acid of several degrees of strength, which he indicates, and supplies tables of their tension at different temperatures (Ann. de Ch., t. xv. p. 173). Of this instrument, which is so convenient in a great many circumstances, he speaks more highly than physicists generally of late, but at the same time remarks that it requires great circumspection in the observer, and that the occasional verification of the instrument by means of the solutions first employed in graduating it is indispensable.

Fig. 44.

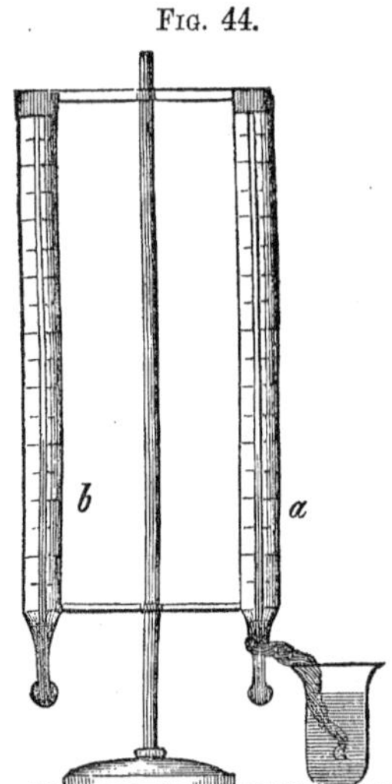

3. The degree of dryness of the air may be judged of by the rapidity of evaporation. Leslie made use of his differential thermometer as a hygrometer, covering one of the bulbs with muslin, and keeping it constantly moist by means of a wet thread from a cup of water placed near it. The evaporation of the moisture cools the ball, and occasions the air in it to contract. This instrument gives useful information in regard to the rapidity of evaporation, or the drying power of the air, but does not indicate directly the quantity of moisture in the air. The *wet-bulb hygrometer*, more commonly used, acts on the same principle, but consists of two similar and very delicate mercurial thermometers, the bulb of one of which (*a*) is kept constantly moist, while the bulb of the other (*b*) is dry. The wet thermometer always indicates a lower temperature than the dry one, unless when the air is fully saturated with moisture, and no evaporation from the moist bulb takes place. In making an observation, the instrument is generally placed, not in absolutely still air, but in an open window where there is a slight draught.

The indications of the wet-bulb hygrometer, or psychrometer, are discovered by simple inspection. It is, therefore, a problem of the greatest importance to deduce from them the dew point, or the tension of the vapour in the air, by an easy rule. Could this inference be made with certainty, the wet-bulb hygrometer is so commodious that it would supersede all others. I shall place

[1] The present and following methods of hygrometry, and all the experimental data required, have lately received a full and critical revision from M. Regnault, of the greatest value. See his "Etudes sur l'Hygrometrie," Annales de Chimie, &c., 1835, 3 sér. t. xv. p. 129.

below a formula for this purpose, which has been used for several years in the north of Europe, and the same as it has been recently modified.[1]

4. The most simple mode of ascertaining the absolute quantity of vapour in the air is to cool the air gradually, and note the degree of temperature at which it begins to deposit moisture, or ceases to be capable of sustaining the whole quantity of vapour which it possesses. The air is saturated with vapour for this particular degree of temperature, which is called its *dew-point.* The saturating quantity of vapour for the degree of temperature indicated may then be learned by reference to a table of the tension of the vapour of water at different temperatures.[2] It is the absolute quantity of vapour which the air at the time of the observation possesses. The dew-point may be ascertained most accurately by exposing to the air a thin cup of silver or tin-plate containing water so cold as to occasion the condensation of dew upon the metallic surface. The water in the cup is stirred with the bulb of a small thermometer, and as the temperature gradually rises, the degree is noted at which the dew disappears from the surface of the vessel. The temperature at which this occurs may be taken as the dew-point. Water may generally be cooled sufficiently in summer to answer for an experiment of this kind by dissolving pounded sal-ammoniac in it.

The dew-point may be observed much more quickly by means of the elegant hygrometer of the late Mr. Daniell. (Daniell's Meteorological Essays, p. 147). This instrument (see figure 45) consists of two glass balls, *a* and *b*, connected by a syphon, and containing a quantity of ether, from which the air has been expelled by the same means as in the cryophorus of Dr. Wollaston (page 75). One of the arms of the syphon tube contains a small thermometer, with its scale, which should be of white enamel; the bulb of the thermometer descends into the ball, *b*, at the extremity of this arm, and is placed, not in the centre of the ball, but as near as possible to some point of its circumference. A zone of this ball is gilt and burnished, so that the deposition of dew may easily be perceived upon it. The other ball, *a*, is covered with muslin. When an observation is to be made, this last ball is moistened with ether, which is supplied slowly by a drop or two at a time. It is cooled by the evaporation of the ether, and becomes capable of condensing the vapour of the included fluid, and thereby occasions evaporation in the opposite ball, *b*, containing the thermometer. The temperature of the ball, *b*, should be thus

[1] The psychrometer was first suggested by Gay-Lussac (Annales de Chimie, &c., t. xxi. p. 91), and its application particularly studied by Dr. E. H. August, of Berlin, (*Ueber die Fortschritte der Hygrometrie*), 1830, and Dr. Apjohn (Philosophical Magazine, 1838, &c.) To obtain the tension of vapour in the atmosphere from the two temperatures observed, the following formula is given by Dr. August, neglecting some very small quantities:—

$$x = f' - \frac{0.568\,(t - t')}{640 - t'} . h;$$

where t and t' are temperatures (Centigrade) of the dry and wet thermometers, f' the tension of vapour in air saturated at the temperature t', h the height of the barometer, and $640 - t'$ the latent heat of aqueous vapour. Some of the numerical data are modified by M. Regnault, and the formula becomes:—

$$x = f' - \frac{0.429\,(t - t')}{610 - t'} . h;$$

Or,

$$x = f' - \frac{0.480\,(t - t')}{616 - t} , h.$$

The last co-efficient 0.480 he finds to give a coincidence almost perfect between the calculated and true results, when the air is not more than four-tenths saturated. Otherwise the first coefficient 0.429 is least objectionable. (Annales, &c., xv. pp. 202 and 226).

[2] A table by M. Regnault for this purpose will be given in an Appendix.

Fig. 45.

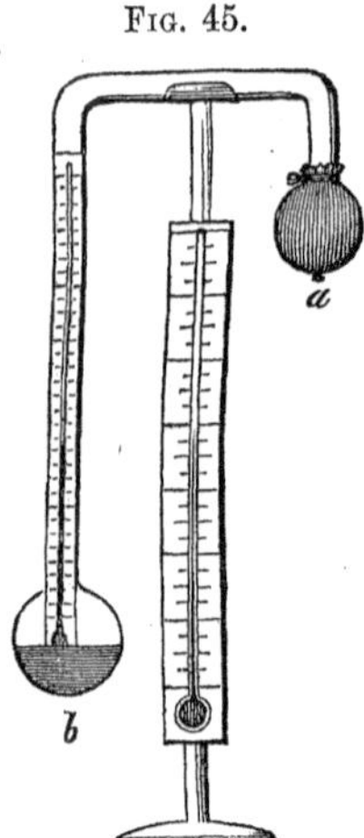

reduced in a gradual manner, so that the degree of the thermometer at which dew begins to be deposited on the metallic part of the surface of the ball may be observed with precision. The temperature of *b* being thereafter allowed to rise, the degree at which the dew disappears from its surface may likewise be noted. It should not differ much from the temperature of the deposition, and will probably give the dew-point more correctly; although, strictly speaking, the mean between the two observations should be the true dew-point. It is convenient to have a second thermometer in the pillar of the instrument, for observing the temperature of the air at the time.

M. Regnault proposes a modification of Daniell's hygrometer, under the name of the *Condenser-hygrometer*, (Annales de Chimie, et Ph. t. xv. Pl. 2), which appears to be the most perfect instrument of the class. It consists of a thimble, *a b c*, (figure 46), made of silver, very thin, and perfectly polished, 1-8 inch in depth, and 8-10ths of an inch in diameter, which is fitted tightly upon a glass tube, *c d*, open at both ends. The tube has a small lateral tubulure, *t* The upper opening of the tube is closed by a cork, which is traversed by the stem of a very sensible thermometer occupying its axis; the bulb of the thermometer is in the centre of the silver thimble. A very thin glass tube, *f g*, open at both ends, traverses the same cork, and descends to the bottom of the thimble. Ether is poured into the tube as high as *m n*, and the tubulure *t* is placed in communication by means of a leaden tube with an aspirator jar six or eight pints in capacity, filled with water. The aspirator jar is placed near the observer, while the condenser-hygrometer is kept as far from his person as is desirable.

Fig. 46. Fig. 47.

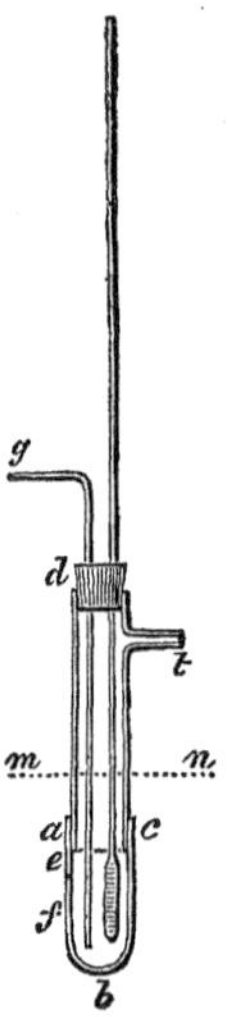

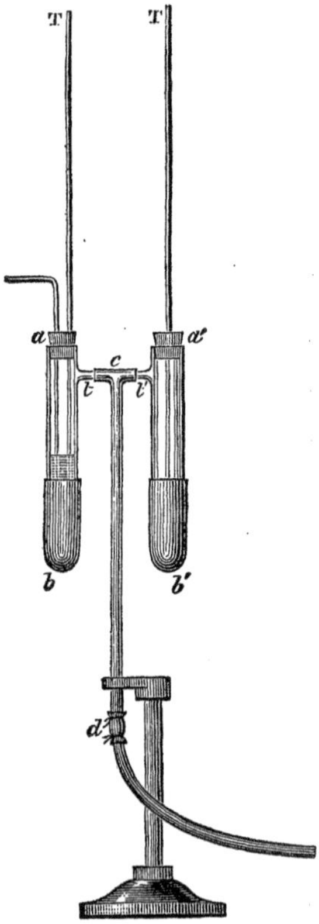

On allowing water to run from the aspirator jar, air enters by the tube *g f*, passing bubble by bubble through the ether, which it cools by carrying away vapour; the refrigeration is the more rapid, the more freely the water is allowed to flow; and the whole mass of ether presents a sensibly uniform temperature, as it is briskly agitated by the passage of the bubbles of air. The temperature is sufficiently lowered in less than a minute to determine an abundant deposit of dew. The thermometer is then observed through a little telescope; suppose that it is read off at 50°. This temperature is evidently somewhat lower than what corresponds exactly to the air's humidity. By closing the stopcock of the aspirator the passage of air is stopped, the dew disappears in a few seconds, and the thermometer again rises. Suppose that it marks 52°: this degree is above the dew-point. The stopcock of the aspirator is then opened very slightly, so as to determine the passage of a very small stream of air bubbles through the ether. If the thermometer continues, notwithstanding, to rise, the stopcock is opened further, and the thermometer brought down to 51°.8: by shutting the stopcock slightly, it is easy to stop the falling range, and make the thermometer remain stationary at 51°.8 as long as is desired. If no dew forms after the lapse of a few seconds, it is evident that 51°.8 is higher than the dew-point. It is brought down to 51°.6, and maintained there by regulating the flow. The metallic surface being now observed to become dim after a few seconds, it is concluded that 51°.6 is too low, while 51°.8 was too high. A still greater approximation even may be made, by now finding whether 51°.7 is above or below the point of condensation. These operations may be executed in a very short time, after a little practice; three or four minutes being found sufficient, by M. Regnault, to determine the dew-point to within about $\frac{1}{10}$th of a degree Fahr. A more considerable fall of temperature may be obtained by means of this than the original instrument of Daniell, with the consumption of a much less quantity of ether; indeed, that liquid may be dispensed with entirely, and alcohol substituted for it. The thermometer, T, to observe the temperature of the air during the experiment, is placed in a second similar glass tube and thimble *a′ b′*, also under the influence of the aspirator, but containing no ether.

In evaporating by means of hot air, as in drying goods in the ordinary bleachers' stove, which is heated by flues from a fire carried along the floor, it should be kept in mind that a certain time must elapse before air is saturated with humidity. Mr. Daniell has observed that a few cubic inches of dry air continue to expand for an hour or two, when exposed to water at the temperature of the air. At high temperatures, the diffusion of vapour into air is more rapid; but still it is not at all instantaneous. Hence, in such a drying stove, means ought to be taken to repress rather than to promote the exit of the hot air; otherwise a loss of heat will be occasioned by the escape of the air, before it is saturated with humidity. The greatest advantage has been derived from closing such a stove as perfectly as possible at the top, and only opening it after the goods are dried and about to be removed, in order to allow of a renewal of the air in the chamber between each operation. In evaporating water by heated air, the vapour itself carries off exactly the same quantity of heat as if it were produced by boiling the water at 212°, while the air associated with it likewise requires to have its temperature raised, and therefore occasions an additional consumption of heat. Hence water can never be evaporated by air in a drying stove with so small an expenditure of fuel as in a close boiler.

When bodies to be dried do not part with their moisture freely, but in a gradual manner, as is the case with roots, and most organic substances, the hot air to dry them may be greatly economised by a particular mode of applying it, which is practised in the madder-stove. The principle of this drying stove is illustrated by the annexed figure, in which *a b* represent a tight chamber, having two openings, one near the roof, by which hot air is admitted into the chamber, and another at the bottom, by which the air escapes into the tall chimney *c*. The chamber contains a series of stages, from the floor to the roof, on the lowest of which, sacks, half filled with the damp madder roots, are first placed. In proportion as the roots dry, the bags are raised from stage to stage, till they arrive at the highest stage, where they are exposed to the air when hottest and most desiccating. As the dried roots are removed from the top, new roots are introduced below, and passed through in the same manner. Here the dry and

Fig. 48.

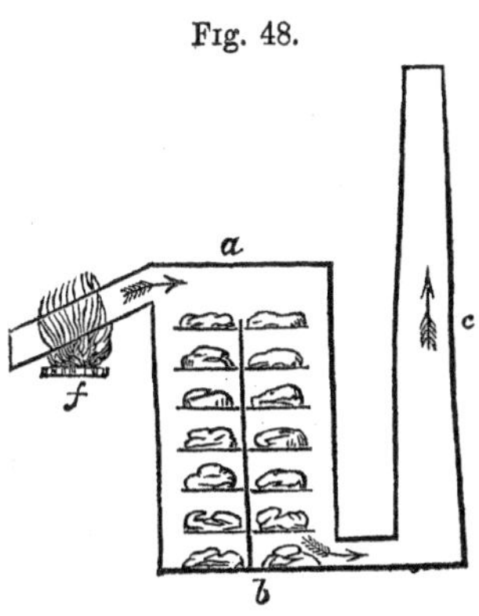

FIG. 49.

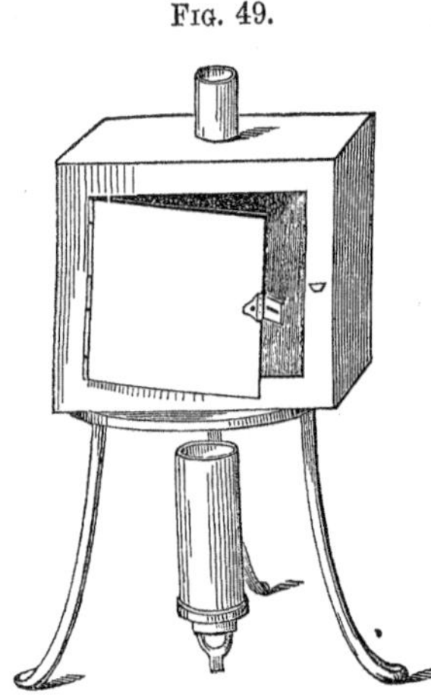

hot air, after taking all the moisture which the roots on the highest stage will part with, descends, and is still capable of abstracting a second quantity of moisture from the roots on the next, and so on, as it proceeds, till it passes away into the chimney absolutely saturated with moisture, after having reached the bottom of the chamber.

It is frequently an object to dry a small quantity of a substance most completely (such as an organic substance for analysis) at some steady temperature, such as 212°. This is effected very conveniently by means of a little oven, (figure 49), consisting of a double box of copper or tin-plate, about six inches square, with water between the casings, which is kept in a state of ebullition by means of a gas flame, or spirit lamp.

NATURE OF HEAT.

It is convenient to adopt the material theory of heat in considering its accumulation in bodies, and in expressing quantities of heat and the relative capacities of bodies for heat. Indeed, every thing relating to the absorption of heat suggests the idea of its substantial existence; for heat, unlike light, is never extinguished when it falls upon a body, but is either reflected and may be farther traced, or is absorbed and accumulated in the body, and may again be derived from it without loss. But the mechanical phenomena of heat, which resemble those of light, may be explained with equal if not greater advantage by assuming an undulatory theory of heat, corresponding with the undulatory theory of light. A peculiar imponderable medium or ether is supposed to pervade all space, through which undulations are propagated that produce the impression of heat. A hot radiant body is a body possessing the faculty to originate or excite such undulations in the ether or medium of heat, which spread on all sides around it, like the waves from a pebble thrown into still water. Sound is propagated by waves in this manner, but the medium in which they are generally produced, or the usual vehicle of sound, is the air; and all the experiments on the reflection and concentration of heat, by concave reflectors, may be imitated by means of sound. Thus, if a watch instead of the lamp be placed in the focus of one of a pair of conjugate reflecting mirrors (fig. 20, p. 54), the waves of air occasioned by its beating emanate from the focus, strike against the mirror, and are reflected from it, so as to break upon the face of the opposite mirror, are concentrated into its focus, and communicate the impression of sound to an ear placed there to receive it. The transmission of heat from the focus of one mirror to the focus of the other may easily be conceived to be the propagation of similar undulations through another and different medium from air, but coexisting in the same space.

In adopting the material theory of heat, we are under the necessity of assuming that there are different kinds of heat, some of which are capable of passing through glass, such as the heat of the sun, while others, such as that radiating from the hand, are entirely intercepted by glass. But on the undulatory theory the different properties of heat are referred to differences in the size of the waves, as the differences of colour are accounted for in light. Heat of the higher degrees of intensity, however, admits of a kind of degradation, or conversion into heat of lower intensity, to which we have nothing parallel in the case of light. Thus when the calorific rays of the sun, which are of the highest intensity, pass through glass, and strike a black wall, they are absorbed, and appear immediately afterwards radiating from the heated wall, as heat of low intensity, and are no longer capable of passing through glass.

It is as yet an unsolved problem to reverse the order of this change, and convert heat of low into heat of high intensity. The same degradation of heat or loss of intensity, is observed in condensing steam in distillation. The whole heat of the steam, both latent and sensible, is transferred without loss in that process, to perhaps fifteen times as much condensing water; but the intensity of the heat is reduced from 212° to perhaps 100° Fahr. The heat is not lost; for the fifteen parts of water at 100° are capable of melting as much ice as the original steam. But by no quantity of this heat at 100° can temperature be raised above that degree: no means are known of giving it intensity.

If heat of low is ever changed into heat of high intensity, it is in the compression of gaseous bodies by mechanical means. Let steam of half the tension of the atmosphere, produced at 180°, in a space otherwise vacuous, be reduced into half its volume, by doubling the pressure upon it, and its temperature will rise to 212°. If the pressure be again doubled, the temperature will become 250°, and the whole latent heat of the steam will now possess that high intensity. When air itself is rapidly compressed in a common syringe, we have a remarkable conversion of heat of low into heat of very high intensity.

It may be imagined that the elevation of temperature produced in the friction of hard bodies has a similar origin; that it results from the conversion of heat of low intensity, which the bodies rubbed together possess, into heat of high intensity. But it would be necessary further to suppose that a supply of heat of low intensity to the bodies rubbed can be endlessly kept up, by conduction or radiation, from contiguous bodies, as there is certainly no limit to the production of heat by means of friction.

Count Rumford, by boring a cylinder of cast iron, raised the temperature of several pounds of cold water to the boiling point. Sir H. Davy succeeded in melting two pieces of ice in the vacuum of an air-pump, by making them rub against each other, while the temperature of the air-pump itself and the surrounding atmosphere was below 32°. M. Haldot observed that when the surface of the rubber was rough, only half as much heat appeared as when the rubber was smooth. When the pressure of the rubber was quadrupled, the proportion of heat evolved was increased seven-fold. When the rubbing apparatus was surrounded by bad conductors of heat, or by non-conductors of electricity, the quantity of heat evolved was diminished. (Nicholson's Journal, xxvi. 30).

According to Pictet, a piece of brass, rubbed with a piece of cedar wood, produced more heat than when rubbed with another piece of metal; and the heat was still greater when two pieces of wood were rubbed together. He also finds that solids alone produce heat by friction; no heat appears to arise from the friction of one liquid upon another liquid, or upon a solid, nor by the friction of a current of air or gas upon a liquid or solid.

One other point only connected with the nature of heat remains, to which there is at present occasion to allude—the existence of a repulsive property in heat. Such a repulsive power in heated bodies is inferred to exist from the appearance of extreme mobility which many fine powders assume, such as precipitated silica, on being heated nearly to redness. Professor Forbes also attributes to such a repulsion the vibrations which take place between metals unequally heated, and the production of tones, to which allusion has already been made. But this repulsive power was rendered conspicuous, and even measurable, by Dr. Baden Powell, in the case of glass lenses, of very slight convexity, pressed together. On the application of heat, a separation of the glasses, through extremely small but finite spaces, was indicated by a change in the tints which appear between the lenses, and which depend upon the thickness of the included plate of air. This repulsion between heated surfaces appears to be promoted by whatever tends to the more rapid communication of heat. (Phil. Trans. 1834, p. 485).

CHAPTER II.

LIGHT.

THE mechanical properties of light constitute the science of optics, and belong, therefore, to physics, and not to chemistry. But it may be useful, by a short recapitulation, to recal them to the memory of the reader.

1. The rays of light emanate with so great velocity from the sun, that they occupy only 7½ minutes in traversing the immense space which separates the earth from that luminary. They travel at the rate of 192,500 miles in a second, and would, therefore, move through a space equal to the circumference of our globe in 1-8th of a second. They are propagated continually in straight lines, and spread or diverge at the same time; so that their density diminishes in the direct proportion of the squares of their distance from the sun. Hence, if the earth were at double its present distance from the sun, it would receive only one-fourth of the light; at three times its present distance, one-ninth; at four times its present distance, one-sixteenth, &c.

2. When the solar rays impinge upon a body, they are reflected from its surface, and bound off, as an elastic ball striking against the same surface in the same direction would do; or they are absorbed by the body upon which they fall, and disappear, being extinguished; or lastly, they pass through the body, which in that case is transparent or diaphanous. In the first case, the body becomes visible, appearing white, or of some particular colour, and we see it in the direction in which the rays reach the eye. In the second case, the body is invisible, no light proceeding from it to the eye; or it appears black, if the surrounding objects are illuminated. In the third case, if the body be absolutely transparent, it is invisible, and we see through it the object from which the light was last reflected. But light is often greatly affected in passing through transparent bodies.

3. If light enter such media, of uniform density, perpendicularly to their surface, its direction is not altered; but in passing obliquely out of one medium into another, it undergoes a change of direction. If the second medium be denser than the first, the ray of light is bent, or *refracted*, nearer to the perpendicular; but in passing out from a denser into a rarer medium, it is refracted from the perpendicular. Thus, when the ray of light *r*, passing through the air, falls obliquely upon a plate of glass at the point *a*, instead of continuing to move in the same straight line *a b*, it is bent towards the perpendicular at *a*, and proceeds in the direction *a c*. The ray is bent to the side on which there is the greatest mass of glass. On passing out from the glass into the air, a rarer medium, at the point *c*, the ray has its direction again changed, and in this case *from* the perpendicular, but still towards the mass of glass. The amount of refraction, generally speaking, is proportional to the density of a body, but combustible bodies possess a higher refracting power than corresponds to their density. Hence the diamond, melted phosphorus, naphtha, and hydrogen gas, exhibit this effect upon light in a greater degree than other transparent bodies. Dr. Wollaston had recourse to this refracting power as a test of the purity of some substances. Thus, genuine oil of cloves had a refracting power expressed by the number 1535, while that of an impure specimen was not more than 1498.

FIG. 50.

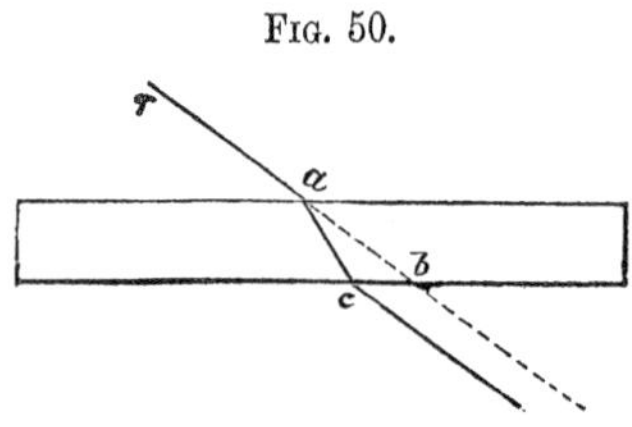

4. In passing through many crystallized bodies, such as Iceland spar, a certain portion of light is refracted in the usual way, and another portion undergoes an

extraordinary refraction, in a plane parallel to the diagonal which joins the two obtuse angles of the crystal. Such bodies are said to *refract doubly*, and exhibit a double image of any body viewed through them.

5. Reflected and likewise doubly refracted light assume new properties. *Common* light, by being reflected from the surface of glass, or any bright surface non-metallic, is more or less of it converted into what is called *polarized* light. If it be reflected at one particular angle of incidence, 56°.45′, it is *all* changed into polarized light; and the further the angle of reflection deviates from 56°, on either side, the less is polarized, and the more remains common light. 56° is the maximum polarizing angle for glass; 52°.45′ for water. The light is said to be *polarized*, from certain properties which it assumes, which seem to indicate that the ray, like a magnetic bar, has sides in which reside peculiar powers. One of these new properties is, that when it falls upon a second glass plate, it is not reflected in the same way as common light. If the plane of the second reflector is *perpendicular* to the first, and the ray fall at an angle of 56°, it is not reflected at all, it *vanishes;* but if *parallel,* it is entirely reflected. Polarized light appears to possess some most extraordinary properties, in regard to vision, of useful application. It is said that a body which is quite transparent to the eye, and which appears upon examination to be as homogeneous in its structure as it is in its aspect, will yet exhibit, under polarized light, the most exquisite organization. As an example of the utility of this agent in exploring mineral, vegetable, and animal structures, Sir D. Brewster refers to the extraordinary structure of the minerals apophyllite and analcime; to the symmetrical and figurate disposition of siliceous crystals in the epidermis of equisetaceous plants, and to the wonderful variations of density in the crystalline lenses, and the integuments of the eyes of animals, which polarized light renders visible. (Rep. of the British Association, vol. i. Report upon Optics, by Sir. D. Brewster.)

6. *Decomposition of light.*— When a beam of light from the sun is admitted

Fig. 51.

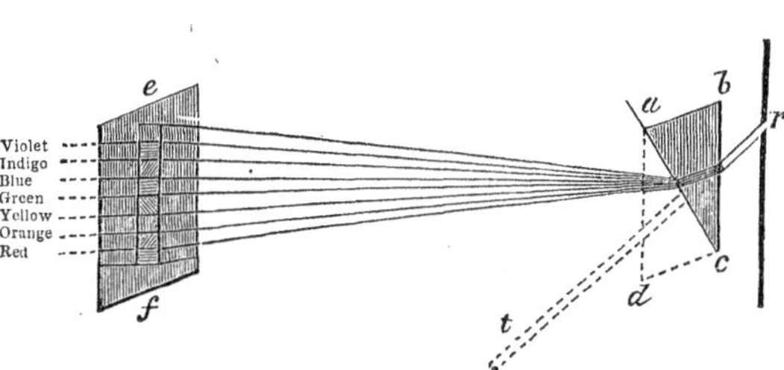

into a dark room, by a small aperture *r* in a window-shutter, and is intercepted in its passage by a wedge or solid angle of glass *a b c*, it is refracted as it enters, and a second time as it issues from the glass; and instead of forming a round spot of white light, as it would have done if allowed to proceed in its original direction *r t*, it illuminates with several colours an oblong space of a white card *e f*, properly placed to receive it. The solid wedge of glass is called a prism, and the oblong coloured image on the card, the solar spectrum. Newton counted seven bands of different colours in the spectrum, which, as they succeed each other from the upper part of the spectrum represented in the figure, are violet, indigo, blue, green, yellow, orange, and red. The beam of light admitted by the aperture in the window-shutter has been separated in passing through the prism into rays of different colours, and this separation obviously depends upon the rays being unequally refrangible. The blue rays are more considerably refracted or deflected out of their course, in passing through the glass, than the yellow rays, and the yellow rays than the red. Hence the violet end is spoken of as the most refrangible, and the red as the least refrangible end of the spectrum.

The coloured bands of the spectrum differ in width, and are shaded into each other; and it is not to be supposed that there are really rays of seven different colours. Sir D. Brewster has established, in his analysis of solar light, that there are rays of three colours only, blue, yellow, and red, which were well known to artists to be the three primary colours of which all others are compounded.

A certain quantity of white light, and a portion of each of the primary rays, may be found at every point from the top to the bottom of the spectrum. But each of the primary rays predominates at a particular part of the spectrum. This point is, for the blue rays, near the top of the spectrum; for the yellow rays, somewhat below the middle; and for the red rays, near the bottom of the spectrum. Hence, there exist rays of each colour of every degree of refrangibility; but the great proportion of the yellow rays is more refrangible than the red, and the great proportion of the blue more refrangible than either the yellow or red. The compound spectrum which we observe is in fact produced by the superposition of three simple spectra, a blue, a yellow, and a red spectrum. The distribution of the rays in each of these simple spectra is represented by the shading in the annexed figures. Of the seven different coloured bands into which Newton divided the spectrum, not one is a pure colour. The orange is produced by a predominance of the yellow and red rays; the green, by the yellow and blue rays, and the indigo and violet are essentially blue, with different proportions of red and yellow.[1]

Fig. 52.

Blue spectrum. Yellow spectrum. Red spectrum.

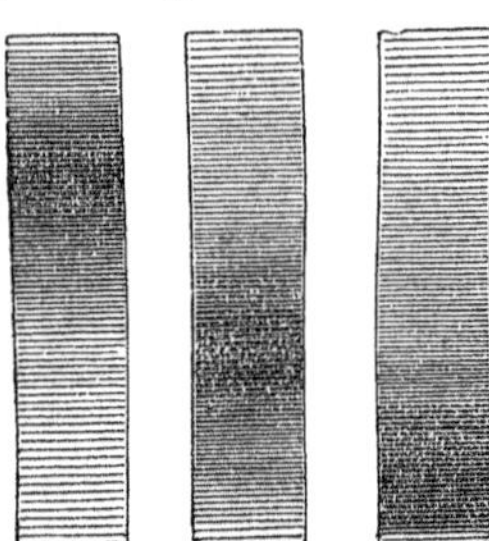

By placing a second prism *a d c*, in a reversed position, in contact with the first prism, the colours disappear, and we have a spot of white light, as if both prisms were absent. The three coloured rays of the spectrum, therefore, produce white light by their union.

On examining the solar spectrum, Dr. Thomas Young observed that it is crossed by several dark lines; that is, that there are interruptions in the spectrum, where there is no light of any colour. Fraunhofer subsequently found that the lines in the spectrum of solar light were much more numerous than Dr. Young had imagined, while the spectrum of *artificial* white flames contains all the rays which are thus wanting. One of the most notable is a double dark line in the yellow, which occurs in the light of the sun, moon, and planets. In the light of the fixed stars, Sirius and Castor, the same double line does not occur; but one conspicuous dark line in the yellow, and two in the blue. The spectrum of Pollux, on the contrary, is the same as that of the sun. Now a very recent discovery of Sir D. Brewster has given these observations an entirely chemical character. He has found that the white light of ordinary flames requires merely to be sent through a certain gaseous medium (nitrous acid vapour) to acquire more than a thousand dark lines in its spectrum. He is hence led to infer that it is the presence of certain gases in the atmosphere of the sun which occasions the observed deficiencies in the solar spectrum. We may thus have it yet in our power to study the nature of the combustion which lights up the suns of other systems. Dr. Miller, by subjecting the spectrum to the absorptive influences of chlorine, iodine, bromine, perchloride of manganese, and other coloured vapours, brought into view numerous dark bands not previously observed. The spectra of coloured flames were also marked by peculiar lines.

The rays of *heat* are distributed very unequally throughout the luminous spectrum; most heat being found associated with the red or least refrangible luminous rays, and

[1] Sir David Brewster, On a New Analysis of the Solar Light, indicating three primary colours, forming coincident spectra of equal length. Edinburgh Phil. Trans. vol. xii. p. 123.

least with the violet rays. Indeed, when the solar beam is decomposed by a prism of a highly diathermanous material, such as rock salt, the rays of heat are found to extend, and to have their point of maximum intensity considerably beyond the visible spectrum, on the side of the red ray. Hence, although there are calorific rays of all degrees of refrangibility, the great proportion of them are even less refrangible than the least refrangible luminous rays. It is to be observed that the least refrangible rays are absorbed in greatest proportion in passing through bodies which are not highly diathermanous; such as crown-glass, and water. Hence prisms of these substances, allowing only the more refrangible rays of heat to pass, give a spectrum which is hottest in the red, or perhaps even in the yellow ray, and possesses little or no heat beyond the border of the red ray. The inequality in refrangibility existing between the rays of heat and of light is decisive of the fact that they are peculiar rays, that can be separated, although associated together in the sunbeam. Indeed, Melloni finds that light from both solar and terrestrial sources is divested of all heat by passing successively through water, and a glass coloured green by oxide of copper, being incapable as it issues from these media of affecting the most delicate thermoscope.

The light of the sun is capable of inducing certain chemical changes which depend neither upon its luminous nor calorific rays, but upon the presence of what are called *chemical rays.* Thus, under the influence of light, chlorine gas is capable of decomposing water, combining with its hydrogen, and liberating oxygen; the chlorine in the freshly precipitated chloride of silver appears to be liberated, and the colour of the salt changes from white to black from the formation of a subchloride. Photographic impressions are obtained on paper by means of this and other salts of silver, particularly the bromide and iodide, which are still more sensitive to light. A polished plate of silver, covered with the thinnest film of iodide, is employed to receive the image in the daguerréotype. The moist chloride of silver is darkened more rapidly by the violet than by the red rays of the spectrum; but this change is produced upon it even when carried a little way out of the visible spectrum on the side of the violet ray. The rays found in that situation are, therefore, more refrangible than any other kind of rays in the spectrum. Their characteristic effect is to promote those chemical decompositions in which oxygen is withdrawn from water and other oxides; and hence they are sometimes named *de-oxidizing* rays. These rays were likewise supposed to communicate magnetism to steel needles exposed to them; but this opinion is no longer tenable.

CHAPTER III.

SECTION I.

CHEMICAL NOMENCLATURE AND NOTATION.

THERE are fifty-nine substances at present known, which are simple, or contain one kind of matter only. Their names are given in the following table, together with certain useful numbers which express the quantities by weight, according to which the different elements combine with each other. The letter or symbol annexed to the name is employed to represent these particular quantities of the elements, or the chemical equivalents.

TABLE OF ELEMENTARY SUBSTANCES,

WITH THEIR CHEMICAL EQUIVALENTS.

*** For the *authorities* for the *numbers* in this table, see note at page 104.

Names of Elements.	Symbols.	Equivalents.		
		Hydrogen =1.	Oxy.=100. H.=12·5.	
Aluminum	Al	13·69	171·17	$Al_2\,O_3$, alumina. $Al_2\,Cl_3$, chloride of aluminum. $Al_2\,O_3$, $3SO_3$, sulphate of alumina.
Antimony (Stibium)........	Sb	129·03	1612·90	SbO_3, oxide of antimony. SbO_5, antimonic acid.
Arsenic	As	75	937·50	AsO_3, arsenious acid. AsO_5, arsenic acid.
Barium	Ba	68·64	858·01	BaO, baryta. BaCl, chloride of barium.
Bismuth	Bi	70·95	886·92	BiO, oxide of bismuth. BiO, NO_5, nitrate of bismuth. BiCl, chloride of bismuth.
Boron	B	10·90	136·20	BO_3, boric or boracic acid. BFl_3, fluoboric acid.
Bromine.............	Br	78·26	978·30	BrO_5, bromic acid. BrH, hydrobromic acid.
Cadmium...........	Cd	55·74	696·77	CdO, oxide of cadmium. [mium. CdS, sulphide or sulphuret of cad-
Calcium	Ca	20	250·00	CaO, lime. CaCl, chloride of calcium.
Carbon...............	C	6	75·00	CO, carbonic oxide. CO_2, carbonic acid. CS_2, sulphide or sulphuret of carbon.
Cerium	Ce	46	575	CeO, oxide of cerium. Ce_2O_3, sesquioxide of cerium.
Chlorine............	Cl	35·50	443·75	ClO_5, chloric acid. ClO_7, perchloric acid. ClH, hydrochloric acid.
Chromium	Cr	28·15	351·82	CrO_3, chromic acid. Cr_2O_3, sesquioxide of chromium. Cr_2O_3, $3SO_3$, sulphate of chromium.
Cobalt	Co	29·52	368·99	CoO, oxide of cobalt. Co_2O_3, sesquioxide of cobalt.
Copper (Cuprum).	Cu	31·66	395·70	Cu_2O, suboxide of copper. CuO, oxide of copper. CuO, SO_3, sulphate of copper.
Didymium	D	49·6	620	
Fluorine.............	F	18·70	233·80	HF, hydrofluoric acid. BF_3, fluoboric acid.
Glucinum	Gl	26·50	331·26	Gl_2O_3, glucina. Gl_2Cl_3, chloride of glucinum.
Gold (Aurum)......	Au	98·33	1229·16	Au_2O. oxide of gold. Au_2O_3, sesquioxide of gold.
Hydrogen...........	H	1	12·50	HO, water. HO_2, binoxide of hydrogen.
Iodine................	I	126·36	1579·50	IO, iodic acid. HI, hydriodic acid.
Iridium	Ir	98·68	1233·50	IrO, protoxide of iridium. Ir_2O_3, sesquioxide of iridium.
Iron (Ferrum).....	Fe	28	350·00	FeO, protoxide of iron. Fe_2O_3, sesquioxide of iron. [of iron. Fe_2O_3, $3SO_3$, sulphate of sesquioxide
Lanthanum.........	Ln	48	600	LnO, oxide of lanthanum.

Name of Elements.	Symbols.	Equivalents. Hydrogen =1.	Oxy.=100. H.=12·5.	
Lead (Plumbum)..	Pb	103·56	1294·50	PbO, oxide of lead. $PbCl$, chloride of lead.
Lithium	Li	6·43	80·37	LiO, oxide of lithium. $LiCl$, chloride of lithium.
Magnesium.........	Mg	12·67	158·35	MgO, magnesia. $MgCl$, chloride of magnesium.
Manganese..........	Mn	27·67	345·90	MnO, protoxide of manganese. MnO_2, binoxide of manganese. MnO_3, manganic acid. Mn_2O_7, permanganic acid.
Mercury (Hydrargyrum)	Hg	100·07	1250·9	Hg_2O, suboxide (black oxide). HgO, oxide (red oxide). Hg_2Cl, subchloride (calomel). $HgCl$, chloride (sublimate).
Molybdenum.......	Mo	47·88	598·52	MO_3, molybdic acid.
Nickel................	Ni	29·57	369·68	NiO, protoxide of nickel. Ni_2O_3, sesquioxide of nickel.
Niobium.............	...	......	......	
Nitrogen or azote.	N (or Az)	14	175·00	NO_5, nitric acid. NO_2, binoxide of nitrogen. NH_3, ammonia.
Osmium	Os	99·56	1244·49	OsO_4, osmic acid.
Oxygen	O	8	100·00	
Palladium..........	Pd	53·27	665·90	PdO, protoxide of palladium. PdO_2, peroxide of palladium.
Pelopium...........	...	......	......	
Phosphorus........	P	32·02	400·3	PO_5, phosphoric acid. PO_3, phosphorous acid. PH_3, phosphuretted hydrogen.
Platinum...........	Pt	98·68	1233·50	PtO, protoxide of platinum. PtO_2, binoxide of platinum.
Potassium (Kalium)	K	39·00	487·50	KO, potassa. KCl, chloride of potassium.
Rhodium...........	R	52·11	651·39	RO, protoxide of rhodium. R_2O_3, sesquioxide of rhodium.
Ruthenium	Ru	52·11	651·39	Ru_2O_3, sesquioxide of ruthenium.
Selenium...........	Se	39·57	494·58	SeO_3, selenic acid. SeH, hydroselenic acid.
Silicium	Si	21·35	266·82	SiO_3, silicic acid, or silica. SiF_3, fluosilicic acid.
Silver (Argentum)	Ag	108·00	1350·00	AgO, oxide of silver. $AgCl$, chloride of silver.
Sodium (Natronium).....	Na	22·97	287·17	NaO, soda. $NaCl$, chloride of sodium.
Strontium..........	Sr	43·84	548·02	SrO, strontium. $SrCl$, chloride of strontium.
Sulphur	S	16	200·00	SO_3, sulphuric acid. SH, hydrosulphuric acid.
Tantalum or Columbium	Ta	92·30	1153·72	TaO, oxide of tantalum. TaO_3, tantalic acid.
Telurium............	Te	66·14	801·76	TeO_3, telluric acid. TeH, hydrotelluric acid.
Thorium............	Th	59·59	744·90	ThO, oxide of thorium. $ThCl$, chloride of thorium.
Tin (Stannum).....	Sn	58·82	735·29	SnO, protoxide of tin. SnO_2, binoxide of tin.
Titanium...........	Ti	24·29	303·66	TiO_2, titanic acid. $TiCl_2$, bichloride of titanium.

Name of Elements.	Symbols.	Equivalents.		
		Hydrogen =1.	Oxy.=100. H.=12·5.	
Tungsten (Wolfram)........	W	94·64	1183·00	WO_3, tungstic acid.
Uranium	U	60	750	UO, oxide of uranium (urane of Peligot.) U_2O_3, uranic acid.
Vanadium...........	V	68·55	856·89	VO_3, vanadic acid.
Yttrium..............	Y	32·20	402·51	YO, yttria. YCl, chloride of yttrium.
Zinc	Zn	32·52	406·59	ZnO, oxide of zinc. $ZnCl$, chloride of zinc.
Zirconium..........	Zr	33·62	420·20	Zr_2O_3, zirconia. Zr_2Cl_3, chloride of zirconium.

*** The numbers in the preceding table are, with several exceptions, those of Berzelius. The equivalent of carbon has lately been reduced, with the general concurrence of chemists, from 76·44, on the oxygen scale, to 75, and hydrogen made 12·5 exactly, chiefly from the experiments of M. Dumas on the combustion of carbon and hydrogen gas by means of oxygen and oxide of copper, in his refined arrangement for organic analysis (Ann. de Chimie, 3 sér. t. i. p. 5). For nitrogen, M. Pelouze obtained, by two analyses of sal-ammoniac, the numbers 175·58 and 174·78; M. Marignac obtained for the same element the number 175·25, from the analysis of nitrate of silver; and Dr. T. Anderson has been led to nearly the same result, by an analysis of the nitrate of lead. These results permit the adoption of 175 as the equivalent of nitrogen: the old number was 177·04.

The equivalents of chlorine, potassium, and silver, the most fundamental numbers in the table, which were determined by Berzelius with remarkable precision, have received small corrections from M. Marignac. Seven experiments were made by the latter chemist on the decomposition of chlorate of potash by heat, in each of which from 800 to 1100 grains of the salt were employed, which gave him from 39.155 to 39.167 per cent. of oxygen; he adopts 39.161, the actual result of two experiments. Berzelius had obtained, thirty years before, 39.15. Pelouze has also obtained identically the same result (Poggendorff's Annalen, lviii. 171). On the other hand, 100 parts of silver required for precipitation from solution of nitrate, 69.062 parts of chloride of potassium (mean of six experiments); the maximum was 69.067, and the minimum 69.049; while the precipitated chloride of silver amounted after fusion to 132.84 parts, as the mean of five experiments, of which the maximum was 132.844, and the minimum 132.825 parts (Marignac). These experiments, from which the equivalents are deduced, obtain the unqualified approbation of Berzelius, who gives the numbers reduced to equivalents as they appear below. (Rapport Annuel sur le Progrès de la Chimie, par J. Berzelius, Paris, 1845, p. 32).

	Marignac.	Berzelius (old numbers).
Chlorine	443.20	442.65
Potassium	488.94	489.92
Silver	1349.01	1351.61

Finally, M. Maumené has investigated the same three important equivalents; decomposing the chlorate of potash by heat, and by guarding against certain minute sources of inaccuracy, raising the proportion of oxygen from 100 salt to 39.209; also decomposing the fused chloride of silver by hydrogen gas, and analyzing the oxalate and acetate of silver. The experiments of this chemist appear to be executed with a degree of exactness which can scarcely be exceeded, and lead to conclusions of the highest interest, as they give numbers which approach so closely to multiples of 6.25, the half equivalent of hydrogen, that the differences may be safely considered as falling within the unavoidable errors of observation, and the multiple numbers assumed as the true numbers for the three equivalents in question, (Annales, &c. 1846, 3 sér. xviii. 41). The results are:—

	Maumené.	Multiple Numbers.
Chlorine	443.669	443.75 = 6.25 × 71
Potassium	487.004	487.50 = " × 78
Silver	1350.322	1350.00 = " × 216

The following short table contains numbers lately obtained by M. Pelouze, for several elements, differing sensibly from the numbers of Berzelius, for which they are substituted, and multiples of 6.25, to which they all closely approximate.

	Berzelius.	Pelouze.	Multiples of 6.25.
Sodium	290.90	287.17	287.50 = 6.25 × 46
Barium	856.88	858.03	856.25 = " × 137
Strontium	547.29	548.02	550.00 = " × 88
Silicium	277.29	266.82	268.75 = " × 43
Phosphorus	392.29	400.30	400.00 = " × 64
Arsenic	940.08	937.50	937.50 = " × 150

The equivalent of sodium was determined from the quantity of chloride of sodium required to precipitate 200 parts of silver from the nitrate. Barium, strontium, silicium, phosphorus, and arsenic, in a similar manner, also by the quantity of silver which their chlorides precipitated.

The equivalent of calcium is taken at 250, after Dumas; MM. Erdmann and Marchand have confirmed this equivalent; Berzelius himself reduces his first number from 256.02 to 251.94. Sulphur and mercury are also after Erdmann and Marchand; Berzelius has, on recalculating his old results, reduced the number for sulphur from 201.17 to 200.8.

The equivalent of iron was lately found 349.80 by MM. Swanberg and Norlin, and their results confirmed by Berzelius, who now obtains 350.27 and 350.369 (instead of 339.21, the old equivalent): an intermediate number 350 is adopted in the table.

The number for zinc is that of M. Axel Erdmann, who took unusual pains in purifying the metal: it is 412.63 according to M. Favre, and 414 according to M. Jacquelain; the number of Berzelius is 403.23.

The number for uranium is that adopted by M. Peligot; it has been found 746.36 by M. Wertheim, and 742.875 by Ebelmen.

The number for gold is that lately deduced by Berzelius from an analysis of the double chloride of gold and potassium (Poggendorff's Annalen, lxv. 314); it replaces his former number 1243.01. Those of cerium and ruthenium are by Hermann (Annuaire de Chimie, 1835, p. 130). M. Rammelsberg has adopted for the former metal 574.7, and M. Beringer 577; the number of M. Hermann is intermediate. Ruthenium, the new metal from native platinum, is considered by its discoverer, Prof. Haus, to be isomorphous with, and to have the same equivalent as, rhodium, from the composition of the double sesqui-chloride of ruthenium and potassium, $2 K Cl + Ru_2Cl_3$.

No data exist for fixing the equivalents of the metallic elements lately discovered, whose names appear in the table, namely, didymium found with lanthanum in cerite (Mosander); niobium and pelopium in the tantalite of Bavaria (H. Rose).

Names of Elements.	Symbols.	Equivalents.		
		Hydrogen =1.	Oxy. =100. H. =12·5.	
Aridium	Ar			
Donarium	Do	79.72	997.4	Do_2O_3, sesquioxide of donarium.
Erbium	E			
Ilmenium	Il	60.4	753.	IlO_2, ilmenic acid.
Norium	No			
Terbium	Tb			

[The elements given in the above table have been made known since the publication of this part of the original work. Aridium by Ulgren, Donarium by Bergemann, Erbium and Terbium by Mosander, Ilmenium by Hermann, and Norium by Svanberg. The equivalents, as far as ascertained, are given on the authority of the discoverers. — R. B.]

In the class of simple substances are placed all those bodies which are not known to be compound, on the principle that whatever cannot be decomposed or resolved by any process of chemistry into other kinds of matter, is to be considered as simple. They are the only bodies the names of which are at present independent of any rule. An attempt was, indeed, made on the first introduction of a systematic nomenclature, to make the names of several of them significant; but some confusion in regard to their derivatives was found to be the consequence of this, and many of them being familiar substances, were almost of necessity allowed to retain the names they bear in common language: such as, sulphur, tin, silver, and the other metals known in the arts. To newly discovered elements, however, such names were applied as were suggested by any striking physical property they possessed, or remarkable circumstance in their history. The names of the newer metals, platinum, potassium, vanadium, &c., have a common termination, which serves to distinguish them as metals. Another class of elementary bodies, resembling each other in certain particulars, is marked in a similar manner; namely, that composed of chlorine, iodine, bromine, and fluorine.

The names of *compound bodies* are contrived to express their composition, and the class to which they belong, and are founded on a distribution of compounds into three orders, namely, first, compounds of one element with another element; as, for instance, oxygen with sulphur in sulphuric acid, or oxygen with sodium in soda, which are called binary compounds. Secondly, combinations of binary compounds with each other, as of sulphuric acid with soda in Glauber's salt, and the salts generally, which are termed ternary compounds. And thirdly, combinations of salts with one another, or double salts, such as alum, which are quaternary compounds.

1.—Of the compounds of the first order, the greater number known to the original framers of the chemical nomenclature contained oxygen as one of their two constituents; and hence an exclusive importance was attached to that element. Its compounds with the other elementary bodies may be divided by their properties into: (*a*) the class of neutral bodies and bases; and (*b*) the class of acids.

(*a*). To members of the first class the generic term *oxide* was applied, the first syllable of oxygen, with a termination (*ide*) indicative of combination; to which the name of the other element was joined to express the specific compound. Thus a compound of oxygen and hydrogen is *oxide of hydrogen;* of oxygen and potassium, *oxide of potassium;* of which compounds, the first, or water, is an instance of a neutral oxide; and the second, or potash, of a base or alkaline oxide. But the same elementary body often combines with oxygen in more than one proportion, forming two or more oxides; to distinguish which the Greek prefix (*proto*, πρῶτος, first) is applied to the oxide containing the least proportion of oxygen; *deuto* (δεύτερος, second) to the oxide containing more oxygen than the protoxide; and trito (τριτος, third) to the oxide containing still more oxygen than the deutoxide; which last oxide, if it contains the largest proportion of oxygen with which the element can unite to form an oxide, is more commonly named the *peroxide;* from *per*, the Latin particle of intensity. Thus, the three compounds of the metal manganese and oxygen are distinguished as follows:—

Names.	Composition.	
	Manganese.	Oxygen.
Protoxide of manganese	100	28.91
Deutoxide of manganese	100	43.36
Peroxide of manganese	100	57.82

As the prefix *per* implies simply the highest degree of oxidation, it may be applied to the second oxide where there are only two, as in the oxides of iron, the second oxide of which is called, indifferently, the deutoxide or peroxide of iron. M. Thenard, in his Traité de Chimie, avoids the use of the term deutoxide, and confines the application of peroxide to such of these oxides as, like the peroxide of manganese, do not combine with acids. He applies the names *sesquioxide* and *binoxide* to oxides, which are capable of combining with acids, and contain respectively, once and a half and twice as much oxygen as the protoxides of the same metal. He has thus the protoxide, sesquioxide, and peroxide of manganese, the protoxide and sesquioxide of iron, the protoxide and binoxide of tin, &c. This distinction is useful, and will be adopted in the present work. Certain inferior oxides, which do not combine with acids, are called *suboxides;* such as the suboxide of lead, which contains less oxygen than the oxide distinguished as the protoxide of the same metal.

The compounds of chlorine and several other elements are distinguished in the same manner as the oxides. Such elements resemble oxygen in several respects, particularly in the manner in which their compounds are decomposed by electricity.

Chlorine, for example, like oxygen, proceeds to the positive pole, and is therefore classed with oxygen as an electro-negative substance, in a division of elements grounded on their electrical relations. Thus, with the other elementary bodies,

Oxygen	forms	oxides,
Chlorine	"	chlorides,
Bromine	"	bromides,
Iodine	"	iodides,
Fluorine	"	fluorides,
Sulphur	"	sulphides (or sulphurets),
Phosphorus	"	phosphides (or phosphurets),
Carbon	"	carbides (or carburets),
Nitrogen	"	nitrides,
Hydrogen	"	hydrides,
Cyanogen ($N C_2$)	"	cyanides,
Sulphion ($S O_4$)	"	sulphionides.

As cyanogen and sulphion, although compound bodies, comport themselves in their combinations like electro-negative elements, their compounds are named in the same manner as the oxides.

When several chlorides of the same metal exist, they are distinguished by the same numerical prefixes as the oxides. Thus we have the protochloride and the sesquichloride of iron; the protochloride, and the bichloride of tin. The compounds of sulphur greatly resemble the oxides, but they have been generally named sulphurets, and not sulphides or sulphurides. Berzelius, indeed, applies the term sulphuret to such binary compounds of sulphur only as are basic and correspond with basic oxides; while sulphide is applied by him to such as are acid, or correspond with acid oxides. Hence, he has the *sulphuret of potassium*, and the *sulphide of arsenic* and *sulphide of carbon*. Compounds of chlorine are distinguished by him into chlorurets and chlorides, on the same principle; thus he speaks of the *chloruret of potassium*, and of the *chloride of phosphorus*. But these distinctions have not served any important purpose, while besides conducing to perspicuity it is an object of some consequence in a systematic point of view to allow the termination *ide*, already restricted to electro-negative substances, to apply to all of them without exception.

The combinations of metallic elements among themselves are distinguished by the general term *alloys*, and those of mercury as *amalgams*.

(*b*). The binary compounds of oxygen which possess acid properties are named on a different principle. Thus the acid compound of titanium and oxygen is called *titanic acid;* of chromium and oxygen, *chromic acid;* or the name of the acid is derived from that of the substance in combination with oxygen, with the termination *ic*. Where the same element was known to form two acid compounds with oxygen, the termination *ous* was applied to that which contained the least proportion of oxygen, as in *sulphurous* and sulphuric acids. On the discovery of an acid compound of sulphur which contained less oxygen than that already named sulphurous acid, it was called *hyposulphurous acid*, (from the Greek ὑπο, under); and another new compound, intermediate between the sulphurous and sulphuric acids, was named *hyposulphuric acid.* On the same principle, an acid containing a greater proportion of oxygen than that already named chloric acid, was named *hyperchloric acid*, (from the Greek ὑπερ, over;) but now more generally *perchloric acid*. The names of the different acid compounds of oxygen and sulphur, which have been referred to for illustration, with the relative proportions of oxygen which they contain, are as follows:

Names.	Composition. Sulphur.	Oxygen.
Hyposulphurous acid	100	50
Sulphurous acid	100	100
Hyposulphuric acid	100	125
Sulphuric acid	100	150

The same system is adopted for all analogous acids. An acid of chlorine, containing more oxygen than chloric acid, is named perchloric acid, and other similar compounds, which all contain an unusually large proportion of oxygen, are distinguished in the same manner; as periodic acid, and permanganic acid. The perchloric acid is also sometimes called *oxichloric;* but this last term does not seem so suitable as the first.

Another class of acids exists in which sulphur is united with the other element in the place of oxygen. The acids thus formed are called *sulphur acids.* The names of the corresponding oxygen acids are sometimes applied to these, with the prefix *sulph,* as *sulpharsenious* and *sulpharsenic* acids, which resemble arsenious and arsenic acids respectively in composition, but contain sulphur instead of oxygen.

Lastly, certain substances, such as chlorine, sulphur and cyanogen, form acids with hydrogen, which are called hydrogen acids, or *hydracids.* In these acid compounds the names of both constituents appear, as in the terms *hydrochloric acid, hydrosulphuric acid,* and *hydrocyanic acid.* Thenard has proposed to alter these names to *chlorhydric, sulphohydric,* and *cyanhydric acids,* which in some respects are preferable terms.

2.—Compounds of the second order, or salts, are named according to the acid they contain, the termination *ic* of the acid being changed into *ate*, and *ous* into *ite.* Thus a salt of sulphuric acid is a *sulphate;* of sulphurous acid, a *sulphite;* of hyposulphurous acid, a *hyposulphite;* of hyposulphuric acid, a *hyposulphate;* and of perchloric acid, a *perchlorate;* and the name of the oxide indicates the species—as sulphate of oxide of silver, or sulphate of silver; for the oxide of the metal being always understood, it is unnecessary to express it, unless when more than one oxide of the same metal combines with acids, as sulphate of protoxide of iron, and sulphate of sesquioxide of iron. These salts are often called protosulphate and persulphate of iron, where the prefixes proto and per refer to the degree of oxidation of the iron. The two oxides of iron are named *ferrous oxide* and *ferric oxide* by Berzelius, and the salts referred to, the ferrous sulphate, and the ferric sulphate. The names stannous sulphate and stannic sulphate express in the same way the sulphate of the protoxide of tin, and the sulphate of the peroxide of tin. But such names, although truly systematic, and replacing very cumbrous expressions, involve too great a change in chemical nomenclature to be speedily adopted. Having found its way into common language, chemical nomenclature can no longer be altered materially without great inconvenience. It must be learned as a language, and not be viewed and treated as the expression of a system. A *super*-sulphate contains a greater proportion of acid than the sulphate or neutral sulphate; a *bi*-sulphate twice as much, and a *sesqui*-sulphate once and a half as much as the neutral sulphate; while a *sub*-sulphate contains a less proportion than the neutral salt; the prefixes referring in all cases to the proportion of acid in the salt, or to the *electro-negative* ingredient, as with oxides. The excess of base in sub-salts is sometimes indicated by Greek prefixes expressive of quantity, as *di*-chromate of lead, *tris*-acetate of lead; but this deviation is apt to lead to confusion. If a precise expression for such subsalts were required, it would be better to say, the bibasic subchromate of lead, the tribasic subacetate of lead. But the names of both acid and basic salts are less in accordance with correct views of their constitution, than the names of any other class of compounds.

Combinations of water with other oxides are called *hydrates:* as hydrate of potassa, hydrate of boracic acid.

3.—In the names of quarternary compounds or of double salts, the names of the constituent salts are expressed, thus:—*Sulphate of alumina and potash* is the compound of sulphate of alumina and sulphate of potash; the name of the acid being expressed only once, as it is the same in both of the constituent salts. The name alum, which has been assigned by common usage to the same double salt, is likewise received in scientific language. The *chloride of platinum and potassium* expresses, in the same way, a compound of chloride of platinum and chloride of potassium. An oxichloride, such as the *oxichloride of mercury*, is a compound of the oxide with the chloride of the same metal.

The first ideas of a chemical nomenclature are due to Guyton de Morveau, whose views were published in 1782; but the chief merit of the construction of the valuable system in use is justly assigned to Lavoisier, who reported to the French Academy on the subject, in the name of a committee, in 1787. It has not been materially modified or expanded since its first publication. The present, or Lavoisierian nomenclature, does not furnish precise expressions for many new classes of compounds, the existence of which was not contemplated by its inventors, and many of its names express theoretical views of the constitution of bodies which are doubtful, and not admitted by all chemists. But its deficiencies are supplied, and the composition of bodies more accurately represented, in certain written expressions, or chemical formulæ, which are also employed to denote particular substances, and which form a valuable supplement to the nomenclature still generally used. These formulæ are constructed on the simplest principles, and besides supplying the deficiencies of the nomenclature, they at once exhibit to the eye the composition of bodies, and afford a mechanical aid in observing relations in composition, of the same kind as the use of figures in the comparison of arithmetical sums.

Symbols of the elements.—Each elementary substance is represented by the initial letter of its Latin name, as will be seen by reference to the table of elementary substances, page 102; but when the names of two or more elements begin with the same letter, a second in a smaller character is added for distinction; thus oxygen is represented by the letter O, the metal osmium by Os, fluorine by F, and iron (ferrum) by Fe; small letters, it is to be observed, never being significant of themselves, but employed only in connexion with the large letters as distinctive adjuncts. These symbols represent, at the same time, certain relative quantities of the elements, the letter O expressing not oxygen indefinitely, but 100 parts by weight of oxygen, and Fe 350 parts by weight of iron, or any other quantities of these two substances which are in the proportion of these numbers: 8 parts of oxygen, for instance, and 28 of iron. It will immediately be explained that the elementary bodies combine with each other in certain proportional quantities only, which are expressed by one or other indifferently of the two series of numbers placed against the names of the elements in the table referred to. These quantities are conveniently spoken of as the combining proportions, the equivalent quantities, or *the equivalents* of the elements. The symbol, or letter, of itself representing *one* equivalent of the element, several equivalents are represented by repeating the symbol, or by placing figures before it; thus Fe Fe, or 2 Fe, and 3 O, mean two equivalents of iron and three of oxygen. Or small figures are placed either above or below the symbol, and to the right; thus Fe^2, and O^3, or $Fe_2\,O_3$, are of the same value as the former expressions, but are used only when symbols are placed together in the formulæ of compounds. Two equivalents of an element are sometimes expressed by placing a dash through, or under its symbol, but such abbreviations will not be made use of in the present work.

Formulæ of compounds.—The collocation of symbols expresses combination: thus Fe O represents a compound of one equivalent or proportion of iron, and one of oxygen, or the protoxide of iron; SO_3, a compound of one equivalent of sulphur, and three of oxygen—that is, one equivalent of sulphuric acid; and sulphate of iron itself, consisting of one equivalent of each of the preceding compounds, may be represented as follows:

$$Fe\,O \quad S\,O_3, \text{ or}$$
$$Fe\,O + S\,O_3, \text{ or}$$
$$Fe\,O, \quad S\,O_3.$$

The sign plus (+) or the comma, being introduced in the second and third formulæ, to indicate a distribution of the elements of the salt into its two proximate constituents, oxide of iron, and sulphuric acid, which is not so distinctly indicated in the first formula. It may often be advantageous to make use of both the comma and the plus sign in the same formula, and then it would be a beneficial practice to use them as in the following formula for the double sulphate of iron and potash:

$$Fe\,O,\ SO_3+KO,\ SO_3,$$

in which the comma is employed to indicate combination more intimate in degree, or of a higher order than the plus sign, namely, of the oxide with the acid in each salt, while the combination of the two salts themselves is expressed by the sign +.

The small figures in the preceding formulæ affect only the symbol or letter to which they are immediately attached. Larger figures placed before and in the same line with the symbols apply to the *compound* expressed by the symbols. Thus $3\ S\ O_3$, means three equivalents of sulphuric acid; 2 Pb O, two equivalents of oxide of lead. But the interposition of the comma or plus sign prevents the influence of the figure extending farther, thus

$$2\ Pb\ O,\quad Cr\ O_3,\ \text{or}$$
$$2\ Pb\ O + Cr\ O_3,$$

is two proportions of oxide of lead, and one of chromic acid, or the sub-chromate of lead. To make the figure apply to symbols which are separated by the comma or plus sign, it is necessary to enclose all that is to be affected within brackets, and place the figure before them. Thus,

$$2\ (Pb\ O,\ Cr\ O_3)$$

means two proportions of neutral chromate of lead. The following formulæ of two double salts with their water of crystallization, exhibit the application of these rules:—

Iron-alum, or the sulphate of peroxide of iron and potash:

$$KO,\ SO_3+Fe_2\ O_3,\ 3\ SO_3+24\ HO$$

Oxalate of peroxide of iron and potash:

$$3\ (K\ O,\ C_2\ O_3)+Fe_2\ O_3,\ 3\ C_2\ O_3+6\ HO.$$

It will be found to conduce to perspicuity, to avoid either connecting two formulæ of different substances not in combination, by the sign plus, or allowing them to be separated merely by a comma, as the plus and comma *between* symbols or formulæ are conventionally understood to unite the formulæ into one, and to express combination; and indeed it is advisable to write every complete formula apart, and in a line by itself, if possible.

The only other circumstance to be attended to in the construction of such formulæ is the *arrangement* of the symbols or letters, which is not arbitrary. In naming a binary compound, such as oxide of iron, chloride of potassium, &c., we announce first the oxygen or element most resembling it in the compound; that is, the electro-negative ingredient; but in the formulæ of the same bodies, it is the other or the electro-positive element which is placed first, as in Fe O, and K Cl. In the formulæ of salts, it is likewise the basic oxide or electro-positive constituent which is placed first, and not the acid. Thus the sulphate of potash is $K\ O,\ S\ O_3$, and not $S\ O_3,\ KO$. Information respecting the constitution of a compound may often be expressed in its formula, by attending to this rule. Thus sulphuric acid of specific gravity 1.780, contains two proportions of water to one of acid, but by giving to it the following formula:

$$H\ O,\ S\ O_3+HO,$$

we express that one proportion only of water is combined as a base with the acid, and that the second proportion of water, the formula of which follows that of the acid, is in combination with this sulphate of water.

The above system of notation is complete, and sufficiently convenient for representing all binary compounds, and compounds belonging to the organic department of the science, in the formulæ of which the ultimate elements only are expressed.

But when salts and double salts are expressed, the formulæ sometimes become inconveniently long. They may often be greatly abbreviated, and made more distinct, by expressing each equivalent of oxygen in an oxide or acid, by a point placed over the symbol of the other element, thus:

Protoxide of iron, $\dot{\text{F}}\text{e}$.

Sulphuric acid, $\dddot{\text{S}}$.

Crystallized sulphate of protoxide of iron, $\text{Fe}\,\dddot{\text{S}}, \dot{\text{H}}+6\dot{\text{H}}$.

Alum, $\dot{\text{K}}\,\dddot{\text{S}}, \dot{\text{Al}}\ddot{\text{Al}}\,\dddot{\text{S}}_3+24\dot{\text{H}}$.

Felspar, $\dot{\text{K}}\,\dddot{\text{Si}}, \dot{\text{Al}}\ddot{\text{Al}}\,\dddot{\text{Si}}_3$.

Oxalate of peroxide of iron and potash, $3\dot{\text{K}}\,\dddot{\text{C}}\text{C}+\dot{\text{Fe}}\ddot{\text{Fe}}, 3\dddot{\text{C}}\text{C}+6\dot{\text{H}}$.

Such formulæ are more compact, and more easily compared with each other, the relation between the mineral felspar and alum without its water of crystallization, being seen at a glance on thus placing their formulæ together; the one having the symbol for silicium, the other that for sulphur, but everything else remaining the same. This abbreviated plan also exhibits more distinctly the relation between the equivalents of oxygen in the different constituents of a salt, which is always important.

It is to be observed, that the oxygen expressed by the points placed over a letter is brought under the influence of the small figure attached to that letter: as, for example, $\dddot{\text{S}}_3$ in the preceding formula of alum, means three equivalents of sulphuric acid; so that this sign has the same value as if it were written $3\,\dddot{\text{S}}$.

Equivalents of *sulphur* are likewise sometimes expressed by commas placed over other symbols, as the trito-sulphuride of arsenic by $\overset{,,,}{\text{As}}$; but such compounds are not of constant occurrence like the oxides, and do not create the same necessity for a new and arbitrary symbol. A compound body, such as cyanogen, which combines with a numerous series of other bodies, is often for brevity expressed by the initial letter or letters of its name, as—

Cyanogen .... Cy,
Ethyl .... E;

and the organic acids are sometimes expressed by a letter in the same way, but with the minus sign (—) placed over it: thus—

Acetic acid, by $\overline{\text{A}}$,
Tartaric acid, by $\overline{\text{T}}$.

But arbitrary characters of this kind will always be explained on the occasion of their introduction.

SECTION II.—COMBINING PROPORTIONS.

All analyses prove that the composition of bodies is fixed and invariable: 100 parts of water are uniformly composed of 11.1 parts by weight of hydrogen, and 88.9 parts of oxygen, its constituents never varying either in nature or proportion. This and other substances may exist in an impure condition, from an admixture of foreign matter, but their own composition remains the same in all circumstances. It is this constancy in the composition of bodies which gives to chemical analyses all their value, and rewards the vast care necessarily bestowed upon their execution.

An examination of the composition of a class of bodies, such as the oxides, containing an element in common, shows that any one element unites with very different quantities of the other elements. Thus in each of the five oxides of which the composition is given on page 112, the oxygen and other constituents appear in a different relation to each other:

Composition of Oxides.

Water.	Oxide of Copper.	Oxide of Zinc.	Oxide of Lead.	Oxide of Silver.
Oxygen ... 88.9	Oxygen 20.2	Oxygen ... 19.1	Oxygen ... 7.2	Oxygen 6.9
Hydrogen 11.1	Copper 79.8	Zinc 80.9	Lead 92.8	Silver 93.1
100	100	100	100	100

But the relation between the oxygen and the other constituent in these oxides will be seen more distinctly by stating their composition in such a way as to have the oxygen expressed by the same number in every case, or made equal to 100 parts. Thus:

Composition of Oxides.

Water.	Oxide of Copper.	Oxide of Zinc.	Oxide of Lead.	Oxide of Silver.
Oxygen . 100	Oxygen..... 100	Oxygen ... 100	Oxygen ... 100	Oxygen..... 100
Hydrogen 12.5	Copper...... 396	Zinc........ 406	Lead 1294	Silver...... 1350
112.5	496	506	1394	1450

From which it follows, that—

12.5 parts of hydrogen,
396 parts of copper,
406 parts of zinc,
1294 parts of lead,
1350 parts of silver,
combine with 100 *parts of oxygen.*

These numbers prove to be in some degree characteristic of the substances to which they are here attached, for when the composition of the *sulphides* of the same substances is examined, it is found that exactly corresponding quantities of hydrogen, copper, &c. likewise combine with one and the same quantity of sulphur, although not with 100 parts of that element as of oxygen. The conclusion from an examination of the sulphides is, that—

12.5 parts of hydrogen,
396 parts of copper,
406 parts of zinc,
1294 parts of lead,
1350 parts of silver,
combine with 200 *parts of sulphur.*

An examination of the *chlorides* of the same five elements likewise proves, that—

12.5 parts of hydrogen,
396 parts of copper,
406 parts of zinc,
1294 parts of lead,
1350 parts of silver,
combine with 443.75 *parts of chlorine.*

Hydrogen, copper, &c., are indeed found to unite in the proportions repeated above, with a certain or constant quantity of all other elements; as, for example, with 978 *bromine*, with 1580 *iodine*, &c.

On extending the inquiry to other substances, it appears that for each of them a number may be found which expresses in like manner the proportion in which that substance unites with 100 parts of oxygen, 200 of sulphur, 443.73 of chlorine, &c. These numbers constitute the combining proportions, or equivalent quantities of bodies, which are introduced in the table of the names of the elements at the beginning of this chapter, and which are the quantities understood to be expressed by the chemical symbols of these bodies.

Any series of numbers may be chosen for the combining proportions, provided the true relation between them is preserved, as in the first series of numbers given in the same table, which are all 12½ times less than the numbers of the second series. Hydrogen is reduced from 12.5 to 1, oxygen from 100 to 8, sulphur from 200 to 16: altered in the same proportion, copper becomes 31.66, zinc 32.52, lead 103.56, and silver 108. This series, or the hydrogen scale, is recommended by the circumstance that its numbers are smaller and more easily recollected than those of the other, or oxygen scale. The equivalents of several of the most important elements are now also generally allowed to be the exact multiples of the equivalent of hydrogen, so that the equivalent of the latter element being 1, the equivalents of the former are accurately expressed by entire numbers; — carbon by 6, oxygen by 8, nitrogen by 14, sulphur by 16, and iron by 28.

Having reference to the oxygen series, it is said, in general terms, that the combining proportion of a simple substance represents the quantity of that substance which combines with 100 parts of oxygen to form a protoxide. On the hydrogen scale, which I shall adopt, the definition of a chemical equivalent, or combining proportion becomes as follows:— *The combining proportion of a simple substance represents the quantity of that substance which unites with* 8 *parts of oxygen to form a protoxide.*

The first law of combination is, that "bodies unite with each other in their combining proportions only, or in multiples of them, and in no intermediate proportions." This law may be illustrated by the compounds of nitrogen and oxygen, which are five in number, and are composed as follows:—

Protoxide of nitrogen...................... nitrogen 14, oxygen 8.
Deutoxide of nitrogen .... nitrogen 14, oxygen 16.
Nitrous acid................................ nitrogen 14, oxygen 24.
Peroxide of nitrogen....................... nitrogen 14, oxygen 32.
Nitric acid................................. nitrogen 14, oxygen 40.

The first compound consists of a single combining proportion of each of its constituents. But in the other compounds, a single proportion of nitrogen is united with quantities of oxygen which correspond exactly with two, three, four, and five combining proportions of that element. In the greater number of binary compounds one of the constituents at least is present in the proportion of a single equivalent, like the nitrogen in this series, while the other constituent, generally the oxygen in oxides, and the electro-negative element in other compounds, is present in a multiple of its combining proportion. But the number of equivalents which may enter into a compound is subject to considerable variation, as will appear from the following examples:—

One eq.	of oxygen	+	One eq.	of hydrogen,	forms	water.
Two "	oxygen	+	One "	hydrogen,	form	peroxide of hydrogen.
One "	oxygen	+	Two "	copper,	forms	suboxide of copper.
One "	sulphur	+	Three "	oxygen,	"	sulphuric acid.
Two "	sulphur	+	Two "	oxygen,	form	hyposulphurous acid.
Two "	iron	+	Three "	oxygen,	"	peroxide of iron.
Two "	sulphur	+	Five "	oxygen,	"	hyposulphuric acid.
Two "	manganese	+	Seven "	oxygen,	"	hypermanganic acid.

Representing the constituents of a binary compound by A and B, the last being the oxygen or electro-negative constituent, the most frequent combinations are— A+B, A+2B, A+3B, and A+5B. The combination of 2A+3B, is not unfrequent, but 2A+B, A+4B, A+7B, 2A+2B, or 2A+5B, are of comparatively rare occurrence. Combination between two elements is not known to occur in more complicated ratios than the preceding, if the compounds of carbon and hydrogen be excepted, which are numerous, and exhibit great diversity of composition, like the compounds of organic chemistry generally, to which they properly belong.

Combination likewise takes place among bodies which are themselves compound, in proportional quantities, which are fixed and determined by the law, that "the combining number of a compound body is always the sum of the combining numbers of its constituents." Thus oil of vitriol, which contains water and sulphuric acid, is composed of these bodies in the proportion of—

Water	9
Sulphuric acid	40

in which the combining proportion of the water (9) is the sum of the equivalents of its constituents; namely, of oxygen 8, and of hydrogen 1; and that of sulphuric acid (40), of those of sulphur 16, and of oxygen 24; there being three proportions of oxygen in sulphuric acid. The combining proportion of oxide of zinc is 40.52, the sum of oxygen 8, and zinc 32.52; and the compound of this oxide with sulphuric acid, or the salt, sulphate of zinc, consists of—

Oxide of zinc	40.52
Sulphuric acid	40.
	80.52

Of potash, the combining proportion is 47; or oxygen 8, added to potassium 39; and to this proportion of potash the usual proportion of sulphuric acid is attached in the sulphate of potash, which is composed of—

Potash	47
Sulphuric acid	40
	87

Of these salts themselves, the combining proportions ought to be the sums obtained by the addition of the numbers of their constituents; and accordingly the double sulphate of zinc and potash consists of—

Sulphate of zinc	80.52
Sulphate of potash	87
	167.52

Of nitric acid the constituents are one eq. of nitrogen 14, and five of oxygen 40, making together 54, which is the combining proportion of that acid, and is found to unite with 9 water, with 40.52 oxide of zinc, and with 47 potash; or with the same quantities of these oxides as combine with 40 sulphuric acid. Carbonic acid is composed of one proportion of carbon 6, and two proportions of oxygen 16, so that its combining number is 22; in which proportion it unites with 47 potash, to form carbonate of potash. The equivalent quantities of all other acids and bases correspond in like manner with the numbers deducible from their composition. Indeed, the law is found to hold in compounds of every class and character, and whether they contain few or many equivalents of their elements.

Compound bodies likewise unite among themselves in multiples of their combining proportions, as well as in single equivalents. Thus 47 potash combine with 52.15 chromic acid, and with double that quantity, or 104.30, chromic acid, to form the yellow and red chromates of potash; the first containing one equivalent, and the second two equivalents of acid. The occurrence of multiple proportions was well illustrated by Dr. Wollaston in the carbonate and bicarbonate of potash. A quantity of the latter salt being divided into equal parts, one-half was exposed to a red heat, by the effect of which the salt lost some carbonic acid and became neutral carbonate; and both portions being afterwards decomposed by an acid, the salt in its original condition was found to afford a measure of carbonic acid gas exactly double of that yielded by the portion exposed to the high temperature. By experiments equally simple and convincing, he proved that in the three salts formed by oxalic acid and potash, the quantities of acid which combine with the same quantity of alkali are rigorously among themselves as the numbers 1, 2, and 4. The composition of all other super and sub-salts is found to be in conformity with the same law, one of the constituents being always present in the proportion of two or more equivalents.

The combining proportions of compound bodies depend entirely, therefore, upon those of their constituents, or upon the equivalents of the elementary bodies. The mode of determining these fundamental equivalents generally consists, as may be anticipated, in ascertaining the quantity of any element which exists united with 8 parts of oxygen in the protoxide of that element, which quantity is viewed as a single equivalent. Thus, of hydrogen and lead, the protoxides are water and litharge, in which respectively 8 oxygen are associated with 1 hydrogen and 103.56 lead, which numbers are therefore single equivalents of these elementary substances. But the difficulty still remains to know what is a protoxide; for the rule is not followed in all cases to consider that oxide of an element as the protoxide which contains the least proportion of oxygen. When only one oxide is known, it is presumed to be a protoxide, and composed of single equivalents, unless it corresponds in properties with a higher degree of oxidation of some other element; and of several oxides of the same element, that containing least oxygen is viewed as the protoxide, unless a higher oxide has better claims to be considered as such. Hence magnesia and oxide of zinc being the only oxides of magnesium and zinc known, are protoxides; and water, litharge, potash, soda, lime, and protoxide of iron, which are all the lowest oxides of different metals, are admitted without objection to be protoxides, and become standards of comparison for this class of bodies; while alumina, the only oxide of aluminum, differing entirely from the protoxide of iron, but closely resembling the peroxide of that metal, is considered a peroxide of similar constitution, or to contain three equivalents of oxygen and two of metal. Now in alumina 24 oxygen, or three equivalents, are united with 27.38 aluminum, one-half of which number, or 13.69, is therefore the equivalent of aluminum. The true protoxide of aluminum, if it is capable of existing, still remains to be discovered. The green oxide of chromium, which was till lately the lowest degree of oxidation known of that metal, was notwithstanding considered a peroxide, being analogous to alumina and the peroxide of iron. On the other hand, the second degree of the oxidation of copper, or the black oxide, and not the first degree of oxidation of that metal, must be viewed as the protoxide, or as composed of single equivalents, from its correspondence with the protoxide of iron and a large class of admitted protoxides. The lower degree of oxidation of copper or the red oxide, which contains only half the proportion of oxygen in the black oxide, comes therefore to be considered a suboxide; a compound of two equivalents of metal and one of oxygen. For reasons somewhat similar, the higher of the two grades of oxidation of mercury, or the red oxide of that metal, is now generally received as the protoxide, and the ash-coloured oxide reputed a suboxide. These suboxides of mercury and copper are capable of combining with acids, but they are the only suboxides which possess that property. It is the character of metallic protoxides to form salts with acids; and of several oxides of the same metal, the protoxide is always the most powerful base.

Bodies likewise *replace* each other in combination, in equivalent quantities. Thus in the decomposition of water by chlorine, which occurs in certain circumstances, 35.5 parts of chlorine unite with 1 hydrogen or one equivalent of that body, to form hydrochloric acid, and displace at the same time and liberate 8 parts of oxygen. Hence the number 35.5 represents the combining proportion of chlorine, which is equivalent in combination to, or can be substituted for, 8 oxygen. Again, in decomposing hydriodic acid, 35.5 chlorine unite with 1 hydrogen, and liberate 126.36 iodine, which proportion of iodine may again acquire 1 hydrogen, by decomposing sulphuretted hydrogen, and set free 16 sulphur. Hence 126.36 and 16 are the equivalent quantities of iodine and sulphur, which take the place of 35.5 chlorine or 8 oxygen in combination with 1 hydrogen.

When 32.52 grains of zinc are introduced into a solution of nitrate of copper, they dissolve, acquiring 8 oxygen and 54 nitric acid, and become nitrate of zinc, while 31.66 parts of metallic copper are deposited, which had previously been in the state of nitrate, and in combination with the above-mentioned quantities of oxygen and nitric acid, and the solution remains otherwise unaltered. Zinc throws down nearly all the metals from their solutions in acids in the same manner, and if the quantity of this substance introduced into the solutions, and dissolved, be a combining proportion, as in the instance given, the quantities of the metals precipitated will also be combining proportions of those metals. The quantity of zinc employed may be varied, but the quantity of other metal precipitated will still be, to the quantity of zinc dissolved, in the ratio of the combining numbers of the two metals. Lead, copper, tin, or any other metal, when it acts like zinc as a precipitant, likewise throws down equivalent quantities of other metals, and takes their place in the pre-existing compound. This substitution of one metal for another, in a saline compound, without any change in the character of the compound, shows how justly the combining proportions of bodies are termed their equivalent quantities or equivalents. The metal displaced, and that substituted for it, have evidently the same value in the construction of the compound, and are truly equivalent to each other.

The equivalent proportions of such oxides as are bases are ascertained by finding what quantity of each saturates the known combining proportion of an acid. Thus, to saturate 40 parts, or a combining proportion of sulphuric acid, the following proportions of different bases are requisite, and are equivalent in producing that effect:

Magnesia	20.67
Lime	28
Soda	31
Protoxide of manganese	35.67
Potassa	47
Strontia	51.84
Baryta	76.64
Protoxide of lead	111.56
Oxide of silver	116

The addition of these bodies to sulphuric acid in the above proportions destroys its sour taste and other properties as an acid, of which one of the most characteristic is that of reddening certain vegetable blue colours, such as litmus. The acid is said to be neutralized or saturated, and the product or compound formed is a neutral salt, which does not alter the blue colour of litmus. Of the bases mentioned, magnesia has the greatest saturating power, and oxide of silver the least; the proportion of these bases necessary to saturate the same quantity of sulphuric acid being 20.67 of the former, and 116 of the latter.

Conversely, the equivalent proportions of acids are the quantities which neutralize the known equivalent of any base or alkali. Thus 47 parts of potassa, or a combining proportion, is deprived of its alkaline properties,—of which the most obvious are its caustic taste and power to restore the blue colour of reddened litmus,—by the following proportions of different acids, and a neutral compound or salt produced in every case:—

Sulphurous acid	32
Sulphuric acid	40
Hydrochloric acid	36.5
Nitric acid	54
Chloric acid	75.5
Hyperchloric acid	91.5
Iodic acid	166.36
Hyperiodic acid	182.36

It thus appears that the acids differ as widely among themselves in their equivalent quantities as the bases do. The equivalent of either an acid or base thus deduced from its neutralizing power is always the same as that indicated by its composition, namely the sum of the equivalent numbers of its constituents. As the bases which saturate acids fully are all protoxides, it also necessarily follows that 100 parts of oxygen are always contained in the proportion of base which neutralizes the equivalent of an acid.

The equivalents of both acids and bases are likewise observed in those decompositions in which one acid is substituted for another acid in combination, or one base for another base. Thus an equivalent of sulphuric acid is found to disengage the equivalent quantity exactly of sulphurous acid from the sulphite of soda, of nitric acid from the nitrate of potash, or of hydrochloric acid from the chloride of sodium, and to replace it in combination with the base, forming in every case a neutral sulphate. An equivalent of potash separates in like manner an equivalent of magnesia, of lime, of barytes, or of protoxide of lead, from its combination with an acid. The proportion of acid or base necessary to produce a certain amount of decomposition may therefore be calculated from a knowledge of the equivalents of bodies; and such knowledge comes to be of the most frequent and valuable application for practical purposes.

But the substitution of equivalent quantities of different bodies for one another is most strikingly exhibited in the decompositions which follow the mixture of certain neutral salts. An equivalent of sulphate of magnesia being mixed with an equivalent of nitrate of barytes, the two bases exchange acids, the original salts disappear completely, and two new salts are produced—the sulphate of barytes, which is insoluble and precipitates, and the nitrate of magnesia, which remains in solution; as represented in the following diagram, in which the equivalent quantities are expressed:—

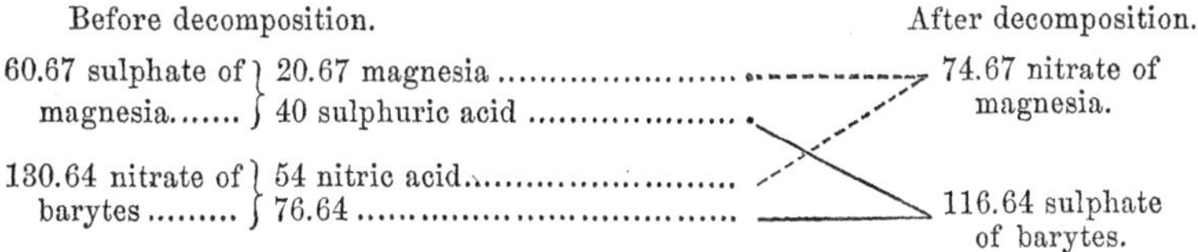

After a double decomposition of this kind, the liquid remains *neutral*, or there is no redundancy of either acid or base; because each of the new salts is composed of a single equivalent of acid and of base, like the salts from which they are formed. If one of the salts be added in a larger proportion than its equivalent quantity, the excess does not interfere with the decomposition, and remains itself unaffected, the decomposition proceeding no farther than the equivalents present. Hence the general observation, that neutral salts continue neutral after decomposition, in whatever proportions they may be mixed.

But the modes of fixing the equivalent numbers which have been stated are inapplicable to several elementary bodies; such as nitrogen, phosphorus, carbon, boron, and some metals of which the protoxides are not bases, and are uncertain. Nitrogen enters into nitric acid, of which acid it is known that the equivalent is 54, and that it contains five equivalents or 40 parts of oxygen, and consequently 14 parts of nitrogen. It is doubtful, however, whether 14 represents one or two equivalents of nitrogen. But the equivalent of ammonia likewise contains 14 nitrogen, and a less proportion is never found in the equivalent of any other compound into which that element enters. The number 14 is, therefore, *the least combining proportion* of nitrogen, and must on that account be taken as one equivalent. The equivalent of phosphorus can be shown on the same principle to be 32, that of arsenic 75, and that of antimony 129, as given in the tables, and not the halves of these numbers, as often estimated. These three bodies agree with nitrogen in their chemical relations, and the numbers recommended represent the quantities which replace 14 of nitrogen in analogous compounds. The equivalent of carbon may be deduced from the known equivalent of its compound, carbonic acid : but the equivalents of boron and silicium cannot be fixed upon with the same certainty, owing to the doubt which hangs over the equivalents of boracic and silicic acids.

Of the facts which involve the principle of combination in definite and equivalent proportions, the last mentioned appears to have been the first observed and explained. Wenzel, of Freiberg in Saxony, so far back as 1777, made an analysis of a great variety of salts with surprising accuracy, which enabled him to perceive that the neutrality which is observed after the reciprocal decomposition of neutral salts depends upon this,—that the quantities of different acids which saturate an equal weight of one base will also saturate equal weights of any other base.

Richter of Berlin confirmed and extended the observations of Wenzel, attaching proportional numbers to the acids and bases, and remarking for the first time that the neutrality does not change during the precipitation of metals by each other, and also that the proportion of oxygen in the equivalents of bases is the same in all, and may be represented by 100 parts. But the first foundations of a complete system of equivalents, embracing both simple bodies and their compounds, were laid by Dalton, at the same time that he announced his atomic theory. (New System of Chemical Philosophy, 1807). The observation that the equivalent of a compound body is the sum of the equivalents of its constituents, and the discovery of combination in multiple proportions, are peculiarly his. Dr. Wollaston afterwards adapted the more important equivalents to the common sliding rule of Gunter, by means of which, proportions can be observed without the trouble of calculation. This instrument, which is known under the name of *the scale of chemical equivalents*, contributed largely to the diffusion of the knowledge of the proportional numbers, but is not itself of much practical value.

The numerical accuracy of the equivalents assigned to bodies depends entirely upon the exactness of the chemical analyses from which they are deduced. The generally received series of numbers, which is adopted in this work, was drawn up by Berzelius from data supplied in a great measure by himself. The consideration of the laws of Wenzel and Richter, which were long overlooked or misunderstood, was revived by him, and by a series of analytical researches unrivalled for their extent and accuracy he first impressed upon chemistry the character of a science of number and quantity, which is now its highest recommendation. Several of Berzelius's numbers received a valuable confirmation from Dr. Turner, whose inquiries were especially directed to test an hypothesis respecting them proposed and ably maintained by Dr. Prout; namely, that the equivalents of all the elements are multiplies of the equivalent of hydrogen, and consequently if that equivalent be made equal to 1, all the others will be whole numbers. (Phil. Trans. 1833, p. 523). Dr. Penny took a part in the same inquiry, (Ibid. 1839, p. 13). More lately laborious researches have been undertaken with the same object by Dumas, Marignac, Pelouze, and others, whose results are quoted under the table of equivalents. It appears to be definitively settled that the equivalents of the elements are not, without exception, multiples of the equivalent of hydrogen. The number for chlorine (35·5) is conclusive against that hypothesis. At the same time, the accurate determinations of the equivalents of chlorine, silver, and potassium, by Maumíné, lend positive support to the opinion that these and all other equivalents are multiples of half the equivalent of hydrogen. So do the recent determinations of carbon and hydrogen in reference to oxygen, and those of nitrogen, sodium, iron, and calcium. The number for lead also, upon the determination of which extraordinary pains have been bestowed by Berzelius at different times, namely 103·56, is favourable to the same view. Now these are the equivalents upon which, above all others, our knowledge is most precise and certain.

Might not, therefore, the equivalent of hydrogen be divided by two, by which chlorine would become 71 and lead 207, hydrogen being 1? The multiple relation would not, however, be established by dividing the equivalent of hydrogen, for, as is justly observed by Berzelius, the chemical reasons which are adduced for the division of the equivalent of hydrogen apply with equal force to the equivalent of chlorine, and the one cannot be divided without dividing the other. The equivalent of chlorine would, therefore, still remain a multiple of half the equivalent of hydrogen.

SECTION III.—ATOMIC THEORY.

The laws of combination, and the doctrine of equivalents, which have just been considered, are founded upon experimental evidence only, and involve no hypothesis. The most general of these laws were not however suggested by observation, but by a theory of the atomic constitution of bodies, in which they are included, and which affords a luminous explanation of them. The partial verification which this theory has received in the establishment of these laws adds greatly to its interest, and is a strong argument in favour of its truth. It is the atomic theory of Dalton, the essential part of which may be stated in a few words.

Although matter appears to be divided and comminuted in many circumstances to an extent beyond our powers of conception, it is possible that it may not be indefinitely divisible; that there may be a limit to the successive division or secability of its parts; a limit which it may be difficult or impossible to reach by experiment, but which nevertheless exists. Matter may be composed of ultimate particles or atoms, which are not farther divisible, and each of which possesses a certain absolute and possibly appreciable weight. Now the question arises, is the atom in every kind of matter of the same weight, or do atoms of different kinds of matter differ in weight? Are the ultimate particles, for instance, to which charcoal and sulphur are reducible, of the same or different weights? Let their weights be supposed to be different, to be in the proportion of the equivalent numbers of sulphur and charcoal, which thus become *atomic weights*, and so of the atoms of other elementary bodies, and the whole laws of combination follow by the simplest reasoning. The atoms of the elementary bodies may be represented to the eye by spheres or by circles in which their symbols are inscribed to distinguish them, as in the following examples, with their relative weights.

Name.	Atom.	Weight of Atom.
Oxygen	(O)	8
Hydrogen	(H)	1
Nitrogen	(N)	14
Carbon	(C)	6
Sulphur	(S)	16
Lead	(Pb)	103.56

Chemical combination takes place between the atoms of bodies, which then come into juxtaposition; and in decomposition the simple atoms separate again from each other, in possession of their original properties. The atom or integrant particle of a compound body is an aggregation of simple atoms, and must therefore have a weight equal to the sum of their weights; as will be obvious from the exhibition of the atomic constitution of a few compounds.

	Atom.	Weight.
Water (oxide of hydrogen)	(H)(O)	1 + 8 = 9
Protoxide of nitrogen	(N)(O)	14 + 8 = 22
Deutoxide of nitrogen	(N)(O)(O)	14 + 16 = 30
Sulphuric acid	(S)(O)(O)(O)	16 + 24 = 40
Oxide of lead	(Pb)(O)	103·56 + 8 = 111·56
Sulphate of lead	{ (Pb)(O) / (S)(O)(O)(O) }	111·56 + 40 = 151·56

It is unnecessary to make any assumption as to the nature, size, form, or even actual weight of the atoms of elementary bodies, or as to the mode in which they are grouped or arranged in compounds. All that is known or likely ever to be known respecting them is their relative weight. The atom of oxygen is eight times heavier than that of hydrogen, but their actual weights are undetermined. To afford the means of expressing the relative weights of these and other atoms, a number which is entirely arbitrary is assigned to one of them, namely 8 to the atom of oxygen, and then the weight of the atom of hydrogen can be said to be 1, of nitrogen 14, of carbon 6, of sulphur 16, and of lead 103·56. A single atom of water contains one atom of oxygen (8), and one of hydrogen (1), and must therefore weigh 9; an atom of oxide of lead contains one atom of oxygen and one of lead, which weigh together 111·56; an atom of sulphuric acid, one atom of sulphur and three atoms of oxygen, which weigh together 40; and an atom of sulphate of lead, including one of each of the preceding compound atoms, must weigh 111·56+40, or 151·56.

The equivalent quantities being now represented by atoms, it necessarily follows that bodies can combine in these quantities or multiples of them only, and not in intermediate proportions, for atoms do not admit of division. In a series of several compounds of the same elements, such as the oxides of nitrogen, which was formerly referred to in illustration of combination in multiple proportions (page 113), one atom of nitrogen combines with one, two, three, four and five atoms of oxygen, and a simple ratio between the quantities of oxygen in these compounds is the consequence. The equivalent of the compound body also is the sum of the equivalents of its constituents, for the weight of a compound atom is the weight of its constituent atoms.

By the juxtaposition, separation, and exchange of one atom for another in compounds, all kinds of combination and decomposition in equivalent quantities may be produced, while the substitution of ponderable masses for the abstract idea of equivalents renders the whole changes most readily conceivable.

This theory being adopted as a useful, while it is at the same time a highly probable representation of the laws of combination, its terms atom or atomic weight may be used as synonymous with equivalent, equivalent quantity, and combining proportion.

M. Dumas is disposed to modify the atomic theory so far as to allow the divisibility of the atoms or ultimate masses in which a body enters into combination, and to suppose that they are groups of more minute atoms, into which they may be divided by physical, but not by chemical forces. He distinguishes the atoms which correspond with equivalents as *chemical atoms*, and allowing them to represent truly and constantly the least quantities in which bodies combine, still supposes that under the influence of heat, and perhaps other physical agencies, these molecules may be subdivided into atoms of an inferior order, of which, for example, two, four, or a thousand, are included in a single chemical atom. (Leçons sur la Philosophie Chimique, professées au Collège de France, par M. Dumas, page 233). But surely such a view is entirely subversive of the atomic theory. It is principally founded on the assumed existence of a similarity between atoms in their capacity for heat, and in their volume while in the gaseous state.

SPECIFIC HEAT OF ATOMS.

The quantity of heat necessary to raise the temperature of equal weights of different bodies a single degree, varies according to their nature, and may be expressed by numbers which are the capacities for heat or specific heats of these bodies (page 49). This difference appears in the numbers for several simple bodies placed together in the first column of the following table, among which no relation can be perceived. But if the comparison is made between the capacity for heat not of *equal* weights, but of *atomic* weights or equivalent quantities of the same bodies, as in the second and third columns of the table, then the numbers for several bodies are found to be nearly the same, and those of others to bear a simple relation to each other.

SPECIFIC HEAT.

	I. Of equal weights. Specific heat of same weight of water being 1.	II. Of atoms. Specific heat of atom of water being 1.	III. Of atoms. Specific heat of atom of lead being 1.	IV. Atomic weights.
Lead	0·0293	0·3372	1·0000	103·56
Tin	0·0514	0·3358	0·9960	58·82
Zinc	0·0927	0·3321	0·9850	32·52
Copper	0·0949	0·3340	0·9908	31·66
Nickel	0·1035	0·3404	1·0095	29·57
Cobalt	0·10696	0·3508	1·040	29·52
Iron	0·1100	0·3315	0·9831	28
Platinum	0·0314	0·3443	1·0211	98·68
Sulphur	0·1880	0·3359	0·9963	16
Mercury	0·0330	0·3714	1·1015	100·07
Tellurium	0·05155	0·3788	1·123	64·14
Gold	0·0298	0·3292	0·9765	98·33
Arsenic	0·081	0·6768	2·0074	75
Silver	0·0557	0·6694	1·9855	108
Phosphorus	0·385	1·3415	3·9789	32
Iodine	0·10824	1·5197	4·506	126·36
Carbon	0·2411	0·1698	0·4766	6
Bismuth	0·03084	0·2190	0·6494	70·95

Of the first twelve substances, which are all metals, with the exception of sulphur, the capacities of the atoms approach so closely, that they may be considered as identical; their capacities appearing to be all nearly one-third of that of the atom of water, in the second column; and nearly coinciding with the capacity of the atom of lead, one of their number in the third column. The weights of the atoms themselves are added in a fourth column, for convenience of reference. The twelve substances in question, taken in the proportions of their atomic weights, will, therefore, undergo an equal change of temperature on assuming an equal quantity of heat. The two metals which follow in the table, namely, arsenic and silver, appear to have an equal capacity for heat, which is double that of lead and the class which coincides with it, while the capacity of phosphorus is four times, and that of iodine four and a half times greater than that of lead and its class. The capacity of the atom of bismuth appears to be two-thirds, and that of carbon to be one-half of the capacity of that of lead. The general results may therefore be stated as follows:—

			Weight of Atom.
Specific heat of atom of	*lead*	1	103·56
" "	tin	1	58·82
" "	zinc	1	32·52
" "	copper	1	31·66
" "	nickel	1	29·57
" "	cobalt	1	29·52
" "	iron	1	28
" "	platinum	1	98·68
" "	sulphur	1	16
" "	mercury	1	100·07
" "	tellurium	1	64·14
" "	gold	1	98·33
" "	arsenic	2	75
" "	silver	2	108
" "	phosphorus	4	32
" "	iodine	$4\frac{1}{2}$	126·36
" "	bismuth	$\frac{2}{3}$	70·95
" "	carbon	$\frac{1}{2}$	6

Messrs. Dulong and Petit, whose researches supplied the greater portion of these valuable results, drew a more general conclusion from them, namely that all atoms, or at least all simple atoms, have the same capacity for heat, and that those atomic weights which are inconsistent with that supposition, ought to be altered and accommodated to it. The specific heat of a body would thus afford the means of fixing its atomic weight. Some of the alterations in the atomic weights, which would follow the adoption of this law, might be advocated upon other grounds — such as halving the atomic weight of silver, doubling that of carbon, and adding one-half to that of bismuth. But the equivalent of phosphorus would require to be divided by four, while that of arsenic, which it so closely represents in compounds, is divided only by two; changes which are inadmissible.

It must be concluded, then, that elementary atoms have not necessarily the same capacity for heat, although a simple relation appears always to exist between their capacities. The capacities of the three gaseous elements, oxygen, hydrogen, and nitrogen, may likewise be adduced in support of such a relation, provided they are the same for equal volumes of the gases, agreeably to the observations of Dulong. But this relation can only be looked for between bodies while under the same physical condition, and perhaps agreeing in other circumstances also, for the capacity for heat of the same body is known to vary under the different forms of solid, liquid, and gas; and, indeed, while the body is in the same state, its capacity appears not to be absolutely constant, but to increase perceptibly to elevated temperatures (page 49).

The capacities of compound atoms have also been submitted to a sufficiently extensive examination to determine that simple relations subsist among them. In two classes of analogous combinations, the capacities of the atoms for heat were found by M. Neumann, of Königsberg, to approach so closely, that they may be admitted to be the same, the differences being sufficiently accounted for by the errors of observation unavoidable in such delicate researches.

	OF EQUAL WEIGHTS. Specific heat of same weight of water being 1.	OF ATOMIC WEIGHTS. Specific heat of atom of water being 1.
Carbonate of lime	0·2044	0·1148
Carbonate of barytes	0·1080	0·1181
Carbonate of iron	0·1819	0·1156
Carbonate of lead	0·0810	0·1200
Carbonate of zinc	0·1712	0·1187
Carbonate of strontia	0·1445	0.1184
Dolomite (carbonate of lime and magnesia)	0·2111	0·1121
	Mean	0·1162

A small class of sulphates presented a similar result:—

	OF EQUAL WEIGHTS.	OF ATOMIC WEIGHTS.
Sulphate of baryta	0·1068	0·1384
Sulphate of lime	0·1854	0·1412
Sulphate of strontia	0·1300	0·1326
Sulphate of lead	0·0830	0·1398
	Mean	0·1380

The numbers in the second column of both tables deviate very little from their mean, but there is no obvious relation between the two means. Identity in capacity for heat is, therefore, to be looked for in compound atoms of the same nature, and which closely agree in their chemical relations, like the numbers of each group, but not between compound atoms which are differently constituted.

Our information on this subject has been greatly extended of late by the valuable researches of M. Regnault.[1] The *atomic heat* of bodies, as it is named by this chemist, is obtained by multiplying the observed specific heat of each body by its equivalent, the latter being taken upon the oxygen scale. Now this product is found to vary for the metallic elements as the numbers 38 to 42, a greater difference than can result from errors of observation; so that the law of atoms is not verified in an absolute manner. But if it is considered that the atomic weights of the simple substances in question vary at the same time from 200 to 1400, the law must be adopted, as at least closely approximating to the truth. The law would probably represent the results of observation in a perfectly rigorous manner, if the specific heat of each body could be taken at a determinate point of its thermometrical scale, and the specific heat be further disencumbered of all the foreign influences which modify the observation, — such as the state of softness, with the assumption of a certain portion of the latent heat of fusion, which many bodies exhibit before melting entirely, — and the heat absorbed to produce dilatation, which is very great in gases, much more feeble in solid and liquid bodies, but which can in no case be neglected (Regnault). An increase of the density of copper also, produced by hammering it, is found by Regnault to effect a sensible diminution of its specific heat: the latter recovers its original value in the metal after being heated.

The same element, in different conditions as to crystalline form, hardness, and aggregation, may vary greatly in its specific heat, as is observed of carbon both by Regnault, and by Delarive and Marcet. (Annales, &c., t. lxxv. p. 242). The results of the former are as follows:—

SPECIFIC HEAT OF VARIETIES OF CARBON.

Animal charcoal	0·26085
Wood charcoal	0·24150
Coke of coal	0·20307
Charcoal from anthracite	0·20146
Graphite, natural	0·20187
Graphite of iron furnaces	0·19702
Graphite of gas retorts	0·20360
Diamond	0·14687

The calorific capacity of this body is the more feeble in proportion as its state of aggregation is greater: it is an instance of a body which may exist with calorific capacities extending through a very wide range.

The following metallic protoxides of the formula MO,[2] protoxide of lead, red oxide of mercury, protoxide of manganese, oxide of copper, and oxide of nickel, have an atomic heat varying from 70·01 to 76·21, of which the mean is 72.03; these numbers being the observed specific heats of the oxides multiplied by their atomic weights: the same product averages about 40 in the elements. The atomic heat of magnesia is 63·03, and of oxide of zinc, 62·77, expressed in the same manner, which agree very closely together, but differ considerably from the other protoxides.

The protosulphurets, of the formula MS, correspond nearly with the protoxides,— the protosulphurets of iron, nickel, cobalt, zinc, lead, mercury, and tin, varying from 71·34 to 78·34; with a mean of 74·51, while the mean of the protoxides is 72·03.

Sesquioxides, of the formula M_2O_3, give for the product of their specific heats by their atomic weights, numbers between 158·56 and 180·01; with an average of 169·73: they are sesquioxide of iron, sesquioxide of chromium, arsenious acid, oxide

[1] On the specific heat of simple and compound bodies: Annales de Chimie, &c., t. lxxiii. p. 5, and 3rd sér. t. i. p. 129.

[2] M representing 1 eq. of metal.

of antimony, and oxide of bismuth, represented as Bi_2O_3, with an equivalent of 1003·6. But the number of alumina (Al_2O_3) was different, being in the form of corundum 126·87, and the saphire 139·61. Two corresponding sulphurets gave numbers somewhat higher than the oxides, namely, sulphuret of antimony 186·21, and sulphuret of bismuth 195·90, of which the mean is 191·06.

Two oxides, of the formula MO_2, namely, binoxide of tin, and artificial titanic acid, gave the first 87·23, and the second 86·45. The bisulphuret of iron (pyrites) gave 96·45; the bisulphuret of tin 135·66; the sulphuret of molybdenum 123·46; and bisulphuret of arsenic (AsS_2) 174·51.

Oxides, of the form MO_3, gave the following results: tungstic acid 118·38, molybdic acid 118·96, silicic acid 110·48, boric acid 103·52.

The subsulphuret of copper, Cu_2S, gave 120·21; and the sulphuret of silver, usually represented AgS, gave 115·86.

The following chlorides, to which M. Regnault is disposed to assign the common formula M_2Cl, gave results comprised between 156·83 and 163·42, with a mean of 158·64 — chloride of sodium, chloride of potassium, chloride of silver, subchloride of copper, and subchloride of mercury. The corresponding iodides ranged from 162·30 to 169·38, exclusive of the iodide of silver, which was 180·45. Of corresponding bromides, bromide of potassium was 166·21, bromide of silver 173·31, and bromide of sodium 175·65.

Protochlorides of the formula MCl, namely, chlorides of barium, strontium, calcium, magnesium, lead, mercury, zinc, and tin, were comprised between 114·72 and 119·59; with a mean of 117·03. The protochloride of manganese was somewhat lower, 112·51.

Of volatile bichlorides (MCl_2), bichloride of tin gave 239·18, and chloride of titanium 227·63; of which the mean is 233·40. The two corresponding chlorides of arsenic and phosphorus, MCl_3, gave, the first 399·26, and the second 359·86: mean 379·51.

The numbers for iodide of lead and iodide of mercury (MI) also closely approximate, the first being 122·54, and the second 119·36: mean 120·95. The fluoride of calcium (MF) gave 105·31.

The principal results obtained by M. Regnault for the salts are thrown together in the following table. The equivalents given in the general formula are those of the table at the beginning of this chapter.

Name of the salt.	General formula, (M=1 eq. of metal.)	Product of the specific heats by the atomic weights.	Mean.
Nitrate of potassa	$MO+NO_5$	302·49	301·72
Nitrate of soda	"	297·13	
Nitrate of silver	"	305·55	
Nitrate of barytes	"	248·83	
Metaphosphate of lime	$MO+PO_5$	248·64	
Chlorate of potassa	$MO+ClO_5$	321·04	
Arseniate of potassa	$MO+AsO_5$	317·30	
Pyrophosphate of potassa	$2MO+PO_5$	395·79	389·01
Pyrophosphate of soda	"	382·22	
Phosphate of lead	"	302·14	
Phosphate of lead	$3MO+PO_5$	397·96	
Arseniate of lead	$3MO+AsO_5$	409·37	
Sulphate of potassa	$MO+SO_3$	207·40	206·80
Sulphate of soda	"	206·21	
Sulphate of baryta	"	164·54	166·15
Sulphate of strontia	"	164·01	
Sulphate of lead	"	165·39	
Sulphate of lime	"	168·49	
Sulphate of magnesia	"	168·30	

Name of the salt.	General formula, (M=1 eq. of metal.)	Product of the specific heats by the atomic weights.	Mean.
Chromate of potassa	$MO+CrO_3$	229·83	
Bichromate of potassa	$MO+2CrO_3$	358·67	
Biborate of potassa	$MO+2BO_3$	321·27	311·07
Biborate of soda	"	300·88	
Biborate of lead	"	258·60	
Borate of potassa	$MO+BO_3$	219·52	216·06
Borate of soda	"	212·60	
Borate of lead	"	165·54	
Carbonate of potassa	$MO+CO_2$	187·04	184·35
Carbonate of soda	"	181·65	
Carbonate of lime (Iceland spar)	$MO+CO_2$	131·61	134·40
Carbonate of lime (arragonite)	"	131·56	
Ditto (white saccharoid marble)	"	136·20	
Ditto (grey saccharoid marble)	"	132·45	
Ditto (white chalk)	"	135·57	
Carbonate of baryta	"	135·99	
Carbonate of strontia	"	133·58	
Carbonate of iron	"	138·16	

The results of M. Regnault on the specific heat of compound bodies are of great interest with regard to the question of the division of the atomic weights of certain elements, to which reference has been made. They establish an equally close relation between the specific heat of analogous compounds as exists among elementary bodies. The general law is announced by M. Regnault in the following manner:—"In all compound bodies, of the same atomic composition and similar chemical constitution, the specific heats are in the inverse proportion of the atomic weights." This law comprehends, as a particular case, the law of Dulong and Petit for similar bodies, and appears to be verified by experiment within the same limits as the latter.

RELATION BETWEEN THE ATOMIC WEIGHTS AND VOLUMES OF BODIES IN THE GASEOUS STATE.

Several of the elementary bodies are gases, such as oxygen, hydrogen, nitrogen, and chlorine, and the proportions in which they combine can be determined by *measure*, with equal, if not greater facility than by weight. Now a relation of the simplest nature is always found to subsist between the measures or volumes in which any two of the gaseous elementary bodies unite. This arises from the circumstance that the specific gravities of gases either correspond exactly with their atomic weights, or bear a simple relation to them. The atom of chlorine is 35½ times heavier than that of hydrogen; and chlorine gas is also 35½ times heavier than hydrogen gas, so that the combining measures of these two gases, which correspond with single equivalents, are necessarily equal. The atom of nitrogen, and its weight as a gas, being both 14 times greater than the atom and weight of hydrogen gas, their combining volumes must be the same. The atom of oxygen is eight times heavier than that of hydrogen, but oxygen gas is 16 times heavier than hydrogen gas, so that taken in equal volumes these two gases are in the proportion by weight of two equivalents of oxygen to one of hydrogen. Hence, in the combination of single equivalents of these elements to form water, half a volume or measure of oxygen gas unites with a whole volume or measure of hydrogen gas. One volume of nitrogen also unites with half a volume of oxygen, and with a whole volume of the same gas, to form respectively the protoxide and deutoxide of nitrogen.

The exact ratio of one to two in which oxygen and hydrogen gases combine by measure, was first observed by Humboldt and Gay-Lussac in 1805. The subject was pursued by the latter chemist, who established the simple ratios in which gases generally combine, and published the laws observed by him, or his Theory of Volumes, shortly after the announcement of the Atomic Theory by Dalton. They afforded new and independent evidence of the combination of bodies in definite and also in multiple proportions, equally convincing as the observed proportions by weight in which bodies unite. Gay-Lussac likewise observed that the product of the union of two gases, if itself a gas, sometimes retains the original volume of its constituents, no contraction or change of volume resulting from their combination:— thus one volume of nitrogen and one volume of oxygen form two volumes of deutoxide of nitrogen; one volume of chlorine and one volume of hydrogen form two volumes of hydrochloric acid gas; and that when contraction follows combination, which is the most common case, the volume of the compound gas always bears a simple ratio to the volumes of its elements. Thus two volumes of hydrogen, and one of oxygen, form two volumes of steam; one volume of nitrogen and three of hydrogen gas form two volumes of ammoniacal gas; one volume of hydrogen and one-sixth of a volume of sulphur-vapour form one volume of sulphuretted hydrogen gas. In these and all other statements respecting volumes, the gases compared are supposed to be in the same circumstances as to pressure and temperature.

The uniformity of properties observed among gases in compressibility and dilatability by heat, has appeared to many chemists to indicate a similarity of constitution, and to favour the idea that they all contain the same number of atoms in the same volume. May not equal volumes of oxygen and hydrogen gases, for instance, be represented by an equal number of atoms of oxygen and hydrogen respectively placed at equal distances from each other, and the difference of sixteen to one in the densities of the two gases arise from the atom of oxygen being really sixteen times heavier than that of hydrogen? Equal volumes of gases would then contain an equal number of atoms, and one, two, or three volumes would be an equivalent expression to one, two, or three atomic proportions, the terms *volume* and *atom* becoming of the same import, or expressing equal quantities of bodies. But such a view is obviously inapplicable to compound gases, as their volume has a variable relation to that of their elements; and its adoption would require grave alterations to be made in the atomic weights of several of the elements themselves, to accommodate those weights to the observed densities of the bodies in the gaseous state. This will be seen from the following table, in which the volume or fractional part of a volume placed against each element always contains the same number of its presently received atoms. These volumes are, therefore, the equivalent volumes of the elements, and may be viewed as representing the bulk of their atoms in the gaseous state, *the combining measure of hydrogen being taken as two volumes.*

	ATOMS.	
	Volume.	Weight.
Hydrogen	2	1
Nitrogen	2	14
Chlorine	2	35·5
Bromine	2	98·26
Iodine	2	126·36
Mercury	2	100·07
Oxygen	1	8
Phosphorus	1	32
Arsenic	1	75
Sulphur	$\frac{1}{3}$	16

Of the first six bodies enumerated, equivalent weights occupy each two volumes. It was, indeed, the observation of this equality between the atom and volume in these gases, that led to the supposition of that relation being general. But the atoms of oxygen, phosphorus, and arsenic, occupy only one volume, and would require to be doubled to fill the same volume as the preceding class; or the latter rather preserved fixed, and the former class divided by two. The present atom of sulphur affords only one-third of a volume of vapour, and must, therefore, be multiplied by six to afford two volumes.

It will be found conducive to perspicuity to apply the expression *combining measure* to the volume or volumes of a gas which enter into combination. The combining measure of oxygen being one volume, the combining measure of hydrogen and its class will be two volumes; or the atom of oxygen gives one, and the atom of hydrogen two volumes of gas. Volumes of the gases may be represented by equal squares with their relative weights inscribed, the numbers having reference to the number assigned to the oxygen volume. If that number be 8, or the atomic weight of oxygen, as in column 1 of the table which follows, then the number to be inscribed in each of the two volumes forming the combining measure of hydrogen will be 0·5, or half its atomic weight, the combining measure itself having the full atomic weight of hydrogen, namely 1. So, of other gases, the combining measure has the whole atomic weight, which is divided among the component volumes. But there is the reason for preferring the number 1105·6 to 8 for the standard oxygen volume, that the weight of a volume of *air* being taken as 1000, that of an equal volume of oxygen is 1105·6; and consequently the corresponding number for the volume of hydrogen, 69·3, expresses the relation in weight of that gas also to air, and so do the corresponding numbers for all the other gases. The numbers on this scale, which express the relative weights of a volume of each gas, and are inscribed in the squares of column 2, are indeed the common *specific gravities of the gases.*

	I. Atomic weight.	Combining measure.	II. Combining measure.
Air			1000
Oxygen	1	8	1105·6
Phosphorus	32	32	4422
Hydrogen	1	0·5 .... 0.5	69·3 .... 69·3
Chlorine	35·5	17·75 .... 17·75	2453 2453

The double squares, which represent the combining measures of hydrogen and chlorine, are divided into volumes by *dotted* lines, to show that the division is imaginary, the partition of a combining measure, like that of an atom which it represents, being impossible. The specific gravities of gases being merely the relative weights of equal volumes, may be expressed by the numbers in the squares of the first column; and the specific gravity of oxygen being accordingly made 8, the specific gravity of any other gas will either be the same number as its atomic weight, or an aliquot part of it. Or if the specific gravity of oxygen be made 1 or 1000, the relation of densities to atomic weights will still be very obvious. (See page 84).

The combining measures of *compound* gases, although variable, have still a constant and simple relation to each other—such as 1 to 1, 1 to 2, or 2 to 3; their elements in combining suffering either no condensation, or a definite and very simple change of volume. Hence the density of a compound gas may often be calculated with more precision from the densities of its constituents, and a knowledge of the change of volume, if any, which occurred in combination, than it can be determined by experiment.

To deduce on this principle the specific gravity of steam. Water consists of single equivalents of oxygen and hydrogen, of which the combining measure of the first is one, and that of the second two volumes. These *three* volumes weigh 1105·6 + 69·3 + 69·3 = 1244·2, and they form *two* volumes of steam; of which one volume must, therefore, weigh 1244·2 divided by two, or 622·1, which is, consequently, the calculated specific gravity of steam, referred to that of air as 1000. The relations in volume of the gases before and after combination may be thus exhibited:—

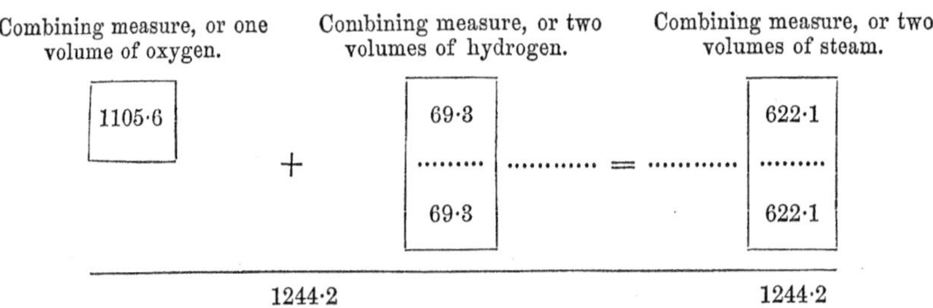

It thus appears necessary to inscribe 622·1 in each volume of steam, to make up 1244·2, the known weight of the two volumes.

In the formation of hydrochloric acid equal measures of chlorine and hydrogen unite without condensation, so that the product possesses the united volumes of its constituent gases.

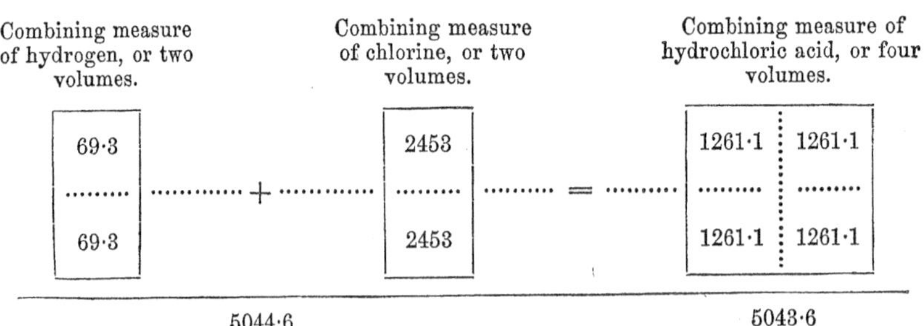

The specific gravity or weight of a single volume of hydrochloric acid is, therefore, obtained by dividing 5044·6 by 4, and is 1261·1.

The specific gravity of the vapour of an elementary body which there are no means of ascertaining experimentally, may sometimes be calculated from the known density of a gaseous compound containing it. The density of carbon vapour may be thus deduced from the observed density of carbonic oxide gas. Assuming that the combining measure of carbon is double that of oxygen, as is true of hydrogen and several other elementary bodies, then carbonic oxide, which like water consists of single equivalents of its constituents, will resemble steam in its constitution also, and be composed of one volume of oxygen gas, and two volumes of carbon vapour condensed into two volumes. The weight of a single volume of carbonic oxide being 972·7, two volumes (1945·4) may be resolved, as shown in the diagram below, into one volume of oxygen, 1105·6, and two volumes of carbon-vapour, 839·8, (1945·4 — 1105·6 = 839·8) each of which it follows must weigh 419·9, or 420.

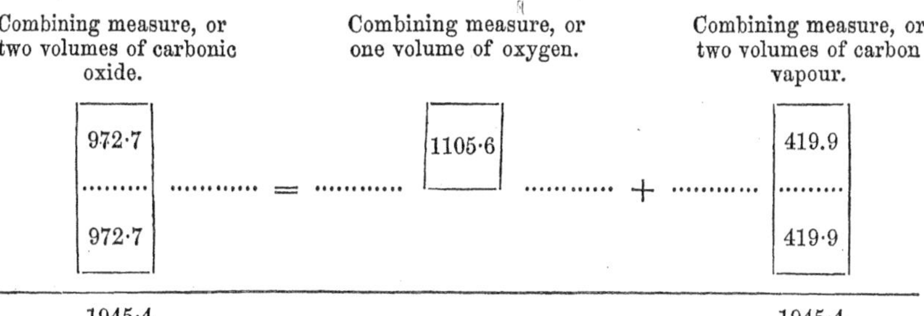

But the density 420 thus assigned to carbon vapour will only be true if it corresponds with hydrogen in its combining measure; but the combining measure of carbon vapour may as well be one-half that of hydrogen, like that of phosphorus, or one-sixth, like that of sulphur, and then the density will be double or six times that supposed. The important conclusion, however, that the density of carbon vapour is either 420, or some multiple or sub-multiple of that number, is quite certain.

The following Table comprises nearly all the accurate information which chemists at present possess respecting the specific gravities of gaseous bodies. The bodies placed first in the table are generally considered as belonging to the inorganic, and those in the latter part to the organic department of the science. They are all experimental results, with the exception of two or three cases which are calculated. The specific gravity of carbon-vapour is assumed here as six-sixteenths of that of oxygen (1105·6).

TABLE OF SPECIFIC GRAVITY OF GASES AND VAPOURS.

Names of Substances.	Proportion of an eq. in 1 volume.	Specific Gravity.			Observers.
		Air = 1.	Oxyg. = 1.	H. = 1.	
Sulphur	3 S	6617	5983·9	96	D.
Oxygen	O	1105·63	1000	16	R.
Phosphorus	P	4355	3938·3	64	D.
Arsenic	As	10600	9586·6	150	M.
Hydrogen	$\frac{H}{2}$	69·26	62·6	1	R.
Carbon (hypothetical)	$\frac{C}{2}$	414·61	375	6	Calcul.
Nitrogen	$\frac{N}{2}$	971·37	878·5	14	R.
Chlorine	$\frac{Cl}{2}$	2421·6	2189·9	35·5	G-L.
Bromine	$\frac{Br}{2}$	5540	5009·7	78	M.
Iodine	$\frac{I}{2}$	8716	7882	126	D.
Mercury	$\frac{Hg}{2}$	6976	6308·5	100·07	D.
Water	$\frac{HO}{2}$	622	562·6	9	R.
Carbonic oxide	$\frac{CO}{2}$	971·2	875	14	Calc.
Protoxide of nitrogen	$\frac{NO}{2}$	1520.4	1375	22	C.
Carbonic acid	$\frac{CO_2}{2}$	1524·5	1378·6	22	B. D.
Chlorocarbonic acid	$\frac{COCl}{2}$	3399	3564.8	49·5	
Sulphide of carbon	$\frac{CS_2}{2}$	2644·7	2391·6	38	G-L.
Hydrosulphuric acid	$\frac{HS}{2}$	1191·2	1077·3	17	G. T.

Names of substances.	Proportion of an eq. in 1 volume.	Specific Gravity.			Ob-servers.
		Air = 1.	Oxyg. = 1.	H. = 1.	
Hypochlorous acid	$\frac{ClO}{2}$	2998·4	2693·4	43·5	
Cyanogen	$\frac{NC_2}{2}$	1806·4	1633·7	26	G-L.
Sulphurous acid	$\frac{SO_2}{2}$	2193	1983·1	32	H-D.
Sulphuric acid (anhydrous)..	$\frac{SO_3}{2}$	3000	2713	40	M.
Chlorosulphuric acid	$\frac{SO_2\ Cl}{2}$	4665	4219	67·5	R.
Chloride of sulphur...........	$\frac{SCl_2}{2}$	3685	3332·7	51·5	D.
Arsenious acid	AsO_3	13850	12526	198	M.
Sulphate of water at 848° ...	$\frac{HO,\ SO_3}{2}$	1680	1519	24·5	B.
Chloride of mercury	$\frac{HgCl}{2}$	9800	8862·3	135·5	M.
Bromide of mercury	$\frac{Hg\ Br}{2}$	12160	10996·6	178	M.
Iodide of mercury	$\frac{HgI}{2}$	15630	14134·6	226	M.
Bichloride of tin	$\frac{SnCl_2}{2}$	9199·7	8389·4	...	D.
Bichloride of titanium........	$\frac{TiCl_2}{2}$	6876	6181·9	...	D.
Sulphuret of mercury	$\frac{HgS}{3}$	5510	4982·9	77·3	M.
Penta-chloride of phosphorus	$\frac{PCl_5}{8}$	3680	3329	52·375	C'.
Fluoride of silicium...........	$\frac{SiFl_3}{3}$	3600	3255·5	...	D.
Chloride of silicium...........	$\frac{SiCl_3}{3}$	5939	5370·7	...	D.
Hydrochloric acid	$\frac{HCl}{4}$	1247·4	1128	18·25	B. A.

Names of Substances.	Proportion of an eq. in 1 volume.	Specific Gravity.			Ob-servers.
		Air = 1.	Oxy. = 1.	H. = 1.	
Hydrobomic acid	$\frac{HBr}{4}$	2731	2469·7	39·5	
Hydriodic acid	$\frac{HI}{4}$	4443	4017·8	63·5	G-L.
Hydrocyanic acid	$\frac{HCy}{4}$	947·6	856·9	13·5	G-L.
Chloride of cyanogen	$\frac{Cy\ Cl}{4}$	2111	1908·9	30·75	G-L.
Deutoxide of nitrogen	$\frac{NO_2}{4}$	1038·8	939·3	15	B'.
Peroxide of nitrogen	$\frac{NO_4}{4}$	1720	1555·4	23	C.
Ammonia	$\frac{NH_3}{4}$	596·7	539·6	8·5	B. & A.
Phosphuretted hydrogen	$\frac{PH_3}{4}$	1214	1097·8	17·25	D.
Hydride of arsenic	$\frac{AsH_3}{4}$	2695	2437	39	D.
Terchloride of phosphorus...	$\frac{PCl_3}{4}$	4875	4408·5	69·75	D.
Terchloride of arsenic	$\frac{AsCl_3}{4}$	6300·6	5697·7	91·5	D.
Chloride of bismuth	$\frac{BiCl}{4}$	11160	10092·1	...	J.
Iodide of arsenic	$\frac{AsI_3}{4}$	16100	14560	226·5	M.
Subchloride of mercury	$\frac{Hg_2\ Cl}{4}$	8350	7551	136	M.
Subbromide of mercury	$\frac{Hg_2\ Br}{4}$	10140	9170	180	M.
Fluoride of boron	$\frac{BF_3}{4}$	2312·4	2091·2	...	J-D.
Chloride of boron	$\frac{BCl_3}{4}$	3942	3564·8	...	D.
Carburetted hydrogen	$\frac{C_2\ H_4}{4}$	559·6	506·1	8	

Names of substances.	Proportion of an eq. in 1 volume.	Specific Gravity.			Ob-servers.
		Air = 1.	Oxyg. = 1.	H. = 1.	
Methylene (?)	$\frac{C_2H_2}{4}$	490	443	7	
Olefiant gas	$\frac{C_4H_4}{4}$	985·2	891	14	T. S.
Oil gas	$\frac{C_8H_8}{4}$	1892	1711	28	F.
Cetene	$\frac{C_{32}H_{32}}{4}$	8007	7240·8	112	D. P.
Oléene	$\frac{C_{12}H_{12}}{4}$	2875	2600·8	42	F''.
Elœene	$\frac{C_{18}H_{18}}{4}$	4071	3681·5	63	F'.
Amilene	$\frac{C_{20}H_{20}}{4}$	5061	4576·6	70	C''.
Naphthaline	$\frac{C_{20}H_8}{4}$	4528	4072	64	D.
Paranaphthaline	$\frac{C_{30}H_{12}}{4}$	6741	6096	96	D.
Benzene (benzole)	$\frac{C_{12}H_6}{4}$	2770	2505	39	M.
Terebene	$\frac{C_{20}H_{16}}{4}$	4765	4309	68	D.
Citrene	$\frac{C_{20}H_{18}}{4}$	4891	4422·8	68	C''.
Retinaphtha	$\frac{C_{14}H_8}{4}$	3230	2921	46	P. W.
Retinile	$\frac{C_{18}H_{12}}{4}$	4242	3836	60	P. W.
Retinole	$\frac{C_{32}H_{16}}{4}$	7110	6429·7	104	P. W.
Sweet oil of wine	$C_{10}H_8$	9476	8569·3	136	R.
Volatile sweet oil of ether ...	$\frac{C_8\ H_9}{2}$	3965	3582	57	M.
Mesitylene	$\frac{C_{12}H_8}{4}$	2805	2836·5	40	C''.

Names of Substances.	Proportion of an eq. in 1 volume.	Specific Gravity.			Observers.
		Air = 1.	Oxy. = 1.	H. = 1.	
Wood-spirit	$\frac{C_2H_4O_2}{4}$	1120	1012·8	16	D. P.
Methylic ether	$\frac{C_2H_3O}{2}$	1617	1462·3	23	*Id.*
Methylic ether (monochlorinated)	$\frac{C_2H_2ClO}{2}$	3903	3529·5	57·5	R.
Methylic ether (bichlorinated)	$\frac{C_2HCl_2O}{6}$	2115	1912·6	30·66	R.
Methylic ether (perchlorinated)	$\frac{C_2Cl_3O}{3}$	4670	4223·2	62·75	*Id.*
Formic acid at 321°·8 F.	$\frac{C_2H_2O_4}{4}$	1610	1456	23	B.
Sulphide of methyl	C_2H_3S	6367	5557·8	94	*Id.*
Chloride of carbon	$\frac{C_2Cl_4}{4}$	5330	4820	77	R.
Chloride of carbon (another)	$\frac{C_4Cl_4}{4}$	5820	5263·1	83	*Id.*
Chloride of carbon (another)	$\frac{C_4Cl_6}{4}$	8157	7376·5	118·5	*Id.*
Chloride of methyl	$\frac{C_2H_3Cl}{4}$	1731	1565·4	25·25	D. P.
Chloride of methyl (monochlorinated)	$\frac{C_2H_2Cl_2}{4}$	3012	2724	42·5	R.
Fluoride of methyl	$\frac{C_2H_3F}{4}$	1186	1072·5	16·5	D. P.
Iodide of methyl	$\frac{C_2H_3I}{4}$	4883	4415·8	70·5	*Id.*
Sulphate of methyl	$\frac{C_2H_3O,\ SO_3}{2}$	4565	4128·1	63	*Id.*
Nitrate of methyl	$\frac{C_2H_3O,\ NO_5}{2}$	2653	2399·6	38·5	*Id.*
Formiate of methyl	$\frac{C_2H_3O,\ C_2HO_3}{4}$	2084	1884·5	30	*Id.*
Acetate of methyl	$\frac{C_2H_3O,\ C_4H_3O_3}{4}$	5563	2317·7	37	*Id.*

Names of substances.	Proportion of an eq. in 1 volume.	Specific Gravity.			Ob-servers.
		Air = 1.	Oxyg. = 1.	H. = 1.	
Methylal	$\frac{C_6H_8O_4}{4}$	2625	2374	38	M'.
Alcohol	$\frac{C_4H_6O_2}{4}$	1613	1458·7	23	G-L.
Mercaptan	$\frac{C_4H_6S_2}{4}$	2326	2103·4	31	B.
Ether	$\frac{C_4H_5O}{2}$	2586	2338·5	37	G-L.
Sulphuret of ethyl	$\frac{C_4H_5S}{2}$	3100	2803·4	45	R.
Chloride of ethyl	$\frac{C_4H_5Cl}{4}$	2299	2006·6	42·25	T.
Chloride of ethyl (monochlorinated)	$\frac{C_4H_4Cl_2}{4}$	3478	3145·2	49·5	R.
Chloride of ethyl (bichlorinated)	$\frac{C_4H_3Cl_3}{4}$	4530	4096·5	66·75	R.
Chloride of ethyl (trichlorinated)	$\frac{C_4H_2Cl_4}{4}$	5799	5244·1	84	*Id.*
Chloride of ethyl (quadrichlorinated)	$\frac{C_4HCl_5}{4}$	6975	6307·6	101·25	*Id.*
Iodide of ethyl	$\frac{C_4H_5I}{4}$	5475	4951·2	77·5	G-L.
Nitrous ether	$\frac{C_4H_5O,\ NO_3}{4}$	2626	2374·7	37·5	D. B'.
Chlorocarbonic ether	$\frac{C_4H_5O,\ C_2O_3Cl}{4}$	3829	3462·6	54·25	D.
Sulphurous ether	$\frac{C_4H_5O,\ SO_2}{2}$	4780	4323	69	E. & B.
Oxalic ether	$\frac{C_4H_5O,\ C_2O_3}{2}$	5087	4600·3	73	D. B'.
Silicic ether (tribasic)	$\frac{3C_4H_5O,\ SiO_3}{3}$	7210	6521	104	E.
Boric ether (tribasic)	$\frac{3C_4H_5O,\ BO_3}{4}$	5140	4649	72	E. & B.
Acetic ether	$\frac{C_4H_5O,\ C_4H_3O_3}{4}$	3067	2773·5	44	*Id.*

Names of Substances.	Proportion of an eq. in 1 volume.	Specific Gravity.			Observers.
		Air = 1.	Oxy. = 1.	H. = 1.	
Benzoic ether	$\frac{C_4H_5O,\ C_{14}H_5O_2}{4}$	5409	4899	71	*Id.*
Succinic ether	$\frac{C_4H_5O,\ C_4H_3O_3}{2}$	6220	5624·8	87	A.
Pyromucic ether	$\frac{C_4H_5O,\ C_{10}H_3O_5}{4}$	4859	4394·1	70	M.
Œnanthic ether	$\frac{C_4H_5O,\ C_{14}H_{13}O_2}{2}$	10508	9502·5	150	L. & P.
Dutch liquid	$\frac{C_4H_3Cl,\ HCl}{4}$	3443	3113·5	49·5	G-L. D.
Bromide of olefiant gas	$\frac{C_4H_3Br,\ HBr}{4}$	6485	5864·5	94	R.
Chloral	$\frac{C_4HCl_3O_2}{4}$	5130	4639·1	73·75	D.
Chloroform	$\frac{C_4HCl_3}{4}$	4199	3797·2	65·78	D.
Aldehyde	$\frac{C_4H_4O_2}{4}$	1532	1385·4	22	L.
Alcarsin	$\frac{C_4H_6As}{2}$	7184	6496·6	105	B.
Acetic acid at 482° F.	$\frac{C_4H_4O_4}{4}$	2080	1879·8	30	C".
Chloracetic acid	$\frac{C_4HCl_3O_4}{4}$	5300	4792·9	81·75	D.
Acetone	$\frac{C_6H_6O_2}{4}$	2019	1825·8	29	*Id.*
Benzoic acid	$\frac{C_{14}H_6O_4}{4}$	4270	3861·4	61	D. M.
Hydride of salicyl	$\frac{C_{14}H_6O_4}{4}$	4276	3867·1	61	P.
Eugenic acid	$\frac{C_{20}H_{12}O_5}{4}$	6400	5787·6	86	D.
Camphor	$\frac{C_{20}H_{16}O_2}{4}$	5468	4945·7	76	*Id.*
Urethane	$\frac{C_6NH_7O_4}{4}$	3096	2800	44·5	*Id.*

After the name of each substance in the preceding table is given the formula of its equivalent, which is divided by the number of volumes of vapour which the equivalent gives and the combining measure contains. The equivalent thus divided therefore expresses the composition of a single volume of the vapour. The first column of numbers contains the specific gravities referred to air as 1000; the second, in which the specific gravities are expressed with reference to that of oxygen as 1000, is obtained by dividing the former specific gravities by 1105·6, the specific gravity of oxygen gas. In the third column, the specific gravities are referred to hydrogen as 1; and consequently the number for any vapour expresses how many times that vapour is heavier than hydrogen. The numbers of this column only are obtained by calculation from the equivalents, and are therefore the theoretical densities: if divided by 16 they give corresponding theoretical densities on the scale of oxygen equal to 1; or if divided by 14·416 (the number of times which air is heavier than hydrogen) they give the theoretical densities on the scale of air equal to 1. The letter or letters in the last column refer to the name of the observer on whose authority the experimental specific gravities of the first and second columns of numbers are given.[1]

An extraordinary variation in the specific gravity of acetic acid at different temperatures was observed by M. Dumas, which is confirmed by M. Cahours and M. Bineau, (Annales de Chimie, &c. 3e sér. t. xviii. p. 226), and the anomaly found to extend to certain acids allied to the acetic; namely formic, butyric, and valerianic acids. Thus the vapour of acetic acid ($H\,O, C_4\,H_3\,O_3$), has a specific gravity of 3200 at 125° Centig., 2480 at 160° C., 2220 at 200°, 2090 at 230°, 2080 at 250°, and retains the last specific gravity, which corresponds with the theoretical density of four volumes from one equivalent, at higher temperatures; the observation being made up to 338° C. This vapour has, indeed, been observed with a density so great as 3950, under reduced pressure, and at a low temperature, namely 69° Fahr. The variation is probably accounted for by considering the acid to be bibasic at low temperatures, with a double equivalent and double density, and to assume progressively the molecular form and single density of the monobasic acid, as the temperature rises. The acid undergoes no permanent or constitutional alteration at the highest of the temperatures specified, but condenses again in possession of all its usual properties.

Butyric acid has a density of 3680 at 177° C., which falls to 3070 at 261° C., and remains the same at 330° C. Valerianic acid gave similar results, but the variation was less excessive (Cahours).

Formic acid vapour was observed by M. Bineau with a specific gravity as high as 3230, under a pressure of about one-fiftieth of an atmosphere, and at the temperature of 51° F., while it rarefied to 1610 at 416° Fahr., under the usual atmospheric pressure. The two sorts of molecular groups of this acid correspond respectively with the specific gravities, 1590 and 3180; in the first case the ordinary equivalent ($C_2\,H\,O_3 + H\,O$) gives four, and in the second two equivalents of vapour.

The acetic and other acids of this class were formerly supposed to give three volumes of vapour, but it is doubted whether the proportions of three and six volumes exist at all, or that the vaporous molecule of compound bodies is ever divisible except by 2, 4, or 8. Three compounds of silicium form exceptions to this rule—the chloride $Si\,Cl_3$, and the corresponding fluoride and ether, which give three volumes. From this circumstance, and the analogy which subsists between silicic acid, and the titanic acid and binoxide of tin, it has been proposed to diminish the

---

[1] A signifies Felix d'Arcet; B, Bunsen; B', Berard; BA, Biot and Arago; BD, Berzelius and Dulong; C, Colin; C', Cruikshanks; C'', Cahours; D, Dumas; DB, Dumas and Boussingault; DB', Dumas and P. Boullay; DP, Dumas and Peligot; E, Ebelmen; E and B, Ebelmen and Bouquet; F', Fremy; G-L, Gay-Lussac; GT, Gay-Lussac and Thenard; L, Liebig; LP, Liebig and Pelouze; M, Mitscherlich; M', Malaguti; P, Piria; PW, Peletier and Walter; R, Regnault; TS, Theodore de Saussure. The table itself is that given by M. Baudrimont in his excellent Traité de Chimie, somewhat modified and extended.

equivalent of silicium one-third, representing silicic acid by $Si\ O_2$; and, in consequence, the chloride and fluoride of silicium and silicic ether would possess, in the state of vapour, a molecule divisible by 2. Two chlorinated compounds of methyl and the sulphuret of mercury are the only other substances of which the equivalents are divided in the table by 6 or 3.

The specific gravity of the vapour of oil of vitriol $H\ O,\ S\ O_3$, was found to vary from 2500 at 630° Fahr., to 1680 at 928° Fahr. This substance should have a density of 1640 on the hypothesis of the union of the anhydrous acid and water without condensation; a number which corresponds sufficiently well with observations of the density made at temperatures above 750° Fahr. But the vapours of the acids are not the only bodies which present such anomalies; the oils of aniseed and fennel, which are perfectly neutral, offer similar results. Thus the vapour of the oil of aniseed varies in specific gravity from 5980 at 473° Fahr. to 5190 at 640° Fahr.; its theoretical density being 5180. The greater part, however, of the compound ethers, and a large number of the volatile oils, particularly the pure hydrocarbon oils, furnish, at from 60 to 80 degrees above the boiling point, numbers which accord closely with theory.

The specific gravity of the pentachloride of phosphorus, taken by M. Mitscherlich at 335° Fahr., is represented by 4850, which led to the conclusion that the molecule of this compound gives six volumes of vapour. But M. Cahours finds that the density of this vapour varies with the temperature, from 4990 at 374° to 3656 at 621°: about 554° the density is 3680, which corresponds with eight volumes of vapour.

From these tables, it appears that a simple relation always subsists between the combining measures of different bodies in the gaseous state:

That the combining measure of a few bodies is the same as that of oxygen, or *one volume;* of a large number, double that of oxygen, or *two volumes;* and of a still larger number, four times that of oxygen, or *four volumes;* while combining measures of other numbers of volumes, such as *three* and *six*, or of fractional portions of one volume, such as *one-third*, are comparatively rare:

That the specific gravity of a gas may be calculated from its atomic weight, or the atomic weight from the specific gravity, as they are necessarily related to each other. Thus, to find the specific gravity of a vapour like that of phosphorus, of which the combining measure is one volume, or the same as that of oxygen. The specific gravities of two bodies, of which the *volumes* of the atoms are the same, must obviously be as the *weights* of these atoms. Hence, 8 and 32 being the atomic weights of oxygen and phosphorus, and 1105·6, the known specific gravity of oxygen, the specific gravity of phosphorus vapour is obtained by the following proportion—

$$8 : 32 :: 1105{\cdot}6 : 4422$$

= sp. gr. of phosphorus vapour.

Secondly, to find the specific gravity of a vapour like that of fluorine, of which the combining measure is assumed to be two volumes, or double that of oxygen. The atomic weight of fluorine 18·70,

$$8 : 18{\cdot}70 :: 1105{\cdot}6 : 2584{\cdot}34 =$$

twice the specific gravity of fluorine, being the weight of two volumes, and the specific gravity required is 1292·17.

These cases are examples of a general rule, that the specific gravity of a body in the state of vapour is obtained by multiplying the atomic weight of the body by 1105·6, the specific gravity of oxygen, and dividing by 8. The number thus found must then be divided by the number of volumes which are known to compose the combining measure of vapour.

The specific gravities thus calculated are generally more accurate than those obtained by direct experiment, from the circumstance that the operation of taking the specific gravity of a gas is generally less susceptible of precision, than the chemical analyses on which the atomic weights are founded. The densities of vapours, taken only a few degrees above their condensing points, are generally a little greater than the truth, owing to a peculiarity in their physical constitution which was formerly explained (page 81). Of such bodies, therefore, the theoretical is a necessary check upon the experimental density.

SECTION IV.—RELATION BETWEEN THE CRYSTALLINE FORM AND ATOMIC CONSTITUTION OF BODIES—ISOMORPHISM.

Bodies on passing from the gaseous or liquid to the solid state generally present themselves in crystals, or regular geometrical figures, which are the larger and more distinct the more slowly and gradually they are produced. Their formation is readily observed in the spontaneous evaporation of a solution of sea-salt, or in the slow cooling of a hot and saturated solution of alum, which salts assume the forms of the cube and regular octohedron. The crystalline form of a body is constant, or subject only to certain geometrical modifications which can be calculated, and is most serviceable as a physical character for distinguishing salts and minerals. Between bodies of similar atomic constitution, a relation in form has been observed of great interest and beauty, which now forms a fundamental doctrine of physical science, like the subjects of atomic weights and volumes just considered.

Gay-Lussac first made the remark that a crystal of potash-alum transferred to a solution of ammonia-alum continued to increase without its form being modified, and might thus be covered with alternate layers of the two alums, preserving its regularity and proper crystalline figure. M. Beudant afterwards observed that other bodies, such as the sulphates of iron and copper, might present themselves in crystals of the same form and angles, although the form was not a simple one like that of alum. But M. Mitscherlich first recognised this correspondence in a sufficient number of cases to prove that it was a general consequence of similarity of composition in different bodies. To the relation in form he applied the term *isomorphism*, (from ἴσος, equal, and μορφή, shape), and distinguished bodies which assume the same figure as *isomorphous*, or (in the same sense) as *similiform* bodies. The law at which he arrived is as follows:—"The same number of atoms combined in the same way produce the same crystalline form; and crystalline form is independent of the chemical nature of the atoms, and determined only by their number and relative position."

This law has not been established in all its generality, but perhaps no fact is certainly known which is inconsistent with it, while an indisposition which certain classes of elements have to form compounds at all similar in composition to those formed by other classes, limits the cases for comparison, and makes it impossible to trace the law, throughout the whole range of the elements, in the present state of our knowledge respecting them.

The relation of isomorphism is most frequently observed between salts, from their superior aptitude to form good crystals. Thus the arseniate and phosphate of soda are obtained in the same form, and are exactly alike in composition, each salt containing one proportion of acid, two of soda, and one of water as bases, together with twenty-four atoms of water of crystallization. With a different proportion of water of crystallization, namely, with fourteen atoms, and the other constituents unchanged, the crystalline form is totally different, but is again the same in both salts. For every arseniate, there is a phosphate corresponding in composition, and identical in form; the isomorphism of these two classes of salts is indeed perfect. The arsenic and phosphoric acids contain each five proportions of oxygen to one of arsenic and phosphorus respectively, and are supposed to be themselves isomorphous, although the fact cannot be demonstrated, as the acids do not crystallize. The elements, phosphorus and arsenic, are also known to be isomorphous: and the isomorphism of their acids and salts is referred to the isomorphism of the elements themselves; isomorphous compounds in general appearing to arise from isomorphous elements uniting in the same manner with the same substance.

The isomorphism of the sulphate, seleniate, chromate, and manganate of the same base is likewise clear and easily observed; each of the acids in these cases containing three proportions of oxygen to one of selenium, sulphur, chromium, and manganese, themselves presumed to be isomorphous.

Of bases, the isomorphism of the class consisting of magnesia, oxide of zinc, oxide of cadmium, and the protoxides of nickel, iron and cobalt, is well marked in the salts which they form with a common acid, and is particularly observable in the double salts of these oxides, such as the sulphate of magnesia and potassa, sulphate of zinc and potassa, sulphate of copper and potassa, which have all six atoms of water and a common form. The sulphates themselves of these bases differ, most of them affecting seven atoms of water of crystallization, while the sulphate of copper affects five; but those with the seven may likewise be crystallized in favourable circumstances with five atoms of water, and then assume the form of the copper salt, thus exhibiting a second isomorphism like the arseniate and phosphate of soda.

The sesquioxides of the same class of metals with alumina and the sesquioxide of chromium, which consist of two atoms of metal and three of oxygen, also afford an instructive example of isomorphism, particularly in their double salts. The sulphate of the sesquioxide of iron with sulphate of potassa and twenty-four atoms of water, forms a double salt having the octohedral form of sulphate of alumina and potassa, or common alum, the same astringent taste, with other physical and chemical properties so similar, that the two salts can with difficulty be distinguished from each other. The salt is called iron alum, and there are corresponding manganese and chrome alums, neither of which contains alumina, but the sesquioxide of manganese and sesquioxide of chromium in its place, with the proportions of acid and water which exist in common alum. In all these salts another substitution may occur without change of form; namely, that of soda or oxide of ammonium for the potassa in the sulphate of potassa, giving rise to the formation of what are called soda and ammonia alums.

Certain facts have been supposed to militate against the principles of isomorphism, which require consideration.

1. It appears that the corresponding angles of crystals reputed isomorphous are not always exactly equal, but are sometimes found to differ two or three degrees, although the errors of observation in good crystals rarely exceed 10′ or 20′ of a degree. But it has been shown by Mitscherlich that a difference may exist between the inclinations of two series of similar faces in different specimens of the same salt, of 59′; while it is also known that the angles of a crystal alter sensibly in their relative dimensions with a change of temperature (page 34). The angles of crystals are, therefore, affected in their values within small limits by causes of an accidental character, and absolute identity in crystalline form may require the concurrence of circumstances which are not found together in the ordinary modes of producing many crystals, which are still truly isomorphous.

The following table exhibits the inequalities which have been observed between the angles of certain isomorphous crystals:—

	Rhomboidal form.	
Carbonate of	manganese (diallogite)	103°
"	lime (calc-spar)	105° 5′
"	lime and magnesia (dolomite)	106° 15′
"	magnesia (giobertite)	107° 25′
"	iron (spathic iron)	107°
"	zinc (smithsonite)	107° 40′
	Square prismatic with rhomboidal base.	
Carbonate of	lime (arragonite)	116° 5′
"	lead (ceruse)	117°
"	strontia (strontianite)	117° 32′
"	baryta (witherite)	118° 57′

Sulphate of baryta .... 101° 42′
" lead (anglesite) .... 103° 42′
" strontia (celestine) .... 104° 30′

2. It appears that the same body may assume in different circumstances two forms which are totally dissimilar, and have no relation to each other. Thus sulphur on crystallizing from solution in the bisulphide of carbon or in oil of turpentine, at a temperature under 100°, forms octohedrons with rhombic bases, but when melted by itself and allowed to cool slowly, it assumes the form of an oblique rhombic prism on solidifying at 232°. These are incompatible crystalline forms, as they cannot be derived from one common form. Carbon occurs in the diamond in regular octohedrons, and in graphite or plumbago in six-sided plates, forms which are likewise incompatible. Sulphur and charcoal have each, therefore, two crystalline forms, and are said to be *dimorphous*, (from δις, twice, and μορφή, shape). Carbonate of lime is another familiar instance of dimorphism, forming two mineral species, calc-spar and arragonite, which are identical in composition, but differ entirely in crystalline form. G. Rose has shown that the first or second of these forms may be given to the granular carbonate of lime formed artificially, according as it is precipitated at the temperature of the air, or near the boiling point of water. Of its two forms, carbonate of lime most frequently affects that of calc-spar: but carbonate of lead, which assumes the same two forms, and is therefore isodimorphous with carbonate of lime, chiefly affects that of arragonite, and is very rarely found in the other form. Had these carbonates, therefore, been each known only in its common form, their isomorphism would not have been suspected,—an important observation, as the want of isomorphism between certain other bodies may be caused by their being really dimorphous, although the two forms have not yet been perceived. Crystallization in three forms is not unknown: thus titanic acid is found in three distinct forms, as the minerals rutile, brookite, and anatase.

3. The observation of the isomorphism of bodies is of the greatest value as an indication that they possess a similar constitution, and contain a like number of atoms of their constituents. But it must be admitted that the most perfect coincidence in form is likewise observed between certain bodies which are quite different in composition. Thus bisulphate of potassa is dimorphous, and crystallizes in one of the two forms of sulphur (Mitscherlich). Nitrate of potassa in common nitre has the form of arragonite, and occurs also, there is reason to believe, in microscopic crystals in the form of calc-spar. Nitrate of soda, again, has the form of calc-spar. Permanganate of baryta and the anhydrous sulphate of soda likewise crystallize in one form. Between the first pair, sulphur and bisulphate of potassa, the absence of all analogy in composition is sufficiently obvious, notwithstanding their isomorphism. Between nitrate of potassa and carbonate of lime, and between permanganate of baryta and sulphate of soda, there is no similarity of composition, on the ordinary view which is taken of the constitution of these salts, but both of these pairs have been assimilated, in speculative views of their constitution proposed by Mr. Johnston (Philos. Mag. third series, vol. xii. page 480) in regard to the first pair, and by Dr. Clark (Records of General Science, vol. iv. page 45) in regard to the second, which merit consideration, although the hypotheses cannot be both correct, as they are based upon incompatible data. To these may be added, the sulphate of baryta with perchlorate and permanganate of potassa: BaO, SO_3 with KO, ClO_7 and $KO, Mn_2 O_7$. The sulphide of antimony with sulphate of magnesia: $Sb S_3$ with MgO, SO_3+7HO. Borax with augite, labradorite and anorthite, quartz and chabasite, mohsite and eudialite, anatase and apophyllite, zircon and wernerite, manganite and prehnite. Copper pyrites, $Cu Fe S_2$, has also the same form as braunite or sesquioxide of manganese, $Mn_2 O_3$. Leucite and analcime both belong to the regular system, and are aluminous silicates of similar composition; but, while the first contains one equivalent of potassa, the other contains one equivalent of soda $+ 2HO$.

The nitrite of lead has the same octohedral figure as the nitrate of lead, with two atoms of oxygen less in its acid.

Of examples of identity of crystalline form without any well-established relation in composition, many others might be quoted, if occurrence in the simple forms of the cube and regular octohedron should be allowed to constitute isomorphism. For example: carbon, sea-salt, arsenious acid, galena, the magnetic oxide of iron, and alum, all occur in octohedrons, although they are no way related in composition. But these simple forms are so common, that they can be held as affording no proof of isomorphism, unless in cases where it is to be expected from admitted similarity of composition, as between the different alums, or between chrome iron and the magnetic oxide of iron, $Cr_2 O_3$, FeO and $Fe_2 O_3$, FeO.

But notwithstanding the occurrence of such apparently fortuitous coincidences in form, isomorphism must still be considered as the surest criterion of similarity of composition which we possess. Truly isomorphous bodies generally correspond in a variety of other properties besides external form. Arsenic and phosphorus resemble each other remarkably in odour, although the one is a metal and the other a non-metallic body, while the corresponding arseniates and phosphates agree in taste, in solubility, in the degree of force with which they retain water of crystallization, and in various other properties. The seleniate and sulphate of soda, with ten atoms of water, which are isomorphous, are both efflorescent salts, and correspond in solubility, even so far as to agree in an unwonted deviation from the usually observed increasing rate of solubility at high temperatures, both salts being more soluble in water at 100° than at 212°. In fact, isomorphism appears to be always accompanied by many common properties, and to be the feature which indicates the closest relationship between bodies.

It will afterwards appear that the more nearly bodies agree in composition, they are the more likely to act as solvents of each other, or to be miscible in the liquid form. An attraction for each other of the same character is probably the cause of the easy blending together of the particles of isomorphous bodies, and of the difficulty of separating them after they are once dissolved in a common menstruum; such isomorphous salts as the permanganate and perchlorate of potassa may, indeed, crystallize apart from the same solution, owing to a considerable difference of solubility; and potassa-alum may be purified, in a great measure, by crystallization, from iron-alum, which is more soluble, and remains in the mother-liquor; but most isomorphous salts, such as the sulphates of iron and copper, or the iodide and chloride of potassium, when once dissolved together, do not crystallize apart, but compose homogeneous crystals, which are mixtures of the two salts in indefinite proportions. This intermixture of isomorphous compounds is of frequent occurrence in minerals, and was quite inexplicable, and appeared to militate against the doctrine of combination in definite proportions, till the power of isomorphous bodies to replace each other in compounds was recognized as a law of nature. Thus, in garnet, which is a silicate of alumina and lime, Al_2O_3, $SiO_3 + 3CaO\ SiO_3$, the alumina is found often wholly or in part replaced by an equivalent quantity of peroxide of iron; while the lime, at the same time, may be exchanged wholly or in part for protoxide of iron, or for magnesia, without the proper crystalline character of the mineral being destroyed. Hence the composition of mineral species is most properly expressed by general formulæ, where a letter, such as R, expresses an equivalent of metal which may be calcium, magnesium, manganese, iron, &c. :—

The Pyroxenes by 3RO, $2SiO_3$.
The Epidotes by 3RO, $2Al_2O_3$, $3SiO_3$.

⁂ The various forms of crystals were first happily described by Professor Weiss, of Berlin, by reference to "crystalline axes," which are three straight lines passing through the same point, and terminating in the surfaces or angles of the crystal. The simplest case is that in which the three axes cross each other at right angles, and are equal in length, as represented (fig. 53); c being the vertical, and a and b the two horizontal **axes**. A crystal is formed by applying planes in three principal ways to these axes.

1. By applying six planes so that each shall be perpendicular to one axis and parallel to the other two, the hexahedron, or, as it is more commonly termed, the cube (fig. 54), is formed. Here the axes terminate in the centre of each of the six faces of the crystal.

FIG. 53.

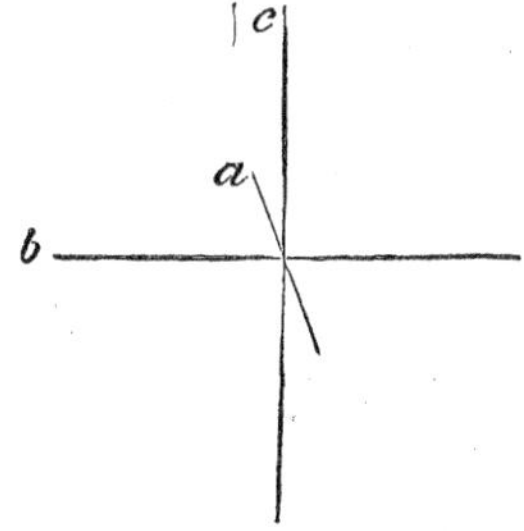

FIG. 54.

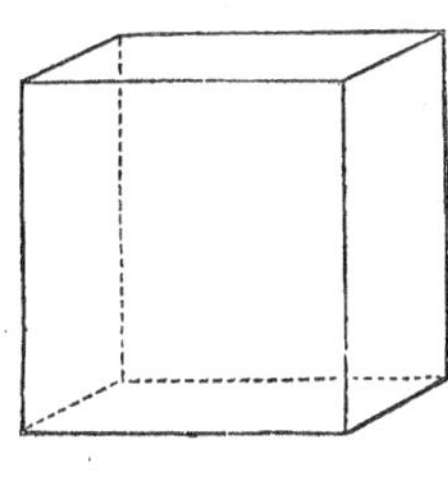

2. By applying one plane to an extremity of each of the three axes, as to the points a, b, and c (fig. 63), and seven planes in the same manner to other extremities, the regular octohedron is produced, of which the eight faces or planes are all equilateral triangles (fig. 55). The axes here terminate in the angles of the crystal.

3. The plane may be applied to the extremities of two axes, and be parallel to the third, which will require twelve planes to close the figure, and give rise to the rhombic dodecahedron (fig. 56).

FIG. 55. FIG. 56.

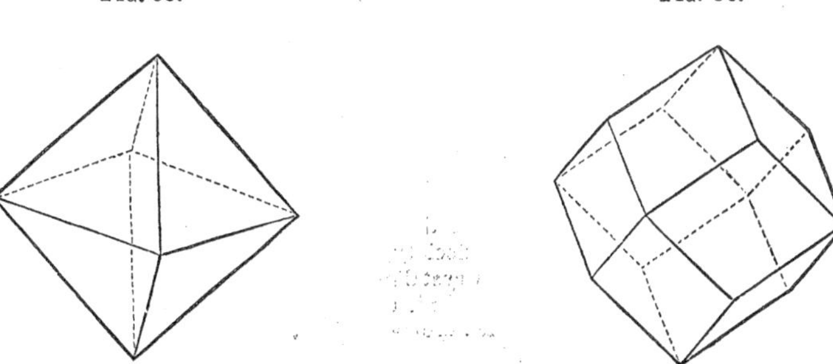

In these three principal forms, the planes are applied to the axes at equal distances from the centre. They may also cut the axes at unequal distances from the centre, giving rise to four other less usual forms.

A body in crystallizing may assume any of these forms, the only thing constant being the crystalline axes. Hence common salt crystallizes both in the cube and octohedron, although most usually in the former figure; and the magnetic oxide of iron both in the octohedron and rhombic dodecahedron. A body may even assume several of these forms at the same time; that is, may present at once faces of the cube, octohedron, and dodecahedron. Of the octohedral crystals of alum, for instance, the solid angles are always found to be cut or truncated by planes which belong to the cube of the same axes (fig. 57); and the edges of the octohedron in the same salt are sometimes removed or bevelled by the faces of the dodecahedron (fig. 58). Fig. 59 represents a combination of all these three forms; and similar or even more complicated combinations are often found in nature.

Fig. 57.

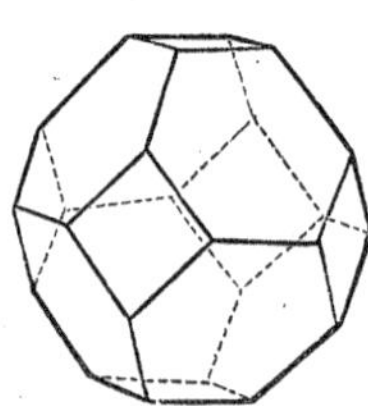

Fig. 58.

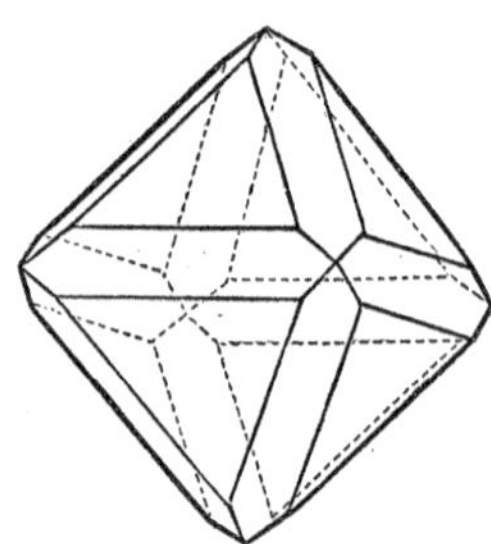

Fig. 59.

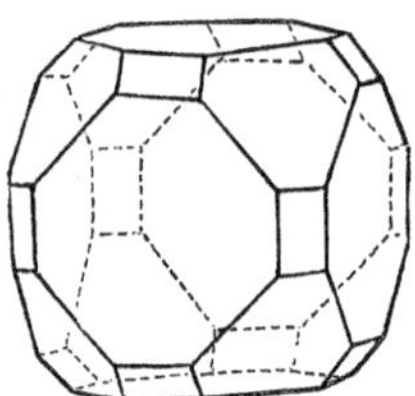

The groups of forms thus associated, by being deducible from the same axes, constitute what is called a "system of crystallization." Six such systems are enumerated by Weiss, to some one of which every crystalline body belongs.

1. The octohedral or regular system of crystallization, with the three principal axes at right angles to each other, and equal in length. It is that already described.

2. The square prismatic, with the axes at right angles, but two only of them equal in length.

3. The right prismatic, with the axes at right angles, but unequal in length.

4. The rhombohedral, with the axes equal, and crossing at equal but not right angles.

5. The oblique prismatic, with two of the axes intersecting each other obliquely, while the third is perpendicular to both, and unequal in length.

6. The doubly-oblique prismatic, with all three axes intersecting each other obliquely, and unequal.

By the apposition of planes to these different sets of crystalline axes, in the same modes as to the axes of the regular system, series of forms are produced, having a general analogy in all the systems, but specifically different.

For additional information on the subject of crystallography, which, although highly important to the chemical inquirer, is not exactly a department of chemistry, reference may be made to the Essay of Dr. Whewell, in the Phil. Trans. for 1825; to an Essay by Dr. Leeson, in the Memoirs of the Chemical Society, vol. iii.; the German Elements of Crystallography of G. Rose; the Systems of Crystallography of Professor Miller and Mr. J. J. Griffin; and to a short work lately published, entitled "Elements de Crystallographie, par M. J. Müller, traduits de l'Allemand par Jerome Nickles," which appears to be well adapted to the wants of the chemist. A full list of isomorphous substances is given by M. Gmelin in his invaluable *Handbuch der Chemie*, vol. i. p. 83.

CLASSIFICATION OF ELEMENTS.

The extent to which the isomorphous relations of bodies have been traced, will appear on reviewing the groups or natural families in which the elements may be arranged, and observing the links by which the different groups themselves are connected; these classes not being abruptly separated, but shading into each other in their characters, like the classes created by the naturalist for the objects of the organic world.

I. *Sulphur Class.*—This class comprises four elementary bodies: oxygen, sulphur, selenium, tellurium. The three last of these elements exhibit the closest parallelism in their own properties, in the range of their affinities for other bodies, and in the properties of their analogous compounds. They all form gases with one atom of hydrogen, and powerful acids with three atoms of oxygen, of which the salts, the sulphates, seleniates, and tellurates are isomorphous; and the same relation undoubtedly holds in all the corresponding compounds of these elements.

Oxygen has not yet been connected with this group by a certain isomorphism of any of its compounds; but a close correspondence between it and sulphur appears, in their compounds with one class of metals being alkaline bases of similar properties, forming the two great classes of oxygen and sulphur bases, such as oxide of potassium and sulphide of potassium; and in their compounds with another class of elements being similar acids, giving rise to the great classes of oxygen and sulphur acids, such as arsenious and sulphursenious acids. They farther agree in the analogy of their compounds with hydrogen, particularly of binoxide of hydrogen and bisulphide of hydrogen, both of which bleach, and are remarkable for their instability; and in the analogy of the oxide, sulphide, and telluride of ethyl, and of alcohol and mercaptan, which last is an alcohol with its oxygen replaced by sulphur. This class is connected with the next by manganese, of which manganic acid is isomorphous with sulphuric acid, and consequently manganese with sulphur.

II. *Magnesian Class.*—This class comprises magnesium, calcium, manganese, iron, cobalt, nickel, zinc, cadmium, copper, hydrogen, chromium, aluminum, glucinum, vanadium, zirconium, yttrium, thorinum. The protoxides of this class, including water, form analogous salts with acids. A hydrated acid, such as crystallized oxalic acid or the oxalate of water, corresponding with the oxalate of magnesia in the number of atoms of water with which it crystallizes, and the force with which the same number of atoms is retained at high temperatures; hydrated sulphuric acid (HO, SO_3+HO) with the sulphate of magnesia (MgO, SO_3+HO). The isomorphism of the salts of magnesia, zinc, cadmium, and the protoxides of manganese, iron, nickel, and cobalt, is perfect. Water (HO) and oxide of zinc (ZnO) have both been observed in thin regular six-sided prisms; but the isomorphism of these crystals has not yet been established by the measurement of the angles. Oxide of hydrogen has not, therefore, been shown to be isomorphous with these oxides, although it greatly resembles oxide of copper in its chemical relations. Lime is not so closely related as the other protoxides of this group, being allied to the following class. But its carbonate, both anhydrous and hydrated, its nitrate, and the chloride of calcium, assimilate with the corresponding compounds of the group; while to its sulphate or gypsum, CaO, SO_3+2HO, one parallel and isomorphous compound, at least, can be adduced, a sulphate of iron, FeO, SO_3+2HO (Mitscherlich), which is also sparingly soluble in water, like gypsum. Glucina is isomorphous with lime from the isomorphism of the minerals euclase and zoisite (Brooke).

The salts of the sesquioxide of chromium, of alumina, and glucina, are isomorphous with those of sesquioxide of iron ($Fe_2\,O_3$), with which these oxides correspond in composition; and the salts of manganic and chromic acids are isomorphous, and agree with the sulphates. The vanadiates are believed to be isomorphous with the chromates. Zirconium is placed in this class, because its fluoride is isomorphous with that of aluminum and that of iron, and its oxide appears to have the same constitution as alumina; and yttrium and thorium, solely because their oxides, supposed to be protoxides, are classed among the earths.

III. *Barium Class.*—Barium, strontium, lead. The salts of their protoxides, baryta, strontia, and oxide of lead, are strictly isomorphous, and one of them at least, oxide of lead, is dimorphous, and assumes the form of lime, and the preceding class in the mineral plumbocalcite, a carbonate of lead and lime (Johnston). But certain carbonates of the second class are dimorphous, and enter into the present class, as the carbonate of lime in arragonite, carbonate of iron in junckerite, and carbonate of magnesia procured by evaporating its solution in carbonic acid water to dryness by the water-bath (G. Rose), which have all the common form of carbonate of strontia. Indeed, these two classes are very closely related.

IV. *Potassium Class.*—The fourth class consists of potassium, ammonium, sodium, silver. The term ammonium is applied to a hypothetical compound of one atom of nitrogen and four of hydrogen (NH_4), which is certainly, therefore, not an elementary body, and probably not even a metal, but which is conveniently assimilated in name to potassium, as these two bodies occupy the same place in the two great classes of potassa and ammonia salts, between which there is the most complete isomorphism. Potassium and ammonium themselves are, therefore, isomorphous. The sulphates of soda and silver are similiform, and hence also the metals sodium and silver; but their isomorphism with the preceding pair is not so clearly established. Soda replaces potassa in soda-alum, but the form of the crystal is the common regular octohedron; nitrate of potassa has also been observed in microscopic crystals, having the arragonitic form of nitrate of soda,[1] which is better evidence of isomorphism, although not beyond cavil, as the crystals were not measured. There are also grounds for believing that potassa replaces soda in equivalent quantities in the mineral chabasite, without change of form. The probable conclusion is, that potassa and soda are isomorphous, but that this relation is concealed by dimorphism, except in a very few of their salts.

This class is connected in an interesting way with the other classes through the second. The subsulphide of copper and the sulphide of silver appear to be isomorphous, (see *sulphide of silver*, under silver, in this work), although two atoms of copper are combined in the one sulphide, and one atom of silver in the other, with one atom of sulphur; their formulæ being—

$$Cu_2\,S \text{ and } Ag\,S.$$

Are then *two* atoms of copper isomorphous with *one* atom of silver? In the present state of our knowledge of isomorphism, it appears necessary to admit that they are.

The fourth class will thus stand apart from the second, which is represented by copper, and also from the other classes connected with the second, in so far as one atom of the present class is equivalent to two atoms of the other classes in the production of the same crystalline form. This discrepancy may be at once removed by halving the atomic weight of silver, and thus making both sulphides to contain two atoms of metal to one of sulphur. But the division of the equivalents of sodium, potassium, and ammonium, which would follow that of silver, and the consideration of potassa and soda as suboxides, are assumptions not to be lightly entertained.

It was inferred by M. Mosander, that lime with an atom of water is isomorphous with potassa and soda, because $CaO + HO$ appears to replace KO or NaO in mesotype, chabasite, and other minerals of the zeolite family. The isomorphism of natrolite and scolezite is so explained: $NaO, Al_2O_3, 2SiO_3, 2HO$ with $CaO, Al_2O_3, 2SiO_3, 3HO$. On the other hand, it is strongly argued by M. T. Scheerer, that one equivalent of magnesia is isomorphous with three equivalents of water, from the equality of the forms of cordierite and a new mineral aspasiolite, the first containing MgO, and the second 3HO in its place; and from a review of a considerable number of alumino-magnesian minerals. One equivalent of oxide of copper, however, is supposed to be replaced by two equivalents of water.[2]

V. *Chlorine Class.*—Chlorine, iodine, bromine, fluorine. These four elements form a well-defined natural family. The three first are isomorphous throughout their whole combinations—chlorides with bromides and iodides, chlorates with bromates and iodates, perchlorates with periodates, &c.; and such fluorides also as can be compared with chlorides appear to affect the same forms. The fluoride of calcium of apatite, $CaF, 3(3CaO, PO_5)$, is also replaced by the chloride of calcium. It is connected with the second class through perchloric acid; the perchlorates being strictly isomorphous with the permanganates. But the formulæ of these two acids are—

---

[1] Frankenheim, in Poggendorff's *Annalen*, vol. xl. page 447. See also a paper by Professor Johnston on the received equivalents of potassa, soda, and silver; Phil. Mag. third series, vol. xii. p. 324.

[2] Poggendorff's *Annalen der Physik und Chemie*, t. lxviii. p. 319. Also, Millon and Reiset's Annuaire de Chimie, 1847, 8vo. Paris, pp. 52 and 234.

$Cl\ O_7$ and $Mn_2\ O_7$,

one atom of chlorine replacing two atoms of manganese. Or, this class has the same isomorphous relation as the preceding class to the others: and such I shall assume to be its true relation. Although halving the atomic weight of chlorine, which would give two atoms of chlorine to perchloric acid, is not an improbable supposition, still it would lead to the same strange conclusion as follows the division of the equivalent of sodium,—namely, that chlorine enters into its other compounds, as well as into perchloric acid, always in the proportion of two atoms; for that element is never known to combine in a less proportion than is expressed by its presently received equivalent. Cyanogen ($C_2\ N$), although a compound body, has some claim to enter this class, as the cyanides have the same form as the chlorides.

VI. *Phosphorus Class.*—Nitrogen, phosphorus, arsenic, antimony, and bismuth; also composing a well-marked natural group, of which nitrogen and bismuth are the two extremes, and of which the analogous compounds exhibit isomorphism. These five elements all form gaseous compounds with three atoms of hydrogen; namely, ammonia, phosphuretted hydrogen, arsenietted hydrogen, &c. The hydriodates of ammonia and of phosphuretted hydrogen are not, however, isomorphous. Arsenious acid and the oxide of antimony, both of which contain three atoms of oxygen to one of metal, are doubly isomorphous. Arsenious acid also is capable of replacing oxide of antimony in tartrate of antimony and potassa or tartar emetic, without change of form; and arsenic often substitutes antimony in its native sulphide. The native sulphide of bismuth ($Bi\ S_3$) is also isomorphous with the sulphide of antimony ($Sb\ S_3$). Nitrous acid (NO^3), which should correspond with arsenious acid and oxide of antimony, likewise acts occasionally as a base, as in the crystalline compound with sulphuric acid of the leaden chambers. The complete isomorphism of the arseniates and phosphates has already been noticed. But phosphoric acid forms two other classes of salts, the pyrophosphates and metaphosphates, to which arsenic acid supplies no parallels.

This class of elements is connected with the others by means of the following links:—Bisulphide of iron is usually cubic, or of the regular system; but it is dimorphous, and, in spirkise, it passes into another system, and has the form of arsenide of iron; $Fe\ S_2$, or rather $Fe_2\ S_4$, being isomorphous with $Fe_2\ As\ S_2$. Again, bisulphide of iron, in the pentagonal-dodecahedron of the regular system, is isomorphous with cobalt-glance, $Fe_2\ S_4$ with $Co_2\ As\ S_2$: so that one equivalent of arsenic appears to be isomorphous with 2S. This is also supported by the isomorphism of the sulphide of cadmium and sulphide of nickel ($Cd\ S$ and $Ni\ S$, or $Cd_2\ S_2$ and $Ni_2\ S_2$), with the arsenide of nickel ($Ni_2\ As$). Tellurium has also been observed in the same form as metallic arsenic and antimony. The phosphorus class approximates also to the chlorine class; nitrogen and chlorine both forming a powerful acid with five equivalents of oxygen, nitric acid, and chloric acid; but of the many nitrates and chlorates which can be compared, no two have proved isomorphous. Nor do the metaphosphates appear at all like the nitrates, although their formulæ correspond.

Nitrogen, it must be admitted, is but loosely attached to this class. It is greatly more negative than the other members of the class, approaching oxygen in that character, with which, indeed, nitrogen might be grouped, N being equivalent to 2O. For while phosphuretted hydrogen is the hydride of phosphorus, or has hydrogen for its negative and phosphorus for its positive constituent, ammonia is undoubtedly the nitride of hydrogen, or has nitrogen for its negative and hydrogen for its positive constituent. The one should be written PH_3, and the other H_3N—a difference in constitution which separates these bodies very widely. An important consequence of classing nitrogen with oxygen is, that, in the respective series of compounds of these elements, cyanogen becomes the analogue of carbonic oxide, C_2N being equivalent to CO, or rather C_2O_2.

VII. *Tin Class.* — Tin, titanium. Connected by the isomorphism of titanic acid (TiO_2) in rutile with peroxide of tin (SnO_2) in tin-stone. Titanium is connected with iron and the second class. Ilmenite and other varieties of titanic iron which have the crystalline form of the sesquioxide of that metal, — namely, that of specular iron, and also of corundum (alumina), — are mixtures of a sesquioxide of titanium (Ti_2O_3) with sesquioxide of iron (H. Rose).

VIII. *Gold Class.* — Gold, which is isomorphous with silver in the metallic state. Gold will thus be connected, through silver, with sodium and the fourth class.

IX. *Platinum Class.* — Platinum, iridium, osmium. From the isomorphism of their double chlorides. The double bichloride of tin and chloride of potassium crystallizes in regular octohedrons, like the double bichloride of platinum and potassium, and other double chlorides of this group; which, although not alone sufficient to establish an isomorphous relation between this class and the seventh, yet favours its existence (Dr. Clark). The alloy of osmium and iridium (IrOs) is isomorphous with the sulphide of cadmium (CdS) and sulphide of nickel (NiS) (Breithaupt).

X. *Tungsten Class.* — Tungsten, molybdenum, tantalum, niobium, and pelopium. From the isomorphism of the tungstates and molybdates, the salts of tungstic and molybdic acids, WO_3 and MoO_3. Tantalic acid is isomorphous with tungstic acid: tantalite (FeO, TaO_3) with wolfram (FeO, WO_3). So are molybdic and chromic acids, the tungstate of lime, tungstate of lead, molybdate of lead, and chromate of lead (in the least usual of its two forms), being all of the same form. This establishes a relation between molybdic, chromic, sulphuric, and other analogous acids (Johnston, Phil. Mag. 3d series, vol. xii. p. 387). Niobium and pelopium are introduced into this class as they replace tantalum in the tantalites of Bavaria.

XI. *Carbon Class.* — Carbon, boron, silicium. These elements are placed together, from a general resemblance which they exhibit without any precise relation. They are not known to be isomorphous among themselves, or with any other element. They are non-metallic, and form weak acids with oxygen, — the carbonic, consisting of two of oxygen and one of carbon, and the boric and silicic acids, which are generally viewed as composed of three of oxygen to one of boron and silicium. Silicic acid may, perhaps, replace alumina in some minerals, but this is uncertain.

Of the elements which have not been classed, no isomorphous relations are known. They are mercury, which in some of its chemical properties is analogous to silver, and in others to copper, cerium, didymium, lanthanum, lithium, rhodium, ruthenium, palladium, and uranium. Ruthenium, however, is believed to be isomorphous with rhodium, from the correspondence in composition of their double chlorides. Didymium and lanthanum are also probably isomorphous with cerium, as they appear to replace that metal in cerite.

According to the original law of Mitscherlich, that isomorphism depends upon equality in the number of atoms, and similarity in their arrangement, without reference to their nature, the elements themselves should all be isomorphous. Most of the metals crystallize in the simple forms of the cube or regular octohedron, which are not sufficient to establish this relation. But the isomorphism of a large proportion, if not the whole, of the elements may be inferred from the isomorphism of their analogous compounds. Thus, from the facts just adduced, it appears that the members of the following large class of elements are linked together from the isomorphism of one or more of their compounds. This large class may be subdivided into smaller classes, between the members of which isomorphism is of more frequent occurrence, and which are then to be viewed as isomorphous groups.

ISOMORPHOUS ELEMENTS.

1. Sulphur
   Selenium
   Tellurium

2. Magnesium
   Calcium
   Manganese
   Iron
   Cobalt
   Nickel
   Zinc
   Cadmium
   Copper
   Chromium
   Aluminum
   Glucinum
   Vanadium
   Zirconium

3. Barium
   Strontium
   Lead

4. Tin
   Titanium

5. Platinum
   Iridium
   Osmium

6. Tungsten
   Molybdenum
   Tantalum

With two atoms of the preceding elements.

7. Sodium
   Silver
   Gold
   Potassium
   Ammonium

Chlorine
Iodine
Bromine
Fluorine
Cyanogen

9. Phosphorus
   Arsenic
   Antimony
   Bismuth

The tendency of discovery is to bring all the elements into one class, either as isomorphous atom to atom, or with the relation to the others which sodium, chlorine, and arsenic exhibit.

But must not isomorphism be implicitly relied upon in estimating atomic weights, and the alterations which it suggests be adopted without hesitation in every case? Chemists have always been most anxious to possess a simple physical character by which atoms might be recognised; and equality of volume in the gaseous state, equality of specific heat, and similarity in crystalline form, have all in their turn been upheld as affording a certain criterion. The indications of isomorphism certainly accord much better than those of the other two criteria with views of the constitution of bodies derived from considerations purely chemical, and are indeed invaluable in establishing analogy of composition in a class of bodies, by supplying a precise character which can be expressed in numbers, instead of that general and ill-defined resemblance between allied bodies, which chemists perceived by an acquired tact rather than by any rule, and which was heretofore their only guide in classification. Admitting that isomorphism is a certain proof of similarity of atomic constitution within a class of elements and their compounds, it may still be doubted whether the relation of the atom to crystalline form is the same without modification throughout the whole series of the elements, or whether all atoms agree exactly in this or any other physical character.

Crystalline form and the isomorphous relation may prove not to be a reflection of atomic constitution, or immediately and necessarily connected with it, but to arise from some secondary property of bodies, such as their relation to heat; in which a simple atom may occasionally resemble a compound body, as we find sulphur isomorphous in one of its forms with bisulphate of potassa; while we find another simple atom, potassium, isomorphous through a long series of compounds with the group of five atoms which constitute ammonium. The occurrence of dimorphism also, both in simple and compound bodies, gives to crystalline form a less fundamental character.

Is it probable that sulphur and carbonate of lime could be made to appear in sets of crystals which are wholly unlike, merely by a slight change of temperature, if form were the consequence of an invariable atomic constitution? Crystalline form, then, may possibly depend upon some at present unknown property of bodies, which may have a frequent and general, but certainly not an invariable relation to their atomic constitution. There may be nothing truly inconsistent with the principles of isomorphism in one atom of a certain class of elements having the same crystallographic value as two atoms of another class, the relation which has been assumed to exist between the sodium, chlorine, and phosphorus classes, and the others, particularly when the classes stand apart, and differ in their properties from all the others, as those of sodium and chlorine do.

SECTION V.—ALLATROPY.

Many solid, and a few liquid bodies admit of a variation of properties, and may present different appearances at the same temperature.

Dimorphism, or the assumption of two incompatible crystalline forms by the same body, in different circumstances, has already been noticed as occurring with sulphur, carbon, carbonates of lime and lead, bisulphate of potassa, and chromate of lead. It is also observed in the biphosphate of soda, and in a considerable number of minerals. The sulphate of nickel (NiO, SO_3+7HO) is *trimorphous;* the other salts of similar composition, such as sulphate of magnesia and sulphate of zinc, have been found in two only of these forms. Dimorphous crystals may differ in density, the densities of calc-spar and arragonite, the forms of carbonate of lime being 2.719 and 2.949, and indeed all resemblance in properties between the crystals may be lost, as in diamond and graphite, the two forms of carbon. The particular form assumed by sulphur and carbonate of lime, which may be made to crystallize in either of their forms at will, is found to depend upon the degree of temperature at which the solid is produced; carbonate of lime being precipitated, on adding chloride of calcium to carbonate of ammonia, in a powder, of which the grains have the form of calc-spar or of arragonite, according as the temperature of the solution is 50° or 150° (G. Rose, Phil. Mag. 3d series, vol. xii. p. 465). A large crystal of arragonite, when heated by a spirit-lamp, decrepitates, and falls into a powder composed of grains of calc-spar. Native carbonate of iron is isodimorphous with carbonate of lime; as spathic iron its specific gravity is 3.872, as junckerite 3.815. The crystals of sulphur produced at the higher of two temperatures become opaque when kept for some days in the air, and pass spontaneously into the other form; while the crystals produced at the lower temperature are disintegrated and changed into the other form by a moderate heat. These observations are important, as establishing a relation between dimorphism and solidification at different temperatures.

A considerable variation of properties is likewise often observable in a solid which is not crystalline, or of which the crystalline form is indeterminate. This fact has been designated *allatropy* by Berzelius (from ἀλλοτροπος, of a different nature): dimorphism, or diversity in crystalline form, is, therefore, a particular case of allatropy. Sulphide of mercury obtained by precipitating corrosive sublimate by hydrosulphuric acid, is black; but the same body, when sublimed by heat, or produced by agitating mercury in a solution of the persulphide of potassium, forms cinnabar, of which the powder is the red pigment vermilion; while vermilion itself, if heated till sulphur begins to sublime from it, and then suddenly thrown into cold water, becomes black; although, if allowed to cool slowly, it remains red. Yet it is of the same composition exactly in the black and red states. The iodide of mercury newly sublimed is of a lively yellow colour, and may remain so for a long time; but it generally begins to pass into a fine scarlet on cooling, and may be made to undergo this change of colour in an instant by strongly pressing it: these, however, are two different crystalline forms. The precipitated sulphide of antimony may be deprived of the water it contains, at the melting point of tin, without losing its peculiar orange colour; but, when heated a little above that temperature, it shrinks, and assumes the black colour and metallic lustre of the native sulphide, without any loss of weight. Again, the black sulphide, when heated strongly and thrown into water, loses its metallic lustre, and acquires a good deal of the appearance of the precipitated sulphide. Chromate of lead, which is usually yellow, if fused and thrown into cold water, gives a red powder. The nitrates of lead are sometimes white, and sometimes yellow; and crystals of sulphate of manganese are often deposited from the same solution, some of which are pink, and others colourless, although identical in composition.

Such differences of colour are permanent, and not to be confounded with changes which are peculiar to certain temperatures: thus oxide of zinc is of a lemon-yellow colour, when strongly heated, but milk-white at a low temperature; the oxide of mercury is much redder at a high than at a low temperature, and bichromate of potassa, which is naturally red, becomes almost black when fused by heat. Even bodies in the gaseous state are liable to transient changes of this kind, the brown nitrous fumes being nearly colourless below zero, and on the other hand deepening greatly in colour at a high temperature.

The condition of *glass* is a remarkable modification of the solid form assumed by many bodies. Matter in this state is not crystallized, and on breaking, presents curved and not plain surfaces, or its fracture, in mineralogical language, is *conchoidal*, and not *sparry*. The indisposition to crystallize, which causes solidification in the form of glass, is more remarkable in some bodies, such as phosphoric and boracic acids, and their compounds, than in others. The biphosphate and binarseniate of soda have the closest resemblance in properties, yet when both are fused by a lamp, the first solidifies on cooling into a transparent colourless glass, and the second into a white opake mass composed of interlaced crystalline fibres. The phosphate at the same time discharges sensibly less heat than the arseniate in solidifying, retaining probably a portion of its heat of fluidity, or latent heat in a state of combination, while a glass. None of the compounds of silicic acid and a single base, such as soda or lime, or simple silicate, becomes a glass on cooling from a state of fusion, with the exception of the silicate of lead containing a great excess of oxide: they all crystallize. But a *mixture* of the same silicates, when fused, exhibits a peculiar viscosity or tenacity, appears to have lost the faculty of crystallizing, and constantly forms a glass. The varieties of glass in common use are all such mixtures of silicates. Glass is sometimes devitrified when kept soft by heat for a long time, owing to the separation of the silicates from each other, and their crystallization; and the less mixed glasses are known to be most liable to this change. It is probable that all bodies differ, when in the vitreous and in the crystalline form, in the proportion of combined heat which they possess, as has been observed of melted sugar (page 61) in these two conditions.

Arsenious acid, when fused or newly sublimed, appears as a transparent glass of a light yellow tint; but left to itself, it slowly becomes opake and milk white, the change commencing at the surface and advancing to the centre, and often requiring years to complete it, in a considerable mass. The arsenious acid is no longer vitreous, being changed into a multitude of little crystals, whence results its opacity; and it has altered slightly at the same time in density and in solubility. But the passage from the vitreous to the crystalline state may take place instantaneously, and give rise to an interesting phenomenon observed by H. Rose. The vitreous arsenious acid seems to dissolve in dilute and boiling hydrochloric acid without change, but the solution on cooling deposits crystals which are of the opake acid, and a flash of light, which may be perceived in the dark, is emitted in the formation of each crystal. This phenomenon depends upon and indicates the transition, for it does not occur when arsenious acid already opaque is substituted for vitreous acid, and dissolved and allowed to crystallize in the same manner.

A still greater change than those described, is induced upon certain bodies by exposure to a high temperature, without any corresponding change in their composition. Several metallic peroxides, such as alumina, sesquioxide of chromium and binoxide of tin, cease to be soluble in acids after being heated to redness. The same is true of a variety of salts, such as many phosphates, tungstates, antimoniates, and silicates. Many of these bodies contain water in combination, when most readily dissolved by acids, which constituent is dissipated at a high temperature, but in general before the loss of solubility occurs, so that the contained water alone is not the cause of the solubility. Berzelius remarked an appearance often observable when such bodies are under the influence of heat, and in the act of passing from the soluble to the insoluble state. They suddenly glow or become luminous, rising in temperature above the containing vessel, from a discharge of heat. The rare mineral gadolinite, which is a silicate of yttria, affords a beautiful example of this change. When heated it appears to burn, emits light, and becomes yellow, but undergoes no change in weight. Fluorspar, and many other crystalline substances, exhibit a feeble phosphorescence when heated, which has no relation to this change, and is to be distinguished from it.

The circumstance most certain respecting this change in bodies, which affects so deeply their chemical properties, is that the bodies do not contain a quantity of heat, after the change, which they must have possessed before its occurrence in a combined or latent form. No ponderable constituent is lost, but there is this loss of heat. A change of arrangement of the particles, it is true, must occur at the same time in some of these bodies, such as is observed when sulphite of soda is converted by heat into a mixture of sulphate of soda and sulphuret of sodium, without change of weight; but it would be difficult to apply an explanation of this nature to oxides, such as alumina and binoxide of tin, which contain only two constituents, and still more so to an element such as carbon. The loss of heat observed will afford all the explanation necessary, if heat be admitted as a constituent of bodies equally essential as their ponderable elements. As the oxide of chromium possesses more combined heat when in the soluble than in the insoluble state, the first may justly be viewed as the higher *caloride,* and the body in question may have different proportions of this as well as of any other constituent. But it is to be regretted that our knowledge respecting heat as a constituent of bodies is extremely limited; the definite proportion in which it enters into ice and other solids in melting, and into steam and vapours, has been studied, and also the proportion emitted during the combustion of many bodies, which has likewise proved to be definite. But the influence which its addition or subtraction may have on the chemical properties of a body is at present entirely matter of conjecture. The phenomena under consideration seem to require the admission of heat as a true constituent which can modify the properties of bodies very considerably; otherwise a great physical law must be abandoned, namely, that "no change of properties can occur without a change of composition." But if heat be once admitted as a chemical constituent of bodies, then a solution of the present difficulties may be looked for, for nothing is more certain than that "a change in composition will account for any change in properties." Heat thus combined in definite proportions with bodies, and viewed as a constituent, must not be confounded with the specific heat of the same bodies, or their capacity for sensible heat, which may have no relation to their combined heat.

SECTION VI. — ISOMERISM.

In such changes of properties as have already been described, the individuality of the body is never lost. But numerous instances have presented themselves of two or more bodies possessing the same composition, which are unquestionably different substances, and not mutually convertible into each other. Different bodies thus agreeing in composition, but differing in properties, are said to be *isomeric,* (from ἰσος, equal, and μερος, part), and their relation is termed *isomerism.* The discovery of such bodies excited much interest, and they have received a considerable share of the attention of chemists. But the result of a careful study of the bodies associated by similarity of composition, though differing in properties, has been upon the whole unfavourable to the doctrine of isomerism. Isomeric bodies have in general been proved by the progress of discovery to agree in the relative proportion of their constituents only, and to differ either in the aggregate number of the atoms composing them, or in the mode of arrangement of these atoms; and although new cases of isomerism are constantly arising, others are removed as they come to admit of explanation. This is what was to be expected, for isomerism in the abstract is improbable; a difference in properties between bodies, without a difference in their composition, appearing to be an effect without a sufficient cause. Hence, the term isomerism is now generally employed in a limited sense, to indicate simply the identity in composition of two or more bodies as expressed in the proportion of their constituents in 100 parts. Several classes of such isomeric bodies may be formed.

The members of the most numerous class of isomeric bodies differ in atomic weight. Thus we know at present three gases, three or four liquids, and as many solids, which all consist exactly of carbon and hydrogen, in the proportion of one atom to one atom, or, in weight, of 86 parts of carbon and 14 of hydrogen, very nearly. These agree in ultimate composition, but differ completely in every other respect. But a representation of their chemical constitution explains at once the cause of the differences they present, as is obvious in the following formulæ of four well characterized members of this isomeric group:—

	Equivalents and combining measure.
Olefiant gas	C_4H_4 or 4 volumes.
Gas from oil	C_8H_8 or 4 volumes.
Naphthene	$C_{16}H_{16}$ or 4 volumes.
Cetene	$C_{32}H_{32}$ or 4 volumes.

It thus appears that the atom of cetene contains twice as many atoms of carbon and hydrogen as the atom of naphthene, four times as many as the atom of the gas from oil, and eight times as many as the atom of olefiant gas; while as the atom of all these bodies affords the same measure of vapour, or four volumes, they must differ as much in density as they do in the number of their constituent atoms. It is not surprising, therefore, that they all possess different and peculiar properties. Several groups of bodies might be selected from the Table at page 130, which have a similar relation to each other, the number of their atoms being different, although their relative proportion is the same: such as—

Oil of lemons.......... $C_{10}H_8$
Oil of turpentine.......... $C_{20}H_{16}$

and,

Naphthaline.......... $C_{20}H_8$
Paranaphthaline.......... $C_{30}H_{12}$

A still more remarkable case is presented by alcohol and the ether from wood-spirit, in which there is identity of condensation as well as of composition, with different equivalents. The vapours of these two liquids have in fact the same specific gravity, and contain, under equal volumes, equal quantities of carbon, hydrogen, and oxygen. But we know that they are of a different type, alcohol being the hydrated oxide of ethyl, and ether of wood-spirit the oxide of methyl, so that their constitution and rational formulæ are quite different:—

Alcohol.......... C_4H_5O+HO.
Ether of wood-spirit.......... C_2H_3O.

In another class of isomeric bodies, the atomic weight may be equal, as well as the elementary composition. A pair belonging to this class are known, which coincide besides in the specific gravity of their vapours. The composition and atom of both the formiate of the oxide of ethyl (formic ether) and the acetate of oxide of methyl, may be represented by $C_6H_5O_4$: the density of both their vapours is 2574: and what is very remarkable, these bodies in their ordinary liquid state almost coincide in properties, the density of formic ether being 0·916, and that of the acetate of methylene 0·919, (density of water being $=1.000$), while the first boils at 133°, and the last at 136·4°. But when acted on by alkalies, their products are entirely different, the one affording formic acid and alcohol, and the other acetic acid and wood-spirit. Each of the isomeric bodies in question contains, indeed, two different binary compounds, and their constitution is truly represented by different formulæ: —

Formiate of oxide of ethyl........................ $C_4H_5O + C_2HO_3$
Acetate of oxide of methyl .... $C_2H_3O + C_4H_3O_3$

in which the same atoms are seen to be very differently arranged. The term *metameric* has been applied to bodies so related.

The last class of isomeric bodies are of the same atomic weights, but their constitution or molecular arrangement being unknown, their isomerism cannot at present be explained. It can scarcely be doubted, however, that their molecular arrangement is really different.

One pair of such isomeric bodies will illustrate the coincidences observed not at all unfrequently among organic substances. The racemic and tartaric acids, of which the composition is the same, exhibit a similarity of properties, and a parallelism in their chemical characters, that are truly astonishing. These acids are found together in the grape of the Upper Rhine. They differ considerably in solubility, the racemic being the least soluble, so that they may be separated from each other by crystallization; and the racemic acid contains an atom of water of crystallization, which is not found in the crystals of tartaric acid. They form salts which correspond very closely in their solubility and other properties. The bitartrate and biracemate of potassa are both sparingly soluble salts: the tartrates and racemates of lime, lead, and barytes, are all alike insoluble. Both acids form a double salt with soda and ammonia, which is an unusual kind of combination. But what is most surprising, crystals of these double salts not only coincide in the proportion of their water and other constituents, and in the composition of their acids, but also in external form, having been observed by Mitscherlich to be isomorphous. A nearer approach to identity could scarcely be conceived than is exhibited by these salts, which are, indeed, the same both in form and composition. The crystallized acids are both modified in an unusual manner by heat, and form three classes of salts, as phosphoric acid does. The formulæ of both acids in their ordinary class of salts is $C_8H_4O_{10}+$ two atoms of base (Fremy); but by no treatment can the one acid be transmuted into the other. Lastly, every organic acid produces a new acid by destructive distillation, which is peculiar to it, and is termed its pyr-acid. Now racemic and tartaric acid, when destroyed by heat, agree in giving birth to one and the same pyr-acid.

The allatropy of elements has been supposed to throw light upon the multiplication of series of compounds arising from one radical, and the isomerism of certain compounds. Fused sulphur passes through several allatropic conditions as its temperature is raised, in which it is imagined that the equivalent of the element may be doubled, tripled, and even quadrupled by a coalition of so many single atoms and the formation of compound atoms, which are distinguished as α sulphur, β sulphur, δ sulphur, γ sulphur, &c. In the different series of the oxygen acids of sulphur, containing one, two, three, and four equivalents of sulphur, the different allatropic varieties of sulphur are imagined to exist. Silicium in its combustible and incombustible allatropic conditions may thus give rise to different silicic acids, and allatropic borons and tungstens to the isomeric boric and tungstic acids.

SECTION VII. — ARRANGEMENT OF THE ELEMENTS IN COMPOUNDS.

The names of some compounds imply that they contain other compounds, and indicate a certain atomic constitution, while the names of other compounds express no particular arrangement of their constituent atoms, but leave it to be inferred that the atoms are all directly combined together. Thus sulphate of soda implies the continued existence of sulphuric acid and soda in the salt, while nitric acid, or binoxide of hydrogen, supposes no partition of the compound to which it is applied. But it is to be remembered that the original framers of the nomenclature were guided more by facilities of an etymological nature, in constructing such terms, than by views of the constitution of compounds.

Of a binary compound containing single atoms of its constituents, there cannot be two modes of representing the constitution; but where one of the constituents is present in the proportion of two or more atoms, several hypotheses can always be formed of their mode of aggregation. In a series of binary combinations of the same elements, such as that of nitrogen and oxygen, NO_1, NO_2, NO_3, NO_4, NO_5, the simplest view has generally been taken, namely, that it is the elements themselves which unite. But in particular cases the chemist is often involuntarily led into another opinion. Thus binoxide of nitrogen is so often a product of the decomposition of nitric acid, that the acid appears more like a compound of that oxide of nitrogen with oxygen, than a compound of nitrogen itself with oxygen. When the binoxide of hydrogen was first discovered by Thénard, he was led by the whole train of its properties to view it as a compound of water and oxygen, into which it is resolved with so much facility, and to name it accordingly *oxygenated water*, which it may be, and not a direct combination of hydrogen and oxygen; or its formula be $HO+O$, and not HO_2. The periodide of potassium, and the other analogous compounds obtained by dissolving iodine in metallic iodides, were first termed *ioduretted iodides* from similar considerations, and the hyposulphites, obtained by dissolving sulphur in sulphites, *sulphuretted sulphites*. It may be doubted whether chemists would return with advantage to any of these expressions, the views of composition which they indicate being uncertain, and not offering a sufficient inducement to depart from the more systematic designations. The binoxide of hydrogen, for instance, may be easily resolved into water and oxygen, not because water pre-exists in it, but because water is a compound of great stability, and is formed when binoxide of hydrogen is decomposed. Nitric acid, also, is as likely to be a compound of quadoxide of nitrogen with an additional atom of oxygen, as of binoxide of nitrogen with three atoms of the same element.

Certain compound bodies, however, have been observed to act the part of a simple body in combination, and can be traced through a series of compounds. The following substances, for instance, may be represented with considerable probability as compounds of carbonic oxide, as in the formulæ :—

CO, carbonic oxide.
$CO+O$, carbonic acid.
$CO+Cl$, chloroxicarbonic acid.
$2CO+O$, oxalic acid.

Carbonic oxide is said to be the *radical* of this series, a name applied to any compound which is *capable of combining with simple bodies*, as carbonic oxide appears to do with oxygen and chlorine in these compounds. Messrs. Liebig and Wöhler first proved by decisive experiments that such a radical exists in the benzoic combinations, which may be represented thus :—

$C_{14}H_5O_2+O$, benzoic acid.
$C_{14}H_5O_2+H$, essential oil of bitter almonds.
$C_{14}H_5O_2+Cl$, chloride of benzoyl, &c.

Cyanogen was the first recognised member of the class of compound radicals, of which the number known to chemists is constantly increasing, and which appear to pervade the whole compounds of organic chemistry. In combining with simple bodies, radicals act the part of other simple bodies, such as metals, chlorine, oxygen, &c., which they replace in compounds.

With the elements themselves compound radicals may be divided into two great classes :—

The *Basyl* class, consisting of metals the oxides of which are bases, hydrogen, the corresponding compound radicals, ammonium, ethyl, &c. These are electro-positive bodies.

The *salt-radical* class—chlorine, sulphur, oxygen, &c., with cyanogen, and other compound radicals which combine with metals and other members of the former class, and form salts or compounds partaking of the saline character. Such radicals are also termed *salogens;* they are electro-positive.

Constitution of salts.— Of the supposed combinations of binary compounds with binary compounds, the most numerous and important class are salts. Sulphate of soda is commonly viewed as a direct combination of sulphuric acid and soda, each preserving its proper nature in the compound; and so are all similar compounds of an acid oxide with a basic oxide. An oxygen acid is allowed to exist in them, and they are particularly distinguished as "oxygen-acid salts." But an opinion was promulgated long ago by Davy, that these salts might be constituted on the plan of the binary compounds of the former class, and their hydrated acids on the plan of a hydrogen acid; a view which is supported by many analogies, and has latterly had a preference given to it by some of our leading chemical authorities. It is, therefore, deserving of serious consideration.

One class of acids, the hydrogen acids, and the salts which they produce with alkalies, are unquestionably binary compounds, and were assumed by Davy as the types of acids and salts in general. Hydrochloric acid is composed of two elements, chlorine and hydrogen, and with soda it forms water and chloride of sodium, thus:—

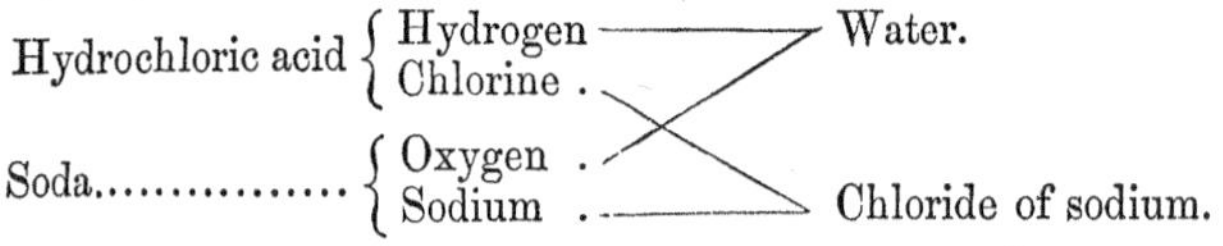

Hydrochloric acid { Hydrogen ——— Water.
Chlorine .

Soda............... { Oxygen .
Sodium .——— Chloride of sodium.

the hydrogen of the acid being replaced by sodium in the salt formed. Hydrocyanic is another hydrogen acid, of which cyanide of sodium is a salt. In general terms, a *radical* (which may be either simple or compound, like chlorine or cyanogen) forms an *acid* with hydrogen, and a *salt* with sodium or any other metal.

Hydrated sulphuric acid, which consists of sulphuric acid and an atom of water, $HO+SO_3$, is represented as a hydrogen acid by transferring the oxygen of the water to the sulphuric acid to form a new radical, SO_4, which is supposed to be in direct combination with the remaining atom of hydrogen, as $H+SO_4$. In sulphate of soda, the oxygen of the soda is in the same manner transferred to the acid, or the formula of the salt is changed from $NaO+SO_3$ to $Na+SO_4$. To SO_4, the salt-radical of sulphates, the name *sulphion* has been applied, from the circumstance that, in the voltaic decomposition of a sulphate, SO_4, travels to the positive pole, and the metal or hydrogen to the negative pole. Its compounds, or the sulphates, become *sulphionides.* The hydrated acid and its soda salt are thus named and denoted on the two views of their constitution—

I. ON THE ACID THEORY:

Hydrated sulphuric acid, sulphate of oxide of hydrogen, or hydric sulphate.................... $HO+SO_3$
Sulphate of soda, sulphate of oxide of sodium, or soda sulphate $NaO+SO_3$

II. ON THE SALT-RADICAL THEORY:

Sulphionide of hydrogen.................... $H+SO_4$
Sulphionide of sodium.................... $Na+SO_4$

which last formulæ are strictly comparable with those of an admitted hydrogen acid and its salt, such as—

Hydrochloric acid or chloride of hydrogen.................... $H+Cl$
Chloride of sodium.................... $Na+Cl$

or as—

Hydrocyanic acid or cyanide of hydrogen.................... $H+C_2N$
Cyanide of sodium.................... $Na+C_2N$

which thus appear compounds of three different radicals, chlorine (Cl), cyanogen (C_2N), and sulphion (SO_4), with the same elementary bodies, hydrogen and sodium. Sulphion is known only in combination, and has not been obtained in a separate state like chlorine and cyanogen. The body, sulphuric acid, SO_3, which may be separated from some sulphates, and can exist by itself, is looked upon as a product of the decomposition of these salts, and not to pre-exist in them, so that a secondary character is assigned to it.

Hydrated nitric acid, or aqua fortis, becomes a hydrogen acid by the creation of a nitrate radical, nitration. It is the nitrationide of hydrogen instead of the nitrate of water—

$$H+NO_6, \text{ instead of } HO+NO_5.$$

The nitrate of potassa becomes the nitrationide of potassium, and so of all other nitrates.

It is evident that the same view is applicable to hydrated oxygen acids in general, which may be made hydrogen acids, by assuming the existence of a new salt-radical for each, containing an atom more of oxygen than the oxygen acid itself, and capable of combining directly with hydrogen and the metals. The class of oxygen acid salts is thus abolished, and they become binary compounds like the chlorides and cyanides. Even oxygen acids themselves can no longer be recognized. It is not sulphuric acid (SO_3), but what was former viewed as its compound with water, that is the acid, and it is a hydrogen acid. The properties which characterize acids are undoubtedly only observed in the hydrates of the oxygen acids. Thus the anhydrous sulphuric acid does not redden litmus, and exhibits a disposition to combine with salts, such as chloride of potassium and sulphate of potassa, rather than with bases. The liquid carbonic acid has little affinity for water, does not combine directly with lime, but dissolves in alcohol, ether, and essential oils, like certain neutral bodies. It is only when associated with water that the bodies referred to exhibit acid properties, and then hydrogen acids may be produced.

On this view, it is obvious that the acid and salt are really bodies of the same constitution, hydrochloric acid being the chloride of hydrogen, as common salt is the chloride of sodium, and sulphuric acid and sulphate of soda being the sulphionides of hydrogen and of sodium. The acid reaction and sour taste are not peculiar to the hydrogen compound, and do not separate it from the others; the chloride, sulphionide, and nitrationide of copper being nearly as acid and corrosive as the chloride, sulphionide, and nitrationide of hydrogen, and clearly bodies of the same character and composition: they are all equally salts in constitution. The term "acid" is not absolutely required for any class of bodies included in the theory, and might, therefore, be dropped, if it were not that an inconvenience would be felt in having no common name for such bodies as anhydrous sulphuric acid SO_3, anhydrous nitric acid NO_5, sulphurous acid SO_2, carbonic acid CO_2, &c. To these substances, which first bore the name, it should now be confined. In considering the generation of salts, three orders of bodies would be admitted, as in the following tabular exposition of a few examples:—

I. The Acid.	II. The Salt-radical.	III. The Salt.
SO_3	SO_4	SO_4+H or a metal.
NO_5	NO_6	NO_6+H or a metal.
	NC_2	NC_2+H or a metal.
	Cl	$Cl+H$ or a metal.

The first term of the series, or "the acid," is wanting in the last two examples; and that is the peculiarity of those bodies which constituted the original class of hydrogen acids and their salts: while, to the old class of oxygen acid salts, both an acid and a salt-radical can be assigned, as in the first two examples.

The peculiar advantages of the salt-radical theory are—

First: That, instead of two, it makes but one great class of salts, assimilating in constitution bodies which certainly resemble each other in properties. Chloride of sodium and sulphate of soda are both neutral, and possess a common character, which is that of a soda salt; but they are separated widely from each other on the view of their constitution which is expressed in their names.

Secondly: It accounts for a remarkable law which is observed in the construction of salts; namely, that bases always combine with as many atoms of acid as they themselves contain of oxygen; a protoxide, which contains one atom of oxygen, combining and forming a neutral salt with one atom of an oxygen acid; while an oxide which contains two atoms of oxygen to one of metal, like binoxide of palladium, forms a neutral salt with two atoms of acid; and an oxide of three atoms of oxygen to two of metal, like sesquioxide of iron, forms a neutral salt with three atoms of acid. The acid and oxygen are thus always together in the exact proportion to form the salt-radical, there being always an atom of oxygen for every atom of acid in the salt. This will appear more distinctly in the following formulæ, which exhibit the composition of the neutral sulphates of a metal in four different states of oxidation, an atom of metal being represented by R:—

FORMULÆ OF NEUTRAL SULPHATES.

I. As consisting of Oxide and Acid.	II. As consisting of Metal and Salt-radical.	
$RO+SO_3$	$R+SO_4$	as in sulphate of soda.
R_2O+SO_3	R_2+SO_4	as in sulphate of suboxide of mercury.
RO_2+2SO_3	$R+2SO_4$	as in sulphate of binoxide of palladium.
$R_2O_3+3SO_3$	R_2+3SO_4	as in sulphate of sesquioxide of iron.

The acid is seen in the first column to be always in the proper proportion to form a sulphionide of the metal in the second column; and these sulphionides correspond exactly with known chlorides, such as RCl, R_2Cl, RCl_2, R_2Cl_3.

Thirdly: It offers a more simple and philosophical explanation of the action of certain metals upon acid solutions, and of the decomposition of such solutions in other circumstances. Thus when zinc is introduced into hydrochloric acid (chloride of hydrogen), it is allowed on both views, that the metal simply displaces the hydrogen which is evolved, and that chloride of zinc is formed in the place of chloride of hydrogen. In the same way, when zinc is introduced into diluted sulphuric acid, which contains the sulphionide of hydrogen on the binary theory, hydrogen is simply displaced and evolved as before, and the sulphionide of zinc is formed in the place of the sulphionide of hydrogen. The metal in question appears to be incapable of decomposing pure water by displacing its hydrogen at the temperature of the air; but this fact does not interfere with the preceding explanation, as zinc may have a greater affinity for sulphion than for oxygen, and, therefore, be capable of decomposing the sulphionide, but not the oxide of hydrogen. If the acid solution, however, contains sulphate of water, as it does on the old view, then zinc does and does not decompose water; decomposing it when in combination, but not when free. It becomes necessary to assume that the presence of the acid enhances the affinity of the metal for the oxygen of the water, in a manner which cannot be clearly explained; for the solubility of oxide of zinc in the acid, to which the influence of the acid is often ascribed, accounts for the continuance of the action, by providing for the removal of the oxide, rather than for its first commencement. The phenomena of the decomposition of an acid solution in the voltaic circle, are also most simply explained on the salt-radical theory. Oxide of hydrogen and sulphionide of hydrogen, are both binary "electrolytes," which are decomposed in the voltaic circle in the same manner, although not with equal facility; the common element, hydrogen, proceeding from both to the negative electrode, and oxygen in the one case and sulphion in the other to the positive electrode. The sulphion finds water there, and resolves itself into sulphionide of hydrogen and free oxygen. The decomposition of the sulphionide of sodium or any other salt may be explained in the same simple manner; while on the other view, it must be assumed that a simultaneous transference between the electrodes of acid and alkali with the oxygen and hydrogen of water takes place; and the effect of the acid in promoting the decomposition of the water remains unaccounted for.

When a metallic oxide is dissolved in an acid solution, as oxide of zinc in diluted sulphuric acid, the reaction which occurs is thus explained on the binary theory:

Sulphionide of hydrogen ...	Hydrogen...........	Water.
	Sulphion	
Oxide of zinc..	Oxygen	
	Zinc	Sulphionide of zinc;

as in the reaction between the same oxide and hydrochloric acid (page 156).

The chief objections to the salt-radical theory, are—

First: The creation of so many hypothetical radicals; namely, one for every class of oxygen-acid salts. But it is to be remembered that the great proportion of oxygen acids, such as acetic, oxalic, &c. are equally of an ideal character, and cannot be exhibited in a separate state.

Secondly: The peculiarities of the salts of phosphoric acid which are supposed to be inimical to the new view. That acid forms three different and independent classes of salts, containing respectively one, two, and three, equivalents of base to one of acid. On the binary theory, these three classes of salts must contain three different salt-radicals, combined respectively with one, two, and three equivalents of hydrogen or metal. The three phosphates of water and the corresponding phosphionides of hydrogen would be represented as follows:—

I.	II.	III.
$HO+PO_5$	$2HO+PO_5$	$3HO+PO_5$
$H+PO_6$	$2H+PO_7$	$3H+PO_8$

Such salt-radicals and such compounds with hydrogen startle us, from their novelty, but it may be questioned whether they are really more singular than the anormal classes of phosphates, containing several equivalents of base, for which they are substituted, but which we have been more accustomed to contemplate. All the salt-radicals known in a separate state, such as chlorine and cyanogen, combine with one equivalent only of hydrogen, or are monobasylous, but it would be unfair to assume in the present imperfect state of our knowledge that other salt-radicals may not exist, capable of combining with two or three equivalents of hydrogen, as the phosphate-radicals are supposed to do. The existence of at least one such radical is highly probable, as will afterwards appear.

In conclusion, it may be stated that neither view of the constitution of the oxygen-acid salts, (which alone are affected by this discussion), rests on demonstrative evidence; they are both hypotheses, and are both capable of explaining all the phenomena of the salts. But to whichever of them a speculative preference is given, we can scarcely avoid using the language of the acid theory, in the present state of chemical science.

[Additional objections may be urged against the salt-radical theory:

As long as it is applied to salts constituted according to the law that "bases always combine with as many atoms of acid as they themselves contain of oxygen," the subject is without difficulty, but when it is applied to anhydrous compounds containing more than one equivalent of acid, it fails, or necessitates the creation of as many hypothetical salt-radicals as there are examples of this kind. Thus, the anhydrous sulphates of potassa and soda, the chromates, &c. are not mere combinations of one equivalent of the base with one and more equivalents of the acid, but become compounds of a metal with a greater number of salt-radicals. The neutral chromate of potassa, KO, CrO_3, is on the salt-radical theory; K, CrO_4, the bichromate; KO, $2CrO_3$, is K, Cr_2O_7; and the terchromate, KO, $3CrO_3$, is K, Cr_3O_{10}; or potassium combined with three different and new substances, each requiring a new and distinctive name. Moreover, the theory is involved in the same difficulty when the attempt is made to apply it to those salts which are exceptions to the above law, or in which the number of atoms of oxygen in the base does not correspond with the number of atoms of acid. The following example may be taken from the salts of tartaric acid, which, considered as bibasic, has the formula $C_8H_4O_{10}$, for which the conventional symbol $\bar{T}$ may be substituted, and we shall then have four of the salts represented below, on the old and new views of their constitution:

KO, HO. $\bar{T}$=K, H. $\bar{T}O_2$, ... Cream of tartar.
KO, NaO. $\bar{T}$=K, Na. $\bar{T}O_2$ .... Rochelle salt.
KO, Fe_2O_3. $\bar{T}$=K, Fe_2. $\bar{T}O_4$ .... Tartarized iron.
KO, SbO_3. $\bar{T}$=K, Sb. $\bar{T}O_4$ .... Tartar emetic.

In the first two formulæ the elements are readily transposed to suit either view; but in the two latter a new hypothetical salt-radical appears, endowed with new powers, viz. the capability of combining respectively with two atoms of metallic-radical K, Sb, and with three atoms of radical K, Fe_2.

This theory explains very readily the reaction which takes place when water is decomposed under the influence of readily oxidated metals and hydrated acids, by the supposition that the metal replaces the hydrogen of the combined water. But there exist acids of which we have no known hydrate, equivalent for equivalent, carbonic, chromic, &c. acids. These being destitute of combined water do not admit of similar substitution; no hydrogen being combined, no replacement can take place.

Any theory, to be perfect, must include all known cases; and hence, if this hypothesis is not applicable to all oxygen salts, to the same extent as former views, it fails in its promised advantages. It has not yet been carried out or exhibited in detail by its advocates, which would seem to show they are aware of its difficulties, and are not yet prepared to obviate them. One of the points requiring explanation is the supposition in some of the examples quoted, that potassium and oxygen, two elements occupying the extremes of the electro-chemical series, can be placed in contact with each other without combining, a supposition requiring a subversion of chemical affinity which does not correspond with known facts.

It is not evident why "oxygen-acid salts alone are affected by this discussion." The compounds of sulphur, selenium, &c. are very analogous in character; and as sulphur-acids, combine only with sulphur bases, the same transfer of sulphur will be here required as of oxygen in the former salts, giving rise to as many new sulphur salt-radicals as those of oxygen.—R. B.]

Without deciding definitively in favour of one or other of the rival theories, it is well to keep in view that the great class of salts includes compounds which differ essentially in their capacity of analytical decomposition. A certain number of salts contain salt-radicals which can be isolated, others oxygen-acids which can be isolated, while others have yet afforded neither salt-radical nor acid in a separate state. Hence, they may be classed as—

1. Salts of isolable salt-radicals: chlorides, cyanides, sulphocyanides, &c.
2. Salts of isolable acids: sulphates, nitrates, carbonates, &c.
3. Salts which contain neither an isolable salt-radical nor an isolable acid: acetates, hyposulphites, &c. Even admitting that all salts have the same constitution, the capability of breaking up in such different ways must affect their modes of decomposition in different circumstances, and produce differences in properties which render such distinctions important.

It has become further necessary to recognize three classes of oxygen-acid salts, which in the language of the acid theory contain one, two, and three equivalents of base to one of acid.

1. *Monobasic salts.*—The great proportion of acids, such as sulphuric, nitric, &c. neutralize but one equivalent of base, or more correctly combine in the proportion of one equivalent of acid to each equivalent of oxygen in the base, and form, therefore, *mono*basic salts. (See formulæ of the neutral sulphates, page 158). But this is not inconsistent with an acid forming two series of salts with the same base or class of isomorphous bases. Thus there appear to be two well-marked classes of sulphates of the magnesian oxides, which agree in having one equivalent of base, but differ essentially in the proportions of combined water which they affect. In one series the sulphate is combined with one, three, five, or seven equivalents of water. Copperas (a sulphate of iron), Epsom salt (a sulphate of magnesia), blue vitriol (a sulphate of copper), and most of the well-known magnesian sulphates, belong to this class, which may be called the copperas class of snlphates. All the members of it are very soluble in water, and form double salts with sulphate of potassa. The other series affect two, four, and six equivalents of water. They are less known, but appear to be of sparing solubility, and to be incapable of forming double salts with sulphate of potassa. Gypsum or sulphate of lime belongs to this class, which may, therefore, be called the gypsum class of magnesian sulphates. Sulphate of iron is said to crystallize from solution in sulphuric acid with two equivalents of water, with the crystalline form and sparing solubility of gypsum. Dr. Kane obtained a sulphate of copper with four equivalents of water, by exposing the anhydrous salt to the vapour of hydrochloric acid, which appears to be the second term in this series; and Mitscherlich still maintains the existence of a peculiar sulphate of magnesia containing six equivalents of water of crystallization, which will constitute the third term. It is evident that the cause of such double classes of salts is as deeply seated as that of dimorphism, and hence, possibly, the magnesian sulphate itself, which exists in the two classes, is not the same in its constitution with reference to heat.

2. *Bibasic salts.*—That class of phosphates which received the name of pyrophosphates, was the first in which one equivalent of acid was found to neutralize two equivalents of base; their formulæ being $2RO, PO_5$. The classes of tartrates and racemates which have long been known to chemists, are also bibasic salts. It is the character of a bibasic acid to unite at once with two different bases of the same natural family, which accounts for the formation of Rochelle salt, the tartrate of potassa and soda, of which the formula is $KO, NaO + C_8H_4O_{10}$. It has also been shown that gallic acid is bibasic, the gallate of lead being thus composed: $2PbO + C_7HO_3$. Now if we attempt to make this a monobasic salt by dividing the equivalents both in base and acid by two, an equivalent of gallic acid would come to contain half an equivalent of hydrogen, which Liebig considers as conclusive against the division of its atomic weight. Itaconic, comenic, euchronic, fulminic, and several other organic bibasic acids, might be named. The compound acids formed by the union of two others, and called copulated acids, such as hyposulphobenzoic acid, are usually of this class.

3. *Tribasic salts.*—The tribasic phosphates of the formula $3RO, PO_5$, have likewise proved to be the type of a class of salts. One equivalent of arsenic acid neutralizes three equivalents of base; so, it is probable, does one atom of phosphorous acid. Tannic acid also saturates three atoms of base, the formula of the tannate of lead being $3PbO + C_{18}H_5O_9$ (Liebig). There is the same necessity to admit that citric acid is tribasic, and the formula of a citrate $3RO + C_{12}H_5O_{11}$, as there is to allow that gallic acid is bibasic. Most of the citrates contain two equivalents of fixed base and one of water, but the citrate of silver contains three equivalents of oxide of silver. Cyanuric, meconic, camphoric, and several other organic acids, are tribasic.

Two of the three atoms of base in this class of salts may be different, as is observed in certain citrates, cyanurates, and phosphates, or the whole three may be different, as in the phosphate called microcosmic salt, which contains at once soda, oxide of ammonium, and water as bases.[1] Two or more of the bases may likewise be isomorphous, or at least belong to the same natural family as soda and oxide of ammonium, water, and magnesia.

Salts usually denominated Subsalts.—The preceding classes of salts, and many other bodies also, are capable of combining with a certain proportion of water, generally vaguely spoken of as water of crystallization. The compounds of the present class appear to be salts which have assumed a fixed metallic oxide in the place of this water. They may, therefore, be truly neutral in composition, the excess of oxide not standing in the relation of base to the acid. It appears that the formulæ of the nitrates named are as follows:—

Nitrate of water (acid of sp. gr. 1.42)........HO, NO_5+3HO.
Nitrate of copper (prismatic)....................CuO, NO_5+3HO.
Nitrate of copper (rhomboidal)..................CuO, NO_5+6HO.
Subnitrate of copper.............................$CuO, NO_5+3(CuO, HO)$.

I have distinguished as *constitutional* the three atoms of water which exist in these and all the magnesian nitrates, and which are replaced by three atoms of hydrated oxide of copper in the subnitrate of copper, which is therefore a nitrate of copper, with the addition of constitutional (not basic) oxide of copper; a view which is expressed by the arrangement of the symbols in its formula.

The subnitrates of zinc and lead, and probably also those of nickel and cobalt, have a similar composition (Gerhardt). A similar correspondence is observed between the crystallized neutral sulphate of copper, and the subsulphate of copper, containing four equivalents of oxide of copper, and five of water to one of acid:—

Sulphate of copper, $CuO, SO_3, HO+4HO$.
Subsulphate of copper, $CuO, SO_3, (CuO, HO)+2\,(CuO, HO)+2HO$.

Three equivalents of water in the neutral salt appear to be replaced by three equivalents of hydrated oxide of copper in the subsalt. The remaining 2HO of the latter salt are expelled by a moderate heat, while the other 4HO in combination with oxide of copper, are extricated by a much higher temperature, and their separation attended by a palpable decomposition of the salt, as it affords a portion of soluble neutral salt afterwards to water. The remark is made by M. Gerhardt, that the number of such subsalts is greatly exaggerated, which is quite in accordance with my own observations; few salts combining with an excess of oxide in more than one or two proportions. Most subsalts are entirely insoluble in water, but when they possess a certain degree of solubility, they may afford other analogous subsalts by double decomposition. Thus a solution of bisubnitrate of lead, PbO, NO_5+PbO, HO, on the addition of neutral chromate of potassa allows the red bisubchromate of lead, PbO, CrO_3+PbO, to precipitate. M. Gerhardt, who observed this fact, considers that it assimilates the nitrates and pyrophosphates, and indicates that the latter are ordinary subsalts. But this is really a coincidence of small importance, while nitric acid affords no bibasic hydrate, nor a bibasic salt of soda, as phosphoric acid does.

Water, oxide of copper, oxide of lead, and the hydrates of these metallic oxides, appear to be the bodies most disposed to attach themselves to salts in this manner. The strong alkalies, potassa and soda, are never found in such a relation, or discharging any other function than that of base to the acid of the salt. These views of subsalts, in which their constitutional neutrality is preserved, have been extended to organic compounds. Many neutral organic bodies appear to be capable of combining with metallic oxides, particularly with oxide of lead—such as sugar, amidin, dextrin, orcin, and they generally combine with several atoms of the oxide. Thus in the compound of orcin and oxide of lead, $C_{18}H_7O_3+5PbO$, the orcin is combined

[1] Inquiries respecting the Constitution of Salts; of oxalates, nitrates, phosphates, sulphates, and chlorides. Phil. Trans. 1837, page 47.

with five atoms of constitutional oxide of lead, which actually replace five atoms of constitutional water, which orcin in its ordinary state contains.

Constitutional water is sometimes replaced by a *salt*, which never happens with basic water. Thus cane sugar may be represented as $C_{12}H_{11}O_{11}$, or rather $C_{24}H_{22}O_{22}$; of which one atom of water may be replaced by chloride of sodium, and the compound formed, $C_{24}H_{21}O_{21}+NaCl$. It is to be observed that constitutional water is superadded to a salt, and such an element is removed and replaced without affecting the structure of the body to which it is attached. The replacing substance may also be a compound of a very different character from water; for besides metallic oxides and salts, ammonia and certain anhydrous acids appear to be capable of attaching themselves to salts, in the same manner as constitutional water.

A different view of the constitution of subsalts is advocated by M. Millon, who assumes the existence of poly-atomic bases, or that two, three, four, and even six equivalents of water or a metallic oxide, may together constitute a single equivalent of base, and unite as such with a single equivalent of acid to form a neutral salt (Annales de Chim. et de Phys., xviii. 333).

Salts of the type of red chromate of potassa.—Several salts unite with anhydrous acids. Thus both chloride of sodium and chloride of potassium absorb and combine with two atoms of anhydrous sulphuric acid without decomposition, when exposed to the vapour of that substance. Sulphate of potassa also combines with one atom of anhydrous sulphuric acid. All these compounds are destroyed by water. But the red chromate of potassa, generally called bichromate of potassa, which consists of chromate of potassa together with one atom of chromic acid, is possessed of greater stability, as is likewise the compound of chloride of sodium or potassium with two atoms of chromic acid. Another compound containing one atom of potassium and three atoms of chromic acid, known as the terchromate of potassa, may be viewed as a combination of chromate of potassa with two atoms of chromic acid, and represented by KO, CrO_3+2CrO_3. The bichromate of potassa will then be KO, CrO_3+CrO_3, and the chromate containing chloride of potassium, $KCl+2CrO_3$. The biniodate of potassa (iodate of water and potassa) may be rendered anhydrous, and, when so, is a salt of the same class.

Double salts.—Salts combine with each other, but by no means indiscriminately. With a few exceptions, which may be placed out of consideration for the present, the combining salts have always the same acid—sulphates combining with sulphates, chlorides with chlorides. Their bases or their metals, however, must belong to different natural families. Thus it may be questioned whether a salt of potassa ever combines with a salt of soda, certainly never with a salt of ammonia. Salts of the numerous metals including hydrogen, belonging to the magnesian family, do not combine together. Thus sulphate of magnesia does not form a double salt with sulphate of lime, with sulphate of zinc, or with sulphate of water; while on the other hand salts of this family are much disposed to combine with salts of the potassium family — sulphate of soda, for instance, forming double salts with sulphate of lime, sulphate of zinc, and sulphate of water. We have thus the means of distinguishing between a double salt, and the salt of a bibasic or tribasic acid. The bisulphate and binoxalate of potassa saturated with soda, form sulphates and oxalates of potassa and soda, which separate from each other by crystallization, although the acid salts are themselves double salts of water and potassa. But the acid fulminate of silver, or the acid tartrate of potassa (bitartrate), affords only one salt when saturated with soda, in which isomorphous bases exist, and which, therefore, is a salt of one acid, and not a compound of two salts. The great proportion of the salts which are named *super*, *acid* and *bi*-salts, contain a salt of water, and are double salts — such as the supercarbonate of soda (HO, CO_2+NaO,CO_2), the bisulphate of potassa (HO, SO_3+KO, SO_3), and the binacetate of soda: but a few of them are bibasic or tribasic salts, containing one or two atoms of water as base — such as the salt called bitartrate of potassa, or biphosphate of potassa ($2HO, KO+PO_5$).

From these observations must be excepted double salts formed by fusion, and many salts formed in highly acid solutions, which are scarcely limited in variety of composition; carbonate of potassa fusing with the carbonate or sulphate of soda, and sulphate of baryta crystallizing in combination with sulphate of water, from solution in sulphuric acid. Such salts are decomposed by water, and are otherwise deficient in stability, compared with the soluble double salts, to which alone the preceding remarks apply.

There is no parallelism between the constitution of a double salt and that of a simple salt itself, or foundation for the statements which are sometimes made, that one of the salts which compose a double salt has the relation to the other of an acid to a base, and that one salt is electro-negative to the other. The resolution of a double salt into its constituent salts by electricity, has never been exhibited, and is not to be expected, from what is known of electrolytic action; while no analogy whatever subsists between a double salt and a simple salt on the binary view of the constitution of the latter. Besides, the supposed analogy is destroyed by what is known of the derivation of double salts. Sulphate of magnesia acquires an atom of sulphate of potassa in the place of an atom of water, which is strongly attached to it, in becoming the double sulphate of magnesia and potassa. In the same way, the sulphate of water has an atom of water also replaced by sulphate of potassa, in becoming the bisulphate of potassa; relations which appear in the rational formulæ of these salts:

Sulphate of magnesia.................................. $\dot{M}g\dddot{S}(\dot{H})+6\dot{H}$
Sulphate of magnesia and potassa..................... $\dot{M}g\dddot{S}(\dot{K}\dddot{S})+6\dot{H}$
Sulphate of water (acid of sp. gr. 1.78)............ $\dot{H}\dddot{S}(\dot{H})$
Bisulphate of potassa.................................. $\dot{H}\dddot{S}(\dot{K}\dddot{S})$

It thus appears that a provision exists in sulphate of magnesia itself for the formation of a double salt, and that the molecular structure is unaltered, notwithstanding the assumption of the sulphate of potassa as a constituent. The derivation of the acid oxalates likewise throws much light on the nature of double salts. The oxalate of potassa contains an atom of constitutional water, which is replaced by hydrated oxalic acid (the crystallized oxalate of water), in the formation of the binoxalate of potassa (double oxalate of potassa and water), or by the oxalate of copper in the formation of the double oxalate of potassa and copper, as exhibited in the following formulæ, in which the replacing substances are enclosed in brackets to mark them as before:

Oxalate of potassa................................. $\dot{K}\dddot{C}C, (\dot{H})$
Binoxalate of potassa.............................. $\dot{K}\dddot{C}C, (\dot{H}\dddot{C}CH_2)$
Oxalate of potassa and copper..................... $\dot{K}\dddot{C}C, (\dot{C}u\dddot{C}C\dot{H}_2)$

Now the anomalous salt, quadroxalate of potassa, is derived in the same way from the binoxalate, as the binoxalate itself is derived from the neutral oxalate, two atoms of water being displaced by two atoms of hydrated oxalic acid, thus:

Binoxalate of potassa.............................. $\dot{K}\dddot{C}C, \dot{H}\dddot{C}C, (2H)$
Quadroxalate of potassa........................... $\dot{K}\dddot{C}C, \dot{H}\dddot{C}C, (2\dot{H}\dddot{C}CH_2)$

These examples illustrate the derivation of double salts by *substitution.* The structure of the salts, too, exemplifies what may be called *consecutive* combination. The basis of the last mentioned salt, for instance, is oxalate of potassa, which is in direct combination with oxalate of water. A compound body is thus produced which seems to unite *as a wholė* with two atoms of hydrated oxalic acid. This is very different from the direct combination of all the elements which compose the salt.

In the formation of many other classes of double salts, no substitution is observed, but simply the attachment of two salts together, often of an anhydrous with a hydrated salt, in which case the last often carries its combined water along with it, and sometimes acquires an additional proportion. Thus in the formula of the double chloride of potassium and copper, $KCl+CuCl, 2HO$, the formulæ of its constituent salts reappear without alteration; and in that of alum, sulphate of potassa is found with the hydrated sulphate of alumina annexed, of which the water is increased from eighteen to twenty-four atoms. In these and all other double salts, the characters of the constituent salts are very little affected by their state of union. If one of them has an acid reaction, like sulphate of alumina or chloride of copper, it retains the same character in combination; and nothing resembling a mutual neutralization of the salts by each other is ever observed. No heat is evolved in their formation. (Memoirs of the Chemical Society, ii. 51).

The compounds of chlorides with chlorides, and of iodides with iodides, are numerous, and were viewed by Bonsdorf as simple salts, in which one of the chlorides is the acid, and the other the base. But such an opinion can no longer be entertained, the chlorides themselves being unquestionably salts, and their compounds, therefore, double salts.

The combinations of such salts with each other as contain different acids are not so well understood, the theory of their formation having hitherto been little attended to. They are in general decomposed by water, and easily, if the solubility of one of their constituents is considerable, as is observed of the compounds of iodate of soda with one and with two proportions of chloride of sodium, of the biniodate of potassa with the sulphate of potassa, of the oxalate of lime with the chloride of calcium.

The compound cyanides, which form a considerable class of salts, must be excepted from all the preceding general statements in regard to double salts. Cyanides of the same family combine together, as cyanide of iron with cyanide of hydrogen; the compound cyanide also generally consists of three and not of two simple cyanides; and lastly, the properties of compound cyanides are very different from those of the simple cyanides which are supposed to compose them. The simple cyanide of potassium, for instance, is highly poisonous, while the double cyanide of potassium and iron is as mild in its action upon the animal economy as sulphate of soda. But the compound cyanides may be removed from the class of double salts, on a speculative view of their constitution which their anomalous character led me to propose. It is to be premised that the supposed double proto-cyanide of iron and potassium (yellow prussiate of potassa) affords no hydrocyanic acid whatever when distilled with an excess of sulphuric acid at a temperature not exceeding 100°; which suggests the idea that it does not contain cyanides or cyanogen. Assuming the existence of a new compound radical, N_3C_6, which has three times the atomic weight of cyanogen, and may be called *prussine*, and which is also *tribasylous* or capable of combining with three atoms of hydrogen or metal, like the radical of the tribasic class of phosphates, then the compound cyanides assume a constitution of extreme simplicity. We have one atom of prussine combined always with three atoms of hydrogen or metal in the following salts: in the proto-cyanide of iron and potassium with one of iron and two of potassium; in the compound called ferro-cyanic acid, with one of iron and two of hydrogen; in Mosander's salts, with one of iron, one of potassium and one of barium, calcium, &c.; with two of iron and one of potassium in the salt which precipitates on distilling the yellow prussiate of potassa with sulphuric acid at 212°. To many of these, parallel combinations might be adduced from the tribasic phosphates. Prussides likewise combine together, producing double prussides, such as

Percyanide of iron and potassium
(red prussiate of potassa)....... $Fe_2, N_3C_6+K_3, N_3C_6$
Prussian blue........................ $Fe_2, N_3C_6+Fe_3, N_3C_6$
Basic prussian blue................ $Fe_2, N_3C_6+Fe_3, N_3C_6+Fe_2O_3$

Formation of salts by substitution. — Chemists have come to pronounce less decidedly on theories of the constitution of salts and the arrangement of elements in these and other compounds, since their attention has been fixed upon the formation of compounds, by the subtitution of one element for another, without injury to the original form or type, and often to give a preference to empirical over rational formulæ, while their opinions on chemical constitution were suspended. The elementary composition of oil of vitriol, or the hydric sulphate, is expressed by SO_4H; the sulphate type, and other neutral sulphates, are formed by replacing the hydrogen by a metal; the zinc sulphate, SO_4Zn; the soda sulphate, SO_4Na. M. Gerhardt, assuming as a law that the equivalent of all compound bodies gives two volumes of vapour, divides the equivalents of the following elements by two — nitrogen, phosphorus, chlorine, hydrogen, and all the metals; and is thereby enabled to construct substitution formulæ, which are often remarkable for their simplicity. This will appear in the following selected formulæ: —

Formulæ by M. Gerhardt.

($O=8$, $S=16$; *the other symbols = half the usual equivalents.*)

I. NITRATES.

Hydric nitrate	NO_3H	Monobasylous salts.
Magnesia nitrate	NO_3Mg	
Potassa nitrate	NO_3K	

II. SULPHATES.

Hydric sulphate	SO_4H_2	Bibasylous salts.
Magnesia sulphate	SO_4Mg_2	
Potassa sulphate	SO_4K_2	
Potassa bisulphate	SO_4KH	

III. TRIBASIC PHOSPHATES.

Hydric phosphate	PO_4H_3	Tribasylous salts.
Subphosphate of soda	PO_4Na_3	
Phosphate of soda	PO_4Na_2H	
Biphosphate of soda	PO_4NaH_2	

The preceding groups are symbolized without any division of the equivalents used; but M. Gerhardt departs from this practice, when necessary, in the *unitary* system of notation which he recommends: —

Anhydrous alum	$SO_4\,(K_{\frac{1}{2}}Al_{\frac{3}{2}})$
Pyrophosphate of soda	$PO_{\frac{7}{2}}\,(Na_2)$
Subphosphate of soda + HO	$PO_{\frac{9}{2}}\,(Na_3H)$

Although a rational formula, strictly speaking, expresses no more than a decomposition, — and the rational formulæ of a compound may truly, therefore, be as numerous as the modes of decomposition of which it is susceptible, — still much would undoubtedly be lost by abandoning such formulæ for formulæ which are entirely empirical; unless, indeed, it is found that the uniform practice of exhibiting the leading constituent, in the proportion of a single equivalent, should bring together different bodies under common formulæ, which are types of useful classification, as M. Gerhardt maintains.

Salts of Ammonia. — Ammonia is a gaseous compound of one equivalent of nitrogen and three of hydrogen, of which the solution in water is caustic and alkaline, and which neutralizes acids perfectly, as potassa and soda do. But all its oxygen-acid salts contain, besides ammonia, an equivalent of water which is essential to them, and inseparable without the destruction of the salt; and with this additional constituent they are isomorphous with the salts of potassa. Hydrochloric acid also unites with ammonia without losing its hydrogen, and the compound or hydrochlorate of ammonia, which is isomorphous with the chloride of potassium, contains, therefore, an equivalent of hydrogen, besides chlorine and ammonia. On the now generally received theory of these salts, the ammonia with this hydrogen, or that of the water in the oxygen-acid salts, constitutes a hypothetical basyl, *ammonium* (NH_4), to which allusion has already been made as being isomorphous with potassium. This view of the constitution of the salts of ammonia will be made obvious by a few examples: —

	ON THE AMMONIUM THEORY.
Hydrochlorate of ammonia, HN_3, HCl ...	Chloride of ammonium, NH_4, Cl
Sulphate of ammonia, NH_3, HO, SO_3 ...	Sulphate of oxide of ammonium, NH_4O, SO_3
Nitrate of ammonia, NH_3 HO, NO_5 ...	Nitrate of oxide of ammonium, NH_4O, NO_5

The application of this theory to the compounds of ammonia with hydrosulphuric acid and sulphur is particularly felicitous. These compounds may be thus represented, and placed in comparison with their potassium analogues, NH_4 being equivalent to K: —

Sulphide of ammonium........................	NH_4S	... KS
Sulphide of ammonium and hydrogen (bihydrosulphate of ammonia.....................	NH_4S, HS	... KS, HS
Tritosulphide of ammonium....................	NH_4S_3	... KS_3
Pentasulphide of ammonium..................	NH_4S_5	... KS_5

Ammonium is supposed to present itself in a tangible form, and in possession of metallic characters, in the formation of what is called the *ammoniacal amalgam.* When mercury alloyed with one per cent. of sodium is poured into a saturated cold solution of sal ammoniac (chloride of ammonium), it undergoes a prodigious increase of bulk, expanding sometimes from one volume to two hundred volumes, without becoming in the least degree vesicular, and acquiring a butyraceous consistence, while its metallic lustre is not impaired. A small addition is at the same time made to its weight, estimated at from 1 part in 2000 to 1 in 10,000, which certainly consists of ammonia and hydrogen in the proportions of ammonium. The sodium, it is supposed, combines with the chlorine of chloride of ammonium, and the liberated ammonium with mercury, so that the metallic product is an amalgam of ammonium. It speedily resolves itself again spontaneously into running mercury, ammonia, and hydrogen, unless the temperature be reduced so far as to freeze it. After all, however, neither isolation nor the metallic character is essential to ammonium as an alkaline radical, other basyls being now admitted, such as ethyl and benzoyl, which have no claim to such characters.

Other classes of ammoniacal salts may be formed in which the fourth equivalent of hydrogen in ammonium is replaced by a metal of the magnesian family, — by copper in particular, which most resembles hydrogen. Thus anhydrous chloride of copper absorbs a single equivalent of ammonia with great avidity and the evolution of much heat, which cannot afterwards be separated from it by the agency of heat. The compound appears to be strictly analogous to chloride of ammonium, but contains an equivalent of copper in the place of hydrogen. Its formula is NH_3Cu, Cl, and it may be named the chloride of *cuprammonium.* This salt and many others are likewise capable of combining with more ammonia, which is retained less strongly, and has the relation of constitutional water to the salt. The constitution of these combinations will be more minutely considered in other parts of the work.

Amidogen and amides. — The existence of another compound of nitrogen and hydrogen (NH_2), containing an equivalent less of hydrogen than ammonia, is recognized in an important series of saline compounds, although it has not been isolated. These compounds are called *amides*, and hence the name amidogen applied to their radical. When potassium is heated in ammoniacal gas, the metal is converted into a fusible green matter, which is the amide of potassium, while an equivalent of hydrogen is disengaged. Amidogen exists also in the white precipitate of mercury formed on adding ammonia to corrosive sublimate, the product being a double chloride and amide of mercury ($HgCl + HgNH_2$).

Amides are produced in an interesting way, by the abstraction of the elements of water from compounds of ammonia with oxygen acids. Thus, on decomposing oxalate of ammonia by heat, the acid losing a proportion of oxygen, and the ammonia a proportion of hydrogen, *oxamide* sublimes, which consists of $NH_2 + 2CO$. When ammoniacal gas and anhydrous sulphuric acid vapour are mixed together, a saline substance is produced which dissolves in water, but is not sulphate of ammonia, the solution affording no indications of sulphuric acid. It is believed to be a hydrated *sulphamide*, or to be constituted thus, $NH_2, SO_2 + HO$; a compound which it will be observed contains neither ammonia nor sulphuric acid. Similar products result from the action of ammonia on dry carbonic acid and all the other anhydrous oxygen salts. The difference between these compounds and the true salts of ammonia affords an argument in favour of the ammonium theory of the latter.

ANTITHETIC OR POLAR FORMULÆ.

Formulæ for compounds may be constructed to exhibit the attraction of the ultimate elements for each other without involving any contested theory of the constitution of compounds, and which indeed might supersede the consideration of such views, were it not that the nomenclature, which it would be inconvenient to alter greatly, is founded upon the latter. A certain amount of information is given in the ordinary formulæ by the arrangement of the symbols, the symbol of the basylous or positive constituent being placed before the symbol of the halogenous or negative constituent, as in HO for water, SO_3 for sulphuric acid. To carry out this principle farther, and make its application more perspicuous, I have suggested the writing of a formula in two lines, placing all the negative constituents in the upper, and the positive in the lower line: —

Potassa....... $\frac{O}{K}$ Water....... $\frac{O}{H}$ Sulphuric acid.... $\frac{O_3}{S}$ Ammonia....... $\frac{N}{H_3}$

Cyanogen.... $\frac{N}{C_2}$ Olefiant gas $\frac{H_4}{C_4}$ Carbonic oxide... $\frac{O_2}{C_2}$ Hydric oxalate $\frac{O_3O}{C_2H}$

From their construction these formulæ are named *antithetic*, the two orders of constituents being placed opposite or against each other; or *polar*, from exhibiting the opposite attractive forces of the elements. Several decompositions already referred to, and others, may be made more intelligible by their aid.

Decomposition of ammoniacal salts. — In the decomposition of oxalate of ammonia and formation of oxamide, the change consists in the abstraction of two equivalents of water from the constituents of the salt: the formulæ being —

$$\text{Oxalate of ammonia.....................} \quad \frac{NOO_3}{H_3HC_2} - \frac{O_2}{H_2} = \frac{NO_2}{H_2C_2} \text{ oxamide.}$$

The interesting observation has lately been made by M. Dumas, that by distillation with anhydrous phosphoric acid, four equivalents of water are separated from oxalate of ammonia, and cyanogen formed. Supposing that the formation of oxamide precedes this last decomposition, we have —

$$\text{Oxamide.................................} \quad \frac{NO_2}{H_2C_2} - \frac{O_2}{H_2} = \frac{N}{C_2} \text{ cyanogen.}$$

It is seen, that although we cannot say that water exists either in oxalate of ammonia or in oxamide, still 4O is negative and 4H positive in the first of these substances, and 2O negative with 2H positive in the second, the relation which these elements bear to each other in water. The polar relation of these elements, therefore, does not require to be subverted, when they are led to unite and take the form of water, under the influence of the attraction of phosphoric acid for that oxide. It is manifestly a law of decomposition that those decompositions take place most readily which permit the elements to continue in their original polar condition and position in the formulæ; the explanation being, that such decompositions are promoted by the peculiar attractions of the ultimate elements for each other as they exist in the original compound; or the compound molecule is broken up in the direction in which it naturally divides.

The decomposition by phosphoric acid of other salts of ammonia containing acids related to the alcohols, illustrates the same constancy of polar relation in the elements before and after the change. Thus, formiate of ammonia gives hydrocyanic acid by the abstraction of four equivalents of water:—

Formiate of ammonia...... $\frac{N\ O}{H_3H}\frac{HO_3}{C_2} - \frac{O_4}{H_4} = \frac{NH}{C_2}$ hydrocyanic acid.

Here the hydrogen of hydrocyanic acid is represented as negative, and it can certainly be replaced by chlorine, a negative element, and the chloride of cyanogen formed:—

Hydrocyanic acid............ $\frac{NH}{C_2}$ Chloride of cyanogen......... $\frac{NCl}{C_2}$

With a metallic oxide, however, hydrocyanic acid gives a cyanide, and then the hydrogen appears positive—

Hydrocyanic acid............ $\frac{N}{C_2H}$ Cyanide of silver............ $\frac{N}{C_2Ag}$

But hydrocyanic acid is in the lowest degree feeble in its powers as an acid, or as cyanide of hydrogen, and its hydrogen appears to be just on the limit between the basylous and halogenous character and position.

Acetate of ammonia distilled with phosphoric acid also loses four equivalents of water, like all the ammoniacal salts in question, and gives the cyanide of methyl:—

Acetate of ammonia $\frac{N\ O}{H_3H}\frac{O_3H_3}{C_4} - \frac{O_4}{H_4} = \frac{H_2H}{C_2}\frac{N}{C_2}$ cyanide of methyl.

The chloracetate of ammonia in losing 4HO gives a liquid body of the composition C_4Cl_3N:—

Chloracetate of ammonia.............. $\frac{N\ O}{H_3H}\frac{O_3Cl_3}{C_4} - \frac{O_4}{H_4} = \frac{Cl_2}{C_2}\frac{ClN}{C_2}$

Here the single negative H of hydrocyanic acid is also under the positive attraction of the C_2 of the hydrocarbon, C_2H_2, a cross attraction, which forms a bond of union between the hydrocyanic acid and hydrocarbon, and supports the equilibrium.

Why is ammonia a base?—Of ammonia and hydrochloric acid the antithetic formulæ are—

$$\frac{N}{H_3} \text{ and } \frac{Cl}{H}$$

There can be little doubt but that when these bodies are united, the highly negative chlorine shares, or assumes entirely, the positive attraction of the third equivalent of hydrogen in ammonia, which there is reason to believe is less powerfully attracted or neutralized by the negative nitrogen than the other two equivalents of hydrogen. We thus obtain the following formula:—

Hydrochlorate of ammonia........................ $\frac{N\ Cl}{H_2\ H_2}$

Now the acid character of hydrochloric acid, which is neutralized in the salt, depends upon the former substance being a compound in which a powerful salt-radical, chlorine, is united with a weak basyl, hydrogen. With a powerful basyl, such as potassium, chlorine gives a neutral salt, the chloride of potassium. But it is probable that the subchloride of hydrogen, H_2Cl, if it could exist in a separate state, would be an equally neutral salt, for hydrogen belongs to the magnesian class of elements, two atoms of which appear to be equivalent to one atom of the potassium class, or H_2Cl to be equivalent to KCl, and possibly isomorphous with it. One atom of nitrogen there are also grounds for believing to be equivalent in composition to two atoms of oxygen, or N=2O. Hence the compound $\frac{N}{H_2}$ has a character of saturation, or polar neutralization, like $\frac{O_2}{H_2}$ or two equivalents of water. In ammonia, therefore, the third basylous atom of hydrogen may well be considered as unsaturated, and to be what imparts a basylous or positive character and activity to the compound. In metallic oxides which are bases, we have also the positive property of the metal imperfectly saturated by the weak negative body oxygen, and the positive attraction therefore in excess.

In the oxygen acids, on the contrary, there is an excess of negative attraction from the predominance of the oxygen element, and it is remarkable that in the more powerful acids, such as sulphuric, nitric, and chloric, one equivalent of this oxygen is but feebly united, and its negative attraction free to act, like the positive attraction of the third equivalent of hydrogen in ammonia. Hence ammonia and anhydrous sulphuric acid readily combine :—

$$\frac{N}{H_2H} + \frac{OO_2}{S} = \frac{N\ O\ O_2}{H_2\,H\ S}$$

From the action of the affinities exhibited in the last formula, a stable equilibrium results; but it is not intended to express that amidogen, water, and sulphurous acid, exist ready formed in the compound. Indeed, in no case do the formulæ express actual formation of subordinate compounds, or anything more than what are considered to be the predominating set of attractions among all the possible attractions which the elements have for each other, and all of which they continue to exert in some degree.

In sulphate of oxide of ammonium, the affinities of equilibrium are those of the elements of amidogen, suboxide of hydrogen, and sulphuric acid :—

Constitution of Sulphate of Ammonia. Sulphate of Ammonia.

$$\frac{N}{H_3} + \frac{O}{H} + \frac{O_3}{S} \quad = \quad \frac{N\ O\ O_3}{H_2\,H_2\,S}$$

In this and all the other oxygen-acid salts of ammonia, the highly alkaline oxide H_2O appears, and constitutes the point of attachment for the acid. Other sources of stability in the sulphate of ammonia are — first, the attraction of N for its third atom of hydrogen, which is never entirely relinquished, although the latter is more under the influence of the O of the water; and, secondly, the attraction of the O_3 of the sulphuric acid for the basylous H_2: for these cross attractions prevent the division of the compound into subordinate compounds under the influence of the predominating affinities first enumerated. This salt may be taken as a fair example of the assumed mode of formation of compounds, in which the affinities of the elementary atoms only are operative, to the entire exclusion of the affinities usually assigned to subordinate groups of elements acting as compound radicals or quasi-elements.

Why are arsenic and phosphoric acids tribasic?—Phosphoric acid, PO_5, may be considered, from its properties and mode of formation, as phosphorous acid, PO_3 + two equivalents of oxygen less strongly combined; and in the same way arsenic acid, AsO_5, as arsenious acid, AsO_3 + two equivalents of oxygen. Now, when united with a base, which we shall suppose a metallic protoxide, RO, these two surplus equivalents of oxygen in the phosphoric acid, added to the single equivalent of oxygen in the base, convert an equivalent of the latter into an acid of the formula RO_3. Two more equivalents of base are required—one to neutralize this RO_3, and the other to neutralize the phosphorous acid, PO_3; making three equivalents of base to every single equivalent of phosphoric acid. The general formula for a so-called tribasic phosphate is, therefore—

$$\text{Phosphate} \ldots\ldots \frac{O\;O_3}{R\;R} + \frac{O\;O_3}{R\;P}$$

and resembles a double sulphate, $RO, SO_3 + RO, SO_3$.

$$\text{Tribasic subphosphate of lime } (3CaO, PO_5) \ldots\ldots \frac{O\;O_3}{Ca\;Ca}\frac{O\;O_3}{Ca\;P}$$

Phosphoric acid appears farther to have the power, when heated strongly, of assuming the two equivalents of oxygen referred to into a more intimate state of combination, possibly with the loss of a portion of combined heat, and gives the class of monobasic metaphosphates. The general formula of a metaphosphate is—

$$\text{Metaphosphate} \ldots\ldots \frac{O\;O_5}{R\;P}$$

A pyrophosphate, or so-called bibasic phosphate, is, on this view, a compound of a common phosphate and metaphosphate :—

$$\text{Pyrophosphate} \ldots\ldots \frac{O\;O_3}{R\;R}\frac{O\;O_3}{R\;P} + \frac{O\;O_3}{R\;P}$$

Hence the equivalent of a pyrophospate contains four equivalents of base and two of phosphoric acid—the reason why so many double pyrophosphates appear to exist.

Phosphoric acid is thus supposed to resemble those conjugate organic acids which combine with two equivalents of base, because they possess the elements of two different acids.

ATOMIC VOLUME OF SOLID BODIES.

Since the existence of simple relations between the combining volumes of gaseous bodies was ascertained by Gay-Lussac, various attempts have been made to establish similar relations between the measures, as well as the weights, in which bodies, in the liquid and solid form, enter into combination. If the atoms of all elements had, in the solid form, the same bulk, their specific gravities would be regulated by their atomic weights, and be in the same proportion. It was early observed by M. Dumas, that a close approximation to this simple ratio holds among the specific gravities of a considerable number of isomorphous bodies; but it is by no means general. The subject has received its fullest investigation from Professor Schroeder of Mannheim[1], Dr. Hermann Kopp[2] of Giessen, and Messrs. Playfair and Joule.[3] Much information has been collected, and many curious relations in the specific gravities of particular bodies pointed out; but the general deductions drawn can, in general, claim only a certain degree of probability. Much of the uncertainty arises from the specific gravity of a body in the solid form being often variable between rather wide

---

[1] Die Molecularvolume der chemischen Verbindungen im festen und flussigen Zustande: Mannheim, 1843.

[2] Bemerkungen zur Volumtheorie, Braunschweig, 1844; Annales de Chimie et de Physique, 2e Sér. T. lxxv. and 3e Sér. T. iv. p. 462.

[3] Memoirs of the Chemical Society of London, vol. ii. p. 401; vol. iii. pp. 57 and 199. Also, a paper on the Constitution of Aqueous Solutions of Acids and Alkalies, by Mr. J. J. Griffin; ibid. p. 155.

limits. Thus platinum, in a pulverulent state, reduced from its oxide and from the double chloride of platinum and ammonium respectively, is found to have the specific gravity 17·766 in the first case, and 21·206 in the second (Playfair and Joule); and the effect of compression upon the malleable metals is generally very sensible. As the rate of dilatation of different solids and liquids by heat is very dissimilar, it is obvious their relations in density may also be disturbed or disguised by temperature.

At present, I shall confine myself to a summary of the results of M. Kopp on this subject, which partake least of a speculative character. The *atomic volume*, which I substitute for the *specific volume* of Dr. Kopp, in the following tables, is the volume or measure of an equivalent or atomic proportion of the different substances enumerated. The *calculated density* is obtained by dividing the atomic weight by this volume. Thus an equivalent of mercury, 1266 parts by weight, has the volume 93 assigned to it. Now 1266, divided by 93, gives 13·6 as the "calculated" specific gravity, which coincides with the specific gravity of mercury actually observed by Kupffer and others. The atomic volume for oxygen will afterwards appear to be 16, or a multiple of that number, and is the modulus of the scale.

TABLE I.

Atomic Volume and Specific Gravity of Elements.

Substances.		Atomic Weight.	Primitive Atomic Volume.	Calculated Sp. Grav.	Observed Specific Gravity.
Antimony......	Sb	806	120	6·72	6·70 Karsten; 6·6 Breithaupt; 6·85 Muschenbroeck.
Arsenic	As	470	80	5·87	5·70, 5·96 Guibourt; 5·62 Karsten; 5·67 Herapath.
Bismuth	Bi	1330	135	9·85	9·88 Thenard; 9·83 Herapath; 9·65 Karsten.
Bromine	Br	489	160	3·06	2·99 Loewig; 2·97 Balard.
Cadmium	Cd	697	81	8·60	8·66 Herapath; 8·63 Karsten, Kopp; 8·60 Stromeyer.
Chlorine	Cl	221	160	1·38	1·33 Faraday.
Chromium	Cr	352	69	5·10	5·10 Thomson.
Cobalt	Co	369	44	8·39	8·49 Brunner; 8·51 Berz.; 8·71 Lampadius.
Copper	Cu	396	44	9·00	8·96 Berzelius; 9·00 Muschenb.; 8·72 Karsten.
Cyanogen	Cy	165	160	1·03	About 0·9 Faraday.
Gold	Au	1243	65	19·1	19·26 Brisson.
Iridium.........	Ir	1233	57	21·6	19·5 Mohs; 23·5 Breithaupt; 21·8 Hare.
Iodine	I	789	160	4·93	4·95 Gay-Lussac.
Iron	Fe	339	44	7·70	7·6, 7·8 Broling; 7·79 Karsten.
Lead	Pb	1294	114	11·35	11·33 Kupffer; 11·39 Karsten; 11·35 Herapath.
Manganese....	Mn	346	44	7·86	8·03 Bachmann; 8·01 John.
Mercury	Hg	1266	93	13·6	13·6 Kupffer, Karsten, Cavallo.
Molybdenum .	Mo	599	69	8·68	8·62, 8·64 Bucholz.
Nickel	Ni	370	44	8·41	8·40 Tourte; 8·38 Tupputi; 8·60 Brunner.
Osmium........	Os	1244	57	21·8	Native; 19·5 (?) Thenard.
Palladium	Pd	666	57	11·7	11·3 Wollaston; 12·1 Lowry.
Phosphorus ...	P	196	111	1·77	1·77 Berzelius.
Platinum	Pt	1233	57	21·6	21·0 Borda; 21·5 Berzelius; 23·5 (?) Cloud.
Potassium.....	K	490	583	0·84	0·86 Gay-Lussac, Thenard; 0·87 Sementini.
Rhodium	R	651	57	11·4	11·0 Wollaston; 11·2 Cloud.
Selenium	Se	495	115	4·30	4·30, 4·32 Berzelius; 4·31 Boullay.
Silver	Ag	1352	130	10·4	10·4 Karsten.
Sodium.........	Na	291	292	0·99	0·97 Gay-Lussac and Thenard.
Sulphur........	S	201	101	1·99	1·99, 2·05 Karsten; 1·99 Breithaupt.
Tin	St	735	101	7·28	7·28 Herapath; 7·29 Kupffer, Karsten.
Titanium	T	304	57	5·33	5·3 Wollaston; 5·28 Karsten.
Tungsten	W	1183	69	17·1	17·2 Allan and Aiken; 17·4 Bucholz.
Zinc	Zn	403	58	6·95	6·92 Karsten; 6·86, 7·21 Berzelius.

It will be observed that certain analogous substances possess the same atomic volume:—bromine, chlorine, cyanogen, and iodine; chromium, molybdenum, and tungsten; cobalt, copper, iron, manganese, and nickel; iridium, osmium, palladium, platinum, and rhodium.

There are also analogous substances of which the atomic volume of one is double that of the other. The volume of an equivalent of silver is double that of gold, and the volume of potassium double that of sodium.

When a substance enters into combination, it either occupies its own volume, or assumes a new volume, which last may remain constant through a class of compounds. Hence the volumes in the preceding table are described as the primitive atomic volumes. The metals enumerated possess the following atomic volumes in their salts:—

	Atomic volume in Salts.
Ammonium	218
Barium	143
Calcium	60
Magnesium	40
Potassium	234
Sodium	130
Strontium	108

The other metals are supposed to retain their primitive volumes in combination.

In explaining the atomic volume of carbonates, it is supposed by Dr. Kopp that the salt-radical CO_3 enters into its combinations with the atomic volume 151.

In the nitrates, the salt-radical NO_6 is supposed to have the atomic volume 358.

In one class of sulphates, SO_4 is supposed to have the atomic volume 236; in another, the atomic volume 186.

In the chromates, the atomic volume of CrO_4 is 228; and, in the tungstates, that of WO_4 is 244.

The atomic volume of chlorine is 196 in one class of chlorides, and 245 in another.

On combining the atomic volumes of the metals contained in the salts with these suppositions for their salt-radicals, the atomic volume of the compound is obtained, and the following calculated specific gravities:—

TABLE II.—*Atomic Volume and Specific Gravity of Salts.*

CARBONATES.

CARBONATES.	Atomic Weight.	Formula.	Calculated Atomic Volume.	Calculated Sp. Gr.	Observed Specific Gravity.
Cadmium	1073	$Cd+CO_3$	$81+151=232$	4·63	4·42 Herapath; 4·49 K.
Iron	715	$Fe+CO_3$	$144+151=195$	3·67	3·33 Mohs; 3·87 Naum.
Lead	1670	$Pb+CO_3$	$114+151=265$	6·30	6·43 Karsten; 6·47 Breithaupt.
Manganese	722	$Mn+CO_3$	$44+151=195$	3·70	3·55, 3·59 Mohs.
Silver	1728	$Ag+CO_3$	$130+151=281$	6·15	6·08 Karsten.
Zinc	779	$Zn+CO_3$	$58+151=209$	3·73	4·44 Mohs; 4·4, 4·5 Naumann.
Baryta	1233	$Ba+CO_3$	$143+151=294$	4·19	4·30 Karsten; 4·24 Breit.; 4·30 Mohs.
Lime	632	$Ca+CO_3$	$60+151=211$	3·00	Arragonite 3·00 Breit; 2·93 Mohs. Calc. spar 2·70 Kar.; 2·72 Beudant.
Magnesia	534	$Mg+CO_3$	$40+151=191$	2·80	2·31 Breithaupt; 3·00, 3·11 Mohs; 2·88, 2·97 Naum.
Potassa	866	$K+CO_3$	$234+151=385$	2·25	2·26 Karsten.
Soda	667	$Na+CO_3$	$130+151=281$	2·37	2·47 Karsten.
Strontia	923	$Sr+CO_3$	$108+151=259$	3·56	3·60 Mohs; 3·62 K.
Dolomite	1166	$Mg+CO_3$ $Ca+CO_3$	$40+151$ $60+151$ $=402$	2·90	2·88 Mohs.
Mesiline	1250	$Mg+CO_3$ $Fe+CO_3$	$40+151$ $44+151$ $=386$	3·24	3·35 Mohs.

NITRATES.

Nitrates.	Atomic Weight.	Formula.	Calculated Atomic Volume.	Calculated Sp. Gr.	Observed Specific Gravity.
Lead	2071	$Pb+NO_6$	114+358=472	4·40	4·40 Karsten; 4·77 Breithaupt; 4·34 Kopp.
Silver	2129	$Ag+NO_6$	130+358=488	4·36	4·36 Karsten.
Ammonia	1004	$Am+NO_6$	218+358=576	1·74	1·74 Kopp.
Baryta	1634	$Ba+NO_5$	143+358=501	3·20	3·19 Karsten.
Potassa	1267	$K+NO_6$	234+358=592	2·14	2·10 Karst.; 2·06 Kopp.
Soda	1068	$Na+NO_6$	130+358=488	2·19	2·19 Marx.; 2·20 Kopp. 2·26 Karsten.
Strontia	1324	$Sr+NO_6$	108+358=466	2·84	2·89 Karsten.

SULPHATES: FIRST CLASS.

Sulphates.	Atomic Weight.	Formula.	Calculated Atomic Volume.	Calculated Sp. Gr.	Observed Specific Gravity.
Copper	997	$Cu+SO_4$	44+236=280	3·56	3·53 Karsten.
Silver	1953	$Ag+SO_4$	130+236=366	5·34	5·34 Karsten.
Zinc	1004	$Zn+SO_4$	58+236=294	3·42	3·40 Karsten.
Lime	857	$Ca+SO_4$	60+236=296	2·90	2·96 Naumann; 2·93 Karsten.
Magnesia	759	$Mg+SO_4$	40+236=276	2·75	2·61 Karsten.
Soda	892	$Na+SO_4$	130+236=366	2·44	2·46 Mohs; 2·63 K.

SULPHATES: SECOND CLASS.

Sulphates.	Atomic Weight.	Formula.	Calculated Atomic Volume.	Calculated Sp. Gr.	Observed Specific Gravity.
Lead	1895	$Pb+SO_4$	114+186=300	6·32	6·30 Mohs; 6·17 Karst.
Baryta	1458	$Ba+SO_4$	143+186=329	4·43	4·45 Mohs; 4·20 Karst.
Potassa	1091	$K+SO_4$	234+186=420	2·60	2·62 Karst.; 2·66 Kopp.
Strontia	1148	$Sr+SO_4$	108+186=294	3·90	3·95 Breit.; 3·59 Karsten.

CHROMATES AND TUNGSTATES.

Chromates and Tungstates.	Atomic Weight.	Formula.	Calculated Atomic Volume.	Calculated Sp. Gr.	Observed Specific Gravity.
Lead	2046	$Pb+CrO_4$	114+228=342	5·98	5·95 Breith.; 6·00 Mohs.
Potassa	1241	$K+CrO_4$	234+228=462	2·69	2·64 Karst.; 2·70 Kopp.
Lead	2877	$Pb+WO_4$	114+244=358	8·04	8·0 Gmel.; 8·1 Leonh.
Lime	1839	$Ca+WO_4$	60+244=304	6·05	6·04 Kars.; 6·03 Meiss.

CHLORIDES: FIRST CLASS.

Chlorides.	Atomic Weight.	Formula.	Calculated Atomic Volume.	Calculated Sp. Gr.	Observed Specific Gravity.
Lead	1736	Pb+Cl	114+196=310	5·60	5·68, 5·80 Karsten; 5·24, 5·34 Monro.
Silver	1794	Ag+Cl	130+196=326	5·50	5·50, 5·57 Karsten; 5·55 Boul.; 5·13 Herap.
Barium	1299	Ba+Cl	143+196=339	3·83	3·86 Boul.; 3·70 Karst.
Sodium	733	Na+Cl	130+196=326	2·25	2·26 Mohs; 2·15 Kopp; 2·08 Karsten.

CHLORIDES: SECOND CLASS.

CHLORIDES.	Atomic Weight.	Formula.	Calculated Atomic Volume.	Calculated Sp. Gr.	Observed Specific Gravity.
Copper	1234	2Cu+Cl	88+245=333	3·70	3·68 Karsten.
Mercury	1708	Hg+Cl	93+245=338	5·05	5·14 Gmel.; 5·43 Boul.; 5·40 Karsten.
Mercury	2974	2Hg+Cl	186+245=431	6·90	6·99 Karsten; 6·71 Herapath; 7·14 Boullay.
Ammonium	669	Am+Cl	218+245=463	1·44	1·45 Watson; 1·50 Kopp; 1·53 Mohs.
Calcium	698	Ca+Cl	60+245=305	2·29	2·21, 2·27 Boullay; 1·92 Karsten.
Potassium	932	R+Cl	234+245=479	1·94	1·94 Kopp; 1·92 Karsten.
Strontium	989	Sn+Cl	108+245=353	2·80	2·80 Karsten.

In explaining the specific gravity of oxides, it is necessary to make three assumptions for the specific volume of oxygen. In the first small class of oxides, the oxygen is contained with the atomic volume 16; in the second and large class, with the atomic volume 32; and, in the third class, with the atomic volume 64. The metals are supposed to retain their primitive atomic volumes.

TABLE III.—*Atomic Volume and Specific Gravity of Oxides.*

FIRST CLASS.

OXIDES.	Atomic Weight.	Formula.	Calculated Atomic Volume.	Calculated Sp. Gr.	Observed Specific Gravity.
Antimony	1006	Sb+2O	120+32=154	6·53	6·53 Boullay; 6·70 Karst.
Chromium	1003	2Cr+3O	138+48=186	5·39	5·21 Wöhler.
Tin	935	Sn+2O	101+32=133	7·03	6·96 Mohs; 6·90 Boullay; 6·64 Herapath.

SECOND CLASS.

OXIDES.	Atomic Weight.	Formula.	Calculated Atomic Volume.	Calculated Sp. Gr.	Observed Specific Gravity.
Antimony	1913	2Sb+3O	240+96=336	5·69	5·78 Boullay; 5·57 Mohs.
Bismuth	2960	2Bi+3O	270+96=366	8·09	8·17 Karst.; 8·21 Herap.; 8·45 Royer and Dum.
Cadmium	797	Cd+O	81+32=113	7·05	6·95 Karsten.
Cobalt	1038	2Co+3O	88+96=184	5·64	5·60 Boullay; 5·32 Herap.
Copper	496	Cu+O	44+32= 76	6·53	6·43 Karst.; 6·13 Boul.; 6·40 Herapath.
Iron	978	2Fe+3O	88+96=184	5·31	5·23 Boullay; 5·25 Mohs.
Lead	1394	Pb+O	114+32=146	9·55	9·50 Boullay; 9·28 Herap.; 9·21 Karsten.
Lead	1494	Pb+2O	114+64=178	8·40	8·90 Herap.; 8·92 Karst.
Lead	2889	2Pb+3O	228+96=324	8·91	8·94 Muschenbroek; 8·60 Karst.; 9·20 Boullay.
Manganese	446	Mn+O	44+32— 76	5·87	4·73 Herapath.
Mercury	1366	Hg+O	93+32=125	10·9	11·0 Boullay; 11·1 Herapath; 11·2 Karsten.
Molybdenum	799	Mo+2O	69+64=133	6·01	5·67 Bucholz.
Tin	835	Sn+O	101+32=133	6·28	6·67 Herapath.
Titanium	504	Ti+2O	57+64—121	4·16	4·18 Klaproth; 4·20, 4·25 Breithaupt.
Zinc	503	Zn+O	58+32= 90	5·48	5·43 Mohs; 5·60 Boullay; 5·73 Karsten.
Ilmenite	942	{Fe, Ti}+3O	{44, 57}+96=197	4·78	4·73, 4·79 Breithaupt; 4·75, 4·78 Kupffer.

THIRD CLASS.

Oxides.	Atomic Weight.	Formula.	Calculated Atomic Volume.	Calculated Sp. Gr.	Observed Specific Gravity.
Copper	892	2Cu+O	88+ 64=152	5·87	5·75 Karsten, Boyer and Dumas; 6·05 Herapath.
Mercury	2632	2Hg+O	186+ 64=250	10·05	10·69 Herap.; 8·95 Karst.
Molybdenum...	899	Mo+3O	69+192=261	3·44	3·46 Bergman, Thomson; 3·49 Berzelius.
Silver...........	1452	Ag+O	130+ 64=194	7·48	7·14 Herapath; 7·25 Boullon; 8·26 Karsten.
Tungsten	1483	W+3O	69+192=261	5·68	5·27 Herapath; 6·12 Berzelius; 7·14 Karsten.

Dr. Kopp has endeavoured to determine the atomic volume of the constituents of many other classes of compounds. The specific gravity of the compounds of sulphur and arsenic with the metals, of water with oxides and salts, of chlorine with the non-metallic elements, are explained in a similar manner on a small number of suppositions. He also shows with considerable success that in those isomorphous substances, of which the crystalline form is only similar, and not absolutely identical, as the carbonates (p. 140), the observed difference between the atomic volumes corresponds with the difference between the crystalline forms. The variation in the atomic volume is thus manifested by a variation in the crystalline form.

CHAPTER IV.

CHEMICAL AFFINITY.

In the preceding section, compound bodies have been viewed as already formed, and existing in a state of rest. The arrangement, weights, and other properties of their atoms, have also been examined with the relations and classification of the compounds themselves. But chemistry is more than a descriptive science; for it embraces, in addition to views of composition, the consideration of the action of bodies upon each other, which leads to the formation and destruction of compounds. Certain bodies, when placed in contact, exhibit a proneness to combine with each other, or to undergo decomposition, while others may be mixed most intimately without change. The actual phenomena of combination suggest the idea of peculiar attachments and aversions subsisting between different bodies, and it was in this figurative sense that the term *affinity* was first applied by Boerhaave to a property of matter. A specific attraction between different kinds of matter must be admitted as the cause of combination, and this attraction may be conveniently distinguished as *chemical affinity.*

The particles of a body in the solid or liquid state exhibit an attraction for each other, which is the force of *cohesion,* and even different kinds of matter have often an attraction for each other, which is probably of the same nature, although distinguished as *adhesion.* This force retains bodies in contact which are once placed in sufficient proximity to each other. It is exhibited in the adhesion of two smooth pieces of lead pressed together, or perfectly flat pieces of plate-glass, which sometimes cannot again be separated. The action of glue, wax, mortar, and other cements, in attaching bodies together, depends entirely upon the same force. In detaching glue from the surface of glass, the latter is sometimes injured, and portions of it are torn off by the glue, the adhesive attraction of the two bodies being greater than the cohesion of the glass. The property of water to adhere to solid surfaces and wet them, its imbibition by a sponge, the ascent of liquids in narrow tubes, and other phenomena of capillary attraction, and the rapid diffusion of a drop of oil over the surface of water, are illustrations of the same attraction between a liquid and a solid, and between different liquids. But this kind of attraction is deficient in a character which is never absent in true chemical affinity — *it effects no change in the properties of bodies.* It may bind different kinds of matter together, but it does not alter their nature.

The tendency of different gases to diffuse through each other till a uniform mixture is formed, is another property of matter, — the effect of a force wholly independent of chemical affinity. It is certain that this physical property is not lost in liquids, and that it contributes to that equable diffusion of a salt through a menstruum, which occurs spontaneously, and without agitation to promote it. (Jerichau, in Poggendorff's Annalen, xxxiv. 613; or Dove and Moser's Repertorium der Physik, i. 96, 1837.)

Solution. — The attraction between salt and water, which occasions the solution of the former, differs in several circumstances from the affinity which leads to the production of definite chemical compounds. In solution, combination takes place in indefinite proportions, a certain quantity of common salt dissolving in, or combining with any quantity of water however large; while a certain quantity of water, such as 100 parts, can dissolve any quantity of that salt less than 37 parts, the proportion which saturates it. Water has a constant solvent power for every other soluble salt; but the maximum proportion of salt dissolved, or the saturating quantity, has no relation to the atomic weight of the salt, and indeed varies exceedingly with the temperature of the solvent. The limit to the solubility of a salt seems to be immediately occasioned by its cohesion. Water, in proportion as it takes up salt, has its power to disintegrate and dissolve more of the soluble body gradually diminished; it dissolves the last portions slowly and with difficulty, and at last, when saturated, is incapable of overcoming the cohesion of more salt that may be added to it. The solubility in water of another body in the liquid state is not restrained by cohesion, and is in general unlimited. Thus alcohol, and also soluble salts above the temperature at which they liquefy in their water of crystallization, dissolve in water in any proportion. Generally speaking, also, those salts dissolve in largest quantity which are most fusible, or of which the cohesion is most easily overcome by heat, as the hydrated salts; and among anhydrous salts, the nitrates, chlorates, chlorides, and iodides, which are all remarkable for their fusibility. In this species of combination, bodies are not materially altered in properties; indeed, are little affected except in their cohesion.

The union also between a body and its solvent differs in a marked manner from proper chemical combination in the relation of the bodies to each other which exhibit it. Bodies *combine* chemically with so much the more force as their properties are more opposed, but they *dissolve* the more readily in each other, the more similar their properties. Thus, metals combine with non-metallic bodies, acids with alkalies; but to dissolve a metal, another metal must be used, such as mercury; oxidated bodies dissolve in oxidated solvents, as the salts and acids in water; while liquids which contain much hydrogen are the best solvents of hydrogenated bodies —an oil, for instance, of a fat or a resin; alcohol and ether dissolving the essential oils and most organic principles, but few salts of oxygen acids. The force which produces solution differs, therefore, essentially from chemical affinity in being exerted between analogous particles, in preference to particles which are very unlike; and resembles more, in this respect, the attraction of cohesion.

A more accurate idea of the varying solubility of a salt at different temperatures may be conveyed by a curve constructed to represent it, than by any other means. The perpendicular lines in the following diagram, indicate the degrees of temperature which are marked below them, and the horizontal lines, quantities of salt dissolved by 100 parts by weight of water. The proportion of any salt dissolved at a particular temperature may be learned by carrying the eye along the perpendicular line expressing that temperature, till it cuts the curve of the salt, and then horizontally to the column of parts dissolved.[1]

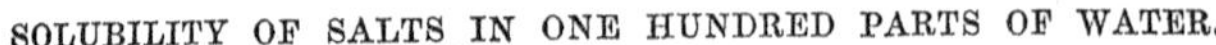

SOLUBILITY OF SALTS IN ONE HUNDRED PARTS OF WATER.

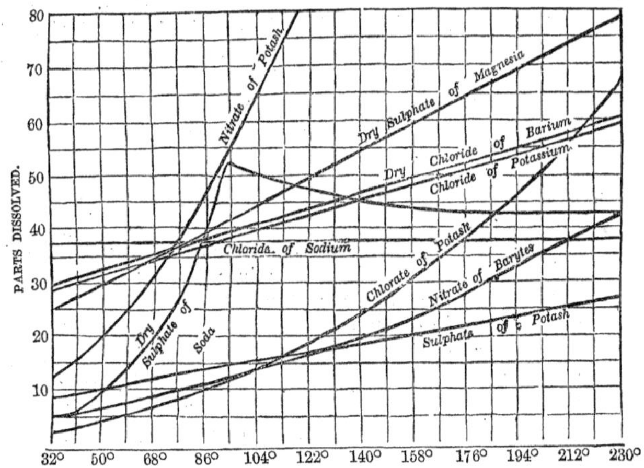

It will be observed that the perpendicular lines advance by 9°, the first being 32°, and the last 230°. The solubility of nitrate of potassa increases from 13 parts in 100 water at 32°, to 80 parts at 118°, or very rapidly with the temperature. Sulphate of soda is seen by the form of its curve to increase in solubility from 5 parts at 32° to 52 parts at 92°, but then to diminish in solubility with farther elevation of temperature. In this salt, sulphate of magnesia and chloride of barium, the solubility is expressed in parts of the anhydrous, and not the hydrated salt. The lines of chloride of barium and chloride of potassium are parallel, showing a remarkable relation between the solubilities of these two salts, which does not appear in any others. The line of chloride of sodium is observed to cut all the lines of temperature at the same height, 100 parts of water dissolving 37 parts of that salt at all temperatures.

Chemical affinity acts only at insensible distances, and has no effect in causing bodies to approach each other which are not in contact, differing in this respect from the attraction of gravitation, which acts at all distances, however great, although with a diminishing force. Hence, the closest approximation of unlike particles is necessary to develope their affinities, and produce combination. Sulphur and copper in mass have no effect upon each other, but if both be in a state of great division, and rubbed together in a mortar, a powerful affinity is brought into play, the bodies themselves disappear, and sulphuret of copper is produced by their union, with the evolution of much heat. The affinity of bodies is, therefore, promoted by every thing which tends to their close approximation; in solids, by their pulverization and intermixture, this attraction residing in the ultimate particles of bodies; in gases, by their spontaneous diffusion through each other, which occasions a more complete intermixture than is attainable by mechanical means; and between liquids, or between a liquid and solid, by the adhesive attraction which liquids possess, which must lead to perfect contact, and also by a disposition of liquid bodies to intermix, of the same physical character as gaseous diffusion. Elevation of temperature has

[1] An extensive and very careful series of experiments on the solubility of salts in water at different temperatures has been made by M. Poggiale, Ann. de Chim. et de Phys. 3e Sér. T. viii. p. 463; and the *Rapport Annuel* of Berzelius, Paris, 1846, p. 18.

certainly often a specific action in increasing the affinity of two bodies, but it also often acts by producing a perfect contact between them, from the fusion or vaporization of one or both bodies. Hence, no practice is more general to promote the combination of bodies than to heat them together.

If the affinity between two gases is sufficiently great to begin combination, the process is never interrupted, but is continued from the diffusion of the gases through each other till complete, or at least till one of the gases is entirely consumed. Thus, when hydrochloric acid and ammoniacal gases, in equal measures, are introduced into a jar containing at the same time a large quantity of air, the formation of hydrochlorate of ammonia proceeds, the gases appearing to search out each other, till no portion of uncombined gas remains. The combination of two liquids, or of a liquid and a solid, is also facilitated in the same manner by the mobility of the fluid, and proceeds without interruption, unless, perhaps, the product of the combination be solid, and by its formation interpose an obstacle to the contact of the combining bodies. But the affinities of two solids which are not volatile are rarely developed at all, owing to the imperfection of contact. Even the action of very powerful affinities between a solid and a liquid or a gas, is often arrested in the outset from the physical condition of the former. Thus, the affinity between oxygen and lead is certainly considerable, for the metal is rapidly converted into a white oxide when ground to powder and agitated with water in its usual aerated condition; and in the state of extreme division in which lead is obtained by calcining its tartrate in a glass tube, the metal is a pyrophorus, and combines with oxygen when cold with so much avidity as to take fire and burn the moment it is exposed to the air. Iron also, in the spongy and divided state in which it is procured, by reducing the peroxide by means of hydrogen gas at a low red heat, absorbs oxygen with equal avidity at the temperature of the air, and takes fire and burns. But notwithstanding an affinity for oxygen of such intensity, these metals in mass oxidate very slowly in air, particularly lead, which is quickly tarnished indeed, but the thin coating of oxide formed does not penetrate to a sensible depth in the course of several years. The suspension of the oxidation may be partly due to the comparatively small surface which a compact body exposes to air, and which becomes covered by a coat of oxide, and protected from farther change; but partly also to the effect of the conducting power of a considerable mass of metal in preventing the elevation of temperature consequent upon the oxidation of its surface. For metals oxidate with increased facility at a high temperature, such as the lead pyrophorus quickly attains from the oxidation of the great surface which it exposes, compared with its weight. The heat from the oxidation of the superficial particles of the compact metal, however, is not accumulated, but carried off and dissipated by the conducting power of the contiguous particles, so that elevation of temperature is effectually repressed. It thus appears that the state of aggregation of a solid may oppose an insuperable bar to the action of a very powerful affinity.

The affinity of two bodies, one or both of which are in the state of gas, is often promoted in an extraordinary manner by the contact of certain solid bodies. Thus, oxygen and hydrogen gases may be mixed and retained for any length of time in that state without exhibiting any affinity for each other, and the gaseous mixture may, indeed, be heated in a glass vessel to any temperature short of redness without showing any disposition to combine. But if a clean plate of platinum be introduced into the cold mixture, the gases in contact with the metallic surface instantly unite and form water; other portions of the mixture come then in contact with the platinum, and combine successively under its influence, so that a large quantity of the gaseous mixture may be quickly united. The temperature of the platinum also rises, from the heat evolved by the combination occurring at its surface, and the influence of the metal increasing with its temperature, combination proceeds at an accelerated rate, till the platinum becoming red hot, may cause the combination to extend to a distance from it, by kindling the gaseous mixture. Platinum acts in this manner with greatest energy when in a highly divided state, as in the form of spongy platinum, owing to the greater surface exposed, and the rapidity with which it is heated. The metal itself contributes no element to the water formed, and is in no respect altered. It is an action of the metallic surface, which must be perfectly clean, and is retarded or altogether prevented by the presence of oily vapours and many other combustible gases, which soil the metallic surface. Mr. Faraday is disposed to refer the action to an adhesive attraction of the gases for the metal, under the influence of which they are condensed and their particles approximated within the sphere of their mutual attraction, so as to combine. This opinion is favoured by the circumstance that the property is not peculiar to platinum, but appears also in other metals, in charcoal, pounded glass, and all other solid bodies; although all of them, except the metals, act only when their temperature is above the boiling point of mercury. But, on the other hand, at low temperatures, the property appears to be confined to a few metals only which resemble platinum in their chemical characters; namely, in having little or no disposition to combine with oxygen gas, and in not undergoing oxidation in the air. The action of platinum may, therefore, be connected with its chemical properties, although in a way which is quite unknown to us. The same metal disposes carbonic oxide gas to combine with oxygen, but much more slowly than hydrogen; and it is remarkable that if the most minute quantity of carbonic oxide be mixed with hydrogen, the oxidation of the latter under the influence of the platinum is arrested, and not resumed till after the carbonic oxide has been slowly oxidated and consumed, which thus takes the precedence of the hydrogen in combining with oxygen. This extraordinary interference of a minute quantity of carbonic oxide gas, which cannot from its nature be supposed to soil the surface of the platinum like a liquefiable vapour, seems to point to a chemical, perhaps to an electrical explanation of the action of the platinum, rather than to the adhesive attraction of the metal. The oxidation of alcohol at the temperature of the air, and also at a low red heat, is promoted in the same manner by contact with platinum.

Order of affinity. — The affinity between bodies appears to be of different degrees of intensity. Lead, for instance, has certainly a greater affinity than silver for oxygen, the oxide of the latter being easily decomposed when heated to redness, while the oxide of the former may be exposed to the most intense heat without losing a particle of oxygen. Again, it may be inferred that potassium has a still greater affinity for oxygen than lead possesses, as we find the oxide of lead easily reduced to the metallic state when heated in contact with charcoal, while potassa is decomposed in the same manner with great difficulty. But the order of affinity is often more strikingly exhibited in the decomposition of a compound by another body. Thus, sulphuretted hydrogen gas is decomposed by iodine, which combines with the hydrogen, forming hydriodic acid, and liberates sulphur. The affinity of iodine for hydrogen is, therefore, greater than that of sulphur for the same body. But hydriodic acid is deprived of its hydrogen by bromine, and hydrobromic acid is formed; and this last is decomposed in its turn by chlorine, and hydrochloric acid produced. It thus appears that the order of the affinity of the elements mentioned *for hydrogen* is, chlorine, bromine, iodine, sulphur. The order of decompositions, in the precipitation of metals by each other from their saline solutions, also indicates the degree of affinity. Thus, from the decomposition of the nitrates of the following metals, the order of their affinity *for nitric acid and oxygen* may be inferred to be as follows: — zinc, lead, copper, mercury, silver; zinc throwing down lead from the nitrate of lead, and all the other metals which follow it; lead throwing down copper; copper, mercury; and mercury, silver; while nitrate of zinc itself is not affected by any other metal, and nitrate of silver is decomposed by all the metals enumerated. Bodies were first thus arranged according to the degree of their affinity for a particular substance, inferred from the order of their decompositions, by Geoffroy and Bergman, and tables of affinity constructed, of which the following is an example: —

Order of Affinity of the Alkalies and Earths for Sulphuric Acid.

Baryta.
Strontia.
Potassa.
Soda.
Lime.
Ammonia.
Magnesia.

Baryta is capable of taking sulphuric acid from strontia, potassa, and every other base which follows it in the table, — the experiment being made upon sulphates of these bases dissolved in water; while sulphate of baryta is not decomposed by any other base. Lime separates ammonia and magnesia from sulphuric acid, but has no effect upon the sulphates of soda, potassa, strontia, and baryta; and in the same manner any other base decomposes the sulphates of the bases below it in the column, but has no effect upon those above it. Tables of this kind, when accurately constructed, may convey much valuable information of a practical kind, but it is never to be forgotten that they are strictly tables of the order of decomposition and of the comparative force or order of affinity in one set of conditions only. This will appear by examining how far decomposition is affected by accessory circumstances in a few cases.

Circumstances which affect the order of decomposition. — Volatility in a body promotes its separation from others which are more fixed, and consequently facilitates the decomposition of compounds into which the volatile body enters. Hence, by the agency of heat, water is separated from hydrated salts; ammonia, from its combinations with a fixed acid, such as the phosphoric; and a volatile acid from many of its salts: as sulphuric acid from the sulphate of iron, carbonic acid from the carbonate of lime, &c. Ammonia decomposes hydrochlorate of morphia at a low temperature, but, on the other hand, morphia decomposes the hydrochlorate of ammonia at the boiling point of water, and liberates ammonia, owing to the volatility of that body. The fixed acids, such as the silicic and phosphoric, disengage in the same way at a high temperature those acids which are generally reputed most powerful, and by which silicates and phosphates are decomposed with facility at a low temperature. Many such cases might be adduced in which the order of decomposition is reversed by a change of temperature. The volatility of one of its constituents must, therefore, be considered an element of instability in a compound.

Decomposition from unequal volatility is, of course, checked by pressure, and promoted by its removal and by every thing which favours the escape of vapour; such as the presence of an atmosphere of a different sort into which the volatile constituent may evaporate. Carbonate of lime is decomposed easily at a red heat, provided a current of air or of steam is passing over it which may carry off the carbonic acid gas, but the decomposition ceases when the carbonate is surrounded by an atmosphere of its own gas; and the carbonate may even be heated to fusion, in the lower part of a crucible, without decomposition. Here the occurrence of decomposition depends entirely upon the existence of a foreign atmosphere into which carbonic acid can diffuse. Nitrates of alumina, and peroxide of iron in solution, are decomposed by the spontaneous evaporation of their acid, even at the temperature of the air; and so is an alkaline bicarbonate when in solution, but not when dry. A change in the composition of the gaseous atmosphere may affect the order of decomposition, as in the following cases: —

When steam is passed over iron at a red heat, a portion of it is decomposed, oxide of iron being formed and hydrogen gas evolved. From this experiment it might be inferred that the affinity of iron for oxygen is greater than that of hydrogen. But let a stream of hydrogen gas be conducted over oxide of iron at the very same temperature, and water is formed, while the oxide of iron is reduced to the metallic state. Here the hydrogen appears to have the greater affinity for oxygen. But the result is obviously connected with the relative proportion between the hydrogen and steam which are at once in contact with the metal and its oxide at a red heat. When steam is in excess, water is decomposed, but when hydrogen is in excess, oxide of iron is decomposed; and why? because the excess of steam in the first case is an atmosphere into which hydrogen can diffuse, and the disengagement of that gas is therefore favoured; but in the second case the atmosphere is principally hydrogen, and represses the evolution of more hydrogen, but facilitates that of steam. The affinity of iron and hydrogen for oxygen at the temperature of the experiment is so nearly balanced, that the one affinity prevails over the other, according as there is a proper atmosphere into which the gaseous product of its action may diffuse. This affords an intelligible instance of the influence of mass or quantity of material, in promoting a chemical change; the steam or the hydrogen, as it preponderates, exerting a specific influence, in the capacity of a gaseous atmosphere.

The remarkable decomposition of alcohol by sulphuric acid, which affords ether, is another similar illustration of decomposition depending upon volatility, and affected by changes in the nature of the atmosphere into which evaporation takes place. Alcohol or the hydrate of ether is added in a gradual manner to sulphuric acid somewhat diluted, and heated to 280°. In these circumstances, the double sulphate of ether and water is formed; water, which was previously combined as a base to the acid, being displaced by ether, and set free together with the water of the alcohol. The first effect of the reaction, therefore, is the disengagement of watery vapour, and the creation of an atmosphere of that substance which tends to check its farther evolution. But the existence of such an atmosphere offers a facility for the evaporation of ether, which accordingly escapes from combination with the acid and continues to be replaced by the water, the affinity of sulphuric acid for water and for ether being nearly equal, till ether forms such a proportion of the gaseous atmosphere as to check its own evolution, and to favour the evolution of watery vapour. Then the sulphate of ether comes in its turn to be decomposed as before, and ether evolved. Hence, both ether and water distil over in this process, the evolution of one of these bodies favouring the separation and disengagement of the other. In this description, the evolution of water and ether are for the sake of perspicuity supposed to alternate, but it is evident that the result of such an action will be the simultaneous evolution of the two vapours in a certain constant relation to each other.

Influence of insolubility. — The great proportion of chemical reactions which we witness are exhibited by bodies dissolved in water or some other menstruum, and are affected to a great extent by the relations of themselves and their products to their solvent. Thus carbonate of potassa dissolved in water is decomposed by acetic acid, and carbonic acid evolved, the affinity of the acetic acid prevailing over that of the carbonic acid for potassa. But if a stream of carbonic acid gas be sent through acetate of potassa dissolved in alcohol, acetic acid is displaced, or the carbonic acid prevails, apparently from the insolubility of the carbonate of potassa in alcohol. The insolubility of a body appears to depend upon the cohesive attraction of its particles, and such decompositions may therefore be ascribed to the prevalence of that force.

Formation of compounds by substitution. — It is remarkable that compounds are in general more easily formed by substitution, than by the direct union of their constituents; indeed, many compounds can be formed only in that manner. Carbonic acid is not absorbed by anhydrous lime, but readily by the hydrate of lime, the water of which is displaced in the formation of the carbonate. In the same manner, ether, although a strong base, does not combine directly with acids, but the salts of ether are derived from its hydrate or alcohol, by the substitution of an acid for the water of the alcohol. In all the cases, likewise, in which hydrogen is evolved during the solution of a metal in a hydrated acid, a simple substitution of the metal for hydrogen occurs.

Combination takes place with the greatest facility of all when *double decomposition* can occur. Thus carbonate of lime is instantly formed and precipitated, when carbonate of soda is added to nitrate of lime, nitrate of soda being formed at the same time and remaining in solution.

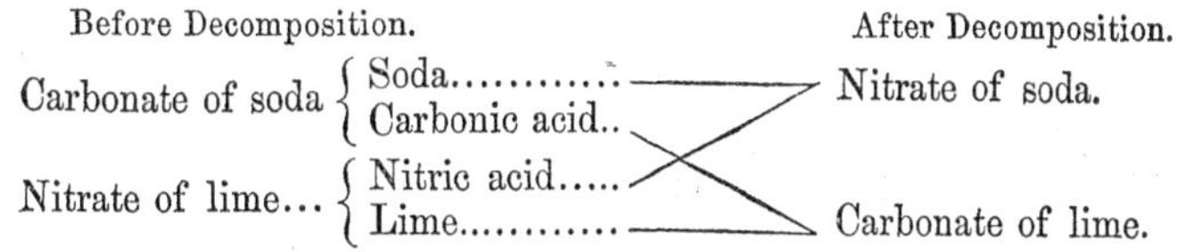

Here a double substitution occurs, lime being substituted for soda in the carbonate, and soda for lime in the nitrate. Such reactions may therefore be truly described as double substitutions as well as double decompositions. They are most commonly observed on mixing two binary compounds or two salts. But reactions of the same nature may occur between compounds of a higher order, such as double salts, and new compounds be thus produced, which cannot be formed by the direct union of their constituents. Thus the two salts, sulphate of zinc and sulphate of soda, when simply dissolved together, at the ordinary temperature, always crystallize apart, and do not combine. But the double sulphate of zinc and soda is formed on mixing strong solutions of sulphate of zinc and bisulphate of soda, and separates by crystallization; the sulphate of water with constitutional water (hydrated acid of sp. gr. 1·78) being produced at the same time, and remaining in solution. The reaction which occurs may be thus expressed:

Before Decomposition.		After Decomposition.
$\left.\begin{array}{l} HO, SO_3+(NaO, SO_3) \\ ZnO, SO_3+(HO) \end{array}\right\}$	=	$\left\{\begin{array}{l} HO, SO_3+HO \\ ZnO, SO_3+NaO, SO_3 \end{array}\right.$

in which the constituents of both salts before decomposition inclosed in brackets, are found to have exchanged places after decomposition, without any other change in the original salts.[1] The double sulphate of lime and soda can be formed artificially only in circumstances which are somewhat similar. It is produced on adding sulphate of soda to acetate of lime, the sulphate of lime, as it then precipitates, carrying down sulphate of soda in the place of constitutional water (Liebig).

Different hydrates of the same body, such as peroxide of tin, differ sensibly in properties, and afford different compounds with acids, unquestionably because these compounds are formed by substitution. The constant formation of phosphates containing one, two, or three atoms of base, on neutralising the corresponding hydrates of phosphoric acid with a fixed base, likewise illustrates in a striking manner the derivation of compounds, on this principle. Many insoluble substances, such as the earth silica, possess a larger proportion of water, when newly precipitated, than they retain afterwards, and in that high state of hydration they may exhibit affinities for certain bodies which do not appear in other circumstances. Hydrated silica dissolves in water at the moment of its separation from a caustic alkali; and alumina dissolves readily in ammonia, when produced in contact with that substance by the oxidation of aluminum. The usual disposition to enter into combination, which silica and alumina then exhibit, is generally ascribed to their being in the *nascent state;* a body at the moment of its formation and liberation, in consequence of a decomposition, being, it is supposed, in a favourable condition to enter anew into combination. But their degree of hydration in the nascent state may be the real cause of their superior aptitude to combine.

Double decompositions take place without the great evolution of heat which often accompanies the direct combination of two bodies, and with an apparent facility or absence of effort, as if the combinations were just balanced by the decompositions which occur at the same time. It is, perhaps, from this cause that the result of double decomposition is so much affected by circumstances, particularly by the insolubility of one of the compounds. For it is a general law, to which there is no excep-

[1] On Water as a Constituent of Sulphates, Phil. Mag. 3d series, vol. vi. p. 417.

tion, that two soluble salts cannot be mixed without the occurrence of decomposition, if one of the products that may be formed is an insoluble salt. On mixing carbonate of soda and nitrate of lime, the decomposition seems to be determined entirely by the insolubility of the carbonate of lime, which precipitates. When sulphate of soda and nitrate of potassa are mixed, no visible change occurs, and it is doubtful whether the salts act upon each other, but if the mixed solution be concentrated, decomposition occurs, and sulphate of potassa separates by crystallization owing to its inferior solubility.

It may sometimes be proved that double decomposition occurs on mixing soluble salts, although no precipitation supervenes. Thus, on mixing strong solutions of sulphate of copper and chloride of sodium, the colour of the solution changes from blue to green, which indicates the formation of chloride of copper and consequently that of sulphate of soda also. Now it is known that hydrochloric acid will displace sulphuric acid from the sulphate of copper at the temperature of the experiment, while sulphuric acid will, on the other hand, displace hydrochloric from chloride of sodium. It hence appears that in the preceding double decomposition, those acids and bases unite which have the strongest affinity for each other, and the same thing may happen on mixing other salts. But where the order of the affinities for each other of the acids and bases is unknown, the occurrence of any change upon mixing salts, or the extent to which the change proceeds, is entirely matter of conjecture.

It was the opinion of Berthollet, founded principally upon the phenomena of the double decompositions of salts, that decompositions are at all times dependent upon accidental circumstances, such as the volatility or insolubility of the product, and never result from the prevalence of certain affinities over others; and consequently that in accounting for such changes, the consideration of affinity may be neglected. He supposed that when a portion of base is presented at once to two acids, it is divided equally between them, or in the proportion of the quantities of the two acids, and that one acid can come to possess the base exclusively, only when it forms a volatile or an insoluble compound with that body, and thereby withdraws it from the solution, and from the influence of the other acid. His doctrine will be most easily explained by applying it to a particular case, and expressing it in the language of the atomic theory. The reaction between sulphuric acid and nitrate of potassa is supposed to be as follows. On mixing eight atoms of the acid with the same number of atoms of the salt, the latter immediately undergoes partial decomposition, its base being equally shared between the two acids which are present in equal quantities; and a state of statical equilibrium is attained in which the bodies in contact are—

(*a*) Four atoms sulphate of potassa.
Four atoms nitrate of potassa.
Four atoms sulphuric acid.
Four atoms nitric acid.

The nitrate of potassa, it is supposed, is decomposed to the extent stated, and no farther, however long the contact is protracted. But let the whole of the free nitric acid now be distilled off by the application of heat to the mixture, and a second partition of the potassa of the remaining nitrate of potassa is the consequence; the free sulphuric acid decomposing the salt till the proportion of the two acids uncombined in the mixture is again equal, when a state of equilibrium is attained. The mixture then consists of—

(*b*) Six atoms sulphate of potassa.
Two atoms nitrate of potassa.
Two atoms sulphuric acid.
Two atoms nitric acid.

On removing the free nitric acid as before, a third partition of the potassa of the remaining nitrate of potassa between the two acids on the same principle takes place, of which the result is —

(*c*) Seven atoms sulphate of potassa.
One atom nitrate of potassa.
One atom sulphuric acid.
One atom nitric acid.

The proportion of the two acids free being always the same. The repeated application of heat, by removing the free nitric acid, will cause the sulphuric to be again in excess, which will necessitate a new partition of the potassa of the remaining nitrate of potassa, till at last the entire separation of the nitric acid will be effected, and the fixed product of the decomposition be—

(*d*) Eight atoms sulphate of potassa.

Here the affinity of the sulphuric and nitric acids for potassa is supposed to be equal; and the complete decomposition of the nitrate of potassa by the former acid, which takes place, is ascribed to the volatility of the latter acid, which, by occasioning its removal in proportion as it is liberated, causes the fixed sulphuric acid to be ever in excess.

Complete decompositions in which the precipitation of an insoluble substance occurs, were explained by Berthollet in the same manner. On adding a portion of baryta to sulphate of soda, the baryta decomposes the salt, and acquires sulphuric acid, till that acid is divided between the two bases in the proportion in which they are present, and at this point decomposition would cease, were it not that the whole sulphate of baryta formed is removed by precipitation. But a new formation of that salt is the necessary consequence of that equable partition of the acid between the two bases in contact with it, which is the condition of equilibrium; and the new product precipitating, more and more of it is formed, till the sulphate of soda is entirely decomposed, and its sulphuric acid removed by an equivalent of baryta.

According to these views of Berthollet, no decomposition should be complete unless the product be volatile or insoluble, as in the cases instanced. But such a conclusion is not consistent with observation, as it can be shown that a body may be separated completely from a compound, and supplanted by another body, although none of the products is removed by the operation of either of the causes specified, but all continue in solution and in contact with each other. Thus the salt borax, which is a biborate of soda, is entirely decomposed by the addition to its solution of a quantity of sulphuric acid not more than equivalent to its soda, although the liberated boracic acid remains in solution; for the liquid imparts to blue litmus paper a purple or wine-red tint, which indicates free boracic acid, and not that characteristic red tint, resembling the red of the skin of the onion, which would inevitably be produced by the most minute quantity of the stronger acid, if free. But if the borax were only decomposed in part in these circumstances, and its soda equally divided between the two acids, then free sulphuric, as well as boracic acid, should be found in the solution. The complete decomposition of the salt can be accounted for in no way but by ascribing it to the higher affinity of sulphuric acid for soda, than that of boracic acid for the same base.

According to the same views, on mixing together two neutral salts containing different acids and bases, and which do not precipitate each other, each acid should combine with both bases, so as to occasion the formation of four salts. Again, four salts, of which the acids and bases are all dissimilar, should react upon each other in such a way as to produce sixteen salts, each acid acquiring a portion of the four bases; and certain acids and bases, dissolved together in certain proportions, could have but one arrangement in which they would remain in equilibrio. Hence the salts in a mineral water would be ascertained by determining the acids and bases present, and supposing all the bases proportionally divided among the acids. But this conclusion is inconsistent with a fact observed in the preparation of factitious mineral waters, namely, that their taste depends not only on the nature of the salts, but also upon the order in which they are added. (Dr. Struve, of Dresden.) Before we can determine how the acids and bases are arranged in a mineral water, or what salts it contains, it may therefore be necessary to know the history of its formation. Instead of supposing the bases equally distributed among the acids in mixed saline solutions, it is now more generally assumed that the strongest base may be exclusively in possession of the strongest acid, and the weaker bases be united with the weaker acids; a mode of viewing their composition which agrees best with the medical qualities of mineral waters. It thus appears that the doctrines of Berthollet, by which the resulting actions between bodies in contact are made to depend upon their relative quantities or masses, and the physical properties of the products of their combination, to the entire exclusion of the agency of proper affinities between the bodies, cannot be admitted as a true representation of the actual phenomena of combination.

CATALYSIS, OR DECOMPOSITION BY CONTACT.

An interesting class of decompositions has of late attracted considerable attention, which, as they cannot be accounted for on the ordinary laws of chemical affinity, have been referred by Berzelius to a new power, or rather new form of the force of chemical affinity, which he has distinguished as the *Catalytic force*, and the effect of its action as *Catalysis* (from κατα, downwards, and λυω, I unloosen). A body in which this power resides, resolves others into new compounds, merely by contact with them, or by an action of presence, as it has been termed, without gaining or losing anything itself. Thus an acid converts a solution of starch (at a certain temperature), first into gum and then into sugar of grapes, although no combination takes place between the elements of the acid and those of the starch, the acid being found free, and undiminished in quantity, after effecting the change. The same mutations are produced in a more remarkable manner by the presence of a minute quantity of a vegetable principle, *diastase*, allied in its general properties to gluten, which appears in the germination of barley and other seeds, and converts their starch into sugar and gum, which, being soluble, form the sap that rises into the germ, and nourishes the plant. This example of the action of a catalytic power in an organic secretion is probably not the only one in the animal and vegetable kingdoms, for it is not unlikely that it is by the action of such a force that very different substances are obtained from the same crude material by different organs. In animals, this crude material, which is the blood, flows in the uninterrupted vessels, and gives rise to all the different secretions; such as milk, bile, urine, &c., without the presence of any foreign body which could form new combinations. A beautiful instance of an action of catalysis was traced by Liebig and Wöhler in the chemical changes which the bitter almond exhibits. The application of heat and water to the almond, by giving solubility to its emulsin or albuminous principle, enables it to act upon an associated principle, amygdalin, of a neutral character, which then furnishes bodies so unlike itself as the volatile oil of almonds, hydrocyanic and formic acids. The action of yeast in fermentation is a more familiar illustration of a similar power. The presence of that substance, although insoluble, is sufficient to cause the resolution of sugar into carbonic acid gas and alcohol, a decomposition which can be effected by no other known means. Changes of this kind, although most frequent in organic compounds, are not confined to them. The binoxide of hydrogen is a body of which the elements are held together by a very slight affinity. It is not decomposed by acids, but alkalies give its elements a tendency to separate, slow effervescence occurring with the disengagement of oxygen, and water being formed. Nor do soluble substances alone produce this effect; other organic and inorganic bodies, also — such as manganese, silver, platinum, gold, fibrin, &c., which are perfectly insoluble — exert a similar power. The decomposition, in these instances, takes place by the mere presence of the foreign body, and without the smallest quantity of it entering into the new compound; for the most minute researches have failed in discovering the slightest alteration in the foreign body itself. The liquid persulphide of hydrogen, and a solution of the nitrosulphate of ammonia of Pelouze, are decomposed in the same way, and by contact of nearly all the substances which act upon peroxide of hydrogen. One remarkable difference, indeed, is observable, namely, that alkalies impart stability to nitrosulphate of ammonia, while acids decompose it, or the reverse of what happens with both the binoxide and bisulphide of hydrogen (Phil. Mag. 3d Series, vol. x. p. 489).

The phenomena referred to catalysis are of a recondite nature, and much in need of elucidation. The influence of platinum, formerly noticed, in disposing hydrogen and oxygen to unite, is probably connected with the catalytic power of the same metal, but is at present equally inexplicable. It would be unphilosophical to rest satisfied by referring such phenomena to a force of the existence of which we have no evidence. The doctrine of catalysis must be viewed in no other light than as a convenient fiction, by which we are enabled to class together a number of decompositions not provided for in the theory of chemical affinity, as at present understood, but which, it is to be expected, will receive their explanation from new investigations. It is a provisional hypothesis, like the doctrine of isomerism, for which the occasion will cease as the science advances.

SECTION II. — CHEMICAL POLARITY.

Illustrations from magnetical polarity. — The ideas of induction and polarity, which now play so important a part in physical theories, were originally suggested by the phenomena of magnetism, which still afford the best experimental illustrations of them. A bar magnet exhibits attractive power which is not possessed in an equal degree by every particle composing the bar, but is chiefly localized in two points at or near its extremities. The powers, too, residing at these points are not one and the same, or similar, but different, indeed contrary, in their nature; and are distinguished by the different names of Austral magnetism and Boreal magnetism. The opposition in the mode of action of these powers is so perfect, that they completely negative or neutralize each other when residing in the same particle of matter in equal quantity or degree, as they are supposed really to exist in iron before it is magnetized; and they only signalize their presence when displaced and separated to a distance from each other, as they are in a magnet. A body possessing any such powers residing in it, which are not general but local, and not the same but opposite, is said (in the most general sense) to possess *polarity*.

In the theory of magnetism, it is found necessary to consider a magnet as composed of minute indivisible particles or filaments of iron, each of which has individually the properties of a separate magnet. The displacement or separation of the two attractive powers takes place only within these small particles, which are called the magnetic elements, and must be supposed so minute that they may be the ultimate particles or atoms themselves of the iron. A magnetic bar may, therefore, be represented as composed of minute portions, *a b* in fig. 60 representing one such portion; the right hand extremities of each of which possess one species of magnetism, and the left hand extremities the other. The unshaded ends being supposed to possess austral, and the shaded ends boreal magnetism, then the ends of the bar itself, of which these sides of the elementary magnets form the faces, possess respectively austral and boreal magnetism, and are the austral and boreal poles of the magnet. Such, then, is the polar condition of a bar of iron possessing magnetism, of which the attractive and repulsive powers residing at the extremities are the results. Of the existence of such a structure the breaking of a magnet into two or more parts affords a proof, for it forms as many complete mag-

Fig. 60.

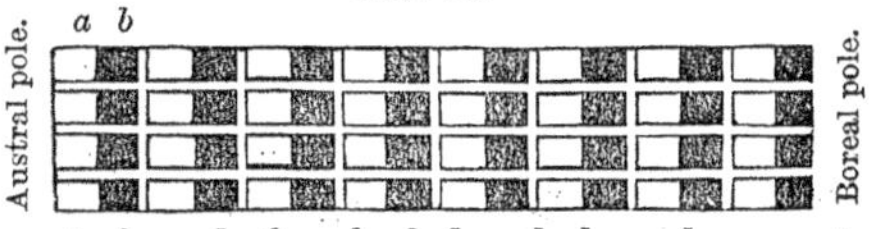

a b

Austral pole.

Boreal pole.

nets as there are parts, new poles appearing at all the fractured extremities. A magnetic element, it is to be remembered, is itself insecable, like a chemical atom, so that the division must take place between magnetic elements.

Fig. 61.

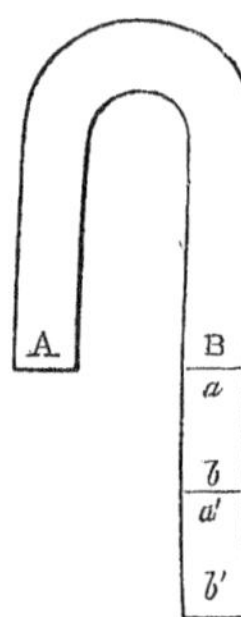

When to the boreal pole B of a magnet (fig. 61), which may be of the horse-shoe form, a piece of soft iron, *a b*, wholly destitute of magnetic powers, is presented, a similar displacement of the magnetic forces of its elements occurs as in the magnet itself; or *a b* becomes a magnet by induction, and may attract and induce magnetism in a second bar *a' b'*; both of which continue magnetic so long as the first remains in the same position, and under the influence of A B. These induced magnets must have the same polar molecular structure as the original magnet, but their magnetism is only temporary, and is immediately lost when they are removed from the permanent magnet. The displacement of the magnetisms in these induced magnets commences at the extremity *a* of *a b*, in contact with B, which extremity has the opposite magnetism of B, (the different kinds of magnetism being mutually attractive,) and is the austral pole of *a b*; and *b* is its boreal pole. Of *a' b'*, again, the upper extremity *a'* in contact with *b'* is the austral, and the lower extremity *b'* the boreal pole, or *b* and *b'* have the same kind of magnetic power as the pole B of the original magnet, from which they are dependent. A third bar of soft iron placed at *b'* is likewise polarized, and the series of induced magnets may be still farther extended, but the attractive powers developed in the different members of the series become less and less with their distance from the pole B of the original magnet.

Fig. 62. Fig. 63.

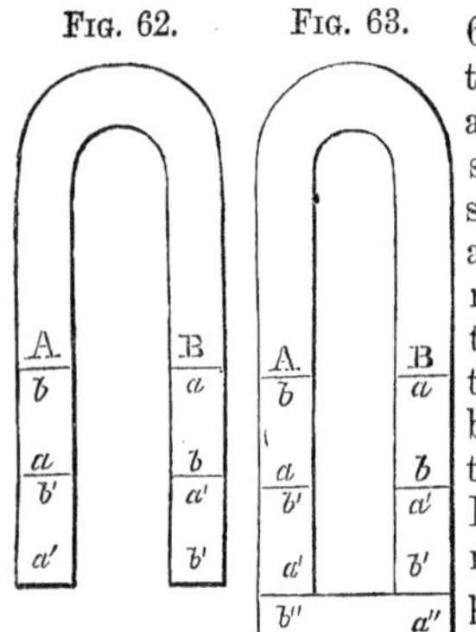

A similar set of bars may be connected with A (fig. 62), which become temporary magnets also according to the same law, the lower extremities of this set being austral. On now uniting the lower extremities of both sets by another bar of soft iron *a'' b''*, (fig. 63), either set renders *a'' b''* a magnet, having its austral pole at *a''* and its boreal pole at *b''*; and acting together, they communicate a degree of magnetism to the uniting bar greater than either set possessed before they were united. By this connexion, also, the inductive actions of each set of bars are brought to bear upon the other, and the attractive forces at all their poles are thereby greatly increased. In the most favourable conditions as to the size and connexion of the temporary magnets with relation to the primary magnet, each of the former, however numerous, acquires powers equal to those of the original magnet. This general enhancement of power in the induced magnets has been acquired, therefore, by completing the circle of them between A and B.

Fig. 64.

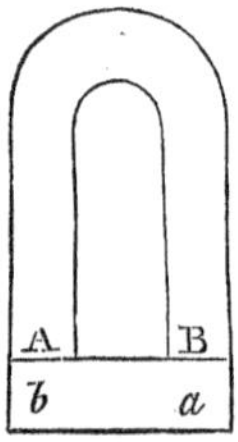

It is also important to observe, with a view to the future application of the remark, that a single bar of soft iron, or *lifter*, as *b a*, (fig. 64), connecting the poles of a magnet A B, not only acquires at *b* and *a* equal though opposite powers to the contiguous poles of the magnet, but also reacts by induction on these poles themselves in a gradual manner, and increases their magnetism. The original magnetic forces of A and B are therefore increased, by the opportunity to act inductively, which the connecting bar affords them. The threads of steel filings which are taken up by a magnet, (see figure 65) illustrate the inductive action of magnetism, for each grain of steel is a complete magnet, and the threads a series of connected magnets. It will be observed also that these threads diverge from each other; because, while unlike poles are in contact in each thread, which attract, like poles

FIG. 65

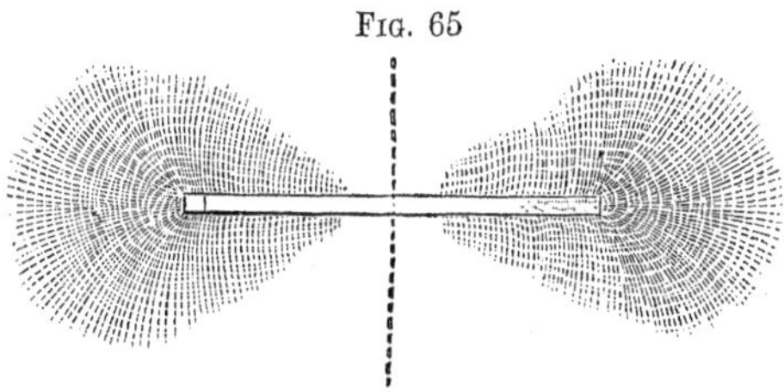

are in contact of adjoining threads which repel. This repulsion of polar chains by each other, there will be occasion again to refer to.

Atomic representation of a double decomposition.—Chemical polarity, although less adapted for exhibition, is still more simple than magnetic polarity in its nature, while it is of a more fundamental character, and appears to be the basis of all other polarities whatever. In a binary compound,—such as chloride of potassium,—there reside two attractive powers, opposite in their nature; namely, the halogenous affinity of the salt-radical chlorine, and the basylous affinity of the metal potassium. The atomic theory gives form to the molecule of chloride of potassium: one atom, Cl, being the seat of the halogenous, chlorous, or negative affinity (as we shall also call it); and the other atom, K, the seat of the basylous or positive affinity. A binary saline molecule is thus entirely similar to a magnetic element. We have to deal with two affinities only,—the chlorous and basylous. Atoms possessing different affinities attract each other; while atoms possessing the same affinity repel each other.

The two binary compounds, hydrochloric acid (chloride of hydrogen) and oxide of lead, when brought into contact, mutually decompose each other, forming chloride of lead and water: $H\,Cl$ and $Pb\,O = Pb\,Cl$ and $H\,O$. At the instant of acting upon each other, the two compound molecules must have a certain relative position. Under (1), the basylous hydrogen of the hydrochloric acid is presented to the basylous lead of the oxide of lead, atoms which repel each other. In (2) and (3), on the contrary, a basylous atom of one molecule is presented to a halogenous atom in the other, H to O in (2), and Cl to Pb in (3). These are attractive pairs; but, before they can enter into new combinations, they must be released from the atoms with which they are already combined; which can be effected in (4), the only disposition of the polar molecules in which both attractive poles are together, and the actual decompositions and combinations possible: Cl is in contact with Pb at the same time that H is in contact with O, allowing the simultaneous formation of $Pb\,Cl$ and $H\,O$. This is no more than the expression of a double decomposition in the language of the atomic theory.

(1)	(2)	(3)
Cl	Cl	H
H	H	Cl
Pb	O	Pb
O	Pb	O

(4)

Cl	Pb
H	O

It is further to be observed, that, in the original polar molecules (4), although approximation and combination are promoted by the attraction of the contiguous unlike poles, they are opposed by the mutual repulsion of the like poles; Cl repelling O, and Pb repelling H. This unfavourable influence of the repulsions is reduced to a minimum in the arrangement of several pairs of the hydrochloric acid and oxide of lead molecules to form one circle. In (5), four pairs of the polar molecules are symmetrically placed; HCl alternately with PbO, and the attractive poles of the different molecules together. Affinities tending to a simultaneous formation of chloride of lead and water are equally favoured in this arrangement, as in (4); while the mutual repulsion of the like atoms,—such as the H and Pb, or the Cl and O of the adjoining molecules A and B—is less, as these like atoms are more distant from each other in the circular arrangement. It is obvious that the

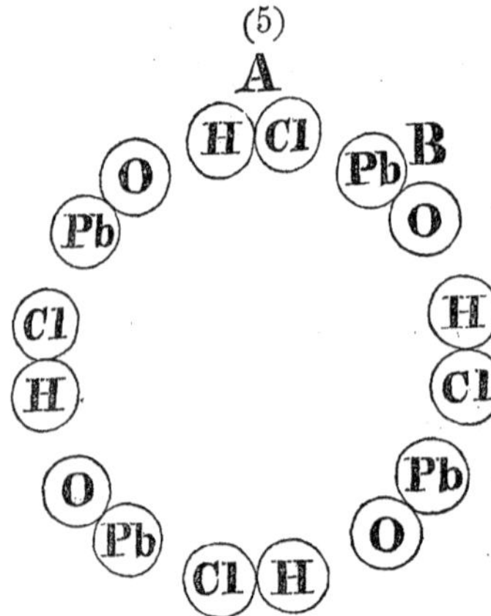

repelling atoms will be more distant the larger the circle, or the more nearly a segment of it approaches to a straight line. This arrangement of many pairs in a circle, being a condition of equilibrium, is a necessary one, and must take place in all double decompositions occurring in a liquid where the binary molecules are free to move. The formation of such polar circuits explains the ready occurrence of double decompositions; but it is of still more importance, as being the simplest and most intelligible exhibition of a voltaic circle.

Action of an acid upon two metals in contact.—When a plate of zinc is plunged into hydrochloric acid, a chemical change of a simple nature ensues; the metal dissolves, combining with the chlorine of the acid and displacing its hydrogen, the gas-bubbles of which form upon the zinc plate, increase in size, detach themselves, and rise through the liquor to its surface. The solution of zinc, when effected by its substitution for hydrogen, as in this experiment, is attended by a train of extraordinary phenomena, which become apparent when a second metal, such as copper, silver, or platinum, is placed in the same acid fluid, and allowed to touch the zinc, the second metal being one upon which the fluid exerts no solvent action, or a less action than upon zinc.

Fig. 66.

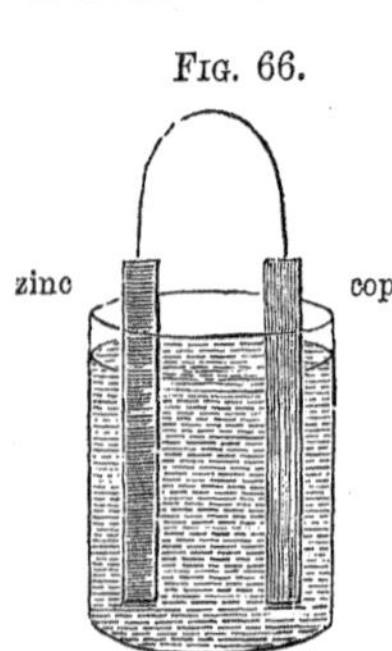

The zinc plate being connected by a metallic wire with a copper plate, as represented in fig. 66, and both dipped together in the hydrochloric acid, the zinc only is acted upon, and dissolves as rapidly as before; but much of the hydrogen gas now appears upon, and is discharged from the surface of the copper plate, and not fron the zinc. The hydrogen, being produced by the solution of the zinc, thus appears to travel through the liquid from that metal to the copper. But no current or movement in the liquid is perceptible, nor any phenomenon whatever to indicate the actual passage of matter through the liquid in that direction. The transference of the hydrogen must take place by the propagation of a decomposition through a chain of particles of hydrochloric acid extending from the zinc to the copper, and may be conceived by the diagram on the margin, in which each pair of associated circles marked *cl* and *h* represents a particle of hydrochloric acid. The chlorine *cl* of particle 1 in contact with the zinc combining with that metal, its hydrogen *h* combines, the moment it is set free, with the chlorine of particle 2, as indicated by the connecting bracket below, and liberates the hydrogen of that particle, which hydrogen forthwith combines with the chlorine of particle 3, and so on through a series of particles of any extent till the decomposition reaches the copper plate, when the last liberated atom of hydrogen (that of particle 3, in the diagram) not having hydrochloric acid to act upon, is evolved and rises as gas in contact with the copper plate.

Fig. 67.

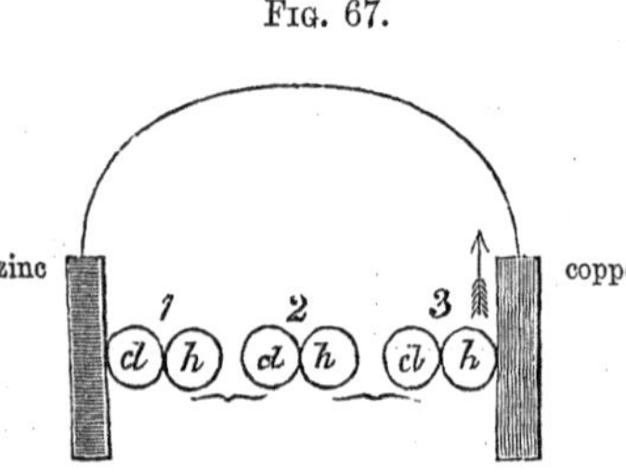

It is to be observed that this succession of decompositions and recombinations leading to the discharge of the hydrogen at the copper, does not occur at all unless that plate be in metallic connexion with the zinc, by means of a wire, as in the figure, or by the plates themselves touching without or within the acid fluid. This would seem to indicate that while the decomposition travels from the zinc to the copper through the acid, some force or influence is propagated at the same time through the wire, from the copper back again to the zinc. That something does pass through the wire in these circumstances is proved by its being heated, and by its temporary assumption of certain electrical and magnetic properties. Whether anything material does pass, or it is merely a vibration or vibratory impulse, or a certain induced condition that is propagated through the molecules of the wire, of which the electrical appearances are the effects, cannot be determined with certainty. But a power to effect decomposition, the same in kind as that occurring in the acid jar, and which acts in the same sense or direction, is propagated through the wire, and appears to be fundamental to all the other phenomena.

Fig. 68.

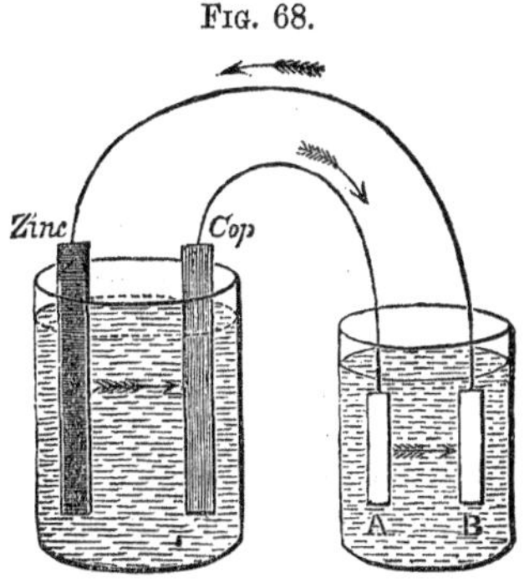

Let the wire, supposed to be of platinum, connecting the zinc and copper plates, be divided in the middle, and the extremities A and B of the portions attached to the copper and zinc plates respectively be flattened into small plates, and then dipped at a little distance from each other in a second vessel containing hydriodic acid. Iodine will soon appear at A, although that element is incapable of combining with the substance of the platinum, and hydrogen gas will appear at B. If the connecting wire and the small plates A and B were of zinc or of copper, the hydriodic acid would be decomposed precisely in the same manner, but the iodine as it reached A would unite with the metal and form an iodide. Supposing a decomposing force to have originated in the zinc plate, and to have circulated through the hydrochloric acid in the jar to the copper plate, and onwards through the wires and the hydriodic acid back to the zinc, as indicated by the direction of the arrows, then the hydrogen of the hydriodic acid has followed the same course, and been discharged against the metallic surface to which the arrow points.

The solution of the zinc in hydrochloric acid which developes these powers, acting at a distance, is not itself impeded, but on the contrary is promoted by exerting such an influence: for, placed alone in the acid, that metal scarcely dissolves at all, if pure and uncontaminated with other metals, or if its surface has been silvered with mercury; but it dissolves with rapidity when a copper plate is associated with it in the same jar, in the manner described. Hence the decomposing power which appears between A and B cannot be viewed as actually a portion of that which causes the solution of the zinc in the hydrochloric acid, for that force has suffered no diminution in its own proper sphere of action.

This combination of metals and fluids is known as the *simple voltaic circle.*

To explain the phenomena of the voltaic circle, the existence of a substantial principle, the electric fluid, has been assumed, of such a nature that it is readily communicable to matter, and capable of circulating through the voltaic arrangement, carrying with it peculiar attractive and repulsive forces which occasion the decompositions observed. A vehicle was thus created for the chemical affinity which is found to circulate. But it is generally allowed that this form of the electrical hypothesis has not received support from observations of a recent date, particularly from the great discoveries of Mr. Faraday, which have completely altered the aspect of this department of science, and suggest a very different interpretation of the phenomena. All electrical phenomena whatever are found to involve the presence of matter, or there is no evidence of the independent existence of electricity apart from matter; so that these phenomena may really be exhibitions of the inherent properties of matter. The idea of anything like a circulation of electricity through the voltaic circle appears to be abandoned. Electrical induction, by which certain forces are propagated to a distance, is found to be always an action of contiguous particles upon each other, in which it is unnecessary to suppose that any thing passes from particle to particle, or is taken from one particle and added to another. The change which a particle undergoes takes place within itself, and it is looked upon as a temporary development of different powers in different points of the same particle. The doctrine of polarity has thus come to be introduced into the discussion of electrical phenomena.[1]

One reason for retaining the theory of an electric fluid or fluids is, that it affords the means of expressing in distinct terms those strictly physical laws which are reputed electrical; and for many purposes such an hypothesis is unquestionably useful, if not absolutely necessary; but it has nothing to recommend it in the description of the chemical phenomena of the voltaic circle. These admit of a perfectly intelligible statement, when viewed as an exhibition of ordinary chemical affinity, acting in particular circumstances, without any electrical hypothesis.

Polarity of the arrangement.—It is to be assumed that the zinc and hydrochloric acid are both composed of particles, or molecules, which are susceptible of a polar condition. Of hydrochloric acid, the chemical atom is the polar molecule, and it therefore consists of an atom of chlorine and an atom of hydrogen associated together. The polar molecule of zinc may be supposed, for a reason which will afterwards appear, to consist of a pair likewise of associated atoms, which, however, are in this body both of the same element. The powers appearing in a polar molecule of zinc and of hydrochloric acid are the same. One pole of each molecule has the basylous attraction, or affinity, which is characteristic of zinc, or *zincous* attraction, and may be called the zincous pole; while the other has the halogenous attraction, or affinity, which is characteristic of chlorine, or *chlorous* attraction, and may be called the chlorous pole.

Zinc and acid in contact may therefore be represented (fig. 69) by trains of associated pairs of atoms. In the molecule of hydrochloric acid B, which is next the zinc, the chlorine atom forms the chlorous pole, and is turned towards the zinc, the fluidity of the acid allowing its molecule to take that position, which may be indicated by inscribing *cl* in the circle which represents the chlorine atom. The other atom of the molecule B, or the hydrogen, is the opposite, or zincous pole, and is marked *z*. Of the two atoms forming the polar molecule A of the zinc, the exterior atom which is in contact with the acid has thereby zincous attraction developed in it, and becomes the zincous pole, while the interior becomes the chlorous pole, as indicated in both by the inscribed letters. This polar condition of the zinc must be supposed the necessary and immediate consequence of its contact with the polar acid.

Fig. 69.

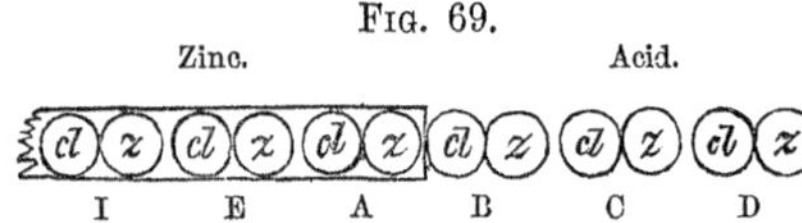

But each of these particles throws a train of particles of its own kind into a similar state of polarity: A, the contiguous particles E and I of the zinc, and B, the contiguous particles C and D of the acid. For *cl* of A becoming a chlorous pole, developes near it in an opposite, or zincous pole in *z* of E, and a chlorous pole in *cl*, the more remote extremity of E; in the same manner as the austral pole of a magnet developes, by induction, a boreal and austral pole in a piece of soft iron applied to it. And as the induced magnet, thus formed, will react upon a second piece of iron, and render it also magnetic, so the polarized particle E renders I

[1] For Mr. Faraday's views, the eleventh and subsequent series of his Researches, in the Philosophical Transactions for 1836, and the following years, may be referred to. He has favoured the scientific world with a reprint of the whole series: Faraday's Experimental Researches in Electricity: R. and J. E. Taylor, London, 1839. The subject is also systematically treated in the work of the late Professor Daniell, entitled an Introduction to the Study of Chemical Philosophy, which may be consulted with advantage.

similarly polar. The polar arrangement of the particles C and D of the acid is produced by B in the same manner. But as in a series of induced magnets (fig. 61, page 188), the magnetism acquired diminishes with the distance from the pole of the original magnet, so in trains of chemically polarized molecules, such as A, E, I and B, C, D, the amount of polarity developed in each molecule will diminish with the distance from the sources of induction A and B; I being polarized to a less degree than E, and D than C.

In the electrical theory of the voltaic circle as modified by Mr. Faraday, the zinc and hydrochloric acid are equally supposed to have a polarizable molecule. The polarity is also developed in these molecules by their approximation or contact. The molecule of hydrochloric acid is supposed to contain the positive and negative electricities, which possess contrary powers, like the two magnetisms, and are in combination and neutralize each other, in the non-polar condition of the molecule. But the contact of zinc causes the separation of the two electricities in the acid molecule, its atom of chlorine next the zinc becoming negative, and its atom of hydrogen positive. The electricities of the zinc molecule are separated at the same time, the side of the molecule next the acid becoming positive, and the distant side negative. The positive and negative sides of the two different molecules are thus in contact, the different electricities, like the different magnetisms, attracting each other. Hence, one side of each molecule is said to be positive instead of zincous, and the other side to be negative instead of chlorous. Polarity of the molecule is supposed in both views, but on one view the polar forces are the two electricities, on the other two chemical affinities. The difference between the two views is little more than nominal, for in both the same powers and properties are ascribed to the acting forces. The electricities are supposed to be the cause of the chemical affinities, but it may with equal justice be assumed that chemical affinities are the cause of the phenomena reputed electrical. One set of forces only is necessary for the explanation of the phenomena of combination, and the question is, whether are these forces electrical or chemical? Shall electricity supersede chemical affinity, or chemical affinity supersede electricity? If the electricities should be retained, in discussing the voltaic circle, their names might well be changed, the positive called zincous electricity, and the negative chlorous electricity, which express (as will appear more clearly afterwards), the nature of the chemical affinities with which these electricities are invested, and of which they are indeed constituted the sole depositories. The propagation of the effects to a distance is supposed to take place by the polarization of chains of molecules, on the electrical as well as chemical theory of the voltaic circle; so that the explanations which follow, although expressed in the language of the chemical theory, are the same in substance as those which are given on the electrical theory as now understood.

If the attractions of the respective zincous and chlorous poles of A and B which are in contact, rise to a certain point, the atom z of A is detached from the mass of metal, and combines with the atom cl of B, which last atom is disengaged at the same time from its hydrogen. Chloride of zinc is produced and dissolves in the liquid, while hydrogen is disengaged and rises from the surface of the metal; or we have the ordinary circumstances of the solution of an isolated mass of zinc in hydrochloric acid.

SIMPLE VOLTAIC CIRCLE.

Circle with the connecting wire unbroken.—When the zinc is pure, or its surface amalgamated with mercury, the zincous and chlorous attractions of the touching poles of A and B are not sufficiently intense to produce these effects, and combination does not occur. Let a copper plate F G H (fig. 70), be then introduced into the acid, and connected by a metallic wire H K I with the zinc. The particles of the acid assume chlorous and zincous poles as before; so also do those of the zinc, and the chain of polar molecules is now continued through the zinc and wire to the

Fig. 70.

Connecting wire.

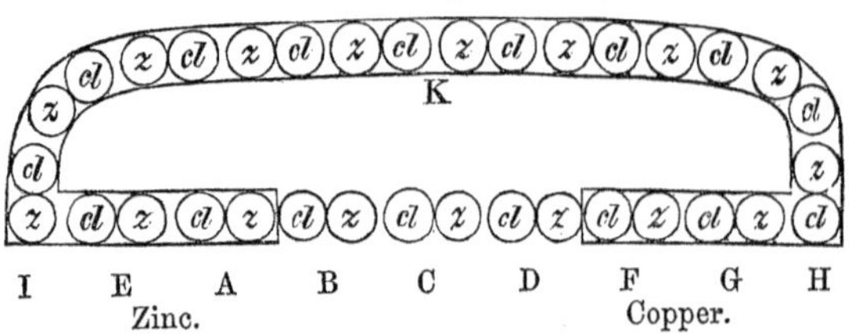

copper, the exterior particle F of which, it will be observed, comes thereby to present a chlorous pole to the acid. The contiguous particle D of acid is thus exposed to a second induction from the chlorous polarity of the copper, which increases the zincous polarity of the side of D next F, and, therefore, co-operates in enhancing the polar conditions already assumed by the chain of acid particles extending between the two metals. An endless chain or circle of polar molecules symmetrically arranged is thus formed, such as exists in a magnet of which the poles are united by a lifter, in which every particle in the chain has its own polar condition elevated by induction, and at the same time does itself react upon and elevate the polar condition of every other particle in the chain. The result of this is that the primary attraction of the zinc atom *z* of A, for the chlorine, *cl* of the hydrochloric acid B, is increased, and attains that degree of intensity at which the resistance to the impending combination is overcome, and the *z* and *cl* of A and B unite. But in a circle of polar molecules, in which the condition of any one molecule determines and is determined by that of every other, the intensity of the polar condition is necessarily the same in every element of the circle. The chemical polarity, therefore, of the other particles forming the chain, must increase to an equal degree as with A and B, when the circle is completed, and the same change must now occur in all of them that has occurred in A and B. The pole of B next C is intensely zincous, while that of C next B is intensely chlorous, whence the chlorine and hydrogen *cl* and *z* of these two particles combine together. At the same time, and for the same reason, the hydrogen *z* of C unites with the chlorine *cl* of D; and so on, through a chain of particles of hydrochloric acid of any length, till the copper is reached, when the last acid particle, D in the figure, yields its hydrogen *z* to the chlorous pole of the copper *cl*. But *the hydrogen not being capable of combining permanently with the copper, is liberated as gas upon the surface of that metal.*

Some internal change of a similar character appears to take place in the chain of polarized molecules extending through the metals themselves—a series of molecular detachments and re-attachments, among the atoms of their polar molecules, like the decompositions and recompositions in the acid, causing evolution of heat and other phenomena, generally reputed electrical, which the zinc and copper plates and the connecting wire exhibit.

Amalgamation of the zinc plate.—The polar molecule of the metals has been assumed to contain two atoms (like that of the acid), with the view of assimilating these intestine changes in the solid to those occurring in the fluid portion of the voltaic circuit, and also because it appears to account for the advantage of amalgamating the zinc surface. In the amalgamated plate, it is not zinc itself, but a chemical combination of mercury and zinc, which is presented to the acid, in which mercury is the negative element, and which might, therefore, be called a hydrargyride of zinc. That combination likewise is *fluid*. It must constitute the polar molecule, which will then consist of an atom of mercury as chlorous pole, and an atom of zinc as zincous pole, and not of two atoms of zinc. Such metallic molecules being capable of movement from their fluidity will place themselves, in forming a polar chain, with their unlike poles together, as the fluid acid particles arrange themselves. So that in an amalgam of zinc, of which A, E, and I, are polar molecules (fig. 70), all the atoms marked *cl* are mercury, and those marked *z* are zinc. It thus follows that, when by contact with an acid the amalgam is polarized, it presents a face of zinc only to the acid. If the mercury were exposed to the acid, that metal would completely derange the result, acting locally like a copper plate, as will afterwards be explained. The previous combination of the zinc (with mercury) likewise prevents that metal from yielding easily to the chlorine of hydrochloric acid; and the zinc of the amalgam is, therefore, not dissolved, till the affinities are enhanced by the introduction of a copper plate into the acid, and the formation of a voltaic circle.

It would thus appear that zinc, associated with copper, dissolves more readily in the acid than when alone, because the attraction or affinity of the zinc for chlorine is increased by the completion of a circle of similarly polar molecules, in the same manner as the magnetic intensity at one of the poles of a magnet is increased on completing the circle of similarly polarized elements, by connecting that pole by means of soft iron with the other pole (Fig. 64, page 188).

Although the terms of the electrical hypothesis are at present avoided, still it will be convenient to denominate the zinc, being the metal which dissolves in the acid, the active or *positive* metal, and the copper, which does not dissolve, the inactive or *negative* metal of the voltaic circle.

Looking to the condition of the two connected metals in the acid, it will be observed that the surface of the zinc presented to the acid has zincous affinity, or is zinco-polar, but the surface of the copper presented to the acid has, on the contrary, chlorous affinity, or is chloro-polar. Such a condition of the copper is necessary to the propagation of the induction; and the advantage of copper or platinum as the negative metal in a voltaic arrangement depends upon there being little or no impediment to either of these metals assuming the chlorous condition, that can arise from the peculiar affinity of the metals named for the chlorine of the acid; an affinity which tends to cause them to be superficially zincous instead of chlorous. If the second metal were zinc, the surface of it would be disposed to dissolve in the acid, and becoming on that account zincous, would induce a polarization in the intermediate acid in an opposite sense from that induced by the first plate of zinc; which counter polarizing actions would mutually neutralize each other. The acid between the two zinc plates would be like a piece of iron connecting two like magnetic poles, which itself is not then polarized.

But if one of the two zinc plates were less disposed to dissolve in the acid than the other, from the physical condition of its surface, from the acid being weaker there, or from any other cause, then the plate so situated might become negative to the other, and a voltaic circle of weak power be established, in which both metals were zinc.

Impurity of the zinc.—If zinc is alone in the acid, and every superficial particle of the metal equally disposed dissolve, then the zinc everywhere exposes a surface in a state of zincous polarity; and a polar circle in the liquid, starting from one particle of the zinc and returning upon another, cannot be established, as this requires that a part of the zinc surface be chlorous. But if the zinc contains on its surface a single particle of copper, a chlorous pole is created, upon which an inductive circle starting from an adjoining particle of zinc, A, (fig. 71), and passing through the liquid, may return as shown in the figure. It is the formation of such circles that causes impure zinc, which is contaminated by other metals, to dissolve so much more quickly in an acid than the pure metal. Why such circles are not formed when the positive metal in combination with the zinc is mercury, which forms a fluid alloy, has already been accounted for; and the nature of the evil which might otherwise attend the amalgamation of the zinc is now evident.

Fig. 71.

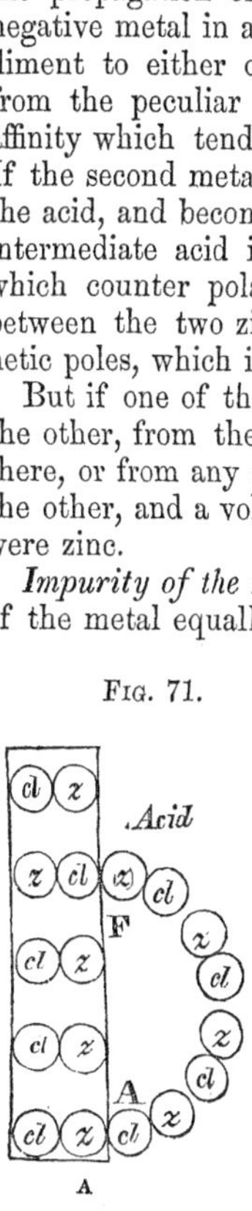

The whole chain of polar molecules in the voltaic circle admits of a natural division into two segments, the acid or liquid segment B C D (fig. 70), and the metallic segment, A K F, each of which has a pair of poles, the unlike poles of the two segments being opposed te each other. The pole at B of the acid portion is chlorous, and is opposed to the zincous pole at A of the metallic segment; while the pole of the liquid segment at D is zincous, and is opposed to the chlorous pole or the metallic segment at F. The distribution of polarity in these two segments is, therefore, the same as in two magnets with their unlike or attracting poles in contact.

Such, then, is the action of affinity by induction, which the mere introduction of zinc and copper in contact into the same acid liquid is sufficient to develope, and which accounts for the discharge of the hydrogen upon the surface of the copper in such an arrangement, the remarkable phenomenon by a description of which this subject was introduced.

Circle with the connecting wire broken.—It remains for us to apply the same principles to explain the additional phenomena of the second case described, in which the connecting wire, supposed to be of platinum, between the zinc and copper plates, is divided, and the broken extremities introduced into hydriodic acid (fig. 70, page 194).

Broken at any point, as at K, (Fig. 70), it is evident that if the polarized condition be still sustained, the portion of the metallic segment connected with the copper plate will terminate with a zincous pole at K, and that connected with the zinc with a chlorous pole; which may be indicated respectively by K and L, in fig. 72. When

Fig. 72.

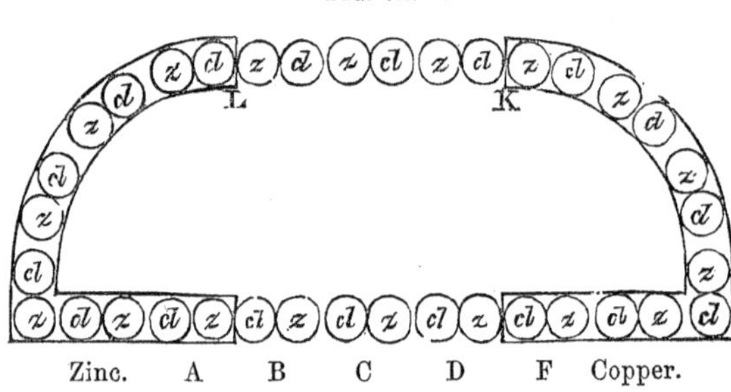

hydriodic acid is interposed between K and L, the breach is repaired by the polarization of a chain of particles of that acid. The extremity K, being zincous, induces chlorous polarity in the side of the hydriodic acid particle which it touches; in consequence of which the iodine atom (the analogue of chlorine) of the hydriodic acid molecule is presented to that pole, and liberated there when decomposition occurs. The extremity L of the zinc or positive metal element is chlorous, and therefore induces zincous polarity in the particle of hydriodic acid which it touches, and hydrogen (the analogue of zinc) is liberated there. The polarity in an induced circle must necessarily be of equal intensity at every point in it, and being sufficient at A to cause the decomposition of the hydrochloric acid, must also decompose the hydriodic acid between K and L; otherwise it is never established at A, nor anywhere else.

In the present arrangement, the voltaic circle is broken into four segments, or has four polar elements, every terminal pole of which is in contact with a pole of a different name; and the whole arrangement may be compared to a circle of four magnets with the attractive poles in contact.

These elements are:—First, the zinc plate or positive metal, A L, of which the end at A, in the hydrochloric acid (fig. 73), has zincous affinity, and the end faced with platinum at L, in the hydriodic acid, chlorous affinity.

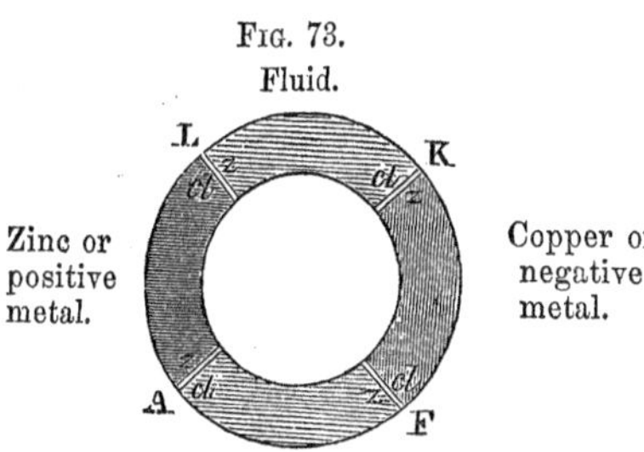

FIG. 73.

Secondly, the body of hydrochloric acid, A F, between the zinc and copper plates, of which the surface at A, in contact with the positive metal, has chlorous, and that at F, in contact with the negative metal, zincous affinity.

Thirdly, the copper or negative metal F K, of which the end at F in the hydrochloric acid has chlorous affinity, and that faced with platinum at K in the hydriodic acid, zincous affinity.

And fourthly, the body of hydriodic acid, K L, between the zincous and chlorous poles of the negative and positive metals, of which the surface K, in contact with the negative metal, is chlorous, and the surface L, in contact with the positive metal, zincous.

In every voltaic circle employed to produce decomposition these four elements are to be looked for. Hereafter, in adverting to any one of these elements, it will be sufficient to confine our notice to its terminal polarities or affinities, without recurring to the polarized condition of the element itself, upon which its terminal affinities depend.

COMPOUND VOLTAIC CIRCLE.

In both the arrangements described there is only one source of polarizing force, namely, the action between the zinc and acid at A. But a circle of a similar nature may be constructed embracing within itself two or more of such primary sources of polarizing power, and the intensity of the polar condition of the whole circle be thereby greatly increased.

Figure 74 represents such a circle, in which there are two zinc plates, both supposed to be in contact with hydrochloric acid, namely at A and at C, and a copper plate attached to each of these zincs. The polar condition of such a circle will easily be observed. By the contact of the acid and zinc at A, a zincous pole is established there in the first zinc plate, and a chlorous pole in the acid, which are so inscribed in the diagram. These occasion the formation of a chlorous pole at D in the first copper, the united zinc and copper A D forming together one polar element; and a zincous pole at B in the acid, the column A B of acid being the second polar element. The further effect of the induction is to produce a chlorous pole at B in the second copper, of which the corresponding zincous pole is at C, in the second zinc; the united zinc and copper B C forming together a third polar element. And, as a last consequence of the inducing force originating at A, the column of acid between C and D becomes a fourth polar element of the circle, having a chlorous pole at C and a zincous pole at D. Now it will be observed that the chemical affinity between the acid and zinc at C tends to produce the same polar conditions at that point as are already established there from the effect of induction. The extremity of the zinc plate at C is in fact zincous, both primarily and by induction; and the acid in contact with it chlorous, likewise both primarily and by induction; and generally, throughout the whole circle, the polar conditions determined by the second chemical action at C are the same as those determined by the first action at A.

FIG. 74.

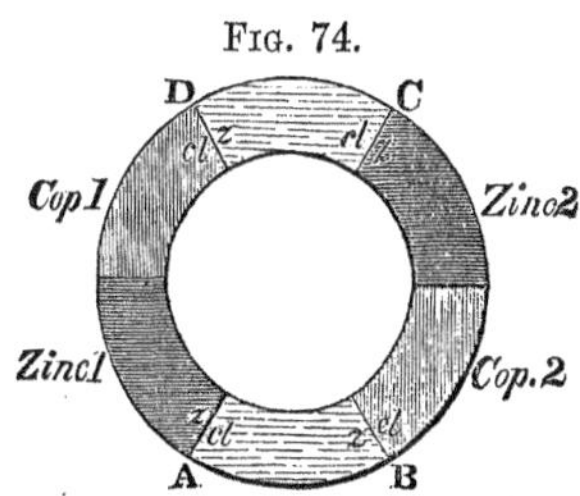

In the last arrangement, the inductive actions are in the same direction, and favour each other; but a circle may be constructed in which the inductions, being

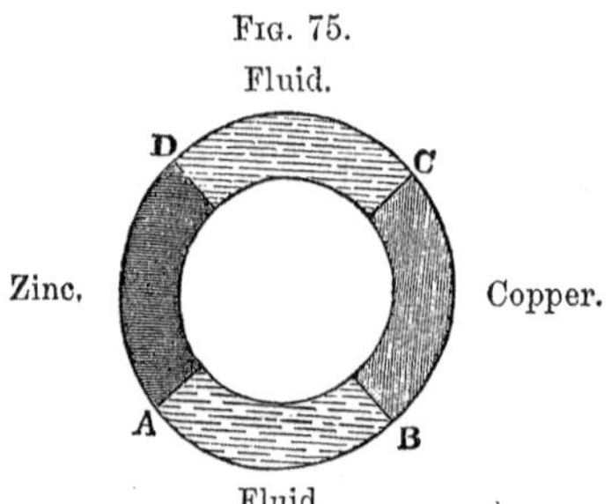

Fig. 75.

in opposite directions, oppose and neutralize each other. Thus if A D (fig. 75) be entirely zinc, both its extremities being exposed to acid, will tend equally to be zincous. In the same way, if B C be entirely copper, the condition of both its extremities will be chlorous, from the action of the acid on the two ends of the zinc; and, consequently, the elements of such a circle could have no polarity.

A circle is represented in fig. 76, containing three sources of polarizing force. It consists of three alternations of copper and zinc symmetrically arranged, and forming three polar elements F A, B C, and D E, with three acid columns between these alternations, which form three additional polar elements, A B, C D, and E F. The number of alternations of copper and zinc with acid may obviously be increased to any extent, and the chemical action of the acid on the zinc in each alternation is found to increase in a marked manner up to the number of 10 or 12 alternations. This increase of the affinity is undoubtedly owing to the favouring inductive action which the chemical actions at the different points have upon each other. Such a compound circle may be compared to a number of magnets disposed in a circle with their attracting poles together, of which each would have its magnetic intensity exalted by induction from all the rest. When such a circle is broken at any point, all chemical action and polarization cease till contact is again made, and the circuit completed. The polarization, too, being the result of a circular induction involving so many lines or chains of particles, cannot, when once established, be more or less at any one point in the circuit than at others. The resulting chemical action must, therefore, be every where equal in the circle, and consequently the same quantity of zinc be dissolved, and hydrogen evolved in each acid.

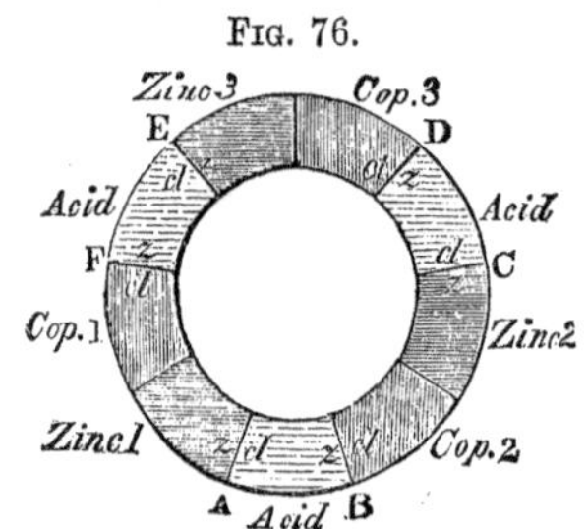

Fig. 76.

If any metallic element of this compound circle be broken, and a polarizable liquid be interposed between the metallic extremities so as to complete the circuit, decomposition occurs in that liquid as in the simple interrupted circle (fig. 72). But the polarizing influence of the compound circle being of high intensity, more numerous and difficult decompositions are effected by means of it than by the simple circle. The compound voltaic circle is indeed a decomposing instrument of great efficiency.

If, in this arrangement, the position of one of the metals in the series be reversed, so that a zinc is where a copper should be, then, by the action of the acid on that zinc, polarization in the wrong direction is occasioned, which greatly diminishes the general polarity of the circle, reducing it in an arrangement of ten alternations to one-fourth, according to Mr. Daniell.

Voltaic battery.—In the first of the two annexed diagrams (see fig. 77) is represented a compound circle, such as is employed to produce decomposition, and called a voltaic battery, consisting of three acid jars, each of which contains a zinc and copper plate, and which are termed active cells, as they are sources of polarizing power, from the action of acid upon zinc which takes place in them.

In the second diagram (see fig. 78), the same arrangement is repeated, with the addition of a third jar, termed the decomposing cell, which contains any binary polar liquid, with two platinum plates immersed in it. Each copper, it will be seen, is connected by a wire with the following zinc; and, in the first diagram, the copper in the third cell C'' is immediately connected with the zinc in the first cell Z by a wire,

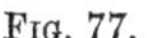

Fig. 77.

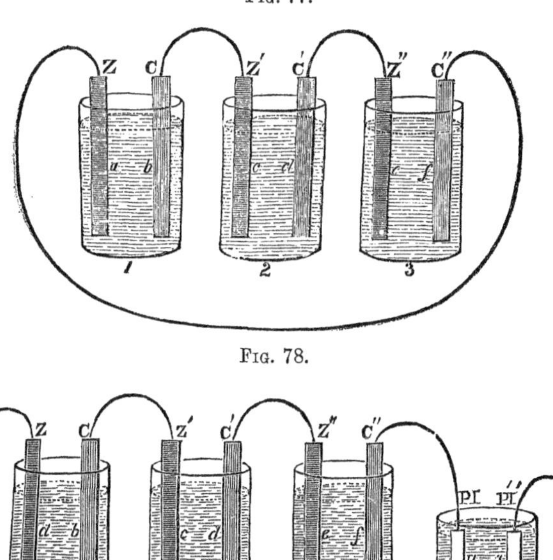

Fig. 78.

and the circuit thus completed. The polar elements in the circle of the first diagram, it will be found, are six in number; namely, the three acid columns between the metals in the cells *a b*, *c d*, and *e f*; and the three pairs of zinc and copper plates, each of which pairs forms a single polar element, of which the surface of the zinc is the zincous, and the surface of the copper the chlorous pole. In the second diagram, one of these metallic elements Z C'' is divided, and a polar liquid *g h*, in the cell of decomposition, interposed between the broken extremities Pl and Pl'. To ascertain the polar condition of the extremities, or the terminal platinum plates in the decomposing cell, it is to be observed that Pl' with Z forms one polar element, of which Z being a zincous pole, Pl' must be a chlorous pole. Again, Pl with C'' forms one polar element, of which C'' being a chlorous pole, Pl must be a zincous pole. Now, the platinum plates Pl and Pl', which are thus zincous and chlorous, are disposed in the decomposing cell, in regard to one another,—the first to the left, and the second to the right, as the zincous and chlorous plates (the zinc and copper) also are arranged in the active cells. It will be convenient to distinguish by names the poles which these terminal platinum plates constitute, as they are much more frequently referred to, and of greater consequence than any other poles in the voltaic battery, when used as an instrument of decomposition, as it constantly is. The chlorous plate Pl', which is in connexion with a zinc plate Z, may be called the *chloroid* (like chlorine), and the zincous plate Pl, which is connected with a copper plate C'', may be called the *zincoid* (like zinc),—names which express the virtual properties of each plate, or the particular attractive power and affinity which each of them acquires from its place in the circle.

When hydrochloric acid is the polar liquid interposed between these plates, chlorine is of course attracted by the surface of the zincoid, and discharged there; and hydrogen by the face of the chloroid, and discharged upon that plate. On the electrical hypothesis, the same plates are variously denominated:—

The *zincoid* as the positive pole, the positive electrode, the anode, and the zincode.

The *chloroid* as the negative pole, the negative electrode, the cathode, and the platinode.

The cell of decomposition thus interpolated in the voltaic circle is an obstacle to induction, and reacts on the whole series, reducing the chemical action and evolution of hydrogen in each of the active cells by at least one-third. In that retarding cell itself, the amount of decomposition is necessarily the same as in the other cells. Mr. Daniell found the chemical action reduced to one-tenth in a series of eight active and two such retarding cells; and entirely stopped by three retarding to seven active cells.

OF THE SOLID ELEMENTS OF THE VOLTAIC CIRCLE.

The elements of a Voltaic Circle are obviously of two different kinds—the metals or solid portions, through the substance of which chemical induction is propagated without decomposition; and the liquids in the cells, which yield to the induction and suffer decomposition. In reference to the first, it is to be observed that, as only iron and one or two other metals of the same natural family are susceptible of magnetic polarity, so the susceptibility of chemical polarity which appears in the voltaic battery is not possessed by solids in general, but is confined to the class of bodies to which zinc belongs,—the metals, all of which possess it, with the addition of carbon in the form of charcoal, and certain metallic sulphides, more particularly the sulphide of silver when heated. Weak solutions of the alkaline sulphides, containing an excess of sulphur, also admit of a feeble polarity without undergoing decomposition. The non-metallic elements, with their compounds, the oxides and salts of the metals, are destitute of this power, and cannot, therefore, be used as solid elements of the circle. A body available for this purpose is termed a *conductor* on the electrical hypothesis, a name which may be retained as it is not at variance with the function assigned to the metals in the circle viewed as a chemico-polar arrangement. Two different metals are combined in a circle, one of which is acted on by the liquid, and, therefore, called the active or the positive metal; while the other is not acted upon, and is, therefore, called the inactive or the negative metal; and it has already been stated, that the more easily acted on by the liquid, or the more highly positive the one metal, and the less easily acted upon, or more negative the other metal, the more proper and efficacious is the combination. In the following table several of the metals are arranged in the order in which they appear positive or negative to each other, when acted on by the acid fluids commonly employed in the voltaic battery. Each metal is positive to any one below it in the table, and negative to any one above it.

Most positive.

Potassium.
Sodium.
Manganese.
Zinc.
Cadmium.
Iron.
Nickel.
Cobalt.
Lead.
Tin.
Bismuth.
Copper.
Silver.
Mercury.
Palladium.

Carbon.
Platinum.
Rhodium.
Iridium.
Gold.
Most negative.

Zinc, which stands high in the list, is the only metal which can be used with advantage in the voltaic battery, as the positive metal. Although closely approaching zinc in the strength of its affinities, iron is ill adapted for the purpose, from the impossibility of amalgamating its surface, the irregularity of its structure, and certain peculiarities of this metal in reference to chemico-polarity. Platinum forms an excellent negative metal, from the weakness of its affinities, and is generally used for the plates in the cell of decomposition. Silver also is highly negative, but copper is the only negative metal which from its cheapness can be used in the construction of active cells of considerable magnitude.

Voltaic protection of metals.—But although the difference between two metals in point of affinity be very small, yet their association in the same acid always gives a decided predominance to the affinity of the more positive, by causing the surface of the other to become chlorous, and therefore wholly inactive in an acid fluid. A negative metal may thus be protected from the solvent action of saline and acid liquids, by association with a more positive metal; iron, for instance, by zinc, as in articles of *galvanized iron*, which are coated with the former metal. The process is analogous to the making of tin-plate. The surface of the iron (generally sheet iron) is first cleaned from all adhering oxide by a dilute acid; then immersed in a weak solution of tin, with fragments of metallic tin, according to the improved practice of Messrs. Morewood and Rogers, by which the iron is covered by a film of tin, to which zinc is capable of adhering more uniformly than to an iron surface. The article so prepared is then passed once through a bath of melted zinc, of which the surface is covered by the fused chloride of zinc and ammonium, to protect the metal from oxidation. It thus acquires a smooth and beautifully crystallized coating of zinc. Copper is protected by either zinc or iron, as was remarkably illustrated in the attempt made by Sir H. Davy to defend the copper sheathing of ships from corrosion in sea-water, by means of his *protectors*. These were small masses of iron or zinc fixed upon the ship's copper, at different points under the water line. They completely answered the purpose of protecting the copper, but unfortunately gave rise to a deposition of earthy matter upon that metal to which barnacles and sea-weeds attached themselves, and thereby diminished the facility of the ship's motion through the water. The more recent substitution, by Mr. Muntz, of an alloy of 60 parts of copper and 40 of zinc, for pure copper, has proved more successful. In acting as a protecting positive metal, zinc necessarily undergoes corrosion, but more slowly than might be expected. On zinced articles which are exposed to the air only, and not immersed in water, a film of suboxide of zinc soon appears, which forms a hard covering, and protects the metal below from further change.

On the other hand, the injurious effect of association with a *negative* metal is often accidentally illustrated, as in the corrosion of the ends of iron railings, which are fixed in their sockets by lead, a more negative metal. In dye-coppers, an iron steam-pipe with a rose of lead or copper is quickly destroyed. Some kinds of cast iron undergo a rapid corrosion, when exposed to sea-water, the carbon acting as a negative body and ultimately remaining in the form of plumbago after all the metal has disappeared.

A weak voltaic circle may even be formed of a single positive metal in an acid, as the zinc A B (fig. 79), provided the surfaces of the metal exposed to the acid at A and B are in different conditions as to purity or mechanical structure, and therefore unequally acted upon by the acid; whereupon the part least disposed to dissolve becomes negative to the other. A zinc plate may also be unequally acted on and

Fig. 79.

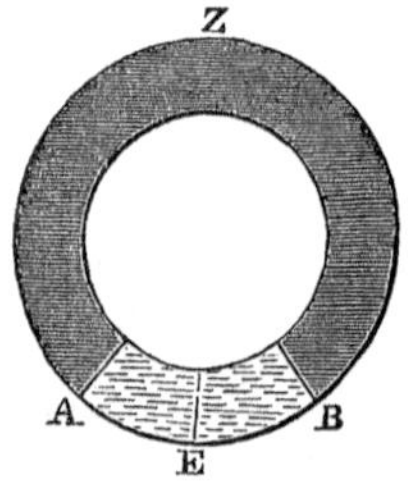

thrown into a polar state, from the liquid in which it is immersed varying in composition and activity at different points of the metallic surface. A circle may thus be formed of one metal A Z B, with two liquids A E and E B, which merge into each other, and form together one polar element A B.

The two metals in a circle have generally been exhibited in metallic contact, and forming together one polar element, but they may be separated, as are the zinc and copper plates A D and C B in the diagram (fig. 80), by two fluids, provided these fluids are such as a strong acid at A B, and as iodide of potassium at D C, the first of which acts very powerfully on zinc, while the other acts very feebly upon that metal (unless associated with copper); so that of the consequent opposing inductions, that originating at A greatly exceeds and overpowers that of D. It is likewise necessary that the fluid D C be of easy decomposition, so as to yield to the polar power of the single circle. In this arrangement, however, it is obvious that the zinc itself forms a complete polar segment, of which A is the zincous, and D the chlorous pole; and the copper also an entire polar segment of which B is the chlorous, and C the zincous pole.

Fig. 80.

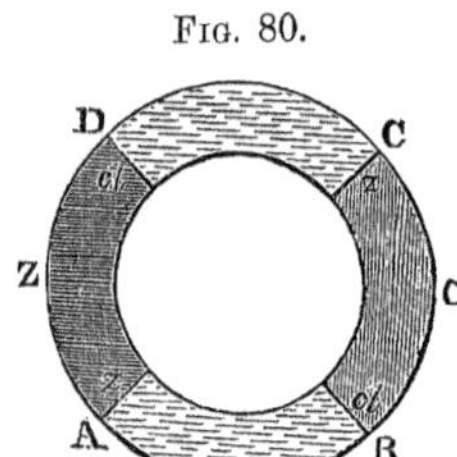

The preceding table exhibits the relation which the metals enumerated assume to each other, in the acid and saline solutions usually employed as exciting fluids. But the relation of any one metal to another is not the same in all exciting fluids. Thus when tin and copper are placed in acid solutions, the former is most rapidly corroded and becomes the positive metal, according to its position in the series, but if they are put into a solution of ammonia which acts most upon the copper, then the latter becomes the positive metal. Copper is positive to lead in strong nitric acid, which oxidizes the former most freely, whereas in dilute nitric acid, by which the lead is most rapidly dissolved, the lead is positive.

LIQUID ELEMENTS OF THE VOLTAIC CIRCLE.

With the view of simplifying the statement of the circular decompositions which occur in the voltaic circle, the exciting fluid has hitherto always been supposed to be hydrochloric acid (chloride of hydrogen), and this compound is a fair type of the class of bodies which possess a polar molecule, and are available for the purpose of bringing these changes into play. The exciting fluid is always a *saline body* in the general sense; that is, a binary compound of a salt-radical or halogen, such as chlorine, with a basyl, such as hydrogen or a metal. The chloride of copper, chloride of sodium, chloride of ammonium, or the chloride of any other basyl, may be substituted for hydrochloric acid, although not all with the same advantage; and the chlorides of basyls may be replaced by their iodides, sulphionides (sulphates), nitrationides (nitrates), and salts of other acids, as exciting fluids, provided they have the condition of liquidity, which gives mobility to their particles, and permits that disposition of them which is assumed in a polar chain. The liquids which yield in the cell of decomposition are of the same nature, possessing always a binary polar molecule, although the liquid which forms the best exciting fluid is not always the most easily decomposed in the decomposing cell.

The positive metal which is exposed to the exciting fluid always acts in one way, displacing the basyl and combining with the halogen of that body; in the manner the zinc has been seen to liberate hydrogen and combine with chlorine, when hydrochloric acid is the exciting fluid. The positive metal is thus substituted for a similar basyl in a pre-existing saline compound. That metal may dissolve in another manner, by uniting directly, for instance, with free chlorine or iodine in solution, but then no polar chain is formed. Particles of chlorine may extend from the zinc to the associated negative metal, but not possessing a binary molecule they have no occasion to throw themselves into a polar chain in order to act upon the zinc, as the molecules of hydrochloric acid require to do in the same circumstances. The particles of these free elements appear to be incapable of that polar condition, having chlorous affinity on one side and zincous on the other, of which both the solid and liquid constituents of the voltaic circle must be susceptible. Judging from the uniformity in composition of exciting liquids, their capacity to form polar chains depends on their consisting of an atom of basyl and an atom of salt-radical, which are respectively the locus of zincous and chlorous affinity or polarity. Such molecules may be looked upon as in a state of tension when forming a part of a polar chain, each about to divide into its chlorous and zincous atoms. Mr. Faraday had established that all exciting liquids are binary compounds of single equivalents of salt-radical and basyl, or *proto-compounds*, such as hydrochloric acid itself, proto-chloride of tin, &c. Other saline bodies which are *per-compounds*, such as bichloride of tin, are not exciting or polar, because, as may be supposed, they are not naturally resolvable into a chlorous and zincous atom, but into a chlorous atom and *another salt;* the bichloride of tin, for instance, into chlorine and proto-chloride of tin. Certain compounds, which are deficient in the saline character and not polarizable, such as chloride of sulphur, and the liquid chlorides of phosphorus and carbon, have been enumerated as exceptions to this rule. None of these bodies, however, is really a proto-compound.

The zinc or positive metal, too, always forms a proto-compound in dissolving, which is a saline body. The order of the chemical changes in the exciting fluid therefore is as follows: — The zinc in decomposing a binary compound and forming a binary compound liberates an atom of its own class; which atom repeats the same actions; supplying at the same time another atom of the same kind to act in the same manner, and that another, from the zinc to the copper plate. The combining bodies are always a basyl and a salt-radical, and therefore only two kinds of attraction or affinity are at work throughout the chain, those of a basyl and a salt-radical, the zincous and chlorous affinities. Hence, in the present subject of chemical polarity, we have to deal with but two attractive forces, the zincous and the chlorous, as in magnetism with but two magnetic forces, the austral and the boreal.

On the electrical hypothesis, a body which is thus decomposed in the active cells, or in the cell of decomposition, is called an *electrolyte* (decomposable by electricity), and this kind of decomposition is distinguished as *electrolysis*. The two elements of an electrolyte, which travel or are transferred in opposite directions, in its decomposition have been named *ions* (from 'ιων, going); the halogen which travels to the positive metal or terminal, the *anion* (going upwards), and the basyl, which is transferred to the negative metal, or terminal, the *cation* (going downwards). Strictly chemical expressions equivalent to the former would be *zincolyte* and *zincolysis*, the decompositions throughout the circle being referred to the affinity of zinc or the positive metal.

The characters of the two constituents of an electrolyte may be shortly noticed. The class of basyl constituents is composed of the metals in their order as positive metals, beginning with potassium, and terminating with mercury, platinum, and the less oxidable metals. Ammonium has a claim to be introduced high in this list, and should probably be accompanied by the analogous basyl of the aniline class of bases and of the vegeto-alkalies, although in respect to the decomposition of their salts in the voltaic circle, we have little precise information. Hydrogen likewise finds a place near copper in this class.

At the head of the halogen constituents of electrolytes may be placed iodine and the other members of the chlorine family. These are followed by the halogens of the sulphates, nitrates, carbonates, acetates, and other oxygen-acid salts. Sulphur must be allowed to follow the last, as the salt-radical of the soluble sulphides, and the lowest place be assigned to oxygen, as the salt-radical of the soluble metallic oxides; of oxide of potassium, for instance, and of water. It is unusual to speak of oxygen as a salt-radical, and of caustic potassa and water as salts, but the binary theory of salts recognizes no essential difference between the chloride, sulphionide, and oxide of a basyl, the oxide being connected with the more highly saline compounds through the sulphide, and the list of salt-radicals forming a continuous descending series from iodine to oxygen.

The facility of decomposition of different electrolytes appears to depend more upon the high place of their salt-radical, than upon the nature of their other constituent. The iodides, for instance, as iodide of potassium and hydriodic acid, are the most easily decomposed of all salts, yielding to the polar influence of the single circle. Then follow the chlorides, — chloride of lead, fused by heat, yielding to a very moderate power. After these the salts of strong oxygen acids, such as sulphates and nitrates either of strong bases, such as potassa and soda, or of weak bases, such as oxide of copper and water (the hydrated acids are such salts). The carbonates and acetates, which have much weaker salt-radicals, are still less easily decomposed, and finally oxides are decomposed with great difficulty. Water itself is polarized with such extreme difficulty, and decomposed when alone to so minute a degree, even by a powerful battery, as long to have left its claim uncertain to be considered an electrolyte, when in a state of purity.

Widely as the more characteristic halogens and basyls differ, still the classes pass by imperceptible gradations into each other, and form portions of one great circular series. Mercury and the more negative metals, although clearly basyls, appear at times to assume the salt-radical relation to the highly positive metals; such a character is evinced in mercury, by the energy with which it unites with sodium and potassium, and by its function in the amalgamated zinc plate of the voltaic circle. So that the salt-radical or basyl character of a body is not absolute, but always relative to certain other bodies.

The addition of a salt or acid, even in minute quantity, to water in the cell of decomposition, causes the copious evolution of oxygen and hydrogen gases at the zincoid and chloroid, and is therefore often spoken of as facilitating, by its presence, the decomposition of the water, in some way which cannot be explained. But the phenomena are unattended with difficulty on the binary theory of saline bodies. When sulphate of soda exists in the water of the decomposing cell, it may be sulphionide of sodium which is decomposed, SO_4, the sulphate radical being evolved at the zincoid, and sodium at the chloroid. But the sodium having a strong affinity for oxygen reacts upon the water at the pole, forming soda and liberating hydrogen, which therefore appear together; while SO_4 having, as a high salt-radical, a powerful affinity for hydrogen, likewise decomposes water, and thus evolves oxygen, which, with a free acid, appears at the zincoid. A solution of chloride of sodium is decomposed in the same manner, its elements chlorine and sodium being attracted to the zincoid and chloroid respectively, but neither of these elements appearing as such. Both decompose water, and thus produce oxygen with hydrochloric acid at the zincoid, and soda with hydrogen at the chloroid. It has indeed been ascertained that the polar influence which apparently effects two decompositions in these circumstances, namely, that of water into oxygen and hydrogen, and of a salt into its acid and alkali, is no more in quantity than is necessary to decompose one of these bodies, the circulating power being measured by the quantity of fused chloride of lead decomposed in another part of the circuit (Daniell). There can be little doubt, then, that only one binary compound is immediately decomposed, and that the two sets of products which appear at the terminals are the results of secondary decomposition. Indeed, the decomposition of salts in the voltaic circle is supposed to afford considerable support to the salt-radical theory of these bodies (page 156.)

Certain salts form a polar chain, or conduct, without undergoing decomposition, in a way which cannot at present be explained, particularly the iodide of mercury and fluoride of lead, both fused by heat. According to recent observations of M. Matteucci many other fused salts conduct to a greater extent than is indicated by their decomposition.

Secondary decompositions.—The products of voltaic action are frequently of the secondary character just described, the original products being lost from their reaction upon the liquid in which they are produced, or upon the substance of the metallic terminals. Thus, salts of the vegetable acids often afford carbonic acid, and salts of ammonia nitrogen, instead of oxygen, at the positive terminal or zincoid; the oxygen liberated having reacted upon the combustible constituents of these bodies. Nitrates, again, may afford nitrogen, or nitric oxide, at the negative terminal or chloroid, in consequence of the oxidation of the hydrogen evolved there. The nascent condition of the liberated elements favours such secondary actions. When the zincoid is composed of a positive metal, such as zinc itself or copper, the chlorous element is absorbed there, combining with the metal. The decomposition of a salt is also then much easier, the action of the circle being greatly assisted by the proper affinity of the matter of the zincoid for a chlorous body. Indeed, when two pieces of the same metal communicate by means of one of its salts, the phenomena are the same as if the metallic circuit were complete (Faraday). Insoluble sulphides, chlorides, and other compounds of a positive metal acting as the zincoid, have thus been slowly produced in a single circle with a weak exciting fluid; which products have exhibited distinct crystalline forms, resembling natural minerals, not otherwise producible by art. The hydrogen evolved upon a platinum chloroid, immersed in the solution of a copper or iron salt, may also reduce these metals upon the surface of the platinum, in the form of brilliant octahedral crystals. In the active cells themselves a secondary decomposition is apt to occur, the hydrogen evolved decomposing the salt of zinc which accumulates in the liquid, and occasioning a deposition of that metal upon the copper plate; an occurrence which may determine an opposite polarity, and cause the action of the circle to decline. But on disconnecting the zinc and copper plates, the foreign deposit upon the latter is quickly dissolved off by the acid. The inconvenience of this secondary decomposition in the exciting cells is avoided by dividing the cell into two compartments, by a porous plate of earthenware interposed between the zinc and copper plates. The salt of zinc formed about that metal is prevented from diffusing to the copper, by the diaphragm, although it allows, from its porosity, a continuity of liquid polar molecules between the metals.

Two polar liquids separated by a porous diaphragm.—The liquids on either side of the porous division may also be different, provided they have both a polar molecule. Thus, in fig. 81, the polar chain is composed of molecules of hydrochloric acid, extending from the zinc to the porous division at *a*; and of molecules of chloride of copper, from *a* to the copper plate. When the Cl of molecule 1 unites with zinc, the H of that molecule unites with the Cl of molecule 2 (as indicated by the connecting bracket below), the H of molecule 2 with the Cl of molecule 3, the Cu of molecule 3 with the Cl of molecule 4, and the Cu of this molecule, being the last

FIG. 81.

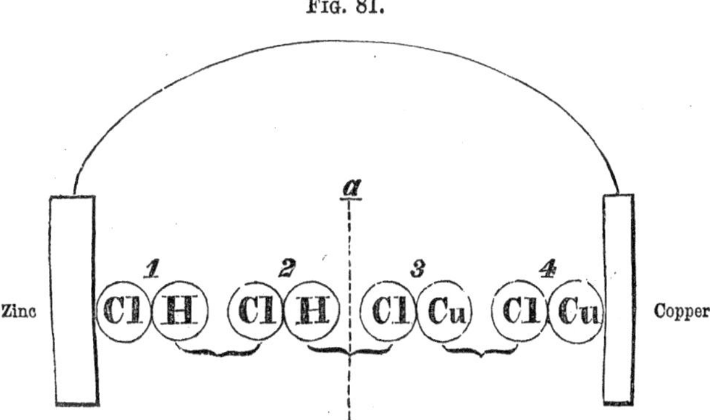

in the chain, is deposited upon the copper plate. Dilute sulphuric acid, in contact with an amalgamated zinc plate, and the same acid fluid saturated with sulphate of copper, in contact with the copper plate, are a combination of fluids of most frequent application. In such an arrangement, the formation of small gas bubbles upon the negative plate, which makes its contact with the acid fluid imperfect, is avoided; and the surface of that plate is kept clean and entirely metallic by the constant deposition of fresh copper upon it. The copper is deposited in a coherent state, and forms a plate, which may be stripped off from the original copper after attaining any desired degree of thickness, — and presents an exact impression of the surface of the latter. In the operation of *electrotyping*, the article to be copied is so placed in a copper solution as the negative plate of a voltaic pair, being first made conducting, if not metallic and already so, by rubbing its surface over with fine plumbago. With a negative plate of platinum, undiluted nitric acid may be used in the place of the acid solution of copper in the last arrangement, with oil of vitriol, diluted with four or five times its bulk of water, about a positive plate of amalgamated zinc. The polar molecules will be, on the binary theory of salts, NO_6+H, in the former, and SO_4+H, in the latter fluid. The hydrogen is also here entirely suppressed at the negative plate, uniting with the fifth equivalent of oxygen in nitric acid to form water, which is attended with the evolution of peroxide of nitrogen, NO_4. The solution of the zinc, with such an arrangement of fluids, appears to give the most intense polarization that can be attained.

Application of the voltaic circle to chemical synthesis. — The liquid in the decomposition cell may be divided by a porous diaphragm placed between the platinum plates, which form the zincoid and the chloroid in a similar manner, and the synthetical results of the voltaic action be had more readily apart from each other. With a solution of chlorate of potassa between the plates, it is found that the oxygen, instead of being evolved at the positive pole as gas, is communicated to the chlorate of potassa there, and converts it into perchlorate (Berzelius). In a solution of chloride of potassium, even when rendered acid by sulphuric acid, chlorate, and afterwards perchlorate of potassa were found at the positive pole (Kolbe). A concentrated solution of chloride of ammonium evolves hydrogen at the negative pole; but neither oxygen nor chlorine at the positive pole. But the surface of the platinum plate representing the latter pole is covered with small, yellow, oily drops of chloride of nitrogen, which, as soon as the two poles are brought into contact, decompose with explosion (Kolbe). A solution of the yellow prussiate of potassa is converted into the red prussiate by the action of the oxygen at the positive pole (Smee). Dr. Kolbe oxidized the cyanide of potassium in the same manner, and converted it into cyanate of potassa, but did not succeed in obtaining a percyanate: nor did he succeed in forming a fluorate of potassa from the fluoride of potassium by the same means (Mem. of the Chem. Soc., vol. iii. p. 287). The decomposition of a concentrated neutral solution of valerianate of potassa in the cold gave a gaseous carbo-hydrogen, C_8H_8, of double the density of olefiant gas, and what appeared to be a new ether, containing C_2H_2 less than amylic ether. Such transformations from the series of one alcohol to that of another are of great importance, and the attaining them by voltaic action highly interesting. Six pairs of Bunsen's carbo-zinc battery were employed in these decompositions, and the action continued for several days (Kolbe, Memoirs of the Chemical Society, vol. iii. p. 378).

Transference of the ions. — With a double diaphragm cell, in which the liquid between the poles was divided into three portions, Messrs. Daniell and Miller were enabled to make some singular observations on the transfer of the ions and their accumulation at the poles. With a neutral salt of the potassium family (such as sulphate of soda), for one equivalent of salt decomposed, half an equivalent of free acid is added to the division of the cell containing the positive pole, and half an equivalent of free alkali to the division containing the negative pole — the amount of transference which the polar decomposition requires: but, with a salt of the magnesian family (such as sulphate of zinc), while the acid travels as usual to the positive pole and accumulates there, no corresponding transference of oxide of zinc takes place in the opposite direction. This seems to imply that water travels, as base, instead of oxide of zinc. All the magnesian salts retain one equivalent of water very strongly; and, in the polar chain, probably assume this water as their base, so as to become equivalent to hydrated acids in solution. In the decomposition of salts of oxide of ammonium, the ammonia also appears passive, and does not move towards the negative pole, although the acid of the salt travels as usual towards the positive pole. The water, which is essential to the salts of oxide of ammonium, appears to be here again, the base which travels; and in a polar chain extending through a salt of ammonia, such as the sulphate of ammonia, we have probably sulphate of water as the polar molecule; the ions being SO_4 and H; not SO_4 and NH_4.[1]

Voltaic endosmose.—It was first observed by Mr. Porrett, that in the decomposition cell, divided into two chambers by a permeable diaphragm of wet bladder or porous earthenware, the liquid tends to pass from the chamber containing the positive terminal plate into that containing the negative terminal, so as to rise at times several inches in the latter above its level in the former (Annals of Philosophy, 1816). This accumulation of liquid at the negative pole is only considerable with liquids of an inferior conducting power, that is, of difficult decomposition, and is greatest in pure water.

The transfer takes place of a large quantity of water with the hydrogen of the negative pole, as if the ions were O on the one side, and H + Water on the other. In a polar molecule, such as this implies, we must have an aggregation of many atoms of water forming one compound polar atom. Let us suppose six atoms of water associated H_6O_6; the polar molecule will be $H_6O_5 + O$, in which H_6O_5 is the basyl, and O the salt-radical. Taking advantage of the graphical representation of such a compound molecule by a polar formula (page 168), in which the letters exhibit the relative position of the constituent atoms, we have—

$$\text{Positive Pole.}\ \left|\ \begin{array}{cccccc} 1 & 2 & 3 & 4 & 5 & 6 \\ O & O & O & O & O & O \\ \hline H & H & H & H & H & H \end{array}\ \right|\ \text{Negative Pole.}$$

The oxygen 1 is alone attracted by the positive metal or pole with which it is in contact, while hydrogen (1) being so far relieved from the attraction of its own oxygen, comes under the influence of oxygen 2, 3, 4, 5, and 6. As the salt-radical O (1) separates, we have thus the temporary formation of the basylous atom—

$$\frac{O\ \ O\ \ O\ \ O\ \ O}{H\ H\ H\ H\ H\ H},\ \text{or}\ \frac{O_5}{H_6}.$$

But instead of involving six atoms of water, as in this illustration, the compound polar molecule may embrace hundreds or thousands. It will always be represented by $H_nO_{n-1} + O$; H_nO_{n-1} being the basylous atom which is transferred to the negative pole, and O the salt-radical atom which is transferred to the positive pole. It appears to be by a polarization of this sort that, in bad conductors, mass compensates for conducting power; as in the return current of the electric telegraph through the earth, where the resistance is found to be even less than in the metallic wires; indeed, quite inappreciable.

It is found by Mr. J. Napier that the passage of a salt without decomposition, such as sulphate of copper, from the positive to the negative division of the decomposition cell, may take place independently of the water in which it is dissolved, and to a greater proportional amount (Mem. Chem. Soc. ii. 28). This unequal movement of the salt and water proves that the phenomenon is not simply a flowing of

[1] Professors Daniell and Miller, "On the Electrolysis of Secondary Compounds," in the Philosophical Transactions, 1844.

the liquid towards the negative pole; and it allows us to suppose that an aggregate polar molecule may be formed of many atoms of a salt, as well as of water. It is only in dilute saline solutions that the voltaic endosmose is perceptible.

VOLTAIC CIRCLES WITHOUT A POSITIVE METAL.

If we dip together into an acid fluid two platinum plates, one clean, and the other coated with a film of zinc or highly positive metal, we have the speedy solution of the positive metal by the usual polar decomposition, and hydrogen transferred to the opposite platinum plate. It appears that hydrogen, sulphur, phosphorus, and various other oxidable substances, will originate a polar decomposition in water or a saline fluid, when associated with platinum, in the same manner as the zinc is in the last experiment; and circles may thus be formed without a positive metal. The non-metallic but oxidable elements enumerated cannot be substituted in mass for zinc or the positive metal, because they are non-conductors; but in the thinnest films they are not so, if we may judge from experiments of this kind, and become quite equivalent to metals. Farther, with chlorine or any other strongly halogenous element dissolved in water, and placed in contact with one of the platinum plates, while the other is clean, we may have a polarization originating with the chlorine, and causing the transfer of the oxygen or salt-radical of the interposed water, or saline fluid, to the clean platinum. Nothing like this is witnessed in the voltaic combination of two metals; it is equivalent to an action in which the copper or negative metal originated the polarization by its affinity for the hydrogen or basylous constituent of the polar liquid.

1. With hydrogen gas dissolved in the acid fluid of one chamber of the divided cell, and air or oxygen in the other, polarization occurs on uniting the platinum plates, attended with the oxidation of the hydrogen and disappearance of both gases (Schönbein). Viewing this arrangement as a simple circle, consisting of a liquid and metallic segment (page 194), we have to consider particularly the composition of the terminal polar molecules at either end of the metallic segment — platinum with hydrogen must form the one at the positive pole, and platinum with oxygen the other at the negative pole: —

$$\text{(1)} \qquad \underset{-\ \ +}{\text{Pt H}} \ \ldots\ldots\ldots\ldots\ldots\ldots\ldots\ldots \underset{\text{acid}}{} \ldots \ \underset{-\ \ +}{\text{O Pt}}$$

These are equivalent to the external molecules of the two metals, zinc and copper, in the usual voltaic arrangement, which are composed in that case of two atoms of zinc on the one side, and two atoms of copper on the other (fig. 70, page 194): —

$$\text{(2)} \qquad \underset{-\ \ +}{\text{Zn Zn}} \ \ldots\ldots\ldots\ldots\ldots\ldots\ldots\ldots \underset{\text{acid}}{} \ldots \ \underset{-\ \ +}{\text{Cu Cu}}$$

The peculiar superiority of platinum, as the single metal, in arrangements of the present class, depends upon its strictly intermediate character between basyls and halogens, so that it lends itself to form a polar binary molecule equally with hydrogen or oxygen in (1), — with both basyl and salt-radical.

The intermediate liquid (the acid) must be a binary compound as usual. Here the positive hydrogen combines with the salt-radical of that binary compound, and sends its hydrogen or basyl to the second or opposite plate; while the oxygen at that plate decomposes the binary liquid also, sending back oxygen or salt-radical to the hydrogen of the first plate. There are, therefore, two concurring polarizations in every polar chain, tending to bring about simultaneously the same combinations and decompositions throughout the circle: hydrogen enters into combination on the one side, and oxygen on the other, in one and the same polar chain. The union of concurring primary zincous and chlorous polarizations, exhibited in such an arrangement, offers a new means of increasing polar intensity, entirely different from the multiplication of couples in the compound circle, of which the application will be fully observed afterwards in the nitric acid battery of Mr. Grove. The temporary combination of hydrogen with copper, the former as the basylous and the latter as the halogenous element of one polar molecule, which it is necessary to assume in explaining the circular polarity of the ordinary voltaic circle (page 194), is quite in accordance with the relation of hydrogen to platinum in the present circles.

2. A circle of still higher power is formed with chlorine gas, dissolved in the negative chamber, against hydrogen in the positive chamber of the divided cell. Here the terminal polar molecules of the metallic segment are:—

$$(3) \qquad \underset{-\ +}{\text{Pt H}} \ \ldots\ldots\ldots\ldots\ldots\ldots\ldots\ldots \ \underset{-\ +}{\text{Cl Pt}}$$

3. *Inflammation of mixed hydrogen and oxygen by platinum.*—There is every reason to believe that the remarkable action of clean platinum, both in the form of a plate and of platinum sponge, in disposing a mixture of oxygen and hydrogen in the gaseous state to unite, is the same in nature as its action upon these elements liquefied and in solution in water. In the former, as in the latter case, a polar chain must arrange itself in the platinum mass, of which one terminal molecule is platinide of hydrogen, and the other oxide of platinum (3). A less certain point is, whether the chain is completed by the interposition of a binary molecule of water already formed, between the polar H and O; or these atoms come immediately into contact, and close the circle, without the intervention of any compound polar molecule.

4. *Gas-battery.*—The gas-battery of Mr. Grove belongs to this class of voltaic arrangements. It is essentially an apparatus in which a supply of both negative and positive gas is kept over the liquid at each plate, to supply loss by absorption. A simple circle consists of a bottle (fig. 82), containing a dilute acid, with two tubes filled with oxygen and hydrogen respectively, and placed in two openings in the bottle. The platinum plates contained in these tubes are made rough by adhering reduced spongy platinum, which enables them also to retain the better on their surface a portion of the acid fluid into which they dip. The two plates are connected by a wire above the tubes, which is represented in the figure as carried round a magnetic needle, to obtain evidence of polarization in the wire. Here, as in (2), the gases only act when in contact with the platinum surface and taking a part in the terminal polar molecule, and also when covered by liquid, which is necessary to complete the polar chain between the terminal polar molecules on each side. The gases in the tubes are supplementary, and do not take a part in the polar chain. The modifications of this battery, where, instead of hydrogen gas, sulphur or phosphorus, vaporized in nitrogen gas, or a gaseous hydrocarbon, is placed at the positive pole, are of the same character, and only act by supplying a film of an oxidable body, such as sulphur, or phosphorus, to the surface of the platinum, capable of forming the positive element of a polar molecule with that metal. This, again, must be covered by the binary acid fluid, in order to communicate by a polar chain with the oxygen of the terminal molecule of platinum and oxygen in the negative chamber of the divided cell. (Grove, on the Gas Voltaic Battery: Philosophical Transactions, 1843 and 1845).

Fig. 82.

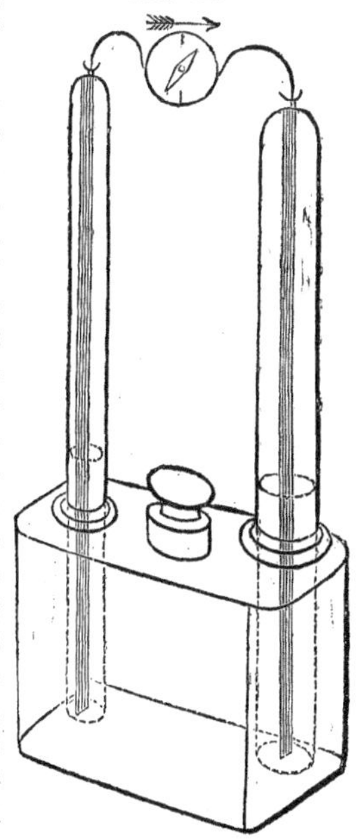

5. Closely resembling these circles is that in which one of the platinum plates is covered by a film of peroxide of lead or peroxide of manganese. The platinum plate may be so prepared by making it the negative terminal for a short time in a

solution of acetate of lead or of protosulphate of manganese. In an acid fluid, which is capable of dissolving the protoxide of lead or manganese, polarization occurs, the excess of oxygen of the attached peroxide forming with platinum a polar molecule, in which the oxygen is the chlorous element. This decomposes the saline molecule of the acid, or water, causing the transference of the salt-radical or oxygen to the clean platinum plate, where it may be evolved as gas. This most nearly resembles the case with chlorine—water at one platinum plate, which causes the evolution of oxygen at the other platinum plate; the only source of polarizing power in the circle being a chlorous affinity.

6. By much the most powerful voltaic arrangement of this class is that in which one chamber of the divided cell is charged with a solution of sulphide of potassium, and the other chamber with strong nitric acid.[1] Here we have two concurring sources of polarization in one polar chain, namely, the affinity of sulphur for oxygen, tending to transmit hydrogen in one direction, and the easy decomposition of nitric acid into $N\,O_4$ and O, supplying oxygen to the surface of the platinum, which sends a chlorous element in the opposite direction. The terminal polar molecules of the metallic segment of the circle are—

$$(4)\qquad \underset{-\;+}{\mathrm{Pt\ S}} \;\ldots\ldots\ldots\ldots\ldots\ldots\ldots\ldots\; \underset{-\;+}{\mathrm{O\ Pt}}$$

With a single pair of plates so charged, water may be decomposed. The action is equally powerful with chlorine substituted for the nitric acid. Such combinations of fluids may be greatly varied: all that is necessary is an oxidable substance at one plate, and an oxidizing substance at the other. In the first class are protosalts of iron, tin and manganese, sulphides, sulphites, hyposulphites; in the second, chlorine, nitric, chromic and manganic acids, and persalts of iron and tin. Taking protoxide of iron against peroxide, as an example of these cases, the terminal molecules of the metallic segment may be represented as—

$$(5)\qquad \underset{-\;+}{\mathrm{Pt\ Fe}} \;\ldots\ldots\ldots\ldots\ldots\ldots\ldots\ldots\; \underset{-\;+}{\mathrm{O\ Pt}}$$

It is true we have no evidence of the actual separation of the iron or of the oxygen upon the platinum surface; still there is reason to believe such a polarity to be established, assisted by secondary affinities; the oxygen of the protoxide of iron passing over to an adjoining double molecule of protoxide, and converting it into peroxide, to allow the metal to join in a polar molecule with the platinum. At the same time, the peroxide of iron at the negative plate may become protoxide, while its oxygen is engaged in forming a polar molecule with the platinum. But the intensity of polarization with the salts of iron against each other is feeble compared with that of chlorine or nitric acid against an alkaline sulphide. In all these cases the polar circle must be completed by a saline compound in the liquid or liquids, which may serve as the means of connecting the terminal molecules described of the platinum plates, and by metallic polar molecules through the wire connecting the platinum plates.

It was supposed by M. Becquerel that a circle of the present description may be formed in which the affinities are those of an acid for an alkali: the acid and alkaline solutions being separated by porous baked clay, which leaves them in free liquid contact, although their actual mixture proceeds with extreme slowness. Sulphuric acid and potassa, however, are generally admitted to be nearly or altogether incapable of producing this effect, while acids which part readily with oxygen, such as iodic, chloric, chromic, or nitric acid, with an alkali, produce a powerful effect. The polarization may be referred to the oxygen of the acids, in these last cases, at the negative terminal, and is a chlorous affinity. It may possibly be often assisted by

[1] Mr. A. R. Arnott, on "Some New Cases of Voltaic Action;" Memoirs of the Chem. Soc. i. 142.

minute quantities of ammonia, organic or other oxidizable matter, at the positive terminal in the alkaline solution. (Becquerel, Elements d'Electro-Chimie, 1843).

Theoretical considerations.—The facility with which circular decompositions take place, and the necessity of their occurrence in the action of binary compounds, which was explained under the atomic exhibition of a double decomposition at page 189, are undoubtedly the key to the great stimulus to chemical activity, which the voltaic arrangement affords. Reverting to the original illustration of the action of hydrochloric acid upon zinc, it may be observed that zinc has a strong attraction for chlorine, and would combine at once with that element if the latter were free, without foreign aid of any kind. But with the chlorine of hydrochloric acid the case is different. That chlorine is already combined and strongly retained by its own hydrogen: to enable the chlorine to enter into a new combination we must relieve it from this attraction, by engaging otherwise the affinity of the hydrogen. The contrivance of the voltaic circle is to present another halogen to the hydrogen, and thus divert its affinity from the chlorine—the latter being thereby left free to combine with the zinc. This requires a train of similar decompositions passing round a circle to the zinc, illustrated in diagram 70 of page 194; and which ends in relieving the external combining atom of zinc from the attraction of even the contiguous atom of the same kind; thus dissolving the attraction of aggregation in the metal, and resigning the external atom of zinc entirely to the attraction of the equally relieved chlorine. It is entirely, therefore, because the agent applied to the zinc is a binary compound, and not a free element, that this circular mode of action is necessary.

It is to be remarked in explanation of the facility with which the mutual combinations and decompositions in a circular chain occur, that they do not necessarily consume any power or occasion waste of force. They may be compared to the movement of a nicely balanced beam on its pivot, or the oscillation of a pendulum, in which the motion is equal in two opposite directions, and requires only the minimum of effort to produce it.

Farther, it is not to be supposed that zinc dissolves by a circular action of affinity, only when a negative metal is attached to it, and a voltaic circle purposely constructed. For this positive metal never appears to dissolve in hydrochloric acid in any other manner; the formation of little polar circles in the fluid, starting from one point of the metallic mass and returning upon another, being always required for its solution (page 195). In the solution of zinc, therefore, by a binary saline body, such as hydrochloric acid, the circular or voltaic polarization is the necessary, as well as the most effective mode of action of chemical affinity.

The molecular condition of conductors, such as carbon and the metals, in a voltaic circle, appears to be that of polymeric combination. Their atoms must be feebly basylous and chlorous to each other; the distinction possibly depending upon inequality in their proportions of combined heat, and maintain the relation of combination. Again, many of these binary molecules are associated together like the many similar atoms of carbon, or of hydrogen, which we find associated in the polymeric hydrocarbons. The whole must be held together by their chemical affinities, and the aggregation of the mass be the final resultant of the same attractions. The determination of the polar condition in two metals, by the mere application of heat or cold to their junction, requires the assumption of the sali-molecular structure of metals; and the other proposition, that affinity passes into aggregation, is equally necessary to account for the polar (or electrical) effects which are produced by friction or abrasion, as they appear to extend to the division of chemical molecules.

The cumulative nature of chemical combination is well illustrated in such compounds as the acid hydrates—in dilute sulphuric acid, for instance, where we find an atom of acid uniting with more and more atoms of water, with a decreasing affinity, but without any assignable limit to their number. It is worthy of remark that the acids are bodies with chlorous or negative atoms, and their peculiar affinity in excess. The polar formula for sulphuric acid (page 168) is $\frac{O_3}{S}$; or three negative to one positive atom. By the apposition of a single binary molecule of water, sulphate of water is produced, $\frac{O_3\ O}{S\ H}$, in which the excessive proportion of chlorous atoms and affinity in the compound is in some degree diminished, the formula of the latter presenting four negative to two positive atoms. The apposition of more and more molecules of water is determined by this excess of chlorous affinity, which it tends to neutralize; the constant difference, or excess of two chlorous over the number of basylous atoms, becoming proportionally less with large numbers of added molecules of water. All the magnesian bases appear to assume water to assist in neutralizing their acid in the same manner, and retain one equivalent of this water in general very strongly. In the formation of a polar chain through a solution of a sulphate of this class, we have had reason to suppose that the sulphuric acid applies itself, for the time, to the water rather than the metallic oxide as its base (page 206). The phenomena of voltaic endosmose were also found to favour the idea of the polarization of highly aggregated molecules, in which the binary molecule was represented by a single atom of chlorine or salt-radical, against a single atom of hydrogen or metal associated with a large number of atoms of water, which constituted together the basylous atom. The application of polar formulæ to the explanation of voltaic decompositions of all kinds would, I believe, more correctly express the molecular changes that occur, than the usual assumption of the binary division of the compound body, in an absolute manner, into a basylous atom and a fictitious group forming a halogen body.

GENERAL SUMMARY.

1. In a closed voltaic circle, a certain number of lines or chains of polarized molecules is established, each chain being continuous round the circle. Hence the polar condition of the circle must be every where the same. The same number of particles of exciting fluid are simultaneously polar upon the surface of every zinc plate in the active cells, and also upon the surface of the zincoid in the cell of decomposition, and the consequent chemical change, or decomposition occurring, is of the same amount in all the cells in the same time. This equality in condition and results is essential to a circular polarization, such as exists in the voltaic circle.

The number of polar chains that can be established at the same time in a particular voltaic arrangement, is obviously affected by several circumstances: —

(1) By the size of the zinc plate: the number of particles of zinc that may be simultaneously acted upon by the exciting fluid being directly proportional to the extent of metallic surface exposed.

(2) By the nature and accidental state of the exciting liquid, some electrolytes being more easily acted on by the positive metal than others; while the state of dilution, temperature, and other circumstances, may affect the facility of decomposition of any particular electrolyte.

(3) The adhesion of the gas bubbles of hydrogen to the copper plate, at which they are evolved, interferes much with the action of a battery; partly by reducing the surface of copper in contact with acid, and partly by acting as a zincous element, and originating an opposite polarization in the battery (page 209). By taking up the hydrogen, by means of a solution of sulphate of copper in contact with the copper plate, Mr. Daniell increased the amount of circulating force six times.

(4) The chemical action in a cell is also diminished by increasing the distance from each other in the exciting fluid of the positive and negative metals.

(5) The lines of chemico-polar molecules in the exciting fluid should be repulsive of each other, like lines of magneto-polar elements, as illustrated in the mutual repulsion and divergence of the threads of steel filings which attach themselves to

the pole of a magnet (fig. 65, page 189). That the lines of induction do diverge greatly in the acid, starting from the zinc as a centre, is placed beyond doubt by many experiments of Mr. Daniell. A small ball of zinc suspended in a hollow copper globe filled with acid, is the arrangement in which this divergence is least restrained, and was found to be the most effective form of the voltaic circle. When the copper, too, is a flat plate, and wholly immersed in the acid, the back is found to act as a negative surface, as well as the face directly exposed to the zinc, showing that the lines of induction in the acid expand, and open out from each other, some bending round the edge of the copper plate and terminating their action, after a second flexure, on its opposite side. To collect these diverging lines, the surface of the copper may be increased with advantage to at least four times that of the zinc.

(6) The polar chains of molecules, in the connecting wires and other metallic portions of the circle must be equally repulsive of each other. Hence the small size of the negative plates in the active cells, and of the platinum plates in the cell of decomposition, and the thinness of the connecting wires, are among the circumstances which diminish the number of polar chains that can be established, and impair the general efficiency of a battery.

2. The effect of multiplying the active cells in a battery is not to increase the number of polar chains, or *quantity* of decomposition, but to increase the *intensity* of the induction in each chain; although this increase in intensity generally augments the quantity also, in an indirect manner, by overcoming more or less completely such obstacles to induction as have been enumerated.

3. The intensity of the induction, also, is much greater with some electrolytes than others. Thus a single pair of zinc and platinum plates excited by dilute sulphuric acid, decomposes iodide of potassium, proto-chloride of tin, and fused chloride of silver, but not fused nitre, chloride or iodide of lead, or solution of sulphate of soda. With the addition, however, of a little nitric acid to the sulphuric, the same single circle decomposes all these bodies, and even water itself. Here we have a primary chlorous induction from the oxygen of the nitrous acid, in addition to the basylous induction of the zinc (page 208). The former action also is attended by the suppression of the hydrogen, so that the evolution of that gas upon the negative plate is avoided.

4. The division of the connecting wire, and the separation of its extremities to the most minute distance from each other, is sufficient to stop all induction and the propagation of the polar condition in an arrangement with the usual good conducting fluids. In a powerful voltaic battery consisting of seventy large Daniell cells, no induction was observed to pass when the terminal wires were separated not more than the one-thousandth of an inch, even with the flame of a spirit-lamp or rarefied air between them. Absolute contact of the wires was necessary to establish the circulation. But after contact was made, and the wires were heated to whiteness, they might be separated to a small distance without the induction being interrupted: the space between them was then filled with an arch of dazzling light, containing detached particles of the wire in a state of intense ignition, which were found to proceed from the zincoid to the chloroid, — the former losing matter, and the other acquiring it. So highly fixed a substance as platinum is carried from the one terminal to the other in this manner; but the transference of matter is most remarkable between charcoal points, which may be separated to the greatest distance, and afford the largest and most brilliant arch of flame. A similar, although it may be an excessively minute detachment of matter, is found to accompany the electric spark in all circumstances. Hence, the electric spark always contains matter. In a powerful water battery, however, of a thousand couples, where the conducting power of the liquid is low, good sparks are obtained on approaching the terminals (Gassiot).

5. When terminal wires of a voltaic circle are grasped in the hands, the circuit may be completed by the fluids of the body, provided the battery contains a considerable number of cells, and the induction is of high intensity: the nervous system is then affected, the sensation of the electric shock being experienced.

6. The conducting wire becomes heated precisely in proportion to the number of polar chains established in it, and consequently in proportion to the size of the zinc plate; and this to the same degree from the induction of a single cell as from any number of similar cells. Wires of different metals are unequally heated, according to the resistance which they offer to induction. The following numbers express the heat evolved by the same circulation in different metals, as observed by Mr. Snow Harris:—

	Heat evolved.	Resistance.
Silver	6	1
Copper	6	1
Gold	9	1½
Zinc	18	3
Platinum	30	5
Iron	30	5
Tin	36	6
Lead	72	12
Brass	10	3

The conducting powers of the metals are inversely as these numbers; silver being a better conductor than platinum in the proportion of 5 to 1. The conducting power of all of them is found to be diminished by heat.

7. As a portion of the voltaic circle, the conducting wire acquires extraordinary powers of another kind, which can only be very shortly referred to here, belonging as they properly do to physics.

(1) Another wire placed near and parallel to the conducting wire, has the polar condition of its molecules disturbed, and an induction propagated through it in an opposite direction to that in the conducting wire.

(2) If the conducting wire be twisted in the manner of a corkscrew so as to form a hollow spiral or helix, it will be found in that form to represent a *magnet*, one end of the helix being a north, and the other a south pole; and, if moveable, will arrange itself in the magnetic meridian, under the influence of the earth's magnetism. Its poles are attracted by the unlike poles of an ordinary magnet, and it imparts magnetism to soft iron or steel by induction. Two such helices attract and repel each other by their different poles, like two magnets. Indeed, an ordinary magnet may be viewed as a body having a helical chain of its molecules in a state of permanent chemico-polarity.

(3) If a bar of soft iron bent into the form of a horse-shoe, with a copper wire twisted spirally round it, be applied like a lifter to the poles of a permanent magnet, at the instant of the soft iron becoming a magnet by induction, the molecules of the spiral become chemico-polar; and when contact is broken with the permanent magnet, and the soft iron ceases to be a magnet, the wire exhibits a polarity the reverse of the former. By a proper arrangement, electric sparks and shocks may be obtained from the wire, while the soft iron included within it is being made and unmade a magnet. The magneto-electric machine is a contrivance for this purpose, and is now coming to supersede the old electric machine, as a source of what is termed electricity of tension. Magnetic and electric effects are thus reciprocally produced from each other.

(4) When the pole of a magnetic needle is placed near the conducting wire, the former neither approaches nor recedes from the latter, but exhibits a disposition *to revolve round it.* The extraordinary and beautiful phenomena of electrical rotation are exhibited in an endless variety of contrivances and experiments. As the magnetic needle is generally supported upon a pivot, it is free to move only in a horizontal plane, and consequently when the conducting wire is held over or under it (the needle being supposed in the magnetic meridian), the poles in beginning to describe circles in opposite directions round the wire, proceed to move to the right and left of it, and thus deviate from the true meridian. The amount of deviation in degrees

is proportional to the quantity of circulating induction, and may be taken to represent it, as is done in a useful instrument, the galvanometer, to be afterwards described. It was in the form of these deflections, that the phenomena exhibited by a magnet, under the influence of a conducting wire, first presented themselves to Oersted in 1819.

8. Thermo-electrical phenomena are produced from the effect of unequal temperature upon metals in contact. If heat be applied to the point *c* (fig. 83), at which two bars of bismuth and antimony *b* and *a* are soldered together, on connecting the free extremities by a wire, the whole is found to form a weak voltaic circle, with the induction from *b* through the wire to *a*. Hence in this thermo-polar arrangement the bismuth is the negative metal, and may be compared to the copper in the voltaic cell. If cold instead of heat be applied to *c*, a current also is established, but in an opposite direction to the former. Similar circuits may be formed of other metals, which may be arranged in the following order, the most powerful combination being formed of those metals which are most distant from each other in the following enumeration: bismuth, platinum, lead, tin, copper or silver, zinc, iron, antimony. When heated at their points of contact, the current proceeds through the wire from those which stand first to the last. According to Nobili, similar circuits may be formed with substances of which the conducting power is lower than that of the metals.

Fig. 83.

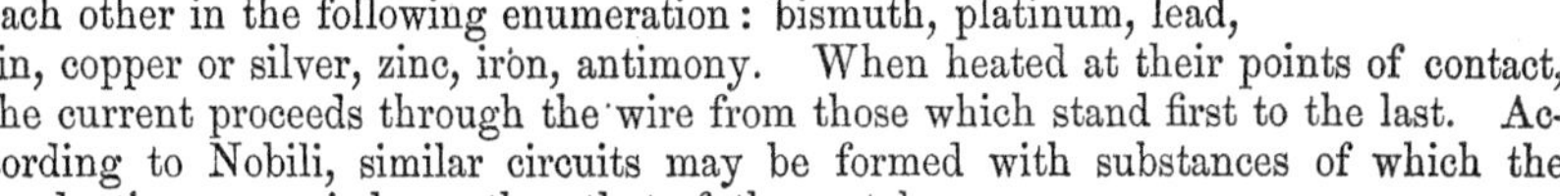

Several pairs of bismuth and antimony bars may be associated as in fig. 84, and the extreme bars being connected by a wire, form an arrangement resembling a compound voltaic circle. Upon heating the upper junctions, and keeping the lower ones cool, or on heating the lower ones and keeping the others cool, an induction is established in the wire, more intense than in the single pair of metals, but still very weak. The conducting wire strongly affects a needle, causing a deflection proportional to the inequality of temperature between the ends of the bars. Melloni's thermo-multiplier is a delicate instrument of this kind, which is even more sensitive to changes of temperature than the air-thermometer, and has afforded great assistance in exploring the phenomena of radiant heat (page 55).

Fig. 84.

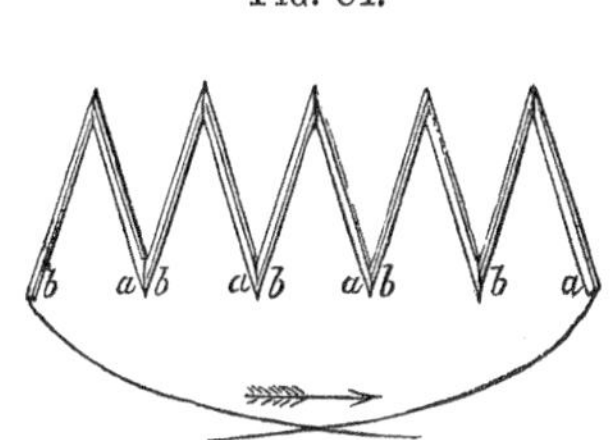

In such a compound bar, also, unequal temperature may be produced, by making it the connecting wire of a single and weak voltaic circle; whereupon the metals become cold at their junction, if the induction is from the bismuth to the antimony, and hot at the same point if the induction is in the opposite direction. These are the converse of the preceding phenomena, in which electrical effects were produced by inequality of temperature.

9. The friction of different bodies is another source of electrical phenomena. One, at least, of the bodies rubbed together must not be a conductor, and in general two non-conductors are used. When a silk handkerchief or a piece of resin is rubbed upon glass, both are found, after separation, in a polar condition, and continue in it. The rubbing surface of the glass becomes and remains zincous, and that of the resin or silk is chlorous; and a molecular polarization is at the same time established through the whole mass of both the glass and resin, reaching to their opposite surfaces, which exhibit the other polarity. The powers thus appearing on the two rubbing surfaces, being manifestly different, were distinguished by the names of the bodies on which they are developed; that upon the glass as *vitreous* electricity (basylous affinity), and that upon the resin as *resinous* electricity (halogenous affinity).

In comparing the chemico-polarity excited by friction with that of the voltaic circle, we observe that the former is of high intensity but small in quantity, or affecting only a small number of trains of molecules. Also that the polar condition is more or less permanent, depending upon the insulation, and attended with a disturbance of the polar condition of surrounding bodies to a considerable distance, giving rise to electrical attractions and repulsions, or statical phenomena. If both the excited vitreous and resinous surfaces have a conducting metal, such as a sheet of tin-foil, applied to them, and each sheet have a wire proceeding from it, the wires and tin-foil are polarized similarly to the glass and the resin which they cover; and a saline body placed between the extremities of the wires, which are respectively a zincoid and chloroid, is polarized also, and decomposed. But the amount of decomposition, which is a true measure of the *quantity* of polar chains, is extremely minute compared with the amount of polarization in the voltaic circle. Thus, Mr. Faraday has calculated that the decomposition of one grain of water by zinc, in the active cell of the voltaic circle, produces as great an amount of polarization and decomposition in the cell of decomposition, as 950,000 charges of a large Leyden battery, of several square feet of coated surface; an enormous quantity of power, equal to a most destructive thunder-storm. The polarization from friction is therefore singularly intense, although remarkably deficient in quantity, or in the number of chains of polar molecules.

The kinds of matter susceptible of this intense polarization are so many and so various, such as glass, minerals, wood, resins, sulphur, oils, air, &c., as to make it difficult to suppose that the polar molecule is of the same chemical constitution in all of them, as it is in the electrolytes of the voltaic circle. Indeed, it must be admitted that all matter whatever may be forced into a polar condition by a most intense induction.

Electrical induction at a distance, Mr. Faraday has shown to be always an action of contiguous particles, chains of particles of air, or some other "dielectric," extending between the excited body which is inducing, and the induced body. His investigations of this subject led to the remarkable discovery that the intensity of electric induction at a constant distance from the inducing body is not always the same, but varies in different media, the induction through a certain thickness of shell-lac, for instance, being twice as great as through the same thickness of air. Numbers may be attached to different bodies which express their relative inductive capacities:—

Specific inductive capacity of		air	1
"	"	glass	1.76
"	"	shell-lac	2
"	"	sulphur	2.24

The inductive capacity of all gases is the same as that of air, and this property, it is remarkable, does not alter in these bodies with variations in their density.

10. Mr. Faraday has lately made the important discovery that a ray of polarized light passing through a transparent liquid or solid, is deflected, and takes a spiral direction, or has a motion of rotation communicated to it by the approximation of the pole of a powerful electro or natural magnet; the pole of the latter being so placed that the ray is in the direction of the lines of attraction of the magnet. The amount of the deflection of the ray varies in different transparent bodies, and is approximatively expressed for oil of turpentine by 11.8, heavy borate of lead glass 6.0, flint-glass 2.8, rock-salt 2.2, water 1, alcohol and ether less than water (Phil. Trans. 1846).

11. Operating with electro-magnets of the highest power, Mr. Faraday has obtained results of a fundamental nature respecting the magnetic capacity of different kinds of matter. The magnetic field being represented as in fig. 85, where N and S are the two poles, the dotted line N S connecting these poles, or line of magnetic force, is conveniently termed the axial direction, and the line *e r*, perpendicular to the former, the equatorial direction. When a bar of bismuth, two inches long, 0.33 inch wide, and 0.2 thick, was delicately suspended by a thread of

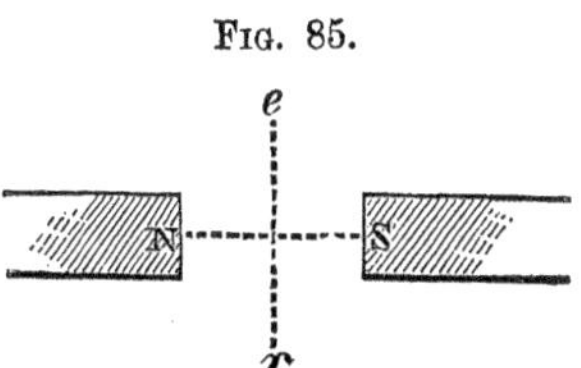

Fig. 85.

untwisted silk, and placed between the magnets, it arranged itself in the direction of *e r*, or equatorially. All kinds of solid, liquid, and even gaseous matter have a certain amount of tendency to place themselves, like the bismuth bar, across the axial or proper magnetic direction. This equatorial tendency is, however, overcome and negatived by the smallest proper magnetic property which bodies may possess, as this is the axial polarity, and causes the substance to set with its greatest length in the direction N S. Besides iron, nickel and cobalt, the usual magnetic metals, platinum, palladium and titanium, proved to be axial bodies. So are all the salts containing iron, nickel, or cobalt, as base. Even bottle glass is comparatively very magnetic, from the iron it contains; so is crown (window) glass, but not flint glass. The solutions of these salts are also magnetic. Crystals of the yellow ferrocyanide and red ferricyanide of potassium are not magnetic, but set equatorially. The iron, it will be remembered, belongs to the acid in these last salts. The salts of the oxides of the following metals proved magnetic, and Mr. Faraday is disposed to infer that the metals themselves are so — manganese, cerium, chromium. Paper and many other organic and mineral substances often contain enough of iron to make them fall into the same class.

The bodies which place themselves equatorially are named *diamagnetic*. The endless list of them is also headed by metals, which appear to possess this power in different degrees of intensity according to the following order: —

DIAMAGNETIC METALS.

Bismuth	Cadmium
Antimony	Mercury
Zinc	Silver
Tin	Copper

The other non-magnetic metals are diamagnetic in a less degree. This property is not sensibly impaired by heating the metals up to their fusing points. The property may be experimentally illustrated by pointed pieces of rock crystal, glass, phosphorus, sealing-wax, caoutchouc, wood, beef, bread, &c. (Phil. Trans. 1846).

Hot air and flame are more diamagnetic than cold or cooler air, so that a stream of the former spreads itself equatorially in ascending between magnetic poles. Of many gases and vapours tried by Mr. Faraday, oxygen was found to be the least diamagnetic; and this element appears to lower the equatorial tendency of the gases into which it enters as a constituent. Nitrogen is more highly diamagnetic than carbonic acid or hydrogen. In an atmosphere of carbonic acid gas (instead of air) between the magnetic poles, streams of hydrogen gas, coal gas, olefiant gas, muriatic acid, and ammonia, passed equatorially, and are therefore more diamagnetic. A stream of oxygen, which is so little diamagnetic, had, consequently, "the appearance of being strongly magnetic in coal gas, passing with great impetuosity to the magnetic axis, and clinging about it; and if much muriate of ammonia fume were purposely formed at the time, it was carried by the oxygen to the magnetic field with such force as to hide the ends of the magnetic poles. If, then, the magnetic action were suspended for a moment, this cloud descended by its gravity; but being quite below the poles, if the magnet were again rendered active, the oxygen cloud immediately started up and took its former place. The attraction of iron filings to a magnetic pole is not more striking than the appearance presented by the oxygen under these circumstances" (Faraday, Phil. Mag. xxxi. 415).

VOLTAIC INSTRUMENTS.

Daniell's constant battery.—A cell of this battery consists of a cylinder of copper $3\frac{1}{2}$ inches in diameter, which experience has proved to the inventor to afford the most advantageous distance between the metallic surfaces, but which may vary in height from 6 to 20 inches, according to the power which it is wished to obtain. A membranous bag, formed of the gullet of an ox, is hung in the centre by a collar and circular copper plate, resting upon a rim within and near the top of the cylinder; and in this is suspended by a wooden cross-bar, a cylindrical rod of amalgamated zinc half an inch in diameter. Or a tube of porous earthenware, shut at the bottom, is substituted for the membrane with great convenience. The outer cell is charged with a mixture of 8 measures of water and 1 of oil of vitriol, which has been saturated with sulphate of copper, and portions of the solid salt are placed upon the circular copper plate, which is perforated like a colander, for the purpose of keeping the solution always in a state of saturation. The internal tube is filled with the same acid mixture without the salt of copper. A section of the upper part of one of these cells is here represented: *a b c d* (fig. 86) is the external copper cylinder; *e f g h*, the internal cylinder of earthenware, and *l m* the rod of amalgamated zinc. Upon a ledge *c d*, within an inch or two of the top of the cylinder, rests the cylindrical colander *i k*, which contains the copper salt, and both the sides and bottom of which are perforated with holes. A number of such cells may be connected into a compound circuit, with wires soldered to the copper cylinders, and fastened to the zinc by clamps and screws as shown below, in fig. 87 (Daniell's Int. to Ch. Phil.) Instead of the zinc cylinder a thick plate of laminated zinc is now generally used, which is more regularly amalgamated than the cast cylinder.

Fig. 86.

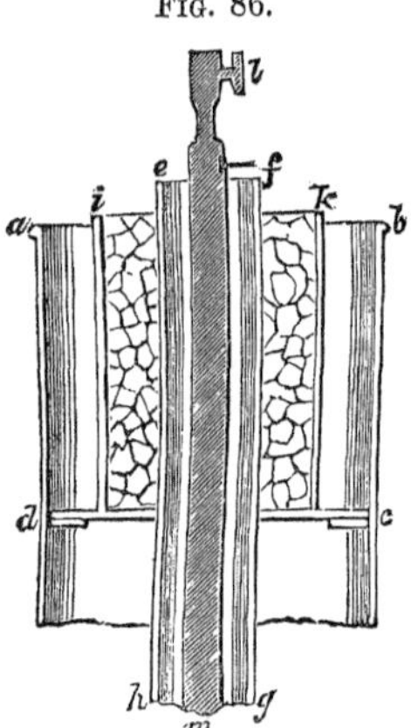

Fig. 87.

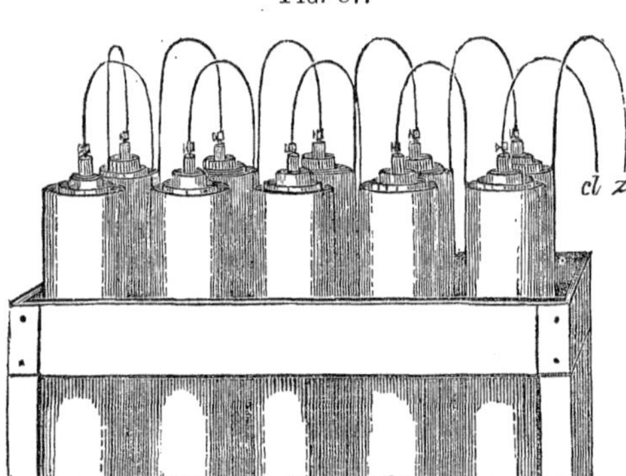

In this instrument the sulphate of zinc, formed by the solution of the zinc rod, is retained in the stoneware cylinder, and prevented from diffusing to the copper surface; while the hydrogen, instead of being evolved as gas on the surface of the latter metal, decomposes the oxide of copper of the salt there, and occasions a deposition of metallic copper on the copper plate. Such a circle will not vary in its action for hours together, which makes it invaluable in the investigation of voltaic laws. It owes its superiority principally to three circumstances:—to the amalgamation of the zinc, which prevents the waste of that metal by solution when the circuit is not completed; to the non-occurrence of the precipitation of zinc upon the copper surface; and to the complete absorption of the hydrogen at the copper surface, the adhesion of globules of gas to the metallic plates greatly diminishing, and introducing much irregularity into the action of a circle.

Grove's nitric acid battery.—In this battery the positive metal is amalgamated zinc, and the negative metal platinum, while the intermediate liquid is of two kinds, dilute sulphuric acid of sp. gr. 1.125 in contact with the zinc, and strong nitric acid

Fig. 88.

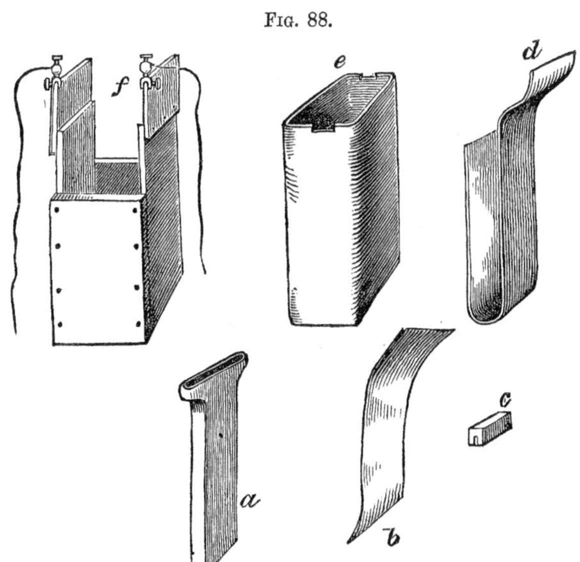

in contact with the platinum. In fig. 88, *a* represents a flat cell of porous earthenware, to contain the nitric acid and platinum plate; *b*, the platinum plate; *d*, the zinc plate, which is doubled up to include the porous cell; *e*, a cell of glazed earthenware to contain the sulphuric acid and zinc plate; *f*, a wooden frame to support the last cell, terminated above by copper plates provided with clamps, by which the terminal wires are attached. Two wooden wedges, such as *c*, are required to fix the upper end of the zinc plate on the one side, and the platinum plate on the other, as in fig. 89. Convenient dimensions for the principal parts are, the the external cell *e*, $4\frac{1}{2}$ inches by $2\frac{3}{4}$ and $1\frac{1}{4}$; porous cell *a*, $4\frac{1}{2}$ by $2\frac{1}{2}$ and $\frac{5}{8}$ inch; platinum plate 5 inches by $2\frac{1}{2}$, and weighing about 10 grains in the square inch.

Fig. 89.

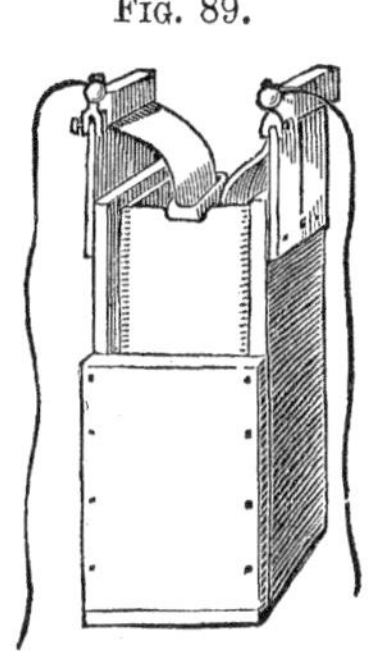

Fig. 90.

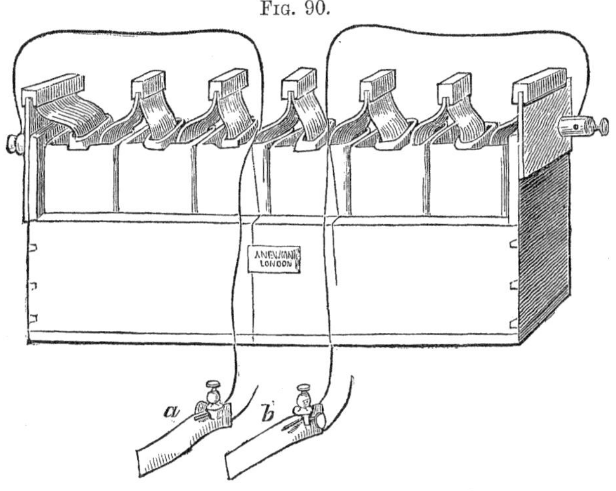

In fig. 90, six of these cells are placed together in a wooden frame, with the upper part of each end of the frame of stout sheet copper, to which the plates and wires can be clamped. The wires from the platinum and zinc ends of the battery, have platinum plates, *a* and *b*, attached to them as terminals. A battery of this size will evolve 8 or 10 cubic inches of mixed oxygen and hydrogen gases in the voltameter per minute. It is equal to several times as many cells of the preceding battery. The polarizing power is very intense, and little more decomposing power is gained by increasing the number of cells beyond five or six.

The carbo-zinc battery of Bunsen, which is much used on the continent, is a modification of the last construction, in which charcoal in contact with the nitric acid is substituted for platinum. The carbon is in the form of a hollow cylinder, and is made by coking pounded coal in a proper iron mould. By soaking the coke in sugar, and calcining a second time, great compactness is given to the cylinder. The latter is so large as to include the porous cell containing the zinc and acid, and is itself placed in a stout glass cylinder, of which the neck is contracted so as to support the coke cylinder (fig. 91). The zinc cylinder *c* is connected by a slip *b* and ring *a* of the same metal with the coke cylinder, of which the upper end is made a little conical to hold the ring. This battery has the advantage of enlarged negative surface, and provides ample space for the nitric acid.

Fig. 91.

For other useful forms of the battery, such as that introduced by Mr. Smee, in which a thin sheet of silver covered by a deposit of platinum (platinized silver) is the negative metal, I must refer to works upon Electricity.

Bird's battery and decomposing cell.—To M. Becquerel we are particularly indebted for the investigation of the decomposing powers of feeble currents, sustained for a long time, the results of which are of great interest, both from the nature of the substances that can be thus decomposed, and from the form in which the elements of the body decomposed are presented, the slow formation of these bodies permitting their deposition in regular crystals (Traité Experimental de l'Electricité et du Magnétisme, par M. Becquerel). Dr. Golding Bird has also added to the number of bodies decomposed by such means, and contrived a simple form of the battery, which, with Becquerel's decomposing cell, renders such decompositions certain and easy (Phil. Trans. 1837, p. 37). The decomposing cell consists of a glass cylinder *a* (fig. 92) within another glass cylinder *b*. The inner cylinder *a* is 4 inches long, and 1½ inch in diameter, and is closed at the lower end by a plug of plaster of Paris 0.7 inch in thickness: this cylinder is fixed by means of wedges of cork within the other, which is a plain jar, about 8 inches deep by 2 inches in diameter. A piece of sheet copper *c*, 4 inches long and 3 inches wide, having a copper conducting wire soldered to it, is loosely coiled up and placed in the inner cylinder with the plaster bottom: a piece of sheet zinc *z*, of equal size, is also loosely coiled, and placed in the outer cylinder; this zinc likewise being furnished with a conducting wire. The outer cylinder is then nearly filled with a weak solution of common salt, and the inner with a saturated solution of sulphate of copper. The two fluids are prevented from mixing by the plaster diaphragm, and care being taken that they are at the same level in both the cylinders, the circle will afford, on joining the wires, a continuous current for weeks, the chloride of

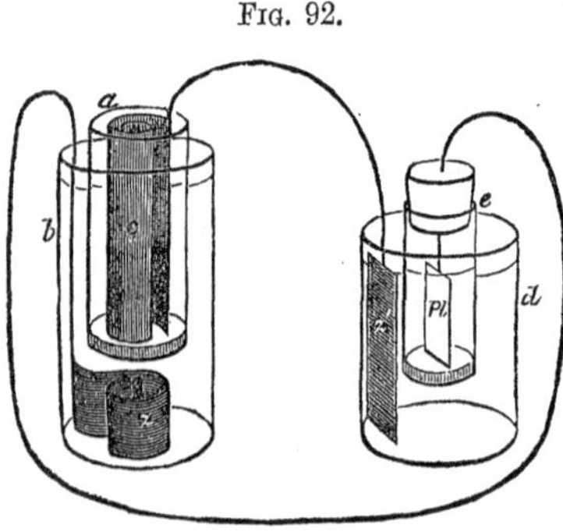

Fig. 92.

sodium and the sulphate of copper being very slowly decomposed. After it has been in action for some weeks, chloride of zinc is found in the outer cylinder: and beautiful crystals of metallic copper, frequently mixed with the ruby suboxide (closely resembling the native copper ruby ore in appearance), with large crystals of sulphate of soda, are found adhering to the copper plate in the smaller cylinder, especially on that part where it touches the plaster diaphragm.

The decomposing cell is the counterpart of the battery itself, consisting, like it, of two glass cylinders, one within the other, the smaller one *e* having a bottom of plaster of Paris fixed into it: this smaller tube may be about ½ inch wide and 3 inches in length, and is intended to hold the metallic or other solution to be decomposed, the external tube *d*, in which the other is immersed, being filled with a weak solution of common salt. In the latter solution a slip of amalgamated zinc-plate *z'*, soldered to the wire coming from the copper plate *c* of the battery, is immersed; and a slip of platinum foil *pl*, connected with the wire from the zinc plate *z* of the battery, is immersed in the liquor of the smaller tube, being held in its place by a cork, through which its wire passes. The whole arrangement is now obviously a pair of active cells, of which *c z'* is one metallic element, and *z pl* the other; and the fluid between *z* and *c* divided by the porous plaster diaphragm, one fluid element, and the fluid between *z* and *pl*, divided by a porous plaster diaphragm, another fluid element; although it will be convenient to speak of the last as the cell of decomposition. With a solution of chlorides or nitrates of iron, copper, tin, zinc, bismuth, antimony, lead or silver, in the smaller tube, Dr. Bird finds the metals to be reduced upon the surface of the platinum, generally but not invariably in possession of a perfect metallic lustre, always more or less crystalline, and often very beautifully so. The crystals of copper rival in hardness and malleability the finest specimens of native copper, and those of silver, which are needles, are white and very brilliant. The solution of fluoride of silicon in alcohol being introduced into the small tube by Dr. Bird, a deposition of silicon upon the platinum was found to take place in 24 hours, which was nearly black and granular, and is described as exhibiting a tendency to a crystalline form. From an aqueous solution of the same fluoride, a deposition of gelatinous silica was observed to take place around the reduced silicon, mixed with which, or precipitated in a zone on the sides of the tube, especially if of small diameter, frequently appear minute crystalline grains of silica or quartz, of sufficient hardness to scratch glass, and appearing translucent under the microscope. With a modification of the decomposing cell described, Dr. Bird succeeded in decomposing a solution of chloride of potassium, and obtained an amalgam of potassium. The inner tube *e* was replaced by a small glass funnel, the lower opening of which was stopped with stucco, and which thus closed retained a weak solution of the alkaline chloride poured into it. Every thing external to this funnel remaining as usual, mercury, contained in a short glass tube, like a thimble, was placed in the funnel, and covered by the liquid, and instead of the platinum plate, a platinum wire, coiled into a spiral at the extremity, was plunged into the mercury, the other end of this wire being connected with the zinc plate *z* of the battery. The circuit having been thus completed, the mercury had swollen in eight or ten hours to double its former bulk, and when afterwards thrown into distilled water, evolved hydrogen, and produced an alkaline solution. A solution of hydrochlorate of ammonia being substituted for that of chloride of potassium, in this experiment, the metal swells to five or six times its bulk in a few hours, and the semi-fluid amalgam of ammonium is formed. These feeble currents thus effect decompositions in the lapse of time, which batteries of the ordinary form, and considerable magnitude, may effect very imperfectly, or fail entirely in producing.

Volta-meter. — The decomposing power of a battery is represented by the quantity of oxygen and hydrogen gases evolved in a cell of decomposition containing dilute sulphuric acid. The volta-meter (figure 93) is simply a cell so charged, and of a proper form to allow of the gases evolved being collected and measured.

FIG. 93.

Galvanometer.—The sensibility of the magnetic needle to the influence of the conducting wire of a voltaic circle brought near it, has been applied to the construction of an instrument which will indicate the feeblest polarization or slightest current in the connecting wire. It consists of a pair of magnetic needles (fig. 94), fixed on one axis with their attracting poles opposite each other, so as to leave them little or no directive power, and render them astatic, which is delicately suspended by a single fibre of unspun silk. The lower needle is enclosed within a circle formed by a hank of covered wire B, of which *p* and *n* are the extremities. When the terminal wires of a battery are connected with the wires, the hank of wire of the galvanometer becomes part of the connecting wire, and the needle is deflected. The inductions proceeding in one direction above the needle and returning in the opposite direction below the needle, conspire to produce the same deflection; and the upper needle having its poles reversed, is deflected in the same direction, by the wire below it, as the lower needle is by the wire above that needle. Every turn of the wire also repeats the influence upon the needle, so that the deflection is increased in proportion to the number of turns or coils in the hank of wire.

FIG. 94.

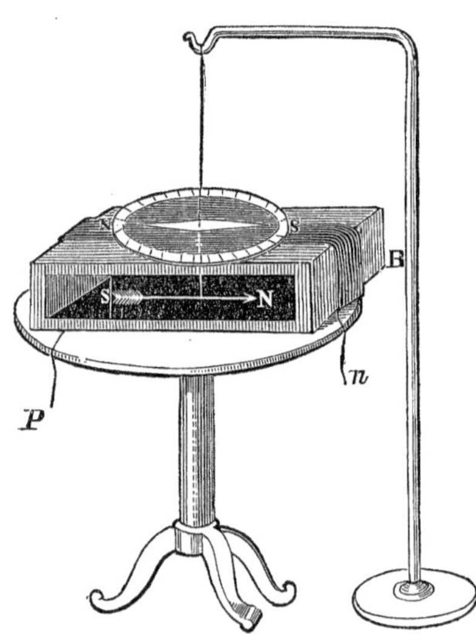

CHAPTER V.

NON-METALLIC ELEMENTS.

Section I.—Oxygen.

Equivalent 8 (*hydrogen* = 1, *or* 100 *as the basis of the Oxygen Scale; density* 1105·6 (*air* = 1000); *combining measure* □ (*one volume.*)

The following thirteen of the sixty-two elementary bodies known,[1] are included in the class of non-metallic elements:—oxygen, hydrogen, nitrogen, carbon, boron, silicon or silicium, sulphur, selenium, phosphorus, chlorine, bromine, iodine, and fluorine. Of these, oxygen, from certain relations which it bears to all the others, and from its general importance, demands the earliest consideration.

The name oxygen is compounded of ὀξὺς, acid, and γεννάω, I give rise to, and was given to this element by Lavoisier, with reference to its property of forming acids in uniting with other elementary bodies. Oxygen is a permanent gas, when uncombined, and forms one-fifth part of the air of the atmosphere. In a state of combination, this element is the most extensively diffused body in nature, entering as a constituent into water, into nearly all the earths and rocks of which the crust of the globe is composed, and into all organic products, with a few exceptions. It was first recognised as a distinct substance by Dr. Priestley in England, in 1774, and about a year afterwards by Scheele in Sweden, without any knowledge of Priestley's experiments. From this discovery may be dated the origin of true chemical theory.

Preparation.—Oxygen gas is generally disengaged from some compound containing it, by the action of heat.

1. It was first procured by Priestley, by heating Red Precipitate (oxide of mercury), which is thereby resolved into fluid mercury and oxygen gas. To illustrate the formation of oxygen in this way, 200 grains of red precipitate may be introduced into the body of a small retort *a* of hard or difficultly fusible glass, and the retort

Fig. 95.

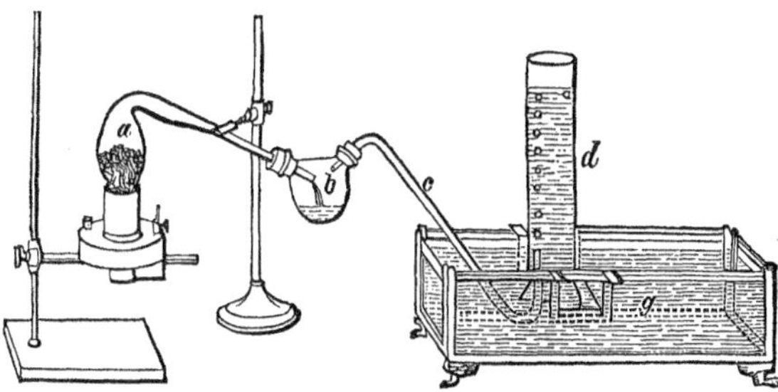

united in an air-tight manner with a small globular flask *b*, having two openings, both closed by perforated corks, one of which admits the beak of the retort, and the

[1] This number includes three elements — erbium, terbium, and ilmenium, of which the existence is doubtful.

other an exit tube *c*, of glass, bent as in the figure. The extremity of the exit tube is introduced into a graduated jar capable of holding 50 or 60 cubic inches, and placed in an inverted position, full of water, upon the shelf of a pneumatic water-trough. Heat is then applied to the retort by means of an Argand spirit lamp powerful enough to raise it to a red heat, and maintain it at that temperature for a considerable time. The first effect of the heat is to expand the air in the retort, bubbles of which issue from the tube *c*, and rise to the top of the jar displacing water; but more gas follows, which is oxygen, and at the same time metallic mercury condenses in the neck of the retort and runs down into the intermediate flask *b*. When the red precipitate in the retort has entirely disappeared, the lamp may be extinguished, and the retort allowed to cool completely. The end of the exit tube *c* being now above the level of the water in the jar, which is nearly full of gas, a portion of the latter, equal in bulk to the air which first left the retort, will return to it, from the contraction of the gas within the retort. The jar will be found in the end to contain 44 cubic inches of gas, which is therefore the measure of oxygen produced in the experiment, and the flask to contain 185 grains of mercury. Now 44 cubic inches of oxygen weigh 15 grains; and a true analysis of the red precipitate has been effected, of which the result is, that 200 grains of that substance consist of—

185 grains mercury.
15 " oxygen, (44 cubic inches).
———
200

But oxygen gas is more generally derived from two other substances—oxide of manganese and chlorate of potassa.

2. When the gas is required in large quantity, and exact purity is immaterial, the oxide of manganese is preferred from its cheapness. This is a black, heavy mineral, found in Devonshire, in Hesse Darmstadt, and other localities, of which upwards of 40,000 tons are consumed annually in the manufactures of the country. It is called an oxide of manganese, because it is a compound of the metal manganese with oxygen. In explanation of what takes place when this substance is heated, it is necessary to state that manganese is capable of uniting with oxygen in several proportions, namely, one equivalent, or 27.67 parts of manganese, with 8, and with 16 parts of oxygen; and two equivalents of manganese with 24 parts of oxygen. These compounds are:—

Protoxide of manganese	Mn+ O.
Sesquioxide	2Mn+3O.
Binoxide, or native black oxide	Mn+2O.

Now the binoxide, however strongly heated, never loses more than one-third of its oxygen, being converted into a compound of the first two oxides: that is, three equivalents of binoxide (131.01 parts) lose two equivalents of oxygen (16 parts), and leave a compound of one eq. of sesquioxide and one eq. of protoxide; a change which may be thus expressed:—

$$3Mn\,O_2 = \begin{cases} 2O \\ Mn^2\,O^3 + Mn\,O. \end{cases}$$

One of the malleable iron bottles in which mercury is imported is readily converted into a retort, in which the black oxide may be heated, by removing its screwed iron stopper, and replacing this by an iron pipe of three feet in length, one end of which has been cut to the screw of the bottle. This pipe may be bent like *a*, figure 96, if the bottle is to be heated in an open fire, or in a furnace open at the top. From 3 to 9 pounds of the oxide may be introduced as a charge, according to the quantity of gas to be prepared, each pound of good German manganese yielding about 1400 cubic inches, or 5.05 gallons of gas. Upon the first application of heat, water comes off, as steam, mixed occasionally with a gas which extinguishes flame; this is owing to the impurity of the oxide. The products may be allowed to escape, till the point of a wood-match, red without flame, applied to the orifice, is rekindled and made to burn with brilliancy; the gas is then sufficiently pure, and means must be taken for collecting it. A small flexible tin tube *b*, of any convenient length, is

FIG. 96.

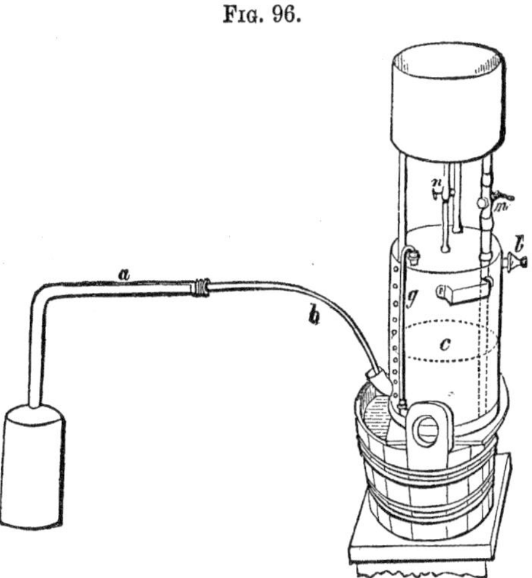

adapted to the iron pipe, by means of a perforated cork, by which the gas is conveyed to a pneumatic trough, and collected in glass jars filled with water, as in the former experiment; or, as this process affords considerable quantities of oxygen, the gas is more generally conducted into the inferior cylinder or drum of a copper gas holder *c*, full of water. The water does not flow out by the recurved tube which forms the lower opening, but is retained in the vessel by the pressure of the atmosphere on the surface of the water in that tube, as water is retained in a bird's drinking-glass. But when the tin tube is introduced into the gas-holder by this opening, water escapes by it, in proportion as gas is thrown into the cylinder and rises in bubbles to the top. The progress of filling the gas-holder may be observed by the glass gauge-tube *g*, which is open at both ends, and connected with the top and bottom of the cylinder, so that the water stands at the same height in the tube as in the cylinder. Convenient dimensions for the cylinder itself are 16 inches in height by 12 in diameter; to fill which a charge of three pounds of manganese may be used. The gauge-tube is so apt to be broken, or to occasion leakage at its junctions with the cylinder, when the latter is large and unwieldy, that it is generally better to forego the advantage it offers, and dispense with this addition to the gas-holder. When applied to a small gas-holder, the ends of the tube are conveniently adapted to the openings of the cylinder, by means of perforated corks, which are afterwards covered by a mixture of white and red lead with a drying oil.

FIG. 97.

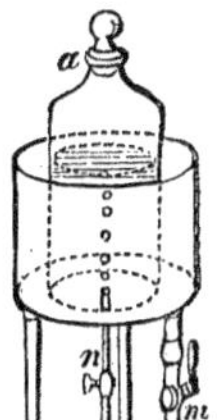

After the cylinder is filled, the lower opening by which the gas was admitted is closed by a good cork, or by a brass cap made to screw over it. The superior cylinder is an open water trough, connected with the inferior cylinder by two tubes provided with stop-cocks, *m* and *n*, one of which, *m*, is continued to the bottom of that vessel, and conveys water from the superior cylinder, while the other tube, *n*, terminates at the top of the inferior cylinder, and affords a passage by which the gas can escape from it, when water is allowed to descend by the other tube. The tube and perforation of the stopcock of *m* should be considerably wider than *n*. A jar *a* is filled with gas by inverting it full of water in the superior cylinder, over the opening of *n*, as exhibited in the figure, and allowing the gas to ascend from the inferior cylinder. Gas may likewise be obtained by the stopcock *l* (fig. 96), water being allowed to enter by *m* at the same time.

Oxygen may likewise be disengaged from oxide of manganese in a flask or retort, by means of sulphuric acid diluted with an equal bulk of water, but this is not a process to be recommended. When the quantity of oxygen required is not very large, it is better to have recourse to chlorate of potassa, which has also the advantage of giving a perfectly pure gas.

3. A well-cleansed Florence oil flask, the edges of the mouth of which have been heated and turned over so as to form a lip, with a bent glass tube and perforated cork fitted to it (fig. 98), forms a convenient retort in which about half an ounce of chlorate of potassa may be heated by means of a gas flame or Argand spirit lamp. The salt melts, although it contains no water, and when nearly red-hot emits abundance of oxygen gas. At one point of the decomposition, the effervescence may become so violent as to burst the flask, especially if the exit tube be narrow, unless the heat be moderated. The chlorate of potassa parts with all the oxygen it possesses, which amounts to 39·2 per cent. of its weight, and leaves a white hard salt, the chloride of potassium.

Fig. 98.

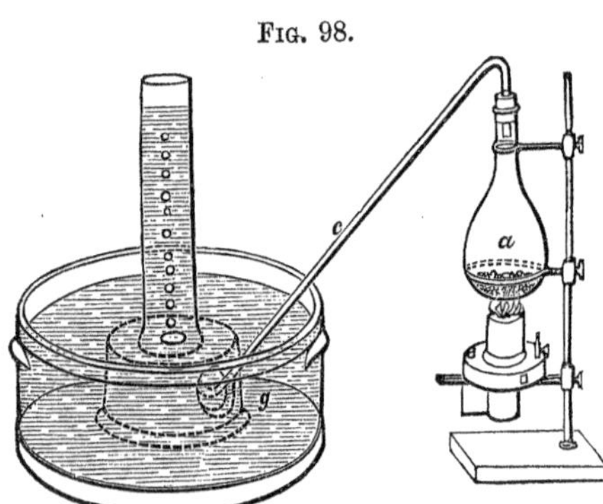

The only inconvenience attending the preceding process is the high temperature required, which would soften a retort or flask of flint glass. It was discovered, however, by M. Mitscherlich, that chlorate of potassa is decomposed at a much lower temperature when mixed with dry powders, upon which it exercises no chemical action, particularly metallic peroxides, such as the binoxide of manganese and the black oxide of copper. Nothing can answer better than the binoxide of manganese, after being made anhydrous by a short exposure to a red heat. Two parts of chlorate of potassa in powder, mixed with one part of the dried oxide, forms a useful "oxygen mixture," which may be made in quantity and preserved for occasional use.

From an atomic statement of the composition of chlorate of potassa, it appears that one equivalent of it (122·5 parts) contains six equivalents of oxygen (48 parts), namely five eq. in the chloric acid and one eq. in the potassa, the whole of which come off, leaving one equivalent of chloride of potassium (74·5 parts):—

$$KO + Cl\,O_5 = \begin{cases} 6O \\ K\,Cl. \end{cases}$$

Half an ounce of chlorate of potassa should yield 270 cubic inches, or nearly a gallon of pure oxygen gas.

4. Another process for oxygen gas, proposed by Mr. Balmain, consists in heating in a retort 3 parts of the bichromate of potassa in powder, with 4 parts of undiluted sulphuric acid: the gas comes off in a continuous stream, and a mixture of sulphate of potassa and sulphate of sesquioxide of chromium remains behind in the retort. The decomposition which takes place is explained in the following formula:—

$$KO,\ Cr_2O_6 \text{ with } 4\ SO_3, \text{ give } KO,\ SO_3 \text{ with } Cr_2O_3\ 3SO_3 \text{ and } 3O.$$

The bichromate of potassa loses one-half of the oxygen contained in the chromic acid, or about 16 per cent. of its weight; one ounce of salt yielding about 200 cubic inches of gas.

[5. When a perfectly pure gas is not required, oxygen may be obtained in large quantity from nitrate of potassa. The same apparatus is used as in the decomposition of black oxide of manganese: the nitre, of which 8 or 10 pounds may be used at once, is to be exposed to a well-regulated heat of a charcoal fire, the draught being urged or diminished in proportion to the rapidity of the flow of gas. A red heat is about the best temperature for the operation. The gas which comes over at this temperature contains about 96 per cent. of oxygen, and when after some time it is found necessary to urge the fire that the flow of gas may be kept up, the per centage diminishes, and may fall as low as 66. Two of the five equivalents of oxygen in the nitric acid are given off in the first part of the operation, and in the latter part the remaining oxygen with the nitrogen.—R. B.]

Properties.—Oxygen gas is colourless, and destitute of odour and taste. It is heavier than air in the ratio of 1105·6 to 1000, according to the latest careful determination, that of M. Regnault.[1]

At the temperature of 60°, and with the barometer at 30 inches, 100 cubic inches of oxygen gas weigh 34·19 grains (Regnault). One cubic inch, therefore, weighs 0·3419 gr., or about one-third of a grain. It has never been liquefied by cold or pressure.

Oxygen is so sparingly soluble in water, that when agitated in contact with that fluid no perceptible diminution of its volume takes place. But when water is previously deprived of air by boiling, and allowed to cool in a close vessel, 100 cubic inches of it dissolve $3\frac{1}{2}$ cubic inches of this gas.

If a lighted wax taper attached to a copper wire be blown out, and dipped into a vessel of oxygen gas, while the wick remains red-hot, it instantly rekindles with a slight explosion, and burns with great brilliancy. If soon withdrawn and blown out, it may be revived again in the same manner, and the experiment be repeated several times in the same gas. Lighted tinder burns with flame in oxygen, and red-hot charcoal with brilliant scintillations. Burning sulphur introduced into this gas in a little hemispherical cup of iron-plate with a wire attached to it, burns with an azure blue flame of considerable intensity. Phosphorus introduced into oxygen in the same manner, burns with a dazzling light of the greatest splendour, particularly after the phosphorus boils and rises through the gas in vapour. Indeed, all bodies which burn in air, burn with increased vivacity in oxygen gas. Even iron wire may be burned in this gas. For this purpose thin harpsichord wire should be coiled about a cylindrical rod into a spiral form. The rod being withdrawn, a piece of thread must be twisted about one end of the wire, and dipped into melted sulphur; the other end of the wire is to be fixed into a cork, so that the spiral may hang vertically. The sulphured end is then to be lighted, and the wire suspended in a jar of oxygen, open at the bottom, such as that represented in fig. 97, page 225, supported

Fig. 99.

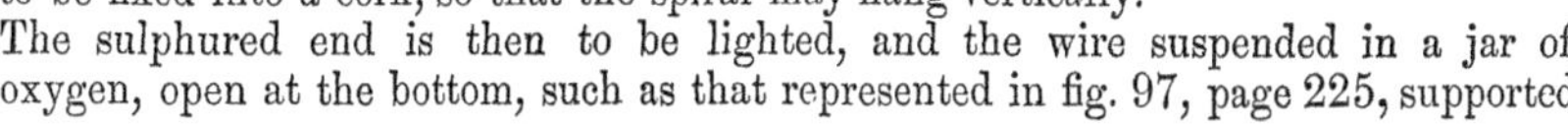

[1] Annales de Chimie, &c., 1845, 3e. ser. t. xiv. p. 211. The mean of three weighings previously made by MM. Dumas and Boussingault, was 1105·7 (ibid. t. viii. p. 201). Baron Wrede found 1105·2. At a much earlier period T. de Saussure obtained Regnault's number, 1105·6. These coincidences in the results of independent observers appear to prove that a close approximation has been made to the true density of this gas: an important datum. The earlier determination of MM. Dulong and Berzelius was 1102·6 (ibid. 1820, 2e. ser. t. xv. p. 386). According to M. Regnault, the weight of 1000 cubic centimeters (1 liter) of oxygen gas, at 32° F., barometer 29·92 inches (760 millimeters), is 1·4298 gramme. Hence, 1000 c. c. being equal to 61·028 English c. inches, and 1 gramme to 15·4440 English grains, 100 cubic inches of oxygen, at the specified temperature and pressure, weigh 36·1390 grains. Calculating with Regnault's coefficient for the expansion of air (page 40), 1 volume of oxygen will become 1·05701 volume, at 60°, and 100 cubic inches of oxygen will weigh 34·1898 grains at that temperature.

upon an earthenware plate. The wire is kindled by the sulphur, and burns with an intense white light, throwing out a number of sparks, or occasionally allowing a globule of fused oxide to fall; while the wire itself continues to fuse and burn till it is entirely consumed, or the oxygen is exhausted. The experiment forms one of the most beautiful and brilliant in chemistry. The globules of fused oxide are of so elevated a temperature, that they remain red-hot for some time under the surface of water, and fuse deeply into the substance of the stoneware plate upon which they fall.

[A portion of cast iron placed upon ignited charcoal and subjected to a stream of oxygen, soon melts and burns brilliantly, throwing off showers of bright sparks on all sides. — R. B.]

Oxygen gas is respirable, and indeed is constantly taken into the lungs from the atmosphere in ordinary respiration. When a portion of dark blood drawn from a vein is agitated with this gas, the colour becomes of a fine vermilion red. The same change occurs in the blood of living animals, during respiration, from the absorption of oxygen gas, which is required to maintain the animal heat. A small animal, also, such as a mouse or bird, lives four or five times longer in a vessel of oxygen than it will in an equal bulk of air. But the continued respiration of this gas in a state of purity is injurious to animal life. A rabbit is found to breathe it without inconvenience for some time, but after an interval of an hour or more the circulation and respiration are much quickened, and a state of great excitement of the general system supervenes; this is by and by followed by debility, and death occurs in from six to ten hours. The blood is found to be highly florid in the veins as well as the arteries, and, according to Broughton, the heart continues to act strongly after the breathing has ceased.

Oxygen may be made to unite with all the other elements except fluorine, and forms *oxides,* while the process of uniting with oxygen is termed *oxidation.* With the same element oxygen often unites in several proportions, forming a series of oxides, which are then distinguished from each other by the different prefixes enumerated under *Chemical nomenclature* (page 106). Many of its compounds are *acids,* particularly those which contain more than one equivalent of oxygen to one of the other element, and compounds of this nature are those which it most readily forms with the non-metallic elements: such as carbonic acid with carbon, sulphurous acid with sulphur, phosphoric acid with phosphorus. But oxygen unites in preference with single equivalents of a large proportion of the metallic class of elements, and forms bodies which are alkaline or have the character of *bases:* such as potassa, lime, magnesia, protoxide of iron, &c. A certain number of its compounds are neither acid nor alkaline, and are therefore called *neutral* bodies: such as the oxide of hydrogen or water, carbonic oxide, and nitrous oxide. The greater number of these neutral oxides are also protoxides.

It has already been stated that in a classification of the elements oxygen does not stand alone, but forms one of a small natural family along with sulphur, selenium, and tellurium. These elements also form acid, basic, and neutral classes of compounds, with the same bodies as oxygen does, of which the sulphur compounds are well known, and always exhibit a well-marked analogy to the corresponding oxides. Oxygen-acids unite with oxygen bases, and form neutral *salts:* so do sulphur-acids with sulphur-bases, selenium-acids with selenium-bases, and tellurium-acids with tellurium-bases.

The combinations of oxygen, like those of all other bodies, are attended with the evolution of heat. This result, which is often overlooked in other combinations, in which the proportions of the bodies uniting and the properties of their compound receive most attention, assumes an unusual degree of importance in the combinations of oxygen. The economical applications of the light and heat evolved in these combinations are of the highest consequence and value, and oxidation alone, of all chemical actions, is practised, not for the value of the products which it affords, and indeed without reference to them, but for the sake of the incidental phenomena attending it. Of the chemical combinations, too, which we habitually witness, those of oxygen are infinitely the most frequent, which arises from its constant presence and interference as a constituent of the atmosphere. Hence, when a body combines with oxygen, it is said to be *burned;* and instead of undergoing oxidation it is said to suffer *combustion;* and a body which can combine with oxygen and emit heat is termed a *combustible.* Oxygen, in which the body burns, is then said to support combustion, and called a *supporter* of combustion.

The heat evolved in combustion is definite, and can be measured. With this view it is employed to melt ice, to raise the temperature of water from 32° to 212°, or to convert water into steam, and its quantity is estimated by the extent to which it produces these effects. The heat from the oxidation of a combustible body is thus found to be as constant as any other of its properties. Despretz obtained, by such experiments, the results contained in the following table:—

HEAT FROM COMBUSTION.

1 pound of	pure charcoal..........................	heats from 32° to 212°,	78	pounds of water.
—	charcoal from wood	—	75	—
—	baked wood..............................	—	36	—
—	wood containing 20 per cent. of water	—	27	—
—	bituminous coal........................	—	60	—
—	turf..	—	25 to 30	—
—	alcohol.....................................	—	67·5	—
—	olive oil, wax, &c......................	—	90 to 95	—
—	ether	—	80	—
—	hydrogen..................................	—	236·4	—

The quantity of heat evolved appears to be connected with the proportion of oxygen consumed, for the greater the weight of oxygen with which a pound of any combustible unites, the more heat is produced. The following results indicate that the heat depends exclusively upon the oxygen consumed, four different combustibles in consuming a pound of oxygen affording nearly the same quantity of heat:—

HEAT OF COMBUSTION.

1 pound of oxygen	with hydrogen	heats from 32° to 212°,	29½	pounds of water.
—	with charcoal	—	29	—
—	with alcohol	—	28	—
—	with ether	—	28½	—

The quantity of combustible consumed in these experiments varied considerably, but the oxygen being the same, the heat evolved was nearly the same also. But when the same quantity of oxygen converted phosphorus into phosphoric acid, exactly twice as much heat was evolved, according to Despretz, as in the former experiments. The superior vivacity of the combustion of these and other bodies in pure oxygen, compared with air, depends entirely upon the rapidity of the process, and the larger quantity of combustible oxidated in a given time. A candle burns with more light and heat in oxygen than in air, but it consumes proportionally faster.

Oxidation is often a very slow process, and imperceptible in its progress—as in the rusting of iron and tarnishing of lead exposed to the atmosphere. The heat being then evolved in a gradual manner is instantly dissipated, and never accumulates. But when the oxide formed is the same, the nature of the change effected is in no way altered by its slowness. Iron oxidates rapidly when introduced in a state of ignition into oxygen gas, and lead, in the form of the lead pyrophorus, which contains that metal in a high state of division, takes fire spontaneously and burns in the air; circumstances then favouring the rapid progress of oxidation.

Oxidation may also go on with a degree of rapidity sufficient to occasion a sensible evolution of heat, but without flame and open combustion. The absorption of oxygen by spirituous liquors in becoming acetic acid, and by many other organic substances, is always attended with the production of heat. The smouldering combustion of iron pyrites and some other metallic ores in the atmosphere, is a phenomenon of the same nature. Most bodies which burn with flame also admit of being oxidated at a temperature short of redness, and exhibit the phenomenon of *low combustion.* Thus, tallow thrown upon an iron plate not visibly red-hot, melts and undergoes oxidation, diffusing a pale lambent flame visible only in the dark (Dr. C. J. B. Williams). If the tallow be heated in a little cup with a wire attached, till it boils and catches fire, and the flame then be blown out, the hot tallow will still continue in a state of low combustion, of which the flame may not be visible, but which is sufficient to cause the renewal of the high combustion, if the cup is immediately introduced into a jar of oxygen gas. A candle newly blown out is sometimes rekindled in oxygen, although no point of the wick remains visibly red, owing to the continuance of this low combustion. When a coil of thin platinum wire, or a piece of platinum foil, is first heated to redness, and then held over a vessel containing ether or hot alcohol, the vapours of these substances, mixed with the air, oxidate upon the hot metallic surface, and may sustain the metal at a red heat for a long time, without the occurrence of combustion with flame. The product, however, of the low combustion of these bodies is peculiar, as is obvious from its pungent odour.

Combustion in air. — The affinity for oxygen of all ordinary combustibles is greatly promoted by heating them, and is indeed rarely developed at all except at a high temperature. Hence, to determine the commencement of combustion, it is commonly necessary that the combustible be heated to a certain point. But the degree of heat necessary to inflame the combustible is in general greatly inferior to what is evolved during the progress of the combustion, so that a combustible, once inflamed, maintains itself sufficiently hot to continue burning till it is entirely consumed. Here the difference may be observed between combustion and simple ignition. A brick heated till it be red-hot in a furnace, and taken out, exhibits ignition, but has no means within itself of sustaining a high temperature, and soon loses the heat which it had acquired in the fire, and on cooling is found unchanged.

The oxidable constituents of wood, coal, oils, tallow, wax, and all the ordinary combustibles, are the same, namely, carbon and hydrogen, which in combining with oxygen, at a high temperature, always produce carbonic acid and water; volatile bodies, which disappear, forming part of the aerial column that rises from the burning body. The constant removal of the product of oxidation, thus effected by its volatility, greatly favours the progress of combustion in such bodies, by permitting the free access of air to the unconsumed combustible. The influence of air in combustion is obvious from the facility with which a fire is checked or extinguished when the supply of air is lessened or withheld, and, on the contrary, revived and animated when the supply of air is increased by blowing up the fire. For the oxygen of the air being consumed in combining with the combustible, a constant renewal of it is necessary. Hence, if a lighted taper, floated by a cork upon water, be covered with a bell jar having an opening at top, such as that in which the iron-wire was burned, the taper will burn for a short time without change, then more and more feebly, in proportion as the oxygen is exhausted, and at last will expire. The air remaining in the jar is no longer suitable to support combustion, and a second lighted taper introduced into it by the opening at top is immediately extinguished.

In combustion, no loss whatever of ponderable matter occurs; nothing is annihilated. The matter formed may always be collected without difficulty, and is found to have exactly the weight of the oxygen and combustible together which have disappeared. The most simple illustrations of this fact are obtained in the combustion of those bodies which afford a solid product. Thus when two grains of phosphorus are kindled in a measured volume of oxygen gas, they are found converted after combustion into a quantity of white powder (phosphoric acid), which weighs 4½ grains, or the phosphorus acquires 2½ grains; at the same time 7½ cubic inches of oxygen disappear, which weigh exactly 2½ grains. In the same way, when iron-wire is burned in oxygen, the weight of solid oxide produced is found to be equal to that of the wire originally employed added to that of the oxygen gas which has disappeared. But the oxidation of mercury affords a more complete illustration of what occurs in combustion. Exposed to a moderate degree of heat for a considerable time in a vessel filled with oxygen, that metal is converted into red scales of oxide, possessing the additional weight of a certain volume of oxygen which has disappeared. But if the oxide of mercury so produced be then put into a small retort, and reconverted by a red heat into oxygen and fluid mercury, the quantity of oxygen emitted is found to be the same as had combined with the mercury in the first part of the operation; thus proving that oxygen is really present in the oxidized body.

The evolution of heat, which is the most striking phenomenon of combustion, still remains to be accounted for. It has been referred to the loss of latent heat by the combustible and oxygen, when, from the condition of gas or liquid, one or both become solid after combustion; to a reduction of capacity for heat, the specific heat of the product being supposed to be less than that of the bodies burned; and to a discharge of the electricities belonging to the different bodies, occurring in the act of combination. But the first two hypotheses are manifestly insufficient, and the last is purely speculative. The evolution of heat during intense chemical combination, such as oxidation, may be received at present as an ultimate fact; but if we choose to go beyond it, we must suppose that the heat exists in a combined and latent state in either the oxygen or combustible, or in both; that each of these bodies is a compound of its material basis with heat, the whole or a definite quantity of which they throw off on combining with each other. Heat, like other material substances, is here supposed not to evince its peculiar properties while in a state of combination with other matter, but only when isolated and free. This view gives a literal character to the expressions — liberation, disengagement, and evolution of heat during combustion. The phenomenon, it is to be remembered, is not confined to oxidation, but occurs in an equal degree in combinations without oxygen, and indeed to a greater or less extent in all chemical combinations whatever.

Fig. 100.

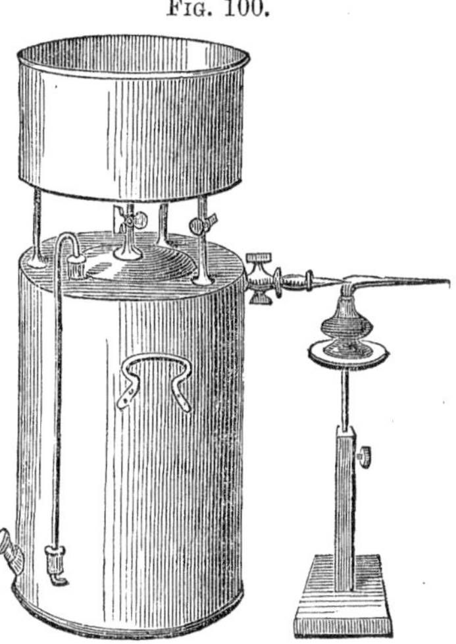

Fig. 101.

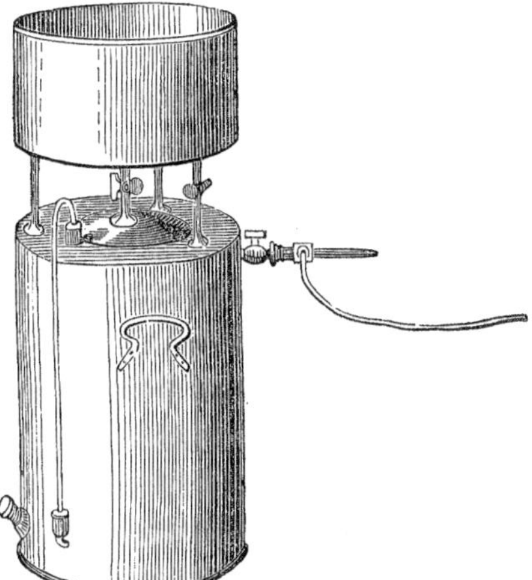

Pure oxygen has not as yet found any considerable application in the arts. But by the chemist it is applied to support combustion with the view of producing intense heat. A jet of this gas from a gas-holder (fig. 100), thrown upon the flame of a spirit-lamp, produces a blow-pipe flame of great intensity, adequate to fuse platinum. Or, if coal-gas be conducted to the oxygen jet (fig. 101), and the gases kindled as they issue together, a flame is produced of equally high temperature. Where a large quantity of oxygen is required, as in this application of it, the gas may be obtained by heating oxide of manganese in a cylinder of cast iron supported over a furnace, like the retort for coal gas. The calcined oxide does not regain its oxygen when afterwards exposed to the air, as was once supposèd, but would still be of some value in the preparation of chlorine.

Ozone.—When electric sparks are taken through perfectly dry oxygen, a small portion of the gas acquires new properties, according to A. de la Rive, and is supposed by Berzelius to pass into an allatropic condition, in which it is named ozone from the peculiar odour it possesses, and which is somewhat metallic in character. The oxygen evolved from the decomposition of water in the voltameter (page 221) has the same odour. But the most ready mode of producing it is to place a few sticks of phosphorus in a quart bottle containing a little water at the bottom of it. While the sticks of phosphorus undergo the low combustion and are luminous, producing fumes of phosphorus acid and absorbing much oxygen, they give rise to the appearance of ozone in the air of the bottle in a manner not at present understood.

This substance has never been obtained in a separate state, but air impregnated with it acts very much as if a trace of chlorine gas were present, which ozone appears to resemble. In ozonized air, paper impregnated with a solution of iodide of potassium immediately becomes brown from the liberation of iodine; also paper containing a solution of sulphate of manganese soon becomes brown or black, from the formation of binoxide of manganese. The same air made to stream through a solution of the yellow-ferrocyanide of potassium converts it into the red ferricyanide. Ozone appears to be a gas not sensibly dissolved by water. It is destroyed by a heat of 140°, by contact with olefiant gas, and such other hydrocarbons as combine with chlorine, by phosphorus, or reduced silver. In the latter case nothing appears except oxide of silver. It passes, I find, through dry and porous stoneware, and is therefore not likely to be merely an electrical grouping of gaseous molecules. Professor Schönbein, who named this substance, and has made it the object of many investigations, considers it to be a volatile peroxide of hydrogen.

SECTION II.

HYDROGEN.

Equivalent 1, *as the basis of the Hydrogen Scale, or* 12·5 (*oxygen*=100); *symbol* H; *density* 69·26 (*air* 1000); *combining measure* [|] (*two volumes*).

Hydrogen gas, which was long confounded with other inflammable airs, was first correctly described by Cavendish, in 1766. It does not exist uncombined in nature; at least the atmosphere does not contain any appreciable proportion of hydrogen. But it is one of the elements of water, and enters into nearly every organic substance. Its name is derived from ὑδωρ, water, and γεννaω, I give rise to, and refers to its forming water when oxidated.

Preparation.—This element, although resembling oxygen in being a gas, appears to be more analogous to a metal in its relations to other elements. By heating oxide of mercury, it is resolved into oxygen and mercury; and several other metallic oxides, such as those of silver and gold, are susceptible of a similar decomposition. But some others are deprived of only a portion of their oxygen by the most intense heat, such as binoxide of manganese; and many, such as the protoxide of lead, are not decomposed at all by simple calcination. By igniting the latter oxide, however, mixed with charcoal, its oxygen goes off in combination with carbon, as carbonic oxide, and the lead is left. The oxide of hydrogen or water is similarly affected. Potassium and sodium brought into contact with it, at the temperature of the air, combine with its oxygen, and are converted into the oxides, potassa and soda; and hydrogen is consequently liberated.

Iron and many other metals decompose water, and become oxides, at a red heat. Hence, hydrogen gas is sometimes procured by transmitting steam through an iron tube filled with iron turnings, placed across a furnace and heated red-hot (fig. 102).

Fig. 102.

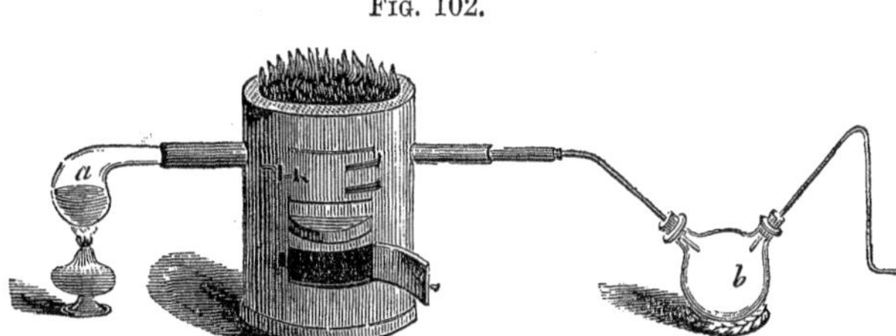

The vapour is obtained by boiling water in the small retort *a*, and the gas produced by its decomposition collected in the usual manner at the pneumatic trough. But it is necessary to have a flask *b* between the iron tube and the trough, to prevent an accident from the water of the trough finding access to the red-hot tube, in the event of condensation of the vapour in *a*.

Some other compounds of hydrogen are decomposed more easily than water, by iron and zinc. The chloride of hydrogen or hydrochloric acid is decomposed by these metals, and evolves hydrogen at the ordinary temperature of the air. But this gas is more generally obtained by putting pieces of zinc or iron into oil of vitriol or the concentrated sulphuric acid, diluted with six or eight times its bulk of water. The hydrogen is then derived from the decomposition of the proportion of water intimately united with the acid, as illustrated in the following diagram, zinc being used, and the quantities expressed:—

Before decomposition.			After decomposition.
49 oil of vitriol, or sulphate of water	Hydrogen	1	1 Hydrogen.
	Oxygen	8	
	Sulphuric acid	40	
32.52 zinc	Zinc	32.52	80.52 Sulphate of oxide of zinc.
81.52		81.52	81.52

Or by symbols:—

$$HO+SO_3 \text{ and } Zn=ZnO+SO_3 \text{ and } H.$$

The zinc dissolves in the acid with effervescence, from the escape of hydrogen gas. It will be observed that the products after decomposition, mentioned in the last column, hydrogen and sulphate of oxide of zinc, are similar to those before decomposition, in the first column, zinc and sulphate of water; and that the change occurring is simply the *substitution* of zinc for hydrogen in the sulphate of water. The large quantity of water used with the acid is useful to dissolve the sulphate of zinc formed.

Zinc is generally preferred to iron, in the preparation of hydrogen, and is previously granulated, by being fused in a stone-ware crucible, and poured into water; if sheet zinc be used, which is better, it is cut into small pieces. The common glass retort may be used in the experiment, or a gas-bottle, such as the half-pound phial (see fig. 103), with a cork having two perforations fitted with glass tubes, one of which descends to the bottom of the bottle, and is terminated externally by a funnel for introducing the acid, whilst the other is the exit tube, by which the hydrogen

Fig. 103.

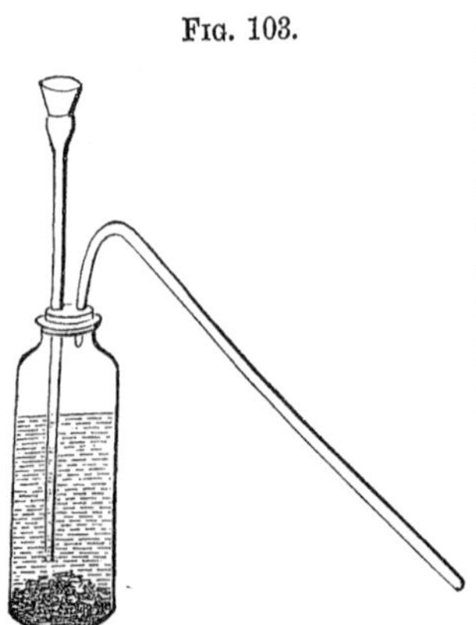

escapes. With an ounce or two of zinc in it, the bottle is two-thirds filled with water, and the undiluted acid added from time to time by the funnel, so as to sustain a continued effervescence. No gas escapes by the funnel tube, as its extremity within the bottle is always covered by the fluid. To produce large quantities, a half-gallon stone-ware jar may be mounted as a gas-bottle, with a flexible metallic pipe fitted to the cork as the exit tube. This gas may be collected, like oxygen, either in jars over the pneumatic trough, or in the gas-holder. The first jar or two filled will contain the air of the gas-bottle, and therefore must not be considered as pure hydrogen. One ounce of zinc is found to cause the evolution of 615 cubic inches of hydrogen gas.

Properties.—Hydrogen gas thus prepared is not absolutely pure, but contains traces of sulphuretted hydrogen and carbonic acid, which may be removed by agitating the gas with lime-water or caustic alkali. It has also a particular odour, which is not essential to hydrogen, as the gas evolved from the amalgam of sodium, acted on by pure water without acid, is perfectly inodorous. An oily compound of carbon and hydrogen, which appears to be the cause of this odour, may be separated in a sensible quantity from the gas prepared by iron, by transmitting it through alcohol. Of the pure gas, water does not dissolve more than $1\frac{1}{2}$ per cent. of its bulk. Hydrogen has never been liquefied by cold or pressure.

Hydrogen is the lightest substance in nature, being sixteen times lighter than oxygen, and 14.4 times lighter than air; 100 cubic inches of it weigh only 2.14 grains. Soap-bubbles blown with this gas ascend in the atmosphere; and it is used, as is well known, to inflate balloons, which begin to rise when the weight of the stuff of which they are made and the hydrogen together, are less than the weight of an equal bulk of air. A light bag is prepared for making this experiment in the chamber, by distending the lining membrane of the crop of the turkey, which may weigh 35 or 36 grains, and when filled with hydrogen, about 5 grains more, or 41 grains; the same bulk of air, however, would weigh 50 or 51 grains; so that the little balloon when filled with hydrogen has a buoyant power of 9 or 10 grains. Larger bags are prepared for the same purpose, of gold-beaters' skin. Sounds produced in this gas were found by Leslie to be extremely feeble; much more feeble, indeed, than its rarity compared with air could account for. Hydrogen may be taken into the lungs without inconvenience, when mixed with a large quantity of air, being in no way deleterious; but it does not, like oxygen, support respiration, and therefore an animal placed in pure hydrogen soon dies of suffocation. A lighted taper is extinguished in the same gas.

Hydrogen is eminently combustible, and burns when kindled in the air with a yellow flame of little intensity, which moistens a dry glass jar held over it; the gas combining with the oxygen of the air in burning, and producing water. If before being kindled the gas is first mixed with enough of air to burn it completely, or with between two and three times its volume, and then kindled, the combustion of the whole hydrogen is instantaneous and attended with explosion. With pure oxygen, instead of air, the explosion is much more violent, particularly when the gases are mixed in the proportions of two volumes of hydrogen to one of oxygen, which are the proper quantities for combination. The combustion is not thus propagated through a mixture of these gases, when either of them is in great excess. The sound in such detonations is occasioned by the concussion which the atmosphere receives from the sudden dilatation of gaseous matter, in this case of steam, which is prodigiously expanded from the heat evolved in its formation.

A musical note may be produced by means of these detonations, when they are made to succeed each other very rapidly. If hydrogen be generated in a gas-bottle (fig. 104), and kindled as it escapes from an upright glass jet having a small aperture, the gas will be found to burn tranquilly; but on holding an open glass tube of about two feet in length over the jet, like a chimney, the flame will be elongated and become flickering. A succession of little detonations is produced, from the gas being carried up and mixing with the air of the tube, which follow each other so quickly as to produce a continuous sound or musical note.

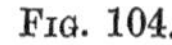

Fig. 104.

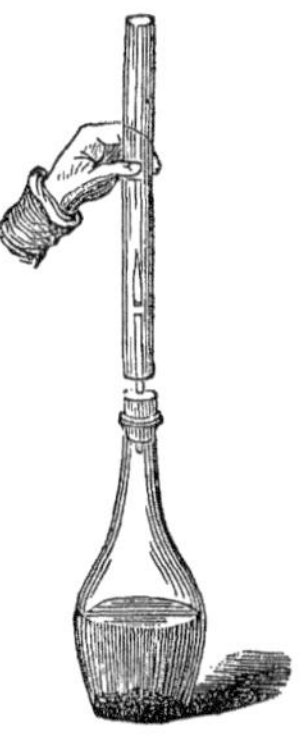

Several circumstances affect the combination of hydrogen with oxygen, which are important. These gases may be mixed together in a glass vessel, and preserved for any length of time without combining. But combination is instantly determined by flame, by passing the electric spark through the mixture, or even by introducing into it a glass rod, not more than just visibly red-hot. Hydrogen, indeed, is one of the more easily inflammable gases. If the mixed gases be heated in a vessel containing a quantity of pulverized glass, or any sharp powder, they begin to unite in contact with the foreign body in a gradual manner without explosion, at a temperature not exceeding 660°. The presence of metals disposes them to unite at a still lower temperature; and of the metals, those which have no disposition of themselves to oxidate, such as gold and platinum, occasion this slow combustion at the lowest temperature. In 1824, Dobereiner made the remarkable discovery that newly prepared spongy platinum has an action upon hydrogen mixed with oxygen, independently of its temperature, and quickly becomes red-hot when a jet of hydrogen is thrown upon it in air, combination of the gases being effected by their contact with the metal. In consequence of this ignition of the platinum the hydrogen itself is soon inflamed, as it issues from the jet. An instrument depending upon this action of platinum has been constructed for producing an instantaneous light. Afterwards, Mr. Faraday observed, that the divided state of the platinum, although favourable, is not essential to this action; and that a plate of that metal, if its surface be scrupulously clean, will cause a combination of the gases, accompanied with the same phenomena as the spongy platinum. This action of platinum is manifested at temperatures considerably below the freezing point of water, and in an explosive mixture largely diluted with air or hydrogen. Spongy platinum, made into pellets with a little pipe-clay, and dried, when introduced into mixtures of oxygen and hydrogen will be found to cause a gradual and silent combination of the gases, in whatever proportions they are mingled, which will not cease till one of them is completely exhausted. The theory of this effect of platinum is very obscure. It belongs to a class of actions depending upon surface, not confined to that metal, and by which other combustible vaporous bodies are affected besides hydrogen.

Fig 105.

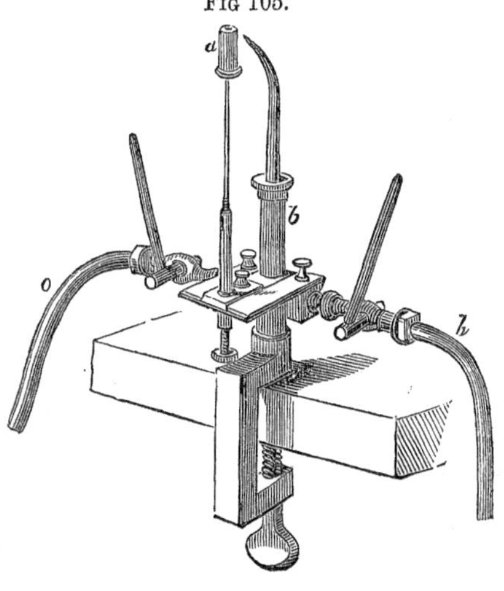

The flame of hydrogen, although so slightly luminous, is intensely hot; few combinations producing so high a temperature as the combustion of hydrogen. In the oxi-hydrogen blow-pipe, oxygen and hydrogen gases are brought by tubes *o* and *h* (fig. 105), from different gas-holders, and allowed to mix immediately before they escape by the same orifice, at which they are inflamed. This is most safely effected by fixing a jet for the oxygen within the jet of hydrogen (fig. 106), so that the oxygen is introduced into the middle of the flame of hydrogen

Fig. 106.

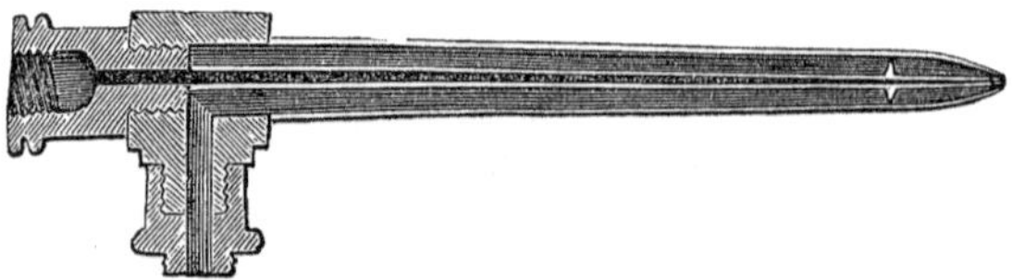

—a construction first proposed by Mr. Maugham, [first made and used by Professor Hare.—R. B.] and adapted to the use of coal-gas instead of hydrogen by Mr. Daniell. (Phil. Mag. 3d ser., vol. ii. p. 57.) Each of the gases may be more conveniently contained in a separate air-tight bag of Macintosh cloth capable of holding from 4 to 6 cubic feet of gas, and provided with press-boards. These require to be loaded with two or three 56lbs., when in use, to send out the gas with sufficient pressure. At this flame the most refractory substances, such as pipe-clay, silica and platinum, are fused with facility, and the latter even dissipated in the state of vapour. The flame itself, owing to the absence of solid matter, is scarcely luminous, but any of the less fusible earths, upon which it is thrown, —a mass of quick-lime, for instance (*a*, fig. 105)—is heated most intensely, and diffuses a light, which, for whiteness and brilliancy, may be compared to that of the sun. With the requisite supply of the gases, this light may be sustained for hours, care being taken to move the mass of lime slowly before the flame, so that the same surface may not be long acted upon; for the high irradiating power of the lime is soon impaired, it is supposed from a slight agglutination of its particles occasioned by the heat. This light, placed in the focus of a parabolic reflector, was found to be visible, in the direction in which it was thrown, at a distance of 69 miles, in one experiment made by Mr. Drummond, when using it as a signal light. The heating effects are even more intense when the gases are forced into a common receptacle, and allowed to escape from under pressure, but there is the greatest risk of the flame passing back through the exit tube and exploding the mixed gases; an accident which would expose the operator to the greatest danger. Mr. Hemming's apparatus, however, may be used without the least apprehension. A common bladder is used to hold the mixture, and the gas before reaching the jet, at which it is burned, is made to pass through his safety tube. This consists of a brass cylinder about six inches long and three-fourths of an inch wide, filled with fine brass wire of the same length, which is tightly wedged by forcibly inserting a pointed rod of metal into the centre of the bundle. The conducting power of the metallic channels through which the gas has then to pass is so great as completely to intercept the passage of flame. A similar safety tube of smaller size is interposed at *b*, in fig. 105, of the first arrangement.

Fig. 107.

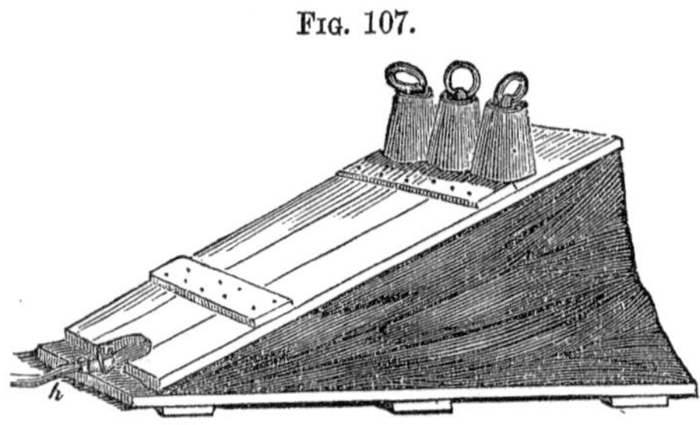

Hydrogen is capable of forming two compounds with oxygen, namely, water, which is the protoxide, and the binoxide of hydrogen.

The most important of the present applications of hydrogen gas is in the oxi-hydrogen blow-pipe. It has been superseded, as a material for inflating balloons, by coal gas, the balloon being proportionally enlarged to compensate for the less buoyancy of the latter gas.

PROTOXIDE OF HYDROGEN.—WATER.

Equivalent 9, *or* 112.5 *on the oxygen scale; formula* H+O, *or* HO; *density* 1; *as steam* 622 (*air* 1000); *combining measure of steam* ☐☐.

Mr. Cavendish first demonstrated, in 1781, that the product of the combustion of hydrogen and oxygen is water. He burned known quantities of these gases in a dry glass vessel, and found that water was formed in quantity exactly equal to the weights of the gases which disappeared. It was afterwards established by Humboldt and Gay-Lussac, that the gases unite rigorously in the proportion of two volumes of hydrogen to one volume of oxygen, and that the water produced by their union occupies, while it remains in the state of vapour, exactly two volumes (page 126). The proportion of the constituents of water by weight was determined with great care by Berzelius and Dulong. Their method was to transmit dry hydrogen gas over a known weight of the black oxide of copper, contained in a glass tube, and heated to redness by a lamp. The gas was afterwards conveyed through another weighed tube containing the hygrometric salt, chloride of calcium. The hydrogen gas in passing over the oxide of copper, combines with its oxygen and forms water, which is carried forward by the excess of hydrogen gas, and absorbed in the chloride of calcium tube. The weight of this water being ascertained, the proportion of oxygen it contains is determined by ascertaining the loss which the oxide of copper has sustained: the difference is the hydrogen.

The apparatus for such an experiment is illustrated in the following diagram (fig. 108). The oxide of copper to be reduced is contained in F, a small flask of

Fig. 108.

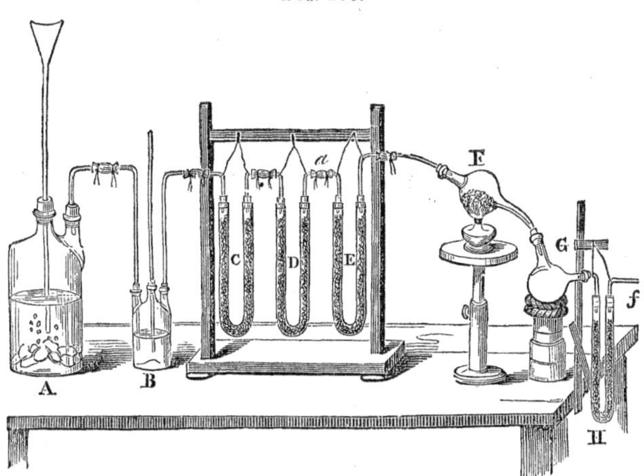

hard glass, having two openings, and heated by a spirit lamp. This flask communicates with another, G, intended to receive the greater part of the water produced in the experiment, which is followed by a bent tube H, containing fragments of pumice soaked in oil of vitriol, intended to receive the last portions. The hydrogen gas for this purpose must be very pure, and thoroughly dry. It is evolved slowly from a gas-bottle A, and passes through a second bottle B, and the bent tube C, both containing a concentrated solution of caustic potassa; and afterwards the bent tube D, containing a solution of chloride of mercury in pumice: and lastly through the bent tube F, containing oil of vitriol in pumice, proceeding thence entirely purified into F, and the excess of hydrogen gas escaping by *f*. Numerous most careful experiments, lately executed in this manner by M. Dumas, prove that water consists exactly by weight of—

Oxygen	88·91	8
Hydrogen	11·09	1
	100·00	9

The oxygen and hydrogen are therefore combined exactly in the proportion of 8 to 1, as appears by the proportions of the last column. This experiment serves not only to determine rigorously the composition of water, but it offers also the best method of ascertaining the composition of such metallic oxides as are de-oxidized by hydrogen.

Properties. — When cooled down to 32°, water freezes, if in a state of agitation, but may retain the liquid condition at a lower temperature, if at rest (page 60); the ice, however, into which it is converted cannot be heated above 32° without melting. Ice is lighter than water, its specific gravity being 0·916; and one of the forms (fig. 109) of its crystal is a rhomboid, very nearly resembling Iceland spar.

Fig. 109.

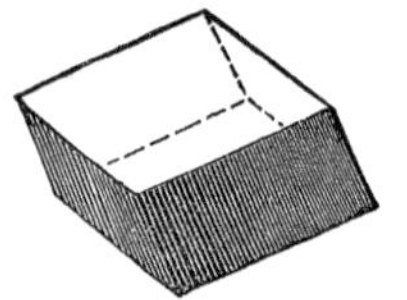
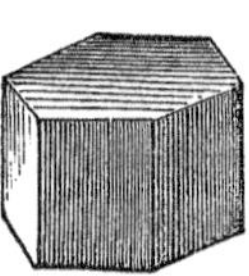
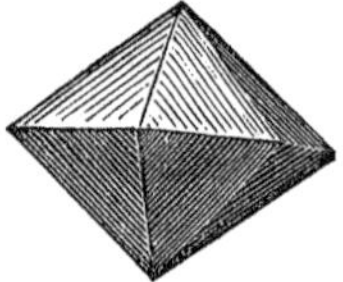

Water is elastic and compressible, yielding, according to Oersted, 53 millionths of its bulk to the pressure of the atmosphere, and, like air, in proportion to the compressing force for different pressures. The peculiarities of its expansion by heat, while liquid, have already been fully described (page 38). Under a barometric pressure of 30 inches, it boils at 212°, but evaporates at all inferior temperatures. Its boiling point is elevated by the solution of salts in it, and the temperature of the steam from these solutions is not constantly 212°, as has been alleged, but that of the last strata of liquid through which the steam has passed. When mixed with air, the vapour of water has a tendency to condense, it is said, in vesicles, which inclose air; forming in this condition the masses of clouds, which remain suspended in the atmosphere from the lightness of the vesicles, the substance of mists and fogs, and "vapour" generally, in its popular meaning. The vesicles may be observed by a lens of an inch in focal length, over the dark surface of hot tea or coffee, mixed with an occasional solid drop which contrasts with them. According to the experiments of Saussure, made upon the mists of high mountains, these vesicles generally vary in size from the 1-4500th to the 1-2780th of an inch, but are occasionally observed as large as a pea. They are generally condensed by their collision into solid drops, and fall as rain; but their precipitation in that form is much retarded in some conditions of the atmosphere. It is proper to add, however, that Prof. J. Forbes and several other eminent meteorologists disbelieve entirely the existence of vesicular vapour.

It was lately discovered by Mr. Grove that the vapour of water is decomposed to a small but sensible extent by an exceedingly high temperature, and resolved into its constituent gases. If a small ball of platinum, of the size of a large pea, with a wire attached to it, be heated in the flame of the oxi-hydrogen blow-pipe to bright whiteness, and till it begins to show symptoms of fusion, and then plunged into hot water, minute bubbles of gas rise with the steam, which consist of a mixture of oxygen and hydrogen. Only a small portion of the steam, not amounting to even one-thousandth part of the whole produced (it is supposed) suffers decomposition. The occurrence of a decomposition in such circumstances, which is unquestionable, appears singular, seeing that oxygen and hydrogen certainly combine at the same, or even a higher, temperature in the flame of the blow-pipe, which is employed to heat the platinum ball. The combustion in the blow-pipe may, indeed, be incomplete, but this is unlikely, for I find that when the mixed gases are exploded in a glass tube, the combustion is so complete that certainly not one part in four thousand, if any portion whatever, escapes combustion. It is a question whether the decomposition of the steam by ignited platinum is not an exhibition of the deoxidizing action of light rather than the effect of heat; the blow-pipe flame itself being scarcely visible, while the decomposing platinum, although necessarily of a lower temperature, is highly incandescent.

A cubic inch of water at 62°, Bar. 30 inches, weighs in air 252.458 grains. The imperial gallon has been defined to contain 10 pounds avoirdupois (70,000 grains) of distilled water at that temperature and pressure. Its capacity is therefore 277.19 cubic inches. The specific gravity of water at 60° is 1, being the unit to which the densities of all other liquids and solids are conveniently referred; it is 815 times heavier than air at that temperature.

In its chemical relations water is eminently a neutral body. Its range of affinity is exceedingly extensive, water forming definite compounds, to all of which the name *hydrate* is applied, with both acids and alkalies, with a large proportion of the salts, and indeed with most bodies containing oxygen. It is also the most general of all solvents. Gay-Lussac has observed that the solution of a salt is uniformly attended with the production of cold, whether the salt be anhydrous or hydrated, and that, on the contrary, the formation of a definite hydrate is always attended with heat; a circumstance which indicates an essential difference between solution and chemical combination (Ann. de Ch. et de Phys. t. lxx. p. 407). Even the dilution of strong solutions of some salts, such as those of ammonia, occasions a fall of temperature. The solvent power of water for most bodies increases with its temperature. Thus at 57° water dissolves one-fourth of its weight of nitre, at 92° one-half, at 131° an equal weight, and at 212° twice its weight of that salt. Solutions of such salts, saturated at a high temperature, deposit crystals on cooling. But the crystallization of some saturated solutions is often suspended for a time, in a remarkable manner, and afterwards determined by slight causes. Thus, if two pounds of crystallized sulphate of soda be dissolved in one pound of water, with the assistance of heat, and the solution be filtered while hot through paper, to remove foreign solid particles, and then set aside in a glass matrass, with a few drops of oil on its surface, it may become perfectly cold without crystallization occurring. Violent agitation even may not cause it to crystallize. But when any solid body, such as the point of a glass rod, or a grain of salt, is introduced into the solution, crystals immediately begin to form about the solid nucleus, and shoot out in all directions through the liquid. The solubility of many salts of soda and lime does not increase with the temperature, like that of other salts.

Water is also capable of dissolving a certain quantity of air and other gases, which may again be expelled from it by boiling the water, or by placing it in vacuo. Rain-water generally affords 2½ per cent. of its bulk of air, in which the proportion of oxygen gas is so high as 32 per cent., and in water from freshly melted snow 34.8 per cent., according to the observations of Gay-Lussac and Humboldt, while the oxygen in atmospheric air does not exceed 21 per cent. Boussingault finds that the quantity of air retained by water, at an altitude of 6 or 8000 feet, is reduced to one-third of its usual proportion. Hence it is that fishes cannot live in Alpine lakes, the air contained in the water not being in adequate quantity for their respiration. The following table exhibits the absorbability of different gases by water deprived of all its air by ebullition: —

100 cubic inches of water at 60° and 30 Bar., absorb of

	Dalton.	Henry.	Saussure.
Hydrosulphuric acid............	100 C. I.	106	253
Carbonic acid.....................	100	100	106
Nitrous oxide.....................	100	77.6	76
Olefiant gas	12.5	14	15.5
Oxygen..............................	3.7	3.55	6.5
Carbonic oxide	1.56	2.01	6.2
Nitrogen............................	1.56	1.47	4.2
Hydrogen	1.64	1.53	4.6

The results of Saussure are probably nearest the truth for hydrosulphuric acid and nitrous oxide, but for the other gases those of Dalton (Manchester Memoirs, 2d ser. p. 287) and Henry (Phil. Trans. 1843, pp. 29, 274) are most to be depended on.

Uses. — Rain received after it has continued to fall for some time may be taken as pure water, excepting for the air it contains. But after once touching the soil, it becomes impregnated with various earthy and organic matters, from which it can only be completely purified by distillation. A copper still should be used for that purpose, provided with a copper or block-tin worm, which is not used for the distillation of spirits, as traces of alcohol remaining in the worm and becoming acetic acid, cause the formation of acetate of copper, which would be washed out and contaminate the distilled water. The use of white lead cement about the joinings of the worm is also to be avoided, as the oxide of lead is readily dissolved by distilled water. The first portions of the distilled water should be rejected, as they often contain ammonia, and the distillation should not be carried to dryness.

Water employed for economical purposes is generally submitted to a more simple process, that of filtration, by which it is rendered clear and transparent by the removal of matter mechanically suspended in it. Such foreign matter may often be removed in a considerable degree by subsidence, on which account it is desirable that the water should stand at rest for a time, before being filtered. The filtration of liquids generally is effected, on the small scale, by allowing them to flow through unsized or filter paper, and that of water, on the large scale, by passing it through beds of sand. The sand preferred for that purpose is not fine, but gravelly, and crushed cinders or furnace clinkers may be substituted for it. Its function, as that also of the paper in the chemist's filter, is to act as a *support* for the finer particles of mud or precipitate which are first deposited on its surface, and form the bed that really filters the water. When the mud accumulates so as to impede the action of the sand filter, the surface of the sand is scraped, and an inch or two of it removed.

Fig. 110 is a section of the water-filter, as it is usually constructed for public

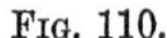
Fig. 110.

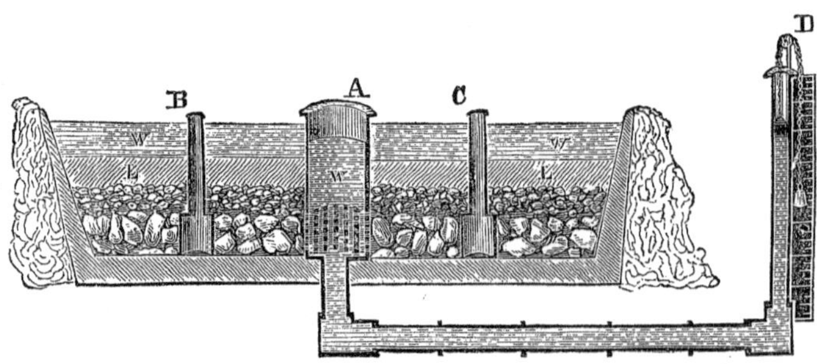

works in Lancashire. An excavation of about six feet in depth, and of sufficient extent, is lined to a considerable thickness with well puddled clay, to make it water-tight. Upon the clay floor is laid first a stratum of large stones, then a stratum of smaller stones, and, finally, a bed of coarse sand or gravel, L L. To allow the air to escape from the lower beds, small upright tubes, open at both ends, B and C, are inserted in these beds, and rising above the surface of the water W W. The filtered water enters, from the lowest bed, into a large open iron cylinder A, the lower part of which is perforated for that purpose. The filtered water stands at the same height in the gauge tube D as in A; this height is observed by means of a float balanced by a weight which traverses a scale of feet and inches at D.

Upward filtration through a bed of sand is sometimes practised, but it has the disadvantage that the filter cannot be cleaned in the manner indicated. Filtering under high pressure, and with great rapidity, has been practised in a very compact apparatus, consisting of a box, not above three feet square, filled with sand. This filter, which becomes speedily choked with the mud it detains, is cleansed by suddenly reversing the direction in which the water is passing through the box, which occasions a shock that has the effect of loosening the sand, and allowing the water to bring away the mud. The action of such a filter, erected at the Hôtel-Dieu of Paris, was favourably reported on by M. Arago (Annal. de Chim. et de Phys. t. lxv. p. 428).

Matter actually dissolved in water is not affected by filtration. No repetition of the process would withdraw the salt from sea-water and make it fresh. Hence the impregnation of peaty matter, which river water generally contains, and to the greatest extent in summer, when the water is concentrated by evaporation, is not removed by filtering. Animal charcoal is the proper substance for discolouring liquids, as it withdraws organic colouring matter, even when in a state of solution.

In the process of clarifying liquors by dissolving in them the white of egg and other albuminous fluids, the temperature is raised so as to coagulate the albumen, which thus forms a delicate net-work throughout the liquid, and is afterwards thrown up as scum in the boiling, carrying all the foreign matter suspended in the liquid along with it.

Gelatine, isinglass, or other "finings," added to wine in a turbid state, produce a precipitate with its tannin, which carries down all suspended matter; and on the settling of this precipitate, or its separation by filtering, the wine is found transparent.

The most usual earthy impurities in water, occasioning its hardness, are sulphate of lime, and the carbonate of lime dissolved in carbonic acid, both of which are precipitated on boiling the water, and occasion an earthy incrustation of the boiler. So far as this precipitation is due to carbonate of lime it may be avoided by adding hydrochlorate of ammonia to the water, by which the lime is converted into chloride of calcium and becomes soluble. Water containing carbonate of lime may be also softened by the addition of lime-water, as recommended by Professor Clark. Thames water requires for this purpose the addition of about one-fourteenth of its bulk of lime-water. This action of lime-water will be explained under carbonic acid.

When waters contain iron, they are termed *chalybeate:* this metal is most frequently in the state of carbonate dissolved in carbonic acid, and rarely in a proportion exceeding one grain in a pound of water. The *sulphurous* waters, which are recognised by their peculiar odour, and by blackening silver and salts of lead, contain hydrosulphuric acid in a proportion not exceeding the usual proportion of air in spring water, and generally no oxygen. *Saline* waters for the most part contain various salts of lime and magnesia, and generally common salt. Their density is always considerably higher than that of pure water. *Sea-water* contains 3½ per cent. of saline matter, and has a density 1.0274. Its composition is interesting, as the sea comes to be the grand depository of all the soluble matter of the globe. A minute analysis of the water of the English Channel, executed by Mr. Schweitzer, is subjoined:—

Sea-water of the English Channel.	Grains.
Water	964.74372
Chloride of sodium	27.05948
Chloride of potassium	0.76552
Chloride of magnesium	3.66658
Bromide of magnesium	0.02929
Sulphate of magnesia	2.29578
Sulphate of lime	1.40662
Carbonate of lime	0.03301
	1000.0000

In addition to those constituents, distinct traces of iodine and of ammonia were detected (Phil. Mag. 3d ser. vol. xv. p. 58). According to Professor Forchammer, the whole quantity of saline matter in water from different parts of the Atlantic varied from 35.7 parts (German sea) to 36.6 parts (tropics) in 1000 parts of the water. The relative proportion of the salts in the water of different seas varied very little (Reports of the British Association, 1846, p. 90).

BINOXIDE OF HYDROGEN.

Equivalent, 17, *or* 212.5 *on Oxygen Scale; formula* $H+2O$ *or* HO_2.

The second compound of hydrogen and oxygen is a liquid, containing twice as much oxygen as water, and is a body possessed of very extraordinary properties. It was discovered by Thenard, in 1818, who prepared it by a long and intricate process.

Preparation.—The formation of the binoxide of hydrogen depends upon the existence of a corresponding binoxide of barium. The latter is obtained by calcining pure nitrate of baryta at a high temperature in a porcelain retort, and afterwards exposing the earth baryta or protoxide of barium, which is left, in a porcelain tube heated to redness, to a stream of oxygen gas, which the protoxide rapidly absorbs, becoming binoxide. Treated with a little water, the binoxide of barium slakes and falls to powder, forming a hydrate, of which the formula is BaO_2+HO. Dilute acids have a peculiar action upon this hydrate, which will be easily understood, if the binoxide of barium is represented as the protoxide united with an additional equivalent of oxygen, or as $BaO+O$. They combine with the protoxide of barium, forming salts of baryta, and the second equivalent of oxygen, instead of being liberated in consequence, unites with the water of the hydrate, the HO of the preceding formula giving rise to $HO+O$ or the binoxide of hydrogen, which dissolves in the water. Although it would be inconvenient to abandon the systematic name binoxide of hydrogen for this compound, still it must be allowed that the properties of the body, as well as its mode of preparation, are more favourable to the idea of its being a combination of water with oxygen, or *oxygenated water*, as it was first named by its discoverer, than a direct combination of its elements. It is recommended by Thenard to dissolve the binoxide of barium in hydrochloric acid considerably diluted with water, and to remove the baryta by sulphuric acid, which forms an insoluble sulphate of baryta. The hydrochloric acid, again free in the liquor, is saturated a second time with binoxide of barium, and precipitated; and after several repetitions of these two operations, the hydrochloric acid itself is removed by the cautious addition of sulphate of silver, and the sulphuric acid of the last salt by solid baryta. Such is an outline of the process, but its success requires attention to a number of minute precautions, which are fully detailed in the Traité de Chemie of the author quoted (vol. i. p. 479, 6th ed.) The weak solution of binoxide of hydrogen, which this process affords, may be concentrated by placing it with a vessel of strong sulphuric acid under the receiver of an air-pump, until the solution attains a density of 1.452, when the binoxide itself begins to rise in vapour without change. It then contains 475 times its volume of oxygen.

M. Pelouze abridges this process considerably by employing hydro-fluoric acid or fluosilicic acid, in place of hydrochloric acid, to decompose the binoxide of barium. By this operation, the baryta separates at once with the acid, in the state of the insoluble fluoride of barium, and nothing remains in solution but the binoxide of hydrogen. After thus decomposing several portions of binoxide of barium successively in the same liquor, the fluoride of barium may be separated by filtration, and the binoxide of hydrogen, which is still dilute, be concentrated by means of the air-pump.

Properties.—Binoxide of hydrogen is a colourless liquid resembling water, but less volatile, having a metallic taste, and instantly bleaching litmus and other organic colouring matters. It is decomposed with extreme facility, effervescing from escape of oxygen at a temperature of 59°, and when suddenly exposed to a greater heat, such as 212°, actually exploding from the rapid evolution of that gas. It is rendered more permanent by dilution with water, and still more so by the addition of the stronger acids, while alkalies have the opposite effect.

The circumstances attending the decomposition of this body are the most curious facts in its history. Many pure metals and metallic oxides occasion its instantaneous resolution into water and oxygen gas, by simple contact, without undergoing any change themselves, affording a striking illustration of catalysis (page 186); and this decomposition may excite an intense temperature, the glass tube in which the experiment is made sometimes becoming red hot. Some protoxides absorb at the same time a portion of the oxygen evolved, and are raised to a higher degree of oxidation, but most of them do not; and certain oxides, such as the oxides of silver and gold, are reduced to the metallic state, their own oxygen going off along with that of the binoxide of hydrogen. The decomposition of these metallic oxides cannot be ascribed to the heat evolved, for oxide of silver is reduced in a very dilute solution of the binoxide of hydrogen, although the decomposition is not then attended with a sensible elevation of temperature. The metallic oxides which are decomposed in this remarkable manner are originally formed by the decomposition of other compounds, and not by the direct union of their elements, which, in fact, exhibit little affinity for each other. In this general character they agree with binoxide of hydrogen itself.

Uses.—The binoxide of hydrogen is a substance which it is exceedingly desirable to possess, with the view of employing it in bleaching, and for other purposes, as a powerful oxidating agent. But the expense and uncertainty of the process for preparing this compound have hitherto prevented any application of it in the arts, or even its occasional use as a chemical re-agent.

SECTION III.

NITROGEN.

Synonyme, AZOTE. *Equiv.* 14, *or* 175 (O=100); *symbol* N; *density* 971.37; *combining measure* [|] .

Dr. Rutherford, of Edinburgh, examined the air which remains after the respiration of an animal, and found that after being washed with lime-water, which removes carbonic acid, it was incapable of supporting either combustion or respiration. He concluded that it was a peculiar gas. Lavoisier afterwards discovered that this gas exists in the air of the atmosphere, forming indeed 4-5ths of that mixture, and gave it the name azote, (from α, privative, and ζωη, life), from its inability to support respiration. It was afterwards named nitrogen by Chaptal, because it is an element of nitric acid. Besides existing in air, nitrogen forms a constituent of most animal and of many vegetable substances. In a natural arrangement of the elements, nitrogen appears to have its place between oxygen and phosphorus (page 147).

Preparation.—Nitrogen is generally procured by allowing a combustible body to combine with the oxygen of a certain quantity of air confined in a vessel. For that purpose a little metallic or porcelain cup may be floated, by means of a cork, on the surface of the water-trough. A few drops of alcohol are then introduced into the cup, or a small piece of phosphorus is placed in it, and being kindled, a tall bell jar is held over the cup, with its lip in the water. The combustion soon terminates, and the water of the trough rises in the jar. Alcohol does not consume the oxygen entirely, a small portion of it still remaining mingled with the nitrogen; a certain quantity of carbonic acid gas is also produced by its combustion. But the combustion of phosphorus exhausts the oxygen completely, and leaves nitrogen unmixed with any other gas.

Fig. 111.

Nitrogen may be likewise conveniently obtained by conducting chlorine gas into diluted ammonia. For delicate purposes of research this gas is best prepared by carrying air through a tube filled with reduced metallic copper in a pulverulent form, and heated to redness, by which the oxygen is entirely absorbed.

Properties.—Nitrogen gas is tasteless and inodorous; has never been liquefied, and is less soluble in water than oxygen. It is a little lighter than air, (specific gravity .9714), which possesses the mean density of 79.1 volumes of nitrogen and 20.9 volumes of oxygen. Nitrogen is a singularly inert substance, and does not unite directly with any other single element, so far as I am aware, under the influence of light or of a high temperature, unless, perhaps, oxygen and carbon. A burning taper is instantly extinguished in this gas, and an animal soon dies in it, not because the gas is injurious, but from the privation of oxygen, which is required in the respiration of animals. Nitrogen appears to be chiefly useful in the atmosphere, as a diluent of the oxygen, thereby repressing to a certain degree the activity of combustion and other oxidating processes. Of the fixation of free nitrogen of plants, there is no evidence. When heated with oxygen, nitrogen does not burn like hydrogen, nor undergo oxidation. But nitrogen may be made to unite with oxygen by transmitting several hundred electric sparks through a mixture of these gases in a tube, with water or an alkali present, and nitric acid is produced. The water formed by the combustion of hydrogen in air, or of a mixture of hydrogen and nitrogen in oxygen, has often an acid reaction, which is due to a trace of nitric acid. But when the hydrogen is mixed with air in excess, so as to prevent great elevation of temperature during the combustion, the oxidation of the nitrogen does not take place (Kolbe). Nitric acid is also a product of the oxidation of a variety of compounds containing nitrogen. Ammonia mixed with air, on passing over spongy platinum at a temperature of about 572°, is decomposed, and the nitrogen it contains is completely converted into nitric acid, by combining with the oxygen of the air. Cyanogen and air, under similar circumstances, occasion the formation of nitric and carbonic acids. (Kuhlman, Phil. Mag. 3d ser., vol. xiv. p. 157). Nitric acid is also largely produced by the oxidation of organic matters during putrefaction in air, when an alkali or lime is present, as in the natural nitre soils and artificial nitre beds.

A suspicion has always existed that nitrogen may be a compound body, but it has resisted all attempts to decompose it, and the evidence of its elementary character is equally good with that of most other bodies reputed simple. Before considering the compounds of nitrogen with oxygen, we may notice the properties of atmospheric air, which is regarded as a mechanical mixture of these gases.

THE ATMOSPHERE.

According to the new and most careful determination of the weight of air by M. Regnault, 100 cubic inches of atmospheric air, deprived of aqueous vapour and the small quantity of carbonic acid it usually contains, weigh 30.82926 grains, at 60° and 30 Bar. Its density at the same temperature and pressure is estimated at 1000, and is conveniently assumed as the standard of comparison for the densities of gaseous bodies, as water is for solids and liquids. Hence, at 62°, air is 810 times lighter than water, and 11,000 times lighter than mercury. The bulk of air varies with its temperature and the pressure affecting it, according to the same laws as other gases (pages 40 and 81).[1]

The mean pressure of the atmosphere at the surface of the sea is generally estimated as equal to the weight of a column of mercury of 30 inches in height, which is about 15 pounds on the square inch of surface, and is equivalent to a column of water of nearly 34 feet in height. The oxygen alone is equal to a column of 7.8 feet of water over the whole earth's surface, from which an idea may be formed of the immense quantity of that element, and how small the effect must be of the oxidating processes observed at the earth's surface in diminishing it. If the atmosphere were of uniform density, its height, as inferred from the barometer, would be 11,000 times 30 inches, or 5.208 miles, but the density of air being proportional to the pressure upon it, diminishes with its elevation, the superior strata being always more rare and expanded than the inferior strata upon which they press.

DENSITY OF THE ATMOSPHERE.

Height above the sea in miles.	Volume.
0	1
2·705	2
5·41	4
8·115	8
10·82	16
13·424	32
16·23	64

At a height of 2.705 miles (11,556 feet) the atmosphere is of half density, by calculation, or one volume is expanded into 2, and the barometer would stand at 15 inches; the density is again halved for every 2.7 miles additional elevation. From calculations founded on the phenomena of refraction, the atmosphere is supposed to

---

[1] I. Weight of 1 Litre of Gases, at 0° C., Bar. 0.76 metre (Regnault).

	In Grammes.
Atmospheric Air	1.293187
Nitrogen	1.256167
Oxygen	1.429802
Hydrogen	0.089578
Carbonic Acid	1.977414

II. Weight of 100 Cubic Inches of Gases; Bar. 29.92 inches.

	At 32° F. In Grains.	At 60° F. In Grains.
Atmospheric Air	32.58684	30.82926
Nitrogen	31.66020	29.95260
Oxygen	36.13896	34.18979
Hydrogen	2.16216	2.04554
Carbonic Acid	50.03856	47.33972

Here the French litre is taken at 61.028 English cubic inches; the gramme at 15.4440 grains; and the volume of air and the other gases, at 60°, 1.05701, their volume at 32° being 1. (Regnault, Compt. Rend. t. 20, p. 975).

extend, in a state of sensible density, to a height of nearly 45 miles. It is certainly limited, probably from the expansibility of the aërial particles having a natural limit (page 81). The atmospheric pressure also varies at the same place, from the effect of winds and other causes, which are not fully understood. Hence the use of the barometer as a weather glass; for wet and stormy weather is generally preceded by a fall of the mercury in the barometer, and fair and calm weather by its rise.

The temperature of the atmosphere is greatest at the earth's surface, and has been observed to diminish one degree for every 352 feet of ascent, in the lower strata. It is believed, however, that the progressive diminution is less rapid at great distances from the earth. But at a certain height, the region of perpetual congelation is attained even in the warmest climates; the summits of the Andes, which rise 21,000 feet, being perpetually covered with snow under the equator. The line of perpetual congelation, which has been fixed at 15,207 feet at 0° latitude, descends progressively in higher latitudes, being 3,818 feet at 60°, and only 1,016 feet at 75°. The decrease of temperature with elevation in the atmosphere is ascribed to two causes. 1. To the property which air has of becoming cold by expansion, which arises from an increase of its latent heat with rarefaction. The actual temperature of the different strata of the atmosphere is indeed believed to be that due to their dilatation, supposing that they had all the same original temperature and density as the lowest stratum. 2. To the circumstance that the atmosphere derives its heat principally from contact with the earth's surface. The sun's rays appear to suffer little absorption in passing through the atmosphere; but there are some observations on the force of solar radiation which are not easily reconciled with that circumstance. A thermometer of which the bulb is blackened, rises a certain number of degrees above the temperature of the air, when exposed to the sun, but the rise is decidedly greater on high mountains than near the level of the sea, and in temperate, or even arctic climates, which is more remarkable, than within the tropics. It is a question how solar radiation is obstructed in the hotter climates (Daniell's Meteorological Essays, 2d edit.)

The blue colour of the sky has been found by Brewster to be due to light that has suffered polarization, which is therefore reflected light, like the white light of clouds. The air of the atmosphere must therefore have a disposition to absorb the red and yellow solar rays, and to reflect the blue rays. At great heights, the blue colour of the sky was observed by Theodore de Saussure to become deeper and deeper, being mixed with black, owing to the absence of white reflecting vapour or clouds. The red and golden tints of clouds appear to be connected with a remarkable property of steam observed by Professor J. Forbes. A light seen at night through steam issuing into the atmosphere from under a pressure of from 5 to 30 pounds on the inch, is found to appear of a deep orange red colour, exactly as if observed through a bottle containing nitrous acid vapour. The steam, when it possesses this colour, is mixed with air, and on the verge of condensation; and it is known that the golden hues of sunset depend upon a large proportion of vapour in the air, and are indeed a popular prognostic of rain (Phil. Mag. 3d ser. vol. xiv. pp. 121, 425, and vol. xv. pp. 25, 419.)

Winds.—The movement of masses of air, or wind, is always produced by inequality of temperature of the atmosphere at different points of the earth's surface, or in different regions of the atmosphere of equal elevation. The primary movement is always an ascending current, the heated and expanded air over some spot rising in a vertical column. Dense and colder air flows towards that point, producing the horizontal current which is remarked by an observer on the earth's surface. Some winds are of a very limited range, and depend upon local circumstances; such are the sea and land breeze experienced upon the coasts of tropical countries. From its low conducting power, the surface of the land is more quickly heated than the sea, so that soon after sunrise the expanded air over the former begins to ascend, and is replaced by colder air from the sea, forming the sea-breeze. But after sunset, the earth's heat, being less in quantity, is more quickly dissipated by radiation than that of the sea, and the air over the land becomes dense and flows outwards, displacing the air over the sea, and producing the land-breeze. It is obvious that these inferior currents must be attended by a superior current in an opposite direction, or that the air in these winds is carried in a perpendicular vortex of no great extent, of which the motion is reversed twice every twenty-four hours. A grand movement of a similar nature is produced in the atmosphere, from the high temperature of the equatorial compared with the polar regions of the globe; the air over the former constantly ascending, and having its place supplied by horizontal currents from the latter, within the lower region of the atmosphere. Hence, if the earth were at rest, the wind would constantly blow at its surface, from the poles to the equator, and in the opposite direction in the upper strata of the atmosphere. But the earth, accompanied by its atmosphere, makes a diurnal revolution upon its axis, in which any point on its surface is always passing to a point in space previously to the east of it, and with a velocity proportional to its circle of latitude on the globe; a velocity which is consequently nothing at the poles, and attains its maximum at the equator. The result of this is, that the lower current or polar stream, in tending to the equator, is constantly passing over parallels of latitude which have a greater degree of velocity of rotation to the east, than the stream itself, which comes thus to be felt as a resistance from the east; and instead of appearing as a wind directly from the north, as it really is, this stream appears as a wind from the east, with a certain northerly declination, which diminishes as the stream approaches the equator, where it flows directly from the east, constituting the great trade-wind which constantly blows across the Atlantic and Pacific Oceans from east to west within the tropics. Our keen east winds in spring have a low temperature, which attests their arctic origin. The upper or equatorial current has its course deflected by similar causes; starting from the equator it has a greater projectile force to the east than the parallels of latitude over which it has to pass, and retaining this motion towards the east it appears, as it passes over them, a west wind or wind from the west. The upper current, flowing in the opposite direction from the trade-wind below, was actually experienced by Humboldt and Bonpland on the summit of the Peak of Teneriffe, and has been indicated at various times by the transport of volcanic ashes by its means.

These currents, instead of flowing in a uniform manner over and under each other, appear often to descend, and to flow side by side, giving rise to hot and cold seasons in their different courses, and the great variability of climate of the temperate zone. On the great oceans, within the temperate zone, westerly winds prevail greatly over easterly, which are supposed by some to be the upper current descending to the surface of the earth. These westerly winds temper the climate of the western seaboard both of Europe and America, which is much milder than the climate of their eastern coasts.

The nature of the movement of the atmosphere in hurricanes has lately received considerable elucidation. It appears that they move in circles, and are great horizontal vortices, which are probably produced by currents of air meeting obliquely, like the little eddies or whirlwinds formed at the corner of streets. The whole vortex also travels, but its movement of translation is slow compared with its velocity of rotation (Colonel Reid on the Law of Storms; also the work of Mr. Espy).

Some hurricanes in the United States have a path of only a few hundred yards in width, but extending for many miles. An interesting theory of the origin of these, and many other local winds, has been proposed by Mr. Espy, and favourably reported upon by M. Babinet, to the French Institute. When a column of air, saturated with vapour at a high temperature, ascends in the atmosphere, it expands by the removal of pressure and becomes colder, as happens with dry air of the same temperature. But on being cooled to a certain point of temperature by its ascent, vapour condenses in the former, and raising the temperature of the column makes it specifically lighter and more buoyant. The ascent of damp air has thus a tendency to perpetuate itself, and may give rise to a most powerful upward aspiration, as is shown by calculation, quite adequate to prostrate trees, and produce the mechanical effects observed; the whole funnel being carried over the surface of the earth by a more general movement of the atmosphere.

Vapour. — The properties of the atmosphere are much affected by the presence of watery vapour in it, which it acquires from contact with the surface of the sea, lakes, rivers, and humid soil. The quantity which can rise into the air is limited by its temperature (page 90), and comes to be deposited again from various causes. The surface of the earth is cooled by radiation, and occasions the precipitation of dew from the air in contact with it. Vapour is also condensed into drops, from various agencics within the atmosphere itself. The following are the principal causes of clouds and rain. 1. The ascent of air in the atmosphere, and its consequent rarefaction, which is attended with cold. A cloud will be observed within the receiver of an air-pump, on the plate of which a little water has been spilt, on making two or three rapid strokes of the pump, which is due to this cause. It is observed in operation in the formation of the clouds and mists which settle on the summits of mountains. The wind passing over the surface of a level country is impeded by a mountain; rising in the atmosphere the stream overcomes thc obstacle, and produces a cloud as it passes over the mountain, which appears stationary on its summit. 2. The mixing of opposite currents of hot and cold air, both saturated with humidity, may occasion rain, from the circumstance, first conjectured by Dr. Hutton, that the currents of air on mixing and attaining a mean temperature are incapable of sustaining the mean quantity of vapour. Thus, supposing equal volumes of air at 60° and 40°, both saturated with vapour, to be mixed, the tension of vapour at the former temperature being the 0.524th of an inch of mercury, and at the latter the 0.263d of an inch, the mean tension is 0.393d of an inch. But the tension of vapour at 50°, the intermediate temperature is only the 0.375th of an inch; and consequently the excess of the former tension, or vapour of the 0.018th of an inch of tension, must condense as rain. But this is an inconsiderable cause of rain compared with the next. 3. Contact of air in motion with the cold surface of earth, or mere proximity, appears to be the most usual cause of its refrigeration, and of the precipitation of rain from it. The mean temperature of January in this country is about 34°, but with a south-west wind the thermometer may be observed gradually to rise in the course of 48 hours to 54°. Now supposing this wind to be saturated with vapour at 54°, and to be cooled to 34°, as it is on its first arrival, the moisture which it must deposit is very considerable, as will appear by the following calculation: —

Tension of vapour at 54°	0.429	inch.
" " at 34°	0.214	"
Condensed	0.215	"

The mean annual fall of rain in London amounts to a column of 23 inches. The quantity collected by a rain-gauge is found to be affected to an extraordinary extent by very moderate differences of elevation. Thus the annual fall of rain in three situations was found, by Professor J. Phillips, to be as follows: —

	Inches.	Height.
Top of York Minster	15·910	242 feet.
Roof of Museum	20.461	73 "
Surface of ground	24.401	0 "

The last stated cause of rain throws some light on this inequality: the air is more cooled near the ground, and therefore deposits most humidity.

The annual fall is greater near the equator, and diminishes in high latitudes. At Granada (lat. 12° N.), it is 126 inches; at Calcutta (lat. 19° 46′), 81 inches; Rome, 39 inches; average of England, 31 inches; St. Petersburgh, 16 inches;

Uleaborg, 13½ inches. The number of rainy days follows a different proportion, the average during the year being about as follows:—

In Northern Europe	180
In Central Europe	146
In Southern Europe	120[1]

When clouds form at temperatures below 32°, the aqueous vapour is converted into an infinity of little needle-like crystals, which often diverge from each other at angles of 60° and 120°, as do also the thin crystals in freezing water. Snow differs very much in the arrangement of these spiculæ (fig. 112), but the flakes are all of the same configuration in the same storm. The figures are essentially referable to a hexagonal star or prism, one of the crystalline forms of ice. *Hail* is also produced by cold, but in circumstances which are entirely different. It occurs only in summer or in warm climates, and when the sun is above the horizon. It seems to be produced in a humid ascending current of air, greatly cooled by rarefaction, which has an upward velocity sufficient to sustain the falling hailstones at the same place till they attain considerable magnitude. The formation of hail is always attended with thunder or signs of electricity; and it has been found that small districts may be protected from its devastations by the elevation of many thunder rods.

Fig. 112.

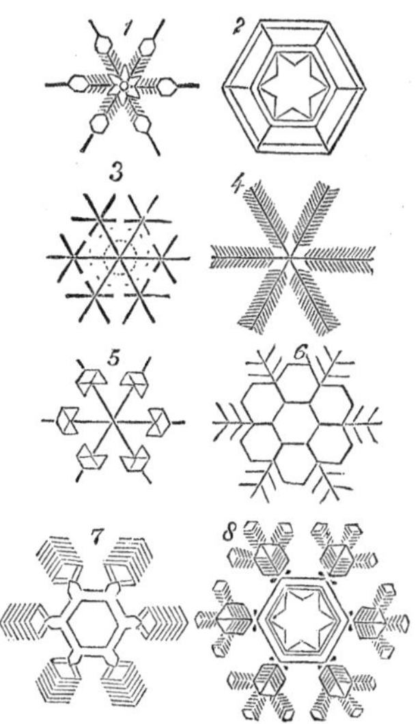

Analysis of air.—A knowledge of the composition of the atmosphere followed that of its constituent gases. Various modes of analysis are practised:—1. A stick of phosphorus introduced into a known measure of air in a graduated tube, effects a complete absorption of the oxygen in 24 hours. On afterwards withdrawing the phosphorus the diminution of volume may be observed, which always indicates 20 or 21 per cent. of oxygen. 2. A known measure of air may be mixed with a slight excess of hydrogen more than sufficient to combine with its oxygen, 100 volumes of air, for example, with 50 volumes of hydrogen, and the mixture exploded in a strong glass tube of proper construction, by means of the electric spark. The diminution in volume of the gases after combustion is observed; and as oxygen and hydrogen unite in the exact ratio of one volume of the first to two volumes of the second, one-third of the diminution represents the volume of oxygen in the measure of air employed. The tube used for this purpose is called the voltaic eudiometer. The syphon eudiometer is a convenient instrument of this kind. It is formed of a straight tube moderately stout, of about 1-4th or 3-8ths of an inch internal diameter, sealed at one end, and about 22 inches long. The closed end of this tube being softened by heat, two stout platinum wires are thrust through the glass from opposite sides of the tubes, so that their extremities in the tube approach within one-tenth of an inch of each other. These are intended for the transmission of the electric spark, and are retained, as if cemented, in the apertures of the glass when the latter cools. One-half the tube next the closed end is afterwards graduated into hundredths of a cubic inch, and the tube is bent in the middle, like a syphon, as represented by *a* in the figure. By a little dexterity, a portion of the gaseous mixture to be exploded is

Fig. 113.

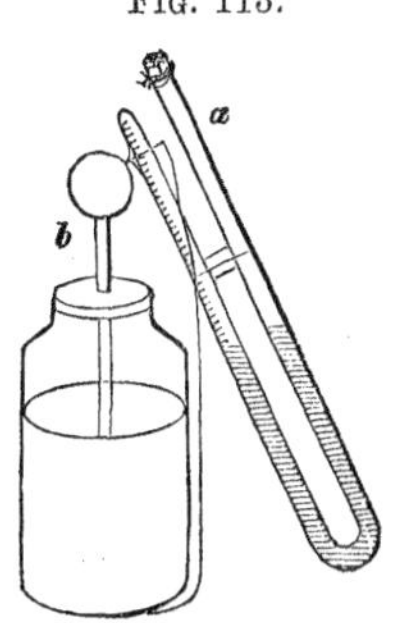

[1] See Müller's Physics and Meteorology, and Kämtz's Meteorology, by Walker.

transferred to the sealed limb of the instrument, at the water or mercurial trough, and the measure noted with the liquid at the same height in both limbs. The mouth of the open limb may then be closed by a cork, which can be fixed down by soft copper wire. A chain being now hung to one platinum wire, the other is presented to the prime conductor of an electric machine, or the knob of a charged Leyden phial *b*, so as to take a spark through the mixture, which is thereby exploded. The risk of the tube being broken by the explosion, which is considerable in the ordinary form of the eudiometer, is completely avoided in this instrument by the compression of the air retained by the cork in the open limb, this air acting as a recoil spring upon the occurrence of the explosion in the other limb. 3. The combustion of the mixed gases may be determined without explosion by means of a little pellet of spongy platinum, and the experiment can then be conducted over mercury in an ordinary graduated tube. 4. Another exact method of removing oxygen from air, recommended by Gay-Lussac, is the introduction into the air of slips of copper moistened with hydrochloric acid, which absorb oxygen with great avidity.

5. A solution in ammonia of the subchloride of copper, or of any salt of the suboxide of that metal, such as the sulphite, absorbs oxygen with great avidity, and may be used in the analysis of air.

6. In the recent careful analyses of air by MM. Dumas and Boussingault (Compt. Rend. 12, 1005) the oxygen was withdrawn, by passing air over reduced metallic copper at a red heat. To obtain the necessary precision in the results, the experiment was conducted in the following manner. In fig. 114, *a b* is a tube of the

FIG. 114.

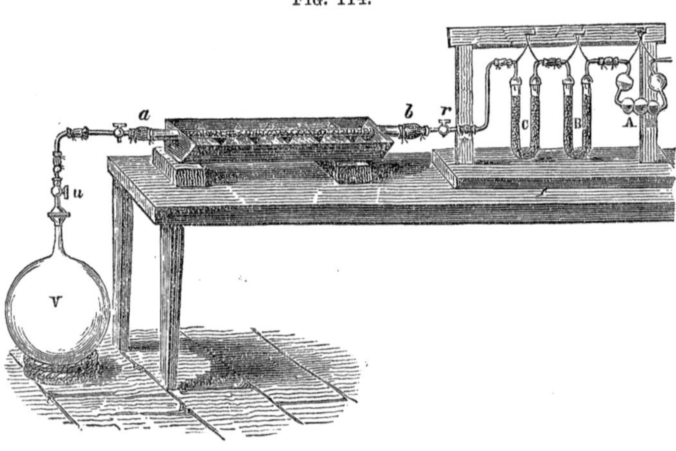

difficultly fusible or hard glass used in organic analysis, which is filled with metallic copper (reduced from the black oxide of copper by hydrogen), and placed in a long trough-furnace of sheet iron, in which it can be heated to redness throughout its whole length. The tube is provided with stopcocks at both ends, and attached by caoutchouc tubes to small glass tubes. By one of these small tubes it communicates with a glass balloon V, of about 1200 cubic inches in capacity, having a stopcock *u*; and by the other *r*, with a series of tubes A, B, and C. Of these A is a series of bulbs containing a concentrated solution of caustic potassa, and is intended for the absorption of the small portion of carbonic acid present in air; the U-shaped tube B contains fragments of pumice impregnated with the same alkaline solution; and the similar tube C is filled with pumice impregnated with oil of vitriol, in order to dry the air.

The balloon V is weighed and applied to the other apparatus in a vacuous state. The tube *a b* containing the metallic copper is also weighed beforehand. The tube and copper being heated to low redness, the stopcocks are partially opened, and air allowed to flow in a gradual manner into V. The oxygen is entirely absorbed by the copper, and the weight of that constituent ascertained by weighing the tube *a b* after the experiment. The nitrogen passes on alone into V, and its weight is found by again weighing that balloon. A great many analyses made in this way gave the following mean results:—

	Air by weight.	Air by volume.
Oxygen	23.10	20·90
Nitrogen	76.90	79.10
	100·00	100·00

Air from distant localities and different elevations has not exhibited any sensible variation in composition.

The theory of the constitution of mixed gases of Dalton supposes that the oxygen and nitrogen of air form independent atmospheres, the one gas not pressing upon or interfering with the other. If each of these atmospheres were of uniform density, their heights would obviously be inversely as the densities of the two gases, the height of the nitrogen column 8, and that of the oxygen 7; and the proportion of the one gas to the other would vary with the elevation. The same variation should occur in the atmosphere in its actual state: the proportion being supposed 21 per cent. at the level of the sea, by a calculation on this principle it should be 20.070 per cent. at a height of 10,000 Parisian feet, and 19.140 per cent. at a height of 20,000 feet. But as the influence of the great polar and equatorial currents is allowed to extend to a greater height in the atmosphere than the last, and than has ever been reached by man, it is not to be wondered at that no diminution in the proportion of oxygen is observable in the accurate analyses of air from the summit of the Faulhorn (8000 feet) which were lately made by Brunner, with the view of testing this hypothesis. (Poggendorff, Handwörterbuch der Chemie, Bd. i. S. 570).

Besides these constituents, the atmosphere always contains a variable quantity of watery vapour and carbonic acid gas. The presence of the latter is observed by exposing to the air a bason of lime-water, which is soon covered by a pellicle of carbonate of lime. Its proportion is ascertained by adding baryta-water of a known strength, from a graduated pipette, to a large bottle of the air to be examined; agitating after each addition, till a slip of yellow turmeric paper is made permanently brown by the baryta-water after agitation, which proves that more of the latter has been added than is neutralized by the carbonic acid of the air. The carbonic acid is in the equivalent proportion (by weight) of the quantity of baryta which has been neutralized.

Another and perhaps more exact method is to draw a large but known volume of dry air through a U tube, containing pumice impregnated with caustic potassa, and to pass it afterwards through a second U tube, containing oil of vitriol. The increase of weight on both tubes weighed together is the proportion of carbonic acid.

Like every subject connected with the atmosphere, the proportion of carbonic acid which it contains was ably investigated by the Saussures. The elder philosopher of that name detected the presence of this gas in the atmosphere resting upon the perpetual snows of the summit of Mont Blanc, so that there can be no doubt that carbonic acid is diffused through the whole mass of the atmosphere. The younger Saussure has ascertained, by a series of several hundred analyses of air, that the mean proportion of carbonic acid is 4.9 volumes in 10,000 volumes of air, or almost exactly 1 in 2000 volumes; but it varies from 6.2 as a maximum to 3.7, as a minimum in 10,000 volumes. Its proportion near the surface of the earth is greater in summer than in winter, and during night than during day upon an average of many observations. It is also rather more abundant in elevated situations, as on the summits of high mountains, than in the plains; a distribution of this gas which proves that the action of vegetation at the surface of the earth is sufficient to keep down the proportion of it in the atmosphere, within a certain limit. (Saussure,

Ann. de Chim. et de Phys. t. xxxviii. p. 411; and t. xliv. p. 5). An enormous quantity of carbonic acid is discharged from the elevated cones of the active volcanoes of America, according to the observations of Boussingault, which may partly account for the high proportion of that gas in the upper regions of the atmosphere. The gas emitted from the volcanoes of the old world, according to Davy and others, is principally nitrogen.

Carbonic acid is a constituent of the atmosphere which is essential to vegetable life, plants absorbing that gas, and deriving from it the whole of their carbon. Extensive forests, such as those of the Landes in France, which grow upon sands absolutely destitute of carbonaceous matter, can obtain their carbon in no other manner. But the oxygen of the carbonic acid is not retained by the plant, for the lignin and other constituent principles of vegetables, contain, it is well known, no more oxygen than is sufficient to form water with their hydrogen, and which, indeed, has entered the plant as water. The oxygen of the carbonic acid must therefore be returned in some form to the atmosphere. The discharge of pure oxygen gas from the leaves of plants was first observed by Priestley, and the general action of plants upon the atmosphere has subsequently been minutely studied by Sir H. Davy and Dr. Daubeny. The decomposition of carbonic acid requires the concurrence of light; and is not therefore sensible during the night. That plants fully compensate for the loss of oxygen occasioned by the respiration of animals and other natural processes is not improbable; but the mass of the atmosphere is so vast that any change in its composition must be very slowly effected. It has, indeed, been estimated that the proportion of oxygen consumed by animated beings in a century does not exceed 1-7200th of the whole quantity.

Other gases and vaporous bodies are observed to enter the atmosphere, but none of them can afterwards be detected in it, with the exception of hydrogen in some form, probably as the light carburetted hydrogen of marshes, of which Boussingault believes that he has been able to detect the presence of a minute but appreciable proportion. (Ann. de Chim. et de Phys. lvii. 148). He also observed concentrated sulphuric acid to be blackened when exposed in a glass capsule to the air, protected from dust, and at a distance from vegetation, which he ascribes to the occasional presence in the air of some volatile carbonaceous compound which is absorbed and decomposed by the acid.

Ammonia (NH_3) also is a minute but essential constituent of air, probably in the form of carbonate. It is brought down by rain, and is the principal source of the nitrogen of plants.

Omitting the aqueous vapour always present in air, but of which the proportion is constantly fluctuating, it may be represented as follows, in 10,000 volumes:—

COMPOSITION OF DRY AIR BY VOLUME.

Nitrogen	7912
Oxygen	2080
Carbonic acid	4
Carburetted hydrogen (CH_2)	4
Ammonia	Trace
	10,000

Of the odoriferous principles of plants, the miasmata of marshes and other matters of contagion, the presence, although sufficiently obvious to the sense of smell, or by their effects upon the human constitution, cannot be detected by chemical tests. But it may be remarked in regard to them, that few or none of the compound volatile bodies we perceive entering the atmosphere, could long escape destruction from oxidation. The atmosphere contains, indeed, within itself, the means of its own purification, and slowly but certainly converts all organic substances exposed to it into simpler forms of matter, such as water, carbonic acid, nitric acid, and ammonia. Although the occasional presence of matters of contagion in the atmosphere is not to be disputed, still it is an assumption, without evidence, that these substances are volatile or truly vaporous. Other matters of infection with which we can compare them, such as the matter of cow-pox, may be dried in the air, and are not in the least degree volatile. Indeed, volatility of a body implies a certain simplicity of constitution and limit to the number of atoms in its integrant particle, which true organic bodies appear not to possess. Again, the source of such bodies being at all times inconsiderable, they would, if vapours, be liable to a speedy attenuation by diffusion so great as to render their action wholly inconceivable. It is more probable that matters of contagion are highly organized particles of fixed matter, which may find its way into the atmosphere, notwithstanding, like the pollen of flowers, and remain for a time suspended in it; a condition which is consistent with the admitted difficulty of reaching and destroying those bodies by gaseous chlorine, and with the washing of walls and floors as an ordinary disinfecting practice. On this obscure subject, I may refer to a valuable paper by the late Dr. Henry upon the application of heat to disinfection, in which it is proved that a temperature of 212° is destructive to such contagious matters as could be made the subject of experiment. (Phil. Mag. 2d ser., vols. x. p. 363; xi. pp. 22, 207 (1832).

With reference to gaseous disinfectants, it may be remarked that sulphurous acid gas (obtained by burning sulphur) is preferable, on speculative grounds, to chlorine. No agent checks more effectually the first development of animal or vegetable life. This it does by preventing oxidation. In the same manner it renders impossible the first step in putrefactive decomposition and fermentation. All animal odours and emanations are most immediately and effectively destroyed by it. The fœtid odour from the boiling solution of cochineal (for instance), which is so persistent in dye-houses, is most completely removed by the admission of sulphurous acid vapour (J. Graham).

The compounds of nitrogen or oxygen are the following: —

Protoxide of nitrogen or nitrous oxide	NO
Binoxide of nitrogen or nitric oxide	NO_2
Nitrous acid	NO_3
Peroxide of nitrogen (hyponitric acid of Thenard)	NO_4
Nitric acid	NO_5

PROTOXIDE OF NITROGEN.

Syn. PROTOXIDE OF AZOTE, NITROUS OXIDE; *Eq.* 22 *or* 275; NO; *density* 1520·4; ☐☐.

This gas was discovered by Dr. Priestley about 1776, and studied by Davy, whose "Researches, Chemical and Philosophical," published in 1809, contain an elaborate investigation of its properties and composition. Davy first observed the stimulating power of nitrous oxide when taken into the lungs, a property which has since attracted a considerable degree of popular attention to this gas.

Preparation. — Protoxide of nitrogen is always prepared from the nitrate of ammonia. Some attention must be paid to the purity of that salt, which should contain no hydrochlorate of ammonia. It is formed by adding pounded carbonate of ammonia to pure nitric acid, which, if concentrated, may be previously diluted with half its bulk of water, so long as there is effervescence; and a small excess of the carbonate may be left at the end in the liquor. The solution should be filtered, and concentrated till its boiling point begins to rise above 250°, and a drop of it becomes solid on a cool glass plate. On cooling, it forms a solid cake, which may be broken into fragments. To obtain nitrous oxide, a quantity of this salt, which should never be less than 6 or 8 ounces, is introduced into a retort, or a globular

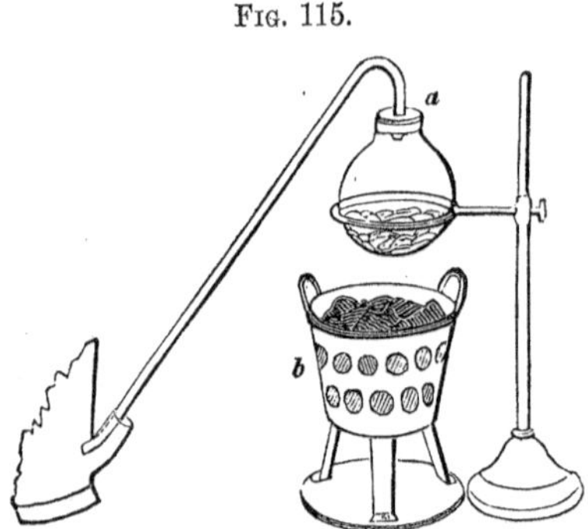

Fig. 115.

flask, called a bolt-head *a*, and heated by a charcoal choffer *b*, the diffused heat of which is more suitable than the heat of a lamp. Paper may be pasted over the cork of the bolt-head to keep it air-tight. At a temperature not under 340° the salt boils and begins to undergo decomposition, being resolved into nitrous oxide and water. As heat is evolved in this decomposition, which is a kind of combustion or deflagration, the choffer must be withdrawn to such a distance from the flask as to sustain only a moderate ebullition. If the temperature is allowed to rise too high, the ebullition becomes tumultuous, and the flask is filled with white fumes, which have an irritating odour; and the gas which then comes off is little more than nitrogen. Nitrous oxide should be collected in a gasometer or in a gas-holder filled with water of a temperature about 90°, as cold water absorbs much of this gas. The whole salt undergoes the same decomposition, and nothing whatever is left in the retort.[1]

Nitrous oxide is likewise produced when the salt called nitro-sulphate of ammonia is thrown into an acid; and also when zinc and tin are dissolved in dilute nitric acid, but the latter processes do not afford the gas in a state of purity.

The nature of the decomposition of the nitrate of ammonia will be best explained by the following diagram, in which an equivalent of the salt, or 80 parts, is supposed to be used. It will be observed that the three equivalents of hydrogen in the ammonia are burned, or combine with three equivalents of the oxygen of the nitric acid, and form water, while the two equivalents of nitrogen in the ammonia and nitric acid combine with the two remaining equivalents of the oxygen of the latter: —

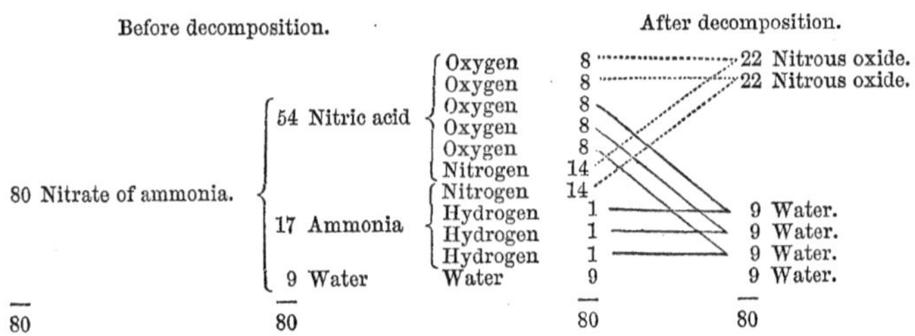

Or in symbols: —

$$NH_3, HO+NO_5=2NO \text{ and } 4HO.$$

From the diagram it appears that 80 grains of the salt yield 44 grains of nitrous oxide and 35 grains of water. One grain of salt yields rather more than one cubic inch of gas.

Properties. — Nitrous oxide possesses the usual mechanical properties of gases, and has a faint agreeable smell. It has been liquefied by evolving it from the decomposition of the nitrate of ammonia in a sealed tube, and possessed in the liquid state an elastic force of above 50 atmospheres at 45°. [It has also been liquefied by mechanical compression (Natterer, Ann. de Phar. 54, 254). Liquid nitrous oxide is colourless, very volatile, boils under the pressure of one atmosphere at

[1] For the preparation and properties of this and other gases, the Elements of Chemistry (1829) of the late Dr. Henry may still be consulted with advantage.

—125° (Regnault, Compt. Rend. t. 28, 333): a drop falling on the hand produces effects similar to a burn; potassium, charcoal, sulphur, and phosphorus float on its surface unaltered, but ignited charcoal burns with brilliancy. Water poured on it, freezes instantly, and the liquid is converted into gas with almost explosive rapidity. Issuing from a jet pipe, part is reduced to a solid state by the sudden evaporation of the rest. The solid is snowlike, and placed on the hand produces the same effects as the liquid (Dumas, Compt. Rend. t. 27, 463). When the liquid is exposed to the cold produced by the vaporization of solid carbonic acid and ether, it freezes at the temperature of —150° (Faraday).—R. B.] The gas is formed by the union of a combining measure, or 2 volumes of nitrogen, with a combining measure, or 1 volume of oxygen, which are condensed into 2 volumes, the combining measure of this gas. The weight of a single volume, or the density of the gas, is therefore by calculation —

$$\frac{971.4 + 971.4 + 1105.6}{2} = 1524.2$$

Cold water agitated with this gas dissolves about three-fourths of its volume of the gas, and acquires a sweetish taste, but, I believe, no stimulating properties. Bodies which burn in air, burn with increased brilliancy in this gas, if introduced in a state of ignition. A newly blown out taper with a red wick may be rekindled in it, as in oxygen. Mixed with an equal bulk of hydrogen, and ignited by flame and the electric spark, it detonates violently. In all these cases of combustion, the nitrous oxide is decomposed, its oxygen uniting with the combustible and its nitrogen being set free. When transmitted through a red-hot porcelain tube, nitrous oxide is likewise decomposed and resolved into oxygen, nitrogen, and the peroxide of nitrogen.

Nitrous oxide was supposed by Davy to combine with alkalies, when generated in contact with them, but these compounds have since been found to contain nitro-sulphuric acid.

This gas may be respired for two or three minutes without inconvenience, and when the gas is unmixed with air, and the lungs have been well emptied of air before respiring, it induces an agreeable state of reverie or intoxication, often accompanied with considerable excitement, which lasts for a minute or two, and disappears without any unpleasant consequences. The gas from an ounce and a half or two ounces of nitrate of ammonia is sufficient for a dose, and it should be respired from a bag of the size of a large ox-bladder, and provided with a wooden tube of an inch internal diameter. The volume of the gas diminishes rapidly during the inspiration, and finally only a few cubic inches remain. An animal entirely confined in this gas soon dies from the prolonged effects of the intoxication.

BINOXIDE OF NITROGEN.

Syn. BINOXIDE, OR DEUTOXIDE OF AZOTE, NITRIC OXIDE; *Eq.* 30 or 375;

NO_2; *density* 1038·8; ⊞.

This gas, which comes off during the action of nitric acid upon most metals, appears to have been collected by Dr. Hales, the father of pneumatic chemistry, but its properties were first minutely studied by Dr. Priestley.

Preparation.—Binoxide of nitrogen is easily procured by the action of nitric acid diluted to the specific gravity 1.2, upon sheet copper clipped into small pieces. As no heat is required, this gas may be evolved like hydrogen from a gas bottle (page 234). Mercury may be substituted for copper, but it is then necessary to apply a gentle heat to the materials. This gas may be collected and retained over water without loss.

In dissolving in nitric acid, the copper takes oxygen from one portion of acid and becomes oxide of copper, which combines with another portion of acid, and forms the nitrate of copper, the solution of which is of a blue colour. The portion of nitric acid which is decomposed losing three equivalents of oxygen and retaining two, appears as nitric oxide gas. This is more clearly shown in the following diagram:—

ACTION OF NITRIC ACID UPON COPPER.

Before decomposition.			After decomposition.
54 Nitric acid	Nitrogen	14	30 Binoxide of nitrogen.
	Oxygen	8	
	Oxygen	8	
	Oxygen	8	
	Oxygen	8	
	Oxygen	8	
32 Copper.........	Copper	32	
54 Nitric acid	Nitric acid............	54	94 Nitrate of copper.
32 Copper	Copper	32	
54 Nitric acid ...	Nitric acid............	54	94 Nitrate of copper.
32 Copper	Copper	32	
54 Nitric acid	Nitric acid............	54	94 Nitrate of copper.
312		312	312

Or in symbols:—

$$4NO_5 \text{ and } 3Cu = 3(Cu\ O,\ NO_5) \text{ and } NO_2.$$

Properties.—This gas is colourless, but when mixed with air it produces ruddy fumes of the peroxide of nitrogen. It is irritating, and causes the glottis to contract spasmodically when an attempt is made to respire it. Nitric oxide has never been liquefied: water at 60°, according to Dr. Henry, takes up only 5 or 6 per cent. of this gas. It is formed of one combining measure of nitrogen or 2 volumes, and two combining measures of oxygen or 2 volumes, united without condensation, so that the combining measure of nitric oxide contains 4 volumes. The weight of one volume, or the density of the gas, is therefore

$$\frac{971.4+971.4+1105.6+1105.6}{4}=1038.5.$$

This gas is not decomposed by a low red heat.

Many combustibles do not burn in nitric oxide, although it contains half its volume of oxygen. A lighted candle and burning sulphur are extinguished by it; mixed with hydrogen, it is not exploded by the electric spark or by flame, but it imparts a green colour to the flame of hydrogen burning in air. Phosphorus and charcoal, however, introduced in a state of ignition into this gas, continue to burn with increased vehemence. The state of combination of the oxygen in this gas appears to prevent that substance from uniting with combustibles, unless, like the two last mentioned, they evolve so much heat as to decompose the nitric oxide. Several of the more oxidable metals, such as iron, withdraw the half of the oxygen from this gas, when left in contact with it, and convert it into nitrous oxide.

No property of nitric oxide is more remarkable than its attraction for oxygen, and it may be employed to separate this from all other gases. Nitric oxide indicates the presence of free oxygen in a gaseous mixture, by the appearance of fumes which are pale and yellow with a small, and reddish brown and dense with a large proportion of the latter gas; and also by a subsequent contraction of the gaseous volume, arising from the absorption of these fumes by water. Added in sufficient quantity, nitric oxide will thus withdraw oxygen most completely from any mixture. But notwithstanding this property, nitric oxide cannot be employed with advantage in the analysis of air or similar mixtures, for the contraction which it occasions does not afford certain data for determining the proportion of oxygen which has disappeared. Nitric oxide is capable of combining with different proportions of oxygen, a combining measure or 4 volumes of the gas uniting, in such experiments, with 1, 2 or 3 volumes of oxygen, and forming nitrous acid, peroxide of nitrogen or nitric acid, or several of these compounds at the same time.

This oxide of nitrogen, like the preceding, is a neutral body, and has a very limited range of affinity. A substance is left on igniting the nitrate of potassa or baryta, which was supposed to be a compound of nitric oxide with potassium, or barium, but Mitscherlich finds it to be either the caustic protoxide itself or the peroxide of the metal. But nitric oxide is absorbed by a solution of the sulphate of iron, which it causes to become black; the greater part of the gas may be expelled again by boiling the solution. All the soluble proto-salts of iron have the same property, and the nitric oxide remains attached to the oxide of iron when precipitated in the insoluble salts of that metal. The proportion of nitric oxide in these combinations is found by Peligot to be definite; one eq. of the nitric oxide to four of the protoxide of iron; or, the nitric oxide contains the proportion of oxygen required to convert the protoxide into sesquioxide of iron. (Ann. de Chim. et de Phys. t. liv. p. 17). Nitric oxide is also absorbed by nitric acid. With sulphurous acid nitric oxide forms a compound which will be more particularly noticed under that acid.

NITROUS ACID.

Syn. AZOTOUS ACID (*Thenard*). *Eq.* 38 or 475; NO_3.

The direct mode of forming this compound is by mixing 4 volumes of binoxide of nitrogen with 1 volume of oxygen, both perfectly dry, and exposing the mixture to a great degree of cold. The gases unite, and condense into a liquid of a green colour, which is very volatile, and forms a deep reddish yellow coloured vapour. Nitrous acid prepared in this way is decomposed at once when thrown into water; an effervescence occurring, from the escape of nitric oxide, and nitric acid being produced, which gives stability to a portion of the nitrous acid. Nitrous acid cannot be made to unite directly with alkalies and earths, probably owing to the action of water first described. But when oxygen gas is mixed with a large excess of nitric oxide, in contact with a solution of caustic potassa, the gases were found by Gay-Lussac always to disappear in the proportions of nitrous acid, which was produced and entered into combination with the potassa, forming a *nitrite* of potassa. Similar nitrites may also be produced by calcining the nitrate of soda till the fused salt becomes alkaline; or by boiling the nitrate of lead with metallic lead. The nitrite of soda may be dissolved and filtered, and the solution precipitated by nitrate of silver; a process which gives the nitrite of silver, a salt possessing a sparing degree of solubility, like that of cream of tartar, but which may be purified by solution and crystallization, and then affords ready means of obtaining the other nitrites by double decomposition (Mitscherlich). Nitrous acid is liberated from the nitrites by acetic acid. When free sulphuric acid is added to a solution of nitrite of silver, the disengaged nitrous acid is immediately resolved into nitric acid and nitric oxide. The subnitrite of lead, on the other hand, may be decomposed by the bisulphate of potassa or soda to obtain a neutral nitrite of one of these bases (Berzelius). The nitrites of potassa and soda are soluble in alcohol, while the nitrates are not so.

Nitrous acid is also capable of combining with several acids, in particular with iodic, nitric, and sulphuric acids. Its combination with the last is obtained by sealing up together liquid sulphurous acid and peroxide of nitrogen (NO_4) in a glass tube. In the course of a few days the tube may be opened: the substances are combined, and form a solid mass, which may be heated up to (200° C.) its point of fusion. At a higher temperature it distils without alteration. In this experiment, sulphurous acid acquires an equivalent of oxygen, and becomes sulphuric acid.

while peroxide of nitrogen loses an equivalent of oxygen, and becomes nitrous acid, but one half only of the latter acid formed unites with sulphuric acid, the composition of the body formed being NO_3+2SO_3. The reaction is expressed as follows:—

$$2SO_3 \text{ and } 2NO_4=NO_3+2SO_3 \text{ and } NO_3.$$

This compound is soluble in strong oil of vitriol without decomposition; but from sulphuric acid somewhat diluted it takes water, and forms a crystalline substance, which often appears in the manufacture of sulphuric acid, as we shall afterwards find. The original solid compound is decomposed by pure water or highly diluted sulphuric acid, and the sulphuric and nitrous acids become free. The tendency of nitrous acid to combine with other acids has already been noticed, as assimilating this compound of nitrogen to arsenious acid and the oxide of antimony (page 147).

PEROXIDE OF NITROGEN.

Syn. HYPONITRIC ACID, NITROUS GAS (*Berzelius*). *Eq.* 46 *or* 575; NO_4; *theoretical density*, 1591.3;

This compound forms the principal part of the ruddy fumes which always appear on mixing nitric oxide with air. As it cannot be made to unite either directly or indirectly with bases, and has no acid properties, any designation for this oxide of nitrogen which implies acidity should be avoided, and the name nitrous acid in particular, which is applied on the continent to the preceding compound. The name peroxide of nitrogen is more in accordance with the rules generally followed in naming such compounds.

Preparation.—When 4 volumes of nitric oxide and 2 of oxygen, both perfectly dry, are mixed, this compound is alone produced, and the six volumes of mixed gases are condensed into 4 volumes, which may be considered the combining measure of peroxide of nitrogen. The weight of 1 volume, or the density of this gas, must therefore be

$$\frac{1038.5\times4+1105.6\times2}{4}=1591.3.$$

The peroxide of nitrogen is also contained in the coloured and fuming nitric acid of commerce, and may be obtained in the liquid condition by gently warming that acid, and condensing the vapour which comes over, by transmitting it through a glass tube surrounded by ice and salt. But it is prepared with most advantage from the nitrate of lead, the crystals of which, after being pounded and well dried, to deprive the salt of hygrometric water, are distilled in a retort of hard glass, or porcelain, at a red heat, and the red vapours condensed in a receiver kept very cold by a freezing mixture. Oxygen gas escapes during the whole process, the nitric acid of the nitrate of lead being resolved into oxygen and peroxide of nitrogen; or $NO_5=NO_4$ and O. As obtained by the last process, which was proposed by Dulong, peroxide of nitrogen is a highly volatile liquid, boiling at 82°, of a red colour at the usual temperature, orange yellow at a lower temperature, and nearly colourless below zero, of density 1·451, and a white solid mass at —40°. It is exceedingly corrosive, and, like nitric acid, stains the skin yellow. The red colour of its vapour becomes paler at a low temperature, but with heat increases greatly in intensity, so as to appear quite opaque when in a considerable body at a high temperature. It is the vapour which Brewster observed to produce so many dark lines in the spectrum of a ray of light which passes through it (page 100). The peroxide is not decomposed by a low red heat, and appears to be the most stable of the oxides of nitrogen. No compound of it is known, unless peroxide of nitrogen be the radical, as some suppose, of nitric acid. But Berzelius is inclined to consider this oxide as itself a compound of nitric and nitrous acids, for

$$NO_5+NO_3=2NO_4.$$[1]

The liquid peroxide of nitrogen is partially decomposed by water, nitric oxide coming off with effervescence, and more and more nitric acid being produced, in proportion to the quantity of water added; but a portion of the peroxide always escapes this action, being protected by the nitric acid formed. In the progress of this dilution the liquid undergoes several changes of colour, passing from red to yellow, from that to green, then to blue, and becoming at last colourless. The peroxide of nitrogen is readily decomposed by the more oxidable metals, and is a powerful oxidizing agent.

NITRIC ACID.

Syn. AZOTIC ACID (*Thenard*). *Eq.* 54 *or* 675; NO_5.

A knowledge of this highly important acid has descended from the earliest ages of chemistry, but its composition was first ascertained by Cavendish, in 1785. He succeeded in forming nitric acid from its elements, by transmitting a succession of electric sparks during several days through a small quantity of air, or through a mixture of 1 volume of nitrogen and 2½ volumes of oxygen, confined in a small tube over water, or over solution of potassa; in the last case, the absorption of the gases was complete, and nitrate of potassa was obtained. A trace of this acid in combination with ammonia has been detected in the rain of thunder-storms, produced probably in the same manner. It was also observed by Gay-Lussac to be the sole product when nitric oxide is added, in a gradual manner, to oxygen in excess over water; the gases then unite, and disappear in the proportion of 4 volumes of the former to 3 of the latter. It is also a constituent of the salt, nitre or salpetre, found in the soil of India and Spain, which is a nitrate of potassa, and also of nitrate of soda, which occurs in large quantities in South America.

[Anhydrous nitric acid was first prepared in 1849, by M. Deville (Compt. Rend. t. 28, p. 257), by treating dry nitrate of silver with dry chlorine. The nitrate of silver is placed in a U-tube, to which a second, having a spherical reservoir at the curved part, is attached. The first tube is immersed in a vessel of water, which can be heated by a spirit-lamp, and the second in a freezing mixture. Chlorine gas is evolved and passed first through a tube containing chloride of calcium, then another filled with pumice moistened with sulphuric acid, that it may be perfectly dried before it reaches the nitrate of silver. All the joints are united by the blow-pipe. The nitrate of silver is heated to 356° F., and a stream of carbonic acid passed through the apparatus to dry the salt, after which it is allowed to cool and the chlorine is transmitted. At common temperatures there is no appearance of action, but when the heat is raised to 203 and then lowered to between 135 and 155°, decomposition takes place, chloride of silver is produced, and crystals of nitric acid begin to appear in the second U-tube at the part not immersed in the freezing mixture, and a small quantity of liquid condenses in the spherical reservoir, while oxygen and chlorine gases escape. To transfer the nitric acid, the stream of chlorine is replaced by carbonic acid, and the freezing mixture taken away; the liquid is now removed from the reservoir and a bulb attached to receive the anhydrous acid. This bulb is immersed in the freezing mixture, and the acid evaporating at ordinary temperature condenses in the bulb, which when filled is to be sealed by the blow-pipe.

Properties. — Anhydrous nitric acid forms transparent colourless crystals, belonging to the right rhombic system. It fuses at a little above 85°, and boils about

---

[1] Traité de Chimie, par J. J. Berzelius, traduite par MM. Esslinger et Hoeffer, Didot, Paris, 1845. An excellent edition of this most valuable system of chemistry.

113°, decomposing slightly at that temperature. In contact with water, it dissolves with the evolution of much heat.

At ordinary temperatures it is liable to spontaneous decomposition, and bursts the bulb by the increased tension of the confined gases (Dumas, Compt. Rend. t. 28, p. 323). — R. B.]

Preparation. — This acid has not until recently been obtained in an insulated state, but in combination with water, as in aqua fortis or the hydrate of nitric acid, or with a fixed base, as in the ordinary nitrates. The hydrate, (which is popularly termed nitric acid,) is eliminated from nitrate of potassa by means of oil of vitriol, which is itself a hydrate of sulphuric acid. That acid unites with potassa, in this decomposition, and forms sulphate of potassa, displacing nitric acid, which last brings off in combination with itself the water of the oil of vitriol. There is a great advantage, first pointed out by Mr. Phillips, in using two equivalents of oil of vitriol to one of nitrate of potassa, which is 98 of the former to 101 of the latter, or nearly equal weights. The acid and salt, in these proportions, are introduced into a capacious plain retort, provided with a flask as a receiver. Upon the application of heat, a little of the nitric acid first evolved undergoes decomposition, and red fumes appear, but soon the vapours become nearly colourless, and are easily condensed in the receiver. During the whole distillation, the temperatnre need not exceed 260°. The mass remains pasty till all the nitric acid is disengaged, and then enters into fusion; red vapours again appearing towards the end of the process. The residuary salt is the bisulphate of potassa, or double sulphate of water and potassa, $HO.SO_3 + KO.SO_3$. The rationale of this important process is exhibited in the following diagram: —

PROCESS FOR NITRIC ACID.

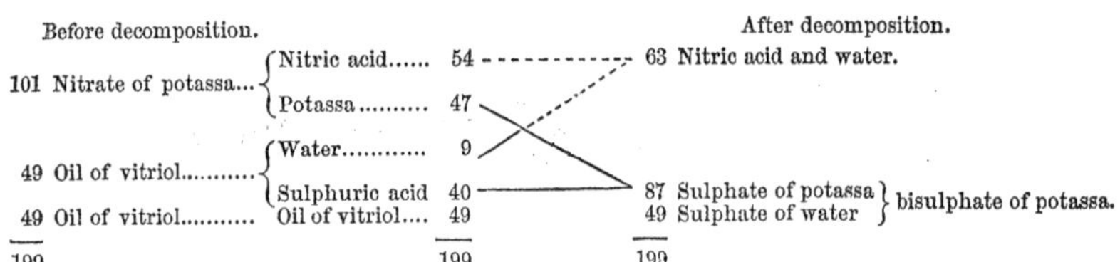

In this operation twice as much sulphuric acid is employed as is required to neutralize the potassa of the nitre, by which means the whole nitric acid is eliminated without loss at a moderate temperature, and a residuary salt is left which is easily removed from the retort.

With half the preceding quantity, or a single equivalent of oil of vitriol, the materials in the retort are apt to undergo a vesicular swelling, upon the application of heat, and to pass into the receiver. Abundance of ruddy fumes are also evolved, that are not easily condensed, and prove that the nitric acid is decomposed. The temperature in this process must also be raised inconveniently high towards the end of the operation, in order to decompose the whole nitre. The peculiarities of the decomposition here arise from the formation of bisulphate of potassa in the operation, the whole sulphuric acid uniting in the first instance with half the potassa of the nitre. Now, it is only at an elevated temperature that the acid salt thus formed can decompose the remaining nitre; — a temperature which is sufficient to decompose nitric acid, as may be proved by transmitting the vapour of the concentrated acid through a tube heated to the same degree.

Ordinary nitric acid for manufacturing purposes is generally prepared by distilling nitrate of soda with an equivalent of sulphuric acid not at its highest degree of concentration in a large cylinder of cast iron (fig. 116, page 261), supported in brickwork over a fire. Both ends of the cylinder are moveable, and generally consist of circular discs of stone. The nitric acid which distils over is condensed in a

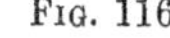

FIG. 116.

FIG. 117.

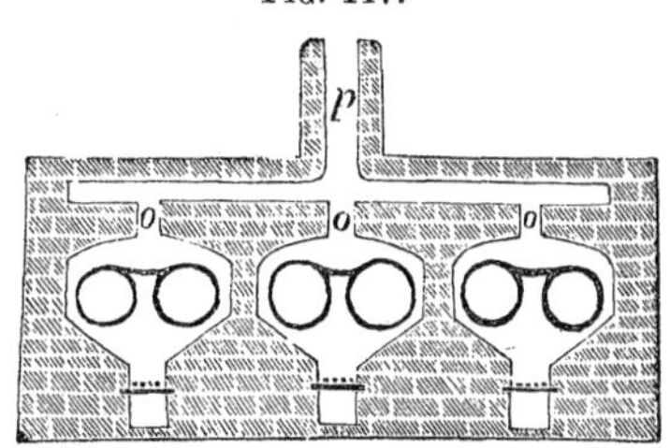

series of large vessels of salt-glaze ware, of the form of Woulf bottles, of which two, A, B, are shown in the figure.

The iron cylinders are generally so supported that two of them are heated by one fire, as in fig. 117, which is a sectional view of three pairs of such retort cylinders. The iron of the vault or roof of the cylinder is most apt to be corroded by the acid vapours, and is therefore protected by a coating of fire-clay or of tiles of the same material cemented together.

Properties. — The acid prepared by the first process is colourless, or has only a straw yellow tint. If the oil of vitriol has been in its most concentrated condition, which is seldom the case, the nitric acid is in its state of highest concentration also, and contains no more than a single equivalent of water. The density of this acid is 1.522 at 58°; but a slight heat disengages a little peroxide of nitrogen from it, and its density becomes 1.521 (Mitscherlich). The density of the strongest colourless nitric acid which Mr. Arthur Smith could prepare was 1.517 at 60°; it boiled at 184°, and came within 1 per cent. of the protohydrate in composition (Chem. Mem. iii. 402). When distilled, it is partially decomposed by the heat, and affords a product of a strong yellow colour. Its vapour transmitted through a porcelain tube, heated to dull redness, is decomposed in a great measure into oxygen and peroxide of nitrogen; and into oxygen and nitrogen gases, when the tube is heated to whiteness. The colourless liquid acid becomes yellow, when exposed to the rays of the sun, and on loosening the stopper of the bottle it is sometimes projected with force, from the state of compression of the disengaged oxygen. Hence to preserve this acid colourless it must be kept in a covered bottle. It congeals at about —40°, but diluted with half its weight of water, it becomes solid at $1\frac{1}{2}$°, and with a little more water its freezing point is again lowered to —45°. Exposed to the air, the concentrated acid fumes, from the condensation by its vapour of the moisture in the atmosphere. It also attracts moisture from damp air, and increases in weight; and when suddenly mixed with 3-4ths of its weight of water, may rise in temperature from 60° to 140°.

Nitric acid has a great affinity for water, and diminishes in density with the proportion of water added to it. A table has been constructed in which the per centage of absolute acid is expressed in mixtures of various densities, which is useful for reference and will be given in an appendix. There appears to be no definite hydrate of this acid between the first (the nitrate of water), and that containing 3 eq. of water additional (A. Smith). The first has no action upon tin or iron. The second is acid of density 1.424, which therefore contains 4 eq. of water. This last hydrate was found by Dr. Dalton to have the highest boiling point of any hydrate of nitric acid: it is 250°, and both weaker and stronger acids are brought to this strength by continued ebullition, the former losing water and the latter acid. The density of the vapour of this hydrate is found to be 1243 by A. Bineau, and it contains 2 volumes of nitrogen, 5 volumes of oxygen, and 8 volumes of steam condensed into 10 volumes, which are therefore the combining measure of this vapour (Ann. de Chim. et de Phys. lxviii. p. 418).

Nitric acid is exceedingly corrosive, and one of the strongest acids, yielding only in that respect to sulphuric acid. The facility with which it parts with its oxygen renders it very proper for oxidating bodies in the humid way, a purpose for which it is constantly employed. Nearly all the metals are oxidized by means of it; some of them with extreme violence, such as copper, mercury, and zinc, when the concentrated acid is used; and tin and iron by the acid very slightly diluted. Poured upon red-hot charcoal, it causes a brilliant combustion. When mixed with a fourth of its bulk of sulphuric acid, and thrown upon a few drops of oil of turpentine, it occasions an explosive combustion of the oil. Sulphur digested in nitric acid at the boiling point is raised to its highest degree of oxidation and becomes sulphuric acid; iodine is also converted by it into iodic acid. Most vegetable and animal substances are converted by nitric acid into oxalic and carbonic acids. It stains the cuticle and nails of a yellow colour, and has the same effect upon wool; the orange patterns upon woollen table-covers are produced by means of it. In the undiluted state it forms a powerful cautery.

In acting upon the less oxidable metals, such as copper and mercury, nitric acid is itself decomposed, and nitric oxide gas produced, which comes off with effervescence. Palladium and silver, when they are dissolved by the acid in the cold, produce nitrous acid in the liquor and evolve no gas, but this is very unusual in the solution of metals by nitric acid. Those metals, such as zinc, which are dissolved in diluted acids with the evolution of hydrogen, act in two ways upon nitric acid; sometimes they decompose it, so as to disengage a mixture of peroxide of nitrogen and nitric oxide, and at other times they decompose both water and nitric acid at once, in such proportions that the hydrogen of the water combines with the nitrogen of the acid to form ammonia, which last combines with another portion of acid, and is retained in the liquor as nitrate of ammonia. The protoxide of nitrogen is also evolved when zinc is dissolved in very feeble nitric acid, which may arise from the action of hydrogen upon nitric oxide. Nitric acid, in its highest state of concentration, exerts no violent action upon certain organic substances, such as lignin or woody fibre and starch, for a short time, but unites with them and forms singular compounds. A proper acid for such experiments is procured with most certainty by distilling 100 parts of nitre, with no more than 60 parts of the strongest oil of vitriol. If paper is soaked for one minute in such an acid, and afterwards washed with water, it is found to shrivel up a little and become nearly as tough as parchment, and when dried to be remarkably inflammable, catching fire at so low a temperature as 356°, and burning without any nitrous odour (Pelouze). Or if the strong undiluted nitric acid of commerce be mixed with an equal weight of oil of vitriol, and cotton-wool be immersed in the mixture for a minute or two and afterwards washed with water, it is converted into gun-cotton, without injury to the cotton fibre (Schönbein).

Nitric acid forms an important class of salts, the nitrates, which occasion deflagration when fused with a combustible at a high temperature, from the oxygen in their acid, and are remarkable as a class for their general solubility, no nitrate being insoluble in water. The nitrate of the black oxide of mercury is perhaps the least soluble of these salts. The nitrates of potassa, soda, ammonia, baryta, and strontia, are anhydrous; but the nitrates of the extensive magnesian class of oxides all contain water in a state of intimate combination, and have a formula analogous to that of hydrated nitric acid, or the nitrate of water itself. Of the four atoms of water contained in hydrated nitric acid of sp. gr. 1.42, one is combined with the acid as base, and may be named *basic* water, while the other three are in combination with the nitrate of water, and may be termed the *constitutional* water of that salt. The same three atoms of constitutional water are found in all the magnesian nitrates, with the addition often of another three atoms of water, as appears from the following formulæ: —

Nitric acid, 1.42	$HO.NO_5+3HO$
Prismatic nitrate of copper	$CuO.NO_5+3HO$
Rhomboidal nitrate of copper	$CuO.NO_5+3HO+3HO$
Nitrate of magnesia	$MgO.NO_5+3HO+3HO$

It is doubtful whether the proportion of constitutional water in any of these nitrates can be reduced below 3 atoms by heat without the loss of a portion of nitric acid at the same time, and the partial decomposition of the salt. The nitrates of the potassa and magnesian classes do not combine together, and no double nitrates are known, nor nitrates with excess of acid. The nitrates with excess of metallic oxide, which are called subnitrates, appear to be formed on the type of the magnesian class: the subnitrate of copper, being $CuO.NO_5+3CuO.3HO$ (Gerhardt), or nitrate of copper with 3 atoms hydrated oxide of copper. The water is strongly retained, and requires a temperature of 300° to expel it. The nitrate of red oxide of mercury is $HgO.NO_5+HgO$ (Kane).

Nitric acid in a solution cannot be detected by precipitating that acid in combination with any base, as the nitrates are all soluble, so that tests of another nature must be had recourse to, to ascertain its presence. A highly diluted solution of sulphate of indigo may be boiled without change, but on adding to it at the boiling temperature a liquid containing free nitric acid, the blue colour of the indigo is soon destroyed. If it is a neutral nitrate which is tested, a little sulphuric acid should be added to the solution, to liberate the nitric acid, before mixing it with the sulphate of indigo. It is also necessary to guard against the presence of a trace of nitric acid in the sulphuric acid. Another test of the presence of nitric acid has been proposed by De Richemont. The liquid containing the nitrate is mixed with rather more than an equal bulk of oil of vitriol, and when the mixture has become cool, a few drops of a strong solution of protosulphate of iron are added to it. Nitric oxide is evolved, and combines with the protosulphate of iron, producing a rose or purple tint even when the quantity of nitric acid is very small. One part of nitric acid in 24,000 of water has been detected in this manner. Free nitric acid also is incapable of dissolving gold-leaf, although heated upon it, but acquires that property when a drop of hydrochloric acid is added to it. But in testing the presence of this acid, it is always advisable to neutralize a portion of the liquor with potassa, and to evaporate so as to obtain the thin prismatic crystals of nitre, which may be recognised by their form, by their cooling nitrous taste, their power to deflagrate combustibles at a red heat, and by the characteristic action of the acid they contain, when liberated by sulphuric acid, upon copper and other metals, in which ruddy nitrous fumes are produced.

[When obtained from nitrate of soda, it may contain iodine. This impure acid yields, on distillation, a sublimate of iodine after all the nitric acid has come over. Neutralized with potassa, mixed with a solution of starch, and sulphuric acid added drop by drop, the liquid assumes a blue colour (Gmelin's Handbook, vol. ii. p. 393). — R. B.]

If nitric acid be rigidly pure, it may be diluted with distilled water, and is not disturbed by nitrate of silver, nor by chloride of barium, the first of which discovers the presence of hydrochloric acid by producing a white precipitate of chloride of silver; the last discovers sulphuric acid by forming the white insoluble sulphate of baryta. The fuming nitric acid may be freed from hydrochloric acid, by retaining it warm on a sand-bath for a day or two, when the chlorine of the hydrochloric acid goes off as gas. To free it from sulphuric, it should be diluted with a little water, and distilled from nitrate of baryta; but the process for nitric acid which has been described gives it without a trace of sulphuric acid, when carefully conducted.

Uses. — Nitric acid is sometimes used in the fumigations required for contagious diseases, particularly in wards of hospitals from which the patients are not removed, the fumes of this acid being greatly less irritating than those of chlorine. For the purpose of fumigation, pounded nitre and concentrated sulphuric acid are used, being heated together in a cup. Nitric acid is par excellence the solvent of metals, and has other most numerous and varied applications not only in chemistry, but likewise in the arts and manufactures.

NITROGEN AND HYDROGEN — AMMONIA.

Eq. 17 *or* 212.5; H_3N; *density* 596.7; ⊞

With hydrogen, nitrogen forms a remarkable gaseous compound — ammonia, which derives its name from sal ammoniac, a salt from which it is generally extracted, and which again was so named from being first prepared in the district of Ammonia, in Libya. Ammonia is produced in the destructive distillation of all organic matters containing nitrogen, which has given rise to one of its popular names, the Spirits of Hartshorn. It is also produced during the putrefaction of the same matters, and finds its way into the atmosphere (page 252). A trace of it is always found in the native oxides of iron, in the varieties of clay, and in some other minerals.

Nitrogen and hydrogen mixed together do not exhibit any disposition to combine, even when heated; but if electric sparks be taken through a mixture of those gases, particularly with the presence of any acid vapour, a sensible trace of a salt of ammonia is produced. Hydrogen, however, if evolved in contact with nitrogen, will in certain circumstances form ammonia. Thus in the rusting of iron in water containing air or nitrogen and carbonic acid, the hydrogen which is then evolved from the decomposition of the water, appears to combine in its nascent state with nitrogen. If, while zinc is dissolving in dilute sulphuric acid, nitric acid be added drop by drop till the evolution of hydrogen gas ceases, the latter will be found to have united with the nitrogen of the nitric acid, and much ammonia to be formed; the oxygen of the nitric acid combining with hydrogen also, to form water, at the same time. If the proportion of nitric acid be relatively small, Mr. Nesbitt finds that it may be entirely converted into ammonia in this manner. When zinc is dissolved in nitric acid alone, which is neither much diluted nor very strong, but in an intermediate condition, the same suppression of hydrogen and formation of ammonia is observed.

Preparation. — In a state of purity, ammonia is a gas, of which the well-known *liquor* or *aqua ammoniæ* is a solution in water. This solution, which is of constant use as a reagent, is prepared by mixing intimately sal ammoniac (hydrochlorate of ammonia) with an equal weight of slaked lime, introducing the mixture into a glass retort or bolt-head, which is afterwards filled up with slaked lime (A, fig. 118), and

Fig. 118.

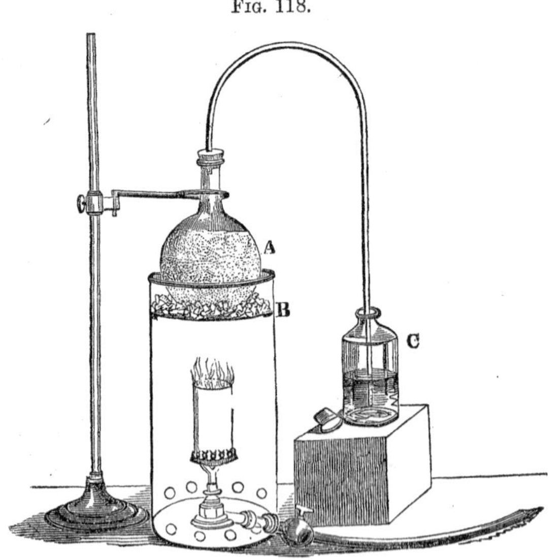

distilling by the diffused heat of a chauffer or sand-pot. If recourse is had to the gas-flame, the heat may be conveniently diffused by placing the burner within a cylinder of sheet iron about 14 inches in height, as represented in the figure, with a perforated stage B, covered with small fragments of pumice-stone, on which the flask A is supported. Ammoniacal gas comes off, which should be conducted into a quantity of distilled water in the bottle C, to condense it, equal to the weight of the salt employed. Chloride of calcium and the excess of lime remain in the retort, and a considerable quantity of water is liberated in the process, and distils over with the ammonia. This reaction is explained in the following diagram:—

PROCESS FOR AMMONIA.

Before decomposition.				After decomposition.	
53.5	Hydrochlorate of ammonia	Ammonia	17	17	Ammonia.
		Hydrogen	1		
		Chlorine	35.5		
28	Lime	Oxygen	8	9	Water.
		Calcium	20	55.5	Chloride of calcium.
81.5			81.5	81.5	

Or in symbols:

$$NH_4Cl \text{ and } CaO = NH_3 \text{ with } HO \text{ and } CaCl.$$

To obtain ammoniacal gas, a portion of the solution prepared by the preceding process may be introduced into a small plain retort, A (fig. 119), by means of the long funnel B; and the short bent tube C being adapted by a perforated cork to the mouth of the retort, the liquid is boiled by a gentle heat, when the gas is first expelled from its superior volatility, and collected in a jar filled with mercury, and inverted over the mercurial trough (fig. 120, page 266). Or the gas may be derived at once from sal ammoniac, mixed with twice its weight of quicklime in a small retort, and collected over mercury.

Fig. 119.

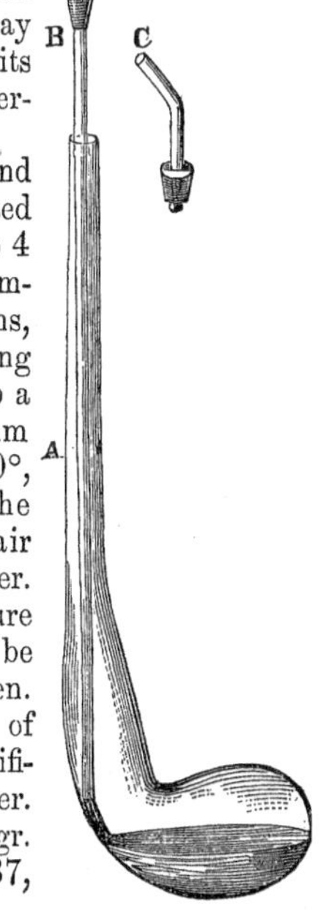

Properties.—Ammonia is a colourless gas, of a strong and pungent odour, familiar in spirits of hartshorn. It is composed of 2 volumes of nitrogen and 6 of hydrogen, condensed into 4 volumes, which form the combining measure of this gas. Ammonia is resolved into its constituent gases, in these proportions, when transmitted through an ignited porcelain tube containing platinum, iron, or copper wire. The two latter metals absorb a little nitrogen (Despretz), and become brittle, but the platinum remains unaltered. By a pressure of 6.5 atmospheres, at 50°, it is condensed into a transparent colourless liquid, of which the sp. gr. is 0.731 at 60°. Ammoniacal gas is inflammable in air in a low degree, burning in contact with the flame of a taper. A small jet of this gas will also burn in oxygen. A mixture of ammoniacal gas with an equal volume of nitrous oxide may be detonated by the electric spark, and affords water and nitrogen. Water is capable of dissolving about 500 times its volume of ammoniacal gas in the cold, and the solution is always specifically lighter, and has a lower boiling point than pure water. According to the observations of Davy, solutions of sp. gr. 0.872, 0.9054, and 0.9692, contain respectively 32.5, 25.37,

Fig. 120.

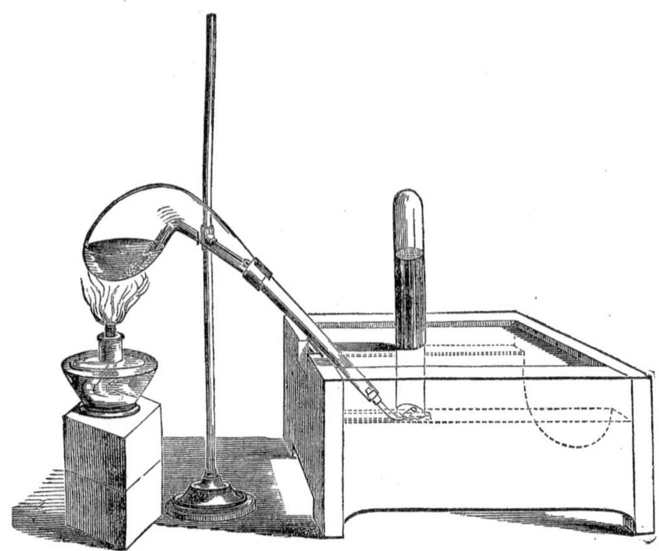

and 9.5 per cent. of ammonia. Mr. Griffin, who has constructed a table of the densities of solutions of ammonia from experiment, finds that no sensible condensation of volume occurs in these mixtures, and that their densities are the mean of those of water 1 and anhydrous liquid ammonia, supposing the latter to be 0.7083 at 62° (Mem. Chem. Soc. iii. 189). Ammoniacal gas is also largely soluble in alcohol.

Solution of ammonia has an acrid alkaline taste, and produces blisters on the tongue and skin. When cooled slowly to —40°, it crystallizes in long needles of a silky lustre. The solution has a temporary action upon turmeric paper, which it causes to be brown while humid; it also restores the blue colour of litmus reddened by an acid, changes the blue colour of the infusion of red cabbage into green, and neutralizes the strongest acids, properties which it possesses in common with the fixed alkalies. It is distinguished as the volatile alkali. When ammonia is free, it may always be discovered, by its odour, by forming dense white fumes with hydrochloric acid, and by producing a deep blue solution with salts of copper.

Ammonia, in solution, is decomposed by chlorine, with the evolution of nitrogen gas and formation of hydrochlorate of ammonia: when ammonia and chlorine, both in the state of gas, are mixed together, the action that ensues is attended with flame. Dry iodine absorbs ammoniacal gas, and forms a brown viscous liquid, which water decomposes, dissolving out hydriodate of ammonia, and leaving a black powder, which is the explosive iodide of nitrogen.

Ammonia forms several classes of compounds with acids and salts (page 166), and exhibits highly curious reactions with many other substances, which do not admit of being discussed so early, but which I shall return to later in the work.

SECTION IV.

CARBON.

Eq. 6 *or* 75; C; *density of vapour* (*hypothetical*) 416 ☐

Carbon is found in great abundance in the mineral kingdom united with other substances, as in coal, of which it is the basis, and in the acid of carbonates: it is also the most considerable element of the solid parts of both animals and vegetables.

It exists in nature, or may be obtained by art, under a variety of appearances, and possessed of very different physical properties. Carbon is a dimorphous body, occurring crystallized in the diamond and graphite in wholly different forms, and when artificially produced forming several amorphous varieties of charcoal which are very unlike each other.

Diamond.—This valuable gem is found throughout the range of the Ghauts in India, but chiefly at Golconda, in Borneo, and also in Brazil. It is always associated with transported materials, such as rolled gravel, or found in a sort of breccia or pudding-stone, composed of fragments of jasper, quartz, and calcedony, so that it is still a question whether the diamond is of mineral or vegetable origin. On removing the crust with which the crystals are covered, they are exceedingly brilliant, refract light powerfully, and are generally perfectly transparent, although diamonds are sometimes black, blue, and of a beautiful rose-colour. The primitive form of diamond is the regular octohedron, or two four-sided pyramids, of which the faces are equilateral triangles, applied base to base (fig. 55, page 143). It is more frequently found in the pyramidal octohedron,—a figure bounded by 24 sides, which presents the general aspect of a regular octohedron, on every facet of which has been placed a low pyramid of three facets; or, each facet of the octohedron is replaced by 6 secondary triangles, and the crystal becomes almost spherical, and presents 48 facets. These facets often appear curved from the effect of attrition. The diamond can always be cleaved in the direction of the faces of the octohedron, which possess that particular brilliancy characteristic of the diamond. It is the hardest of the gems. An edge of its crystal formed by flat planes only scratches glass, but if the edge is formed of curved faces, like the edge of a convex lens, it then, besides abrading the surface, produces a fissure to a small depth, and in the form of the glazier's diamond is used to cut glass. The weight of diamonds is generally estimated by the *carat*, which is about 3.2 grains. The diamond is remarkably indestructible, and may be heated to whiteness in a covered crucible without injury, but it begins to burn in the open air, at about the melting point of silver, charcoal sometimes appearing on its surface, and is entirely converted into carbonic acid gas. When heated to the highest degree between the charcoal points of a strong voltaic battery, the diamond swells up considerably, and divides into portions. After cooling, it is found entirely altered in appearance, having become of a metallic gray, friable, and resembling in every respect the coke from bituminous coal. This experiment appears to show that a high temperature is unfavourable to the existence of diamonds, and that they cannot therefore be originally formed at a very elevated temperature. The diamond is quickly consumed in fused nitre, when the carbonic acid is retained by the potash; this is a simple mode of analyzing the diamond, by which it has been proved to be pure carbon. The diamond is a non-conductor of electricity. Its density varies from 3.5 to 3.55.

Graphite.—This mineral, which is also known as Black Lead and Plumbago, occurs in rounded masses deposited in beds in the primitive formations, particularly in granite, mica-schist, and primitive limestone. Borrowdale in Cumberland is a celebrated locality of graphite, and affords the only specimens which are sufficiently hard for making pencils. It is occasionally found crystallized in plates which are six-sided tables. Graphite may also be produced artificially, by putting an excess of charcoal in contact with fused cast iron, when a portion of the carbon dissolves, and separates again on cooling, in the form of large and beautiful leaflets. In the condition of graphite, carbon is perfectly opaque, soft to the touch, possessed of the metallic lustre, and of a specific gravity from 1.9 to 2.3. It always contains iron and manganese, apparently in the state of oxides, and in combination with silicic and titanic acids, sometimes to the extent of 28 per cent., but in some specimens, as in those from Barreros in Brazil, not more than a trace of those metals is found, which is to be considered an accidental constituent, and not essential to the mineral. Neither in the form of diamond nor graphite does carbon exhibit any indication of fusion or volatility under the most intense heat. *Anthracite* is often nearly pure carbon, but always contains a portion of hydrogen, and is related to bituminous coal, and not to graphite.

Charcoal.—Owing to its infusibility carbon presents itself under a variety of aspects, according to the structure of the substance from which it is derived, and the accidental circumstances of its preparation. The following are the principal varieties: gas carbon, lamp-black, wood charcoal, coke, and ivory black.

1. Gas carbon has the metallic lustre, and a density of 1.76; it is compact, generally of a mammillated structure, but sometimes in fine fibres, and considerably resembles graphite, but is too hard to give a streak upon paper. It is the product of a slow deposition of carbon from coal gas at a high temperature, and is frequently found to line the gas retorts to a considerable thickness, and to fill up accidental fissures in them (Dr. Colquhoun, Ann. of Philos., New Ser., vol. xii. p. 1).

2. Lamp-black is the soot of imperfectly burned combustibles, such as tar or resin. Carbon is deposited in a powder of the same nature, and of the purest form, when alcohol vapour or a volatile oil is transmitted through a porcelain tube at a red heat; and the lustrous charcoal, which is obtained on calcining, in close vessels, starch, sugar, and many other organic substances, which fuse and afford a bright vesicular carbon of a metallic lustre, is possessed of the same characters. The charcoal of the latter sources, however, always retains traces of oxygen and hydrogen. Lamp-black is deficient in an attraction for organic matters in solution, which ordinary charcoal possesses.

3. Wood charcoal. Wood was found by Karsten to lose 57 per cent. of its weight when thoroughly dried at 212°, and 10 per cent. more at 304°. The remaining 33 parts of baked wood afforded, when calcined, 25 of charcoal, while 100 parts of the same wood calcined, without being previously dried, left only 14 per cent. of carbon. It is the absence of this large quantity of water which causes the heat of burning charcoal to be so much more intense than that of wood. When calcined at a high temperature, charcoal becomes dense, hard, and less inflammable. The knots in wood sometimes afford a charcoal which is particularly hard, and is used in polishing metals, but it contains silica. From the minuteness of its pores, the charcoal of wood absorbs many times its volume of the more liquefiable gases; such as ammoniacal gas, hydrochloric acid, hydrosulphuric acid, and carbonic acid, condensing 90 times its volume of the first, and 35 of the last: of oxygen, it condenses 9.25 volumes; of nitrogen, 7.5 volumes; and of hydrogen, 1.75 volumes. It also absorbs moisture with avidity from the atmosphere, and other condensible vapours, such as odoriferous effluvia. From this last property freshly calcined charcoal, when wrapt up in clothes which have contracted a disagreeable odour, destroys it, and has a considerable effect in retarding the putrefaction of organic matter with which it is placed in contact. Water is also found to remain sweet, and wine to be improved in quality, if kept in casks of which the inside has been charred. In the state of a coarse powder, wood charcoal is particularly applicable as a filter for spirits, which it deprives of the essential oil which they contain. It is much less destructible by atmospheric agencies than wood, and hence the points of stakes are often charred, before being driven into the ground, in order to preserve them. Charcoal decomposes the vapour of water at a red heat, giving rise to a gaseous mixture, which was found by Bunsen to consist, in 100 volumes, of hydrogen 56, carbonic oxide 29, carbonic acid 14.8, and light carburetted hydrogen 0.2 volume.

4. The coke of those species of coal which do not fuse when heated is a remarkably dense charcoal, considerably resembling that of wood, and of great value as fuel, from the high temperature which can be produced by its combustion. When burned it generally leaves 2 or 3 per cent. of earthy ashes, while the ashes from wood charcoal seldom exceed 1 per cent. The density of pulverised coke varies from 1.6 to 2.0. Coke and wood charcoal, after being strongly heated, are good conductors of electricity.

5. Ivory black, Bone black, and Animal charcoal, are names applied to bones calcined or converted into charcoal in a close vessel. The charcoal thus produced

is mixed with not less than 10 times its weight of phosphate of lime, and being in a state of extreme division, exposes a great deal of surface. It possesses a remarkable attraction for organic colouring matters, and is extensively used in withdrawing the colouring matter from syrup in the refining of sugar, from the solution of tartaric acid, and in the purification of many other organic liquids. The usual practice, which was introduced by Dumont, is to filter the liquid hot through a bed of this charcoal in grains of the size of those of gunpowder, and of two or three feet in thickness. It is found that the discolouring power is greatly reduced by dissolving out the phosphate of lime from ivory black by an acid, although this must be done in certain applications of it, as when it is used to discolour the vegetable acids. A charcoal possessed of the same valuable property even in a higher degree for its weight, is produced by calcining dried blood, horns, hoofs, clippings of hides, in contact with carbonate of potash, and washing the calcined mass afterwards with water. Even vegetable matters afford a charcoal possessed of considerable discolouring power, if mixed with chalk, calcined flint, or any other earthy powder, before being carbonized. One hundred parts of pipe clay made into a thin paste with water, and well mixed with 20 parts of tar and 500 of coal finely pulverized, have been found to afford, after the mass was dried and ignited out of contact with air, a charcoal which was little inferior to bone black in quality. When charcoal which has been once used in such a filter is calcined again, it is found to have lost much of its discolouring power. This is owing to the deposition upon its surface of a lustrous charcoal, of the lamp-black variety, produced by the decomposition of the organic colouring matters, which has little or no discolouring power. But if the charcoal of the sugar filters be allowed to ferment, the foreign matter in it is destroyed; and if afterwards well washed with water and dried, before being calcined, it will be found to recover a considerable portion of its original power.

Bussy has constructed, from observation, the following table of the efficiency of the different charcoals. (Journ. de Pharm. t. viii. p. 257). These substances are compared with ivory black, as being the most feeble species, although this is superior by several degrees to the best wood charcoal. The relative efficiency, it will be observed, is not the same for two different kinds of colouring matter:—

Species of charcoal same weight.	Relative decolouration of sulphate of indigo.	Relative Decolouration of Syrup.
Blood charred with carbonate of potassa	50	20
Blood charred with chalk	18	11
Blood charred with phosphate of lime	12	10
Glue charred with carbonate of potassa	36	15.5
White of egg charred with the same	34	15.5
Gluten charred with the same	10.6	8.8
Charcoal from acetate of potassa	5.6	4.4
Charcoal from acetate of soda	12	8.8
Lamp-black, not calcined	4	3.3
Lamp-black calcined with carbonate of potassa	15.2	10.6
Bone charcoal, after the extraction of the earth of bones by an acid, and calcination with potassa	45	20
Bone charcoal treated with an acid	1.87	1.6
Oil charred with the phosphate of lime	2	1.9
Bone charcoal, in its ordinary state	1	1

This remarkable action of charcoal in withdrawing matters from solution is certainly an attraction of surface, but it is capable, notwithstanding, of overcoming chemical affinities of some intensity. The matters remain attached to the surface of the charcoal, without being decomposed or altered in nature. For if the blue sulphate of indigo be neutralized and then filtered through charcoal, the whole colouring matter is retained by the latter, and the filtered liquid is colourless. But a solution of caustic alkali will divest the charcoal of the blue colouring matter, and carry it away in solution. The salts of quinine, morphine, and other organic bases and bitter principles, are carried down by animal charcoal used in excess (Warington, Mem. Chem. Soc. iii. 326). Hence this substance is a very general antidote to vegetable poisons, as was proved by Dr. Garrod. Other substances also are carried down by animal charcoal, besides organic matters. Lime from lime water, iodine from solution in iodide of potassium, hydrosulphuric acid from solution in water, soluble subsalts of lead, and metallic oxides dissolved in ammonia or caustic potassa; but it has little or no action upon most neutral salts. The charcoal is apt with time to react upon the substance it carries down, probably from their closeness of contact, reducing the oxides of silver, lead, and copper, for instance, to the metallic state in a short time. Animal charcoal soon disappears when heated in chlorine water, and is converted into carbonic acid; and the affinities of carbon generally are more active in this than in its other forms.

Carbon is chemically the same under all these forms. This element cannot be crystallized artificially by the usual methods of fusion, solution or sublimation, if we except its solution, in cast iron, which gives it in the form of graphite and not of the diamond. It is chemically indifferent to most bodies at a low temperature, but combines directly with some metals by fusion, and forms compounds named *carburets* or *carbides:* in these compounds, however, the metal is most probably the negative constituent. When heated to low redness it burns readily in air or oxygen, forming a gaseous compound carbonic acid, which, when cool, has sensibly the same volume as the original oxygen. With half the proportion of oxygen in carbonic acid, carbon forms a protoxide, carbonic oxide gas. The last gas being supposed similar to steam or to nitrous oxide in its constitution, will be composed of 2 volumes of carbon vapour and 1 volume of oxygen gas condensed into 2 volumes, an assumption upon which the density of carbon vapour, which there are no means of determining experimentally, is usually calculated, and made about 420; the combining measure of this vapour containing 2 volumes (page 129). The density deduced from the equivalent of carbon is more nearly 416.[1] That the equivalent of carbon is exactly 6, as originally maintained by Dr. Prout, has been established beyond doubt by M. Dumas, by the combustion of the diamond in a stream of oxygen gas. Pure carbon then unites with oxygen in the proportion of 3 to 8 exactly, or 6 to 16, to form carbonic acid (p. 272).

Uses.—Several valuable applications of this substance have already been incidentally described. Carbon may be said to surpass all other bodies whatever in its affinity for oxygen at a high temperature; and being infusible, easily got rid of by combustion, and forming compounds with oxygen which escape as gas, this body is more suitable than any other substance to effect the reduction of metallic oxides; that is, to deprive them of their oxygen, and to produce from them the metal with the properties which characterize it.

CARBONIC ACID.

Eq. 22 or 275; CO_2; *density* 1529.0;

This gas was first discovered to exist in lime-stone and the mild alkalies, and to be expelled from the first by heat, and from both by the action of acids, by Dr. Black, and was named by him Fixed Air. He also remarked that the same gas is formed in respiration, fermentation, and combustion; it was afterwards proved to contain carbon by Lavoisier.

---

[1] The number for carbon vapour deduced from the density of oxygen gas, that is, six-sixteenths of that density, is 414.61 (page 130); while six-fourteenths of the density of nitrogen is 416.304, and six times the density of hydrogen, 415.56. The density of nitrogen is probably the least objectionable, and the number deduced from it for carbon (416) therefore the safest.

Fig. 121.

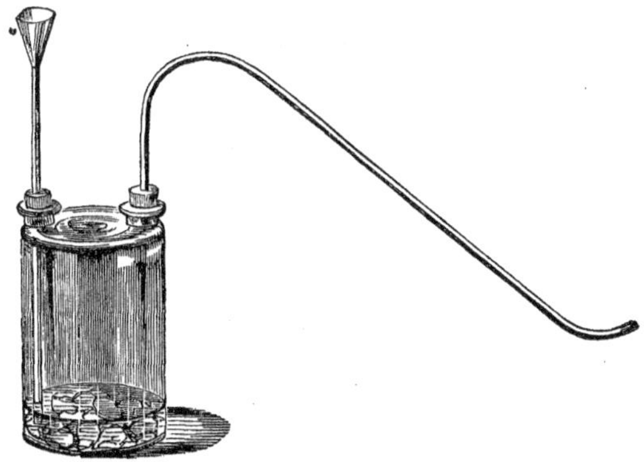

Preparation.—Carbonic acid is readily procured by pouring hydrochloric acid of sp. gr. 1.1, upon fragments of marble contained in a gas-bottle (fig. 121), or by the action of diluted sulphuric acid upon chalk. A gas comes off with effervescence, which may be collected at the water trough, but cannot be retained long over water without considerable loss, owing to its solubility.

From the great weight of carbonic acid a bottle may be filled with this gas by displacing air. The gas being evolved in the gas-bottle A (fig. 122), is first conveyed into a wash-bottle B, containing water, to condense any hydrochloric acid vapour with which the gas may be accompanied; then passing through a U-shaped drying tube C, containing fragments of chloride of calcium, to absorb aqueous vapour, and then conveyed to the lower part of the bottle D. When generated in the close apparatus of Thilorier for the purpose of liquefying it (page 77), this gas is evolved from bicarbonate of soda and sulphuric acid.

Fig. 122.

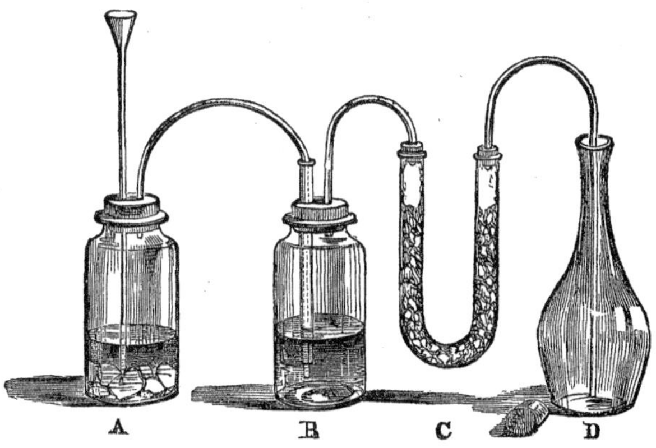

Properties.—This gas extinguishes flame, does not support animal life, and renders lime water turbid. Its density is considerable, being 1529 (Regnault), or a half more than that of air, the gas containing 2 volumes of the hypothetical carbon vapour and 2 volumes of oxygen, condensed into 2 volumes, which form the combining measure. Cold water dissolves rather more than an equal volume of this gas; the solution has an agreeable acidulous taste, and sparkles when poured from one vessel into another. It communicates a wine-red tint to litmus paper, which disappears again when the paper dries; when poured into lime water, it first throws down a white flaky precipitate of carbonate of lime or chalk, which it afterwards redissolves if the solution of the gas be added in excess. The quantity of this gas which water takes up is found to be sensibly proportional to the pressure; a very large volume of the gas is forced into soda, magnesia, and other aërated waters, much of which escapes on removing the pressure from these liquids.

Liquefied by pressure, carbonic acid has an elastic force of 38·5 atmospheres at 32° (Faraday). The specific gravity of liquid carbonic acid, at the same temperature, is 0·83: it dilates remarkably from heat, its expansion being four times greater than that of air, 20 volumes of the liquid at 32° becoming 29 at 86°, and its density varying from 0.9 to 0.6 as its temperature rises from —4° to 86°. (Thilorier, Annal. de Chim. et de Phys. lx. p. 427). It is a colourless liquid, which mixes in all proportions with ether, alcohol, naphtha, oil of turpentine, and bisulphide of carbon, but is insoluble in water and fat oils. At temperatures below —72° it is solid (page 80).

Potassium heated in a small glass bulb blown upon a tube, through which gaseous carbonic acid is transmitted, undergoes oxidation, and liberates carbon, the existence of which in the gas may thus be shown; or, for this experiment, a cleansed and dry Florence oil-flask may be filled, by displacement, with the dried gas (fig. 122), and a pellet of potassium being introduced, combustion may be determined by applying the flame of an Argand spirit-lamp for a few seconds to the bottom of the flask. But burning phosphorus, sulphur, and other combustibles, are immediately extinguished by carbonic acid, and the combustion does not cease from the absence of oxygen only, but from a positive influence in checking combustion which this gas exerts, for a lighted candle is extinguished in air containing no more than one-fourth of its volume of carbonic acid. It is generally believed that any mixture of carbonic acid and air will support the respiration of man, which will maintain the flame of a candle, and therefore a lighted candle is often let down into wells or pits suspected to contain this gas, to ascertain whether they are safe or not. But although air in which a candle can burn may not occasion immediate insensibility, still the continued respiration for several hours of air containing not more than 1 or 2 per cent. of carbonic acid, has been found to produce alarming effects (Broughton). The accidents from burning a pan of charcoal in close rooms are occasioned by this gas. It acts as a narcotic poison upon the system. A small animal thrown into convulsions from the respiration of this gas, may be recovered by sudden immersion in cold water.

Carbonic acid is thrown off from the lungs in respiration, as may be proved by directing a few expirations through lime-water. The air of an ordinary expiration contains, on an average, as observed by Dr. Prout, 3·45 per cent. of its volume of this gas, and the proportion varies from 3.3 to 4.1 per cent.,—being greatest at noon, and least during the night. Carbonic acid is also a product of the vinous fermentation, and is the cause of the agreeable pungency of beer, ale, and other fermented liquors, which become stale when exposed to the air from the loss of this gas. It also exists in all kinds of well and spring water, and contributes to their pleasant flavour, for water which has been deprived of its gases by boiling is insipid and disagreeable. Carbonic acid is also largely produced by the combustion of carbonaceous fuel, and appears to exist in considerable quantity in the earth, being discharged by active volcanoes, and from fissures in their neighbourhood, long after the volcanoes are extinct. The Grotto del Cane in Italy owes its mysterious properties to this gas, and many mineral springs, such as those of Tunbridge, Pyrmont, and Carlsbad, are highly charged with it. It comes thus to be always present in the atmosphere in a sensible, although by no means considerable proportion (page 251).

Composition of carbonic acid.—The composition of this substance, which, like that of water, is one of the fundamental data of chemical analysis, is determined with extreme exactness in the following manner:—A known weight of a very pure form of carbon, such as the diamond, is placed in a little trough or cradle of platinum, which is introduced into a porcelain tube *a b* (fig.123), placed across a furnace. To effect the combustion of the carbon, the end *a* of this tube is made to communicate by means of a glass tube with an apparatus supplying a stream of oxygen gas, perfectly dried by passing through the U tube E, which contains fragments of pumice impregnated with concentrated sulphuric acid. The second and fourth U tubes, A and D, are charged in the same manner. The bulb apparatus B contains a concentrated solution of caustic potassa, and the pumice in the adjoining U tube C is impregnated with the same fluid. These tubes, B and C, containing the alkali, with the tube following them, D, are accurately weighed together in a good balance.

Fig. 123.

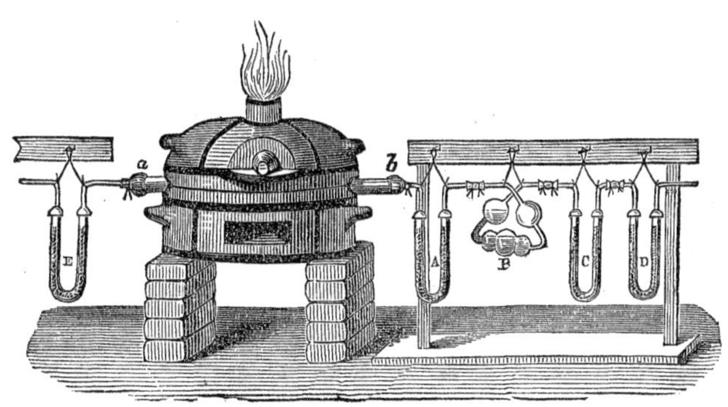

The different parts being connected by short tubes of caoutchouc, as represented in the figure, the apparatus is then filled with oxygen gas, which ought to be slowly disengaged. The tube *a b*, which contains the carbon, is heated to redness, and the latter soon enters into combustion, and is changed into carbonic acid. The gases pass through the series of tubes, A, B, C, D. In A, any trace of moisture is absorbed by the sulphuric acid, which may escape from the inner surface of the tube *a b* when heated, and in B and C the carbonic acid produced is absorbed by the caustic alkali. The excess of oxygen, which passes on uncondensed, takes up a little aqueous vapour in B and C, which tends to diminish the weight of the potassa apparatus; for, although the tension of the vapour of the alkaline solution is small, the latter cannot be used so concentrated as to make the tension insensible. The last U tube D remedies this inconvenience by drying the gases perfectly again before they escape into the atmosphere.

In such a combustion the formation of a little carbonic oxide gas is to be apprehended. This is provided against by filling the part of the tube *a b*, next *b*, with very porous oxide of copper, which is heated to redness during the experiment. In passing through this oxide, any small quantity of carbonic oxide which may exist is necessarily converted into carbonic acid. The oxide of copper is separated by a pad of asbestos from that part of the tube containing the little cradle with the carbon. The evolution of the oxygen is also continued for some time after the combustion of the carbon is complete, in order to sweep the tubes by means of that gas, and carry forward the whole carbonic acid formed into the potassa bulbs where it is absorbed.

On disconnecting the apparatus afterwards, and examining the cradle in which the carbon was placed, to ascertain whether its combustion is complete, a little incombustible earthy matter, not exceeding a few hundredths of a grain, will generally be found remaining, which had existed mechanically diffused through the carbon. The weight of the cradle and residue, deducted from the original weight of the cradle and carbon, gives obviously the exact weight of the carbon consumed; while the original weight of the system of tubes B, C, and D, deducted from their final weight, gives the exact weight of carbonic acid formed. (Cours Elémentaire de Chimie, par M. V. Regnault).

It is found in this way that 6 parts of carbon produce exactly 22 parts of carbonic acid, or carbonic acid contains—

1 eq. carbon	6	27.27
2 eq. oxygen	16	72.73
	22	100.00

Carbonates.—Carbonic acid combines with bases, and forms the class of carbonates. The hydrate of this acid seems incapable of existing in an uncombined state, but it exists in the alkaline bicarbonates, which are double carbonates of water and the alkali. If this hydrate were formed, we may presume that it would be analogous to the crystallized carbonate of magnesia, of which the formula is $MgO,CO_2+HO+2HO$, and also the same with another $2HO$; the salt of magnesia of most acids resembling the salt of water. Carbonate of lime, in the hydrated condition, has a similar formula. Carbonates of potassa, soda, and ammonia, retain a strong alkaline reaction, owing to the weakness of this acid, and the carbonates generally are decomposed with effervescence by all other acids, except the hydrocyanic.

Uses.—Carbonic acid is used in the preparation of aërated waters. The strong vessels in which the impregnation is effected, should be of copper, well tinned, and not of iron, as with the concurrence of water carbonic acid acts strongly upon that metal. It is sometimes desirable to remove carbonic acid from air or other gaseous mixtures, and this is generally done by means of caustic alkali or lime-water. When very dry, or so humid as to be actually wet, the hydrate of lime absorbs this gas with much less avidity than when of a certain degree of dryness, in which it is not so dry as to be dusty, but at the same time not sensibly damp. The dry hydrate may be brought at once to this condition, by mixing it intimately with an equal weight of crystallized sulphate of soda in fine powder; and this mixture, in a stratum of not more than an inch in thickness, intercepts carbonic acid most completely, and may rise in temperature to above 200°, from the rapid absorption of the gas. It is quite possible to respire through a cushion of that thickness, filled with the mixture, and such an article might be found useful by parties entering an atmosphere overcharged with carbonic acid, like that of a coal-mine after the occurrence of an explosion of fire-damp.

Carbonic acid is the highest degree of oxidation of which carbon is susceptible; but another oxide of carbon exists containing less oxygen.

CARBONIC OXIDE.

Eq. 14 *or* 175; CO; *density* 967·8; ☐☐

Priestley is the discoverer of this gas, but its true nature was first pointed out by Cruikshanks, and about the same time by Clement and Desormes.

Preparation.—Carbonic acid is readily deprived of half its oxygen, at a red heat, by a variety of substances, and so reduced to the state of carbonic oxide. The latter gas may therefore be obtained by transmitting carbonic acid over red-hot fragments of charcoal contained in an iron or porcelain tube; or by calcining chalk mixed with 1-4th of its weight of charcoal in an iron retort. It is likewise prepared by gently heating crystallized oxalic acid with five or six times its weight of strong oil of vitriol in a glass retort. The latter process affords a mixture of equal volumes of carbonic acid and carbonic oxide, the elements of oxalic acid being carbon and oxygen in the proportion to form these gases, and this acid being incapable of existing except in combination with water or some other base. Now the sulphuric acid unites with the water of the crystallized oxalic acid, and the latter acid being set free is instantly decomposed. The gas of all these processes contains much carbonic acid, of which it may be deprived by washing it with milk of lime, or a strong solution of potassa.

Another process suggested by Mr. Fownes affords a perfectly pure gas. It consists in heating the crystallized ferrocyanide of potassium in a glass retort or flask A (fig. 124), with four or five times its weight of oil of vitriol. The gas may be

Fig. 124.

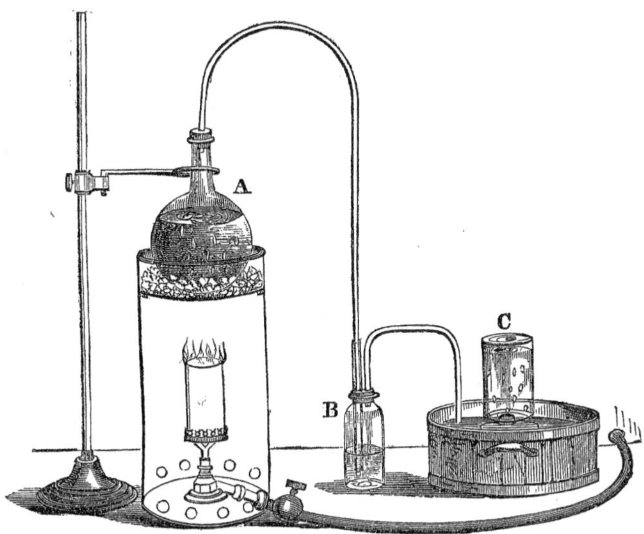

passed through a wash-bottle B, containing a little water, and be collected in the bottle C over the water-trough in the usual manner. One equivalent of ferrocyanide of potassium and 9 equivalents of water are then resolved into 6 equivalents of carbonic oxide, 2 equivalents of potassa, 1 equivalent of protoxide of iron, and 3 equivalents of ammonia: —

$$2K.FeC_6N_3+9HO=6CO+2KO+FeO+3H_3N.$$

Half an ounce of the salt yields 300 cubic inches of carbonic oxide.

Properties. — This gas, as has already been stated, is presumed to contain 2 volumes of carbon vapour, and 1 volume of oxygen, condensed into 2 volumes, so that its combining measure is 2 volumes: its density is 967.79 (Wrede). Carbonic oxide is 14 times heavier than hydrogen, like nitrogen, and coincides remarkably in its rate of transpiration (page 86) and other physical properties with the latter gas. It is very fatal to animals, and when inspired in a pure state almost immediately produces coma. Carbonic oxide is not more soluble in water than atmospheric air, and has never been liquefied. It is easily kindled, and burns with a pale blue flame, like that of sulphur, combining with half its volume of oxygen, and forming carbonic acid, which retains the original volume of the carbonic oxide. This combustion is often witnessed in a coke or charcoal fire. The carbonic acid, produced in the lower part of the fire, is converted into carbonic oxide, as it passes up through the red-hot embers, and afterwards burns above them with a blue flame, where it meets with air.

Carbonic oxide is a neutral body, like water, and combines directly with only a very few substances. It unites with an equal volume of chlorine under the influence of the sun's rays, and forms *phosgene* gas or Chloroxicarbonic Gas. As the gases contract to half their volume on combining, the density of this gas is the sum of carbonic oxide 968, and chlorine 2440, or 3408; its formula is CO.Cl. Chloroxicarbonic gas is colourless, and has a peculiar suffocating odour. In contact with water it is decomposed at the same time with an equivalent of water; hydrochloric and carbonic acids are produced — that is —

$$CO.Cl \text{ and } HO = CO_2 \text{ and } HCl.$$

Carbonic oxide is also absorbed by potassium gently heated, and that metal is employed to separate this gas from a mixture of hydrogen and gaseous carbohydrogens, as in the analysis of coal gas. But carbonic oxide has been supposed to exist in a greater number of compounds, and to be the radical of a series, of which the following substances are members: —

CARBONIC OXIDE SERIES.

Carbonic oxide	CO
Carbonic acid	$CO+O$
Chloroxicarbonic gas	$CO+Cl$
Oxalic acid	$2CO+O$
Oxamide	$2CO+NH_2$
Carbonoxide of potassium	$7CO+3K$
Rhodizonic acid	$7CO+3HO$
Croconic acid	$5CO+H$
Mellitic acid	$4CO+H$

In these compounds carbonic oxide is represented as playing the part of a simple substance, and forming a variety of products by uniting with oxygen, chlorine, hydrogen, and other elements.

Mellitic, croconic, and rhodizonic acids are sometimes enumerated as oxides of carbon, along with carbonic acid, carbonic oxide, and oxalic acid, but the former bodies have not an equal claim to the same early consideration as the latter compounds.

OXALIC ACID.

Eq. 36 *or* 450; C_2O_3. *Oxalate of water*, HO,C_2O_3+2HO.

This acid, discovered by Scheele in 1776, exists in the form of an acid salt of potassa, in a great number of plants, particularly in the species of *Oxalis* and *Rumex*: combined with lime it also forms a part of several lichens. Oxalate of lime occurs likewise as a mineral, *humboldite*, and forms the basis of a species of urinary calculus. This acid is also produced by the oxidation of carbon in combination, in a variety of circumstances, being the general product of the oxidation of organic substances by nitric acid, hypermanganate of potassa, and by fused potassa. Those matters which contain oxygen and hydrogen in the proportion of water furnish the largest quantity of oxalic acid.

This acid has been derived in quantity from lichens, but it is usually prepared by acting upon 1 part of sugar by 5 parts of nitric acid, of 1.42, diluted with 10 parts of water at a gentle heat till no gas is evolved, and evaporating to crystallize. The crystals must be drained, and crystallized a second time, as they are apt to retain a portion of nitric acid. Acting upon 1 part of sugar, with 6.6 parts of nitric acid, of density 1.245, Mr. L. Thompson obtained 1.1 parts of crystallized oxalic acid. One half of the carbon of the sugar appeared to be converted into oxalic acid, and the other half into carbonic acid; the nitric acid being entirely converted into binoxide of nitrogen, by loss of oxygen.

Oxalic acid forms long, four-sided, oblique prisms, with dihedral summits, or terminated by a single face. These crystals contain three equivalents of water, one of which is basic, and the other two constitutional, or water of crystallization. The latter two may be expelled at a temperature above 212°, and the protohydrate rises at the same time in vapour, and condenses as a woolly sublimate. Heated in a retort, the hydrated acid undergoes decomposition about 311°, and is converted into carbonic oxide, carbonic acid, and formic acid, without leaving any fixed residue. Concentrated nitric acid, with heat, converts oxalic acid into water and carbonic acid. When heated with sulphuric acid, oxalic acid yields equal volumes of carbonic oxide and carbonic acid; C_2O_3 being equivalent to $CO+CO_2$ (page 274). No charring, nor evolution of any other gas, occurs, so that the action of concentrated sulphuric acid affords the means of recognising oxalic acid or any oxalate. Crystallized oxalic acid is soluble in 8 parts of water, at 59°, in its own weight of boiling water, and in 4 parts of alcohol, at 59°.

Oxalic acid is a powerful acid, which combines with bases, and forms a well-defined class of salts,—the oxalates: it disengages carbonic acid easily from all its combinations. Added to lime-water, or any soluble salt of lime, oxalic acid forms a white precipitate—the oxalate of lime, which is a highly insoluble salt. Absolute oxalic acid, C_2O_3, has not been isolated, and appears incapable of existing except in combination with water, or some other base.

Composition of oxalic acid.—The analysis of oxalic acid is effected in the following manner:—Ten grains of the crystals, reduced to powder, are exactly weighed and mixed with 200 or 300 grains of oxide of copper, recently calcined, and perfectly dry. This mixture is introduced into a tube of white Bohemian glass, which is not easily fused, open at one end, about 0.4 inch in internal diameter, and 14 or 15 inches long, the other end being drawn out, bent upward, and sealed (*a*, fig. 125).

Fig. 125.

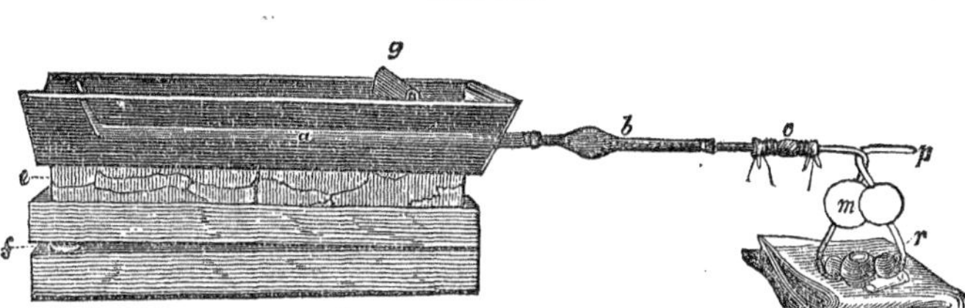

This is placed in a furnace, of a trough form, as represented in the figure, constructed of sheet iron, and heated to low redness by burning charcoal. Immediately connected with the combustion tube, by means of a perforated cork, is a tube of the form *b*, containing fragments of strongly dried chloride of calcium. In this tube the water of the oxalic acid is condensed, and the weight of that constituent is ascertained by weighing the tube, before and after the combustion. Beyond the chloride of calcium tube, and connected with it by a short caoutchouc tube, *c*, is a glass instrument, *p m r*, containing a strong solution of caustic potassa, of density 1.25 to 1.27, for the absorption of the carbonic acid produced by the combustion of the carbon of the oxalic acid by the oxygen of the oxide of copper. This instrument consists of five balls, of which *m* is larger than the others; no more of the potassa ley is put into it than fills the three central balls, leaving a bubble of air in each. One corner is elevated a little by a cork placed under it, and the whole supported on a folded towel: the potassa balls, when filled with ley, commonly weigh from 760 to 900 grains. This apparatus is also weighed before and after the combustion, and the increase ascertained.

The experiment, when properly conducted, gives 4.29 grains water condensed in the chloride of calcium tube, and 6.98 grains of carbonic acid absorbed in the potassa bulbs. But 4.29 grains of water contain 0.47 grain of hydrogen, and 6.98 grains of carbonic acid contain 1.905 grains of carbon. Now, as oxalic acid contains nothing but carbon, hydrogen, and oxygen, we obtain thus, for the composition of 10 grains of oxalic acid:—

Hydrogen	0.476
Carbon	1.905
Oxygen	7.619
	10.000

To learn the relation between the number of equivalents of these constituents of oxalic acid, it is necessary to divide the weight of each of them by its chemical equivalent:—

$$\frac{0.476}{1} = 0.4760. \qquad \frac{1.905}{6} = 0.3175.$$

$$\frac{7.619}{8} = 0.9524.$$

These fractions are in the proportion of 2, 3, and 6; from which it follows, that the formula of the crystallized oxalic acid is $C_2\ H_3\ O_6$ or a multiple of it. Allowing the 3H to be in combination with 3O, as water, we finally obtain the formula $C_2\ O_3 + 3HO$, for the crystallized acid.

CARBON AND HYDROGEN—HYDRIDES OF CARBON.

A large number of compounds of carbon and hydrogen are known; many of them found in the organic kingdom, and others derived from the decomposition of organic compounds. Some of these are liquid bodies, some solid, and others gaseous. At present we shall confine ourselves to the three gaseous compounds which in simplicity of composition resemble inorganic compounds.

PROTOCARBURETTED HYDROGEN.

Syn.[1] *Light carburetted hydrogen, Gas of the Acetates, Marsh-gas, Fire-damp.*

Eq. 16, or 200; $C_2\ H_4$; *density* 559.6; *combining measure* ⊞.

Fig. 126.

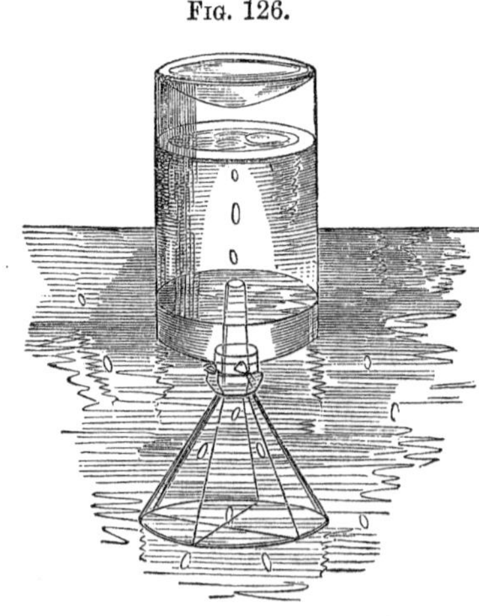

This gas is a constant product of the putrefactive decomposition of wood and other compounds of carbon, under water, and is most readily obtained by stirring the mud at the bottom of stagnant pools, and collecting the gas as it rises in an inverted bottle and funnel (fig. 126). It always contains 10 or 20 per cent. of carbonic acid, which may be separated from it by lime-water, and a small proportion of nitrogen. This gas also issues, in some places, in considerable quantities from fissures in the earth, coming often from subterraneous deposits of coal; and in the working of coal-mines it is found pent up in cavities, and would appear sometimes to be discharged from the fresh surface of the coal in sensible quantity. Hence, this gas is sometimes described as the inflammable

[1] Such systematic designations as have hitherto been applied to this and a few other hydrides of carbon have not in general been clear, and involve the serious error of representing the carbon as the negative element.

air of marshes, and the fire-damp of mines. It is also the most considerable constituent of coal gas, and of the gaseous mixture obtained on passing the vapour of alcohol through an ignited porcelain tube.

Preparation.—This gas is obtained by distilling a mixture of dried acetate of soda, hydrate of potassa and quicklime, in a coated glass retort. Four ounces of cr. acetate of soda may be dried on a sand-bath till anhydrous; the salt is then reduced to powder, and intimately mixed with four ounces of sticks of caustic potassa and six ounces of quicklime, both well pounded. A Florence oil flask, or other flask of hard glass, is coated with a mortar composed of a mixture of Paris-plaster, and half its weight of sand and coal-ashes, A (fig. 127); and provided with a perforated cork

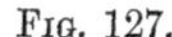
Fig. 127.

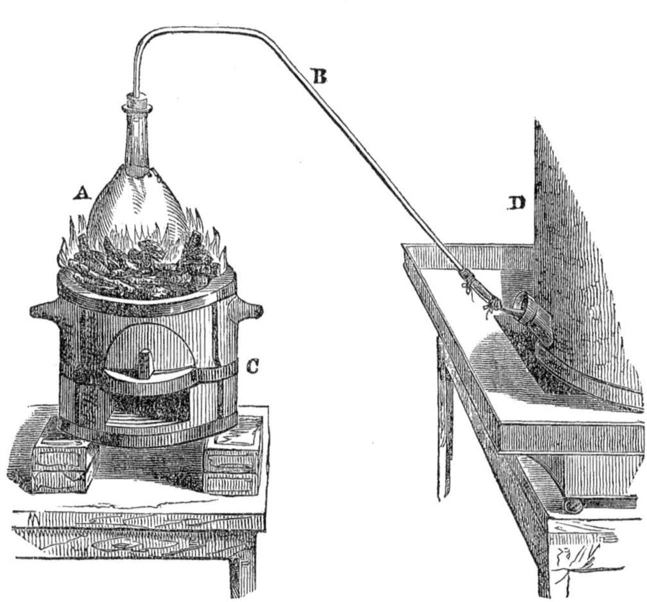

and bent tube B, one extremity of which should descend three or four inches in the neck of the flask. The materials above being introduced into the flask, the latter is placed in an open charcoal furnace C, and strongly heated. The gas comes off, and may be collected in jars over the pneumatic trough, or received in a gas-holder D filled with water.

Properties.—The observed density of protocarburetted hydrogen is 559.6; it is composed of 4 volumes carbon vapour, and 8 volumes hydrogen, condensed into 4 volumes, which are the combining measure of this gas. Hence its specific gravity is by calculation—

$$\frac{416 \times 4 + 69.26 \times 8}{4} = 554.5.$$

It is inodorous, neutral, respirable when mixed with air, not more soluble in water than pure hydrogen, and has never been liquefied. This carburetted hydrogen requires twice its bulk of oxygen to burn it completely, and affords water and an equal bulk of carbonic acid. The oxidation of this gas mixed with oxygen is not determined, at the temperature of the air, by spongy platinum or platinum black. In air it burns, when lighted, with a strong yellow flame. It is a compound of considerable stability, but is decomposed in part when sent through a tube heated to whiteness, and resolved into carbon and hydrogen. This gas is not affected in the dark by chlorine, but when the mixture of these gases, in a moist state, is exposed to light, carbonic and hydrochloric acid gases are produced.

Although instantly kindled by flame, protocarburetted hydrogen requires a high temperature to ignite it. Hydrogen, hydrosulphuric acid gas, and olefiant gas, and carbonic oxide, are all ignited by a glass rod heated to low redness, but glass must be heated to bright redness or to whiteness, to inflame this gas. Sir H. Davy discovered that flame could not be communicated to an explosive mixture of the gas of mines and air, through a narrow tube, because the cooling influence of the sides of the tube prevented the gaseous mixture contained in it from ever rising to the high temperature of ignition. A metallic tube had a greater cooling property, from its high conducting power, and consequently obstructed to a greater degree the passage of flame, than a similar tube of glass; and even the meshes of metallic wire-gauze, when they did not exceed a certain magnitude, were found to be impermeable by flame. Experiments of this kind may be made upon coal-gas, the flame of which will be found incapable of passing through a sheet of iron-wire trellis, containing not less than 400 holes in the square inch. If the gas be allowed to pass through the trellis, and kindled above it, the flame, it will be found, does not return through the apertures to the jet whence the gas issues. Upon these observations, Sir H. Davy founded his invaluable invention of the Safety-lamp, — an instrument now indispensable to the safe working of the most extensive and valuable of our coal-fields.

Safety-lamp. — As left by Davy, this is simply an oil lamp, enclosed in a cage of wire-gauze, the upper part of which is double (fig. 128). Mr. Buddle used iron-wire gauze for the lamp, containing from 784 to 800 holes in the square inch. A crooked wire, which works tightly in a narrow tube passing upwards through the body of the lamp, affords the means of trimming the wick, without undoing the wire-gauze cover of the lamp. When the lamp is carried into an atmosphere charged with fire-damp, a blue flame is observed within the gauze cylinder, from the combustion of the gas, and the flame in the centre of the lamp may be extinguished. The miner should then withdraw, for although the gauze has often been observed to become red-hot, without inflaming the external explosive atmosphere, yet the texture of the gauze may be destroyed, if retained long at so high a temperature. It has always been known, since this lamp was first proposed, that when it is exposed to a strong current of the explosive mixture, the flame may pass too quickly through the apertures of the gauze to be cooled below the point of ignition, and, therefore, communicate with the external atmosphere. But this is easily prevented by protecting the lamp from the draught, and an accident from this cause is not likely to occur in a coal-mine.[1]

Fig. 128.

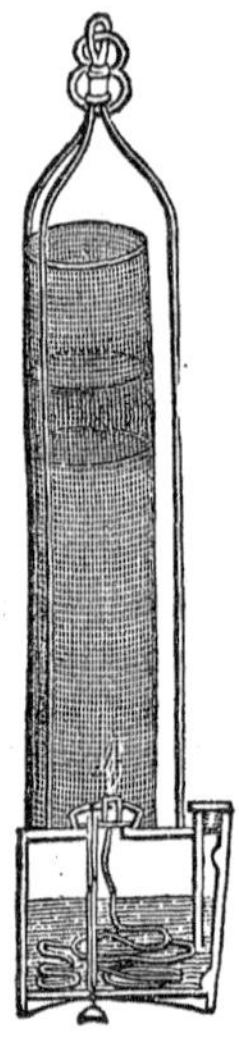

The carburetted hydrogen does not explode when mixed with air in a proportion much above or below the quantity necessary for its complete combustion. With 3 or 4 times its volume of air it does not explode at all, with $5\frac{1}{2}$ or 6 volumes of air it detonates feebly, and with 7 to 8 most powerfully. With 14 volumes of air, the mixture is still explosive, but with larger proportions of air, the gas only burns about the flame of the taper. The large quantity of air which is then mixed with the gas absorbs so much heat as to prevent the temperature of the gaseous atmosphere from rising to the point of ignition.

Coal-gas. — The products of the distillation of coal in an iron retort are of three kinds: a black oily liquid, of a heterogeneous nature, known as coal-tar; a watery

[1] For additional information respecting the safety-lamp, the reader is referred to Davy's Essay on Flame, to Dr. Paris's Life, and Dr. J. Davy's Life of Sir H. Davy, and to the Report of the Parliamentary Committee on Accidents in Mines, 1835.

fluid, known as the ammoniacal liquor, and the elastic fluids which form coal-gas. To purify the gas, it is cooled by transmitting it through iron tubes or shallow boxes, in which it deposits some condensible matter; and it is afterwards exposed to milk of lime, to absorb hydrosulphuric acid gas, which it invariably contains, and frequently afterwards to dilute sulphuric acid or a solution of sulphate of iron, which arrests a little hydrosulphate of ammonia and a trace of hydrocyanic acid. The hydrate of lime is often applied in the state of a damp powder, and not diffused through water.

The process may be illustrated by the arrangement represented in fig. 129. The coal to be distilled is contained in an iron or stoneware retort A, which should not be

Fig. 129.

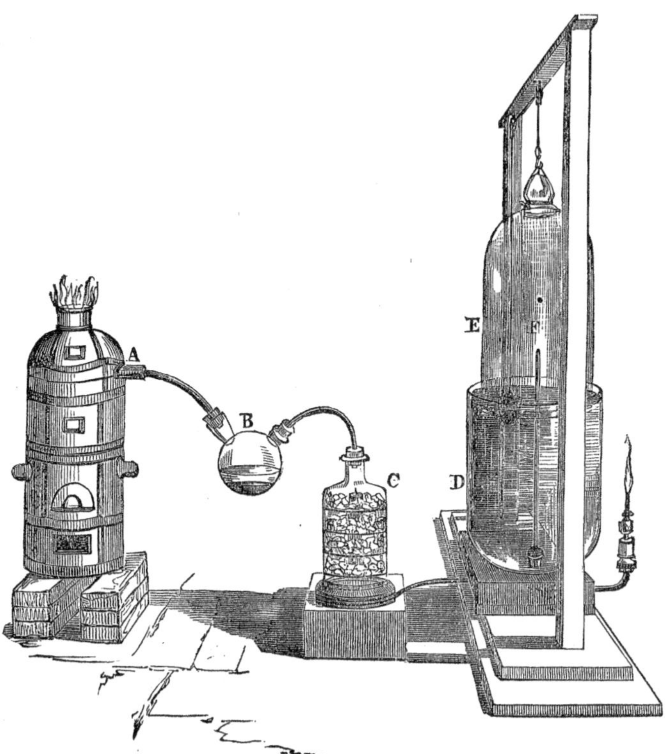

more than half filled if the coal is of a bituminous quality, and is heated by a small charcoal furnace. Tar and a watery fluid containing ammonia condense in B, which represents the condenser. The gas passes on to C, a glass jar, with stages of wire-gauze supporting slaked lime, and forming a lime-purifier. The gas is then conveyed by the tube F into the bell-jar E, filled with water, and inverted over another glass jar D, serving as a water-tank. The jar E, which represents the gasometer, is connected by a string passing over two pulleys above, with an iron weight which balances it. When the gasometer rises and is full, the gas may be allowed to escape by the tube F and the jet and stopcock at the side, by removing or diminishing the counterpoise to the jar E.

Dr. Henry obtained the following results from an examination of the gas from the best cannel coal, at different periods of the distillation:—

COAL GAS IN 100 VOLUMES.

	Density.	Olefiant gas.	Carburetted hydrogen.	Carbonic oxide.	Hydrogen	Nitrogen
At beginning of process...	650	13	82.5	3.2	0	1.3
After five hours.............	500	7	56	11	21.3	4.7
After ten hours..............	345	0	20	10	60	10

Besides the constituents mentioned, coal-gas, when first made, contains small quantities of

Ammonia	Hydrocyanic acid
Hydrosulphuric acid	Bisulphide of carbon
Carbonic acid	Naphtha vapour.[1]

All of these bodies are separated from it in the process of purification, except the two last, namely, naphtha vapour, which is the chief cause of the odour of coal-gas, and bisulphide of carbon, which affords a little sulphurous acid when the gas is burned. The heterogeneous nature of the gaseous mixture is well shown upon introducing a quantity of dry iodine into a bottle of coal-gas, when several liquid and solid compounds of iodine are formed with the different carbohydrogens present. Iodine, on the other hand, is not affected in the slightest degree by fire-damp, but remains with its metallic lustre unchanged in that gas. Indeed, in the ordinary fire-damp no other combustible gas whatever can be found, besides protocarburetted hydrogen (Mem. of Chem. Soc. iii. 7).

The superiority of coal-gas, in illuminating power, depends principally upon the high proportion of olefiant gas and the denser carbohydrogens which it contains. The free hydrogen and carbonic oxide present give no light, and are positively injurious. As the highly illuminating constituents are dense, and contain much carbon, the value of coal-gas is to a certain extent proportional to its density, and to the quantity of oxygen which it requires for complete combustion. In the analysis of coal-gas, the different gases may thus be separated: 1st. Olefiant gas, naphtha vapour, and similar carbohydrogens, by mixing the gas over water, in a dark place, with half its bulk of chlorine, and afterwards washing with caustic potassa; or, by introducing a small pellet of coke charged with fuming sulphuric acid and attached to a platinum wire, into the gaseous mixture, over mercury, and afterwards absorbing the acid vapour by a fragment of hydrate of potassa: 2dly, carbonic oxide, by potassium gently heated in the gas; 3dly, the proportion of protocarburetted hydrogen gas may be determined by detonating the mixture over mercury, in an eudiometer (fig. 113, page 249), with a measured quantity of oxygen, and ascertaining the quantity of carbonic acid formed, which retains the volume of this carburetted hydrogen; 4thly, the free hydrogen, by observing the quantity of oxygen remaining, by means of a stick of phosphorus introduced into the gas, and thereby ascertaining the quantity of oxygen consumed in the last combustion; from this quantity deduct twice the measure of the carburetted hydrogen found, and half the remaining measure of consumed oxygen represents the hydrogen; 5th, the residuary gas after these processes is the nitrogen of the coal-gas.[2]

[1] Dr. Henry's Papers on Coal-Gas are contained in the Philosophical Transactions for 1808, 1820, and 1824.

[2] The tubes and eudiometers for measuring gases require to be very minutely graduated: this is attained with peculiar accuracy and facility by the method recommended by Professor Bunsen. His instrument for graduating glass tubes (fig. 130) consists of a mahogany board A, 5½ feet long, 7 inches wide, and ¾ of an inch thick. In the middle of this board is a groove extending its whole length, 1 inch wide, ½ inch deep, and rounded at bottom as a bed for the reception of the tube. At one part, 5 inches from the end, is placed a brass plate B, 1⅓ foot long and 2 inches wide, in such a position that when screwed down its edge comes one-half over the groove. It is furnished with four screw-nuts, passing through slits

Structure of flame.—The quantity of light obtained from the combustion of coal-gas depends entirely upon the manner in which it is burned, which will appear

---

in the plate, a quarter of an inch long, so as to allow a certain advancement or withdrawal of the plate at pleasure.

Fig. 130.

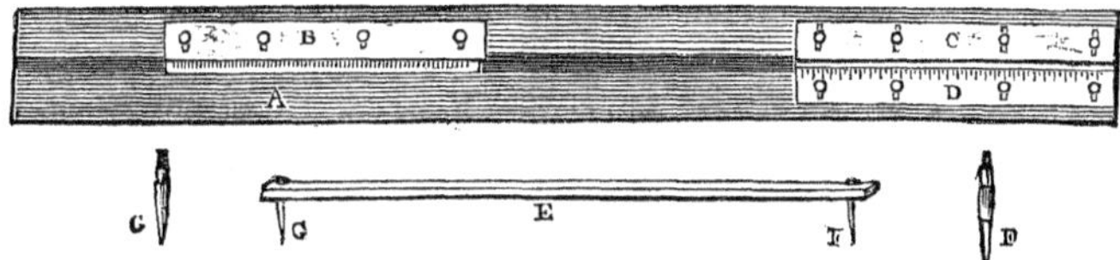

C and D are two similar plates, placed at the other end of the wooden board, C having the same amount of motion as B, and being precisely similar in every respect. D is a brass plate of the same dimensions as B and C, which is cut, at intervals of five millimeters, into notches, every alternate one being one-twentieth and one-tenth of an inch deep. There is also a wooden rod E, 3 feet long, 1 inch broad, and half an inch thick. This is provided with two steel points, placed by screws at half an inch from either end. One of these, F, is in the form of a knife, the other, G, of a bradawl; a screw-driver is also provided, that these points may be attached or removed at pleasure.

When a tube is to be graduated, it is covered with a thin layer of melted wax and turpentine, by means of a camel's hair pencil, and is laid in the groove between C and D, which are then screwed down in their places, so as to retain the tube firmly in its position. A standard tube, previously mathematically divided into millimeters, (the most convenient division,) is now placed in the groove under B, (fig. 131) which is then screwed upon it. The rod, E, is now used, the pointed steel, G, being put into one of the millimeter marks on the standard tube; the knife point, F, falls upon the waxed tube, and is made to produce a line upon it, the length of which is regulated by the distance between the edges of the brass plates C and D. The pointed steel is now removed back one millimeter on the standard tube, and the corresponding mark made on the waxed one; and thus we proceed until the whole of the waxed tube is divided into millimeters. The object of the notches is, that a longer mark may be made at every five millimeters, and a still longer one at every ten, in order to aid the eye in reading. The waxed tube is now removed to a leaden trough containing pounded fluor spar and sulphuric acid, slightly heated, which etches it more successfully than a solution of hydrofluoric acid. Previously, however, to being etched, it is desirable to figure the number of millimeters at the space of every ten; and this is conveniently done by the steel pointer G, after being removed from the rod E. The tube is rubbed with vermilion powder when in use, to make the graduation more legible.

Fig. 131.

We have thus an accurate measure of length etched upon the tube, which should be one of pretty uniform calibre. The next point is to determine the true value of each of the divisional marks: this is done by calibrating it throughout all its length by small portions of mercury,—say equal in bulk to five grains of water. By this means the relative value of each mark is determined, and the proportion which it bears to any given standard. The only possible error is in the assumption that the tube is of even calibre within the space occupied by one measure of mercury; but the quantity of this added is so small, that any such error becomes quite inappreciable. The convenience of this graduator is so great, that a long tube may be beautifully divided in the course of an hour. The standard tube should be made of glass, but the original divisions from which this standard is taken may be those of wood or any other material.

The tubes recommended by Bunsen are 18 or 19 inches in length, about 0.6 inch in internal, and 0.8 inch in external diameter. One of these is converted into a eudiometer, in which the gases are exploded, by inserting near the closed end, by fusion, two platinum wires of the thickness of horse-hair, for the purpose of passing the electric spark. During the explosion the open end of the tube is pressed firmly upon a smooth pad of caoutchouc, placed under the mercury at the bottom of the pneumatic trough. The graduation of these tubes being linear, enables the observer to read off the difference in height between the mercury in the tube and trough, and to make the necessary correction on the volume measured; all exact experiments on gaseous volumes must be made over mercury. This department of chemical analysis has been brought to a high degree of accuracy and perfection by Professor Bunsen. (See Reports of the British Association, 1845, page 148; and Liebig and Poggendorff's Handwörterbuch der Chemie, ii. 1053.)

from the consideration of the structure of luminous flames. The flame of a spirit-lamp, candle, or gas-jet, is hollow, as may be observed by depressing a sheet of wire-trellis upon it, which gives a section of the flame; the seat of the combustion being the margin of the flame, where alone the combustible vapour is in contact with the air. Of volatile carbonaceous combustibles, the flame consists of three parts, which are represented in section (fig. 132):—

FIG. 132.

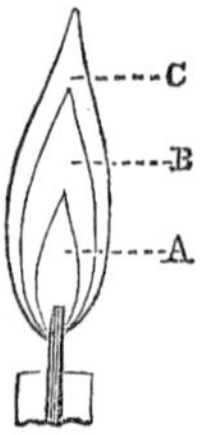

A, cone of vapourized combustible.
B, sphere of partial combustion.
C, sphere of complete combustion.

In B, where the supply of air is insufficient for complete combustion, it is the hydrogen principally which burns, the carbon being liberated in solid particles, which are heated white-hot from the combustion of that gas. The sphere B, indeed, is the luminous portion of the flame, for the light depends entirely upon the deposition of carbon arising from the consecutive combustion of the two elements of the vapour. Gaseous bodies, however strongly heated, emit no light, or at most not more than a sensible glow, and luminous flame has justly been described by Davy as always containing *solid matter heated to whiteness.* The same sphere of the flame, possessing an excess of combustible matter at a high temperature, takes oxygen from metallic oxides, such as arsenious acid, placed in it, and developes their metals. It is, therefore, often referred to as the deoxidizing or reducing flame. In the external hollow cone, C, the deposited carbon meets with oxygen, and is entirely consumed. The hottest point in the whole flame is within this sphere, near the summit of B. This part of the flame, possessing an excess of oxygen at a high temperature, is the proper place for kindling a combustible, and is called the oxidizing flame: its properties are the opposite of those of B.

When coal-gas is mingled with an equal bulk of air before being burned, it is found to lose half its illuminating power. It may be conveniently mixed with a quantity of air sufficient for its complete combustion, by placing over an argand burner, a brass chimney of 5 inches in height provided with a cap of wire-gauze; when kindled above the wire-gauze, the gas burns with a blue flame, not more luminous than that of sulphur. The flame is so feebly luminous because no deposition of carbon occurs in it. The quantity of heat is the same, whether the gas is burned so as to produce much or little light; and where the gas is burned for heat, this mode of combustion has the advantage of giving a flame without smoke. The heat derived from coal-gas burned in this manner is not, however, so intense as that of an argand spirit-lamp.

A result of the circumstances which determine the quantity of light from different flames is, that the larger the flame till it begins to be smoky, the greater the proportion of light obtained from the consumption of the same quantity of gas. It was observed that an argand burner, supplied with 1½ cubic feet of gas per hour, gave as much light as a single candle; with 2 cubic feet per hour the light was equal to 4 candles, and with 3 cubic feet to 10 candles. Hence argands, bat-wings, and other burners, in which a considerable quantity of gas is burned together, are more economical than plain jets. The brightness of ordinary flame, which depends essentially upon the consecutive combustion of hydrogen and carbon, is increased by everything which promotes the rapidity and intensity of the combustion, without deranging the order of oxidation, such as a rapid supply of air, and the substitution of pure oxygen for air, as in Gurney's Bude Light. Not only is there then more light, because there is more combustion in the same time, but the temperature of the flame being greater, the luminous carbon is also heated to a higher degree of whiteness.

BICARBURETTED HYDROGEN.

Syn. Olefiant gas, Elayle; Eq. 28 *or* 350; C_4H_4; *density* 985.2; ☐☐

This gas was discovered in 1796, by certain associated Dutch chemists, who gave it the name of olefiant gas, because it forms with chlorine a compound having the appearance of oil. It is usually prepared by heating together 1 measure of spirits of wine with 3 measures of oil of vitriol, in a capacious retort, till the liquid becomes black and effervescence begins, and maintaining it at that particular temperature. It is collected over water, which deprives it of a portion of ether vapour and sulphurous acid, with which it is accompanied.

A process which yields a purer gas, and in larger volume, is the following. Twenty-eight ounces of water are added to twice their volume of oil of vitriol, in a large globular flask A (fig. 131), which gives an acid of about 1.6 density when

Fig. 133.

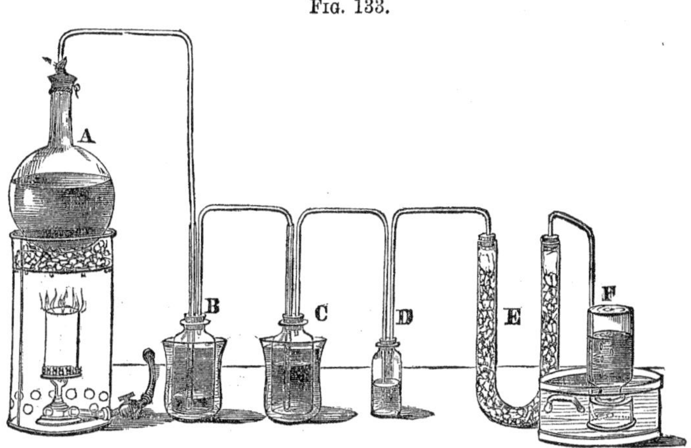

cool. Without waiting to cool, however, 24 ounce measures of spirits of wine are added, and the whole allowed to stand for a night. The flask is supported on a bed of pumice over the gas-flame as already described (page 264), and the latter regulated so as to keep the liquid in a state of moderate ebullition. The gas evolved is passed through two two-pound bottles, B and C, the first of which, B, is empty, or contains only a little water at the beginning, and is intended for the condensation of a considerable portion of alcohol and ether which distil over, while C is half filled with a strong solution of caustic potassa, to absorb the sulphurous and carbonic acids produced. These two wash-bottles are immersed in jars containing cold water. The third wash-bottle, D, contains oil of vitriol, and the U-tube E, pumice soaked in the same fluid to absorb ether-vapour; while the gas is collected at last in bottles, F, over water made sensibly alkaline by caustic potassa.

This gas is formed by a peculiar decomposition of alcohol, in contact with sulphuric acid boiling at 325°, or a little higher, in which the alcohol is resolved into olefiant gas and water, $C_4H_6O_2 = C_4H_4$ and 2HO. This decomposition will be referred to again more particularly under the head of alcohol.

Bicarburetted hydrogen gas contains 2 volumes of carbon vapour and 2 volumes of hydrogen condensed into 1 volume, and is theoretically of the same density as nitrogen and carbonic oxide, or fourteen times heavier than hydrogen. It was condensed by cold and pressure into a transparent liquid, which is not solidifiable (page 79). This gas, when carefully deprived of ether, has a sweet odour, which is peculiar but not strong. Water absorbs about one-eighth of its volume of this gas;

alcohol takes up 2 volumes, oil of turpentine 2.5, and olive oil 1 volume. It is absorbed by fuming sulphuric acid, and by the perchloride of antimony, forming peculiar compounds. The substances named leave certain gaseous impurities uncondensed, which often amount to 15 or 20 per cent., and appear to be principally protocarburetted hydrogen. The gas of the process described above is entirely absorbed by the perchloride of antimony, except about 4 per cent.; but it appears to contain the vapour of some denser carbohydrogen, not absorbed by oil of vitriol, as the specific gravity of the gas so prepared is often as high as that of air, or 1000, instead of 985.2 as observed by Saussure.

This gas burns with a white flame, which is much more brilliant than that of protocarburetted hydrogen. It requires three times its volume of oxygen to burn it completely, and yields twice its volume of carbonic acid gas and twice its volume of aqueous vapour; for one volume of bicarburetted hydrogen contains 2 volumes of carbon vapour, each of which requires 1 volume oxygen and becomes 1 volume carbonic acid, and 2 volumes hydrogen, each of which requires $\frac{1}{2}$ volume oxygen and forms 1 volume steam. This gas is entirely decomposed, when passed through a porcelain tube at a white heat, into carbon, which is deposited, and twice its volume of hydrogen gas.

Bicarburetted hydrogen mixed with twice its volume of chlorine gas is condensed, and forms a liquid compound of an oily consistence, $C_4H_4Cl_2$, from which it was named olefiant gas, or the oil-making gas, and Elayle (from ἐλαιον and ὑλη, the source of an oil), by Berzelius. This substance, which is also known as Dutch liquid, will be described under the derivatives of alcohol.

GAS OF OIL.

Bicarburetted hydrogen of Faraday; Eq. 56 *or* 700; C_8H_8; *density* 1926.4; ☐☐

This gas, which is twice as condensed as olefiant gas, is one of the products of the decomposition of the fixed oils by heat, and exists, therefore, in the gas prepared from oil. It is liquefied when oil gas is greatly compressed, and also by a cold of 0° F. The flame of this gas is very brilliant; it is only sparingly soluble in water, but pretty soluble in alcohol and the fat oils; sulphuric acid dissolves a hundred times its volume. It combines with an equal volume of chlorine, and forms a liquid compound having some analogy to Dutch liquid.

This gas requires 6 volumes of oxygen to burn it, and gives rise to water and 4 volumes of carbonic acid.

CARBON AND NITROGEN—CYANOGEN.

Eq. 26 *or* 325; NC_2; *density* 1819; ☐☐

This compound is a gas, which was first obtained by Gay-Lussac in 1815. It is prepared by heating the cyanide of mercury in a small glass retort, and is collected at the mercurial trough. The cyanide is resolved into running mercury and cyanogen gas, and frequently leaves a black coaly mass in the retort, which Professor Johnston has shown to consist of carbon and nitrogen, in the same proportions as the gas itself.

Cyanogen gas contains 4 volumes of carbon vapour and 2 volumes of nitrogen, condensed into 2 volumes; its density is 1819. When this gas is exploded with twice its volume of oxygen, it affords 2 volumes of carbonic acid gas, and 1 volume of nitrogen; an experiment from which its composition may be deduced. Water at 60° absorbs 4.5 times its volume of this gas, and alcohol 23 volumes. By a pressure of 3.6 atmospheres at 45°, cyanogen is condensed into a limpid liquid, which evaporates again on removal of the pressure. Cyanogen burns with a beautiful purple flame in air or oxygen. The solution of cyanogen in water undergoes spontaneous decomposition. By alkalies the gas is absorbed, and a cyanide and cyanate formed.

Carbon does not burn when heated in nitrogen gas, and appears to be incapable of uniting with that element when alone, or unless when assisted by the presence of a third body, such as potassium, which unites with and gives stability to the compound. Cyanogen is thus produced when nitrogen is sent over fragments of charcoal saturated with potassa, heated white-hot in a porcelain tube placed across a furnace, and obtained as cyanide of potassium. A peculiar form of furnace, in which this remarkable process is conducted on a large scale at Newcastle, with considerable success, is described by Mr. Bramwell (Repertory of Inventions, 3 ser. ix. 280). It consists essentially of a vertical flue in brickwork A B D, (fig. 134), containing charcoal charged with a solution of carbonate of potassa, the middle portion of which, B, is placed within the flue of the adjoining furnace 2 2, by which it is heated intensely, and also obtains a supply of nitrogen, which enters A B D by a number of small openings into the external flue. The passage of gases upwards through the potassa-charcoal is further promoted by the action of air-pumps connected with the tubes G and H. The materials are introduced at the top on removing a lid C, and after descending through the tube are allowed to fall into a cistern of water F, in which the cyanide of potassium is found dissolved. The pipes I and J dip into water, to intercept ammonia or any other volatile product.

Fig. 134.

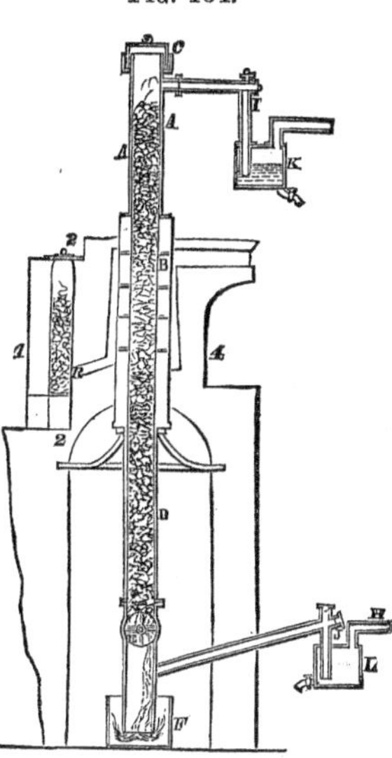

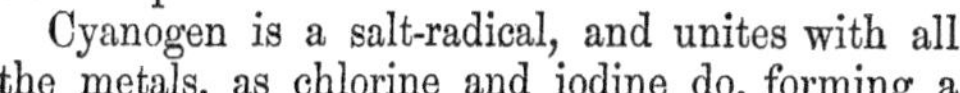

Cyanogen is a salt-radical, and unites with all the metals, as chlorine and iodine do, forming a class of cyanides. It also combines with hydrogen and forms a hydrogen-acid, namely, hydrocyanic or prussic acid. Cyanogen properly belongs to organic chemistry, in which department its numerous combinations will be considered.

Mellon, N_4C_6.—This is another salt-radical, and was formed by Liebig by heating the bisulphide of cyanogen in a glass flask to redness, when it is resolved into sulphur, bisulphide of carbon, and mellon. It is a lemon yellow powder, insoluble in water and alcohol; it unites directly with hydrogen and with potassium, forming hydro-mellonic acid, a hydrogen-acid, and mellonide of potassium, a saline body.

SECTION V.

BORON.

Eq. 10.9 *or* 136.2; B; *density of vapour* (*hypothetical*) 751; ☐☐ .

Boron is an element having some analogy to carbon, but sparingly diffused in nature. It is never found, except in combination with oxygen as boracic acid, of which the salt of soda has long been brought to Europe from India in a crude state, under the name of tinkal, and termed borax when purified. The impure borax or tinkal forms a saline incrustation in the beds of certain small lakes in an upper province of Thibet, which dry up during summer. But the most considerable of the present sources of boracic acid are the hot lagoons of a district in Tuscany, which are charged with the free acid, from the condensation in them of vapours of a volcanic origin. Boracic acid is likewise found in the hot springs of Lipari. It is a constituent also of several minerals, of which datolite and boracite are the most remarkable. Boron was first discovered by Sir H. Davy in 1807, by exposing boracic acid to the action of a powerful voltaic battery, and was afterwards obtained by Gay-Lussac and Thenard in greater quantity, by heating boracic acid with potassium.

Preparation.—Boron is prepared with greatest advantage from a combination of fluoride of boron and fluoride of potassium, which is obtained on saturating hydrofluoric acid with boracic acid, and afterwards adding to it, drop by drop, the fluoride of potassium. This compound, which is of slight solubility, is collected on a filter, and dried at an elevated temperature, but which should not reach a red heat. Equal weights of the compound and of potassium are mixed together in a cylinder or tube of iron, closed at one end, which is gently heated, and the mixture stirred with an iron rod, till the potassium is melted. Heated afterwards more strongly by a spirit-lamp, the mass evolves heat, and becomes red-hot; the potassium combines with the fluorine, and a mixture is obtained of boron and the fluoride of potassium. On treating this with water, the fluoride of potassium dissolves, and the boron remains alone. In washing it farther, instead of pure water, which causes the oxidation of boron, a solution of sal ammoniac should be employed, which does not act upon that body, and the sal ammoniac remaining in the boron may be taken up by alcohol.

Properties.—Thus prepared, boron is obtained in the form of a greenish-brown powder, without the metallic lustre, which becomes hard and assumes a deeper colour, when ignited in vacuo, or in gases which do not combine with it, but undergoes no farther change. Heated in atmospheric air or oxygen it burns with a vivid light, scintillating powerfully, and forms boracic acid. Nitric acid and many other substances also oxidate it easily, and always produce that compound. Fused with carbonate of potassa, it decomposes the carbonic acid, and gives borate of potassa, carbon being liberated. Boron is not known to possess any other degree of oxidation. Boron combines with sulphur, with the disengagement of light, when heated in the vapour of that substance; and it takes fire spontaneously in chlorine, and forms a gaseous chloride of boron, of which the formula is BCl_3, and the density 3942 by observation and 4035 by calculation. This gas is composed of 2 vols. of boron vapour and 6 of chlorine, condensed into 4 vols., which are its combining measure. It may likewise be formed by transmitting chlorine gas over a mixture of boracic acid and charcoal, ignited in a porcelain tube. A corresponding fluoride of boron is evolved from boracic acid, ignited with the fluoride of calcium or fluor-spar, with the formation of borate of lime. The density of this fluoride is 2312.4. Both of these gases are decomposed by water, boracic acid being formed with hydrochloric or hydrofluoric acid.

Boracic or Boric acid.—This acid is prepared by dissolving the salt borax at 212° in two and a half times its weight of water, and adding enough of hydrochloric acid to make the liquid strongly acid to test paper. Chloride of sodium is formed, which continues in solution, while the boracic acid separates in thin shining crystalline plates, on cooling. These plates are drained, and being sparingly soluble, may be washed with a little cold water, and afterwards redissolved in boiling water, and made to crystallize anew. Fused at a red heat in a platinum crucible, these plates give the vitrified acid, of which the density is 1·83. Boracic acid has a weak taste, which is scarcely acid, and it affects blue litmus like carbonic acid, imparting to it a wine-red tint, and not that clear red, free from purple, which the stronger acids produce. It renders yellow turmeric paper brown, like the alkalies. The acid of the carbonates, however, is displaced by boracic acid in the cold, and at a red heat this acid decomposes even the sulphates, from its comparative fixity. The crystals of boracic acid are a hydrate, and contain 3 equivalents of water, of which the formula is $HO.BO_3+2HO$. At 60° it requires 25.66 times its weight of water to dissolve it, but only 2.97 times at 212°. With the assistance of the vapour of water, it is slightly volatile, but alone it is more fixed, and fuses, under a red heat, into a transparent glass. At the white heat of our furnaces boracic acid does not boil; but the tension of its vapour is so considerable at that temperature that it evaporates entirely away in the end. The hydrated acid dissolves in alcohol, and the solution burns with a fine green flame. It communicates fusibility to many substances in uniting with them, and generally forms a glass. On this account borax is much used as a flux.

Borates.—Boracic acid is remarkable for the variety of proportions in which it unites with alkalies; all these borates have an alkaline reaction like the carbonates. The relative proportions of oxygen and boron in boracic acid are known, but the number of equivalents of these elements in this acid is not so certain. Dumas inferred from the density of the chloride that it is a terchloride, and boracic acid, which corresponds, will therefore consist of 3 eq. of oxygen to 1 eq. of boron, and its formula be BO_3. This makes borax the biborate of soda.

SECTION VI.

SILICON OR SILICIUM.

Eq. 21.35 *or* 266.82; Si; *density of vapour (hypothetical)* 1475; ☐☐

Silica or siliceous earth, the oxide of the present element, is the most abundant of all the matters which compose the crust of the globe. It constitutes sand, the varieties of sand-stone and quartz rock, and enters into felspar, mica, and a great variety of minerals, which form the basis of other rocks.

Preparation.—Silica may be decomposed by heating it with potassium, which deprives it of oxygen; but a better process for obtaining silicon is to heat the double fluoride of silicon and potassium, with 8 or 9-10ths of its weight of potassium, with the same precautions as in the preparation of boron. The materials, however, in this case may be heated in a glass tube, as well as in an iron cylinder. The double fluoride employed is prepared by neutralizing fluosilicic acid with potassa. A different process is suggested by Berzelius, which consists in heating potassium in a tube of hard glass with a small bulb blown upon it, which is filled with the vapour of the fluoride of silicon, supplied from the ebullition of that liquid contained in a small retort connected with the glass tube. The potassium burns in this vapour, and at the end, silicon is found, with fluoride of potassium, in the place of the metal (Traité, t. 1, p. 307). But the silicon from all these processes is always in combination with a little potassium, and mixed with a little fluoride of silicon and potassium unreduced. Hence, on applying cold water to the mass, hydrogen gas is disengaged, and potassa formed, and the silicon separates. The potassa thus produced can, with the aid of hot water, dissolve the silicon, which then oxidates and becomes silica, so that cold water only must be employed to wash the silicon, which may be thrown upon a filter. After a time, the liquid which passes has an acid reaction, which arises from its dissolving an acid double fluoride of silicon and potassium, of sparing solubility, which has escaped decomposition, and is mixed with the silicon. The washing is continued so long as the water dissolves anything.

Properties.—The silicon which is thus obtained is, in its pure state, a dull brown powder, which soils the fingers, and when heated in air or oxygen, inflames and burns, but is never more than partially converted into silica. It may be ignited strongly in a covered crucible without loss, and then shrinks in dimensions, acquires a deep chocolate colour, and becomes so dense as to sink in oil of vitriol. By this ignition the properties of silicon are altered to a degree which is very remarkable in a simple substance. It was previously readily soluble in hydrofluoric acid, with evolution of hydrogen, and in caustic potassa, but it is now no longer acted upon by that or any other acid, nor by alkalies. The ignited silicon also refuses to burn in air or oxygen, even when intensely heated by the blowpipe flame. Charcoal, it will be remembered, is more dense and less combustible after being strongly heated; but that substance is not altered by heat to the same extent as silicon. Mixed and heated with dry carbonate of potassa, silicon in any condition is oxidated completely, its action upon the carbonic acid of the salt being attended with ignition, and carbon liberated. Silicon burns when heated in sulphur vapour, and forms a sulphide, which water dissolves, but decomposes at the same time, hydrosulphuric acid and silica being produced, and the last, notwithstanding its usual insolubility, retained in solution. Silicon likewise burns in chlorine; and the chloride of silicon may be otherwise formed by transmitting chlorine over a mixture of charcoal and silica ignited in a porcelain tube. The silica is decomposed by neither charcoal nor chlorine singly, but acting together upon the silica, these bodies produce carbonic oxide and chloride of silicon. This compound is a volatile liquid, of which the formula is $Si\ Cl_3$; that of the sulphide of silicon $Si\ S_3$.

Silica or Silicic Acid, $Si\ O_3$.—This earth, which is the only oxide of silicon, constitutes a number of minerals, nearly in a state of purity, such as rock-crystal, quartz, flint, sandstone, the amethyst, calcedony, cornelian, agate, opal, &c. The first chemical examination of its properties and compounds is due to Bergman.

Preparation.—Silica may be had very nearly, if not absolutely pure, by heating a colourless specimen of rock-crystal to redness and throwing it into water, after which treatment the mineral may be easily pulverized. It is obtained in a state of more minute division, by transmitting the gaseous fluoride of silicon (fluosilicic acid) into water; or by the action of acids upon some of the alkaline compounds of silica. Equal parts of carbonate of potassa and carbonate of soda may be fused in a platinum crucible, at a temperature which is not high; and pounded flint or any other siliceous mineral, thrown by little and little into the fused mass, dissolves in it with an effervescence due to the escape of carbonic acid gas. The addition of the mineral is continued so long as it determines this effervescence. The mass being allowed to cool, is afterwards dissolved in water acidulated with hydrochloric acid, which takes up the silica as well as the alkalies; the liquor is filtered and then evaporated to dryness. The silica may contain a little peroxide of iron or alumina, to dissolve which the saline mass, when perfectly dry, is moistened with concentrated hydrochloric acid, and after two hours the acid mass is washed with hot water. The silica remains undissolved; it may be dried well and ignited.

Properties.—Silica so prepared is a white, tasteless powder, which is rough to the touch, and feels gritty between the teeth. It is extremely mobile when heated, and is thrown out of a crucible, at a high temperature, by the slightest breath of wind. It is absolutely insoluble in water, acids, and most liquids. Finely divided silica, however, decomposes an alkaline carbonate at the boiling point, and is dissolved. Its density is 2.66. The heat of the strongest wind-furnace is not sufficient to fuse silica, but it melts into a limpid colourless glass in the flame of the oxihydrogen blowpipe, and may be drawn out into threads (Girardin). Silica is found frequently crystallized, its ordinary form being a six-sided prism terminated by a six-sided pyramid, as in rock-crystal. Sometimes the prism is very short or disappears entirely, and the pyramid only is seen, as in ordinary quartz.

Silicic acid dissolved by acids.—The conditions of the solubility of silicic acid in other acids are peculiar. Once precipitated, whether gelatinous, like boiled starch, or pulverulent, it is no longer in the least degree soluble either in water or acids. If to a dilute solution of an alkaline silicate, hydrochloric acid be added slowly and drop by drop, the silicic acid is precipitated in proportion as the alkali is neutralized. But, on the contrary, no silicic acid is precipitated, if strong hydrochloric acid in considerable excess be added all at once to the solution of alkaline silicate, or if the latter be poured in a gradual manner into hydrochloric acid whether strong or greatly diluted with water. It thus appears that silicic acid only dissolves in the stronger acids, when presented to them in the *nascent state*, or at the moment of leaving another combination. It appears to enter into combination with the acid which dissolves it; for if the latter is exactly neutralized by adding a strong solution of potassa, drop by drop, the whole of the silica is precipitated.

A pure solution of silicic acid in hydrochloric acid, free from saline matter, is best obtained from the silicate of copper. The latter is prepared by precipitating chloride of copper by the solution of an alkaline silicate; washing the insoluble silicate of copper which falls, by several times mixing it with water and allowing it to subside, so as to get rid of the chloride of potassium present. The silicate of copper is then dissolved in hydrochloric acid, filtered, and hydrosulphuric acid gas made to stream through the liquid, to precipitate the copper. The black insoluble sulphide of copper is removed by filtration, and a perfectly colourless solution of silicic acid is obtained, which may be boiled, to expel the excess of hydrosulphuric acid, without injury. This solution is very acid, and when neutralized by ammonia or potassa it allows gelatinous silica to precipitate.

Hydrates of silicic acid. — When the last solution of silica in hydrochloric acid is evaporated in vacuo over fragments of quicklime, it deposits the protohydrate of silica, SiO_3+HO, in very thin crystalline filaments, grouped in stars, which are colourless, transparent, and possessed of considerable lustre. This is also the composition of the gelatinous silica, precipitated from an alkaline silicate, when allowed to dry in air. The silica has first the appearance of a transparent jelly, which is tenacious, and cracks on drying, forming a mass like gum. When this hydrate is dried at 212°, one half of the water escapes, and another definite hydrate, $2SiO_3+HO$, remains (Doveri). Another hydrate was obtained, by M. Ebelmen, by the spontaneous decomposition of silicic ether, of which the composition is $2SiO_3+3HO$. At 370° C. (698° F.), silicic acid does not retain more than a trace of water. (Doveri: Observations on the Properties of Silica, Annales de Chim. et de Phys. xxi. p. 40, 1847.)

Hydrofluoric acid has an affinity quite peculiar for silica, decomposing it, and carrying off the silicon, in the form of the volatile fluoride of silicon: —

$$3HF \text{ and } SiO_3 = SiF_3 \text{ and } 3HO.$$

The water of springs and wells always contains a little soluble silica, which can only be separated by evaporating the water to dryness. In some mineral waters the proportion of silica is very considerable, and it is often associated with an alkaline carbonate, which silica is capable of decomposing at the boiling point; as in the hot alkaline spring of Reikum in Iceland, and in the boiling jets of the Geyser, which deposit about their crater an incrustation of silica. There can be no doubt likewise that much of the crystalline quartz in nature, besides all the agates, calcedonies, and silicious petrifactions, have been formed from an aqueous solution.

Silicates. — Although silica has no acid reaction, it is certainly an acid, and is indeed capable of displacing the most powerful of the volatile acids at a high temperature. It is capable of uniting with metallic oxides, by way of fusion, in a great variety of proportions. Its compounds with excess of alkali are caustic and soluble, but those with an excess of silica are insoluble, and form the varieties of *glass*, which will be described under the silicate of soda. With alumina it forms the less fusible compounds of porcelain and stoneware, which will be noticed under that earth. A large number of mineral species also are earthy silicates. It seems probable that silicic, like phosphoric acid, forms several classes of salts, of which those containing the largest number of atoms of base are the most easily decomposed by acids. At the same time, some allatropic difference may be suspected between the silicic acid itself, as it exists in these different classes of salts, such as there is between ignited and unignited silicon.

The formula for silicic acid is not very certainly established. Most chemists admit it to be SiO_3, or analogous to sulphuric acid, SO_3, and then the equivalent of silicon is 266.7. But others adopt the formula SiO_2, considering silicic acid analogous to carbonic acid, CO_2; the equivalent of silicon then becomes 177.8. The last view is most in accordance with the density of silicic ether vapour. On the other hand, the composition of two intermediate compounds between the chloride of silicon, $SiCl_3$, and the sulphide of silicon, SiS_3, namely, $SiSCl_2$ and SiS_2Cl, is most simply represented on the first view. (Is. Pierre.)

SECTION VII.

SULPHUR.

Eq. 16 *or* 200; S; *at* 600°, *density of vapour* 6634, *and combining measure* 1-3*d volume; at* 1800°, *density about one-third of above, and combining measure* 1 *volume* □.

This element is exhaled in large quantity from volcanoes, either in a pure state or in combination with hydrogen, and by condensing in fissures forms sulphur veins, from which the greater part of the sulphur of commerce is derived. (See Recherches sur les fumerolles, par MM. Melloni and Piria: Annales de Chim. et de Phys. 2de sér. lxxiv. 331.) It exists also in combination with many metals, as iron, lead, copper, zinc, &c.; and is sometimes extracted from iron pyrites or bisulphide of iron. Sulphur is classed with oxygen; and the higher sulphides resemble peroxides in losing a portion of their sulphur, as some of the latter lose a portion of their oxygen, when strongly heated. Sulphur is likewise extensively diffused, as a constituent of the sulphuric acid, in gypsum and other native sulphates. This element also enters into the organic kingdom, being invariably associated in minute quantity with albuminous or protein compounds.

Properties. — Sulphur is found in commerce in rolls, which are formed by pouring melted sulphur into cylindrical moulds, and also in the form of a fine crystalline powder, the flowers of sulphur, which are obtained by throwing the vapour of sulphur into a close apartment, of which the temperature is below the point of fusion of that substance, and in which the sulphur therefore condenses in the solid form and in minute crystals, just as watery vapour does in the atmosphere below 32°, in the form of snow. The purity of the flowers is more to be depended upon than that of roll-sulphur. Sulphur is insipid and generally inodorous, but acquires an odour when rubbed; it is very friable, a roll of it generally emitting a crackling sound, and sometimes breaking, when held in the warm hand. Its specific gravity is 1.98. It fuses at 234°, forming a transparent and nearly colourless liquid, which is lighter than the solid sulphur. As the temperature is elevated, the liquid becomes more yellow, and passes abruptly into a dark brown at 482°. These allatropic conditions are distinguished by Frankenheim as $S\alpha$ and $S\beta$. In the last state it is so thick and viscous as to flow with difficulty. This change in its degree of fluidity is not occasioned by an increase of density, for fluid sulphur continues to expand with the temperature. Thrown into water, while in this condition, sulphur forms a mass which remains soft and transparent for some time after it is perfectly cool, and may be drawn into threads which have considerable elasticity. From 500° to its boiling point, 788°, when it is distinguished as $S\gamma$, it becomes again more fluid, and if allowed to cool returns through the same conditions, becoming again very fluid, before freezing. Sulphur has considerable volatility, beginning to rise in vapour before it is completely fused. At its boiling point it forms a transparent vapour of an orange colour, and distils over unchanged. The density of this vapour, taken a little above its boiling point, is very considerable, being observed to lie between 6510 and 6617 by Dumas, to be 6900 by Mitscherlich. These results indicate the unusual combining measure of 1-3d of a volume for this vapour, which gives the theoretical density 6634. But sulphur-vapour has lately been shown by M. Bineau to be one of those bodies of which the density changes with the temperature (page 132), and to fall at 1000° C. under ordinary pressure to about one-third of what it is about 450° or 500° C. The anomaly of its density is thus removed, and the combining measure of sulphur-vapour made to be 1 volume, or the same as oxygen.

Sulphur and many other substances may be obtained in distinct crystals, on passing from a state of fusion, by operating in a particular manner. A considerable quantity of sulphur is fused in a stoneware crucible, and allowed to cool till it begins to solidify; the solid crust which covers its surface is then broken, and the portion remaining fluid poured out. On afterwards breaking the crucible, when it has become quite cold, the sulphur is found to have a considerable cavity, which is lined with fine crystals, like a geode in quartz. Sulphur is dimorphous; the form which it assumes at a high temperature, and consequently in its passage from a state of fusion, is a secondary modification of an oblique prism with a rhomboidal base (fig. 135), belonging to the Fifth System of crystallization (page 144). Sulphur is soluble in the sulphide of carbon, the chloride of sulphur and oil of turpentine, and is deposited from solution in these menstrua at a lower temperature, and of its second form, which is an elongated octohedron with a rhomboidal base (fig. 136), belonging to the Third System. Such is likewise the form of the grains of flowers of sulphur, and of the fine transparent crystals of native sulphur; which last appear also to be formed by sublimation.

Fig. 135.

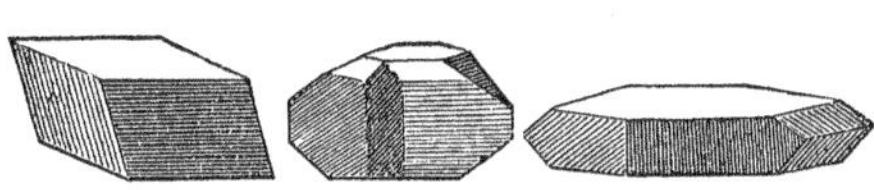

Fig. 136.

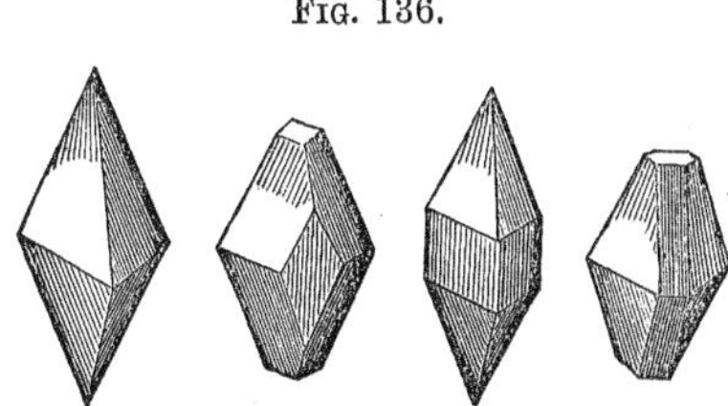

Sulphur is not soluble in water nor in alcohol. It combines readily with most metals; some of them, such as copper and silver in very thin plates, burning in its vapour, as iron does in oxygen gas. When iron and some other metals are mixed in a state of division with flowers of sulphur, and heat applied, the sulphur first melts, and after a few seconds combination ensues with turgescence of the mass, which becomes red-hot. Sulphur unites with bodies generally in the same multiple proportions as oxygen, and sometimes in additional proportions, particularly with potassium, and the metals of the alkalies and alkaline earths. When boiled with caustic potassa or lime, red solutions are formed which contain a large quantity of sulphur, a considerable proportion of which is deposited as a white hydrate of sulphur, upon the addition of an acid. With hydrogen, sulphur unites in single equivalents, and forms hydrosulphuric acid gas, which is the analogue of water in the sulphur series of compounds; and also another compound, the bisulphide of hydrogen, which is deficient in stability, like the binoxide of hydrogen, and is decomposed or preserved by similar agencies.

Sulphur is readily inflamed, taking fire below its boiling point, and burning with a pale blue flame and the formation of suffocating fumes, which are sulphurous acid gas. It exhausts the oxygen of a confined portion of air by its combustion more completely than carbonaceous combustibles, and on that account, and partly also from a negative influence which sulphurous acid has upon the combustion of other bodies, it may be employed in particular circumstances to extinguish combustion; a handful of lump sulphur being dropped into a burning chimney as the most effectual means of extinguishing it. Sulphur unites directly with oxygen only in the proportion of sulphurous acid, but several compounds of the same elements may be formed, which are all acids; namely—

1. Sulphurous acid .... $S\ O_2$
2. Hyposulphurous acid.................................. $S_2\ O_2$
3. Sulphuric acid .... $S\ O_3$
4. Hyposulphuric acid .... $S_2\ O_5$
5. Monosul-hyposulphuric acid .... $S_3\ O_5$
6. Bisul-hyposulphuric acid .... $S_4\ O_5$
7. Trisul-hyposulphuric acid .... $S_5\ O_5$

Uses.—From its ready inflammability sulphur has long been applied to wood matches. But its most considerable applications are in the composition of gunpowder and other deflagrating mixtures, and in the manufacture of sulphuric acid, which there will again be occasion to notice in a more particular manner.

SULPHUROUS ACID.

Eq. 32 *or* 400; SO_2; *density of gas* 2247; *combining measure* □□

Sulphurous acid was distinguished as a particular substance by Stahl, and first recognised as a gas by Dr. Priestley. It was subsequently analyzed with accuracy by Gay-Lussac and by Berzelius.

Preparation.—When sulphur is burned in dry air or oxygen gas, sulphurous acid is the sole product, and the gas is found to have undergone no change in volume. But sulphurous acid is more conveniently prepared in laboratories by several other processes.

FIG. 137.

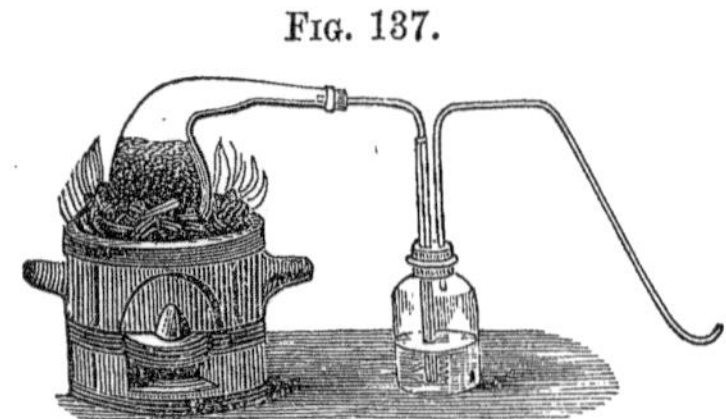

(1.) An intimate mixture of 6 parts of binoxide of manganese and 1 part of flowers of sulphur is heated in a small retort of hard glass (fig. 137;) the gas is carried through a wash-bottle to arrest a little vapour of sulphur which is carried over. Here the sulphur is burnt at the expense of a portion of the oxygen of the binoxide of manganese. Sulphurous acid, which is the product of the combustion, escapes, and protoxide of manganese remains in the retort. (Regnault).

$$S \text{ and } 2\ MnO_2 = SO_2 \text{ and } 2\ MnO.$$

(2.) By heating oil of vitriol upon mercury or copper, either of which becomes an oxide at the expense of one portion of the sulphuric acid, and thereby causes the formation of sulphurous acid. Sheet copper cut into small pieces is put into a flask to which undiluted oil of vitriol is added, and a moderate heat applied. The gas is carried through a bottle, containing a little water to condense the vapour of sulphuric acid, of which a little is carried over, and afterwards through a tube containing chloride of calcium, if it is desired to dry the gas.

(3.) Charcoal, chips of wood, straw, and such bodies, occasion a similar decomposition of sulphuric acid, when heated with it, but the gas is then mixed with a large quantity of carbonic acid. If the sulphurous acid, however, is to be used to impregnate water, or in making alkaline sulphites, the presence of that gas is immaterial. With that object, a quantity of oil of vitriol, equal in volume to 4 ounce measures of water, which for brevity may be spoken of as 4 ounce measures of oil of vitriol, is introduced into a flask with half an ounce of pounded wood-charcoal, and the two substances well mixed with agitation (fig. 138). Effervescence takes place upon applying heat to the flask, from the evolution of gas, which may be conducted in the first instance into an intermediate phial, through the cork of which a stout tube passes, open at both ends, and about 3-8ths of an inch in internal diameter. This phial contains about an ounce of water, into which the wider tube dips, and the tube from the flask descends still lower. The phial serves the purpose of a wash-bottle in condensing any sulphuric acid vapour that may be carried over by the gas, or of intercepting the liquid material in the flask, if thrown out by ebullition, and also of preventing the liquid in the second bottle from passing back, by the glass tube, into the generating flask, on the occurrence of a contraction of the air in that flask, by cooling or any other cause. When that contraction happens in this arrangement, the external air enters the intermediate phial by its open tube. The second bottle is nearly filled with water to be impregnated by the gas.

FIG. 138.

Properties.—Water at 60° is capable of dissolving nearly 50 times its volume of sulphurous acid, which makes it necessary to collect this gas for examination by displacement of air, or in jars filled with mercury in the mercurial trough. Its density is 2247, and it contains 2 volumes of oxygen with 1 volume of sulphur vapour (density 2211), condensed into 2 volumes, which form its combining measure. It may easily be obtained in the liquid state by transmitting the dry gas obtained by the first or second process through a U-shaped tube, surrounded by a freezing mixture of ice and salt, or better, of ice and chloride of calcium. It forms a colourless and very mobile liquid, of sp. gr. 1.45, which boils at 14°. The volatility of this liquid is small at considerably lower temperatures, and it is not applicable with advantage to produce intense cold by its evaporation (Kemp). Sulphurous acid crystallizes from a saturated solution in water, at a temperature of 4 or 5 degrees above 32°, in combination with 72 per cent. of water or 9 equivalents, SO_2+9HO (Pierre, Ann. de Chim. et Phys. 3 ser. 23.416).

Sulphurous acid is not decomposed by a high temperature; but several substances, such as carbon, hydrogen, and potassium, which have a strong affinity for oxygen, decompose it at a red heat. This acid blanches many vegetable and animal colours,—thus violets plunged for a short time into a solution of sulphurous acid become completely white; and the vapours of burning sulphur are therefore employed to whiten straw and to bleach silk, to which they also impart a peculiar gloss. The colours are not destroyed, and may in general be restored by the application of a stronger acid or an alkali. Dry sulphurous acid exhibits no affinity for oxygen, but in contact with a little water these gases slowly combine, and sulphuric acid is formed. From the same affinity for oxygen, sulphurous acid deprives the solution of permanganate of potassa of its red colour, and throws down iodine from iodic acid. It decomposes the solutions of those metals which have a weak affinity for oxygen, such as gold, silver, mercury (with heat), and throws down these bodies in the metallic state. Sulphurous acid is conveniently withdrawn from a gaseous mixture by means of peroxide of lead, which is converted by absorbing this gas into the white sulphate of lead. By nitric acid, sulphurous acid is immediately converted into sulphuric acid.

Sulphites.—The alkaline sulphites have a considerable resemblance to the corresponding sulphates. Their acid is precipitated by the chloride of barium, but the sulphite of baryta is dissolved by hydrochloric acid. When in solution the sulphites gradually absorb oxygen from the air, and pass into sulphates. Sulphurous acid is a weak acid, and its salts are decomposed by most other acids.

Uses.—Besides the application of which sulphurous acid is susceptible in bleaching, it is likewise employed in French hospitals, in the treatment of diseases of the skin. The gas is then applied in the form of a bath. (Dumas, Traité de Chimie appliquée aux Arts, i. 151).

This oxide of sulphur, besides acting as an acid, has been supposed to play the part of a radical, like carbonic oxide, and to pervade a class of compounds, in which hyposulphurous acid and sulphuric acid are included:—

SULPHUROUS ACID SERIES.

Sulphurous acid	SO_2
Sulphuric acid	SO_2+O
Hyposulphurous acid	SO_2+S
Chlorosulphuric acid	SO_2+Cl
Nitrosulphuric acid	SO_2+NO_2
Azotosulphuric acid	$2SO_2+NO_5$

SULPHURIC ACID.

Eq. 40 *or* 500; SO_3; *density of vapour* 2762; □□

Chemists have been in possession of processes for preparing this acid since the end of the fifteenth century. It is of all reagents the one in most frequent use, being the key to the preparation of most other acids; which, in consequence of its superior affinities, it separates from their combinations; and being the acid preferred to others, from its cheapness, for various useful and important purposes in the arts.

Preparation.—Sulphuric acid was first obtained by the distillation of green vitriol or copperas, a native sulphate of iron, and this process is still followed in Bohemia, for the preparation of a highly concentrated acid, known as the Nordhausen acid, from being long produced at Nordhausen in Saxony. The sulphate of iron contains seven equivalents of water, and is first dried, by which its water is reduced considerably below a single equivalent, and then distilled in a retort of stoneware at a red heat. When the experiment is performed on a small scale, the heat of an argand spirit-lamp is sufficient; and in the place of copperas, the sulphate of iron previously peroxidized, the sulphate of bismuth, of antimony, or of mercury, may be employed. The first effect of heat upon the dried sulphate of iron is to cause an evolution of sulphurous acid gas, a portion of sulphuric acid being decomposed in converting the protoxide of iron of that salt into sesquioxide,

$$2\ (FeO.SO_3) = SO_2 \text{ and } SO_3 \text{ and } Fe_2O_3;$$

but the salt used in Bohemia, it appears, is a native sulphate, in which the greater part of the iron is already in the state of sesquioxide, so that little sulphurous acid is lost. Vapours afterwards come over, which condense into a fuming liquid, generally of a black colour, and of a density about 1.9, which is the Nordhausen acid, and contains less than one equivalent of water to two of sulphuric acid. This acid is preferred for dissolving indigo, and for some other purposes in the arts, and is the best source of anhydrous sulphuric acid.

But sulphuric acid is prepared, in vastly greater quantity, by the oxidation of sulphur. When burned in air or oxygen, sulphur does not attain a higher degree of oxidation than sulphurous acid, but an additional proportion of oxygen may be communicated to it by two methods, and sulphuric acid formed.

1. When a mixture of sulphurous acid and air, which must be previously dried, is made to pass over spongy platinum, or a ball of clean platinum wire, at a high temperature, the sulphurous acid is converted into sulphuric acid at the expense of the oxygen of the air. After a time, however, the platinum loses this property, and the process, although interesting in a scientific point of view, does not answer, on account of that change, as a manufacturing method.

2. Sulphurous acid mixed with air may be converted into sulphuric acid, by the agency of nitric oxide, which is the process generally pursued in the manufacture of this acid. The theory of this latter method, which is by no means obvious, has been illustrated by the researches of Clement-Desormes, Davy, De la Provostaye, and others. It is generally considered as depending upon the following reactions:—

1. When binoxide of nitrogen NO_2 mixes with air in excess, it is instantly converted into peroxide of nitrogen NO_4.

2. Peroxide of nitrogen is converted by contact with a small quantity of water into the nitrate of water and nitrous acid.

$$2NO_4 \text{ and } HO = HO.NO^5 \text{ and } NO_3.$$

3. Nitrous acid in contact with a large quantity of water is converted into nitrate of water and binoxide of nitrogen.

$$3NO_3 \text{ and water in excess} = HO.NO_5 + \text{Water and } 2NO_2.$$

Consequently, uniting the last two operations, peroxide of nitrogen is converted by a large quantity of water into nitric acid and binoxide of nitrogen.

4. Sulphurous acid takes oxygen from hydrated nitric acid, and becomes sulphuric acid, disengaging peroxide of nitrogen.

As the peroxide of nitrogen gives nitric acid and binoxide of nitrogen (3), and the last gas is converted by air into peroxide of nitrogen (1), the production of nitric acid may be repeated without end, and more and more sulphurous acid is converted by the latter into sulphuric acid. It thus appears that with a sufficient supply of air or oxygen, a small quantity of nitric acid (or of binoxide of nitrogen) may convert a large quantity of sulphurous acid into sulphuric acid. The binoxide of nitrogen, only acting as a purveyor of oxygen, is re-obtained entire, without loss, at the end of the process. The sulphurous has derived the oxygen necessary to convert it into sulphuric acid, really from the air, but in an indirect manner.

In the manufacture upon the large scale, the sulphurous acid is converted into sulphuric acid, in oblong chambers of sheet-lead, supported by an external framework of wood. Sulphurous acid from burning sulphur, nitric acid vapour, and steam, are simultaneously admitted into the leaden chamber; and the sulphuric acid formed accumulates in the liquid state upon the floor of the chamber. The diagram below represents one of the forms of the chamber, with its appendages.

Fig. 139.

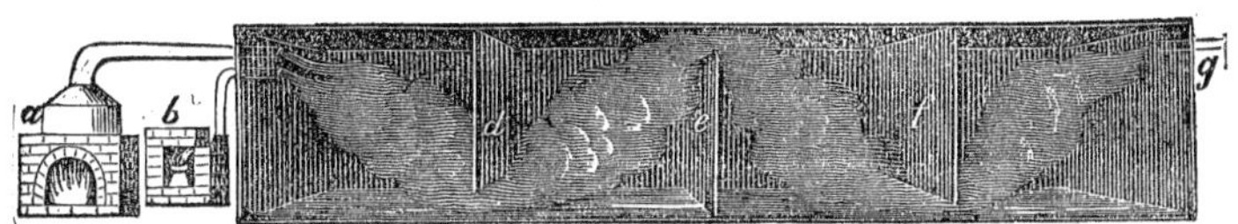

a represents the water boiler with its furnace, for supplying the chamber with steam; *b*, the section of a small chamber in brickwork, or furnace, called the burner, upon the floor of which the sulphur burns, and in which there is a tripod supporting an iron capsule, which contains the materials for nitric acid, namely, oil of vitriol, and either nitre or nitrate of soda. The heat of the burning sulphur evolves the nitric acid from these materials, and consequently the sulphurous acid becomes mixed with nitric acid vapour, which it carries forward with it, by a tube represented in the figure, into the chamber, where these acid vapours meet with the steam admitted near the same point, and the formation of sulphuric acid takes place. The nitric acid vapour is equivalent to binoxide or to peroxide of nitrogen, as the first effect of the sulphurous acid is to reduce the nitric acid to a lower state of oxidation. From 8 to 19 parts of sulphur are consumed in the burner for 1 part of nitrate of soda decomposed there, so that the quantity of nitrous fumes is small compared with the quantity of sulphurous acid thrown into the chamber. The chamber represented is 72 feet in length by 14 in breadth, and 10 in height, and is divided into three compartments, by leaden curtains placed across it, two of which, *d* and *f*, are suspended from the roof, and reach to within six inches of the floor, and one, *e*, rises from the floor to within six inches of the roof: *g* is a leaden conduit tube, for the discharge of the uncondensible gases, which should communicate with a tall chimney, to carry off these gases and to occasion a slight draught through the chamber. The curtains serve to detain the vapours, and cause them to advance in a gradual manner through the chamber, so that the sulphuric acid is deposited as completely as possible, before the vapours reach the discharge tube. When the oxygen of the chamber is exhausted, the admission of acid vapours is discontinued, till the air in it is renewed. But the admission of air to the chamber is generally so regulated, that a continuous current is maintained through the chamber, and the combustion proceeds without interruption. When steam is admitted in proper quantity, as in this method, it is not necessary to begin by covering the floor with water.

The acid may be drawn off from the floor of the chamber of a sp. gr. as high as 1.6. It is further concentrated in open leaden pans, till it begins to act upon the metal, and afterwards in retorts of platinum or glass. It still retains small quantities of nitrous acid and sulphate of lead, from which it can be completely purified by dilution with water and a second distillation. The acid thus obtained, in its most concentrated state, is a definite compound of one eq. acid and one eq. of water, $HO.SO_3$, which last cannot be separated by heat, the hydrate distilling over unchanged. It is the Oil of Vitriol of commerce.

The construction of the leaden chamber is greatly varied; one chamber of great dimensions is often used without any division by curtains; or the vapour is carried successively through a series of three, four, or five connected chambers. The sulphurous acid, also, is often derived from the combustion of bisulphide of iron (iron pyrites), instead of sulphur; a peculiar kiln or flue being employed for burning the former. At the suggestion of Gay-Lussac, the nitrous vapour, as it ultimately leaves the chamber with the air exhausted of oxygen, is absorbed by being made to pass through a column of coke, over which a stream of the concentrated sulphuric acid is flowing. The sulphuric acid, after being charged with nitrous vapours or nitric acid, is transported back to the anterior part of the chamber, and there exposed to the sulphurous acid, as the latter leaves the sulphur burner. This exposure *denitrates* the sulphuric acid, much sulphurous acid becoming sulphuric acid, and peroxide of nitrogen being liberated in the state of vapour. (See Knapp's Chemical Technology, edited by Drs. Ronalds and Richardson, i. 234, Am. ed.).

When the supply of aqueous vapour in the chamber is insufficient, a white crystalline compound appears, known as the crystalline substance of the leaden chambers: it is deposited most frequently in the tube by which two chambers communicate. It contains the elements of 2 eq. sulphuric acid, and 1 eq. nitric acid, $2SO_2+NO_5$; but several other views of the arrangement of its elements may be entertained with equal probability. This substance, which is also termed azoto-sulphuric acid (S_2NO_9), is decomposed by water, and gives sulphuric acid, nitric acid, and binoxide of nitrogen:

$$3(S_2NO_9) \text{ and } 7HO = 6(HO.SO_3) \text{ and } HO.NO_5, \text{ and } 2NO_2.$$

Fig. 140.

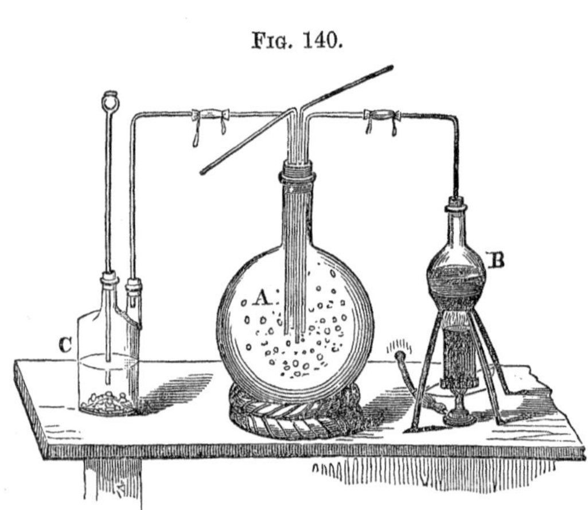

The formation of the crystalline substance, and the general operation of the leaden chamber, may be illustrated by the arrangement in fig. 140. Binoxide of nitrogen evolved by the action of dilute nitric acid on copper in the gas-bottle C, and sulphurous acid evolved by the action of copper clippings on concentrated sulphuric acid in the flask B, are conveyed into a large glass globe, A, containing air. Ruddy fumes of peroxide of nitrogen first appear, but soon the inner surface of the globe is frosted over with the crystalline compound. If steam or water be now introduced, by one of the free tubes, the crystals disappear with effervescence, from escape of gas, sulphuric acid is produced, and the changes are repeated till the air in A is exhausted.

Properties.—Anhydrous sulphuric acid is obtained by gently heating the fuming acid of Nordhausen in a retort, and receiving its vapour in a bottle artificially cooled, which can afterwards be closed by a glass stopper. It condenses in solid fibres, like asbestos, which are tenacious, and may be moulded by the fingers like wax. The density of the solid at 68° is 1.97 : at 77° it is liquid; and a little above that temperature it enters into ebullition, affording a colourless vapour, which produces dense white fumes on mixing with air, by condensing moisture. The dry acid does not redden litmus, an effect which requires the presence of moisture. It combines with sulphur, and produces liquid compounds, which are of a brown, green, and blue colour, and, with one-tenth of its weight of iodine, forms a compound of a fine green colour, which assumes the crystalline form. Heated in the acid vapour, caustic lime or baryta inflames and burns for a few seconds; the vapour is absorbed, and sulphate of lime or baryta formed. The anhydrous acid has a great affinity for water, and when dropped into that liquid, occasions a burst of vapour from the heat evolved. The density of its vapour was found to be 3000 by Mitscherlich, but it is probably 2762, and formed of 3 volumes of oxygen and 1 volume of sulphur vapour condensed into 2 volumes, which constitute its combining measure. This vapour is resolved by a strong red heat into sulphurous acid and oxygen.

When the Nordhausen acid is retained below 32°, well-formed crystals appear in it, which Mitscherlich finds to be a compound of two equivalents of acid and one of water, or $2SO_3+HO$. (Elémens de Chimie, par E. Mitscherlich, t. ii. p. 57). This compound is resolved by heat into the anhydrous acid, which sublimes, and the first hydrate, or oil of vitriol.

The most concentrated oil of vitriol of the leaden chambers ($HO+SO_3$) is a dense, colourless fluid of an oily consistence, which boils at 620°, and freezes at —29°, yielding often regular six-sided prisms of a tabular form. It has a specific gravity at 60° of 1.845. It is a most powerful acid, supplanting all others from their combinations, with a few exceptions, and when undiluted is highly corrosive. It chars and destroys most organic substances. It has a strong sour taste, and reddens litmus even though greatly diluted. Sulphur is soluble to a small extent in the concentrated acid, and communicates a blue, green, or brown tint to it; so are selenium and tellurium. Charcoal also appears to be slightly soluble in this acid, imparting to it a pink tint, which afterwards becomes reddish-brown. The concentrated acid has a great affinity for water, which it absorbs from the atmosphere, and is usefully employed to dry substances placed near it in vacuo. Considerable heat is evolved in its combination with water: when 4 parts by weight of the concentrated acid are suddenly mixed with 1 part of water, the temperature rises to 300°. When diluted with about thirty times its weight of water, sulphate of water $HO.SO_3$, evolves heat, which may be represented by 23 degrees; while $HO.SO_3+HO$, similarly diluted, evolves 14 degrees, or 9 degrees less, and $HO.SO_3+5HO$, 5 degrees only, or 18 degrees less. Hence the first equivalent of water which combines with oil of vitriol appears to evolve as much heat as the following four equivalents (Mem. Chem. Soc., i. 107). In a series of valuable experiments by M. Abria, but which do not admit of being compared with the preceding, he obtained the following results (Annales de Ch. et Ph., 3 sér., xii. 171):—

Quantities of heat disengaged by the combination of sulphate of water,—

With 1 eq. water	64.25	degrees.
2 "	94.69	"
3 "	113.06	"
4 "	124.43	"
5 "	131.66	"
Excess	165.63	"

The anhydrous acid SO_3 disengaged 237.13 degrees in combining with an excess of water. The value of these last degrees, or the unit of heat, is the quantity of heat required to heat up 1 gramme (15.434 grs.) of water 1° Centigrade. Abria concludes that in the combination of anhydrous sulphuric acid with water, the quantities of heat successively disengaged by the different equivalents of water have a multiple relation, and correspond very closely, for the first equivalents, with the numbers—

$$1,\ \tfrac{1}{3},\ \tfrac{1}{6},\ \tfrac{1}{12},\ \tfrac{1}{16},\ \tfrac{1}{24}.$$

The density of sulphuric acid becomes always less by dilution, but not exactly in the ratio of the water added. (Table of Densities of Sulphuric Acid, in Appendix).

Acid of density 1.78 is the second definite hydrate, containing two eq. of water to one of acid. This hydrate forms large and regular crystals, even a little above the freezing point of water, and was observed by Mr. Keir to remain solid till the temperature rose to 45°. If the dilute acid is evaporated at a heat not exceeding 400°, its water is reduced to the proportion of this hydrate. This second eq. of water is expelled by a higher temperature, but the first eq. can only be separated from the acid by a stronger base. Sulphuric acid forms still a third hydrate, of sp. gr. 1.632, containing three eq. of water, the proportion to which the water of a more dilute acid is reduced, by evaporation in vacuo at 212°. It is also in the proportions of this hydrate that the acid and water undergo the greatest condensation, or reduction of volume, in combining. The following, then, are the formulæ of the definite hydrates of this acid, including that derived by Mitscherlich from the Nordhausen acid:—

HYDRATES OF SULPHURIC ACID.

Hydrate in the Nordhausen acid	$HO.2SO_3$
Oil of vitriol, (sp. gr. 1.845)	$HO.SO_3$
Acid of sp. gr. 1.78	$HO.SO_3+HO$
Acid of sp. gr. 1.632	$HO.SO_3+2HO$

The composition of a hydrate of sulphuric acid is ascertained by adding a known weight of oxide of lead to the liquid, in a capsule, and evaporating to dryness. As the sulphuric acid abandons all its water on combining with oxide of lead, and the sulphate of lead may be heated without decomposition, the increase of weight which the oxide on the capsule undergoes is precisely the quantity of dry sulphuric acid in the hydrate examined.

Sulphuric acid acts in two different modes upon metals, dissolving some, such as copper and mercury, with the evolution of sulphurous acid, and others, such as zinc and iron, with the evolution of hydrogen gas. The metal is oxidated at the expense of the acid itself in the one case, and of the water in combination with the acid in the other. The acid acts with most advantage in the first mode when concentrated, and in the second when considerably diluted.

The presence of sulphuric acid in a liquid may always be discovered by means of chloride of barium, which produces with this acid a white precipitate of sulphate of baryta, insoluble in both acids and alkalies.

Sulphates.—Of no class of salts do chemists possess a more minute knowledge than of the sulphates. The sulphates of zinc, magnesia, and other members of the magnesian family, correspond closely with the hydrate of sulphuric acid. Thus of the seven eq. of water which the crystallized sulphate of magnesia possesses, it retains one at 400°, and is then analogous to the sulphate of water of sp. gr. 1.78; the formula of these two salts being,

$$MgO.SO_3+HO,$$
$$HO.SO_3+HO,$$

and the eq. of water in both salts may be replaced by sulphate of potassa, when the sulphate of water forms the salt called the bisulphate of potassa, and the sulphate of magnesia forms the double sulphate of magnesia and potassa, of which the formulæ also correspond:—

$$HO.SO_3+KO.SO_3$$
$$MgO.SO_3+KO.SO_3.$$

In all these sulphates there is one eq. of acid to one of base; but with potassa, sulphuric acid is supposed to form a second salt, in which two of acid are combined with one of base $KO+2SO_3$, and which is said to have lately been obtained in a crystallized state by M. Jacquelin (Annal. de Chim. et de Phys., lxx. 311). This would be a true bisulphate, and would correspond to the red chromate or bichromate of potassa $KO+2CrO_3$; but my own observations have obliged me to call in question the existence of this anhydrous bisulphate (Mem. Chem. Soc., i. 120).

Uses.—Sulphuric acid is employed to a large extent in eliminating nitric acid from nitrate of potassa, and in the preparation of hydrochloric acid and chlorine from chloride of sodium, and also in the processes of bleaching. But the great consumption of this acid is in the formation of sulphates, particularly of sulphate of soda, nearly all the carbonate of soda of commerce being at present procured by the decomposition of that salt.

CHLOROSULPHURIC ACID.

Eq. 67.5 *or* 843.75; SO_2Cl; *density* 4652 ☐☐

Sulphurous acid gas combines with an equal volume of chlorine under the influence of light, and condenses into oily drops, which are denser than water (Regnault, Annales de Chim. et de Phys. lxix. 170, and lxxi. 445). Chlorosulphuric acid in dissolving decomposes 1 eq. of water, and is converted into hydrochloric acid and sulphuric acid,—a reaction which demonstrates the original compound to consist of 1 eq. of sulphurous acid with 1 eq. of chlorine.

The density of the vapour of chlorosulphuric acid was found by experiment to be 4703, which agrees with the theoretical density, 4652. It consists of 2 volumes of sulphurous acid and 2 volumes of chlorine condensed into 2 volumes, which form the combining measure of the vapour. In its condensation, it resembles the vapour of anhydrous sulphuric acid. This body also corresponds exactly in composition with the compound hitherto called chlorochromic acid; CrO_2Cl, chromium being substituted in the latter for the sulphur of the former.

With dry ammoniacal gas, chlorosulphuric acid forms a white powder, which is a mixture of the hydrochlorate of ammonia (sal ammoniac) and *sulphamide*, SO_2+NH_2. It does not combine, as an acid, with bases.

Chlorosulphuric acid may also be represented as a compound of sulphuric acid with a terchloride of sulphur, $3SO_3+SCl_3$. Another compound of the same series has been formed by H. Rose, which is represented by $5SO_3+SCl_3$.

NITROSULPHURIC ACID.

Eq. 62 *or* 775; SNO_4 *or* $SO_2.NO_2$; *not isolable.*

Sir H. Davy made the observation that binoxide of nitrogen is absorbed by a mixture of sulphite of soda and caustic soda, and that a compound is produced, of which the principal characteristic is to disengage abundance of protoxide of nitrogen, upon the addition of an acid to it. He concluded that the nitrous oxide, which then escapes, was previously united with soda, and gave this as an instance of the combination of that neutral oxide with an alkali. As the sulphite of soda became at the same time sulphate, the conversion of the nitric oxide into nitrous oxide appeared to be explained. It was afterwards shown by Pelouze that a new acid is formed in the circumstances of the experiment, to which he has given the name nitrosulphuric, and which may be considered as a compound of sulphurous acid and nitric oxide, or another member of the sulphurous acid series. (Pelouze, in Taylor's Scien. Mem., vol. i. p. 470; or Annal. de Chim. et de Phys. lx. 151).

Preparation.—If a mixture be made over mercury of 2 volumes of sulphurous acid, and 4 volumes of binoxide of nitrogen, which are combining measures of these gases, no change occurs; but on throwing up a strong solution of caustic potassa into the gases, they disappear entirely after some hours, combining with a single equivalent of potassa, and forming together the nitrosulphate of potassa. But it is better to prepare the nitrosulphate of ammonia. A concentrated solution is made of sulphite of ammonia, which is mixed with five or six times its volume of solution of ammonia, and into this binoxide of nitrogen is passed for several hours at a low temperature. A number of beautiful crystals are gradually deposited; they are to be washed with a solution of ammonia, previously cooled, which, besides the advantage of retarding their decomposition, offers that of dissolving less of them than pure water. When the crystals are desiccated, they should be introduced into a well-closed bottle; in this state they undergo no alteration. The same process is applicable to the corresponding salts of potassa and soda. When a strong acid is added to a solution of these salts, for the purpose of liberating the nitrosulphuric acid, the latter, on being set free, decomposes spontaneously into sulphuric acid and protoxide of nitrogen, which comes off with effervescence.

Properties. — The acid of the nitrosulphates is not precipitated by baryta. The nitrosulphate of potassa, when heated, becomes sulphite, and evolves nitric oxide; but the salts of soda and ammonia become sulphates, and evolve nitrous oxide. No nitrosulphates of the metallic oxides, which are insoluble in water, have been formed, or appear capable of existing; for when such salts as chloride of mercury, sulphate of zinc or of copper, sulphate of sesquioxide of iron and nitrate of silver, are added to the nitrosulphate of ammonia, they produce a brisk effervescence of nitrous oxide, with the formation of sulphate of ammonia, or they decompose the nitrosulphate of ammonia as free acids do. Indeed, the only nitrosulphates which have been formed are those of potassa, soda, and ammonia. These are neutral, and have a sharp and slightly bitter taste, with nothing of that of the sulphites.

These salts rival the binoxide of hydrogen in facility of decomposition. The nitrosulphate of ammonia resists 230°, but is decomposed with explosion a few degrees above that temperature, caused by the rapid disengagement of nitrous oxide. Solutions of the nitrosulphates are not stable above the freezing point, but their stability is much increased by an excess of alkali. They are resolved into sulphate and nitrous oxide, by the mere contact of certain substances which do not themselves undergo any change; such as spongy platinum, silver and its oxide, charcoal powder and binoxide of manganese, by acids, even carbonic acids, and by metallic salts.

Azoto-sulphuric acid of De la Provostaye, S_2NO_9.—Liquid sulphurous acid and peroxide of nitrogen, sealed up together in a glass tube, react upon each other, and give rise to a solid compound crystallizing in rectangular square prisms, which has been examined by M. de la Provostaye. A small portion of a blue liquid, possessing an explosive property, which has not been fully examined, is formed at the same time. This substance forms the "crystals of the leaden chamber." It may also be produced, according to Gay-Lussac, by bringing peroxide of nitrogen and oil of vitriol in contact:—

$$2NO_4 \text{ and } 2(HO.SO_3) = HO.NO_5 + HO \text{ and } S_2NO_9.$$

This substance fuses at about 430°, and forms a silky mass on cooling; it may be distilled without decomposition at about 620°. It is decomposed by water, sulphuric acid being formed, and nitrous vapours disengaged. It has been represented as composed of $2SO_2+NO_5$; or as $2SO_3+NO_3$; or $S_2O_5+NO_4$; but nothing certain is known of its molecular arrangement.

Dry binoxide of nitrogen is absorbed by anhydrous sulphuric acid, according to an observation of H. Rose.

HYPOSULPHURIC ACID.

Eq. 72 *or* 900; S_2O_5; *not isolable.*

Preparation. — This acid of sulphur was discovered by Gay-Lussac and Welter, in 1819. To prepare it, a quantity of binoxide of manganese, which must not be hydrated, is reduced to an extremely fine powder, suspended by agitation in water, and sulphurous acid gas is transmitted through the water. When ordinary binoxide of manganese is used, it should be previously treated with nitric acid, to dissolve out the hydrated oxide, and washed. The temperature is apt to rise during the absorption of the gas, but must be repressed, otherwise much sulphuric acid is produced, — the formation of which, indeed, it is impossible to prevent entirely, but of which the quantity is said to be reduced almost to nothing, when the liquid is kept cold during the operation. The binoxide of manganese disappears, and a solution of hyposulphate of the protoxide of manganese is formed; 2 equivalents of sulphurous acid, and 1 of binoxide of manganese, forming one of hydrosulphuric acid and one of protoxide of manganese, or

$$2SO_2 \text{ and } MnO_2 = MnO + S_2O_5.$$

The solution is filtered, and then mixed with a solution of sulphide of barium, which occasions the precipitation of the insoluble sulphide of manganese, with the transference of the hyposulphuric acid to baryta. From this hyposulphate of baryta, the hyposulphates of other metallic oxides may be prepared by adding their sulphates to that salt, when the insoluble sulphate of baryta will precipitate, and the hyposulphate of the metallic oxide added remain in solution. But to procure the hyposulphuric acid itself, the solution of hyposulphate of baryta may be evaporated to dryness, and, being perfectly pure, it is reduced to a fine powder, weighed, and dissolved in water: for 100 parts of it 18.78 parts of oil of vitriol are taken, which, after dilution with three or four times as much water, are employed to decompose this salt of baryta. The liberated hyposulphuric acid solution is filtered, and evaporated *in vacuo* over sulphuric acid, till it attains a density of 1.347, which must not be exceeded, as the acid solution begins then to decompose spontaneously into sulphurous acid, which escapes, and sulphuric acid, which remains in the liquid.

Properties. — This acid has not been obtained in the anhydrous condition. Its aqueous solution has no great stability, being decomposed at its temperature of ebullition. The same solution exposed to air in the cold, slowly absorbs oxygen, according to Heeren, and becomes sulphuric acid. But neither nitric acid, nor chlorine, nor binoxide of manganese, oxidize this acid unless they are boiled in its solution. Its salts are perfectly stable, either when in solution or when dry, and are generally very soluble, having some analogy to the nitrates. A hyposulphite, when heated to redness, leaves a neutral sulphate, and allows a quantity of sulphurous acid to escape, which would be sufficient to form a neutral sulphite with the base of the sulphate. This class of salts was particularly examined by Heeren (Poggendorff's Annalen, v. vii. p. 77). Hyposulphuric acid is imagined to exist in acid compounds produced by the action of sulphuric acid on several organic substances.

The hyposulphate of baryta may be analysed by exposing a portion of it to a red heat, when it gives off sulphurous acid, and leaves pure sulphate of baryta behind. If an equal portion be treated with boiling concentrated nitric acid, the sulphurous acid is converted into sulphuric acid; and if chloride of barium is afterwards added, a quantity of sulphate of baryta is obtained which is exactly double in weight that obtained from the first portion.

HYPOSULPHUROUS ACID.

Eq. 48 *or* 600; S_2O_2, *or* $SO_2 + S$; *not isolable.*

The hyposulphites are better known than hyposulphurous acid itself, which is a body of little stability, quickly undergoing decomposition when liberated by a stronger acid from a solution of any of its salts, and resolving itself into sulphurous acid, hydrosulphuric acid, and sulphur. These salts, long considered as a species of double salts, and called *sulphuretted sulphites*, were first supposed to contain a peculiar acid by Dr. T. Thomson and by Gay-Lussac, — a conjecture afterwards verified by Sir John Herschel, whose early researches upon this acid form the subject of an interesting memoir (Ed. Phil. Journ. vol. i. pp. 8 and 396).

Preparation. — Sulphide of soda is prepared, in the first instance, by saturating a solution of carbonate of soda with sulphurous acid gas, by the apparatus described at page 294). This sulphite, care being taken that it is not acid, is converted into hyposulphite, by digesting it upon flowers of sulphur at a high temperature, but without ebullition. The sulphurous acid assumes 1 eq. of sulphur, and remains in combination with the soda; or, in symbols —

$$NaO + SO_2 \text{ and } S = NaO + SO_2.S.$$

The solution may afterwards be evaporated (ebullition being always avoided, as the hyposulphites are partially decomposed at 212°), and affords large crystals of the hyposulphite of soda. When solution of caustic soda is digested upon sulphur, the latter is likewise dissolved, and a mixture of 1 eq. of hyposulphite of soda with 2 eq. of sulphide of sodium results, of which the last always dissolves an excess of sulphur: —

$$3NaO \text{ and } 4S = NaO + S_2O_2 \text{ and } 2NaS.$$

Exposed to the air, this solution slowly absorbs oxygen, and if it contains a certain excess of sulphur, passes entirely into hyposulphite of soda.

The hyposulphite of lime is also formed by digesting together 1 part of sulphur and 3 of hydrate of lime at a high temperature, when changes of the same nature occur as with sulphur and caustic soda, and the solution becomes red, containing bisulphide of calcium: a stream of sulphurous acid gas is conducted through the solution after it has cooled, and converts the whole salt into hyposulphite, occasioning at the same time a considerable deposition of sulphur. The reaction here may be expressed by the following formula: —

$$2CaS_2 \text{ and } 3SO_2 = 2CaO + 2S_2O_2 \text{ and } 3S.$$

If the waste-lime, in the porous state in which it is removed from the dry-lime purifiers of a gas-work, be exposed to air, it rapidly absorbs oxygen; and, when treated with water, afterwards gives much soluble hyposulphite of lime. This is an economical method of preparing the salt on a large scale (Mem. Chem. Soc. ii. 358).

Zinc and iron also dissolve in the solution of sulphurous acid in water, with little or no effervescence, deriving the oxygen necessary to convert them into oxides, not from water, but from the sulphurous acid, two-thirds of which are thereby converted into hyposulphurous acid, which combines with half the oxide produced; while the other third, remaining as sulphurous acid, unites with the other moiety of the same oxide: —

$$3SO_2 \text{ and } 2Zn = ZnO.S_2O_2 \text{ and } ZnO.SO_2.$$

The hyposulphite obtained by this process is, therefore, mixed with a sulphite.

Properties. — The acid of these salts undergoes decomposition when they are strongly heated, or treated with an acid. It forms soluble salts with lime and strontia, in which respect it differs from sulphurous and sulphuric acids; the hyposulphite of baryta is insoluble. It also forms a remarkable salt with silver, which has no metallic flavour, but tastes extremely sweet. The existence of a hyposulphite in a solution is easily recognised, by its possessing the power to dissolve freshly precipitated chloride of silver, and become sweet. Hyposulphite of soda in solution is apt to become acid by the absorption of oxygen, and then its conversion into sulphate of soda, with deposition of sulphur, proceeds rapidly.

Uses. — The hyposulphite of soda is employed to distinguish between the earths strontia and baryta, — the latter of which it precipitates, and not the former. It is also applied, in certain circumstances, to dissolve the insoluble salts of silver in photography, electro-plating, and the treatment of silver ores.

POLYTHIONIC SERIES.

Three new acids of sulphur have lately been discovered, all containing, like hyposulphuric acid, 5 eq. of oxygen, but evidently more related in constitution and properties to hyposulphurous acid. They were named by Berzelius, from θειον (sulphur); and are composed as follows:—

Trithionic, or monosul-hyposulphuric acid.............. S_3O_5, *or* $S_2O_5 + S$.
Tetrathionic, or bisul-hyposulphuric acid................ S_4O_5, *or* $S_2O_5 + 2S$.
Pentathionic, or trisul-hyposulphuric acid.............. S_5O_5, *or* $S_2O_5 + 3S$.

Hyposulphurous acid becomes the dithionous, and hyposulphuric acid the dithionic acid, as members of the same series; all of which, it will be observed, contain more than 1 equivalent of sulphur, and are therefore polythionic: but the old names of the two acids last referred to are too firmly established to be changed, without a greater necessity for the alteration than appears to exist.

Trithionic or *Monosul-hyposulphuric acid;* eq. 88 *or* 1100, S_3O_5 *or* S_2O_5+S.—This acid was first obtained by M. Langlois (Annal. de Chim. 3 ser. iv. 77). It is the result of the action of sulphur upon the soluble bisulphites, and may be prepared from the bisulphite of baryta. This salt is digested with flowers of sulphur at a temperature not exceeding 122° (50° C.) for several days; the solution first becomes yellow, afterwards loses all colour, and when allowed to cool in this state, deposits a salt in long white silky crystals, which is the trithionate of baryta. By the cautious addition of sulphuric acid to a solution of the new salt, the trithionic acid may be liberated and obtained in solution, while the insoluble sulphate of baryta precipitates. The acid solution may be concentrated in the vacuous receiver of an air-pump, but is rapidly decomposed by heat into sulphurous acid and sulphur. The salt of potassa is easily obtained, either, according to Plessy's method, by passing sulphurous acid into a solution of hyposulphite of potassa; or, according to Langlois, into one of sulphide of potassium: in the latter case hyposulphite of potassa is first formed, and from that the trithionate. (Kessner, Chem. Gaz. vi. p. 369.) The salts of this acid appear to have greater stability than the hyposulphites, and are formed when certain hyposulphites, such as those of zinc, cadmium, and lead, are left to spontaneous decomposition; or even, according to Fordos and Gelis, by the sole effect of the concentration of solutions of these salts. This acid is precipitated black by the salts of the suboxide of mercury, a property which distinguishes the trithionic acid from the two more highly sulphured acids of the same series, which are precipitated yellow by the reagent in question.

Tetrathionic or *Bisul-hyposulphuric acid*; eq. 104 *or* 1300; S_4O_5 *or* $S_2O_5+S_2$.—This acid was discovered by MM. Fordos and Gelis, and is obtained by dissolving iodine in a solution of the hyposulphites, particularly of the hyposulphite of baryta. The reaction in the last case is as follows:—

$$2\ (BaO.S_2O_2) \text{ and } I = BaI \text{ and } BaO.S_4O_5.$$

The new salt, being less soluble than the iodide of barium, is separated by crystallization, and affords the acid when decomposed by a suitable proportion of sulphuric acid. The solution of tetrathionic acid has considerable stability, and may be highly concentrated. The process just described is modified by Kessner, who prepares first the hyposulphite of lead by dissolving 2 parts of hyposulphite of soda in hot water, and pouring this solution into an equally hot dilute solution of 3 parts of acetate of lead. The precipitate is washed with a large quantity of warm water, and mixed (still moist) with 1 part of iodine, and the mass frequently stirred; in the course of a few days the whole is converted into iodide of lead and a solution of tetrathionate of lead. The lead is now removed by sulphuric acid (the use of hydrosulphuric acid being inadmissible), any excess of the latter by carbonate of baryta, and the solution of the tetrathionic acid evaporated. When this acid is saturated with carbonate of soda, or its salt of lead decomposed by sulphate of soda, only products of decomposition are obtained,—sulphur, sulphate, and hyposulphite of soda. (Chem. Gaz. vi., p. 370.) The salts of this acid, therefore, require to be prepared directly, and appear generally to be less stable than the hydrated acid.

Pentathionic or *Trisul-hyposulphuric acid;* = 120 *or* 1500; S_5O_5 *or* $S_2O_5+S_3$.—Several years ago Dr. T. Thomson observed that when hydrosulphuric and sulphurous acids mutually decompose each other in presence of water, the magma of sulphur precipitated is impregnated by a peculiar acid. M. Wackenroder lately found that this acid is an additional number of the present series. To prepare the acid, Wackenroder supersaturates water with sulphurous acid, and then causes hydrosulphuric acid to stream through it till the liquid has the odour and reactions of the latter, evaporating afterwards till the excess of hydrosulphuric acid is expelled. The liquid does not become clear till after clean slips of copper are left in it for some time, to remove the suspended sulphur: copper reduced from the oxide by hydrogen would probably act more rapidly. The addition of chloride of sodium, or saturation with a base, such as an alkaline carbonate, also facilitates the precipitation of the sulphur. In the opinion of Mr. L. Thompson, much of this sulphur, which is supposed to be suspended, is actually in solution.

The clear acid liquid may be concentrated till it attains a density of 1.37; it is inodorous, sour, and a little bitter. It may be preserved at the temperature of the air, without change; but when made to boil it undergoes decomposition, giving off hydrosulphuric acid, followed by sulphurous acid, and leaving behind ordinary sulphuric acid and some sulphur. This acid is decomposed, like the last, by strong bases.

Pentathionic acid was also found by Fordos and Gelis among the products of the decomposition of the chlorides of sulphur by water. The pentothionate of baryta is very soluble, and is easily altered. It was analysed by means of chlorine and the hypochlorites, which transform the whole sulphur into sulphuric acid:

$$S_5O_5 \text{ and } 10\ Cl \text{ and } 10\ HO = 5\ SO_3 \text{ and } 10\ HCl.$$

The pentathionic acid is distinguished from hyposulphurous acid, with which it is isomeric, by the less solubility of the pentathionates, and by the circumstance that the pentathionates have no action upon iodine (Annales de Ch. 3. ser. xxii. 66). The sulphur was supposed by Berzelius to exist in the various polythionic acids, in its different allatropic conditions.

SULPHUR AND HYDROGEN.

HYDROSULPHURIC ACID.

Syn. Sulphuretted hydrogen gas, sulfhydric acid; Eq. 17 *or* 212.5; SH; *density* 1191.2; ☐☐

Sulphur does not combine directly with hydrogen even when heated in that gas, but with that element, notwithstanding, sulphur forms at least two compounds; one of which, hydrosulphuric acid, is a reagent of frequent application and considerable importance.

Preparation.—(1.) Of those metals which dissolve in dilute sulphuric acid, with the displacement of hydrogen, the protosulphides dissolve also in the same acid, but the hydrogen then evolved carries off sulphur in combination, and appears as hydrosulphuric acid gas. The protosulphide of iron, which is commonly employed in this operation, is obtained by depriving yellow pyrites, or bisulphide of iron, of a portion of its sulphur by ignition in a covered crucible; or formed directly by exposing to a low red heat a mixture of 4 parts of coarse sulphur and 7 of iron filings or borings in a covered stoneware or cast-iron crucible. The sulphide of iron, thus obtained, is broken into lumps, and acted upon by diluted sulphuric acid in a gas-bottle (fig. 141), exactly as zinc is treated in the preparation of hydrogen gas. Hydrosulphuric acid is evolved without the application of heat, and should be collected over water at 80° or 90°; or if collected in a gasometer or gasholder, the latter may be filled with brine, in which this gas is less soluble than in pure water. The gas obtained by this process generally contains free hydrogen, arising from an intermixture of metallic iron with the sulphide of iron used. The gas may also be evolved from the action of hydrochloric acid upon the sulphide of iron, but as it is then impregnated with the vapour of the

FIG. 141.

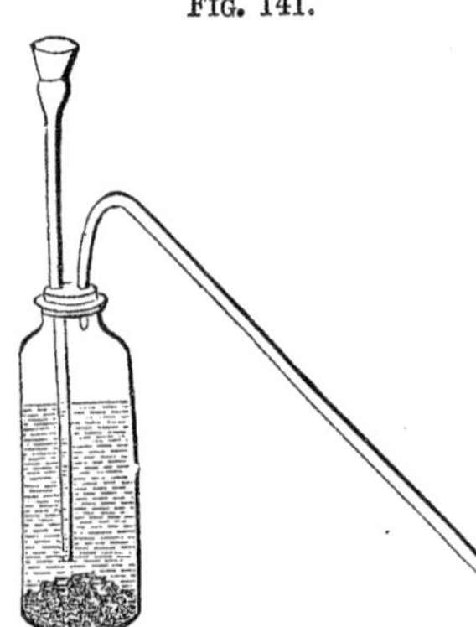

latter acid, and may also, like every gas produced with effervescence, carry over drops of fluid, it should always be transmitted through water in a wash-bottle, before being applied to any purpose as pure gas. The reaction by which hydro-sulphuric acid is usually evolved is expressed in the following equation:

$$FeS \text{ and } HO.SO_3 = HS \text{ and } FeO.SO_3.$$

(2.) Hydrosulphuric acid, without any admixture of free hydrogen, is obtained by digesting in a flask A, used as a retort (fig. 142), with a gentle heat, sulphide

Fig. 142.

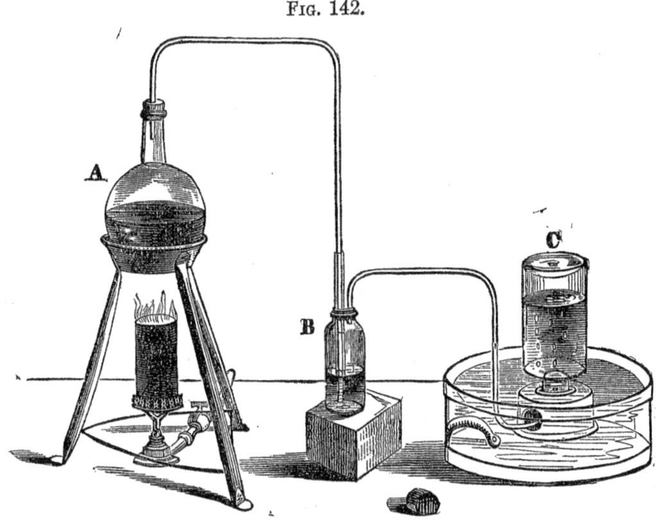

of antimony in fine powder with concentrated hydrochloric acid, in the proportion of 1 ounce of the former to 4 ounce measures of the latter. The gas of this operation is passed through water in a wash-bottle B, and collected over water at 80°, in a bottle C, provided with a good cork. Or, after passing through the wash-bottle, it may be carried over chloride of calcium in a drying tube, and collected over mercury, but is gradually decomposed by that metal, which has a strong affinity for sulphur, and hydrogen is liberated, without any change of volume. The reaction between hydrochloric acid and sulphide of antimony may be thus expressed:

$$3HCl \text{ and } SbS_3 = 3HS \text{ and } SbCl_3.$$

Properties.—Hydrosulphuric acid is a colourless gas, of a strong and very nauseous odour. Its density is 1191.2, by the experiments of Gay-Lussac and Thenard, and its theoretical sp. gr. 17 times that of hydrogen. It consists of 2 volumes of hydrogen and 1 volume of sulphur vapour, condensed into 2 volumes, which form its combining measure. Hydrosulphuric acid is partially decomposed by heat into hydrogen and sulphur; but to obtain complete decomposition it is necessary to pass the gas a great many times through a porcelain tube placed across a furnace, and strongly heated. By a pressure of 17 atmospheres at 50°, it is condensed into a highly limpid colourless liquid, of sp. gr. 0.9, which is of peculiar interest as the analogue of water in the sulphur series of compounds: the solvent powers of this liquid have not been examined. When cooled to —122°, it solidifies, and is then a white crystalline translucent substance, heavier than the liquid (Faraday). The air of a chamber slightly impregnated by this gas may be respired without injury, but a small quantity of the undiluted gas inspired occasions syncope, and its respiration, in a very moderate proportion, was found by Thenard to prove fatal,—birds perishing in air containing 1-1500th, and a dog in air containing 1-800th part of this gas. Its poisonous effects are best counteracted by a slight inhalation of chlorine gas, as the latter may be obtained from a little chloride of lime placed in the folds of a towel wetted with acetic acid. Water dissolves, at 64°, 2½ volumes of this gas, and alcohol 6 volumes. These solutions soon become milky when exposed to air, the oxygen of which combines with the hydrogen of the gas and precipitates the sulphur. Those mineral waters termed sulphureous, such as Harrowgate, contain this gas, although rarely in a proportion exceeding 1½ per cent. of their volume. They are easily recognized by their odour and by blackening silver. It is also found in foul sewers and in putrid eggs. Of deodourizing fluids the solution of nitrate of lead, chloride of zinc, sulphate of iron, and sulphate of manganese, appear to be equally efficacious; the first alone decomposing the free gas, but that salt, and all the others named, decomposing hydrosulphuric acid when in combination with ammonia, the form in which it usually emanates from putrefactive matter.

Hydrosulphuric acid is highly combustible, and burns with a pale blue flame, producing water and sulphurous acid, and generally a deposit of sulphur when oxygen is not present in excess. A little strong nitric acid thrown into a bottle of this gas, occasions the immediate oxidation of its hydrogen, and often a slight explosion with flame, when the escape of the vapour is impeded by closing the mouth of the bottle. Hydrosulphuric acid is immediately decomposed by chlorine, bromine, and iodine, which assume its hydrogen: hence the odour of this gas in a room is soon destroyed on diffusing a little chlorine through it. Tin, and many other metals, heated in this gas, combine with its sulphur with flame, and liberate an equal volume of hydrogen, affording ready means of demonstrating the composition of the gas. Potassium decomposes one half of the gas in that manner, and becomes sulphide of potassium, which unites with the other half without decomposition, forming the hydrosulphate of the sulphide of potassium. The action of other alkaline metals upon hydrosulphuric acid is similar.

This compound has a weak acid reaction, and forms one of the hydrogen-acids. It does not combine and form salts with basic oxides, but it unites with basic sulphides, such as sulphide of potassium, and forms compounds which are strictly comparable with hydrated oxides. When hydrosulphuric acid is passed over lime at a red heat, both compounds are decomposed, and water with sulphide of calcium is formed. The oxides of nearly all the metallic salts, whether dry or in a state of solution, are decomposed by hydrosulphuric acid in a similar manner; but in the salts of those metals of which the protosulphide is dissolved by acids, such as salts of iron, zinc, and manganese, a small quantity of a strong acid entirely prevents precipitation. The sulphides are generally coloured, and many of them are black; hence the effect of hydrosulphuric acid in blackening salts of lead and silver, which renders these compounds so sensitive as tests of the presence of that substance. Hydrosulphuric acid also tarnishes certain metals, such as gold, silver, and brass, so that utensils of which these metals are the basis should not be exposed to this gas.

Bisulphide of hydrogen, HS_2. — When carbonate of potassa is fused with half its weight of sulphur, a persulphide of potassium is formed containing a large excess of sulphur, which affords a solution in water of an orange red colour. The protosulphide of potassium, with hydrochloric acid, gives hydrosulphuric acid and chloride of potassium: HCl and $KS = HS$ and KCl. But when the red solution of persulphide of potassium is poured in a small stream into hydrochloric acid, diluted with two or three volumes of water, while chloride of potassium is formed as before, the hydrosulphuric acid produced combines with another equivalent of sulphur, and forms a yellowish oily fluid, the bisulphide of hydrogen, which falls to the bottom of the acid liquid. Supposing the persulphide of potassium to be a pure bisulphide, then HCl and $KS_2 = HS_2$ and KCl. The result of the combination in this case appears rather capricious; for if the acid and persulphide of potassium be mixed in the other way, — if the acid be added drop by drop to the alkaline sulphide, — then hydrosulphuric acid is evolved, the whole excess of sulphur precipitates, and no persulphide of hydrogen is formed. The oily fluid produced by the first mode of mixing has considerable analogy in its properties to the binoxide of hydrogen, and appears, like that compound, to have a certain degree of stability imparted to it by contact with acids, such as pretty strong hydrochloric acid, while the presence of alkaline bodies, on the contrary, gives its elements a tendency to separate. This decomposition has been taken advantage of to obtain liquid hydrosulphuric acid, by sealing up bisulphide of hydrogen in a Faraday tube (page 77).

Thenard has observed other points of analogy between these compounds. Like binoxide of hydrogen, the bisulphide produces a white spot upon the skin, and destroys vegetable colours, so that it has actually been used in bleaching. The latter compound is also resolved into hydrosulphuric acid and sulphur by all the bodies which effect the transformation of the former into water and oxygen; such as charcoal powder, platinum, iridium, gold, binoxide of manganese, and the oxides of gold and silver, which last, when the bisulphide is dropt upon them, are decomposed in an instant, and even with ignition. The bisulphide of hydrogen undergoes spontaneously the same decomposition, even in well-closed bottles, which are apt, on that account, to be broken. It is soluble in ether, but the solution soon deposits crystals of sulphur. Thenard finds this body not to be uniform in its composition, the proportion of sulphur often exceeding considerably 2 eq. to 1 of hydrogen; but the excess of sulphur is possibly only in solution (Ann. de Ch. 2 ser. xlviii. 79).

SULPHUR AND NITROGEN.

Sulphide of nitrogen; eq. 62 *or* 775; NS_3.—This is a yellow pulverulent solid substance of small stability, and which cannot be formed by the direct union of its elements. The liquid bichloride of sulphur absorbs ammoniacal gas, producing first a flocculent brown matter, $NH_3.SCl_2$, and afterwards, if the action of ammonia is continued, a yellow substance, of which the formula is—

$$2NH_3.SCl_2.$$

Thrown into water this yellow matter undergoes decomposition, producing hydrochlorate and hyposulphite of ammonia, which dissolve, and a yellow powder, which is a mixture of sulphur and the sulphide of nitrogen. This powder is quickly washed with a little water, dried under the receiver of an air-pump, and finally washed several times with ether, which dissolves out the free sulphur, and leaves the sulphide of nitrogen.

The sulphide of nitrogen is a yellow powder, which, a little above 212°, is decomposed in a gradual manner into sulphur and nitrogen, but when sharply heated, violently and with explosion. It is also slowly decomposed by cold water, but much more rapidly at the temperature of ebullition. The composition of sulphide of nitrogen is determined either by boiling a known quantity in fuming nitric acid, which converts the sulphur into sulphuric acid; or, by heating a mixture of this substance and metallic copper in a glass tube, sealed at one end, and arranged as a retort, so that the gas evolved may be collected. The copper and sulphur unite with avidity, and the nitrogen is disengaged as gas.

SULPHUR AND CARBON.

Bisulphide of carbon; sulphocarbonic acid; eq. 38 *or* 475; CS_2.—Charcoal strongly ignited in an atmosphere of sulphur vapour, combines with that element, and forms a compound which holds the same place in the sulphur series that carbonic acid occupies in the oxygen series of compounds. The bisulphide of carbon is a volatile liquid, and may be prepared by distilling, in a porcelain retort, yellow pyrites or bisulphide of iron, with a fourth of its weight of well-dried charcoal, both in the state of fine powder and intimately mixed. The vapour from the retort is conducted to the bottom of a bottle filled with cold water, to condense it. Or sulphur vapour may be sent over fragments of well-dried charcoal in a porcelain or cast iron (not malleable iron) tube, placed across a furnace. The product is generally of a yellow colour, and contains sulphur in solution, to free it from which it is redistilled in a glass retort, by a gentle heat.

Fig. 143.

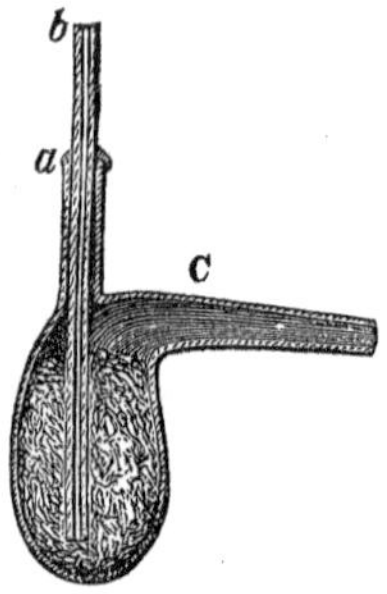

For preparing a larger quantity of bisulphide of carbon, M. Brunner recommends an earthenware retort of the form C (fig. 143), two-thirds filled with dry charcoal, having a tube, *b*, descending through the tubulure *a*, by which fragments of sulphur can be introduced. The retort is raised to a red heat in a furnace (fig. 144), and the vapour which comes over, carried through a condensing tube, *c d*, kept cold by a stream of water, and ultimately conveyed to the lower part of a bottle surrounded by cold water, and also containing a little water, which floats upon the surface of the condensed liquid and prevents its evaporation. The sulphur is gradually introduced into the retort, and, being immediately converted into vapour, produces the bisulphide of carbon in traversing the incandescent charcoal.

Fig. 144.

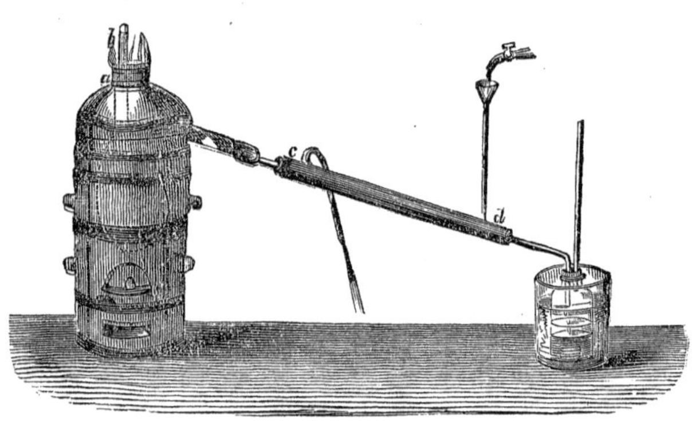

The bisulphide of carbon is a colourless liquid, of high refracting power, and sp. gr. 1.272. Its vapour has a tension of 7.38 Paris inches (Marx) at 50°, and the liquid boils at 110°; a cold of —80° can be produced by its evaporation in vacuo. This compound is extremely combustible, taking fire at a temperature which scarcely exceeds the boiling point of mercury. When a few drops of the liquid are thrown into a bottle of oxygen gas, or nitric oxide, a combustible mixture is formed, which burns, when a light is applied to it, with a brilliant flash of flame, but without a violent explosion. The bisulphide of carbon is insoluble in water, but it is soluble in alcohol. It dissolves sulphur, phosphorus, and iodine. The solution of phosphorus in this liquid is used in electrotyping; objects dipped in the solution and dried are left covered by a film of phosphorus, which enables them to obtain a conducting metallic coating when plunged into a solution of copper.

The observed density of the vapour of bisulphide of carbon is 2644.7 (Gay-Lussac). It consists of 2 volumes carbon vapour (density 416) and 2 volumes sulphur vapour (density 2216), condensed into 2 volumes, which form its combining measure; and is therefore quite analogous in condensation to carbonic acid gas. A complete analysis of the bisulphide of carbon is obtained, by passing it in vapour over a mixture of carbonate of soda and oxide of copper in a combustion tube (page 287) at a red heat: the sulphur is oxidized, and remains in combination with the soda as sulphate of soda, while the carbon is burnt also, and disengaged as carbonic acid gas, accompanied by an equal quantity of carbonic acid liberated from the carbonate of soda by the sulphuric acid formed. The carbon alone of this substance may be advantageously determined as carbonic acid, by a similar combustion with chromate of lead.

The bisulphide of carbon is a sulphur acid, and combines with sulphur bases, such as the sulphide of potassium, forming a class of salts which are called sulphocarbonates. Oxygen bases dissolve it slowly, and are converted into a mixture of carbonate and sulphocarbonate: thus 2 equivalents of potassa with 1 of bisulphide of carbon yield 2 equivalents of sulphide of potassium and 1 of carbonic acid, which combine respectively with bisulphide of carbon and potassa.

Solid sulphide of carbon.—The charcoal left in the tube, after the process for the former compound, is much corroded, and contains a portion of sulphur which cannot be expelled from it by heat. Berzelius considered this sulphur as in chemical combination with the carbon.

SECTION VIII.

SELENIUM.

Eq. 39.28 *or* 491 (*F. Sacc*); Se; *density of vapour unknown.*

This element was discovered in 1817 by Berzelius, in the sulphur of Fahlun employed in a sulphuric acid manufactory in Sweden, and was named by him selenium, from Σελήνη, the moon, on account of its strong analogy to another element, tellurium, which derives its name from *tellus*, the earth. It is one of the least abundant of the elements, but is found in minute quantity in several ores of copper, silver, lead, bismuth, tellurium, and gold, in Sweden and Norway; and in combination with lead, silver, copper, and mercury, in the Hartz. It is extracted from a seleniferous ore of silver of a mine in the latter district, and supplied for sale in little cylinders of the thickness of a goose-quill, and three inches in length, or in the form of small medallions of its discoverer. It has also been found in the Lipari islands associated with sulphur, and can sometimes be detected in the sulphuric acid both of Germany and England. It is separated from its combinations with sulphur and metals by a very complicated process, for which I must refer to the works of Berzelius (Ann. of Phil. vol. xiii. 401; or Ann. de Ch. et de Phys. xi. 160; also Berzelius's Traité, ii. 184, Paris edit. Didot, 1846).

Properties of selenium.—This element is allied to sulphur, and, like that body, exhibits considerable variety in its physical characters. When it cools after being distilled, its surface reflects light like a mirror, has a deep reddish brown colour, with a metallic lustre resembling that of polished blood-stone; its density is between 4.3 and 4.32. When cooled slowly after fusion its surface is rough, of a leaden grey colour, its fracture fine-grained, and the mass resembles exactly a fragment of cobalt. But as selenium does not conduct electricity, and its metallic characters are not constant, it is better classed with the non-metallic bodies. Its powder is of a deep red colour. By heat it is softened, becoming semifluid at 392°, and fusing completely at 482°. It remains a long time soft on cooling, and may then be drawn out like sealing-wax into thin and very flexible threads, which are grey and exhibit a metallic lustre by reflected light, but are transparent and of a ruby red colour by transmitted light. It boils about 1292°, and gives a vapour of a yellow colour, less intense than that of sulphur, but more so than that of chlorine. The density of this vapour has not been ascertained. When heated to the degree of ignition, selenium emits a powerful odour, suggesting that of decaying horse-radish, by means of which the smallest trace of this element may be detected in minerals, when heated before the blow-pipe. The odour was first ascribed to a gaseous oxide of selenium, but it is found by M. Sacc that selenium heated in perfectly dry air is inodorous, and the odour is now referred to the production of a minute quantity of hydroselenic acid.

Selenium combines in two proportions with oxygen, forming selenious acid, which corresponds with sulphurous acid, and selenic acid corresponding with sulphuric acid.

Selenious acid; eq. 55.28 *or* 691; SeO_2.—Selenium does not burn in air, but when strongly heated in the bend of a glass tube *a b c*, (fig. 145), with a current of oxygen passing over it, selenium takes fire and burns with a flame, white at the base, and of a bluish green at the point and edges, but not strongly luminous; selenious acid at the same time condenses in the upper part of the tube as a white sublimate, in long quadrilateral needles. Its vapour has the colour of chlorine. The same acid is the only product of the action of nitric or nitro-muriatic acid upon selenium, and is obtained on slowly cooling the liquor in large prismatic crystals, striated lengthwise, which have a considerable resemblance to nitre. These crystals are hydrated selenious acid. This acid is largely soluble, both in water and alcohol. It is decomposed when in solution, and selenium precipitated by zinc, iron, or sulphite of ammonia, with the assistance of a free acid. The selenite of ammonia is also decomposed by heat, and leaves selenium. The selenious is a strong acid, displacing nitric and hydrochloric acids from their combinations, but is displaced in its turn by the more fixed acids, sulphuric, boracic, &c., at a high temperature. (F. Sacc, Annales de Ch. 3 ser. xxi. 119.)

Fig. 145.

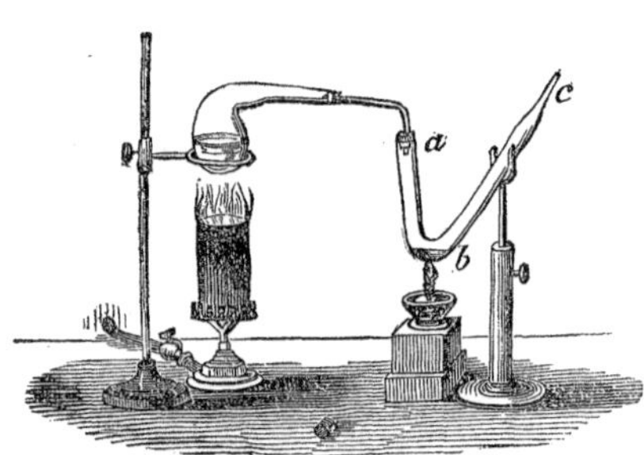

Selenic acid, SeO_3.—Selenium is brought to this superior state of oxidation at a high temperature, by fusion with nitre, a process which affords the seleniate of potassa. The selenic acid is precipitated from that salt by the nitrate of lead; and the insoluble seleniate of lead, after being washed, is diffused through water and decomposed by a stream of hydrosulphuric acid, which converts the lead into insoluble sulphide of lead, and liberates selenic acid. A solution of this acid may be concentrated till its boiling point rises to 536°, but above that temperature it changes rapidly into selenious acid, with disengagement of oxygen. Its density is then 2.60, and it contains little more than a single equivalent of water, and therefore corresponds with the protohydrate of sulphuric acid, or oil of vitriol. Selenic acid has not been obtained in the anhydrous condition. Zinc and iron are dissolved by this acid, with the evolution of hydrogen gas; and with the aid of heat it dissolves copper and even gold, an operation in which it is partially converted into selenious acid. But it does not dissolve platinum. To precipitate its selenium, the acid may be digested with hydrochloric acid, which occasions the formation of selenious acid and the evolution of chlorine, and then sulphurous acid throws down the selenium; for it is singular that selenic acid is not de-oxidized by sulphurous acid, although selenious acid is. The compounds of selenic acid with bases, so much resemble the corresponding sulphates, in their crystalline form, colour, and external characters, that they can only be distinguished from them by the property which the seleniates have of detonating when ignited with charcoal, and causing a disengagement of chlorine when heated with hydrochloric acid. To separate the selenic from the sulphuric acid, Berzelius recommends the saturation of the acids with potassa, and the ignition of the dried salt, mixed with sal-ammoniac; the selenic acid is decomposed by the ammonia and reduced to the state of selenium.

SECTION IX.

PHOSPHORUS.

Eq. 400 *or* 32; *P; density of vapour* 4327; □

This remarkable element appears to be essential to the organization of the higher animals, being found in their fluids, and forming, in the state of phosphate of lime, the basis of the solid structure of the bones. It is also found in most plants, and in a few minerals. Phosphorus was first obtained by Brandt of Hamburgh in 1660, but Kunkel first made public a process for preparing it, which was afterwards improved by Margraff and by Scheele. Its ready inflammability, from which phosphorus derived its name, has always made this substance an object of popular interest; while the singularity, importance, and variety of the phosphoric compounds have drawn to them no ordinary share of the attention of chemists.

Preparation.—Phosphorus is not a substance that can be easily prepared on a small scale, but ever since the time of Godfrey Hankwitz, to whom Mr. Boyle communicated a process for preparing it, phosphorus has been manufactured in London, in considerable quantity and of great purity, for the use of chemists. The earth of bones is decomposed by 2-3ds of its weight of sulphuric acid, and the insoluble sulphate of lime separated by filtration from the soluble phosphoric acid, which passes through with a quantity of phosphate of lime in solution. The acid liquor is then evaporated to the consistence of a syrup, and mixed with charcoal to form a soft paste, which is rubbed well in a mortar, and then dried in an iron pot with constant stirring till the mass begins to be red-hot. It is allowed to cool, and introduced as rapidly as possible into a stoneware retort, previously covered with a coating of fire-clay. The beak of the retort is inserted into a wider copper tube of a few feet in length, the free end of which is bent downwards a few inches from its extremity; and the descending portion introduced into a wide-mouthed bottle, containing enough of water to cover the extremity of the tube to the extent of a line or two. The heat of the furnace in which the retort is placed is slowly raised for three or four hours, and then urged vigorously till phosphorus ceases to drop into the water from the copper tube, which may continue from fifteen to thirty hours, according to the size of the retort. Carbon at a high temperature takes oxygen from the phosphoric acid, and becomes carbonic oxide, so that the phosphorus in distilling over is accompanied all along by that gas.

Wöhler recommends, instead of the preceding process, to calcine ivory black, which is a mixture of phosphate of lime and charcoal, with fine quartzy sand and a little more ordinary charcoal, in cylinders of fire-clay, at a very high temperature. Each cylinder has a bent copper tube adapted to it, one branch of which descends into a vessel containing water. The efficiency of Wöhler's process depends upon the silica acting as an acid, and combining with the lime of the phosphate, at a high temperature, while the liberated phosphoric acid is decomposed by the carbon.

Properties.—At the usual temperature phosphorus is a translucent soft solid of a light amber colour, which may be bent or cut with a knife, and the cut surface has a waxy lustre. Its density is 1.77. Phosphorus melts at 108°, undergoing a remarkable dilatation of 0.0134 of its volume, and becoming transparent and colourless immediately before fusion. It forms a transparent liquid, possessing, like most combustible bodies, a high refracting power. At 217° it begins to emit a slight vapour, and boils at 550°, being converted into a vapour which is colourless, of sp. gr. 4355, according to the experiment of Dumas, which coincides almost with the theoretical density 4327. Its combining measure, like that of oxygen, is 1 volume, allowing its equivalent to be 32. When fused and left undisturbed, it sometimes remains liquid for hours at the usual temperature, particularly when covered by an alkaline liquid, but becomes solid when touched. Phosphorus, when very pure, exhibits, by rapid cooling from a high temperature, a modification analogous to that which sulphur undergoes in the same circumstances, but which is not so easily produced. Light causes it, in all circumstances, to assume a red tint; to avoid which action phosphorus is usually preserved in an opaque bottle. Phosphorus cannot be crystallized from a state of fusion, for this substance passes in a gradual manner from the liquid to the solid condition, a circumstance which is always opposed to crystallization; but from its solution in hot naphtha it may be obtained, in cooling, in rhomboidal dodecahedrons of the regular system. It is quite insoluble in water, but soluble to a small extent, with the aid of heat, in fixed and volatile oils, in bisulphide of carbon, of which 100 parts dissolve 20 of phosphorus; in chloride of sulphur, sulphide of phosphorus, and ether.

[The red substance formed by the action of light appears to be a modification of phosphorus exhibiting chemical and physical characters different from its ordinary condition. Red phosphorus is formed not only by exposure to light, but also by keeping phosphorus at a high temperature (464°—482°) for some time, when it assumes a carmine red colour, thickens, and becomes perfectly opaque. This change takes place in an atmosphere of dry carbonic acid, nitrogen or hydrogen. The unaltered portion of the phosphorus is separated from the red variety by means of bisulphide of carbon, in which this latter is insoluble, and it may be purified to a greater extent by boiling it with a solution of potassa, washing with water, then with very dilute nitric acid, and finally again with water.

Red phosphorus is in the form of a scarlet powder. Its density is 1.964. It remains without alteration in the air; and even when heated gradually in a current of air it does not take fire, requiring a temperature of 500° to combine with oxygen and become luminous, and for complete combustion that of 572°. When heated to the boiling point in a gas which has no action on it, common phosphorus results. Chlorine combines with it at common temperatures without the evolution of light. (Schrœter, Journ. Ph. and Ch. Av. 1851).—R. B.]

Phosphorus undergoes oxidation in the open air, and diffuses white vapours, which have a peculiar odour, suggesting to some that of garlic, and are luminous in the dark; and at the same time the phosphorus becomes covered with acid drops, which arise from the phosphorous acid, produced in these circumstances, attracting the humidity of the air. This slow combustion is attended with a sensible evolution of heat, and may terminate in the fusion of the phosphorus, and its inflammation with combustion at a high temperature. There is a necessity for caution, therefore, in handling phosphorus, a burn from this body in a state of ignition being in general exceedingly severe. It is preserved under the surface of water. The low combustion of phosphorus has been particularly studied. It is not observed a few degrees below 32°, but is sensible at that temperature, and increases perceptibly a few degrees above it. The presence of certain gaseous substances, even in minute quantity, has a remarkable effect in preventing the slow combustion of phosphorus; thus at 66° it is entirely prevented by the presence of,

	Volumes of Air.
1 volume of olefiant gas in	450
1 volume of vapour of sulphuric ether in	150
1 volume of vapour of naphtha in	1820
1 volume of vapour of oil of turpentine in	4444

and the influence of these gases or vapours is not confined to low temperatures, a certain admixture of all of them defending phosphorus from oxidation even at 200°. But on allowing such a gaseous mixture to expand, by diminishing the pressure upon it to a half or a tenth, the phosphorus becomes luminous, and the proportion of foreign gas required to prevent the slow combustion must be greatly increased. The only explanation of this phenomenon which can be offered at present, is that the gases which exert this influence have an attraction for oxygen, and there is reason to believe are themselves undergoing a slow oxidation at the same time. Now when two oxidable bodies are in contact, one of them often takes precedence in combining with oxygen, to the entire exclusion of the other. Potassium is defended from oxidation in air by the same vapours, although to a less degree. (Quarterly Journal of Science, N. S. vol. vi. p. 83). It is curious, that in pure oxygen, phosphorus may remain without oxidating at all, at temperatures below 60°, but an inconsiderable rarefaction of the gas, from diminution of the pressure upon it, will cause the phosphorus to burst into the luminous condition. The dilution of the oxygen with nitrogen, hydrogen, or carbonic acid, produces the same effect. When gradually heated in air, phosphorus generally catches fire, and begins to undergo the high combustion, before its temperature has risen to 140° : of this high combustion, the sole product is phosphoric acid. The inflammability of phosphorus, however, is greatly increased by its impurities, particularly by the presence of the red oxide of phosphorus.

The phosphorus matches now universally employed for procuring a light, are generally the wooden sulphur match, with an additional coating, applied to its extremity, of a paste containing phosphorus, which, when dry, will ignite by friction. The materials added to this paste, to promote the combustion of the phosphorus, are chlorate and nitrate of potassa, or certain metallic oxides, such as the binoxide of manganese or sesquioxide of lead (minium), which abandon readily a portion of their oxygen. The snap, or little detonation which attends the ignition of these matches, is caused by the chlorate of potassa, and is obviated by substituting nitre for that salt; although, to give the proper inflammability, a small proportion of chlorate is found to be indispensable. The phosphorus paste is made by melting phosphorus in a vessel with a certain quantity of water at 120°. The requisite proportion of chlorate or nitrate of potassa is dissolved in this water, and the metallic oxides added, if the latter are used, and then enough of gum to thicken the liquid. The whole are well triturated together, in a mortar, till the globules of phosphorus cease to be visible to the eye; and the mass is coloured blue with Prussian blue, or red with minium. The points of the matches already sulphured are dipped into this paste, so as to cover their extremities, and then cautiously dried in a stove. The gum on drying forms a varnish, which defends the phosphorus from oxidation by the air till the surface is abraded by friction, when the phosphorus first takes fire and communicates its combustion to the sulphur, which again ignites the wood of the match.

Phosphorus is susceptible of four different degrees of oxidation, the highest of which is a powerful acid, while the acid character is not absent even in the lowest. These compounds are:—

Oxide of phosphorus	$2P+O$
Hypophosphorous acid	$P+O$
Phosphorous acid	$P+3O$
Phosphoric acid	$P+5O$

OXIDE OF PHOSPHORUS.

Eq. 72 *or* 900; P_2O.

When burned in air or oxygen, phosphorus generally leaves behind it a small quantity of a red matter, which is an oxide of phosphorus. The same compound is obtained, in larger quantity, by directing a stream of oxygen gas upon melted phosphorus under hot water, and was found by Pelouze to contain 3 equivalents of phosphorus to 2 of oxygen (Annal. de Ch. et de Ph. l. 83).

But this oxide is impure, and the definite oxide appears to have been first obtained by Leverrier (Annal. de Ch. et de. Ph. lxv. 257). His process is to expose to the air small fragments of phosphorus covered by the liquid chloride of phosphorus (PCl_3), in an open bolt-head. Phosphoric acid is formed, and also a yellow matter, which he finds to be a phosphate of the oxide of phosphorus, and which gives a yellow solution with water. This solution is decomposed about 176°, and a flocculent yellow matter subsides, which is a hydrate of the oxide of phosphorus, nearly insoluble in water. This compound abandons its combined water, when dried in vacuo over sulphuric acid, or when cooled below 32°, when the water separates as ice, and oxide of phosphorus remains perfectly pure.

The oxide of phosphorus is a powder of a canary yellow colour, denser than water, and soluble neither in water, alcohol, nor ether. It may be kept in dry air without change. It resists a temperature of 570° without decomposition, but assumes a lively red colour; and does not take fire in the air till heated a little above the boiling point of mercury. This oxide absorbs dry ammoniacal gas, and appears to form feeble combinations with the fixed alkalies. Leverrier assigns to its hydrate the composition P_2O+2HO, and to its phosphate, $2P_2O+3PO_5$.

HYPOPHOSPHOROUS ACID.

Eq. 40 *or* 500; PO; *not isolable. Formula of a Hypophosphite,* MO.PO+2HO.

This acid was discovered in 1816 by Dulong (Annal. de Ch. et de Ph. ii. 141). It was obtained by the action of water upon the phosphide of barium, of which the phosphorus of one portion oxidates and becomes the acid in question, at the expense of the water, while the phosphorus of another portion, combining with the hydrogen of the water, produces phosphuretted hydrogen gas. Rose prepares the same hypophosphite of baryta by boiling phosphorus in a solution of caustic baryta, till all the phosphorus disappears and the vapours have no longer the smell of garlic (H. Rose, sur les Hypophosphites, Annal. de Ch. et de Ph. xxxviii. 258). Wurtz uses sulphide of barium. To separate the hypophosphorous acid from the baryta, diluted sulphuric acid is added, which precipitates the latter. To remove again the excess of sulphuric acid unavoidably added, the acid liquid is saturated with oxide of lead, which forms a soluble hypophosphite of lead and an insoluble sulphate of lead. The latter is separated by filtration, and the lead thrown down from the filtrate by a stream of hydrosulphuric acid gas. The acid remaining in solution may be concentrated with caution to the consistence of a thick syrup, but affords no crystals. More strongly heated, the hydrate of hypophosphorous acid undergoes decomposition, being converted into phosphoric acid, with the evolution of phosphuretted hydrogen and a deposition of phosphorus. The anhydrous acid PO has never been obtained, 3 eq. of water being essential to its composition; namely, 1 eq. as base, and 2 eq., which appear to form elements of the acid itself (Wurtz). Hence the formula of the acid is HO.PO + 2HO; or, believing with Wurtz, that both the oxygen and hydrogen, of 2HO, are negative elements of the acid, like the oxygen in phosphoric acid, the formula is $HO.PH_2O_3$, corresponding with the protohydrate of phosphoric acid $HO.PO_5$.

Hypophosphorous acid is colourless, viscid, and sour to the taste. It withdraws oxygen from the sesquioxide of lead, and some other metallic oxides. When heated with sulphuric acid it changes the latter into sulphurous acid, and also produces a deposit of sulphur, a property by which it is distinguished from phosphorous acid, the complete decomposition of sulphuric acid not being effected by the latter acid. Hypophosphorous acid also decomposes sulphate of copper in solution, producing, when the temperature is only slightly raised, a solid insoluble compound of that metal with hydrogen, the hydride of copper discovered by M. Wurtz, and at the boiling point a deposit of metallic copper with the evolution of hydrogen gas.

The hypophosphites are all soluble in water, and the salts of the magnesian family, such as those of magnesia and cobalt, crystallize well. They are easily obtained by decomposing the hypophosphite of baryta by the soluble sulphates. The dry hypophosphites are permanent in air, but their solutions, evaporated by heat, absorb oxygen. They all contain 2 equivalents of water, which are essential to the constitution of a hypophosphite (Wurtz, Annal. de Ch. et de Ph. 3 sér. vii. 35; and xvi. 190; also, H. Rose, *ib.* viii. 364).

PHOSPHOROUS ACID.

Eq. 56 *or* 800; PO_3. *Formula of a Phosphite*, $2MO.PO_3+HO$.

Preparation. — This acid is the principal product of the slow combustion of phosphorus, but changes after its formation into phosphoric acid, from the further absorption of oxygen from the air. It may be obtained in the anhydrous condition by burning phosphorus with imperfect access of air. Berzelius recommended for this operation a tube of glass, about 10 inches in length and ½ inch in diameter, which is nearly closed at one end, an opening no greater than a large pin-hole being left there, and at a distance of an inch from this extremity the tube is bent at an obtuse angle. A small fragment of phosphorus is introduced into the angle of the tube, and heated till it takes fire. It burns with a pale greenish flame, and the phosphorous acid produced is carried along by the feeble current of air, and condenses in the ascending part of the tube, as a white powder, volatile, but not in the slightest degree crystalline. The phosphorus must not be so much heated as to cause it to sublime unchanged. In contact with air, phosphorous acid is apt to inflame, from the heat occasioned by the condensation of moisture, and is converted into phosphoric acid. The phosphorous acid of the preceding process is immediately soluble in water, while the phosphoric acid, which sometimes accompanies it, remains for a short time undissolved, in the form of white translucent flocks.

Hydrated phosphorous acid is obtained by throwing a few drops of water on the liquid ter-chloride of phosphorus (PCl_3), when that compound evolves hydrochloric acid gas, and gives hydrated phosphorous acid.

$$PCl_3 \text{ and } 3HO = PO_3 \text{ and } 3HCl.$$

The hydrated acid is also obtained by the method of Droquet. Two or three ounces of phosphorus are melted in a cylindrical glass receiver or sealed tube, of 10 or 12 inches in length, and nearly an inch in diameter, and the tube filled up with water. This tube, which will contain a column of fluid phosphorus of 5 or 6 inches in height, is then properly disposed in a bason or bolt-head of warm water, so as to retain the phosphorous fluid. Chlorine gas is conveyed by a quill tube, from a flask in which it is generated, to the bottom of the fluid phosphorus, where combination takes place with ignition, and the chloride of phosphorus is formed. This chloride is dissolved by the water covering the phosphorus, and converted into hydrochloric acid and phosphorous acid. The chlorine must be transmitted very slowly through the phosphorus, as any portion of that gas which reaches the water converts the phosphorous into phosphoric acid; and the absorption of the chlorine by the phosphorus is most complete when it is free from any other gas. When the remaining phosphorus fixes, upon cooling, the acid fluid may be poured off, and concentrated by boiling, till it becomes syrupy and the volatile hydrochloric acid is entirely expelled.

Properties.—In its most concentrated state, the hydrate of phosphorous acid contains three equivalents of water, and crystallizes in transparent prisms. When heated, it is resolved into hydrated phosphoric acid, and pure phosphuretted hydrogen gas, which is not spontaneously inflammable as so prepared. The solution of phosphorous acid absorbs oxygen from the air slowly, if concentrated, but quickly when dilute. Like sulphurous acid, it takes oxygen from the oxide of mercury, when heated with it, and decomposes also the salts of gold and silver. It is one of the more feeble acids.

Phosphites. — The class of phosphites, which has been examined, is bibasic, that is, they contain 2 eq. of base to 1 of phosphorous acid. They also retain 1 eq. of water, the elements of which are proved by Wurtz to enter into the constitution of the acid. Phosphorous acid is thus represented with 5 negative equivalents PHO^4, like phosphoric acid PO_5. Much information respecting the phosphites is contained in the papers of Berzelius. (Annal. de Ch. et de Ph., ii. 151, 217, 329, et x. 278.)

Analysis of phosphorous and hypophosphorous acids. — The composition of both phosphorous and hypophosphorous acid is determined by adding nitric acid to their solutions, by which they are converted into phosphoric acid. But the weight of the resulting phosphoric acid cannot be obtained by simply evaporating its solution to dryness, as that acid retains an indefinite quantity of water in combination. It is necessary to add to the liquid a weighed quantity of oxide of lead, more than sufficient to neutralize the phosphoric acid and what remains of the nitric acid. The whole is then evaporated to dryness in a platinum capsule, and heated sufficiently to expel the nitric acid from the nitrate of lead formed. The water, previously combined with the phosphoric acid, is displaced by the oxide of lead, and escapes, leaving only phosphate of lead with the excess of oxide of lead. This residue is weighed, and the original weight of oxide of lead is deducted from it to obtain the weight of dry phosphoric acid. The composition of phosphoric acid being known (32 phosphorus and 40 oxygen), the quantity of phosphorus in the phosphoric acid of the experiment is obtained by a simple calculation.

Further, if a stream of chlorine gas be transmitted through a solution of hypophosphorous acid, it is converted into phosphoric acid by the oxygen of water which is decomposed. The chlorine uniting with the hydrogen of the water, at the same time, and becoming hydrochloric acid, the quantity of the latter acid produced supplies a measure of the oxygen required to convert the hypophosphorous acid into phosphoric acid.

The composition of phosphorous acid may also be deduced from the analysis of terchloride of phosphorus, which can be made very exactly. One hundred grains of that liquid compound being mixed with water in a flask, it is instantaneously converted into hydrochloric and phosphorous acid; and by the addition of a little nitric acid the latter acid is changed into phosphoric acid. The chloride of silver, precipitated by a solution of nitrate of silver added in excess to the acid liquid, will weigh 310.85 grains, and contains 76.85 grains of chlorine. Hence 100 grains of terchloride of phosphorus contain 76.85 grains of chlorine, and the remaining 23.14 grains is phosphorus. But these numbers are in the proportion 32 phosphorus and 106.5 chlorine, or 1 eq. of the former, and 3 eq. of the latter; giving PCl_3 as the composition of the terchloride of phosphorus. Finally, as phosphorous acid is formed from the terchloride of phosphorus, by replacing the chlorine by an equivalent quantity of oxygen, it follows evidently that the composition of phosphorous acid is PO_3.

PHOSPHORIC ACID.

Eq. 72 *or* 900; PO_5; *forms three hydrates and three classes of salts :*

Formula of a Monobasic phosphate, or Metaphosphate .... $MO.PO_5$

" " *Bibasic phosphate, or Pyrophosphate* .... $2MO.PO_5$

" " *Tribasic phosphate, or Phosphate* .... $3MO.PO_5$

Preparation.—To obtain this acid in a state of purity, a convenient process is to set fire to about a drachm of phosphorus upon a little metallic capsule, placed in the centre of a large stone-ware plate, and immediately cover it by a dry bell jar of the largest size. The phosphorus is converted into white flakes of phosphoric acid, which are retained, with very little loss, within the bell jar, and fall upon the plate like snow.

The process may be made a continuous one, and a large quantity of phosphoric acid prepared by the arrangement of figure 146. The phosphorus is burned within a large glass balloon A, having three tubulures, which has been well dried beforehand. The cork of the upper tubulure is traversed by a long tube, *a b*, open at both ends, and about half an inch in diameter, and which descends to about the centre of the globe. A little capsule of platinum or porcelain *v* is attached, by means of platinum wires, below the lower opening of this tube. To the second tubulure *d* a drying tube C, containing pumice soaked in oil of vitriol, is attached;

and to the third tubulure *g* a somewhat wide bent tube, *g h*, of which the other extremity descends into a well-dried bottle B. This last vessel is placed in communication, by means of the tube *k l*, with any aspirating apparatus, by means of which a continuous current of air is determined, which penetrates by the tube C, where it is dried, and traverses the whole apparatus. A fragment of phosphorus is now dropt upon the capsule *v*, by the tube *a b*, lighted by a hot wire, and the upper opening *a* then closed by a cork. When the combustion is completed, another fragment of phosphorus is added, always taking care to dry the fragment carefully with filter paper before its introduction. The phosphoric acid produced is partly deposited in the globe A, and partly carried forward into the bottle B. It is thus obtained quite anhydrous.

Fig. 146.

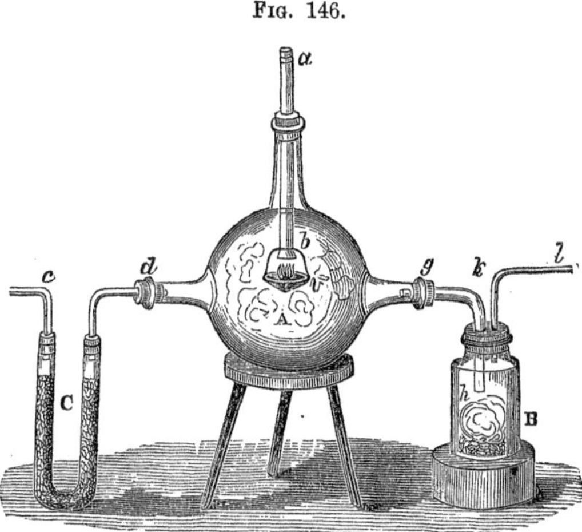

The dry phosphoric acid is distinguished by the same shade of white, absence of crystallization, and perfect opacity, as solid carbonic acid. Exposed for a few minutes to the air, it deliquesces; and when the solid acid is collected in a wine-glass, and a few drops of water are thrown upon it, it is converted into a hydrate with explosive ebullition, from the heat evolved. The anhydrous acid is perfectly fixed, unless in the presence of aqueous vapour, when it sublimes away, probably in the state of a hydrate.

Phosphorus may likewise be oxidated by means of nitric acid. In this operation, the fuming nitric acid should be diluted with an equal bulk of water, to avoid accidents from the violent action of the acid, which may cause the phosphorus to be projected in a state of ignition; the diluted acid is boiled upon the phosphorus, and being afterwards evaporated to dryness, it yields a hydrated phosphoric acid.

Phosphoric acid is also obtained in large quantity from calcined bones, which are reduced to a fine powder and mixed with 4-5ths of their weight of oil of vitriol, previously diluted with 4 or 5 times its bulk of water, as in the preparation of phosphorus (page 313). Carbonate of ammonia is then added to the filtered solution of phosphoric acid, and the resulting phosphate of ammonia being evaporated to dryness and heated to low redness in a platinum crucible, a hydrated phosphoric acid remains, in a fused state, which is known as glacial phosphoric acid, from its resemblance to ice.

To exhibit many of its properties, phosphoric acid must be first dissolved in water, when the compound is found to be marked by an inconstancy and variableness in its characters, most unusual in a strong acid. This arises from the circumstance that it is not actual phosphoric acid which dissolves in water, any more than it is true sulphuric acid which dissolves in water when oil of vitriol is added to that fluid. It is a hydrate of both acids, which is soluble; the phosphate of water in the one case and the sulphate of water in the other. But the phosphoric acid differs from the sulphuric, in a singular and almost peculiar capacity to form three different salts of water, instead of one only; and these three phosphates of water are all soluble without change, and exhibit properties so different, that they might be supposed to contain three different acids. When the dry acid from the combustion of phosphorus is thrown into water, it produces a mixture, in variable proportions, of the three hydrates; but each of them may be had separately, and in a state of purity, by a particular process.

Terhydrate, or tribasic phosphate of water, $3HO + PO_5$. — The common phosphate of soda of pharmacy may be had recourse to for all the hydrates of phosphoric acid; but it should be first dissolved and crystallized anew to purify it. To a warm solution of the pure phosphate of soda in a bason, a solution of acetate of lead in distilled water is added, so long as it occasions a precipitate; the phosphate of soda requires rather more than an equal weight of acetate of lead. The dense insoluble phosphate of lead which precipitates, is washed, and being afterwards suspended in cold water, is decomposed by a stream of hydrosulphuric acid gas sent through it. The liquid may then be warmed, to expel the excess of hydrosulphuric acid, and filtered from the black sulphide of lead: it is very sour, and contains the terhydrate of phosphoric acid. The characters of this acid solution are, to give a yellow precipitate with nitrate of silver, to give a granular crystalline precipitate with ammonia and sulphate of magnesia — the phosphate of magnesia and ammonia, to yield the common phosphate of soda when neutralized with carbonate of soda, to form salts which have invariably 3 eq. of base to 1 of phosphoric acid, and to be unalterable by boiling its solution or keeping it for any length of time. The class of salts which this hydrate forms are the old phosphates, which have long been known, and it is convenient to allow them to be particularly distinguished as the phosphates or the common phosphates.

Deuto-hydrate of phosphoric acid, or bibasic phosphate of water, $2HO + PO_5$.— Dr. Clark first discovered that when the phosphate of soda is heated to redness, it is completely changed, and after being dissolved in water affords crystals of a new salt, which he named the pyrophosphate of soda,—an observation which led to interesting results. (Ed. Journ. of Science, vol. vii. p. 298, 1826; or Annal. de Ch. et de Phys. xli. 276.) If a solution of this salt, which it is not necessary to crystallize, be precipitated by acetate of lead, the insoluble salt of lead washed and decomposed by hydrosulphuric acid, as before, an acid liquor is obtained which contains the deuto-hydrate of phosphoric acid. It must not be warmed to expel the excess of hydrosulphuric acid, but be left in a shallow bason for twenty-four hours to permit the escape of that gas. This acid, when neutralized with carbonate of soda, gives Dr. Clark's pyrophosphate of soda. It also gives a white precipitate with nitrate of silver; all the salts which it forms have uniformly two eq. of base. They were named the *pyrophosphates*, and since that term has come into use, it is not likely to be superseded by the systematic, but rather inconvenient designation of bibasic phosphates. A dilute solution of the deuto-hydrate of phosphoric acid may be preserved for a month without sensible change, but when the solution is exposed for some time to a high temperature, it passes entirely into the terhydrate.

Protohydrate of phosphoric acid.—If the biphosphate of soda be heated to redness, a salt is formed, which treated in a similar manner with the last, gives an acid liquid, containing the protohydrate of phosphoric acid. To prepare the biphosphate itself, a solution of the terhydrate of phosphoric acid is added to a solution of common phosphate of soda, till it is found that a drop of the latter is no longer precipitated by chloride of barium. The biphosphate of soda, which is now in solution, can only be crystallized in cold weather. The glacial phosphoric acid also is in general almost entirely the protohydrate. This hydrate is characterized by producing a white precipitate in solution of albumen, which is not disturbed by the other hydrates, and in solutions of the salts of earths and metallic oxides, precipitates which are remarkable semifluid bodies, or soft solids, without crystallization. All these salts contain only one eq. of base to one of acid, like the protohydrate of the acid itself. The name *metaphosphates* was applied to the class by myself, to mark the cause of the retention of peculiar properties by their acid, when free and in solution; namely, that it was not then simply phosphoric acid, but phosphoric acid *together with* water. (Researches on the Arseniates, Phosphates, and Modifications of Phosphoric Acid, Phil. Trans. 1833, p. 253; or Phil. Mag. 3d ser., vol. iv. p.

401.) This is the least stable of the hydrates of phosphoric acid, being converted rapidly, by the ebullition of its solution, into the terhydrate. If the terms *meta-phosphoric acid* and *pyrophosphoric acid* are employed at all, it is to be remembered that they are applicable to the proto and deutohydrates, and not to the acid itself, which is the same in all the hydrates. But to prevent the chance of misconception, metaphosphate of water and pyrophosphate of water might be substituted for the former terms.

A solution of the terhydrate of phosphoric acid, evaporated in vacuo over sulphuric acid, crystallizes in thin plates, which are extremely deliquescent. The deutohydrate has also been obtained in crystals. When heated to 400°, the terhydrate loses a portion of water, and becomes a mixture of the deuto and protohydrates; and by heating it to redness for some time, the proportion of water may be reduced to one equivalent, or perhaps even less than this; and such is the composition of glacial phosphoric acid. But at that high temperature much of the hydrated phosphoric acid passes off in vapour. The solution of phosphoric acid is not poisonous, nor when concentrated does it act as a cautery, but it injures the teeth from its property of dissolving phosphate of lime. The soluble phosphates, which are not acid, give a precipitate with chloride of barium, which is the phosphate of baryta. This phosphate, in common with all the insoluble phosphates, is dissolved by nitric acid, hydrochloric acid, and even acetic acid, a property by which it is distinguished from sulphate of baryta. A solution of phosphate of lime in phosphoric acid has been prescribed in rickets, a disease which indicates a deficiency of earthy phosphates in the system. The phosphate of soda, also, is given as a mild aperient; its taste is saline, but not disagreeably bitter.

Phosphates.—The formation of three classes of phosphates from the three basic hydrates of phosphoric acid, affords an excellent illustration of the formation of compounds by substitution; the quantity of fixed base, such as soda, with which phosphoric acid combines in the humid way, being entirely regulated by the proportion of water previously in union with the acid, which is simply replaced by the fixed base. Thus, the protohydrate of phosphoric acid combines with no more than one, and the deutohydrate with no more than two equivalents of soda, although a larger quantity of alkali be added to it. The excess of alkali remains free. Again, supposing an equivalent quantity of the terhydrate of phosphoric acid in solution, and one equivalent of soda added to it, one equivalent only of water is displaced, and two retained, and a phosphate formed, containing one of soda and two of water as bases; the salt already adverted to under its old name of biphosphate of soda. Let a second equivalent of soda be added to this salt, and a second basic equivalent of water is displaced, and a tribasic salt produced, containing two of soda and one of water as bases, which is the common phosphate of soda of pharmacy. A third equivalent of soda added to the last salt displaces the last remaining equivalent of basic water, and a tribasic phosphate is formed, of which the whole three equivalents of base are soda, and which has the name of subphosphate of soda. But this last salt can unite with no more soda. The same three salts may be formed by means of the tribasic phosphate of water, in another manner. That acid hydrate decomposes chloride of sodium, but only to a certain extent, expelling hydrochloric acid, so as to acquire one of soda, and becoming $2HO.NaO+PO_5$, or the biphosphate of soda already referred to; the same acid hydrate applied to the carbonate or the acetate of soda, can assume two proportions of soda, displacing twice as much of the weaker carbonic and acetic acids, as of the hydrochloric acid, and so becomes $HO.2NaO+PO_5$, or the common phosphate of soda; and the same acid hydrate applied to the hydrate of soda (caustic soda,) assumes three of soda, and becomes $3NaO+PO_5$, or the subphosphate of soda.

From soluble tribasic phosphates, such as those mentioned, insoluble salts may be precipitated, which are likewise tribasic, by adding solutions of most metallic salts. Thus one equivalent of the common phosphate of soda, added to the nitrate of silver in excess, decomposes 3 equivalents of it, and produces the yellow tribasic phosphate of silver, as explained in the following diagram, in which the name of a substance is understood to express one equivalent of it, and the figures, numbers of equivalents:—

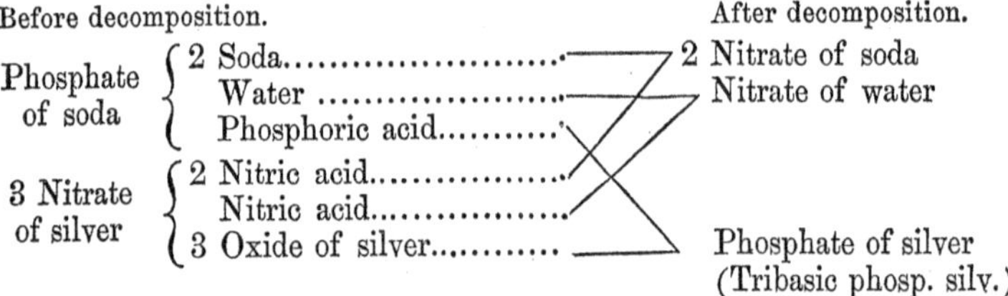

Here, then, is exact mutual decomposition, but it is attended with a phenomenon which does not occur when other neutral salts decompose each other. The liquid does not remain neutral, but becomes highly acid after precipitation; the reason is, that one of the new products is the nitrate of water, or hydrated nitric acid; and consequently the products, although neutral in composition, are not neutral to test paper.

The pyrophosphate of soda, which is bibasic, decomposes, on the other hand, two proportions of nitrate of silver, and gives a pyrophosphate or bibasic phosphate of silver, which is a white precipitate; thus—

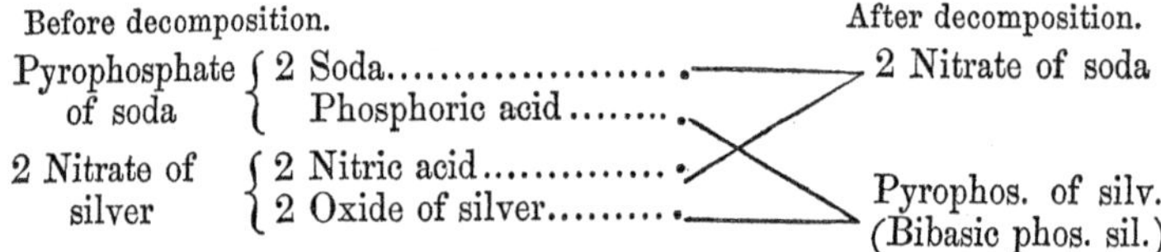

Here there is no salt of water among the products, and consequently the liquid is neutral after precipitation.

The metaphosphate of soda, which is monobasic, like the sulphates, nitrates and other familiar salts, decomposes like these but one proportion of nitrate of silver, and forms a white precipitate; thus—

Before decomposition. After decomposition.

Metaphosph. of soda { Soda .... Nitrate of soda
{ Phosphoric acid ....
Nitrate of silver { Nitric acid ....
{ Oxide of silver .... Metaphosphate of silv. (Monobasic phos. silv.)

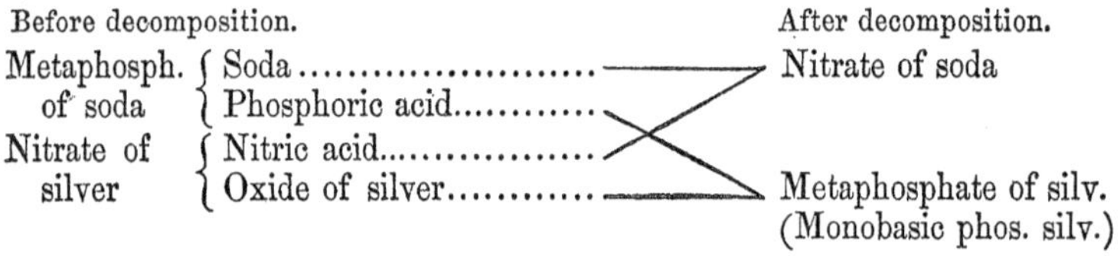

If acetate or nitrate of lead be substituted for nitrate of silver in these decompositions, a tribasic, bibasic, or monobasic salt of lead is obtained in the same manner; and these salts, again, decomposed by hydrosulphuric acid gas, afford respectively the terhydrate, deutohydrate, and protohydrate of phosphoric acid. The statement of the decomposition of the metaphosphate of lead by hydrosulphuric acid will be sufficient to explain how a hydrate of phosphoric acid comes to be formed in all these cases:—

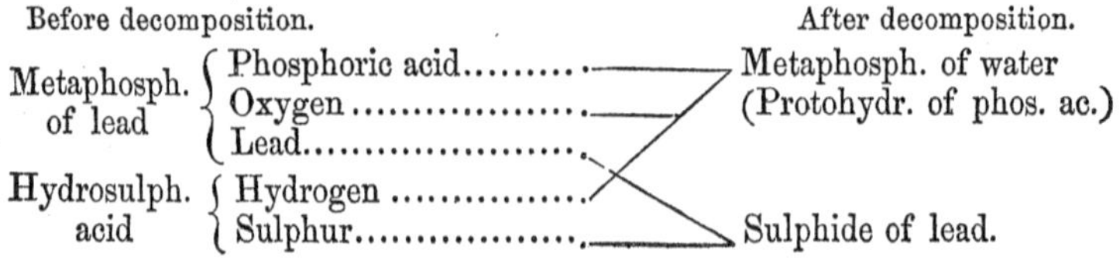

Before decomposition. After decomposition.

Metaphosph. of lead { Phosphoric acid .... Metaphosph. of water (Protohydr. of phos. ac.)
{ Oxygen ....
{ Lead ....
Hydrosulph. acid { Hydrogen ....
{ Sulphur .... Sulphide of lead.

It will be observed that the hydrosulphuric acid forms 1 equivalent of water, at the same time that it throws down the sulphide of lead. In this phosphate of lead, there is only 1 equivalent of oxide of lead, and consequently only 1 equivalent of water is formed; but if there were 2 or 3 equivalents of oxide, there would be 2 or 3 equivalents of water formed and conveyed to the acid; or the phosphoric acid is always left in combination with as many equivalents of water as it previously possessed of oxide of lead. Thus the different hydrates of phosphoric acid are obtained from the decomposition of the corresponding phosphates of lead.

In no decomposition of this kind is there any transition from one class of phosphates into another, because the decompositions are always mutual, and the products of a neutral character. Hence an argument for retaining the trivial names, common phosphates, pyrophosphates, and metaphosphates, for there is no changing, in decompositions by the humid way, from one to the other, and the salts comport themselves so far quite as if they had different acids. The circumstances may now be noticed in which a transition from the one class to the other does occur:—

1st.—Changes without the intervention of a high temperature. When solutions of the metaphosphate and pyrophosphate of water are warmed, they pass gradually into the state of common phosphate, combining with an additional quantity of water; and the metaphosphate of water appears then to become at once common phosphate, without passing through the intermediate state of hydration of the pyrophosphate. The metaphosphate of baryta also, which is an insoluble salt, is gradually dissolved in boiling water, and becomes common phosphate by assuming 2 eq. of basic water. The easy transition from the one class of phosphates to the other, then witnessed, forbids the supposition that they contain different acids, or different isomeric modifications of phosphoric acid. Indeed, it might as well be supposed that in the protoxide and sesqui-oxide of iron, the metal exists in different isomeric conditions, because these oxides possess peculiar properties, and combine in different proportions with the same acid. Iron in its two oxides gives rise to different compounds, because they are formed by substitution; and phosphoric acid in its three hydrates gives rise to different compounds, from the same cause. The degree of oxidation of the iron and the degree of hydration of the acid are anterior conditions, due to the special unexplained affinities with which each element or compound is invested. It is remarkable that pyrophosphates of potassa and of ammonia exist in solution, and perfectly stable, but not in the dry state. These salts do not crystallize. The pyrophosphate of ammonia, indeed, when allowed to evaporate spontaneously, appears to crystallize, but in the act of becoming solid, it passes into common phosphate (the biphosphate of ammonia, $2HO.NH_4O+PO_5$).

2d.—Changes with the intervention of a high temperature. If a single equivalent of phosphoric acid, anhydrous, or in any state of hydration, be calcined at a temperature which may fall short of a red heat (1°), with 1 equivalent of soda or its carbonate, the metaphosphate of soda will be formed; (2°) with 2 equivalents of soda or its carbonate, the pyrophosphate of soda will be formed; and (3°) with 3 equivalents of soda or its carbonate, a common phosphate of soda will be formed. Hence, the formation of none of these classes is peculiarly the effect of a high temperature. Again, a tribasic phosphate, containing one or two equivalents of a volatile base, such as water or ammonia, loses the volatile base, when ignited, and the acid remains in combination with the fixed base. Hence, common phosphate of soda ($HO.2NaO+PO_5$) is converted by heat into pyrophosphate ($2NaO+PO_5$), the original observation of Dr. Clark; and the biphosphate of soda ($2HO.NaO+PO_5$) into metaphosphate of soda ($NaO+PO_5$). The acid remains in combination with the fixed base, and the salt produced may be dissolved in water without assuming basic water.

The metaphosphate of soda is susceptible of a remarkable conversion, by the agency of a certain temperature, and exhibits a change of nature, without a change of composition, such as often occurs in organic compounds, but rarely admits of so satisfactory an explanation. This particular salt, in common with all the other phosphates, combines with water, which becomes attached to the salt, in the state of constitutional water, or water of crystallization. The metaphosphate of soda, so hydrated, when dried at 212°, retains 1 equivalent of water, but that water is not basic, for, on dissolving the salt again, it is found still to be a metaphosphate. But let this hydrated metaphosphate be heated to 300°, and without losing anything, it changes completely, and becomes a pyrophosphate, — the water which was constitutional before, being now basic. The formulæ of the salt in its two states exhibit to the eye the nature of the internal change which occurs in it:

1.—Hydrated metaphosphate of soda.................$NaO.PO_5+HO$,
2.—Pyrophosphate of soda and water.................$NaO.HO+PO_5$.

Phosphates of the form $3MO+2PO_5$. — The recent investigations of Fleitmann and Henneberg establish the existence of two new classes of phosphates, intermediate between the monobasic and bibasic classes. The soda-salt of the preceding formula is produced by fusing together, in a platinum crucible, 100 parts of anhydrous pyrophosphate of soda and 76.87 parts of metaphosphate of soda: the white crystalline mass which results is reduced to powder, and quickly exhausted with water; for, on long digestion, the ordinary phosphates are obtained. The soda-salt is soluble in about twice its weight of cold water, and has a faint alkaline reaction. It gives, by precipitation with nitrate of silver and with phosphate of magnesia, salts corresponding with the soda-salt, and which have not the properties of a mixture of pyrophosphate and metaphosphate.

Phosphates of the form $6MO+5PO_5$. — The soda-salt was obtained by fusing together 100 parts by weight of pyrophosphate of soda and 307.5 of metaphosphate. The solution is by no means stable, but gives, when freshly prepared, a precipitate in nitrate of silver, which is readily soluble in excess of the soda-salt, and possesses the composition, when fused, of $6AgO+5PO_5$. (Liebig's Annalen, lxv. 304.)

Modifications of metaphosphoric acid. — The metaphosphates already described are prepared from the monobasic phosphate of soda in the vitreous condition; this phosphate, when cooled immediately from a state of fusion, remaining a transparent, colourless glass. But if this glassy phosphate be cooled very slowly, a beautiful crystalline mass is obtained. On dissolving it in a small quantity of hot water, the liquid divides into two strata, the more considerable one containing the crystalline salt, and the other a portion of unaltered metaphosphate of soda. The *vitreous metaphosphate*, and all the salts derived from it, are remarkable for not crystallizing, but form liquid or semi-liquid viscid hydrates. But the *crystalline metaphosphate* of soda is described as giving beautiful crystals of the triclinometric system, containing water of crystallization. Its solution is neutral, and has a cooling, pure, saline taste, while the vitreous metaphosphate of soda is insipid. It is rapidly converted into the acid common phosphate by boiling. The corresponding silver-salt is obtained by adding nitrate of silver to a tolerably concentrated solution of the soda-salt. It is white, crystalline, and is represented by the formula $3(AgO.PO_5)+2HO$.

Phosphates were obtained by Mr. Maddrell, by adding the solution of sulphates of magnesia, nickel, copper, soda, lime, baryta, alumina, to an excess of phosphoric acid, evaporating, to expel the sulphuric acid, and heating to upwards of 600°; in the form of a crystalline granular substance, which were all monobasic. They are all anhydrous, insoluble in water and diluted acids, but generally decomposed by concentrated sulphuric acid, and appear to form a class of metaphosphates different from the preceding two. The magnesian metaphosphates of this class have a disposition to combine with the corresponding soda-salt, when any of that base is present in the phosphoric acid with which they are ignited. The double salt of magnesia and soda is represented by $3(MgO.PO_5)+NaO.PO_5$; that of nickel and soda, by $6(NiO.PO_5)+NaO.PO_5$). (Mem. Chem. Soc. iii. 273.)

The only explanation which can be offered of these modifications of the metaphosphoric acid, is, that they are of a polymeric character; such as $MO.PO_5$; $2MO.2PO_5$; $3MO.3PO_5$, or perhaps even higher multiples of $MO.PO_5$. No data, however, appear to exist by which a place in this polymeric series can be ascribed to the respective modifications with any degree of certainty. MM. Fleitmann and Henneberg, who have lately investigated the subject with much ability, are disposed to represent metaphosphoric acid by $6MO.6PO_5$; and certainly with this proportion of base constant and the phosphoric acid variable, the other classes may be consistently represented:—

Common phosphate...............................	$6MO+2PO_5$
Pyrophosphate....................................	$6MO+3PO_5$
Fleitmann and Henneberg's new phosphates	$6MO+4PO_5$ $6MO+5PO_5$
Metaphosphate....................................	$6MO+6PO_5$

The different classes of phosphates are thus represented as all sex-basic salts, with a different polymeric acid in each, P_2O_{10}, P_3O_{15}, &c. But this theory does not embrace the modifications of metaphosphoric acid, nor will it serve to represent several known double phosphates; such, for instance, as the double pyrophosphate of copper and soda, $3(2NaO.PO_5)+2CuO.PO_5$.

Analysis of phosphoric acid and of the phosphates.—Phosphoric acid is produced when the pentachloride of phosphorus is thrown into water:—

$$PCl_5 \text{ and } 5HO=PO_5 \text{ and } 5HCl.$$

It may be inferred with certainty from this decomposition, that phosphoric acid contains 5 equivalents of oxygen, in the same manner as the composition of phosphorous acid is deduced from the decomposition of the terchloride of phosphorus by water (page 318). The affinity of phosphoric acid for water is very intense, the anhydrous phosphoric acid taking water even from oil of vitriol and eliminating anhydrous sulphuric acid, at a high temperature. As hydrated phosphoric acid cannot be made anhydrous by heat, the proportion of dry acid in a solution of the free acid is determined by adding a known weight of oxide of lead, evaporating to dryness, and heating the residue, as in the case of sulphuric acid. The phosphate of lead formed being anhydrous, the increase of weight which the oxide of lead sustains represents exactly the weight of dry phosphoric acid.

In determining the proportion of phosphoric acid in a salt of an alkaline or earthy base, the acid, if not already in the tribasic form, is first brought to that condition by boiling with a little nitric acid. 1. The excess of nitric acid being then neutralized by ammonia, the phosphate is again dissolved in acetic acid. If the solution contains no sulphuric acid nor chlorine, the phosphoric acid may be entirely separated by the addition of nitrate of lead, in the form of an insoluble phosphate of lead, $2PbO.HO.PO_5$, which washes easily, and loses water and becomes pyrophosphate, $2PbO.PO_5$, when calcined (Heintz). This method is based upon the insolubility of phosphate of lead in acetic acid. 2. Phosphoric acid may also be thrown down from the solution of an alkaline phosphate, by adding first carbonate or hydrochlorate of ammonia and then sulphate of magnesia, when, upon stirring the phosphate of magnesia and ammonia,

$$2MgO.NH_4O.PO_5+12HO,$$

falls as a granular precipitate. This phosphate must be precipitated in an alkaline solution, and washed with water containing hydrochlorate of ammonia, as it is very soluble in acids, and even soluble in a sensible degree in pure water. When ignited it loses its volatile constituents, and remains pyrophosphate of magnesia, $2MgO.PO_5$. 3. The phosphoric acid not being in combination with a base which yields a phosphate insoluble in acetic acid, an addition is made to the liquid, which may be acid, of an excess of the acetate of the sesqui-oxide of iron. The phosphate of sesqui-oxide of iron, $Fe_2O_3.PO_5$, immediately separates as a slightly reddish yellow flaky precipitate, which is collected and washed upon a filter. This phosphate is dissolved off the filter by a few drops of hydrochloric acid, then the salt of iron reduced to the state of protoxide by boiling it with sulphite of soda, and afterwards the quantity of iron ascertained by finding how much of a solution of permanganate of potassa of known composition is required to peroxidize the iron. The phosphate of iron being of known composition, the quantity of phosphoric acid is calculated from the iron, 2 eqs. of that metal being present in the phosphate for 1 eq. of phosphoric acid or of phosphorus; that is, 700 parts iron representing 900 parts phosphoric acid (Raewsky and Marguerite). The acetate of sesqui-oxide of iron, which is not permanent, is best prepared extemporaneously from solutions of 100 parts of iron-alum and of 98 parts of acetate of soda in equal quantities of water, of which equal volumes are mixed at the moment the acetate of iron is required.

In describing the various classes of phosphates, with their relations to each other, I have been thus minute, partly because considerable explanatory detail was required, from the extent of the subject, but principally for the sake of the light which the phosphates throw upon the constitution of the class of organic acids, and upon the function of water in many compounds. Indeed, phosphoric acid is one of the links by which mineral and organic compounds are connected. And it may be reasonably supposed that it is that pliancy of constitution which peculiarly adapts the phosphoric, above all other mineral acids, to the wants of the animal economy.

PHOSPHORUS AND HYDROGEN.

Solid hydride of phosphorus, P_2H. — Magnus formed a phosphide of potassium by fusing phosphorus and potassium under naphtha. When this compound is thrown into water, a compound of phosphorus and hydrogen precipitates in the form of a yellow powder. The solid hydride of phosphorus becomes red when exposed to light; it does not shine in the dark, nor take fire below 320° (160° C.) It is insoluble in water and alcohol, and is decomposed by alkalies, with the formation of oxide of phosphorus, free hydrogen, gaseous phosphuretted hydrogen, and a hypophosphite.

Phosphuretted hydrogen gas; eq. 19 *or* 237.5; PH_3. — This gas, which is remarkable for its occasional spontaneous inflammability in air, was discovered by Gengembre in 1783, and has been successively investigated by several chemists. Its true nature was first ascertained by Rose, who proved it to be a compound having the same proportion of hydrogen as ammoniacal gas, with phosphorus in the place of nitrogen. The pure gas is obtained by heating hydrated phosphorous acid, which is resolved into phosphuretted hydrogen and hydrated phosphoric acid: thus —

$$4(3HO+PO_3) \text{ or } 12HO \text{ and } 4PO_3=PH_3 \text{ and } 9HO+3PO_5.$$

The gas so prepared does not inflame spontaneously when allowed to escape into air, but kindles when a light is applied to it, and burns with the white flame of phosphorus. A little air added to the gas, which had no effect at first, has been observed to produce occasionally an explosion after a time. The gas consists of 1 volume of phosphorus vapour and 6 volumes of hydrogen, condensed into 4 volumes, so that it has the same combining measure as ammoniacal gas. Its density is 1185. Phosphuretted hydrogen has a disagreeable alliaceous odour, is but slightly soluble in water, and has no alkaline reaction.

The same gas, in a self-inflammable state, is obtained by boiling phosphorus with water and an excess of lime, or in a strong solution of caustic potassa, in the flask A (fig. 147), at the water-trough B. The first effect is the formation of hypophosphite of lime, with the evolution of phosphuretted hydrogen gas:

$$4P \text{ and } 3CaO \text{ and } 3HO=PH_3 \text{ and } 3CaO+3PO.$$

Phosphuretted hydrogen is again evolved, but mixed with a considerable quantity of free hydrogen, when the hydrated hypophosphite of lime is evaporated to dryness, phosphate of lime being the residuary product.

Fig. 147.

Each bubble of gas on escaping into air takes fire, and produces a beautiful white wreath of smoke, consisting of phosphoric acid. The spontaneous inflammability is due to the presence of a small quantity of the vapour of a liquid compound of phosphorus and hydrogen, and was first explained by M. P. Thénard.

Phosphuretted hydrogen decomposes some metallic solutions, such as those of copper and mercury, and forms metallic phosphides. When the gas is pure, it is entirely absorbed by sulphate of copper and by chloride of lime. With hydriodic acid, phosphuretted hydrogen forms a crystalline compound, which is interesting from its analogy to sal ammoniac. It may be prepared by mixing together its constituent gases over mercury; or more easily by introducing into a small tubulated retort 60 parts of dry iodine with 15 of phosphorus finely granulated, and mixing these bodies intimately with pounded glass; 8 or 9 parts of water are then added to the mixture, and the vapours which immediately come off are allowed to escape by a glass tube open at both ends, adapted to the beak of the retort in which beautiful small crystals of the salt condense, of a diamond lustre. Rose observed that these crystals do not belong to the Regular System, and are, therefore, not isomorphous with sal ammoniac. They are decomposed by water, with evolution of phosphuretted hydrogen.

Phosphuretted hydrogen combines also, like ammonia, with the perchlorides of tin, titanium, chromium, iron, and antimony, forming white saline bodies. The combination with bichloride of tin is decomposed, with escape of the gas in the non-inflammable state, by water, and in the spontaneously inflammable condition by solution of ammonia.

Liquid hydride of phosphorus, PH_2.—This substance, which was discovered by M. Paul Thénard, is obtained by exposing the phosphuretted hydrogen gas, evolved by the action of water, at 140° (60° C.) on the phosphide of calcium Ca_2P, to a freezing mixture in a condensing tube. It is a colourless liquid, of high refracting power, which does not freeze at —4° (—20° C.), but which a temperature of +86° (30° C.) is sufficient to decompose. It is resolved under the influence of light into the gaseous and solid hydrides of phosphorus. The same decomposition is produced by contact with very different substances, such as alcohol, oil of turpentine, hydrochloric acid, and many pulverulent matters.

This compound is one of the most inflammable substances known, taking fire spontaneously in air, and burning with a dazzling flame. The most minute trace of its vapour, diffusing into the different combustible gases, such as hydrogen, carbonic oxide, cyanogen, olefiant gas, &c., communicates to them, as it does to phosphuretted hydrogen, the property of inflaming spontaneously in air or oxygen. (P. Thénard, Annal. de Ch. et Ph., 3me. sér. xiv. 5.)

PHOSPHORUS AND NITROGEN.

Both chlorides of phosphorus absorb ammoniacal gas, and form solid white compounds. The combination of the terchloride contains 2½ equivalents of ammonia, but that of the perchloride was not found equally definite. When exposed to a strong red heat, without access of oxygen, these compounds leave a white amorphous body, which was supposed to be a nitride of phosphorus, PN_2. (Rose: Annal. de Ch. et Ph., liv. 275.) It is most easily prepared by transmitting a stream of dry carbonic acid gas over the ammoniacal compound, in a tube of hard glass, heated by a charcoal fire, so long as vapours of sal ammoniac sublime.

This substance, which is remarkable for its fixity, is not soluble in any menstruum, nor acted upon by dilute acid or alkaline solutions. It is not affected even when heated in an atmosphere of chlorine or sulphur, but is decomposed when heated in hydrogen gas, with the formation of ammonia.

According to M. Gerhardt, the pentachloride of phosphorus absorbs ammonia, with the evolution of some hydrochloric acid, and the formation of a compound $PCl_3.(NH_2)^2$. The nitride of phosphorus also contains hydrogen, and ought to be represented by the formula PN_2H: its formation from the perchloride of phosphorus and ammonia taking place according to the equation:—

$$PCl_5 \text{ and } 2NH_3 = 5HCl \text{ and } PN_2H.$$

This compound, PN_2H, which is named *Phospham* by Gerhardt, is decomposed by fusion with hydrate of potassa, and converted into ammonia, and the ordinary phosphate of potassa. At a high temperature water acts upon phospham, giving rise to ammonia and phosphoric acid.

PHOSPHORUS AND SULPHUR.—SULPHIDES OF PHOSPHORUS.

Phosphorus and sulphur combine in all proportions, with the evolution of much heat, and sometimes with explosion. These elements most safely unite under hot water, of which the temperature, however, must not exceed 160°; for otherwise hydrosulphuric and phosphoric acids may be produced with such rapidity as to occasion an explosion. The compounds obtained in this manner are of a pale yellow colour,—more fusible and more inflammable than phosphorus itself. They were supposed to be indefinite in composition; but Berzelius has shown that they form a series of sulphides of phosphorus corresponding in composition with the oxides, with one sulphide additional. They are represented by the formulæ—

Subsulphide, P_2S .... corresponding with Oxide of Phosphorus, P_2O.
Protosulphide, PS .... " " Hypophosphorous Acid, PO.
Tersulphide, PS_3 .... " " Phosphorous Acid, PO_3.
Pentasulphide, PS_5 .... " " Phosphoric Acid, PO_5.
Persulphide, PS_{12} .... without an oxygen analogue.

These compounds may all be formed directly by fusing sulphur and phosphorus together in the requisite proportions, and are generally crystallizable. The tersulphide was originally obtained by Serullas by the action of hydrosulphuric acid upon the terchloride of phosphorus. They are insoluble in water, alcohol, or ether; but combine readily with alkaline sulphides, and form series of sulphur-salts corresponding with the hypophosphites, phosphites, and phosphates.

SECTION X.

CHLORINE.

Eq. 35.5 *or* 443.75; Cl; *density* 2440; ☐☐.

This substance was discovered by Scheele in 1774, but was believed to be of a compound nature, till Gay-Lussac and Thénard, in 1809, showed that it might reasonably be considered a simple substance. It is to the powerful advocacy of Davy, however, who entered upon the investigation shortly afterwards, that the establishment of the elementary character of chlorine is principally due, and to him it is indebted for the name it now bears, which is derived from χλωρος, yellowish-green, and refers to its colour as a gas, elementary bodies being generally named from some remarkable quality or important circumstance in their history. Chlorine is the leading member of a well-marked natural family, to which also bromine, iodine, and fluorine belong. Phosphorus, carbon, hydrogen, sulphur, and most of the preceding elementary bodies, have little or no action upon each other, or upon the mass of hydrogenous, carbonaceous, and metallic bodies to which they are exposed in the material world; all these substances being too similar in nature to have much affinity for each other. But the class to which chlorine belongs ranks apart, and, with a mutual indifference to each other, they exhibit an intense affinity for the members of the other great and prevailing class — an affinity so general as to give the chlorine family the character of extraordinary chemical activity, and to preclude the possibility of any member of the class existing in a free and uncombined state in nature. The compounds, again, of the chlorine class, with the exception of those of fluorine, are remarkable for solubility, and consequently find a place among the saline constituents of sea water, and are of comparatively rare occurrence in the mineral kingdom; with the single exception of chloride of sodium, which, besides being present in large quantity in sea water, forms extensive beds of rock salt in certain geological formations.

Preparation.—The fuming hydrochloric acid or muriatic acid (as it is also called) of commerce, is a solution in water of hydrochloric gas, a compound of chlorine and hydrogen, from which chlorine gas is easily procured. The liberation of chlorine results from contact of the acid named with binoxide of manganese, and the reaction which then occurs is made most obvious in the following mode of conducting the experiment:—A few ounces of the strongly fuming hydrochloric acid are introduced into a flask *a* (fig. 148), with a perforated cork and tube *b*, upon which a bulb or two have been expanded; and that tube is connected, by means of a short

Fig. 148.

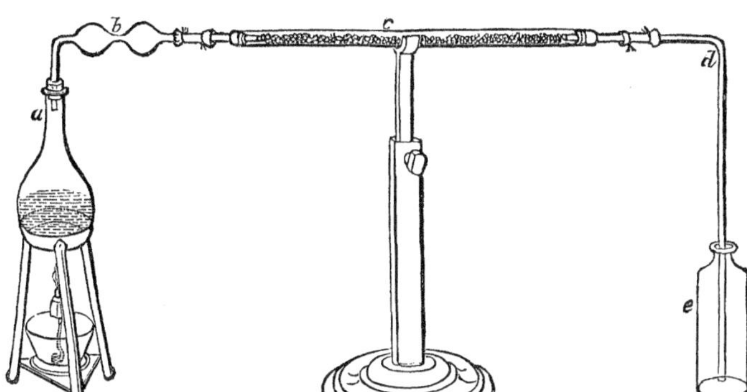

caoutchouc tube, with the drying tube *c*, containing fragments of chloride of calcium, and the last is connected in a similar manner with the exit tube *d*, which descends to the bottom of a dry and empty bottle *e*. Upon applying the spirit-lamp to *a*, the liquid in the flask soon begins to boil, and the hydrochloric gas passes off, depositing, perhaps, a little moisture in the bulbs of *b*, which may be kept cool by wet blotting-paper, and being completely dried in passing through *c*. It is conveyed by *d* to the bottom of the bottle *e*, and finally escapes and produces white fumes in the atmosphere, after displacing the air of that bottle. The hydrochloric gas is obtained in *e* unchanged, and will redden and not bleach a little blue infusion of litmus poured into *e*. But between the tube *c* and *d*, let another tube be now interposed having a pair of bulbs blown upon it *f* and *g* (fig. 149), one of which *f*

Fig. 149.

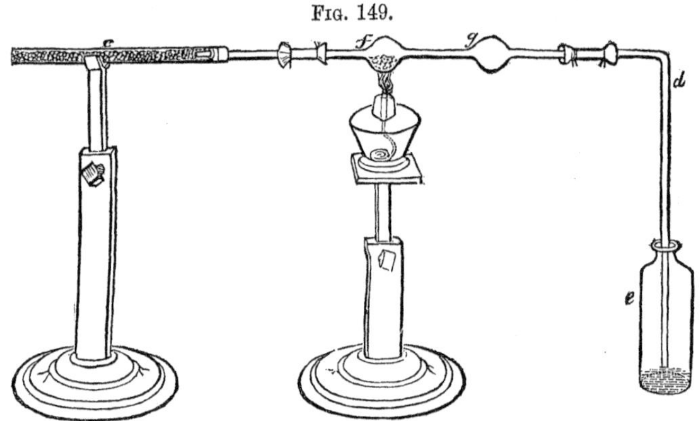

contains a quantity of pounded anhydrous binoxide of manganese; the bottle *e* remaining as before. Then, upon applying heat to the manganese bulb *f*, the hydrochloric gas will be found to suffer decomposition as it traverses that bulb, its hydrogen uniting with the oxygen of the manganese, and forming water, which will condense in drops in *g*, and disengaged chlorine proceeds on to *e*, in which that gas will be perceptible from its yellow tint, and more so by bleaching the infusion of reddened litmus remaining in *e*. If the transmission of hydrochloric acid over the binoxide of manganese be continued for sufficient time, the latter loses all its oxygen, and the metal remains in the state of protochloride. Indeed, only one-half of the chlorine of the decomposed hydrochloric gas is obtained as gas, the other half being retained by the manganese, as will appear by the following diagram:—

PROCESS FOR CHLORINE FROM HYDROCHLORIC ACID AND BINOXIDE OF MANGANESE.

Before decomposition.		After decomposition.
Hydrochloric acid	Chlorine............	Chlorine.
	Hydrogen..........	Water.
Binoxide of mangan.	Oxygen.............	
	Manganese.........	Chloride of manganese.
	Oxygen.............	
Hydrochloric acid	Chlorine............	
	Hydrogen..........	Water.

Or in symbols:—$MnO_2+2HCl=MnCl$ and $2HO$ and Cl.

The most convenient method of preparing chlorine gas is by mixing in a flask A (fig. 150), 1 part of binoxide of manganese with 4 parts of hydrochloric acid,

Fig. 150.

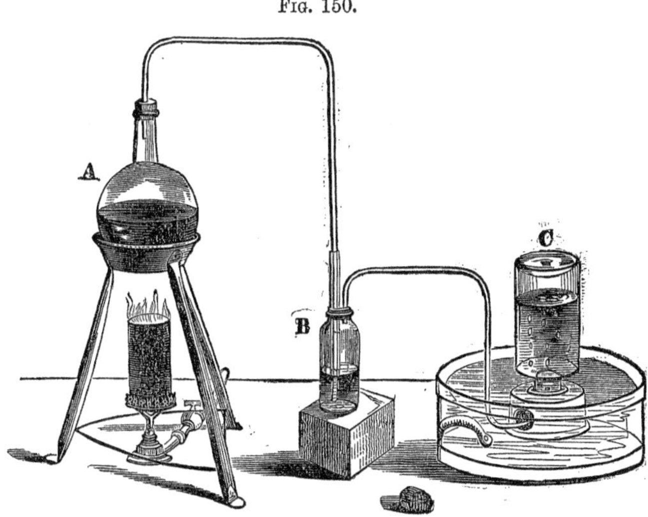

diluted with 1 of water. Effervescence, from escape of gas, takes place in the cold, but is greatly promoted by the application of a gentle heat. The gas is collected in C over water, of which the temperature should not be less than 80° or 90°; otherwise a great waste of the gas occurs from its solution in the water, and also a consequent annoyance to the operator from the escape of the chlorine into the atmosphere, by evaporation from the surface of the water-trough. If the gas is not to be used immediately, but preserved, it should be collected in bottles, into which, when filled with gas, their stoppers greased should be inserted before they are removed from the trough. Before the gas obtained by this process can be considered as pure, it should be transmitted through water in a wash-bottle B, to remove hydrochloric acid. If the gas is to be dried, it must be sent through a tube containing chloride of calcium, of two or three feet in length, some difficulty being experienced in drying this gas in a perfect manner, owing to its low diffusive power. Chlorine cannot be collected over mercury, as it combines at once with that metal.

A somewhat different process for the preparation of chlorine is generally followed on the large scale. About 6 parts of manganese with 8 of common salt are introduced into a large leaden vessel, of a form nearly globular, as represented (fig. 151), and 5 or 6 feet in diameter, and to these is added as much of the unconcentrated sulphuric acid of the leaden chambers as is equivalent to 13 parts of oil of vitriol. The leaden vessel is placed in an iron pan, or has an outer casing, *d e*; and to heat the materials, steam is admitted by *d* into the space between the bottom and outer casing. In the figure, which is a section of the leaden retort, *a* represents the tube by which the chlorine escapes, *b* a large opening for introducing the solid material covered by a lid or water valve, its edges dipping into a channel containing water, *e* a twisted leaden funnel for introducing the acid, *f* a wooden agitator, and *c* a discharge tube, by which the waste materials are run off after the process is finished. A retort of lead cannot be used with safety with binoxide of manganese and hydrochloric acid for chlorine, owing to the action of the acid upon the lead, and

Fig. 151.

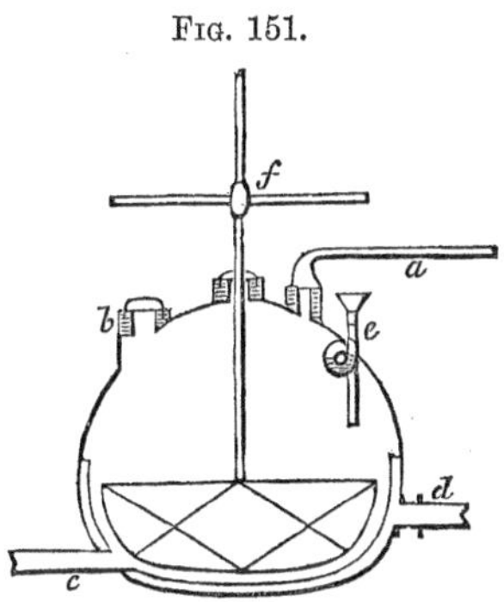

the evolution of hydrogen gas (which produces a spontaneously-explosive mixture with chlorine), or, it is said, of euchlorine. In the reaction which occurs in the leaden retort, it may be supposed either that hydrochloric acid is first liberated from chloride of sodium by sulphuric acid, and afterwards decomposed by binoxide of manganese, as in the preceding experiment; or that sulphates of manganese and soda are simultaneously formed, and chlorine liberated in consequence, as stated in the following diagram, in which the names express (as usual) single equivalents:—

PROCESS FOR CHLORINE FROM CHLORIDE OF SODIUM (COMMON SALT), BINOXIDE OF MANGANESE, AND SULPHURIC ACID.

Before decomposition.		After decomposition.
Chloride of sodium	Chlorine	Chlorine.
	Sodium	
Sulphuric acid	Sulphuric acid	Sulphate of soda.
Binoxide of manganese	Oxygen	
	Protox. manganese	
Sulphuric acid	Sulphuric acid	Sulph. of mangan.

Or in symbols:

$$NaCl \text{ and } 2SO_3 \text{ and } MnO_2 = NaO.SO_3 \text{ and } MnO.SO_3 \text{ and } Cl.$$

A new manufacturing process for chlorine has lately been applied by Mr. C. Tennant Dunlop, in which the use of binoxide of manganese is superseded by nitric acid. One equivalent of nitric acid is found to communicate two equivalents of oxygen to the hydrochloric acid, and thus evolve two equivalents of chlorine. The decomposed nitric acid is evolved in the form of nitrous acid vapour NO_3, and it is an essential part of the process to absorb that vapour by means of sulphuric acid, and to introduce the nitrous acid in this form into the leaden chamber.

Properties.—Chlorine is a dense gas of a pale yellow colour, having a peculiar suffocating odour, absolutely intolerable even when largely diluted with air, and occasioning great irritation in the trachea, with coughing and oppression of the chest. Some relief from these effects is experienced from the inhalation of the vapour of ether or alcohol. The density of chlorine gas is, by experiment, 2470—by theory, 2440. Under a pressure of about 4 atmospheres, chlorine condenses into a limpid liquid of a bright yellow colour, of sp. gr. about 1.33, and which has not been frozen. Water at 60° dissolves twice its volume of this gas, and acquires the yellowish colour, odour, and other properties of chlorine. To form chlorine-water, a stout bottle filled with the gas at the water-trough, may be closed with a good cork, and removed to a basin of cold water: on loosening the cork with the mouth of the bottle under water, a little water will enter it, from the contraction of the gas by cooling; and this water may be agitated in contact with the gas by a lateral movement of the bottle without removing it from the water; on loosening the cork again, more water will be found to enter the bottle, and by repeating the agitation and admission of water, the whole gas (if pure) is absorbed, and the bottle is in the end filled with water, which of course contains an equal volume of chlorine gas. With water near its freezing point, chlorine combines and forms a crystalline hydrate, which Faraday found to contain 10 eqs. of water. Hence chlorine gas cannot be collected at all over water below 40°. Exposed to light, chlorine water soon loses its properties, water being decomposed and hydrochloric acid formed, with the evolution of oxygen gas. But it may be preserved for a long time in an opaque bottle properly closed. When diluted so far that the water does not contain above 1 or $1\frac{1}{2}$ per cent. of its bulk of chlorine, the odour is by no means strong, and such a solution may be employed in bleaching without inconvenience to the workmen, although a combination of chlorine with hydrate of lime, called the chloride of lime, is generally preferred for that purpose.

Chlorine does not in any circumstances unite directly with oxygen, although several compounds of these elements can be formed; nor is it known to combine directly with nitrogen or carbon. Chlorine and hydrogen gases may be mixed and preserved in the dark without uniting, but combination is determined with explosion by spongy platinum or the electric spark, or by exposure to the direct rays of the sun; even under the diffuse light of day, combination of the gases takes place rapidly, but without explosion. Chlorine, indeed, has a strong affinity for hydrogen, and decomposes most bodies containing that element, hydrochloric acid being always formed. In plunging an ignited taper into chlorine gas, its flame is extinguished, but the column of oily vapour rising from the wick is rekindled by the chlorine, and the hydrogenous part of the combustible continues to burn with a red and smoky flame, which expires on removing the taper into air. Paper dipped in oil of turpentine takes fire spontaneously in this gas, and the oil burns, with the deposition of a large quantity of carbon. The affinity of chlorine for most metals is equally great: antimony, arsenic, and several others, showered in powder into this gas, take fire, and produce a brilliant combustion. Chlorine is absorbed by alcohol and many other organic substances, when it generally eliminates more or less hydrogen, as hydrochloric acid, and enters also by substitution into the original compound, in the place of that hydrogen. It bleaches all vegetable and animal colouring matters, and is believed then generally to act in that manner. The colours are destroyed and cannot be revived by any treatment.

A stream of chlorine gas, thrown into a bottle of dry ammoniacal gas, produces a jet of flame from the combustion of the hydrogen of the ammonia. When chlorine is passed through the undiluted solution of ammonia, the same decomposition takes place, and the reaction is a convenient source of nitrogen gas (page 244).

Fig. 152.

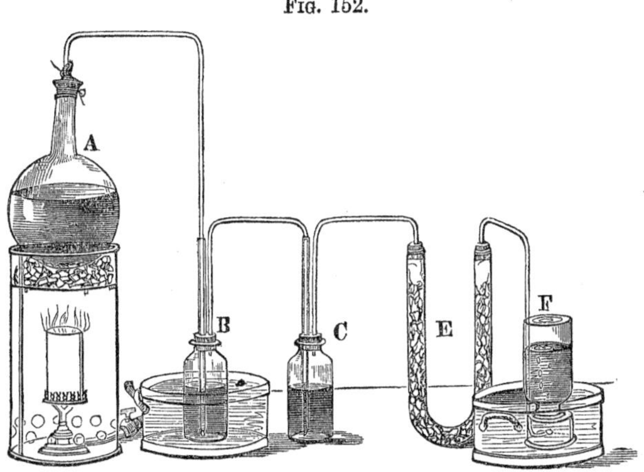

The arrangement represented in fig. 152 may be used for this purpose. It consists of a large globular flask A, in which chlorine is evolved from the usual materials; two wash-bottles, B and C, containing solution of ammonia, the first placed in a bason of cold water to repress its temperature. The nitrogen evolved passes through a U-tube, E, containing fragments of pumice impregnated with a solution of caustic potassa, to absorb any chlorine that may escape the action of the ammonia; and the gas is finally collected in bottles, F, filled with water acidulated with hydrochloric acid, to absorb the vapour of ammonia with which the nitrogen is accompanied.

Chlorine when free is easily recognized by its odour and bleaching power, and by producing both when free and in the soluble chlorides, with nitrate of silver, a white curdy precipitate of chloride of silver, which is soluble in ammonia, but not soluble in cold or boiling nitric acid.

Uses.—Chemistry has presented to the arts few substances of which the applications are more valuable. Chlorine is the discolouring agent of the modern process of bleaching, which, as it is generally conducted with cotton goods, consists of the following operations. The cloth, after being well washed, is boiled first in lime-water and then in caustic soda, which remove from it certain resinous matters soluble in alkali. It is then steeped in a solution of chloride of lime, so dilute as just to taste distinctly, which has little or no perceptible effect in whitening it; but the cloth is afterwards thrown into water acidulated with sulphuric acid, of sp. gr. between 1.010 and 1.020, when a minute disengagement of chlorine takes place throughout the substance of the cloth, and it immediately assumes a bleached appearance. The cloth is boiled a second time with caustic soda, and digested again in dilute chloride of lime and in dilute sulphuric acid, as before. The acid favours the bleaching action, and is required besides to remove the caustic alkali, a portion of which adheres pertinaciously to the cloth. The fibre of the cloth is not injured by dilute sulphuric acid, although digested in it for days, provided the cloth is not allowed to dry with the acid in it, or left above the surface of the liquor. But it is very necessary to wash well after the last *souring*, to get rid of every trace of acid, with which view the cloth may be passed through warm water as a precautionary measure.

Chlorine is had recourse to in disinfecting the wards of hospitals. Mr. Faraday, in fumigating the Millbank Penitentiary, found that a mixture of 1 part of common salt and 1 part of the binoxide of manganese, when acted upon by 2 parts of oil of vitriol previously mixed with 1 part of water (all by weight), and left till cold, produced the best results. Such a mixture, at 60°, in shallow pans of red earthenware, liberated its chlorine gradually but perfectly in four days. The salt and manganese were well mixed, and used in charges of $3\frac{1}{2}$ pounds of the mixture. The acid and water were mixed in a wooden tub, the water being put in first, and then about half the acid: after cooling, the other half was added. The proportions of water and acid are 9 measures of the former to 10 of the latter. (Magazine of Science, 1840, p. 264).

Chlorides.—Chlorine combines with all the metals, and in the same proportions as oxygen. With the exception of the chlorides of silver and lead, and subchlorides of copper and mercury, these compounds are soluble and sapid, and they possess in an eminent degree the saline character. Indeed, common salt, the chloride of sodium, has given its name to the class of salts, and chlorine is the type of salt-radicals or *halogenous* (salt-producing) bodies. Chlorides of metals belonging to different classes often combine together and form double chlorides; the chlorides of the potassium family, in particular, with some chlorides of the magnesian family, as with chloride of copper, with chloride of mercury, with both the chlorides of tin, and with perchlorides generally. A chloride and oxide of the same metal (excepting the potassium family) often combine together, forming *oxichlorides*, which are in general insoluble.

Chlorine is also absorbed by alkaline solutions, and combinations are formed which bleach and exhibit many of the properties of the free element. The chlorine in these compounds, and also in dry chloride of lime, formed by exposing hydrate of lime to chlorine gas, is now generally allowed to exist as hypochlorous acid. They are not permanent compounds, and the chlorine eventually acts upon the metallic oxide, so as to produce a chloride and a chlorate of the metal, as will be afterwards explained.

The following chlorides of the non-metallic elements will now be particularly described:—

Hydrochloric acid	$H\ Cl$	Chloride of boron	$B\ Cl_3$
Hypochlorous acid	$Cl\ O$	Chloride of silicon	$Si\ Cl_3$
Peroxide of chlorine	$Cl\ O_4$	Chloride of sulphur	$S_2\ Cl$
Chloric acid	$Cl\ O_5$	Bichloride of sulphur	$S\ Cl_2$
Hyperchloric acid	$Cl\ O_7$	Terchloride of phosphorus	$P\ Cl_3$
Chloride of nitrogen	$N\ Cl_3$	Pentachloride of phosphorus	$P\ Cl_5$
Chlorocarbonic acid	$CO.Cl$		

HYDROCHLORIC ACID.

Syn. Chlorhydric acid, Muriatic acid; Eq. 36.5 *or* 456.25; ClH; *density* 1269.5;

This acid is one of the most frequently-employed reagents in chemical operations, and has long been known under the names of spirit of salt, marine acid, and muriatic acid (from murias, sea-salt). It was first obtained by Priestley in its pure form of a gas in 1772.

Preparation.—Hydrochloric acid is always obtained by the action of oil of vitriol upon common salt. When the process is conducted on a small scale and in a glass retort, 3 parts of common salt, 5 oil of vitriol, and 5 water, may be taken. The oil of vitriol being mixed with two parts of the water in a thin flask, and cooled, is poured upon the salt contained in a capacious retort A (fig. 153). A flask B, con-

FIG. 153.

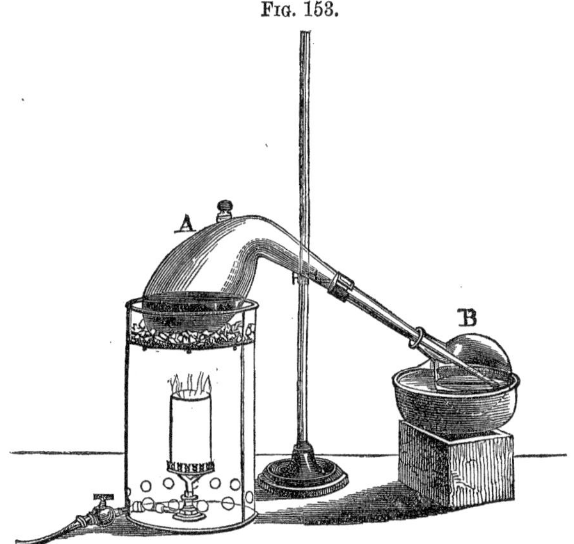

taining the remaining 3 parts of the water, is then adapted to the retort as a condenser. Upon applying heat to the retort, hydrochloric acid gas comes off, and is condensed in the receiver, affording an aqueous solution of the acid, of about sp. gr. 1.170, which contains 34 per cent. of dry acid; while bisulphate of soda remains in the retort. Supposing 2 equivalents of oil of vitriol and 1 of chloride of sodium to be employed, which the preceding proportions represent, then the rationale of the action is as follows:—

PROCESS FOR HYDROCHLORIC ACID.

Before decomposition.				After decomposition.	
58.5	Cloride of sodium	Chlorine	35.5	36.5	hydroc. acid
		Sodium	23		
49	Oil of vitriol	Hydrogen	1		
		Oxygen	8		
		Sulphu. acid	40	71	sulph. of soda
49	Oil of vitriol		49	49	sulph. of water
156.5			156.5	156.5	

Or in symbols: NaCl and $HO.SO_3$=HCl and $NaO.SO_3+HO.SO_3$.

The hydrochloric acid coming off easily and at a low temperature, when 2 eqs. of sulphuric acid are used, is obtained at once pure and free from sulphuric acid.

This process is more economically conducted on the large scale, as for nitric acid (fig. 116, page 261), in a cast-iron cylinder, about 5 feet in length and 2½ in diameter, laid upon its side, which has moveable ends, generally composed of a thin paving-stone cut into a circular disc and divided into two unequal segments. A charge of three or four hundred pounds of salt is introduced into the retort, and after the bottom is heated, sulphuric acid, as it is withdrawn from the leaden chambers, is added in a gradual manner by means of a long funnel, and in proportion not exceeding 1 equivalent for the chloride of sodium. In such circumstances, the lower part of the cylinder exposed to the sulphuric acid is not much acted upon, while the roof of the cylinder is protected from the hydrochloric acid fumes by a coating of fire-clay or thin bricks. The hydrochloric acid gas is conducted by a glass tube into a series of large jars of salt-glaze ware, connected with each other like Wolfe's bottles, and containing water, in which the acid condenses.

Properties. — Hydrochloric acid is obtained in the state of gas by boiling an ounce or two of the fuming aqueous solution in a small retort, or by pouring oil of vitriol upon a small quantity of salt in a retort, and is collected over mercury. It is an invisible gas, of a pungent acid odour, and produces white fumes, when allowed to escape, by condensing the moisture in the air. By a pressure of 40 atmospheres at 50°, it is condensed into a liquid of sp. gr. 1.27. It is quite irrespirable, but much less irritating than chlorine; it is not decomposed by heat alone, nor when heated in contact with charcoal. Hydrochloric acid extinguishes combustion, and is not made to unite with oxygen by heat; but when electric sparks are passed through a mixture of this gas and oxygen, decomposition takes place to a small extent, water being formed and chlorine liberated. It is composed by volume of one combining measure, or two volumes of each of its constituents, united without condensation; so that its combining measure is 4 volumes, and its theoretical density 1269.5. It may be formed directly by the union of its elements.

If a few drops of water or a fragment of ice be thrown up into a jar of hydrochloric acid over mercury, the gas is completely absorbed in a few seconds; or if a stout bottle filled with this gas be closed by the finger and opened under water, an instantaneous condensation of the gas takes place, water rushing into the bottle as into a vacuum. Dr. Thomson found that 1 cubic inch of water absorbs 418 cubic inches of gas at 69°, and becomes 1.34 cubic inch. He constructed the following table, from experiment, of the specific gravity of hydrochloric acid of determinate strengths (First Principles of Cemistry): —

HYDROCHLORIC ACID.

Atoms of Water to 1 of Acid.	Real Acid in 100 of the liquid.	Specific Gravity.	Atoms of Water to 1 of Acid.	Real Acid in 100 of the liquid.	Specific Gravity.
6	40.66	1.203	14	22.700	1.1060
7	37.00	1.179	15	21.512	1.1008
8	33.95	1.162	16	20.442	1.0960
9	31.35	1.149	17	19.474	1.0902
10	29.13	1.139	18	18.590	1.0860
11	27.21	1.1285	19	17.790	1.0820
12	25.52	1.1197	20	17.051	1.0780
13	24.03	1.1127			

To this may be added the following useful table, for which we are indebted to Mr. E. Davy: —

HYDROCHLORIC ACID.

Specific Gravity.	Quantity of Acid per cent.	Specific Gravity.	Quantity of Acid per cent.
1.21	42.43	1.10	20.20
1.20	40.80	1.09	18.18
1.19	38.38	1.08	16.16
1.18	36.36	1.07	14.14
1.17	34.34	1.06	12.12
1.16	32.32	1.05	10.10
1.15	30.30	1.04	8.08
1.14	28.28	1.03	6.00
1.13	26.26	1.02	4.04
1.12	24.24	1.01	2.02
1.11	22.22		

It thus appears that the strongest hydrochloric acid that can be easily formed contains six eqs. of water: this liquid allows acid to escape when evaporated in air, and comes, according to an observation of my own, to contain 12 eqs. of water to 1 of acid. Distilled in a retort, it was found, by Dr. Dalton, to lose more acid than water till it attained the specific gravity 1.094, when its boiling point attained a maximum of 230°, and the acid then distilled over unchanged. Dr. Clark finds by careful experiments that the acid, which is unalterable by distillation, contains 16.4 equivalents of water.

The concentrated acid is a colourless liquid, fuming strongly in air, highly acid, but less corrosive than sulphuric acid; not poisonous when diluted. It is decomposed by substances which yield oxygen readily, such as metallic peroxides and nitric acid, which cause an evolution of chlorine, by oxidating the hydrogen of the hydrochloric acid. A mixture of 1 measure of nitric and 2 measures of muriatic acid forms *aqua regia,* which dissolves the less oxidable metals, such as gold and platinum.

The hydrochloric acid of commerce has a yellow or straw colour, which is generally due to a little iron, but may be occasionally produced by organic matter, as it is sometimes destroyed by light. This acid is rarely free from sulphuric acid, the presence of which is detected by the appearance of a white precipitate of sulphate of baryta on the addition of chloride of barium to the hydrochloric acid diluted with 4 or 5 times its bulk of distilled water. Sulphurous acid is also occasionally present in commercial hydrochloric acid, and is indicated by the addition of a few crystals of protochloride of tin, which salt decomposes sulphurous acid, and occasions, after standing some time, a brown precipitate containing sulphur in combination with tin (Girardin). To purify hydrochloric acid, it may be diluted till its sp. gr. is about 1.1, for which the strongest acid requires an equal volume of water; and with the addition of a portion of chloride of barium, the acid should then be re-distilled. As the acid brings over enough of water to condense it, Liebig's condensing apparatus (fig. 30, page 73) can be used in this distillation. The pure acid thus obtained is strong enough for most purposes, and has the advantage of not fuming in the air. Hydrochloric acid, like chlorine and the soluble chlorides, gives with nitrate of silver a white curdy precipitate, the chloride of silver, soluble in ammonia, but not dissolved by hot or cold nitric acid.

Hydrochloric acid belongs to the class of hydrogen acids or hydracids. On neutralizing this acid with soda or any other basic oxide, no hydrochlorate of soda is formed; but the hydrogen of the acid with the oxygen of the soda forming water, the chlorine and sodium combine, and produce a metallic chloride. Zinc, and the other metals which dissolve in dilute sulphuric acid, with evolution of hydrogen, dissolve with equal facility in this acid, with the same evolution of hydrogen, and a chloride of the metal is then formed.

COMPOUNDS OF CHLORINE AND OXYGEN.

Chlorine and oxygen gases in a free state exhibit no disposition to combine with each other in any circumstances, but this is not inconsistent with their forming a series of compounds, as nitrogen and oxygen, which exhibit a similar indifference to each other, also do. The oxides of chlorine are five in number, namely: —

Hypochlorous acid	ClO
Chlorous acid	ClO_3
Peroxide of chlorine, or Hypochloric acid	ClO_4
Chloric acid	ClO_5
Perchloric acid	ClO_7

Hypochlorous and chloric acids are always primarily formed by a reaction occurring between chlorine and two different classes of metallic oxides; and the chlorous and perchloric acids, again, are derived from the decomposition of chloric acid.

HYPOCLHLOROUS ACID.

Eq. 43.5; ClO; *density of vapour* 2977;

The discovery of this compound in a separate state was made by M. Balard in 1834 (Annal. de Ch. et de Ph. lvii. 225; or Taylor's Scientific Memoirs, i. 269). It was obtained by acting with chlorine upon the red oxide of mercury. If to a two-pound bottle of chlorine gas 300 grains of red oxide of mercury in fine powder be added, with 1½ ounce of water, the chlorine will be found to be rapidly absorbed on agitation. One portion of the chlorine unites with the oxygen of the metallic oxide, and becomes hypochlorous acid, which is dissolved by the water; while another portion forms a chloride with the metal, which chloride unites with a portion of undecomposed oxide, and forms an insoluble oxichloride. The liquid may be poured off and allowed to settle: it is a solution of hypochlorous acid, with generally a little chloride of mercury. This reaction is expressed in the following diagram: —

FORMATION OF HYPOCHLOROUS ACID.

Before decomposition.		After decomposition.	
Chlorine	Chlorine	Hypochlorous acid	
Oxide of mercury	Oxygen		
	Mercury		
Chlorine	Chlorine	Chloride of mercury	combined.
Oxide of mercury	Oxide of mercury	Oxide of mercury	

Or in symbols:

$$2Cl \text{ and } 2HgO = ClO \text{ and } HgCl.HgO.$$

But the oxichloride formed seems not always to contain the same proportion of oxide. The proportion of hypochlorous acid in the liquid may be increased by introducing the same solution into a second bottle of chlorine, with an additional quantity of red oxide of mercury. The oxide of zinc and black oxide of copper, diffused through water, and exposed to chlorine, give rise to a similar formation of hypochlorous acid.

If red oxide of mercury in fine powder be added to chlorine-water so long as the oxide is dissolved, a solution of hypochlorous acid and chloride of mercury is formed, without any insoluble compound: $2Cl$ and $HgO = ClO$ and $HgCl$ (Gay-Lussac).

On the other hand, hypochlorous acid, free from water, and in the liquid state, may be obtained by passing dry chlorine gas in a gradual manner over red oxide of mercury in a glass tube *a b* (fig. 154); care being taken to prevent elevation of temperature, by surrounding the tube with fragments of ice, or immersing it in cold

FIG. 154.

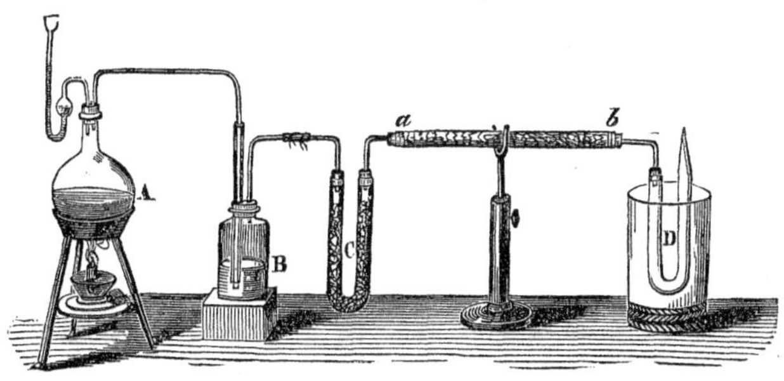

water, as otherwise nothing but oxygen will be disengaged. The chlorine is evolved from the usual materials in the flask A, passed through water in the wash-bottle B to arrest any hydrochloric acid, and afterwards dried over chloride of calcium tube C. Chloride of mercury is formed as in the other processes, and a yellow gas, which is liquefied in the bent tube D, kept cold by a freezing mixture of ice and salt. The oxide of mercury which answers best for this experiment is that precipitated from chloride or nitrate of mercury by potassa, washed and dried at a temperature of about 572° (300° C.) — Regnault's Traité.

Hypochlorous acid is a liquid of an orange-yellow colour, which boils at about 68° (20° C.) Its vapour is of a pale yellow colour, very similar to chlorine. It is composed of 2 volumes of chlorine and 1 volume of oxygen, condensed into 2 volumes, which gives a theoretical density of 2992, while 2977 has been obtained by experiment. It is resolved by a slight elevation of temperature into its constituent gases; a property which allows it to be analyzed, by determining the proportions of the mixed chlorine and oxygen gases. Water dissolves about 200 volumes of this gas, and assumes a fine yellow colour.

Hypochlorous acid is also formed when chlorine is absorbed by weak solutions of alkalies and by hydrate of lime, and, as the acid of the *bleaching chlorides,* possesses considerable interest. It displaces the carbonic acid of alkaline carbonates, but has not much analogy to other acids. Its taste is extremely strong and acrid, but not sour, and its odour penetrating and different from, although somewhat similar to, chlorine. It attacks the epidermis like nitric acid, and is exceedingly corrosive. It bleaches instantly, like chlorine, and is a powerful oxidizing agent. A concentrated solution of it is exceedingly unstable, small bubbles of chlorine gas being spontaneously evolved and chloric acid formed. This decomposition is promoted by the presence of angular bodies, such as pounded glass, and also by heat and light.

Of the elementary bodies, hydrogen has no action upon hypochlorous acid. Sulphur, selenium, phosphorus, and arsenic, act upon it with great energy, and are all of them raised to their highest degree of oxidation, with the evolution of chlorine gas; selenium even being converted into selenic acid, although it is converted into selenious acid only by the action of nitric acid. Iodine is also converted into iodic acid. Iron filings decompose it immediately, and chlorine gas comes off. Copper and mercury combine with both elements of the acid, and form oxichlorides. Many other metals are not acted upon by it, unless another acid be present, such as zinc, tin, antimony, and lead. Silver has a different action upon hypochlorous acid from that of most bodies, combining with its chlorine, and causing an evolution of oxygen gas. Hydrochloric and hypochlorous acid mutually decompose each other, water being formed, and chlorine liberated; if the liquids are both cooled to a very low degree, before mixture, the chlorine is not disengaged, but combines with water to form the hydrate of chlorine, and causes the liquid to become a solid mass. The presence of soluble chlorides is equally incompatible with the existence of hypochlorous acid.

Hypochlorites. — The direct combination of hypochlorous acid with powerful bases is accompanied by heat, which is apt to convert the hypochlorite into a mixture of chlorate and chloride; but by adding the acid in a gradual manner to the alkaline solution, hypochlorites of potassa, soda, lime, baryta, and strontia, may be formed, and may even be obtained in a solid state by evaporation in vacuo, if a considerable excess of alkali be present, which appears to give a certain degree of stability to these salts. They bleach powerfully, and their odour and colour are identically the same as the corresponding decolourizing compounds of chlorine, formed by exposing solutions of the highly basic oxides named to chlorine gas, from which it is impossible to distinguish them by their physical properties. When chlorine, then, is absorbed by a weak solution of potassa, without heat being applied, the hypochlorite of potassa is formed, with chloride of potassium, both of which remain in solution: —

$$2Cl \text{ and } 2KO = KO.ClO \text{ and } KCl.$$

Fig. 155.

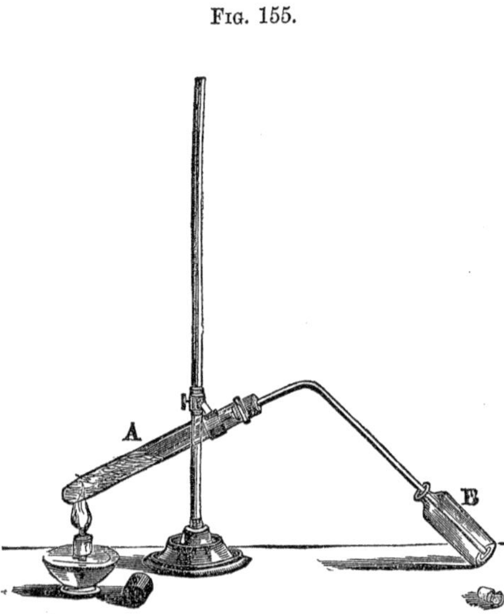

The hypochlorites are salts of a very changeable constitution; a slight increase of temperature, the influence of solar light, even diffused light, converts them into chloride and chlorate.

The *euchlorine* gas of Davy, to which he assigned the composition of hypochlorous acid, has been found to be a mixture of chlorine gas and chlorochloric acid. That mixture is obtained by the action of hydrochloric acid of sp. gr. 1.1 upon chlorate of potassa, aided by a gentle heat. It has a very yellow colour (euchlorine), and explodes feebly when a hot wire is introduced into it, becoming nearly colourless when the chlorochloric acid is decomposed. A tube retort A (fig. 155), is employed for the evolution of this gas, and it is collected in the phial B by displacement.

CHLORIC ACID.

Eq. 75.5 *or* 943.75; $HO.ClO_5$.

When a stream of chlorine gas is transmitted through a strong solution of caustic potassa, the gas is absorbed, and a solution is formed which bleaches at first, but loses that property without any escape of gas, and becomes a mixture of chloride of potassium and chlorate of potassa; the latter of which, being the least soluble, separates in shining tabular crystals. Five equivalents of potassa, (the oxide of potassium) are decomposed by 6 of chlorine, 5 of which unite with the potassium, and form 5 equivalents of chloride of potassium, while the 5 of oxygen form chloric acid with the remaining equivalent of chlorine, as stated in the following diagram, in which the numbers express equivalents:—

ACTION OF CHLORINE UPON POTASSA.

Before decomposition.		After decomposition.	
5 Chlorine	5 Chlorine	5 Chloride of Potassium.	
5 Potassa	5 Potassium		
	5 Oxygen		
Chlorine	Chlorine	Chloric acid	Chlorate of potassa.
Potassa	Potassa	Potassa	

Or in symbols: $6Cl$ and $6KO = KO.ClO_5$ and $5KCl$. Such is the nature of the action of chlorine upon the soluble and highly alkaline metallic oxides, when their solutions are concentrated, or heat applied.

The chlorate of baryta may be formed by transmitting chlorine through caustic baryta in the same manner; and from a solution of the pure chlorate of baryta, chloric acid may be obtained by the cautious addition of sulphuric acid, so long as it occasions a precipitate of sulphate of baryta. The solution may be evaporated by a very gentle heat till it becomes a syrupy liquid, which has no odour, but a very acid taste, is decomposed above 100°, and when distilled at a still higher temperature gives water, then a mixture of chlorine and oxygen gases, and hyperchloric acid; which last acid may be prepared in this way without difficulty. Chloric acid is not isolable, being incapable of existing except in combination with water or a fixed base. This acid first reddens litmus paper, but after a time the colour is bleached, and if the acid has been highly concentrated, the paper often takes fire. It dissolves zinc and iron with disengagement of hydrogen. Chloric acid is decomposed by hydrochloric acid, with escape of chlorine, and by most combustible bodies and acids of the lower degrees of oxidation, such as sulphurous and phosphorous acids, which oxidate themselves at its expense.

This acid, when free or in combination, may be recognized by several properties. It is not precipitated by chloride of barium or nitrate of silver, and its salts have no bleaching power; sulphuric acid causes the disengagement from it of a yellow gas, having a peculiar odour, which bleaches strongly; and its salts, when heated to redness, afford oxygen, and deflagrate with combustibles.

Chlorates.—This class of salts is remarkable for a general solubility, like the nitrates. Those of them which are fusible detonate with extreme violence with combustibles. The chlorate of potassa, of which the preparation and properties will be described under the salts of potassa, has become a familiar chemical product, being largely consumed in the manufacture of deflagrating mixtures. The chlorates were at one time termed *hyperoxymuriates*, and their acid, the existence of which was originally observed by Mr. Chenevix, was first obtained in a separate state by Gay-Lussac.

The composition of chloric acid is ascertained by decomposing a known quantity of chlorate of potassa by heat, and ascertaining the loss of weight which is due to the expulsion of 6 eqs. of oxygen. The chloride of potassium which forms the fixed residue is dissolved, and the chlorine precipitated by nitrate of silver. The chlorine is thus obtained in the form of chloride of silver, of which the composition is known. The relation between the equivalents of chlorine and oxygen is also established by the analysis of the chlorate of potassa (Note, p. 104).

HYPERCHLORIC ACID.

Eq. 91.5 *or* 1143.75; $HO.ClO_7$.

This acid, which is also named perchloric and oxichloric acid, is obtained from chlorate of potassa in different ways. At that particular point of the decomposition of chlorate of potassa by heat, when the evolution of oxygen is about to become very violent, the fused salt is in a pasty state, and contains, as was first observed by

Serullas, a considerable quantity of perchlorate, the oxygen extricated from one portion of chlorate being retained by another portion of the same salt. This salt is rubbed to powder, and dissolved in boiling water, from which the perchlorate is first deposited, on cooling, owing to its sparing solubility. It is stated by M. Millon, that from 50 to 53 per cent. of perchlorate may be obtained by stopping when 9½ litres of gas (580 c. i.) are collected from 100 grammes (1543 grains) of chlorate, instead of 13 litres. (Annal. de Ch. et Ph., 3e sér. vii. 335.) The same salt may also be prepared by throwing chlorate of potassa, in fine powder, and well dried, into oil of vitriol gently heated in an open basin, by a few grains at a time, when the liberated chloric acid resolves itself into peroxide of chlorine and hyperchloric acid, the former coming off as a yellow gas; thus:—

RESOLUTION OF CHLORIC ACID INTO PEROXIDE OF CHLORINE AND HYPERCHLORIC ACID.

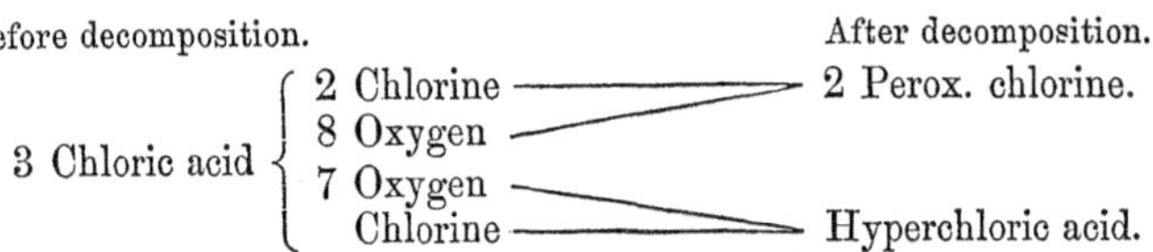

Of the 3 equivalents of potassa, previously in combination with the chloric acid, one remains with hyperchloric acid as hyperchlorate of potassa, and the other two are converted into bisulphate of potassa. The whole reaction between the acid and salt may, therefore, be thus expressed:—

$$3(KO.ClO_5) \text{ and } 4(HO.SO_3) = 2ClO_4 \text{ and } KO.ClO_7 \text{ and } (2HO.SO_3 + KO.SO_3) \text{ and } 2HO.$$

In conducting this operation, the greatest caution is necessary, owing to the explosive property of peroxide of chlorine; for if the order of mixing the substances be reversed, and the acid poured upon the chlorate, or if too much chlorate be added at a time to the acid, a most violent and dangerous detonation may occur. But this reaction is chiefly interesting as affording peroxide of chlorine; for hyperchlorate of potassa may be obtained from chlorate by the action of nitric acid, lately observed by Professor Penny, without danger or inconvenience. The chlorate is tranquilly decomposed in nitric acid gently heated upon it, the chlorine and oxygen at 3 equivalents of peroxide of chlorine being evolved in a state of mixture and not of combination: the saline residue consists of 3 equivalents of nitrate and 1 of perchlorate of potassa, which may be separated by dissolving them in the smallest adequate quantity of boiling water. On cooling, the perchlorate separates in small shining crystals, which may be dissolved a second time to obtain them perfectly pure.

Perchloric acid may be prepared from the last salt by boiling it with an excess of fluosilicic acid, which forms, with potassa, a salt nearly insoluble. After cooling, a clear liquid is decanted and evaporated by the water-bath. To eliminate a small excess of hydrofluoric acid, a little silica in fine powder is added to the liquid, which at a certain degree of concentration carries off the former as fluosilicic acid. After being still further concentrated, the acid liquid may be distilled in a retort by a sand-bath heat. A very dilute acid comes over first, but the temperature of ebullition rises till it attains 392°, after which the receiver should be changed, because what then passes over is a concentrated acid of sp. gr. 1.65. This acid is a colourless liquid which fumes slightly in the air. It may be still farther concentrated by distilling it with 4 or 5 times its weight of strong sulphuric acid, when the greater part of it is decomposed into chlorine and oxygen; but a portion condenses in a mass of small crystals, and also in long four-sided prismatic needles terminated by dihedral summits, which were found by Serullas to be two different hydrates of the acid, the last containing least water and being most volatile. The crystals and the concentrated solution of the acid have a great affinity for water; the acid itself (ClO_7) appears not to be isolable.

Perchloric acid is much the most stable of the oxides of chlorine; it does not bleach, is not altered by the presence of sulphuric acid, and is not decomposed by sulphurous acid or by hydrosulphuric acid. It dissolves zinc and iron with effervescence, and, in point of affinity, is one of the most powerful acids. Perchloric acid is recognized by producing, with potassa, a salt of the same sparing solubility as bitartrate of potassa. It is an interesting acid from its composition, and as being the most accessible of the small class containing periodic and permanganic acids, to which it belongs. The alkaline perchlorates emit much oxygen when heated, and leave metallic chlorides; they do not deflagrate so powerfully with combustibles as the chlorates.

CHLOROUS ACID.

Eq. 59.5 *or* 743.75; ClO_3; *density* 2.646.

This is a gaseous compound of chlorine and oxygen, which is not liquefied at 5° (—15° C.), and is therefore remarkable for its fixity. It was discovered and studied by M. Millon (Annal. de Ch. et Ph., 3 sér. vii.) Chlorous acid is formed by the deoxidation of chloric acid in various circumstances. It is readily obtained from a mixture of three parts of arsenious acid and four of chlorate of potassa, pulverized together, and made into a thin paste with water; twelve parts of ordinary nitric acid diluted with four of water being added, the whole is introduced into a flask, which is filled to the neck with the mixture, and heated cautiously by a water-bath.

Chlorous acid is a gas of a greenish-yellow colour, of which water dissolves five or six times its volume, assuming a golden-yellow tint of considerable intensity. It bleaches litmus and indigo, but does not attack gold, platinum, nor antimony. It is decomposed by heat, in general at 134.6° (57° C.), into perchloric acid, chlorine, and oxygen: $3ClO_3 = ClO_7$ and 2O and 2Cl. Chlorous acid combines with bases, and forms crystallizable salts; the affinity of this and some other anhydrous acids is gradually exerted, and requires time for its action. On pouring a solution of chlorite of potassa into a solution of nitrate of lead, a yellowish-white precipitate of chlorite of lead is obtained, $PbO.ClO_3$, which is easily subjected to analysis by transforming it into sulphate by means of sulphuric acid; or, if the chlorite of lead be fused in a crucible with carbonate of soda, the whole chlorine of the chlorous acid is obtained in the form of chloride of potassium, and may be precipitated from an acid solution by nitrate of silver, and estimated as chloride of silver.

According to M. Millon, the gas which forms when chlorate of potassa is treated with hydrochloric acid (euchlorine), ought to be considered a compound of chloric and chlorous acid, $2ClO_5.ClO_3$. It is named *chlorochloric acid.* Another double acid, which Millon has named *chloroperchloric acid,* is formed when humid chlorous acid is exposed to light, and condenses as a red liquid, $2ClO_7.ClO_3$.

PEROXIDE OF CHLORINE.

Hypochloric acid; eq. 67.5 *or* 843.75: ClO_4.

This substance cannot be obtained in a state of purity without considerable danger. Gay-Lussac recommends, in preparing it, to mix chlorate of potassa in the state of a paste with sulphuric acid previously diluted with half its weight of water and cooled, and to distil the mixture in a small retort by a water-bath. It comes off as a gas, of a yellow colour considerably deeper than chlorine, which cannot be collected over mercury, as it is instantly decomposed by that metal, nor over water, which dissolves it in large quantity. It is composed of 2 volumes of chlorine with 4 volumes of oxygen, condensed into 4 volumes, which gives it a density of 2337.5. This gas is decomposed gradually by light, but between 200° and 212° its elements separate in an instantaneous manner, with the disengagement of light and a violent explosion, which breaks the vessels. Water dissolves about 20 times its volume of this gas: the gas itself is liquefied by cold, and forms a red liquid, which boils at 68° (20° C.) It bleaches damp litmus paper, without first reddening it, and is absorbed by alkaline solutions with the formation of a mixture of a chlorate and chlorite. This compound, then, resembles peroxide of nitrogen, NO_4, and is not a peculiar acid, but may be represented as a compound of chlorous and chloric acids: $2ClO_4 = ClO_3 + ClO_5$.

Peroxide of chlorine has a violent action upon combustibles, kindling phosphorus, sulphur, sugar, and other combustible substances in contact with which it is evolved. Its action upon phosphorus may be shown by throwing a drachm or two of crystallized chlorate of potassa into a deep foot-glass (fig. 156) filled with cold water, to the bottom of which the salt falls without any loss by solution. Oil of vitriol is then conducted to the salt, in a small stream, from a tube funnel, the lower end of which has been drawn out into a jet with a minute opening. A gas of a lively yellow colour is evolved with slight concussions, and immediately dissolved by the water, to which it imparts the same colour. If, while this is occurring, a piece of phosphorus be thrown into the glass, it is ignited by every bubble of gas evolved, and a brilliant combustion is produced under the water, forming a beautiful experiment wholly without danger. If a few grains of chlorate of potassa in fine powder and loaf-sugar be mixed upon paper by the fingers, (rubbing these substances together in a mortar may be attended with a dangerous explosion), and a single drop of sulphuric acid be allowed to fall from a glass rod upon the mixture, an instantaneous deflagration takes place, occasioned by the evolution of the yellow gas, which ignites the mixture. Captain Manby used to fire in this manner the small piece of ordnance, which he proposed, as a life-preserver, to throw a rope over a stranded vessel from the shore; and the same mixture was afterwards employed, with sulphuric acid, in various forms of the instantaneous light-match, all of which, however, are now superseded by other mixtures ignited by friction without sulphuric acid.

Fig. 156.

CHLORINE AND BINOXIDE OF NITROGEN.

Mr. E. Davy appears first to have obtained a gaseous compound of chlorine and binoxide of nitrogen in 1830, and a combination of the same constituents was distilled from *aqua regia* and liquefied by M. Baudrimont in 1843. It is only lately, however, that the nature of the mutual action of nitric and hydrochloric acids has been fully explained by the investigations of M. Gay-Lussac on aqua regia. (Ann. de Ch. et Ph., 3me sér. xxiii. 203; or, Chemical Gazette, 1848, p. 269).

When nitric and hydrochloric acids are mixed, a reaction soon commences if the acids are concentrated; the liquid becomes of a red colour, and effervescence takes place, from the escape of chlorine and a chloro-nitric vapour. On passing this gaseous mixture through a U tube, the angle of which is immersed in a freezing mixture of ice and salt, the chloro-nitric compound condenses as a dark-coloured liquid, and is thus separated from the free chlorine which accompanied it.

Chloro-nitric acid, NO_2Cl_2.—This forms the principal part of the chloro-nitric vapour: it may be represented as a peroxide of nitrogen in which two equivalents of oxygen are replaced by two equivalents of chlorine. A third equivalent of chlorine, due to the third equivalent of oxygen yielded by the nitric acid, is disengaged as gas, and is the agent by which aqua regia dissolves gold, platinum, and other metals having a weak affinity for oxygen, converting them into chlorides: the chloro-nitric acid takes no part in the action. This compound is also formed by the mixture of the two gases in equal volumes, which assume a brilliant orange colour, and suffer a condensation amounting to exactly one-third of their original volume. The theoretical density of this vapour is 1740.2.

Chloro-nitrous acid, NO_2Cl.—This second compound, which corresponds with nitrous acid, NO_3, always appears simultaneously with the other in variable proportions. It is a vaporous liquid of similar properties, of which the vapour density is inferred to be 2259.4. The vapours of both compounds, when conducted into water, are instantly decomposed into hydrochloric acid and peroxide of nitrogen or nitrous acid—a decomposition which affords the means of determining the proportion of chlorine which they contain. The chloro-nitric compounds are also decomposed by mercury, the chlorine combining with the metal and leaving pure binoxide of nitrogen. The solution of the vapours in water decolorizes a solution of permanganate of potassa, owing to the peroxide of nitrogen it contains, but does not bleach indigo because it contains no free chlorine.

CHLORIDE OF NITROGEN.

This is one of the most formidable of explosive compounds, and great caution is necessary in its preparation to avoid accidents. Four ounces of sal ammoniac (which must not smell of animal matter or of nitrate of ammonia), are dissolved in a small quantity of boiling water, filtered, and made up to 3 pounds with distilled water; a two-pound bottle of chlorine is inverted in a basin containing this solution at 80°, being supported by the ring of a retort stand, with its mouth over a small leaden saucer. The chlorine gas is absorbed, and upon the surface of the liquid, which rises into the bottle, an oily substance condenses, which, when it accumulates, precipitates in large drops, and is received in the leaden saucer. During the whole operation, the bottle must not be approached, unless the face is protected by a sheet of wire gauze, and the hands by thick woollen gloves; agitation of the bottle, to make the suspended drop fall, is a common cause of explosion. The leaden saucer, when it contains the chlorine, may be withdrawn from under the bottle, without disturbing the latter, and then no harm can result from the explosion, if it does not occur in contact with glass.

M. Balard finds that this compound may also be produced by suspending a mass of sulphate of ammonia in a strong solution of hypochlorous acid.

The chloride of nitrogen is a volatile oleaginous liquid of a deep yellow colour, and sp. gr. 1.653, of which the vapour is irritating like chlorine, and attacks the eyes. It may be distilled at 160°, but effervesces strongly at 200°, and explodes between 205° and 212°, producing a very loud detonation, and shattering to pieces glass or cast-iron, but producing merely an indentation in a leaden cup. It is resolved into chlorine and nitrogen gases, the instantaneous production of which with heat and light, is the cause of the violence of the explosion. The chloride of nitrogen is decomposed by most organic matters containing hydrogen; and may be safely exploded by touching it with the point of a cane-rod, which has been previously dipped in oil of turpentine.

This compound is represented by NCl_4, but the properties of this compound render its accurate analysis almost impossible, and the correctness of the formula usually assigned to it is very doubtful. M. Millon has shown that it may contain hydrogen, and is possibly a nitride of chlorine with ammonia, Cl_3N+2H_3N. He formed from it corresponding compounds, containing bromine, iodine, and cyanogen, by double decomposition; a bromide, iodide, or cyanide of potassium being introduced into the chloride of nitrogen for that purpose. (Annales de Chim. et de Phys. lxix. 75.)

CHLORIDES OF CARBON.

Sesquichloride of carbon, C_4Cl_6.—The compounds of these elements are not formed directly, but were produced by Mr. Faraday by the action of chlorine upon a certain compound of carbon and hydrogen; the circumstances of their formation were explained with singular felicity by M. Regnault. Chlorine and olefiant gas C_4H_4 combine together in equal volumes, and condense as Dutch liquid (page 286).

Chemists are now generally agreed that the rational formula of this liquid is not C_4H_4+2Cl, but that its elements are thus arranged :—

Dutch liquid........................$C_4H_3.Cl+HCl$.

It is considered a combination of hydrochloric acid HCl, with the chloride of acetyl $C_4H_3.Cl$. When a stream of chlorine gas is transmitted through Dutch liquid, a second eq. of hydrogen is carried off, as hydrochloric acid, and 1 eq. of chlorine left in its place; thus Dutch liquid, $C_4H_3Cl+HCl$ becomes—

$$C_4H_2Cl_2+HCl.$$

This second product, which is a liquid, being submitted to the action of a stream of chlorine, gives rise to a third liquid product, in which the hydrochloric acid of the last formula disappears, and the remaining portion assumes 2 additional eqs. of chlorine, forming—

$$C_4H_2Cl_4.$$

This third liquid is changed by the prolonged action of chlorine into the sesquichloride of carbon, but to hasten the action it is convenient to conduct the operation in the light of the sun; its two remaining eqs. of hydrogen being carried off in the form of hydrochloric acid, and 2 eqs. of chlorine left in their place, which gives the formula

Sesquichloride of carbon..............C_4Cl_6, or $C_4Cl_4+Cl_2$.

This view of the derivation and constitution of the sesquichloride of carbon is confirmed by the density of its vapour, which Regnault found by experiment to be 8157. It should from its formula contain

8 volumes carbon vapour................................	3371
12 volumes chlorine..	29284
	32655

If these form a combining measure of 4 volumes, the most usual of all combining measures, the weight of 1 volume, or density of the vapour, is 8164, which almost coincides with the experimental result.[1]

The sesquichloride of carbon is a volatile crystalline solid, having an aromatic odour resembling that of camphor, fusible at 320° and boiling at 360° (Faraday), of sp. gr. 2, soluble in alcohol, ether, and oils. It was prepared by Mr. Faraday by exposing Dutch liquid to sunlight in an atmosphere of chlorine, which was several times renewed as the chlorine was absorbed.

Protochloride of carbon, C_4Cl_4.—This compound was prepared by Faraday by passing the vapour of the sesquichloride through a glass tube filled with fragments of glass, and heated to redness. A great quantity of chlorine becomes free, and a colourless liquid is obtained, which when purified from sesquichloride of carbon and chlorine as much as possible, boils at 248° (Regnault), has a sp. gr. of 1.5526, and in its chemical relations is very analogous to the sesquichloride of carbon. The density of the vapour of the protochloride decides the nature of its constitution. It was found by Regnault to be 5820, which corresponds to the composition by volume :—

8 volumes carbon vapour...........................	3371
8 volumes chlorine..................................	19523
	22894

$$\text{Density} = \frac{22894}{4} = 5724.$$

[1] Regnault, De l'Action du Chlore sur la liqueur des Hollandais et sur le Chlorure d'Aldéhydène. Ann. de Ch. et de Ph. t. 69, p. 151. Idem, Sur les Chlorures de Carbon, ib. t. 70, p. 104.

It must, therefore, contain 4 eqs. of carbon and 4 of chlorine, and its formula be C_4Cl_4, or it represents olefiant gas C_4H_4 with its whole hydrogen replaced by chlorine. It is interesting to observe how a body retains, after so many mutations, such distinct traces of its origin. From its analysis it might be a compound of single equivalents, C Cl, of the simplest nature, and so it was considered when named protochloride of carbon.

Subchloride of carbon, C_4Cl_2.—Another compound of this class exists, of which a specimen produced accidentally was examined by Messrs. Phillips and Faraday. Regnault has formed it by making the preceding liquid compound pass several times through a tube at a bright red heat. It condenses in the coldest parts of the tube in very fine silky crystals, which may be taken up by ether, and obtained perfectly pure by a second sublimation.

Perchloride of carbon, C_2Cl_4, was obtained by Regnault from the prolonged action of chlorine on hydrochloric ether, wood-spirit, or chloroform, and by M. Kolbe by passing chlorine gas impregnated with the vapour of bisulphide of carbon through a porcelain tube heated to redness. It is a colourless liquid, of density 1.6, boiling at 172° (78° C.) By passing the vapour of this chloride through a tube heated to dull redness, Regnault obtained another chloride of carbon, isomeric with Faraday's sesquichloride, but of which the vapour density was 4.082. Kolbe formed a crystallizable compound of perchloride of carbon and sulphurous acid, which has the formula $2(SO_2)+C_2Cl_4$.

Another chloride of carbon, of the formula $C_{20}Cl_8$, was obtained by M. Laurent, by the action of chlorine upon naphthaline, $C_{20}H_8$, in the form of a crystalline solid, soluble in boiling petroleum.

Chloroxicarbonic gas, CO.Cl.—This gas is formed by exposing equal measures of chlorine and carbonic oxide to sunshine, when rapid but silent combination ensues, and they contract to one half their volume (page 275).

Chloride of boron, B Cl_3.—A gaseous compound of these elements was obtained by Berzelius, by transmitting chlorine over boron heated in a glass tube, and by Dumas by transmitting the same gas over a mixture of boracic acid and carbon ignited in a porcelain tube placed across a furnace (fig. 157). Its density was found to be 4079 by Dumas, and it is considered a terchloride.

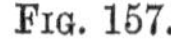

Fig. 157.

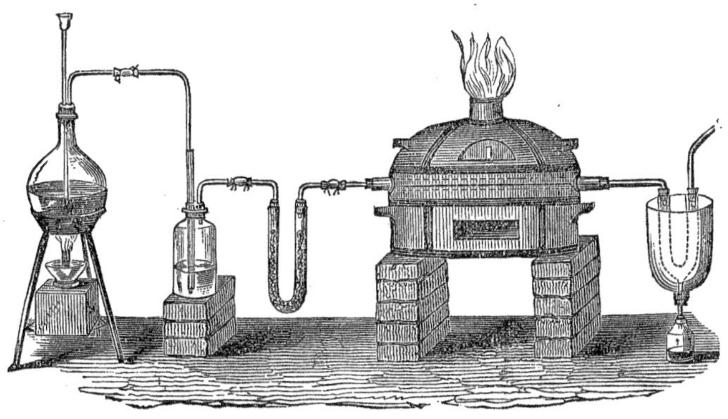

Chloride of silicon; 127.85 *or* 1598.12; $SiCl_3$.—When silicon is heated in a stream of chlorine gas it takes fire, and this compound is formed. It is also obtained in quantity by a process analogous to that of Dumas for the chloride of boron, which it greatly resembles. Silicic acid is not decomposed when heated with carbon, but if chlorine gas be present, then the simultaneous action of the latter element upon the silicon favours the action of the carbon on the oxygen, and carbonic oxide with chloride of silicon results. Precipitated silica (page 290), which is in a highly divided state, is mixed with an equal weight of lamp-black, and made into a stiff paste with a little oil; this is divided into balls, which are rolled in charcoal powder, and then exposed to a strong red heat in a covered crucible. These ignited balls form the mixture of silica and charcoal which is introduced into the porcelain tube (fig. 157), and heated strongly by a charcoal furnace, while chlorine gas, washed by water and dried in a chloride of calcium tube, is carried through the porcelain tube. The chloride of silicon is condensed in a U tube placed in an inverted bell-jar, with an opening at the lower part; a short straight tube is cemented to the lower part of the U tube, and, passing through the tubulure of the jar, terminates in a small, thoroughly dry bottle, where the liquefied chloride of silicon is collected. (Regnault's Traité).

The chloride of silicon is a colourless, highly mobile liquid, of density 1.52; which boils at 138° (59° C.), and fumes in the air. It is instantly decomposed by contact with water, and resolved into hydrochloric acid and silica:—

$$SiCl_3 \text{ and } 3HO = SiO_3 \text{ and } 3HCl.$$

This property affords the means of analyzing the chloride of silicon, as the chlorine of the hydrochloric acid formed may be precipitated by nitrate of silver, and its amount determined. The proportion of oxygen in silicic acid may also be deduced from the same experiment, as the oxygen must necessarily be equivalent to the chlorine in the chloride.

CHLORINE AND SULPHUR.

Chlorine and sulphur appear to combine in several different proportions, some of these compounds being formed only in combination with certain other chlorides. But two compounds of these elements have been obtained in a separate state.

Subchloride of sulphur; 67.5 *or* 843.75; S_2Cl.—This compound was first obtained by Dr. T. Thomson in 1804. To prepare it, a few ounces of flowers of sulphur are introduced into the tubulated retort D (fig. 158), and fused by a lamp

FIG. 158.

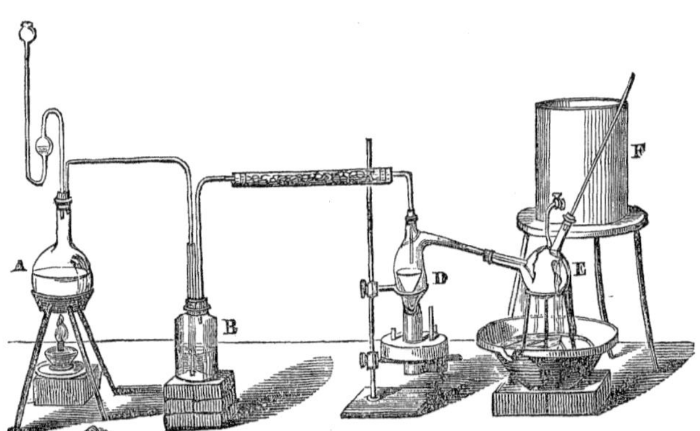

below. Chlorine gas is evolved from hydrochloric acid and binoxide of manganese in the flask A, transmitted through the wash-bottle B containing water, and afterwards dried by chloride of calcium, before the gas reaches the sulphur in D. The chlorine is rapidly absorbed, and a yellowish red dense liquid distils over, and is condensed in the flask with two openings E, which is kept cool by a stream of water from F. It contains an excess of sulphur in solution, but is obtained pure by redistilling the liquid at a moderate temperature (Rose, Annal. de Ch. et de Ph. l. 92). The subchloride of sulphur boils at about 280°, and has a disagreeable odour, somewhat resembling that of sea-weed, but much stronger. Its density in the liquid state is 1.687; the density of its vapour has been found 4668 by experiment. This compound is capable of dissolving a large quantity of sulphur, which may be obtained in crystals from a solution saturated at a high temperature. It is decomposed by water, and hydrochloric acid with acids of sulphur formed.

In one of the processes for *vulcanizing* caoutchouc, the subchloride of sulphur is employed. This compound is dissolved in 50 times its bulk of well rectified coal naphtha, and the articles of caoutchouc immersed in the fluid for one minute, then taken out and dried without heat. The caoutchouc thus acquires a small portion of sulphur, with which it appears to combine, and is improved greatly in elasticity and strength.

Protochloride of sulphur, 51.5 *or* 643.75; SCl. — If chlorine be passed through the former compound, the gas is absorbed in large quantity, and a liquid compound of a deep red colour formed, which contains twice as much chlorine. The new compound dissolves an excess of chlorine, which must be expelled by ebullition. When pure, this chloride boils at 147°.2 (64° C.) Its density in the liquid form is 1.620, and in the state of vapour 3549. It is decomposed like the preceding compound when agitated with water, all its chlorine becoming hydrochloric acid, the quantity of which may be determined by the usual means. Polythionic acids are also formed, with a deposit of sulphur. This compound, of which the formula is SCl, may correspond with hypochlorous acid ClO, or with hyposulphurous acid; but the subchloride of sulphur, S_2Cl, has no analogue among the known compounds of oxygen and chlorine, or of oxygen and sulphur.

When chlorine is passed over the bisulphide of tin, the gas is absorbed, the sulphide fuses, and a compound is formed in yellow crystals, which consists of $SnCl_2 + SCl_2$. The sulphur of the sulphide of titanium and of the sulphides of antimony and arsenic is converted by chlorine in the same manner into bichloride, and the metal itself obtains the same proportions of chlorine as it had of sulphur previously, the new products also remaining in combination with each other (Rose, Annal. de Ch. et de Ph. lxx. 270).

CHLORIDES OF PHOSPHORUS.

Terchloride of phosphorus, PCl_3. — This chloride, which corresponds with phosphorous acid, is obtained by passing chlorine through melted phosphorus, as for chloride of sulphur (fig. 158); a clear and volatile liquid distils over, of sp. gr. 1.45. It is capable of dissolving phosphorus; when mixed with water, it is resolved into hydrochloric and phosphorous acids.

Pentachloride of phosphorus, PCl_5. — Phosphorus takes fire spontaneously in a vessel of dry chlorine, and produces a snow-white woolly sublimate, which is very volatile, rising in vapour below 212°. It is converted by water into hydrochloric and phosphoric acids.

The variation of the vapour-density of this substance observed by M. Cahours, has already been referred to (page 138). This compound is considered by Cahours as a direct combination of the terchloride with 2 eq. chlorine, $PCl_3 + Cl_2$.

Chloroxide of phosphorus, PCl_3O_2. — The vapour of water produces with the pentachloride of phosphorus a compound so named, discovered by M. Wurtz. It is a colourless and very limpid liquid, of density 1.7, which fumes in air. It is decomposed by water.

Chloro-sulphide of phosphorus, PCl_3S_2. — It was discovered by Serullas, and is obtained by the action of hydrosulphuric acid on the pentachloride of phosphorus. It is liquid, boils at 262° (128° C.); is not decomposed by water. The alkaline oxides transform it into a *sulphoxiphosphate*, a metallic chloride being produced at the same time: PCl_3S_2 and $6NaO = 3NaO.PO_3S_2$ and $3NaCl$.

These salts, which correspond with the tribasic phosphates, may be crystallized. The sulphoxiphosphate of soda crystallizes with 24 eq. water, $3NaO.PO_3S_2+24HO$, and has, therefore, a composition exactly similar to the phosphate of soda, $3NaO.PO_5+24HO$, but the form is different. Here, then, sulphur is not isomorphous with oxygen (Wurtz, Annal. de Ch. 3me sér. xx. 472).

SECTION XI.

BROMINE.

Eq. 78.26 *or* 978.30; Br; *density of vapour* 5393; ▢▢.

This element was discovered by M. Balard of Montpellier in 1826. Its name is derived from Βρωμος, mal-odour, and was applied to it on account of its strong and disagreeable odour. Like the other members of the chlorine family, it is found principally in solution, being present in an exceedingly minute but appreciable proportion in sea-water, under the form of bromide of sodium or magnesium, also in the water of the Dead Sea, and in nearly all the saline springs of Europe, of which that of Theodorshall near Kreuznach in Germany is the principal source of bromine, as an article of commerce. Bromine is interesting from its chemical relations, particularly from the extraordinary parallelism in properties with chlorine which it exhibits.

Preparation. — Bromine in combination is discovered by means of chlorine-water, a few drops of which cause the colourless solution of a bromide to become orange-yellow, like nitrous acid, by disengaging bromine, while an excess of chlorine weakens the indication, by forming a chloride of bromine which is nearly colourless. Before the application of this test, the saline water in which bromine is contained must always be greatly concentrated, and, indeed, the greater part of its salts should be separated by crystallization. The bromides are highly soluble, and remain in the crystallizable liquor which is called the mother-ley, or bittern in the case of sea-water. The bromide of magnesium may lose hydrobromic acid during the farther concentration of the mother-ley, by evaporation, on which account Desfosses recommends the addition of hydrate of lime to the liquid, which throws down magnesia, and produces a bromide of calcium which may be evaporated without loss of bromine. Instead of using free chlorine to extricate the bromine, binoxide of manganese and a little hydrochloric acid may be added to the liquid. Upon distilling, bromine is liberated and comes off completely before the liquid boils. The watery vapour which condenses in the receiver along with the bromine contains a portion of chloride of bromine, from which the bromine may be separated by adding baryta to the liquid, and forming a chloride of barium and bromate of baryta; evaporating the liquor to dryness, heating to redness, and treating with alcohol.

Properties. — Bromine condenses in the preceding process as a dense liquid under the water, the sp. gr. of bromine being 2.966. In mass, it is opaque and of a dark brown red, but in a thin stratum, transparent and of a hyacinth red. Its odour is powerful and very like that of chlorine. When cooled 10 or 15 degrees below zero, it freezes, and remains solid at 10°; it then has a leaden gray colour, and a lustre almost metallic. Bromine at the usual temperature is decidedly volatile, and to retard its evaporation it is generally covered by water in the bottle in which it is kept. It boils at 116°.5, and affords a vapour very similar to the ruddy fumes of peroxide of nitrogen. Bromine is soluble to a small extent in water, and gives an orange-coloured solution; it is more soluble in alcohol, and considerably more so in ether.

Bromine bleaches like chlorine, and acts in a similar manner upon the volatile oils and many organic substances containing hydrogen, which element it eliminates in the form of hydrobromic acid. Many metals combine with bromine with ignition, as they do with chlorine; it acts as a caustic on the skin, and stains it yellow, like nitric acid. It forms a compound with starch, which is of a yellow colour; like chlorine it forms a crystalline hydrate with water at 32°, which is of a beautiful red tint.

Hydrobromic acid; 79.26 *or* 990.8; HBr.—This is a gas, in which 2 volumes of each constituent are united without condensation, as in hydrochloric acid, and which has the great attraction for water of that acid. Hydrogen and bromine do not unite at the usual temperature, and a mixture of them is not exploded by flame, but they unite in contact with the flame and form hydrobromic acid. The same acid is more readily prepared by the action of bromine upon certain compounds of hydrogen, such as hydrosulphuric acid, phosphuretted hydrogen, and hydriodic acid. The gas may also be obtained by the mutual action of bromine, phosphorus, and water, and must be collected over mercury.

For the last process, a tube-apparatus, represented fig. 159, is recommended by M. Regnault. It contains a little bromine in the bend *b*, and small portions of phosphorus at *d*, this bend being filled up with fragments of glass, and a very minute quantity of water added. The open end *a* of the tube being closed with a cork, heat is applied to *b*, so as to vapourize the bromine in a gradual manner. A bromide of phosphorus is produced, which is immediately decomposed by the water, while hydrobromic acid is disengaged and escapes by the tube *e*.

Fig. 159.

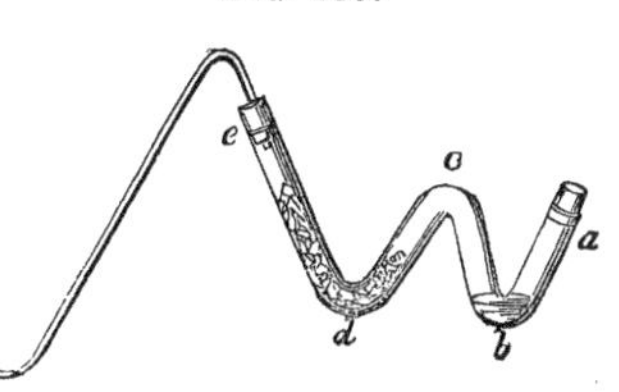

Hydrobromic acid, like all the other bromides, is decomposed by chlorine, which is more powerful in its affinities than bromine, but it is not decomposed by iodine. Its action with metals is precisely similar to that of hydrochloric acid. Hydrobromic acid is not decomposed when heated with oxygen, and water is not decomposed by bromine, so that the affinity of bromine and oxygen for hydrogen may be inferred to be nearly equal. This acid, or a soluble bromide, produces white precipitates with the nitrates of silver, lead, and suboxide of mercury, which are very similar to the chlorides of these metals. The other metallic bromides correspond in solubility with the chlorides. The bromide of silver, like the chloride, is soluble in ammonia.

Bromic acid, BrO_5.—Bromine is dissolved by the strong alkaline bases, and occasions a decomposition exactly similar to that produced by chlorine, in which a bromide of the metal and bromate of the metallic oxide are formed. The bromic acid may be separated from bromate of baryta by sulphuric acid, and its solution may be concentrated to a certain point, like chloric acid, beyond which it undergoes decomposition. It has not been isolated. The chief points of difference between chloric and bromic acid are, that the latter alone is decomposed by sulphurous and phosphorous acids, and by hydrosulphuric acid; and while all the chlorates are soluble, the bromates of silver and suboxides of mercury are insoluble, the former being a white and the latter a yellowish white precipitate. Bromic acid is the only known oxide of bromine.

Chloride of bromine, $BrCl_5$.—Chlorine gas is absorbed by bromine, and a volatile fluid of a reddish yellow colour produced. This chloride appears to dissolve in water, without decomposition, but in an alkaline solution it is converted into chloride and bromate.

Bromide of sulphur.—Bromine combines when mixed with flowers of sulphur, forming a fluid of an oily appearance and reddish tint, much resembling subchloride of sulphur in appearance and properties. This bromide dissolves both sulphur and bromine, and has not been obtained in a state of sufficient purity for analysis.

Bromides of phosphorus, PBr_3 and PBr_5.—If bromine and phosphorus are brought into contact, in a flask filled with carbonic acid gas, a violent action with ignition takes place, of which the products are a volatile crystalline solid and a yellowish liquid. The former, when decomposed by water, affords hydrobromic and phosphoric acids, which proves it to be PBr_5; and the latter affords hydrochloric and phosphorous acids, which proves it to be PBr_3. The liquid bromide does not freeze at 5°, and, like the liquid chloride of phosphorus, is capable of dissolving a large quantity of phosphorus.

Bromide of silicon — Is prepared by a similar process as the chloride of silicon. It is a liquid boiling at 302°, and freezing at 10°. By water it is resolved into hydrobromic acid and silica.

SECTION XII.

IODINE.

Eq. 126.36 *or* 1579.5; I; *density of vapour* 8707;

Iodine was discovered in 1811, by M. Courtois of Paris, in kelp, a substance from which he prepared carbonate of soda. Its chemical properties were examined by Clement, and afterwards, more completely, by Davy and Gay-Lussac, particularly by the latter (Davy, Phil. Trans. for 1814 aud 1815; Gay-Lussac, Annal. de Ch. lxxxviii., xc., et xci.) A trace of iodine has been observed in sea-water (Schweitzer), but it is more abundant in the fuci, ulvi, and other marine plants, and also in sponge, the ashes of which contain iodide of sodium. It is known also to exist in one mineral, a silver ore of Albaradon in Mexico.

Preparation.—The greater part of the iodine of commerce is prepared at Glasgow from the kelp of the west coast of Ireland and western islands of Scotland. The sea-weed thrown upon the beach is collected, dried, and afterwards burned in a shallow pit, in which the ashes accumulate and melt by the heat, being of a fusible material. The fused mass broken into lumps forms kelp, which was prepared and chiefly valued at one time for the carbonate of soda it contains, which varies in quantity from 2 to 5 per cent. It is not all equally rich in iodine. According to the observation of Mr. Whitelaw, the long elastic stems of the fucus palmatus afford most of the iodine contained in kelp, and the kelp prepared from this plant may be recognized by the presence of charred portions of the stems. This being a deep sea plant, iodine is found in largest quantity in the sea-wreck of exposed coasts. A high temperature in the preparation of the kelp, which increases the proportion of alkaline carbonate, diminishes that of the iodine, owing to the volatility of the iodide of sodium at a full red heat. The kelp which contains most iodine generally contains also most chloride of potassium, and it is for these two products that the substance is now valued, more than for its alkali.

The kelp broken into small pieces is lixiviated in water, to which it yields about half its weight of salts. The solution is evaporated down in an open pan, and when concentrated to a certain point, begins to deposit its soda salts, — namely, common salt, carbonate and sulphate of soda, — which are removed from the boiling liquor by means of a shovel pierced with holes like a colander. The liquid is afterwards run into a shallow pan to cool, in which it deposits a crop of crystals of chloride of potassium: the same operations are repeated upon the mother-ley of these crystals until it is exhausted. A dense dark-coloured liquid remains, which contains the iodide, in the form, it is believed, of iodide of sodium, but mixed with a large quantity of other salts; and this is called the iodine ley.

To this ley, sulphuric acid is gradually added in such quantity as to leave the liquid very sour, which causes an evolution of carbonic acid, sulphuretted hydrogen, and sulphurous acid gases, with a considerable deposition of sulphur. After standing for a day or two, the ley so prepared is heated with binoxide of manganese, to separate the iodine. This operation is conducted in a leaden retort *a* (see fig. 160) of a cylindrical form, supported in a sand-bath, which is heated by a small fire below.

The retort has a large opening, to which a capital, *b c*, resembling the head of an alembic, is adapted, and luted with pipe-clay. In the capital itself there are two openings, a larger and a smaller, at *b* and *c*, closed by leaden stoppers. A series of bottles *d*, having each two openings, connected together as represented in the figure, and with their joinings luted, are used as condensers. The prepared ley being heated to about 140° in the retort, the manganese is then introduced, and *b c* luted to *a*. Iodine immediately begins to come off, and proceeds on to the condensers, in which it is collected; the progress of its evolution is watched by occasionally removing the stopper at *c*; and additions of sulphuric acid or manganese are made by *b*, if deemed necessary. The success of the experiment depends much upon its being slowly conducted, and upon the proper management of the temperature, which is more easily regulated when the quantities of materials are considerable, than when the experiment is attempted with small quantities in glass flasks. In the latter circumstances, chlorine is often evolved with the iodine, which escapes in acrid fumes, as the chloride of iodine, and is lost; but this accident can be avoided in the manufacturing process. A little cyanide of iodine often accompanies the iodine, which being more volatile, condenses in the form of white, flexible, prismatic crystals, in the bottle most distant from the leaden retort.

Fig. 160.

In this operation the binoxide of manganese will be in contact at once with hydriodic, hydrochloric, and sulphuric acids; and the iodine of the hydriodic acid may be liberated, from the union with its hydrogen of the oxygen of the manganese, and the formation of water; or hydrochloric acid may be first decomposed by the manganese, and chlorine decompose the hydriodic acid and liberate iodine. If a considerable excess of sulphuric acid be employed, iodine is obtained without the use of binoxide of manganese, the oxygen required by the hydrogen of the hydriodic acid being supplied by the sulphuric acid, a part of which is converted into sulphurous acid. The presence of iodine in the prepared ley may be observed by suddenly mixing it with an equal volume of oil of vitriol, when violet fumes of iodine appear. But the quantity of iodine may be more accurately estimated by means of a solution consisting of 1 part of crystallized sulphate of copper and 2¼ cr. protosulphate of iron, which throws down an insoluble subiodide of copper, almost white. It may also be determined approximatively by precipitation by the ammonio-nitrate of silver.

Properties.—Iodine is generally in crystalline scales of a bluish black colour and metallic lustre. It is obtained, from solution, in modifications of an elongated octohedron with rhomboidal base (fig. 161.) The density of iodine is 4·948; it fuses at 225°, and boils at 347°; but it evaporates at the usual temperature, and more rapidly when damp than when dry, diffusing an odour having considerable resemblance to chlorine, but easily distinguished from it. Iodine stains the skin of a yellow colour, which however disappears in a few hours. Its vapour is of a splendid violet colour, which is seen to great advantage when a scruple or two of iodine is thrown at once upon a hot brick. Hence its name, from Ἰώδης, violet-coloured. The vapour of iodine is one of the heaviest of gaseous bodies, its density being 8716 according to the experiment of Dumas, and 8707.7 according to calculation from its atomic weight.

FIG. 161.

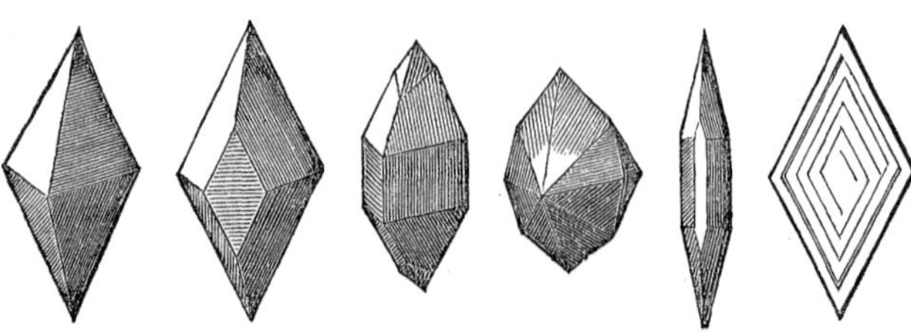

Pure water dissolves about 1-7000th of its weight of iodine, and acquires a brown colour; but when charged with salt, particularly the nitrate or hydrochlorate of ammonia, water dissolves a considerably greater quantity of iodine. The solution of iodine does not disengage oxygen in the light of the sun, and does not destroy vegetable colours, but after a time it becomes colourless, and then contains hydriodic and iodic acids. In other respects, iodine generally comports itself like chlorine, but its affinities are much less powerful. Iodine is soluble in alcohol and ether, with which it forms dark reddish-brown liquids. Solutions of iodides, too, all dissolve much iodine, and become of a deep red colour. A liquid containing 20 grains of iodine and 30 grains of iodide of potassium in 1 ounce of water, is known as Lugol's solution, and preferred to the tincture in medicine, because the iodine is not precipitated from it by dilution with water.

A solution of starch forms a compound with iodine, of a deep blue colour, soluble in pure water but insoluble in acid and saline solutions, the production of which is an exceedingly delicate test of iodine. If the iodine be free, starch produces at once the blue compound, but if the iodine be in combination as a soluble iodide, no change takes place till chlorine is added to liberate the iodine. If more chlorine, however, be added than is necessary for that purpose, the iodine is withdrawn from the starch, chloride of iodine formed, and the blue compound destroyed. Dr. A. T. Thomson, after adding the starch with a drop of sulphuric acid to the liquid containing an iodide, in a cylindrical vessel, allows the vapour only from the chlorine-water bottle to fall upon the solution, and not the chlorine-water itself. In this way, the danger of adding an excess of chlorine is easily avoided, and the test indicates in a sensible manner an exceedingly minute quantity of iodine. The iodide of starch, in water, becomes colourless when heated, but recovers its blue colour if immediately cooled. The soluble iodides give, with the nitrate of silver, an insoluble iodide of silver, of a pale yellow colour, insoluble in ammonia; with salts of lead, an iodide of a rich yellow colour, and with corrosive sublimate, a fine scarlet iodide of mercury.

In ascertaining the quantity of iodine in the mixed chlorides, and iodides of mineral waters and other solutions, Rose recommends the addition of nitrate of silver, which throws down a mixture of chloride and iodide of silver, which is fused and weighed. This is afterwards heated in a tube and chlorine passed over it, by which the iodine is expelled, and the whole becomes chloride of silver. It is weighed again, and a loss is found to have occurred, owing to the equivalent of the replacing chlorine being less than that of the replaced iodine. This loss, multiplied by 1·389, gives the quantity of iodine originally present, which has been expelled by the chlorine. (Handbuch der analyt. Chem. von Heinrich Rose, B. 2, p. 577). Dr. Schweitzer employs a similar method in estimating the quantity of iodine when mixed with bromine, heating the iodide and bromide of silver in an atmosphere of bromine. The difference of weight multiplied by 2.627 gives the proportion of iodine, and multiplied by 1.627 the proportion of bromine. (Phil. Mag., 3d series, xv. p. 57.)

Uses.—Iodine is employed in the laboratory for many chemical preparations, and as a test of starch. It was first introduced into medicine by Coindet of Geneva, who employed it with success, in the treatment of goitre, dissolved in alcohol, in solution of iodide of potassium, or as iodide of sodium; and since that application, most mineral waters to which the virtue of curing goitre was ascribed, have been found to contain iodine. M. Boussingault has adduced striking confirmations of the efficacy of iodine in that disease, in his interesting memoir on the iodiferous mineral waters of the Andes. (Annal. de Chim. et de Phys., liv. 163.) It appears to have a specific action in causing the absorption of glandular swellings, and is also administered as a tonic. Iodine swallowed in the solid state causes ulceration of the mucous membrane of the stomach, and death. But the iodide of potassium or sodium is not poisonous in considerable doses, nor is the iodide of starch hurtful (Dr. A. Buchanan). Iodine and bromine have also found an interesting application to form the film of iodide or bromide of silver, in the silver-plates of the daguerreotype, which is so sensitive to light.

Iodides.—Iodine does not form a hydrate like chlorine, but it combines with another compound body, ammonia; dry iodine absorbing dry ammoniacal gas and running into a brown liquid, which Bineau found to contain 20.4 ammonia to 100 iodine, quantities in the proportion of 3 equivalents of ammonia to 2 of iodine. (Annal. de Chim. et de Phys., lxvii. 226.) This liquid dissolves iodine. Iodine does not combine with dry iodide of potassium, but with the addition of a small quantity of water, it forms what appears to be a ternary compound of iodide of potassium, water and iodine, which is usually fluid, but was obtained in crystals by Bauer. Iodine forms similar compounds with other hydrated metallic iodides. With the metals generally iodine combines, with the same facility, and nearly with as much energy as chlorine does. The iodide of zinc and protiodide of iron, which are very soluble, are formed by simply bringing the metals into contact with iodine, in water. All the iodides are decomposed by bromine, as well as by chlorine.

The compounds of iodine may be shortly described in the following order:—

Hydriodic acid	HI	Iodide of sulphur
Iodic acid	IO_5	Iodides of phosphorus
Periodic acid	IO_7	Chlorides of iodine
Iodide of nitrogen ...	NI_3	Bromides of iodine.

COMPOUNDS OF IODINE.

Hydriodic acid; 127.36 or 1592; HI.—Hydriodic acid cannot be prepared with advantage by treating the iodide of sodium or potassium with hydrated sulphuric acid, as the latter is partially converted into sulphurous acid by hydriodic acid, with the separation of iodine. It may be obtained in the state of gas, by forming an iodide of phosphorus, 9 parts of dry iodine and 1 of phosphorus being introduced into a tube sealed at one end, to be used as a retort, and the mixture covered by pounded glass, and combination determined by a gentle heat; and afterwards decomposing this iodide of phos-

Fig. 162.

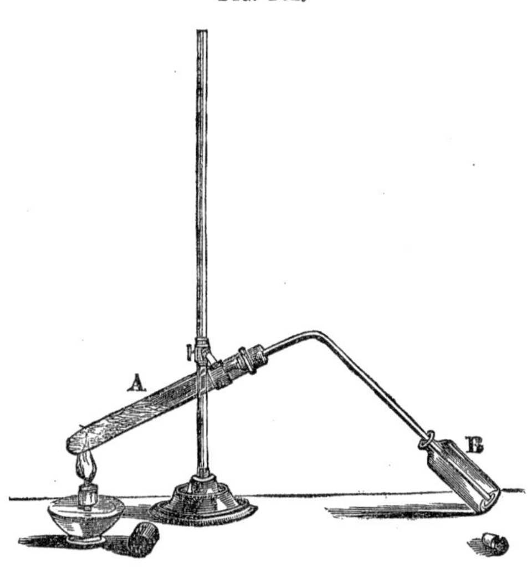

phorus by a few drops of water. Hydriodic acid instantly comes off as gas, and hydrated phosphorous acid remains in the tube:

$$PI_3 \text{ and } 6HO = 3HI \text{ and } 3HO + PO_3.$$

A slight heat may be applied to the tube, when the action abates, to expel the last portions of hydriodic acid; but if the temperature be elevated, the residuary hydrated phosphorous acid is decomposed, with evolution of phosphuretted hydrogen gas, which may, therefore, be obtained by the same operation. This gas is very soluble in water, and soon decomposed over mercury, which combines with its iodine and liberates hydrogen; so that it is collected in a dry bottle, B, by the method of displacement, and the bottle is closed with a glass stopper when full of gas. Hydriodic gas is colourless, of density 4443 by experiment and 4385 by theory, and consists of 2 volumes of iodine vapour and 2 volumes of hydrogen gas united without condensation, or forming 4 volumes, which are, therefore, the combining measure of the gas. In the combination of its constituents by volume, hydriodic acid resembles hydrochloric acid gas and all the other hydrogen acids. Hydriodic acid gas is gradually decomposed by oxygen, with the formation of water: iodine is liberated.

The solution of this acid in water may be obtained by transmitting hydrosulphuric acid gas through water in which iodine is suspended: the iodine combines with the hydrogen of that compound and liberates the sulphur. The liquid may afterwards be warmed to expel the excess of hydrosulphuric acid, and filtered. It is colourless at first, but in a few hours becomes red, owing to the decomposition of hydriodic acid by the oxygen of the air, and solution of the iodine in the acid.

The solution has its maximum boiling point, which lies between 257° and 262°, when of sp. gr. 1.7, according to Gay-Lussac. Nitric and sulphuric acids decompose it, and are decomposed themselves with the formation of water; the starch test then indicates free iodine.

Iodic acid; 166.36 *or* 2079.5; IO_5.—Iodine does not afford a peculiar acid compound with red oxide of mercury and those metallic oxides which yield free hypochlorous acid with chlorine. Nor is it absorbed, like chlorine, by hydrate of lime or alkaline solutions, to form a class of bleaching salts. Such compounds are wanting in the series of oxides of iodine, which is limited to hypoiodic, iodic, and periodic acids. Sementini imagined that he had formed inferior oxides of iodine, but he is evidently mistaken. The iodate of soda combines with iodide of sodium in several proportions, one of which was supposed by Mitscherlich, when he discovered it, to be an iodite of soda; but that this is a double salt of the constitution first mentioned is more probable.

A few grains of iodic acid may easily be prepared by the method of Mr. Connel, which consists in heating the most concentrated nitric acid, free from nitrous vapour, upon a little iodine, in a wide glass tube, and allowing the liquid to cool; the iodine is oxidated at the expense of the nitric acid, and the greater part of the iodic acid is deposited in crystals. When a larger quantity is required, a convenient process is to form, in the first place, an iodate of soda, as suggested by Liebig. An ounce or two of iodine in powder may be suspended in a pound of water, with occasional agitation, and a stream of chlorine be passed through till the whole iodine is dissolved. Carbonate of soda is then added to the liquid, which is of a brown colour and strongly acid, till it becomes slightly alkaline, when a large precipitation of iodine occurs, which may be separated and collected on a filter. This iodine may be suspended in water, and exposed to a stream of chlorine as before.

$$5Cl \text{ and } 5HO \text{ and } I = 5HCl \text{ and } IO_5.$$

The filtered solution contains iodate of soda and chloride of sodium, with a trace of carbonate, which may be neutralized by hydrochloric acid. On afterwards adding chloride of barium to the filtered solution, so long as a precipitate is produced, the whole iodic acid is thrown down as iodate of baryta, which may be collected on a filter and dried. This iodate is anhydrous, and may be decomposed completely, by boiling 9 parts of it for half an hour with 2 parts of oil of vitriol, diluted with 10 or 12 parts of water. The liberated iodic acid dissolves, and being separated from the sulphate of baryta by filtration, is obtained as a crystalline mass when evaporated to dryness by a gentle heat.

This acid is also prepared very easily, according to M. Millon, by digesting iodine in a mixture of nitric acid and chlorate of potassa; the proportions recommended are 4 of iodine, 7.5 chlorate of potassa, 10 of nitric acid, and 40 of water. The iodic acid is afterwards precipitated in the form of iodate of baryta, as in the preceding process, the iodate of baryta then decomposed by sulphuric acid.

Iodic acid crystallizes from a strong solution, as a hydrate, $HO.IO_5$, in large and transparent crystals, which are six-sided tables. This acid is not sublimed, but decomposed into iodine and oxygen, by a high temperature, without any formation of periodic acid. Another definite hydrate of iodic acid was obtained by M. Millon, containing only one-third of an equivalent of water, by maintaining the protohydrate at a temperature of 266° (130° C.), so long as it continued to lose weight. It is also formed when the protohydrate is mixed with an excess of anhydrous alcohol. By drying either of these hydrates at 338° (170° C.), iodic acid is obtained entirely anhydrous (IO_5).

Iodic acid is very soluble in water; and after reddening, bleaches litmus paper. It oxidates all metals with which it has been tried, except gold and platinum. It is deoxidized by sulphurous acid and hydrosulphuric acid, and iodine liberated, but an excess of sulphurous acid causes the iodine again to disappear as hydriodic acid, water being decomposed by the simultaneous action of sulphurous acid and iodine upon its elements. Iodic acid is easily decomposed by heat, disengaging oxygen and vapours of iodine. It is soluble in water, alcohol, and ether.

Iodates.—The salts of iodic acid have a general resemblance to chlorates; when thrown upon burning embers they enliven the combustion, but with less vivacity than chlorates. The iodate of potassa is converted by heat into iodide of potassium and oxygen; so that the composition of iodic acid may be determined from that of iodate of potassa, in the same manner as the composition of chloric acid is determined from that of chlorate of potassa. The iodate of soda, however, loses iodine as well as oxygen, when heated, and a yellow, sparingly soluble, alkaline matter remains, which Liebig supposes to contain the salt of an iodous acid, resolvable into an iodate and iodide by solution in water, but which requires further investigation. The iodates of metallic protoxides, with the exception of the potassa family, are all sparingly soluble or insoluble salts. The iodate of lime contains water, and when heated affords no iodide of calcium, but caustic lime.

Fixed acids, which have little affinity for water, such as iodic acid, appear often to combine in several proportions with oxides of the potassa family. The ordinary biniodate of potassa contains 1 eq. of basic water, but at a high temperature it is made anhydrous, and then a salt remains containing 2 eq. of acid to 1 of potassa. Mr. Penny has crystallized a biniodate and teriodate of soda, both anhydrous.

Iodic acid likewise combines with other acids. These compounds generally precipitate in a crystalline form, when another acid is added to a hot and concentrated solution of iodic acid. Compounds of sulphuric, nitric, phosphoric, and boracic acids, with iodic acid, have been formed. It has been observed by M. Millon, that when the compound with sulphuric acid is submitted to heat, oxygen is evolved, and a hypoiodic acid or peroxide of iodine formed, of which the formula is IO_4. There is formed besides in this decomposition, according to M. Millon, a peculiar double acid, which may be considered a compound of iodous and hypo-iodic acid, having for formula $4IO_4+IO_3$. When vegetable acids are dissolved in iodic acid, they are immediately decomposed by it, carbonic acid being disengaged with effervescence, and iodine precipitated.

Periodic acid, Hyperiodic acid; 182.36 *or* 2279.5; IO_7.—This acid, which was discovered by Magnus and Ammermuller, is formed by transmitting a current of chlorine through a solution of iodate of soda, to which a portion of carbonate is added, and the whole maintained in constant ebullition. On allowing the solution to cool, a basic periodate of soda is deposited in tufts of silky crystals, and the chloride of sodium, formed at the same time, retained in solution. This basic periodate of soda, which is almost insoluble in cold water, is dissolved in nitric acid, and nitrate of silver added, which throws down a basic periodate of silver, also of sparing solubility. The last salt may be washed, and afterwards dissolved in boiling nitric acid, and the solution on cooling yields orange-yellow crystals of neutral periodate of silver. It is remarkable that when these crystals are thrown into water they are decomposed, the whole oxide of silver precipitating with half the periodic acid, as the former basic periodate, while half of the acid is dissolved by the water without a trace of silver, and obtained in a state of purity. This solution when evaporated affords periodic acid in crystals, which are unalterable in the air, and of which the solution in water is not changed by ebullition. The crystals fuse about 266° (130° C.) The solution, treated with hydrochloric acid, affords chlorine and iodic acid, water being formed. Periodic acid is resolved into oxygen and iodine by a high temperature.

Periodates.—Besides neutral salts of this acid, subsalts of the potassa family exist which contain two of base to one of acid. The sparing solubility of the basic salt of soda is the most remarkable character of periodic acid. True subsalts of the potassa family are so extremely unusual, that it is more probable that periodic acid forms a second and bibasic class of salts, to which they belong. (Poggendorff's Annalen, xxviii. 514). The periodates are decomposed by heat like the iodates, but yield more oxygen.

Iodide of nitrogen.—Dry iodine and ammonia unite directly, and form a brown liquid, of which the formula is $3(H_3N).I_2$. But when digested in the solution of ammonia, iodine acts upon that substance as chlorine does, and forms an insoluble black powder, which is powerfully detonating, and analogous to the chloride of nitrogen. The iodide detonates more easily, but less violently, than the chloride, always exploding spontaneously when it dries. Another process is to mix a great excess of ammonia with a saturated solution of iodine in alcohol, and afterwards to add water so long as iodide of nitrogen precipitates. The filter with the humid precipitate should be divided into several pieces, otherwise the whole may explode at once upon drying.

Although named the iodide of nitrogen, this substance contains hydrogen as a constituent, according to the observations of M. Bineau, and may be represented by I_2HN; or ammonia in which 2 eqs. of hydrogen are replaced by 2 eqs. of iodine. The same substance is represented by Millon, as I_3N+2H_3N.

When caustic soda is added to the solution of iodine in alcohol or wood-spirit, a yellow substance of a saffron odour precipitates, which was supposed at one time to be the periodide of carbon, but is really *iodoform*, of which the formula is C_2HI_3. No true iodide of carbon is known.

Iodide of sulphur.—This compound is formed by fusing together 4 parts of iodine and one of sulphur. It has a radiated crystalline structure, but its elements are easily disunited, the iodine escaping entirely from this compound when it is left exposed in the air.

Iodides of phosphorus.—Iodine appears to combine with phosphorus in several proportions, when they are brought in contact and slightly heated. In all these combinations the mass becomes hot without inflaming, if the phosphorus is not at the same time in contact with air. One part of phosphorus with 6, 12, and 20 parts of iodine, forms fusible solids, which may be sublimed without change, but which are decomposed by water, all of them yielding hydriodic acid, and the first affording, besides, phosphorus and phosphorous acid, the second phosphorous acid, and the third phosphoric acid.

Chlorides of iodine.—Chlorine is readily absorbed by dry iodine; when the latter is in excess, a protochloride, ICl, appears to be formed; and when the chlorine is in excess, a terchloride, ICl_3.

Berzelius produced the protochloride by distilling a mixture of 1 part of iodine with 4 parts or more of chlorate of potassa. There is formed in the retort a mixture of iodate and perchlorate of potassa, at the same time that oxygen gas is disengaged, and the chloride of iodine is produced, which condenses in the receiver. This compound is a yellow or reddish liquid, of an oily consistence, of a sharp and peculiar odour, and taste which is feebly acid, but very astringent and rough. It is soluble in water and alcohol; and ether extracts it from its aqueous solution unaltered, so that it is not decomposed by water.

When iodine is saturated with chlorine, it forms a compound which is solid and crystallizable, and of a yellow colour; fusible by heat, but which cannot be sublimed without loss of chlorine. It fumes in air, and has an acrid odour. When this terchloride of iodine is dissolved in water, and the solution saturated with carbonate of soda, chloride of sodium is formed, and some iodate of soda; while at the same time a large quantity of iodine precipitates. By the continued action of chlorine upon iodine in a considerable quantity of water, the liquid becomes at last entirely colourless, and then contains nothing but hydrochloric and iodic acids.

Bromides of iodine.—Iodine likewise forms two bromides, which are both soluble in water. The solution bleaches litmus paper without first reddening it.

SECTION XIII.

FLUORINE.

Eq. 18.70 *or* 233.8; F; *density* (*hypothetical*) 1292; $\square\square$

This elementary body is most frequently found in the mineral kingdom in combination with calcium, as fluoride of calcium, which constitutes the mineral fluor-spar; it exists in small quantity in amphibole, mica, and most of the natural phosphates: a trace of it also occurs in the enamel of the teeth, and in the bones of animals. Of all bodies, fluorine appears to possess the most powerful and general affinities, and to be, therefore, the most difficult to isolate and preserve for the study of its properties. Indeed, we have hitherto learned little more of fluorine than that it exists and may be isolated. Several of its compounds, however, are of less difficult preparation, and well known.

Sir H. Davy made several attempts to isolate fluorine. He exposed the fluoride of silver in a glass tube to gaseous chlorine, at a high temperature, and found that chloride of silver was produced, and fluorine therefore liberated; but it was absorbed and replaced by oxygen, which it disengaged from the silica and soda of the glass. When Davy repeated the same experiment in a platinum vessel, the metal became covered with fluoride of platinum. He proposed afterwards to construct vessels of fluor-spar for the reception of the fluorine, which he expected to disengage from the fluoride of phosphorus by burning it in oxygen gas; but he does not appear to have carried this project into execution. The Messrs. Knox and M. Louyet have announced that they have separated fluorine from the fluorides of silver and mercury, by treating these bodies with chlorine or iodine in vessels of fluor-spar, when fluorine was disengaged in the form of a colourless gas. Gold and platinum did not appear to be acted upon by fluorine, except when it was in the nascent state.

No compound of fluorine and oxygen is yet known, but a compound of fluorine and hydrogen is easily formed, and is of importance from its applications.

HYDROFLUORIC ACID.

Eq. 19.7 *or* 246.3; HF.

Schwankhardt, of Nuremberg, observed in 1670, that it was possible to etch upon glass by means of fluor-spar and sulphuric acid, but it was not till 1771 that Scheele referred this action to a particular acid which sulphuric acid disengaged from fluor-spar. Wenzel first obtained the true hydrofluoric acid, exempt from silica, by preparing it in proper metallic vessels; the acid collected by Scheele being the fluosilicic, and not the hydrofluoric. The preparation and properties of the pure acid were more fully studied by Gay-Lussac and Thénard in 1810. It was then known as fluoric acid, and was supposed, according to the doctrine of the day, to contain oxygen. The idea of its being a hydrogen acid was first suggested, a few years afterwards, by M. Ampère, whose views in theoretical chemistry were often marked by much acuteness and originality. The view of Ampère was generally assented to, and is confirmed by the isomorphism of the fluorides with the chlorides, bromides, and iodides, observed by M. Louyet.

Preparation.—To obtain hydrofluoric acid, a specimen of fluor-spar is selected, free from silicious minerals and galena; this is reduced to an impalpable powder, and distilled in a retort of lead (fig. 163), by a gentle heat, such as that of an oil-bath, with twice its weight of highly concentrated oil of vitriol. The materials become viscid and swell considerably, and an acid vapour distils over, which is even more acrid and suffocating than chlorine, and produces severe sores if allowed to condense upon the hands of the operator. This vapour is received in a bent tube, likewise of lead, used as a receiver, and kept cold by a freezing mixture, in which the hydrofluoric acid condenses without the presence of water. The acid thus obtained may be preserved in vessels of platinum or gold, provided with stoppers of the same metal which fit accurately; or in vessels of lead formed without tin solder, tin being rapidly acted upon by hydrofluoric acid. If a dilute solution of this acid in water is required, the extremity of the leaden tube, from the retort, may be allowed to touch the surface of water in a platinum crucible or capsule, by which the acid vapour is readily condensed; and the dilute acid may be preserved, without much contamination, in a glass bottle which has been previously heated, and coated internally with melted bees'-wax.

Fig. 163.

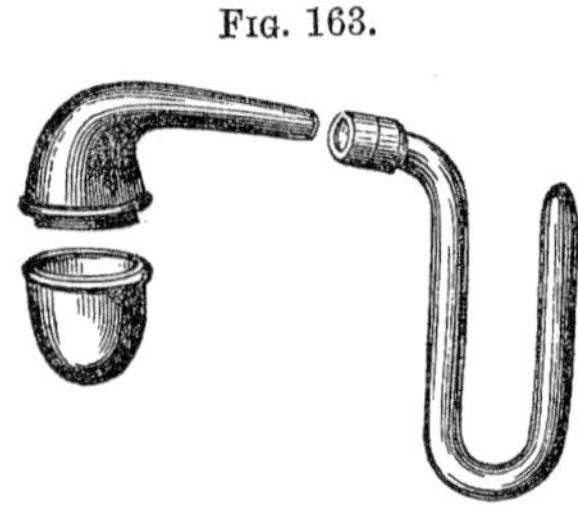

Fluor-spar, which is employed in this operation, is the fluoride of calcium, upon which the action of hydrated sulphuric acid is similar to its action upon chloride of sodium, when hydrochloric acid is produced. Water is decomposed, by the hydrogen and oxygen of which the fluorine and calcium are converted respectively into hydrofluoric acid and lime, the former coming off as vapour, while the latter remains in the retort as sulphate of lime. In symbols—

$$CaF \text{ and } HO.SO_3 = HF \text{ and } CaO.SO_3.$$

Properties.—The acid liquid obtained by the preceding process, which has hitherto been considered as the anhydrous acid, is, according to M. Louyet, a hydrate. Distilled with anhydrous phosphoric acid, it loses water, and gives rise to a colourless gas, fuming in air like hydrochloric acid, which is the true anhydrous hydrofluoric acid. M. Louyet finds this gaseous acid to have no sensible action upon dry glass.

The former product is a colourless, fuming, and very volatile liquid, boiling not much above 60°; and which does not freeze at 4°. Its sp. gr., which is 1.0609, is increased to 1.25 by the addition of a certain quantity of water, for which it has an intense affinity. Hydrofluoric, like hydrochloric acid, dissolves the more oxidable metals with the evolution of hydrogen gas. Mixed with nitric acid, it dissolves ignited silicon and titanium, with disengagement of nitric oxide; but that acid mixture has no action upon the nobler metals, such as gold and platinum, which are dissolved by aqua regia. Several insoluble acid bodies, which are not acted on by sulphuric, nitric, or hydrochloric acid, are dissolved with facility by hydrofluoric acid; such as silica, titanic, tantalic, molybdic and tungstic acids. Water is then formed from the oxygen of these acids and the hydrogen of hydrofluoric acid, and fluorides of silicon or of the metals of the acids enumerated are likewise produced; which fluorides appear to combine with undecomposed hydrofluoric acid, when water is present. This acid destroys glass by acting upon its silica. If a drop of the concentrated acid be allowed to fall upon a glass plate, it becomes hot, enters into ebullition and volatilizes in a thick smoke, leaving the spot with which it was in contact deeply corroded, and covered by a white powder composed of the elements of the glass, excepting a portion of the silica, which has passed off as gaseous fluoride of silicon.

The diluted solution, or the vapour of hydrofluoric acid, is sometimes used to etch upon glass. The purity of the acid being of little moment in this application of it, the sulphuric acid and fluor-spar may be mixed in a stone-ware evaporating basin. The glass is warmed sufficiently to melt bees'-wax rubbed upon it, and thereby covered with a coating of that substance, which is afterwards removed from the parts to be etched, by a pointed rod of lead or tin, employed as a graver. A gentle heat being applied to the basin, acid fumes are evolved, to which the etched surface of the glass is exposed for a minute or two, care being taken not to melt the wax. The wax is afterwards removed by warming the glass, and wiping it with tow and a little oil of turpentine, when the exposed lines are found engraved to a depth proportional to the time they have been exposed to the acid fumes. But in taking impressions upon paper from glass plates engraved in this way, as from a copper-plate, they are too apt to be broken from the pressure applied in printing.

To discover the minute quantity of hydrofluoric acid which exists in many minerals, Berzelius recommends that the substance to be examined be reduced to fine powder and mixed with concentrated sulphuric acid, in a platinum crucible covered by a small plate of glass, waxed and engraved as described. The crucible is then exposed to a gentle heat, insufficient to melt the wax, and, in half an hour, the glass plate may be removed and cleaned. If the mineral submitted to the test contains fluorine, the design will be perceived upon the glass; when the quantity of fluorine, however, is very small, the engraving does not appear immediately, but becomes visible on passing the breath over the glass. The presence of silica in the mineral interferes with this operation, but an indication may then be obtained by heating a fragment of the mineral to redness upon a piece of platinum foil slipped into a glass tube, 8 or 10 inches in length, and open at both ends. The tube is held obliquely with the mineral near the lower end, and so that part of the vapour from the flame passes up the tube. The moisture thus introduced carries away the gaseous fluoride of silicon, and condenses in drops in the upper part of the tube. These drops, when afterwards evaporated, in drying the tube, leave a white spot, which consists of silica, resulting from the decomposition of the fluoride of silicon by the water with which it condensed. (Berzelius).

Fluoride of boron, fluoboric acid; 67·0 *or* 837·5; BF_3.—This compound is gaseous, and is obtained when dry boracic acid is brought in contact with concentrated hydrofluoric acid; when boracic acid is ignited with fluor spar; and most conveniently by heating together in a glass retort, 1 part of vitrified boracic acid in fine powder, 2 of fluor spar, and 12 of concentrated sulphuric acid, although this process does not give it free from fluosilicic acid. The reaction by which the fluoboric acid is then produced may be thus expressed:—

$$3CaF \text{ and } BO_3 \text{ and } 3(HO.SO_3) = 3(CaO.SO_3) \text{ and } 3HO \text{ and } BF_3.$$

Fluoboric acid gas has no action upon glass, and may be collected in glass vessels over mercury. It is colourless, but produces thick fumes when allowed to escape into the atmosphere. Its density, according to Dr. J. Davy, is 2371, and 2312 according to Dumas, who finds 1 volume of this gas to contain $1\frac{1}{2}$ vol. of fluorine. Fluoboric gas is not decomposed by iron and the ordinary metals, even at a bright red heat, but on the contrary, potassium, with the metals of the alkalies and alkaline earths, decomposes it at a red heat; boron is liberated by potassium, and a double fluoride of boron and potassium also formed. Water absorbs fluoboric acid gas with the greatest avidity, taking up, according to J. Davy, 700 times its volume, which increases its bulk considerably, and raises its density to 1.77. Sulphuric acid can dissolve 50 times its volume of the fluoride of boron. The most ready mode of preparing the aqueous solution of this acid is to dissolve crystallized boracic acid in hydrofluoric acid. The acid is extremely caustic and corrosive, charring and destroying wood and organic matters, when concentrated, like sulphuric acid, probably from its avidity for moisture.

A dilute solution of fluoride of boron undergoes spontaneous decomposition, according to Berzelius, depositing one-fourth of its boron in the form of boracic acid, which crystallizes at a low temperature; while a compound of hydrofluoric acid and fluoride of boron remains in solution, which he termed *hydrofluoboric acid.* The fluoride of boron has a great disposition to form double fluorides, and acts upon basic metallic oxides like the following compound.

Fig. 164.

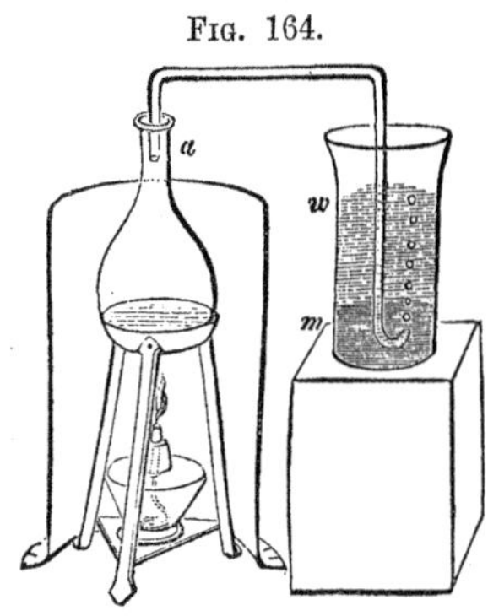

Fluoride of silicon, fluosilicic acid; 77.45 *or* 968.12; SiF_3.—This gas is obtained in the following manner:—Equal parts of fluor spar and broken glass or quartzy sand, in fine powder, are mixed in a glass flask *a* (fig. 164), to be used as a retort, with six parts of concentrated sulphuric acid, and stirred well together. A disengagement of gas immediately takes place, and the mass swells up considerably. After a time, a gentle heat is required to aid the operation. Fluosilicic gas is collected over mercury. In its physical characters it resembles fluoboric gas. It is colourless and fumes in air; it extinguishes bodies in combustion, and does not attack glass. Its density is 3574 according to J. Davy, and 3600 according to Dumas; it contains twice its volume of fluorine.

In transmitting this gas into water, the tube must not dip in the fluid, for it would speedily be choked by the deposition of silica produced by the action of water upon the gas. In the arrangement figured, the extremity of the exit tube is covered by a small column of mercury *m*, in the lower part of the jar, through which the gas passes before it reaches the water *w*. Every bubble of gas exhibits a remarkable phenomenon, as it enters the water, becoming invested with a white bag of silica, which rises to the surface. It often happens, in the course of the operation, that the gas forms tubes of silica in the water, through which it gains the surface without decomposition, if they are not broken from time to time. When water is completely saturated with the fluoride of silicon, it has taken up about once and a half its weight, and is a gelatinous, semi-transparent mass, which fumes in the air. The liquid contains two equivalents of water to one of the original fluoride of silicon: but one-third of the fluoride has been decomposed by the water and converted into hydrofluoric acid and silica. The hydrofluoric acid and fluoride of silicon, in solution, were supposed to be in combination by Berzelius, forming $3HF + 2SiF_3$, which was termed by him *hydrofluosilicic* acid. When this liquid is placed in a moderately warm situation, the whole of it gradually evaporates; the free hydrofluoric acid reacting upon the deposited silica, with formation of water, and fluoride of silicon being revived.

The most remarkable property of the fluoride of silicon is to produce, with neutral salts of potassa, soda and lithia, precipitates which are gelatinous, and so transparent as to be scarcely visible at first in the liquid; and with salts of baryta, a white and crystalline precipitate, which appears in a few seconds. It is often employed to decompose a salt of potassa, for the purpose of isolating its acid. It also serves to distinguish salts of baryta from salts of strontia; the salts of baryta producing with this acid a salt scarcely soluble in water, while the salts of strontia are not precipitated.

Almost all the basic metallic oxides decompose this acid, when they are employed in excess, separating silica, and giving rise to metallic fluorides. When, on the other hand, no more of the base is applied than the quantity required to neutralize the free hydrofluoric acid, combinations are obtained with all bases, which are analogous to double salts; consisting of a metallic fluoride combined with fluoride of silicon, the proportion of the latter containing twice as much fluorine as the former. The formula of one of these compounds, the double fluoride of silicon and potassium, is $2SiF_3 + 3KF$; and those of other metals are similar. The ratio of 2 to 3, in the equivalents of the two fluorides which form these double salts, is unusual. But the double fluorides in question may be represented by single equivalents of fluoride of silicon and metallic fluoride, as was suggested by Dr. Clark, by adopting the low equivalent of silicon 12.6, when silica is made to consist of 1 equivalent of silicon and 2 equivalents of oxygen, and the fluoride of silicon of 1 equivalent of silicon and 2 equivalents of fluorine.

CHAPTER VI.

METALLIC ELEMENTS.

GENERAL OBSERVATIONS.

THE metallic class of elements is considerably more numerous than the non-metallic class, embracing forty-eight elementary bodies. Of these seven only were known to the ancients, and of the remainder, a large proportion are of recent discovery. Their names and their densities, when accurately determined, with the dates and authors of their discovery, are contained in the following table, compiled chiefly from the work of Dr. Turner:—

Table of Metals.

Name.	Density.	Dates and Authors of the Discovery.
Gold	19·257 Brisson, to 19·361	Known to the Ancients.
Silver	10·474, ditto	
Iron	7·778, ditto	
Copper	8·895, Hatchett	
Mercury	13·596, at 32° Regnault	
Lead	11·352, Brisson	
Tin	7·291, ditto	
Antimony	6·702, ditto	1490, described by Basil Valentine.
Bismuth	9·822, ditto	1530, described by Agricola.
Zinc	6·861 to 7·1, ditto	16th century, first mentioned by Paracelsus.
Arsenic	5·884, Turner	1733, Brandt.
Cobalt	8·538, Haüy	
Platinum	20·336 Brisson, to 22·069	1741, Wood, assay-master, Jamaica.
Nickel	8·279, Richter	1751, Cronstedt.
Manganese	7·500	1774, Gahn and Scheele.
Tungsten	17·6, D'Elhuyart	1781, D'Elhuyart.
Tellurium	6·115, Klaproth	1782, Müller.
Molybdenum	7·400, Hielm	1782, Hielm.
Uranium	9·000, Bucholz	1789, Klaproth.
Titanium	5·3, Wollaston	1791, Gregor.
Chromium	5·9,	1797, Vauquelin.
Tantalum		1802, Hatchett.

Table of Metals—continued.

Name.	Density.	Dates and Authors of the Discovery.
Palladium	11·3 to 11·8, Wollaston...	1803, Wollaston.
Rhodium	10·649	
Iridium........	18·680, [21·8, Hare]........	1803, Descotils and Smithson Tennant.
Osmium	10·0	1803, Smithson Tennant.
Cerium.........	....	1804, Hisinger and Berzelius.
Potassium.....	0·865 } Gay Lussac and	1807, Davy.
Sodium	0·972 } Thénard	
Barium	....	
Strontium.....	....	
Calcium........	....	
Cadmium......	8·604, Stromeyer	1818, Stromeyer.
Lithium........	....	1818, Arfwedson.
Zirconium.....	....	1824, Berzelius.
Aluminum	....	1828, Wöhler.
Glucinum	....	
Yttrium........	....	
Thorium	....	1829, Berzelius.
Magnesium ...	....	1829, Bussy.
Vanadium	....	1830, Sefström.
Lantanum	....	1839, Mosander.
Didymium.....	....	Since 1840, Mosander.
Erbium	....	
Terbium	....	
Ruthenium ...	....	1844, Klaus.
Pelopium......	....	1845, H. Rose.
Niobium.......	....	

Of the physical properties of metals and their combinations with each other, the most characteristic is their lustre and power to reflect much of the light which falls upon them,—a property exhibited in a high degree by burnished steel, speculum metal, and the reflecting surface of mercury in glass mirrors. Metals are also remarkable for their opacity, although they have a certain degree of transparency in a highly attenuated state, as fine gold-leaf allows light of a green colour to pass through it. They are peculiarly the conductors of electricity, and also the best conductors of heat. The most dense substances in nature are found among the metals,—gold, for instance, being upwards of nineteen, and laminated platinum twenty-two times heavier than an equal bulk of water. But some of the metals, notwithstanding, are very light, potassium and sodium floating upon the surface of water.

Certain metals possess a valuable property, *malleability*, depending upon a high tenacity with a certain degree of softness; particularly gold, silver, copper, tin, platinum, palladium, cadmium, lead, zinc, iron, nickel, potassium, sodium, and solid mercury. These metals may all be hammered out into plates, or even into thin leaves. In zinc this property is found in the highest degree between 300° and 400°, and in iron at a degree of temperature exceeding a red heat. The same metals are likewise *ductile*, or may be drawn into wires, although the ductility of different metals is not always proportional to their malleability, iron being highly ductile, although it cannot be beaten into very thin leaves. By a peculiar method, Dr. Wollaston formed gold wire so small that it was only 1-5000th of an inch in diameter, and 550 feet of it were required to weigh one grain. He also obtained a wire of platinum not more than 1-30,000th of an inch in diameter, (Phil. Trans. 1813.) The tenacity of different metals is determined by ascertaining the weight required to break wires of them having the same diameter. Iron appears to possess that property in the greatest, and lead in the least degree. It has been observed by M. Baudrimont that the tenacity of wires of iron, copper, and brass, is much injured by annealing them, (Annal. de Chim. et de Phys. lx. 78.) A few of the malleable metals can be *welded*, or portions of them joined into one by hammering them together. Pieces of iron or platinum may be united in this manner at a bright red heat, and fragments of potassium may be made to adhere by pressing them together with the hand at the temperature of the air. Many metals are only malleable in a low degree, and some are actually brittle,—such as bismuth, antimony, and arsenic.

The metals, with the exception of mercury, are all solid at the temperature of the air, but they may be liquefied by heat. Their points of fusion are very different, as will appear from the following table:

Table of the Fusibility of different Metals.

		Fahr.	Different Chemists.
Fusible below a red heat.	Mercury	—39°	
	Potassium	136	Gay-Lussac and Thénard.
	Sodium	190	Gay-Lussac and Thénard.
	Tin	442	Crichton.
	Bismuth	497	Crichton.
	Lead	612	Crichton.
	Tellurium—rather less fusible than lead		Klaproth.
	Arsenic—undetermined.		
	Zinc	773	Daniell.
	Antimony—a little below a red heat.		
	Cadmium	442	Stromeyer.
Infusible below a red heat.	Silver	1873°	Daniell.
	Copper	1996	Daniell.
	Gold	2016	Daniell.
	Cobalt—rather less fusible than iron.		
	Iron, cast	2786	Daniell.
	Iron, malleable	Requiring the highest heat of a smith's forge.	
	Manganese	Requiring the highest heat of a smith's forge.	
	Nickel—nearly the same as cobalt.		
	Palladium.		
	Molybdenum	Almost infusible, and not to be procured in buttons by the heat of a smith's forge.	Fusible before the oxi-hydrogen blow-pipe.
	Uranium	Almost infusible, and not to be procured in buttons by the heat of a smith's forge.	Fusible before the oxi-hydrogen blow-pipe.
	Tungsten	Almost infusible, and not to be procured in buttons by the heat of a smith's forge.	Fusible before the oxi-hydrogen blow-pipe.
	Chromium	Infusible in the heat of a smith's forge, but fusible before the oxi-hydrogen blow-pipe.	
	Titanium	Infusible in the heat of a smith's forge, but fusible before the oxi-hydrogen blow-pipe.	
	Cerium	Infusible in the heat of a smith's forge, but fusible before the oxi-hydrogen blow-pipe.	
	Osmium	Infusible in the heat of a smith's forge, but fusible before the oxi-hydrogen blow-pipe.	
	Iridium	Infusible in the heat of a smith's forge, but fusible before the oxi-hydrogen blow-pipe.	
	Rhodium	Infusible in the heat of a smith's forge, but fusible before the oxi-hydrogen blow-pipe.	
	Platinum	Infusible in the heat of a smith's forge, but fusible before the oxi-hydrogen blow-pipe.	
	Columbium	Infusible in the heat of a smith's forge, but fusible before the oxi-hydrogen blow-pipe.	

The metallic elements are, in general, highly fixed substances, although it is probable that all of them may be dissipated at the highest temperatures. The following metals are so volatile as to be occasionally distilled,—cadmium, mercury, arsenic, tellurium, sodium, potassium, and zinc.

All the metals are capable of uniting with oxygen, but they differ greatly from each other in their affinity for that element. The greater number of them absorb oxygen from dry air at the usual temperature, and undergo oxidation, which is only slight and superficial in many, when they are in mass, but may be complete and perfect in the same metals, when they are highly divided, and in a favourable state for combination, as in the lead and iron pyrophorus exposed to air. The same metals exhibit, at a high temperature, a more intense affinity for oxygen, and combine with the phenomena of combustion.

The metals have been arranged in six groups or sections, differing in their degrees of oxidability: 1. Metals which decompose water even at 32°, with lively effervescence—namely, potassium, sodium, lithium, barium, strontium, calcium. 2. Metals which do not decompose water at 32°, like the metals of the preceding class; they do not decompose it with a lively effervescence, except at a temperature approaching 212°, or even higher, but always much below a red heat. In this class are found magnesium, glucinum, aluminum, zirconium, thorium, yttrium, cerium, and manganese. 3. Metals which do not decompose water except at a red heat, or at the ordinary temperature with the presence of strong acids. This section comprehends iron, nickel, cobalt, zinc, cadmium, tin, chromium, and probably vanadium. Iron is rapidly corroded in water containing carbonic acid, with the evolution of hydrogen. 4. Metals which decompose the vapour of water at a red heat with considerable energy, but which do not decompose water in presence of the strong acids. They are tungsten, molybdenum, osmium, tantalum, titanium, antimony, and uranium. These metals appear to be incapable of decomposing water in contact with acids, because their oxides have but a small basic power, being, indeed, bodies which are ranked among the acids. 5. Metals of which the oxides are not decomposed by heat alone, and which decompose water only in a feeble manner and at a very high temperature. They are also distinguished from the preceding class by their tendency to form basic and not acid oxides. These metals are copper, lead, and bismuth. 6. Metals of which the oxides are reducible by heat alone at a temperature more or less elevated: these metals do not decompose water in any circumstances. They are mercury, silver, palladium, platinum, gold, and probably rhodium and iridium. (Régnault, Annal. de Chim. et de Phys. lxii. 368.) It is to be remarked of nearly all the metals which decompose the vapour of water, and consequently separate hydrogen from oxygen at a certain temperature, that their oxides are reduced, notwithstanding, with great facility by hydrogen gas, and within the same limits of temperature. This anomalous result has already been adverted to in regard to iron (page 181).

Of the non-metallic elements, hydrogen only forms an oxide capable of uniting as a base with acids. It is a general character of the metals, on the contrary, to form such oxides, if tellurium be excepted, which is more analogous in its chemical properties to sulphur than to the metals. Hence, as the former class are principally salt-radicals, the latter are principally basyls.

The protoxides of metals are uniformly and strongly basic, but this feature becomes less distinct in their superior oxides, and passes into the acid character in the high degrees of oxidation of which some metals are susceptible. Thus, of manganese, the protoxide is a strong base; the sesquioxide basic, but in a less degree than the protoxide; the binoxide indifferent; and the still higher oxides are the manganic and permanganic acids, which are respectively isomorphous with sulphuric and perchloric acids. A few metals which have no protoxides, such as arsenic and antimony, are most remarkable for the acids they form with oxygen, and thus more resemble in their chemical history the elements of the non-metallic class. It is, indeed, impossible to draw an exact line of demarcation between the two classes of elements, either with reference to their physical or chemical properties.

Besides combining with oxygen, metals combine with sulphur, chlorine, and with other salt-radicals, whether simple or compound; and hence sulphides, chlorides, and numerous other series of metallic compounds. Of these series the sulphides most resemble the corresponding oxides of the same metals; the chlorides and other series partake more strongly of the saline character. Each metal, or class of metals, effects combination with oxygen in certain proportions, and combines also with sulphur, chlorine, &c. in the same proportions. Hence, given the formulæ of the oxides of a metal, the formulæ of its sulphides, chlorides, &c. may generally be predicated, as they correspond with the former. Thus the oxides of iron being FeO and Fe_2O_3, the sulphides are FeS and Fe_2S_3, and the chlorides $FeCl$ and Fe_2Cl_3; the oxides of arsenic, or arsenious and arsenic acids, being AsO_3 and AsO_5, the sulphides of that metal are AsS_3 and AsS_5, and the chlorides $AsCl_3$ and $AsCl_5$. But sometimes a metal unites with sulphur in more ratios than with oxygen; both iron and arsenic, for example, possessing each a sulphide to which they have no corresponding oxide, namely, iron pyrites and realgar, of which the formulæ are FeS_2 and AsS_2. The potassium family of metals combine also with three and five equivalents of sulphur, without all uniting with oxygen in such high proportions. Again, certain metals of the magnesian and its allied families, such as manganese and chromium, form acid compounds with oxygen, to which no corresponding sulphides exist, such as manganic and chromic acids, MnO_3 and CrO_3. But the circumstance that these acids are isomorphous with sulphuric acid, and the metals they contain isomorphous with sulphur, appears to be a sufficient reason why there should not be similar sulphur acids. The chlorides of a metal generally correspond in number, as they always do in composition, with the oxides; in some cases they are less numerous, but never, I believe, more numerous than the oxides of the same metal.

Combination takes place within a series; that is, oxides combine with oxides, sulphides with sulphides. Those members of the same series which differ greatly in chemical characters being most disposed to combine together, — as oxygen acids with oxygen bases, sulphur acids with sulphur bases. Chlorides also combine with chlorides, to form double chlorides, and iodides with iodides.

Compounds belonging to different series, on the contrary, do not in general combine together, but often mutually decompose each other when brought into contact. Thus hydrochloric acid and potassa do not unite, one belonging to the chlorine and the other to the oxygen series, but form water and chloride of potassium, by mutual decomposition, as explained in the following diagram: —

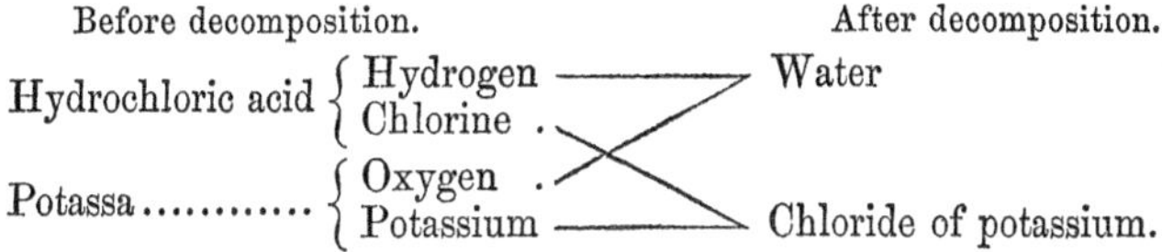

In the same manner, sesqui-oxide of iron, when dissolved in hydrochloric acid, produces water and a perchloride of iron corresponding with the peroxide: —

$$3HCl \text{ and } Fe_2O_3 = 3HO \text{ and } Fe_2Cl_3.$$

And in all cases when a metallic oxide dissolves in hydrochloric acid, without evolution of chlorine, the chloride produced necessarily corresponds with the oxide dissolved. Again, orpiment, or sulph-arsenious acid, does not combine with potassa, when dissolved in that alkaline oxide, the first being a sulphur and the second an oxygen compound, but gives rise to the formation of certain proportions of arsenious acid and sulphide of potassium: —

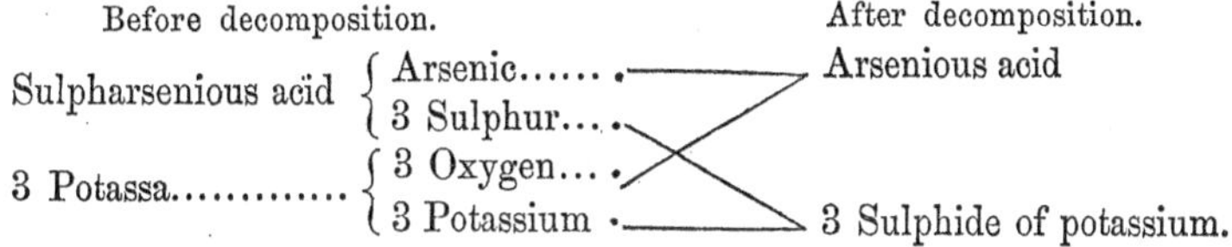

Two pairs of compounds of different series, then, co-exist in the liquid, — an oxygen acid, arsenious acid, which unites with the oxygen base, potassa, and a sulphur base, sulphide of potassium, which unites with undecomposed sulpharsenious acid. Hence the result of dissolving orpiment in potassa is the decomposition of both compounds and formation of two salts of different series, arsenite of potassa and sulpharsenite of sulphide of potassium.

The union of metallic compounds of the oxygen and sulphur series is a rare occurrence. But the red ore of antimony is such a combination, and oxisulphides of mercury also exist. Compounds of metallic oxides with metallic chlorides, and with other highly saline binary compounds, are more frequent; but they are not to be placed in the same category with the compounds of individuals both belonging to the same series, which last are neutral salts. For a metallic oxichloride may generally, if not always, be viewed as a chloride to which a certain proportion of metallic oxide is attached, like constitutional water in a hydrated salt. That metallic oxide is likewise always of the magnesian class, or of a class allied to it. Oxichlorides are then to be associated with those salts of oxygen-acids usually denominated subsalts (page 162); the oxichlorides of lead and of copper,—

$$PbCl + 3PbO \text{ and } CuCl + CuO,$$

with the subacetates and subsulphates of the same metals.

Arrangement of metallic elements.—A distribution of the metals into three classes is generally made, composed respectively of the metals of the alkalies and alkaline earths, the metals of the earths, and the metals proper. The latter class again is subdivided, according to the affinity of the metals contained in it for oxygen, into two groups—the noble and common metals; the oxides of the former, such as gold, silver, &c., abandoning their oxygen at a high temperature, while the oxides of the latter, lead, copper, &c., are undecomposable by heat alone. In treating of the metals, I shall introduce them in the order which appears to facilitate most the study of their combinations, with a general reference to this classification. For subdivisions, I shall avail myself of the natural families into which the elements have been arranged (page 144), which have the advantage of bringing together those metals of which the compounds are most frequently isomorphous. The different metals will therefore be grouped under the following orders:—

I. Metallic bases of the alkalies—three metals:—

	Oxides.
Potassium	Potassa
Sodium	Soda
Lithium	Lithia

II. Metallic bases of the alkaline earths—four metals:—

	Oxides.
Barium	Baryta
Strontium	Strontia
Calcium	Lime
Magnesium	Magnesia

III. Metallic bases of the earths proper—seven metals:—

	Oxides.
Aluminum	Alumina
Glucinum	Glucina
Zirconium	Zirconia
Yttrium	Yttria
Terbium	Terbia
Erbium	Erbia
Thorium	Thorina

IV. Metals proper, of which the protoxides are isomorphous with magnesia—eight metals:—

Manganese	Zinc
Iron	Cadmium
Cobalt	Copper
Nickel	Lead

V. Other metals proper having isomorphous relations with the magnesian family — seven metals: —

Tin	Tungsten
Titanium	Molybdenum
Chromium	Tellurium
Vanadium	

VI. Metals isomorphous with phosphorus — three metals: —

Arsenic	Bismuth
Antimony	

VII. Metals proper, not included in the foregoing classes, of which the oxides are not reduced by heat alone — eight metals: —

Uranium.	Titanium.
Cerium.	Tantalum or Columbium.
Lantanum.	Pelopium.
Didymium.	Niobum.

VIII. Metals proper, of which the oxides are reduced to the metallic state by heat (noble metals)—three metals:—

Mercury.	Gold.
Silver.	

IX. Metals found in native platinum (noble metals) — six metals:—

Platinum.	Osmium.
Palladium.	Rhodium.
Iridium.	Ruthenium.

ORDER I.

METALLIC BASES OF THE ALKALIES.

SECTION I.

POTASSIUM.

Syn. Kalium. *Eq.* 39 *or* 487.5; K.

The alkalies and earths have long been named and distinguished from each other, but they were not known to be the oxides of peculiar metals till a recent period. The terms applied to the new metallic bases are formed from the names of their oxides, as potassium from potash, and calcium from calx, a name sometimes given to lime; while the original names of the oxides are still retained, as those of ordinary objects, and not superseded by appellations indicating their relation to the metals, such as oxide of potassium for potassa, or oxide of calcium for lime.

Preparation.—In 1807, Sir H. Davy made the memorable discovery that potassa is resolved by a powerful voltaic battery into potassium and oxygen. He placed a moistened fragment of hydrate of potassa on mercury, introducing the terminal wire from the zinc extremity of an active battery (the chloroid) into the fluid metal, and touching the potassa with the other terminal wire (the zincoid); bubbles of oxygen gas appeared at the latter wire, and potassium was liberated at the former, and dissolving in the mercury, was protected from oxidation by the air. To effect this decomposition, Davy employed a battery of 200 pairs of four-inch plates; but an amalgam of potassium may be as readily obtained by a more simple voltaic apparatus, in the manner described at page 221. These processes, however, afford potassium only in minute quantity. Soon after the existence of this metal was known, Gay-Lussac and Thénard discovered that potassa is decomposed by iron at a white heat, and they contrived a process by which a more abundant supply of the metal was obtained. It was afterwards noticed by Curaudau, that potassa, like the oxides of common metals, is decomposed by charcoal as well as by iron, which is the basis of the process for potassium now always followed.

This interesting process is described by Mitscherlich, as it is successfully pursued in Germany. Whenever charcoal is used to deprive a metallic oxide of its oxygen, the former must be in a state of minute division, and be intimately mixed with the latter. Carbonate of potassa requires this precaution the more, that it fuses at a red heat, and is thus apt to separate from the charcoal, and sink below it. It is found that the best means to obtain a proper mixture of these substances is to calcine a salt of potassa containing a vegetable acid, which leaves a large quantity of charcoal when decomposed. Crude tartar (bitartrate of potassa) is preferred, and for one operation six pounds of that salt are ignited in a large crucible or melting-pot provided with a lid, so long as combustible gases are disengaged. The crucible is then withdrawn from the fire, and is found to contain a black mass, which is the mixture of charcoal and carbonate of potassa, known as black flux. It is reduced to powder, while still warm, and immediately mixed with about ten ounces of wood-charcoal in small pieces, or in a coarse powder, from which the dust has been separated by a sieve. The use of this additional charcoal is to act as a sponge, and absorb the potassa when liquefied by heat. The mixture is introduced into a bottle of wrought iron, and a mercury bottle (page 224) answers well for the purpose, but must be heated to redness beforehand, to expel a little mercury that remains in it. The mouth of the bottle is enlarged a little by means of a round file, and a straight iron tube of 4 or 5 inches in length fitted into the opening, by grinding. The bottle and tube thus form a retort, which is supported horizontally in a brick furnace, as represented (fig. 165) in which *a* is the iron bottle resting upon two bars of iron *o o*, to which it may also be firmly bound by iron wire. These bars cross the furnace at

Fig. 165.

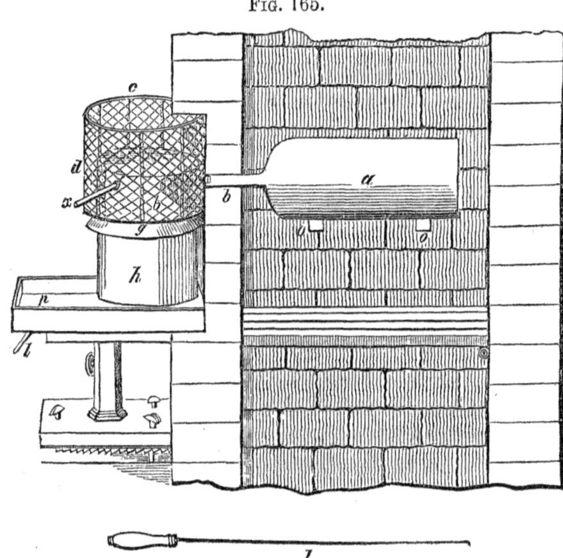

a height of 5 or 6 inches above the grate-bars. A mixture of equal parts of coal and coke makes an excellent fuel for this furnace. The tube *b* of the bottle projects through an aperture in the side-wall of the furnace, and enters a receiver of a peculiar construction required to condense the potassium, which distils over. This receiver is composed of two separate copper cylinders or oval boxes, hard soldered, similar in form and size, which are represented in section (fig. 166), the one, *b n d*, being introduced within the other, *g h k*, and thus forming together a vessel of which *b n d* is the cover. It will also be observed that *b d* is divided into two cells by a diaphragm, *i*, of the same length as the cylinder, and descending with it to within two inches of the bottom, *h*, of *g h k*. A ribbon of copper, *g*, is soldered around *b n d*, so as to form a ledge, which is seen in both figures, and serves as a support for a cage of iron-wire, *c d*, placed over the receiver during the distillation, to hold ice, and also to shed the water from the liquefaction of that ice, which falls into a tray, *p*, below, and flows off by the tube, *l*. The cover has also two short copper tubes, *d* and *b*, of which the copper of *b* is notched so as to clasp firmly by its elasticity the tube *b* from the iron bottle, which is fitted into it. The other tube, *d*, which is exactly opposite to *b*, is fitted with a cork, and the diaphragm, *i*, has a small hole in it to allow of a rod being passed through *b* and *d*. In the same part of the apparatus is a third opening, to which a glass tube, *x*, is fitted by a cork, for the escape of uncondensible gases. The receiver is filled to about one-third with rectified petroleum, a liquid containing no oxygen, so as to come nearly to, but not to cover, the bottom of the partition, *i*. The length of the bottle is 11 inches, its width 4, and the other parts of the apparatus are designed upon the same scale.

FIG. 166.

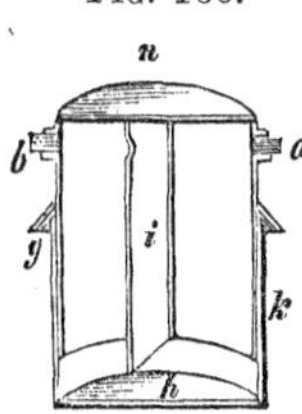

Potassium and carbonic oxide gas are the principal products of the decomposition of the carbonate of potassa, but other substances besides these are found in the receiver; namely, a black mass very rich in potassium, some oxalate and croconate of potassa and free potassa, with a portion of charcoal powder carried over mechanically. Part of these products appears to be formed, after the reduction of the potassium, by the mutual reaction of that metal, carbonic oxide and petroleum. The process is found to succeed best when the iron tube, *b*, is so short that it can be maintained at a red heat through its whole length during the operation, while the receiver is kept at a very low temperature; the potassium then falls from the tube, drop by drop, into the receiver, and does not remain long in contact with carbonic oxide, which is known to combine readily with that metal. One or two other points should always be attended to. The connexion between the tube *b* and the receiver is not made till the iron bottle has been heated to redness, to allow of the escape of a little water, and of a trace of mercury, which had remained in the bottle in the state of vapour, and which come off first. The joining of the tube *b* is not air-tight at first, and allows a little potassium vapour to escape, but this burns and forms potassa, which immediately closes the openings. This tube being always incandescent and the refrigeration properly made, the reduction sometimes proceeds without interruption. But the tube is sometimes obstructed, as appears by the gases ceasing to escape by *x*. Haste must then be made to open the tube *b*, and to clear it by means of a flattened iron rod, *l*, slightly hooked at its anterior extremity. Care has been taken to mark on this rod, with the scratch of a file, how far it has to penetrate into the apparatus to reach the mouth of the bottle, and it must not be introduced farther. The current of air through the furnace is regulated by a register valve in the chimney, and the fire stirred frequently so as to prevent the formation of cavities; the operator being guided in the management of the fire by the rapidity of the current of gas which escapes by the tube *x*. To terminate the operation, the grate bars may be thrown down, by which the fuel will fall into the ash-pit. The quantity of crude tartar mentioned yields about 4 ounces of potassium, which is about 4 per cent. of its weight. The potassium thus obtained, containing a little carbon chemically combined with it, is submitted, together with the black mass found in the receiver, to a second distillation. For this purpose a smaller iron bottle with a bent tube may be employed, the end of which is covered by rectified petroleum in a capacious flask, used as a receiver. (Mitscherlich, Elémens de Chimie, iii. 8).

Properties.—Potassium is solid at the usual temperature, but so soft as to yield like wax to the pressure of the fingers. A fresh surface has a white colour, with a shade of blue, like steel, but is almost instantly covered by a dull film of oxide when exposed to air. The metal is brittle at 32°, and has been observed crystallized in cubes: it is semi-fluid at 70°, and becomes completely liquid at 150°. It may be distilled at a low red heat, and forms a vapour of a green colour. Potassium is considerably lighter than water, its density being 0.865 at 60°.

Potassium oxidates gradually without combustion when exposed to air; but heated till it begins to vaporize, it takes fire and burns with a violet flame. The avidity of this metal for oxygen is strikingly exhibited when a fragment of it is thrown upon water. It instantly decomposes the water, and so much heat is evolved as to kindle the potassium, which moves about upon the surface of the water, burning with a strong flame, of which the vivacity is increased by the combustion of the hydrogen gas disengaged at the same time. A globule of fused potassa remains, which continues to swim about upon the surface of the water for a few seconds, but finally produces an explosive burst of steam, when its temperature falls to a certain point, illustrating the phenomenon of a drop of water on a hot metallic plate (page 64.)

Potassium appears to have the greatest affinity of all bodies for oxygen at temperatures which are not exceedingly elevated. It decomposes nitrous and nitric oxides, and also carbonic oxide gas at a red heat, although potassa is reduced to the metallic state by charcoal at a white heat. It has already been stated that the oxides and fluorides of boron and silicon are decomposed by potassium, and besides these elements, several of the metallic bases of the earths are obtained by means of this metal. It is, indeed, a reducing agent of the greatest value.

COMPOUNDS OF POTASSIUM.

Potassa, or potash; KO; 590 *or* 47.26.—Potassium exposed in thin slices to dry air becomes a white matter, which is the protoxide of potassium or potassa. This compound is fusible at a red heat, and rises in vapour at a strong white heat. It unites with water, with ignition, and forms a fusible hydrate, which is the ordinary condition of caustic potassa.

The hydrate of potassa is obtained in quantity from the carbonate of potassa. Equal weights of that salt and of quicklime are taken, the latter of which is slaked with water, and falls into a powder consisting of hydrate of lime; the former is dissolved in from 6 to 10 times its weight of water, and both boiled together for half an hour in a clean iron pan. The lime abstracts carbonic acid from the potassa, and becomes carbonate of lime; a reaction which may be illustrated by adding lime-water to a solution of carbonate of potassa, when a precipitate of carbonate of lime falls. When the potassa has been deprived entirely of carbonic acid, a little of the clear liquid taken from the pan will be found not to effervesce upon the addition of an acid to it. It is remarkable that the decomposition is never complete if the carbonate of potassa be dissolved in less than the prescribed quantity of water. Liebig has observed that a concentrated solution of potassa decomposes carbonate of lime, and consequently hydrate of lime could not, in the same circumstances, decompose carbonate of potassa. The pan, being covered by a lid, may be allowed to cool; when the insoluble carbonate of lime and the excess of hydrate of lime subside, a considerable quantity of the clear solution of potassa may be drawn off by a syphon, and the remainder may be obtained clear by filtration. In the latter operation a large glass funnel may be employed, to support a filter of washed cotton calico, into which what remains in the pan is transferred. A small portion of liquid, which passes through turbid at first, should be returned to the filter. As the solution of potassa absorbs carbonic acid, it is proper to conduct its filtration with as little exposure to air as possible; on which account the mouth of the the funnel should be covered by a plate, and the liquid which flows from it be immediately received in a bottle, in the mouth of which the funnel may be supported. The bottle in which potassa is preserved should not be of crystal, or of a material containing lead, as the alkali corrodes such glass, particularly when its natural surface has been cut.

To obtain the solid hydrate of potassa, the preceding solution is rapidly evaporated in a clean iron pan or silver basin, till an oily liquid remains at a high temperature, which contains no more than a single equivalent of water. This liquid is poured into cylindrical iron moulds to obtain it in the form of sticks, which are used by surgeons as a cautery, and are the *potassa* or *potassa fusa* of the Pharmacopœia; a form in which it is also convenient to have potassa for some chemical purposes. The sticks generally contain a portion of carbonate of potassa, besides a little oxide of iron and peroxide of potassium, the last of which gives occasion to the evolution of a little oxygen gas when the sticks are dissolved in water. To obtain hydrate of potassa free from carbonate, the sticks are dissolved in alcohol, in which the foreign impurities are insoluble, and the alcoholic solution is evaporated to dryness.

The pure and fused hydrate of potassa is a solid white mass of a structure somewhat crystalline, of sp. gr. 1.706, fusible at a heat under redness. It is a protohydrate, and cannot be deprived of its combined water by the most intense heat. It destroys animal textures. It rapidly deliquesces in damp air, from the absorption of moisture: is soluble in half its weight of water, and also in alcohol. Mixed in powder with a small quantity of water, it forms a second crystalline combination, which is a terhydrate; and its solution in water affords, at a very low temperature, crystals in the forms of four-sided tables and octohedrons, which are a pentahydrate, $KO.HO+4HO$.

The solution of potassa, or potassa ley, has a slight but peculiar odour, characteristic of caustic alkalies, which they acquire from their action upon organic matter, derived from the atmosphere or other sources. The skin and other animal substances are dissolved by this liquid. It is highly caustic, and its taste intensely acrid. It has those properties which are termed alkaline, in an eminent degree. It neutralizes the most powerful acids, restores the blue colour of reddened litmus, changes the blue infusion of cabbage into green, but in a short time altogether destroys these vegetable colours. It acts upon fixed oils, and converts them into soaps, which are soluble in water. It absorbs carbonic acid with great avidity from the air, on which account it should be preserved in well-stopped bottles.

The presence of free potassa or soda, in solutions of their carbonates, may be discovered by nitrate of silver, the oxide of which is precipitated of a brown colour by the caustic alkali, while the white carbonate of silver only is precipitated by the pure carbonated alkali. Potassa, whether free or in combination with an acid as a soluble salt, may be discovered and distinguished from soda and other substances, by means of certain acids, &c., which form sparingly soluble compounds with that alkali. A strong solution of tartaric acid produces a precipitate of bitartrate of potassa, in a liquid containing 1 per cent. of any potassa salt. The precipitate is crystalline, and does not appear immediately, but is thrown down on stirring the liquid strongly, and soonest upon the lines which have been described on the glass by the stirrer. A similar precipitation is occasioned in salts of potassa by perchloric acid. Also by bichloride of platinum, which forms the double chloride of platinum and potassium, in granular octohedrons of a pale yellow colour. In the separation of potassa for its quantitative estimation, the last reagent is preferred, and is added in excess to the potassa solution, together with a few drops of hydrochloric acid, which is then evaporated by a steam heat to dryness. The dry residue is washed with alcohol, which dissolves up everything except the double chloride of platinum and potassium. Ammonia, also, is thrown down by bichloride of platinum; but when the chloride of platinum and ammonium is heated to redness, nothing is left except spongy platinum, while the chloride of platinum and potassium leaves all its potassium in the state of chloride mixed with the platinum. Potassa is likewise separated from acids by means of fluosilicic acid, which throws down a light gelatinous precipitate, the double fluoride of silicon and potassium. Carbazotic acid also produces a yellow crystalline precipitate in solution of potassa.

Salts of potassa, more particularly the chloride, nitrate, and carbonate, communicate to flame a pale violet tint.

Potassa is the base which in general exhibits the highest affinity for acids; it precipitates lime and the insoluble metallic oxides from their solutions in acids. This alkali is employed indifferently with soda for a variety of useful purposes. The principal combinations of potassa with acids will be described after the binary compounds of potassium.

Peroxide of potassium, KO_3.—Heated strongly in air or oxygen, potassium combines with three equivalents of oxygen. The ultimate residue on calcining nitrate of potassa at red heat has been said to be the same compound, but Mitscherlich finds that residue to be potassa. The peroxide of potassium is decomposed by water, being converted into hydrate of potassa, with evolution of oxygen gas.

When potassium is burned with an imperfect supply of air, a grey matter is formed, which Berzelius believed to be a suboxide of potassium. It is not more stable than the peroxide.

Sulphides of potassium.—Sulphur and potassium, when heated together, unite with incandescence, and in several proportions, two of which correspond respectively with the protoxide and peroxide of potassium. The *protosulphide* may be obtained by transmitting hydrogen gas over sulphate of potassa, heated in a bulb of hard glass to full redness, when the whole oxygen of the salt is carried off as water, and the sulphur remains in combination with potassium, forming a fusible compound of a light brown colour. Sulphate of potassa calcined with one-fourth of its weight of pounded charcoal or pit-coal, in a covered Cornish crucible, at a bright red heat, is converted into a black crystalline mass, which is also protosulphide of potassium, with generally a small quantity of a higher sulphide, arising from the combination of the silica of the crucible with potassa of the sulphate. If lamp-black be used instead of charcoal, the sulphide of potassium formed having a great affinity for oxygen, and being in a highly divided state, takes fire when exposed to the air, and forms a pyrophorus. The solution of the protosulphide in water is highly caustic; it is decomposed by acids with effervescence, from the escape of hydrosulphuric acid, but without any deposit of sulphur. Being a sulphur base, it combines without decomposition with sulphur acids.

This sulphide unites directly with hydrosulphuric acid, forming KS.HS; and the compound may be otherwise formed, namely, by transmitting a stream of hydrosulphuric acid through caustic potassa, so long as the gas is absorbed. It is often named the *bihydrosulphate of potassa.* It is analogous in composition to hydrate of potassa (KO.HO) in the oxygen series.

The *trisulphide* is formed when anhydrous carbonate of potassa, mixed with half its weight of sulphur, is maintained at a low red heat so long as carbonic acid gas comes off. Of four proportions of potassa, three become sulphide of potassium, while sulphuric acid is formed, which neutralizes the fourth proportion of potassa: 4KO and $10S = 3KS_3$ and $KO.SO_3$. With carbonate of potassa and sulphur, in equal weights a similar action occurs, at a temperature above the fusing point of sulphur, but five, instead of three, proportions of sulphur then unite with one of potassium, and a *pentasulphide* is formed. With a larger proportion of carbonate of potassa the same sulphide is also produced, provided the temperature does not much exceed the boiling point of sulphur, and the excess of carbonate fuses along with it, without undergoing decomposition. A sulphide obtained by fusing sulphur and carbonate of potassa together has a liver-brown colour, and hence its old pharmaceutic name *hepar sulphuris.* The three sulphides described are deliquescent, and are all soluble in water, the higher sulphides giving red solutions. They may, indeed, be prepared by heating sulphur, in proper proportions, with caustic potassa. A simultaneous formation of hyposulphurous acid then occurs, as already explained (page 304). The preparation, *precipitated sulphur*, is obtained by adding an excess of hydrochloric acid to these solutions, when much sulphur is thrown down, although the potassium be only in the state of protosulphide, for the hydrosulphuric acid, arising from the action of the acid on that sulphide, meets hyposulphurous evolved at the same time from the decomposition of the hyposulphite, with the formation of water and sulphur. The excess of sulphur in the alkaline sulphide also precipitates at the same time. The peculiar whiteness of precipitated sulphur is owing, according to Rose, to its containing a little bisulphide of hydrogen.

Chloride of potassium; eq. 74.5 *or* 931.25; KCl. — Formed by the combustion of potassium in chlorine, or by neutralizing hydrochloric acid by potassa or its carbonate. It is also derived in considerable quantity from kelp (page 352). It crystallizes in cubes and rectangular prisms, resembles common salt in taste, and is considerably more soluble in hot than in cold water. According to the observations of Gay-Lussac, 100 parts of water dissolve of this salt 29.21 parts at 0° C.; 34.53 parts at 19°.35; 43.59 parts at 52°.39; 50.93 parts at 79°.58, and 59.26 parts at 109°.6 C. When pulverised and dissolved in four times its weight of cold water, it produces a depression of temperature of 20½ degrees; while chloride of sodium, dissolved in the same manner, lowers the temperature only 3.4 degrees. Upon the difference between two salts in this property, M. Gay-Lussac founded a method of estimating their proportions in a mixture. Chloride of potassium is principally consumed in the manufacture of alum. Rose observed that chloride of potassium unites with anhydrous sulphuric acid, $KCl+2SO_3$. The same salt unites with terchloride of iodine, $KCl.ICl_3$.

Iodide of potassium; eq. 165.36 *or* 2067; KI. — This salt is obtained by dissolving iodine in solution of potassa till neutral, evaporating to dryness, and heating to redness, to decompose the portion of iodate of potassa formed. M. Freundt recommends to add a little charcoal to the mixed iodide and iodate. Iodide of potassium is more soluble in water than the chloride, and may be obtained in cubes or rectangular prisms, which are generally white and opaque, and have an alkaline reaction from the presence of a trace of carbonate of potassa. Iodide of potassium is also dissolved by alcohol, but in a much less proportion than by water. The dry salt does not combine with more iodine, but in conjunction with a small quantity of water (I believe 4 equivalents) it absorbs the vapour of iodine with great avidity, and runs into a liquid of a deep red, almost black, colour. According to Baup, a saturated solution of iodide of potassium may dissolve so much as two equivalents of iodine, but allows one equivalent to precipitate when diluted. Iodide of potassium, which is often called the *hydriodate of potassa*, is much used in medicine; it is not poisonous even in doses of several drachms. Its solution is also employed as a vehicle for iodine itself, 20 grains of iodine and 30 grains of iodide of potassium being usually dissolved together in 1 ounce of water. The bromide of potassium is capable also of dissolving bromine, but the solution of chloride of potassium has no affinity for chlorine.

Ferrocyanide of potassium. Yellow prussiate of potassa; $K_2.FeCy_3+3HO$; *eq.* 184+27 *or* 2300+337.5. — This important salt is formed when carbonate of potassa is fused at a red heat in an iron pot, with animal matter, such as dried blood, hoofs, clippings of hides, &c., and is the product of a reaction to be hereafter described. This salt occurs in a state of great purity in commerce. It is of a lemon yellow colour, and crystallized in large quadrangular tables, with truncated angles and edges, belonging to the square prismatic system. The crystals contain 3 equivalents of water, which they lose at 212°, are soluble in 4 parts of cold and 2 parts of boiling water, and are insoluble in alcohol. The taste of this salt is saline, and it is not

Fig. 167.

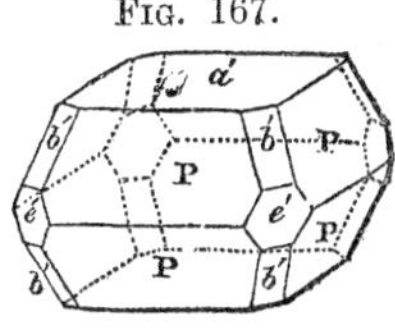

poisonous. By a red heat it is converted, with escape of nitrogen gas, into carburet of iron and cyanide of potassium; but with exposure to air the latter salt absorbs oxygen, and becomes cyanate of potassa. This salt is represented by Liebig as containing a salt-radical, *Ferrocyanogen*, composed of 1 eq. of iron and 3 eq. of cyanogen, or $FeCy_3$. This salt-radical is bibasic, and is in combination with 2 eq. potassium in the salt, as will be seen by reference to its formula. The same salt has been represented by myself as a compound of a tribasic salt-radical *prussine* (3Cy), with Fe+2K. But its reactions with other salts are most easily stated on the former view of its constitution. The iron in this salt is not precipitated by alkalies. When ferrocyanide of potassium is added to salts of lead and various other metallic solutions, it produces precipitates, in which two equivalents of the lead or other metal are substituted, in combination with ferrocyanogen, for the two equivalents of potassium. In salts of sesquioxide of iron, ferrocyanide of potassium produces the well-known precipitate, Prussian blue.

Ferricyanide of potassium, Red prussiate of potassa; $3K.Fe_2Cy_6$; *eq.* 329 *or* 4112.5.—This salt, which, like the last, is a valuable reagent, is formed by transmitting chlorine gas through a solution of the ferrocyanide of potassium, till it ceases to give a precipitate of Prussian blue with a persalt of iron, and no longer. One-fourth of the potassium of the ferrocyanide is converted into chloride, from which the resulting ferricyanide may be separated by crystallization. It forms right rhombic prisms, which are transparent and of a fine red colour. The crystals are anhydrous, soluble in 3.8 parts of cold, and in less hot water. They burn with brilliant scintillations when held in the flame of a candle. The solution of this salt is a delicate test of iron in the state of protoxide, throwing down from its salts a variety of Prussian blue, in which the 3K of the formula are replaced by 3Fe. Liebig views the red prussiate of potassa as containing a salt-radical, *Ferricyanogen*, or ferridcyanogen, Fe_2Cy_6, differing from ferrocyanogen in having twice its atomic weight and in being tribasic.

Cyanide of potassium; eq. 65 *or* 812.5; KCy.—The preparation of this salt is attended with difficulty, owing to the action of the carbonic acid of the air upon its solution, which evolves hydrocyanic acid, and the tendency of the solution itself to undergo spontaneous decomposition, even in close vessels. It may be formed by adding absolute hydrocyanic acid, or a strong solution of that acid, to a solution of potassa in alcohol; a portion of the cyanide falls down as a white crystalline precipitate, which should be washed with alcohol and dried, and an additional quantity is obtained by evaporating the liquid in a retort. But it is prepared with more advantage from the ferrocyanide of potassium, already described. That salt is carefully dried and reduced to a fine powder, 8 parts of which are mixed with 3 parts of carbonate of potassa and 1 part of charcoal, and exposed to a strong red heat in a closed iron crucible, or other convenient vessel. The mass is reduced to powder, placed in a funnel, moistened with a little alcohol, and then washed with cold water. The strong solution of cyanide of potassium which comes through is colourless, and must be rapidly evaporated to dryness in a porcelain basin, and fused at a red heat. The crude salt, obtained by ignition without charcoal, contains a little cyanate of potassa, but this does not interfere with its use for forming and dissolving cyanides of gold and silver, for the processes of voltaic gilding and plating.

Cyanide of potassium crystallizes in colourless cubes, which become opaque and deliquesce in damp air, and are very soluble in water. It bears a red heat without decomposition in close vessels, but with exposure to air absorbs oxygen, and becomes cyanate of potassa (KO.CyO). Its solution smells of hydrocyanic acid, being decomposed by carbonic acid. The action of cyanide of potassium upon the animal economy is equally powerful with that of hydrocyanic acid, and as the dry salt may be preserved in a well-stopped bottle without change, it is preferable to the acid, which is far from stable. Red oxide of mercury dissolves freely in the solution of cyanide of potassium, cyanide of mercury being formed and potassa set free. The purity of the alkaline cyanide may be ascertained from this property; 12 grains of the pure cyanide dissolving 20 grains of finely-pulverised oxide of mercury.

Hydrocyanic acid for medical purposes is conveniently prepared from this cyanide. 24 grains of cyanide of potassium, 56 grains of tartaric acid in crystals, and 1 ounce of water, are agitated together in a stout phial closed by a cork. The liquid is afterwards separated by filtration from the precipitate of bitartrate of potassa; it contains 10 grains of hydrocyanic acid, or rather more than 2 per cent. (Dr. Clark).

Sulphocyanide of potassium; $K.CyS_2$; 1222.2 *or* 97.92. — Sulphocyanogen is a salt-radical consisting of 2 eq. sulphur and 1 eq. cyanogen, which is formed on fusing the ferrocyanides with sulphur. To obtain it in combination with potassium, the ferrocyanide of potassium, made anhydrous by heat and reduced to a fine powder, is mixed with an equal weight of flowers of sulphur in a common cast-iron pot (pitch pot), and kept in a state of fusion for half an hour at a temperature above the melting point of sulphur, but below that at which bubbles of gas escape through the melted mass. No cyanogen is evolved or decomposed, and the residuary matter is a mixture of sulphocyanide of potassium and protosulphocyanide of iron, with the excess of sulphur. Both sulphocyanides dissolve in water, and give a solution which is colourless at first, but soon becomes red from oxidation of the sulphocyanide of iron. To get rid of the iron, carbonate of potassa is added to the boiling solution, so long as a precipitate of carbonate of iron falls, and the liquid is afterwards filtered. This solution gives crystals of sulphocyanide of potassium, when evaporated, which may be freed from any adhering carbonate of potassa by dissolving them in alcohol. The salt crystallizes in long white striated prisms, which are anhydrous, and resemble nitrate of potassa in their appearance and taste. They deliquesce in a damp atmosphere, and are very soluble in hot alcohol, from which the salt crystallizes on cooling. The sulphocyanide of potassium communicates a blood-red colour to solutions of salts of sesquioxide of iron, and is consequently employed as a test of that metal in its higher state of oxidation. The red solution is made perfectly colourless by a moderate dilution with water, when the iron is not present in excess. The sulphocyanide of potassium has been detected in the saliva of man and the sheep.

SALTS OF OXIDE OF POTASSIUM.

Carbonate of potassa; $KO.CO_2$; *eq.* 69 *or* 862.5. — This useful salt is principally obtained from the ashes of plants. Potassa is always contained in a state of combination in clay and other minerals which form the earthy part of soil, and appears to be a constituent of soil essential to vegetation. The alkali is appropriated by plants, and is found in their sap combined with vegetable acids, particularly with oxalic and tartaric acids; also with silicic and sulphuric acids, and as chloride of potassium. When the plants are dried and burned, the salts of the vegetable acids are destroyed, and leave carbonate of potassa: shrubs yielding three, and herbs five times as much saline matters as trees; and the branches of trees being more productive than their trunks—a distribution which may depend upon the potassa existing chiefly in the sap. The whole ashes from wood seldom exceed 1 per cent. of its weight, of which 1-6th may be saline matter. The solution, evaporated to dryness, yields *potashes;* and these, partially purified and ignited, form *pearlash.* The carbonate is mixed in the latter with about 20 per cent. of foreign salts, principally sulphate of potassa and chloride of potassium. The carbonate of potassa is obtained, in a state of greater purity, by dissolving pearlash in an equal weight of water, then separating the solution from undissolved salts, and evaporating it to dryness.

Carbonate of potassa is prepared of greater purity for chemical purposes, by igniting bitartrate of potassa; or better, by burning together 2 parts of that salt and 1 of nitre. In the latter process, the carbon and hydrogen of the tartaric acid are destroyed by the oxygen of the nitric acid, and carbonate of potassa remains mixed with charcoal, from which it may be separated by solution and filtration.

Carbonate of potassa has an acrid, alkaline taste, but is not caustic. It gives a green colour to the blue infusion of cabbage. This salt is highly deliquescent, and soluble in less than an equal weight of water at 60°. It may be crystallized with two equivalents of water. Added to solutions of salts of lime, lead, &c., it throws down insoluble carbonates. It is more frequently used than the caustic alkali, to neutralize acids and to form the salts of potassa.

FIG. 168.

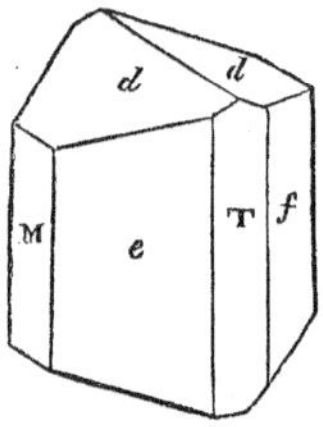

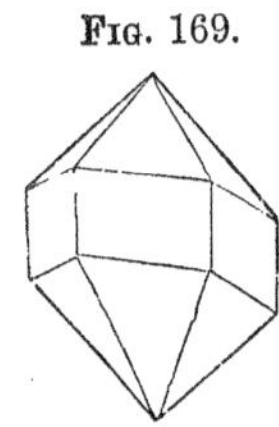

Bicarbonate of potassa; $HO.CO_2 + KO.CO_2$; *eq.* 100 *or* 1250. — Formed by transmitting a stream of carbonic acid gas through a saturated cold solution of the neutral carbonate. It is soluble in four times its weight of water at 60°, and in less water at 212°. The solution has an alkaline taste and reaction, but is not acrid; it does not thrown down magnesia from its soluble salts; it loses carbonic acid when evaporated at all temperatures, and becomes neutral carbonate. The salt contains one proportion of water, which is essential to it, and crystallizes well in prisms of eight sides, having dihedral summits. The existence of a sesquicarbonate of potassa is doubtful.

FIG. 169.

Sulphate of potassa; $KO.SO_3$; *eq.* 87 *or* 1087.5. — This salt precipitates when oil of vitriol is added, drop by drop, to a concentrated solution of potassa. It is generally prepared by neutralizing the residue, composed of bisulphate of potassa, of the nitric acid process (page 260), and crystallizes in double pyramids of six faces, or in oblique four-sided prisms. The crystals are anhydrous, unalterable in air, and they decrepitate strongly when heated; their density is 2.400. The sulphate is one of the least soluble of the neutral salts of potassa: 100 parts of water dissolve 8.36 parts of this salt at 32°, and 0.09666 parts more for each degree above that point.

Hydrated bisulphate of potassa, or sulphate of water and potassa; $HO.SO_3 + KO.SO_3$; *eq.* 136 *or* 1700: the fusible salt remaining, when nitrate of potassa is decomposed in a retort by two equivalents of oil of vitriol. Below 386.6° (197° C.), it is a white crystalline mass. This salt is very soluble in water, but is partially decomposed by that liquid, and deposits sulphate of potassa. It crystallizes from a strong solution in rhombohedral crystals, of which the form is identical with one of the forms of sulphur. But this salt is dimorphous, and crystallizes from a state of fusion by heat in large crystals, which have the form of felspar (Mitscherlich). Its density is 2.163. The excess of acid in this salt acts upon metals and alkaline bases very much as if it were free.

FIG. 170.

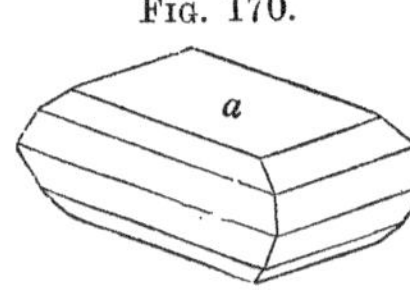

Hydrated sesquisulphate of potassa; $HO.SO_3 + 2(KO.SO_3)$. — A salt in prismatic needles discovered by Mr. Phillips, and which has also accidentally occurred since to Mr. Jacquelin. It is decomposed by water; the circumstances necessary for its formation are unknown.

Sulphate of potassa combines with *hydrated nitric and phosphoric acids*, as well as with hydrated sulphuric acid. On dissolving the neutral salt in nitric acid, a little nitre and hydrated bisulphate of potassa are formed, with a large quantity of a salt in oblique prisms, of which the formula is $HO.NO_5 + 2(KO.SO_3)$. This last salt fuses at 302° (150° C.); its density is 2.38 (Jacquelin). The compound with phosphoric acid is formed by dissolving sulphate of potassa in a syrupy solution of that acid, and crystallizes in oblique prisms of six sides, which fuse at 464° (240° C.), and of which the density is 2.296 (Jacquelin). Its formula is $3HO.PO_5 + 2(KO.SO_3)$. It will be observed that both these compounds agree with Mr. Phillips's sesquisulphate in having 2 eq. sulphate of potassa to 1 eq. of hydrated acid. (Annales de Chimie, lxx).

Nitrate of potassa, Nitre, Saltpetre; $KO.NO_5$; *eq.* 101 *or* 1262.5. — Nitric acid is formed in the decomposition of animal matters containing nitrogen, when they are exposed to air, and are in contact with alkaline substances. It appears to be largely produced in this way in the soil of certain districts of India, from which nitrate of potassa is obtained by lixiviation. Nitrous soils always contain much carbonate of lime, the debris of tertiary calcareous rocks, in which the oxygen and nitrogen of the air unite, according to some, assisted by the porous structure of the rock, and under the influence of an alkaline base, so as to generate nitric acid without the intervention of animal matter. But this conjecture is not founded upon experiment; nor is it a necessary hypothesis, since nitrifiable rocks are never entirely destitute of organic matter. Nitrate of potassa is also prepared in some countries of Europe, by imitating the natural process, in artificial nitre beds, wherein nitrate of lime is formed, and afterwards converted into nitrate of potassa by the addition of wood-ashes to the lixivium.[1]

FIG. 171.

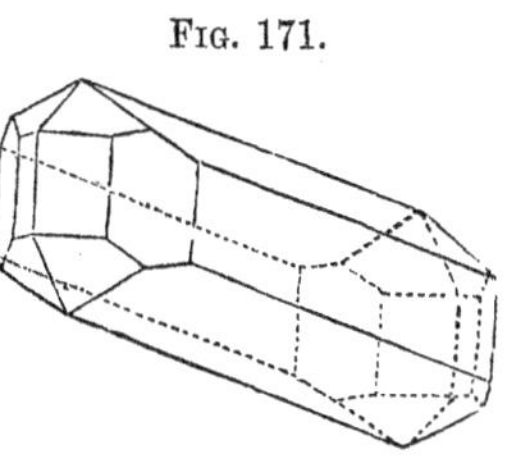

Nitrate of potassa generally crystallizes in long striated six-sided prisms, is anhydrous, unalterable in the air, fusible into a limpid liquid by a heat under redness, in which condition it is cast in moulds, and forms *sal prunelle*. Its density is 1.933 (Dr. Watson). According to Gay-Lussac 100 parts of water dissolve 13.3 parts of this salt at 32°, 29 parts at 64.4°, 74.6 parts at 96.8°, and 236 parts at 206.6°. The taste of the solution is cooling and peculiar; it has considerable antiseptic properties. Nitre is insoluble in absolute alcohol.

From the large quantity of oxygen which nitre contains, and the facility with which it imparts that element to combustibles at a red heat, it is much employed in making gunpowder and other deflagrating mixtures. An intimate mixture of nitre in fine powder with one-third of its weight of wood-charcoal, when touched by a body in ignition, burns with great brilliancy, but without explosion. A mixture of 3 parts of nitre, 2 of dry carbonate of potassa, and 1 of sulphur, forms *pulvis fulminans*, which, heated gently till it enters into fusion, inflames suddenly, and explodes with a deafening report. The violence of the explosion is caused by the reaction between the sulphur and nitre being instantaneous, from their fusion and perfect intermixture, and the consequent sudden formation of much nitrogen gas from the decomposition of nitric acid. Gunpowder contains both sulphur and charcoal, of which the former serves the purpose of accelerating the process of deflagration and supplying heat, while the latter supplies much of the gas, to the formation of which the available force of the explosion is due. Gunpowder yields about 300 times its volume of gas, measured when cold; but its explosive force is greater than this indicates, from the high temperature of the gas, and not less than 1000 atmospheres. The ordinary proportions of gunpowder approach very nearly 1 eq. of nitre, 1 of sulphur, and 3 of carbon, as will be seen by the following comparison:—

Composition of Gunpowder.

	Theoretical Mixture.	English.	Prussian.
Sulphur	11.9	12.5	11.5
Charcoal	13.5	12.5	13.5
Nitre	74.6	75.	75.
	100.0	100.0	100.0

[1] The observations and original experiments upon nitrification, of Professor Kuhlman, are valuable, but do not lead to any general theory of the process. He did not succeed in causing oxygen and nitrogen gases to combine by means of spongy platinum, but he found that under the influence of that substance (1°) all vaporisable compounds of nitrogen, including ammonia, mixed with air, with oxygen, or with an oxidating gas, change into nitric acid or peroxide of nitrogen; and (2°) that all the vaporisable compounds of nitrogen, including nitric acid, mixed with hydrogen or a hydrogenous acid, give rise to ammonia.—(Memoirs of the Academy of Sciences of Lille, 1838, and Liebig's Annalen, xxix. 272, 1839.)

By the combustion of the mixture, carbonic acid and nitrogen gases are formed, with a solid residue of protosulphide of potassium. Thus:—

Deflagration of Gunpowder.

Before decomposition.			After decomposition.
3 Carbon..........	3 Carbon.........	⟶	3 Carbonic acid
Nitrate of potassa	6 Oxygen........	↗	
	Nitrogen.........	⟶	Nitrogen
	Potassium........	↘	
Sulphur............	Sulphur..........	⟶	Sulphide of potassium.

A portion of the potassa is always converted into sulphate of potassa, which must interfere with the exactness of this decomposition. Blasting powder is composed of 20 sulphur, 15 charcoal, and 65 nitre; the proportion of sulphur being increased, by which a more powerfully explosive mixture is obtained, but which is not suitable for fire-arms, as they are injured by an excess of sulphur. The most inflammable charcoal is employed in making gunpowder; which is obtained by calcining branches of about $\frac{3}{4}$ths of an inch in diameter, in an iron retort, for a considerable time, at a heat scarcely amounting to redness, and which has a brown colour without lustre. The granulation of gunpowder increases its explosive force. A charge is thus made sufficiently porous to allow flame to penetrate it, and to kindle every grain composing it at the same time. But still the discharge of gunpowder is not absolutely instantaneous; and it is remarkable that other explosive compounds which burn more rapidly than gunpowder, such as fulminating mercury, are not adapted for the movement of projectiles. Their action in exploding is violent but local; if substituted for gunpowder in charging ordinary fire-arms, they would shatter them to pieces, and not project the ball. It is a common practice to mix with the charge of blasting powder, used in mining, a considerable bulk of sawdust, which renders the combustion of the powder still slower, but productive of a sustained effort, most effectual in moving large masses.

Chlorate of potassa; $KO.ClO_5$; *eq.* 122.5 *or* 1531.25.—This salt is the result of a reaction between chlorine and potassa, which has already been explained (page 341). In the preparation of chlorate of potassa, a strong solution of two or three pounds of carbonate of potassa is made, and chlorine passed through it. The gas is conducted into the liquid by a pretty wide tube, or better by a tube terminated by a funnel, to prevent its being choked by the solid salt which is formed. A stage in the process can be observed before the liquid has discharged much carbonic acid, when bicarbonate, chlorate, and hypochlorite of potassa exist together in solution, and a considerable quantity of chloride of potassium is deposited. The latter salt is removed, and the current of chlorine continued till the liquid, which is often red from hypermanganic acid, becomes colourless or yellow, and ceases to absorb the gas. A considerable quantity of chlorate of potassa is deposited in tubular shining crystals, which are purified by solution and a second crystallization; and more of the same salt is obtained from the liquid evaporated and set aside to crystallize; the separation of the chlorate from chloride of potassium depending upon the solubility at a low temperature of the former salt being greatly less than that of the latter.

The chlorate of potassa may be prepared more economically by exposing to a current of chlorine gas a mixture of 7.6 parts of carbonate of potassa, and 16.8 hydrate of lime in a dry or only slightly damp state. Chlorate of potassa is formed with carbonate of lime and chloride of calcium. The mass is treated with boiling water, which dissolves the chloride of calcium and chlorate of potassa. The latter salt is purified by crystallization. It is stated that other salts of potassa, particularly the sulphate, may be substituted for the carbonate in this process; and that the potassa salt and lime are mixed with hot water when exposed to the chlorine gas.

This salt is anhydrous. It appears in flat crystals of a pearly lustre, of which the forms, according to Brooke, belong to the oblique prismatic system. Its density is 1.989 (Hassenfratz). It has a cooling, disagreeable taste, like that of nitre. According to Gay-Lussac, 100 parts of water dissolve 3⅓ parts of chlorate of potassa at 32°, 6 at 59°, 12 at 95°, 19 at 120.2°, and 60 at 219.2°, the point of ebullition of a saturated solution. This salt fuses readily in a glass retort or tube, enters into ebullition, and discharges oxygen below a red heat. At a certain period in the decomposition, when the mass becomes thick, hyperchlorate of potassa is formed, but ultimately chloride of potassium is the sole residue.

Chlorate of potassa deflagrates with combustibles more violently than the nitrate. A grain or two of it rubbed in a warm mortar with an equal quantity of sulphur, occasions smart explosions, with the formation of sulphurous acid gas. Inclosed with a little phosphorus in paper, and struck by a hammer, it produces a powerful explosion; but this experiment may be attended with danger to the operator from the projection of the flaming phosphorus. A mixture which, when dry, inflames by percussion, and which was applied to lucifer matches, is composed of this salt, sulphur, and charcoal. One of the simplest receipts for this percussion powder consists in washing out the nitre from 10 parts of ordinary gunpowder with water, and mixing the residue intimately, while still humid, with 5¼ parts of chlorate of potassa in an extremely fine powder. This mixture is highly inflammable when dry, and dangerous to preserve in that state. Phosphorus and nitre, however, are now more generally used for these matches (page 315). More chlorate of potassa is employed in the processes of calico-printing, as an oxidizing agent.

Perchlorate of potassa; $KO.ClO_7$; *eq.* 138.5 *or* 1731.25.—Processes for preparing this salt have already been described under perchloric acid (page 342). It is also formed in a strong solution of chlorate of potassa contained in the decomposing cell of a voltaic battery, this salt being deposited in small crystals upon the zincoid, and no oxygen liberated there. It requires 55 parts of water to dissolve it at 59°, but is largely soluble in boiling water. It crystallizes in octohedrons with a square base, which are generally small: they are anhydrous. It deflagrates less strongly with combustibles than the chlorate, loses oxygen at 400°, and is completely decomposed at a red heat, chloride of potassium being left.

Iodate of potassa, $KO.IO_5$; *eq.* 213.36 *or* 2667.—This salt may be formed by neutralizing the chloride of iodine with carbonate of potassa, instead of carbonate of soda (p. 356). It gives small anhydrous crystals, which fuse by heat and lose all their oxygen. Iodic acid likewise forms a biniodate and a teriodate of potassa, according to Serullas. (Annal. de Chim. et de Phys. xliii.) The *biniodate* is obtained by adding an additional proportion of iodic acid to a solution of neutral iodate saturated at a high temperature: it contains an equivalent of water, but may be made anhydrous by a strong heat, according to my own observations. It occurs in prisms with dihedral summits, and requires 75 parts of water at 59° to dissolve it. The *teriodate* is obtained on mixing a strong acid, such as nitric, hydrochloric, or sulphuric, with a hot saturated solution of the neutral iodate, and allowing it to cool slowly. It crystallizes in rhombohedrons, and requires 25 parts of water to dissolve it.

Serullas has observed that the biniodate of potassa has a great disposition to form double salts. A compound with *chloride of potassium*, to which he assigned the formula $KCl+KO.I_2O_{10}$, is obtained on adding a little hydrochloric acid to a solution of iodate of potassa, and allowing the solution to evaporate spontaneously. This salt crystallizes well, but afterwards loses its transparency in the air. It is decomposed by water, and cannot be formed by uniting its constituent salts. Another compound contains bisulphate of potassa: $KO.S_2O_6+KO.I_2O_{10}$. These compounds of iodic acid have also been lately examined by M. Millon.

SECTION II.

SODIUM.

Syn. Natrium. Eq. 23 *or* 287.5; Na.

Davy obtained this metal by the voltaic decomposition of soda, immediately after the discovery of potassium. An intimate mixture of charcoal and carbonate of soda is formed by calcining acetate of soda, from which sodium is commonly prepared, according to the method described for potassium, and with greater facility, owing to the lower affinity of sodium for oxygen.

Sodium is a white metal having the aspect of silver. Its density is 0.972, at 59°, according to Gay-Lussac and Thénard. This metal is so soft, at the usual temperature, that it may be cut with a knife, and yields to the pressure of the fingers; it is quite liquid at 194°. It oxidates spontaneously in the air, although not so quickly as potassium; and when heated nearly to redness takes fire and burns with a yellow flame. Thrown upon water, it oxidates with great vivacity, but without inflaming, evolving hydrogen gas, and forming an alkaline solution of soda. When a few drops only of water are applied to sodium, it easily becomes sufficiently hot to take fire.

As potassium is in some degree characteristic of the vegetable kingdom, so sodium is the alkaline metal of the animal kingdom, its salts being found in all animal fluids. Both of these elements occur in the mineral world; of the two, perhaps potassium is most extensively diffused; felspar, the most common of minerals, containing 12 per cent. of potassa, but from the existence everywhere of a soluble compound of sodium, its chloride, the sources of that element are the more accessible, if not the most abundant.

The anhydrous protoxide of sodium and the peroxide are prepared in the same manner as the corresponding oxides of potassium, which they greatly resemble in properties. The composition of the peroxide of sodium, however, is different, being expressed by the formula $2Na+3O$ (Thénard). It is supposed by M. Millon to be $Na+2O$.

COMPOUNDS OF SODIUM.

Soda; NaO; *eq.* 31 *or* 387.5.—A solution of soda is obtained by decomposing the crystallized carbonate of soda, dissolved in four or five times its weight of water, by means of half its weight of hydrate of lime; the same points being attended to as in the preparation of potassa. A preference is given to this alkali from its cheapness, for most manufacturing purposes, and in the laboratory it may frequently be substituted for potassa, where a caustic alkali is required. On the large scale it is prepared from *salts of soda*, a carbonate containing chloride of sodium and sulphate of soda. The solution of soda is purified from these salts by concentrating it considerably, upon which the foreign salts cease to be soluble in the liquid, and precipitate. (Mr. W. Blythe).

The following table, constructed by Dr. Dalton, exhibits the quantity of caustic soda in solutions of different densities:—

Solution of Caustic Soda.

Density of the Solution.	Alkali per cent.	Density of the Solution.	Alkali per cent.
2·00	77·8	1·40	29·0
1·85	63·6	1·36	26·0
1·72	53·8	1·32	23·0
1·63	46·6	1·29	19·0
1·56	41·2	1·23	16·0
1·50	36·8	1·18	13·0
1·47	34·0	1·12	9·0
1·44	31·0	1·06	4·7

The solid hydrate of soda is obtained by evaporating a solution of soda, precisely in the same manner as the corresponding preparation of potassa. It is soluble in all proportions in water and alcohol.

Soda is distinguished from potassa and other bases by several properties:—1st. All its salts are soluble in water, and it is therefore not precipitated by tartaric acid, chloride of platinum, or any other reagent. 2d. With sulphuric acid it affords a salt which crystallizes in large efflorescent prisms, easily recognised as Glauber's salt. 3d. Its salts communicate a rich yellow tint to flame.

Sulphides of sodium.—These compounds so closely resemble the sulphides of potassium as not to require a particular description. The protosulphide of sodium crystallizes from a strong solution in octohedrons. This salt contains water of crystallization; in contact with air it rapidly passes into caustic soda, and the hyposulphite of the same base.

Chloride of sodium, Sea salt, Common salt, NaCl; *eq.* 58.5 *or* 731.25.—Sodium takes fire in chlorine gas, and combining with that element, produces this salt. The chloride of sodium is also formed on neutralizing hydrochloric acid, by soda or its carbonate, and is obtained thus in the greatest purity. Sea-water contains 2.7 per cent. of chloride of sodium, which is the most considerable of its saline constituents: (analysis of sea-water, page 242). Salt is obtained from that source in warm climates, as at St. Ubes, in Portugal, on the coast of the Mediterranean near Marseilles, and other places where spontaneous evaporation proceeds rapidly; the sea-water being retained in shallow basins or canals, on the surface of which a saline crust forms, with the progress of evaporation, which is broken and raked out. Sea-water is also evaporated artificially, by means of culm, or waste coal, as fuel, on some parts of the coast of Britain, but as much for the sake of the *bittern* as of the common salt it affords. The evaporation is not carried to dryness, but when the greater part of the chloride of sodium is deposited in crystals, the mother liquid, which forms the bittern, is drawn off; it is the source of a portion of the Epsom salt and other magnesian preparations of commerce. Other inexhaustible sources of common salt are the beds of sal-gem or rock salt, which occur in several geological formations posterior to the coal, as at Northwich in Cheshire, in Spain, Poland, and many other localities. These beds appear to have been formed by the evaporation of salt lakes without an outlet, in which the saline matter, continually supplied by rivers, had accumulated, till the water being saturated, a deposition of salt took place upon the bottom of the lake. The Dead Sea is such a lake, and the bottom of it is found to be covered with salt. The salt is sometimes sufficiently pure for its ordinary uses, as it is taken from these deposits, but more generally it is coloured brown from an admixture of clay, and requires to be purified by solution and filtration. Instead of sinking a shaft to the bed of the rock salt, and mining it, the superior strata are often pierced by a bore of merely a few inches in diameter, by which water is admitted to the bed, and the brine formed drawn off by a pump and pipe of copper suspended in the same tubular opening.

Chloride of sodium crystallizes from solution in water in cubes, and sometimes from urine and liquids containing phosphates in the allied form of the regular octohedron. Its crystals are anhydrous, but decrepitate when heated, from the expansion of water confined between their plates. According to Fuchs, pure chloride of sodium has exactly the same degree of solubility in hot and cold water, requiring 2.7 parts of water to dissolve it at all temperatures; but it has been proved by Gay-Lussac, and also by Poggiale, that the solubility of this salt increases sensibly, although not considerably, with the temperature. According to Poggiale 100 parts of water dissolve of chloride of sodium 35.52 parts at 32°; 35·87 parts at 57.2 (14° C.); 39.61 parts at 212° (100° C.); and 40.35 parts at 229.46° (109·7° C.), the temperature of ebullition of a saturated solution (Annales de Ch. 3me Sér. viii. 469). Gay-Lussac also makes the boiling point of a saturated solution 229.5°, but that temperature is too high (I believe) for a solution of pure chloride of sodium. When a saturated solution is exposed to a low temperature between 14° and 5°, the salt crystallizes in hexagonal tables, which have two sides larger than the others. Fuchs found these crystals to contain 6, and Mitscherlich 4 equivalents of water. If their temperature is allowed to rise above 14°, they undergo decomposition, and are converted into a congeries of minute cubes, from which water separates.

The little increase of the solubility of chloride of sodium at a high temperature, makes it impossible to crystallize this salt by cooling a hot solution, but Mr. Arrott finds that with the addition of chloride of calcium to the solution, a greater inequality of solubility at high and low temperatures takes place, and a portion of the chloride of sodium crystallizes from a hot saturated solution on cooling. In the evaporation of brine for salt, certain inconveniences attend the deposition of salt from the boiling solution, which Mr. Arrott proposes to obviate by the presence of chloride of calcium.

Pure chloride of sodium has an agreeable saline taste, deliquesces slightly in damp weather, and dissolves largely in rectified spirits, but is very slightly soluble in absolute alcohol. Its density is 2.557 (Mohs). It fuses at a bright red heat, and at a higher temperature rises in vapour. It is immediately decomposed by oil of vitriol, with the evolution of hydrochloric acid. Besides being used as a seasoning for food, chloride of sodium is employed in the preparation of the sulphate and carbonate of soda. When ignited in contact with clay containing oxide of iron, the sodium of this salt becomes soda, and unites with the silica of the clay, while the chlorine combines with iron, and is volatilized as sesquichloride of iron. On this decomposition is founded the mode of communicating the salt-glaze to pottery: a quantity of salt is thrown into the kiln, where it is converted into vapour by the heat, and condensing upon the surface of the pottery causes its vitrification, which is attended with the formation of hydrochloric acid, and of sesquichloride of iron, if sesquioxide of iron be present. When chloride of sodium and silica, both dry, are heated together, no decomposition takes place; but if steam is passed over the mixture, hydrochloric acid is evolved and silicate of soda formed. These decompositions are represented by the following equations:—

$$SiO_3 \text{ and } 3NaCl \text{ and } Fe_2O_3 = 3NaO.SiO_3 \text{ and } Fe_2Cl_3$$
$$SiO_3 \text{ and } NaCl \text{ and } HO = NaO.SiO_3 \text{ and } HCl.$$

The second reaction has not been applied successfully to the preparation of soda from the chloride of sodium, owing, it is said, to the vitrification of the silicate of soda produced, which covers the undecomposed chloride of sodium, and protects it from the steam. Mr. Tilghman substitutes for the silica precipitated alumina, which is made up into balls with the chloride of sodium, and exposed to steam in a reverberatory furnace at an elevated temperature. Hydrochloric acid escapes, and an aluminate of soda is formed, which may be decomposed, when cold, by dry carbonic acid; the carbonate of soda is dissolved out by water; the alumina is made up again into balls with chloride of sodium, to be ignited and decomposed by steam as before.

The bromide and iodide of sodium crystallize in cubes, and resemble in properties the corresponding compounds of potassium.

Fig. 172.

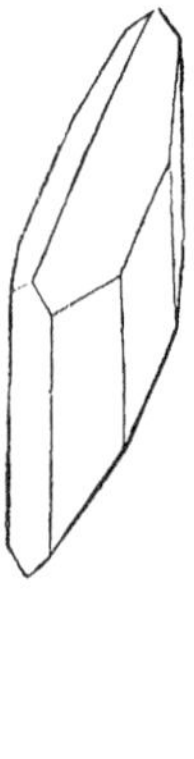

SALTS OF OXIDE OF SODIUM.

Carbonate of soda; $NaO.CO_2+10HO$; *eq.* 53+90, *or* 662.5+1125.—This useful salt is found nearly pure in commerce, in large crystals, which effloresce when exposed to air. These crystals contain 10 equivalents of water, and consist, in 100 parts, of 21.81 soda, 15.43 carbonic acid, and 62.76 water. According to Dr. Thomson, they generally contain about ½ per cent. of sulphate of soda as an accidental impurity: they belong to the oblique prismatic system. Their density is 1.623: 100 parts of water dissolve 20.64 of the crystals at 58.25°, and more than an equal weight at the boiling temperature (Dr. Thomson). In warm weather, the carbonate of soda sometimes crystallizes in another form, which is not efflorescent, and of which the proportion of water is 8 equivalents. The ordinary crystals, by efflorescing in dry air, are reduced to a hydrate of 5 equivalents of water, $NaO.CO_2+5HO$. The same hydrate appears when a solution of carbonate of soda is made to crystallize at 93° (34° C.), in crystals derived from an octohedron with a square base. Again, a solution of this salt evaporated between 158° and 176° (70° and 80° C.), deposits quadrilateral crystals, containing 1 equivalent of water, or 14.77 per cent. Carbonate of soda, therefore, appears to be capable of forming four definite hydrates, containing HO, 5HO, 8HO, and 10HO. The density of the anhydrous salt is 2.509 (Filhol).

The solubility of the carbonate of soda, supposed to be anhydrous, at various temperatures, was observed by M. Poggiale to be as follows:—

100 parts of water at	32° (0° C.)	dissolve	7.08	of carbonate of soda.
100 " "	50° (10° C.)	"	16.66	" "
100 " "	68° (20° C.)	"	25.83	" "
100 " "	86° (30° C.)	"	35.90	" "
100 " "	219.2° (104° C.)	"	48.50	" "

To obtain such determinations of the solubility of a salt at a given temperature, water is kept in contact with a considerable excess of the salt in the state of powder for at least half an hour, at the fixed temperature, with occasional agitation. About two ounces of the solution is then transformed into a light glass flask (fig. 173), and

Fig. 173.

after being accurately weighed, is evaporated either over the gas, or by a small furnace, taking care to hold the neck at an angle of 45°, to avoid drops of fluid being thrown out by the ebullition. After the salt is dry, the heat is still continued, to expel the water of crystallization, the escape of the latter being promoted by blowing air gently into the flask while hot by means of bellows having a bent glass tube attached to the nozzle.

This salt has a disagreeable alkaline taste. When heated, it undergoes the watery fusion; its water is soon dissipated, and a white anhydrous salt remains, which again becomes liquid at a red heat, undergoing then the igneous fusion, and by a greater heat it loses no carbonic acid. A mixture of carbonates of potassa and soda is more fusible than either salt separately.

Carbonate of soda is decomposed at a bright red heat by the vapour of water, which disengages all the carbonic acid, and produces hydrate of soda, $NaO.HO$. The carbon of its acid is also set at liberty by phosphorus at a high temperature, and the phosphate of soda formed. Lime, baryta, strontia, and magnesia, decompose a solution of carbonate of soda, assuming its carbonic acid and liberating soda.

Carbonate of soda is manufactured by a process which will be described immediately under the head of sulphate of soda. Much of the carbonate of commerce is not crystallized, but simply evaporated to dryness, and is then known as *salts of soda*, *soda-salt*, or *soda-ash*. In this form it generally contains chloride of sodium, sulphate of soda, hydrate of soda, and often insoluble matter, and varies considerably in value. The soda which is caustic, and that in combination with carbonic acid alone of the acids, are available in the application of the salt as an alkaline substance. The pure anhydrous carbonate of soda consists of 58.58 soda and 41.42 carbonic acid, and the best soda-salts of commerce contain from 50 to 53 per cent. of available soda. The operation of ascertaining the proportion of alkali in these salts, and in other forms of the carbonate of soda, is a process of importance from its frequent occurrence, and of high interest and value as a general method of analysis of easy execution, and applicable to a great variety of substances. I shall therefore describe minutely the mode of conducting it.

ALKALIMETRY.

The experiment is, to find how many measures of a diluted acid are required to destroy the alkaline reaction of, and to neutralize 100 grains of a specimen of soda-salt. (1.) The acid is measured in the alkalimeter, which is a straight glass tube, or very narrow jar, with a lip (fig. 174), about 5-8ths of an inch in width, and 14 or 15 inches in height, generally mounted upon a foot, which is by no means advantageous, as *a*, (fig. 175), capable of containing at least 1000 grs. of water. It is graduated into 100 parts, each of which holds ten grains of water. In the operation of dividing such an instrument, it is more convenient to use measures of mercury than water,—135.68 grains of mercury being in bulk equal to 10 grains of water, 678.40 grains will be equal to 50 grains of water. A unit measure may be formed of a pipette, *b*, made to hold the last quantity of mercury, into which the metal is poured, the opening at the point of the pipette being closed by the finger, and the height of the mercury in the tube marked by a scratch on the glass made by a triangular file. The bulk of twice that quantity of mercury, or 100 water grain measures, may likewise be marked upon the tube. The former quantity of mercury is then decanted from the tube into the alkalimeter to be graduated, and a scratch made upon the latter at the mercury surface: this is 5 of the 10-grain water measures. Another measure is added, and its height marked; and the same repeated till 20 measures of mercury in all have been added, which are 100 ten-grain water measures. The subdivision of each of these measures into 5 is best made by the eye, and is also marked on the alkalimeter. The divisions are lastly numbered, 0, 5, 10, &c., counting from above downwards, and terminating with 100 on the sole of the instrument. Several alkalimeters may be graduated at the same time, with little more trouble than one, the measured quantities of mercury being transferred from one to the others in succession.

Fig. 174.

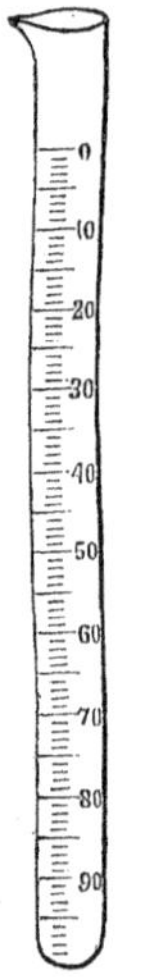

Fig. 175.

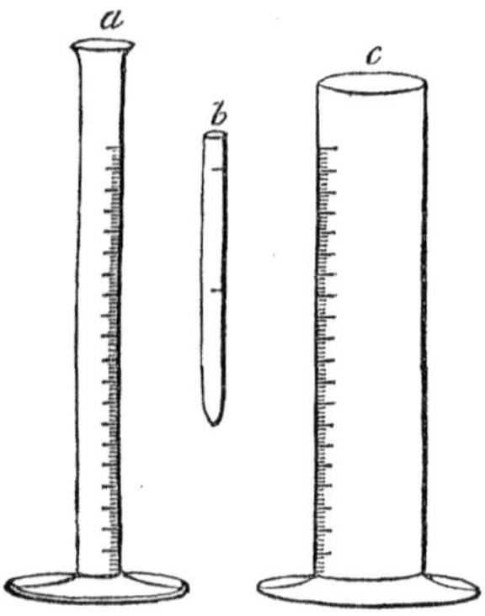

(2.) To form the test acid, 4 ounces of oil of vitriol are diluted with 20 ounces of water; or larger quantities of acid and water are mixed in these proportions. About three-fourths of an ounce of bicarbonate of soda is heated strongly by a lamp for an hour, to obtain pure carbonate of soda; of which 171 grains are immediately weighed; that quantity, or more properly 170.6 grains, containing 100 grains of soda. This portion of carbonate of soda is dissolved in 4 or 5 ounces of hot water, contained in a basin, and kept in a state of gentle ebullition; and the alkalimeter is filled up to 0 with the dilute acid. The measured acid is poured gradually into the soda solution, till the action of the latter upon test-paper ceases to be alkaline, and becomes distinctly acid, and the measures of acid necessary to produce that change accurately observed. The last portions of the acid must be carefully added by a single drop at a time, which is most easily done by using a short glass rod to conduct the stream of acid from the lip of the alkalimeter. It may probably require about 90 measures. But it is convenient to have the acid exactly of the strength at which 100 measures of it saturate 100 grains of soda. A plain cylindrical jar, *c*, of which the capacity is about a pint and a half, is graduated into 100 parts, each containing 100 grain measures of water, or ten times as much as the divisions of the alkalimeter. The divisions of this jar, however, are numbered from the bottom upwards, as is usual in measures of capacity. This jar is filled up with the dilute acid to the extent of 90, or whatever number of the alkalimeter divisions of acid were found to neutralize 100 grains of soda; and *water* is added to make up the acid liquid to 100 measures. Such is the test acid, of which 100 alkalimeter measures neutralize, and are equivalent to, 100 grains of soda; or 1 measure of acid to 1 grain of soda. It is transferred to a stock bottle. The remainder of the original dilute acid is diluted with water to an equal extent, in the same instrument, and added to the bottle. The density of this acid is 1.0995 or 1.0998, which is sensibly the same as 1.1. The protohydrate of sulphuric acid diluted with $5\frac{1}{2}$ times its weight of water, gives this test acid exactly; but as oil of vitriol varies in strength, it is better to form the test acid exactly; but as oil of vitriol varies in strength, it is better to form the test acid in the manner described than to trust to that mixture. Twenty-two measures of the test acid should neutralize 100 grains of cr. carbonate of soda; and $58\frac{1}{2}$ measures, 100 grains of pure anhydrous carbonate of soda.

(3.) In applying the test-acid, it is poured from the alkalimeter, as before, upon 100 grains of the soda-salt to be tested, dissolved in two or three ounces of hot water, the liquid being well stirred by a glass rod after each addition of acid. The salt contains so many grains of soda as it requires measures of acid to neutralize it; and, therefore, so much alkali per cent. The first trial, however, should only be considered an approximation, as much greater accuracy will be obtained on a repetition of it. The experiment is often made in the cold, but it is very advantageous to have the alkaline solution in a basin, in which it is heated and evaporated during the addition of the test-acid. The indications of the test-paper then become greatly more clear and decisive, both from the expulsion of the carbonic acid and the concentration of the solution. With such precautions the proportion of soda may be determined to 0.1 grain in 100 grains of salt, and an alkalimetrical determination, made in a few minutes, is not inferior in precision to an ordinary analysis.

If the soda-salt is mixed with insoluble matter, its solution must be filtered before the test-acid is applied to it. In examining a soda-salt which blackens salts of lead, and contains carbonate of soda with sulphide of sodium and hyposulphite of soda, 100 grains are tested as above, and the whole alkali in the salts thus determined. A neutral solution of chloride of calcium is also added in excess to the solution of a second hundred grains, by which the carbonate of soda is converted into chloride of sodium, while carbonate of lime precipitates. The filtered liquid is still alkaline, and contains all the sulphide of sodium and hyposulphite of soda; the quantity of soda corresponding with which is ascertained by means of the test-acid. This quantity is to be deducted from the whole quantity of alkali observed in the first experiment.

Borax may be analysed by the same test-acid, and will be found, when pure, to contain 16.37 per cent. of soda. The carbonates of potassa may also be examined by the same means; but the per centage of alkali must then be estimated higher than the measures of acid neutralized, in the proportion of the equivalent of soda to that of potassa, which are to each other as 31 to 47.

The test-paper employed in alkalimetry must be delicate. It should be prepared on purpose, by applying a filtered infusion of litmus several times to good letter-paper (not unsized paper), and drying it after each immersion, till the paper is of a distinct but not deep purple colour. If the test-acid be added to the alkaline solution in the cold, the operator must make himself familiar with the difference between the slight reddening of his test-paper by carbonic acid which is disengaged, and the unequivocal reddening which is produced by the smallest quantity of a strong acid. The former is a purple or wine-red tint; the latter a pale or yellow red, without blue, like the skin of an onion.

Method of Gay-Lussac.—The directions for proceeding given by M. Gay-Lussac are recommended by the general utility of the French measures employed for scientific purposes. It is commercial potassa which is supposed to be examined, and its value is expressed in anhydrous oxide of potassium.

The acid employed is the sulphuric, as before, of which 5 grammes at its maximum of concentration, that is, the acid $HO.SO_3$, are taken as a unit. This quantity of acid is diluted with water, so that the mixture occupies fifty cubic centimeters, or one hundred half cubic centimeters.[1] It is capable of neutralizing 4.816 grammes of pure potassa, and one-half cubic centimeter of the dilute acid will consequently indicate 0.04816 gramme of potassa.

To prepare the *normal acid fluid,* as the test-acid is called, it is necessary to have the pure monohydrated sulphuric acid. The acid sold as distilled sulphuric acid is sufficiently free from fixed impurities, but generally contains a little water in excess. By evaporating off one-fourth of this acid, the remaining three-fourths are left of the maximum degree of concentration. One hundred grammes of the monohydrated sulphuric acid are accurately weighed in a small glass bottle. A thin glass flask is also provided, which holds a liter of water when filled to a mark on the neck. The sulphuric acid already weighed is added in a gradual manner to this flask, about half filled with water at first, a circular motion being given to the vessel in order to mix the liquids rapidly. The acid bottle is well rinsed out with water, which is added to the flask; and when the whole cools, more water is added to fill up the flask to the mark on the neck. The normal acid fluid, thus prepared, should be preserved for use in a well-stopped bottle.

Fig. 176.

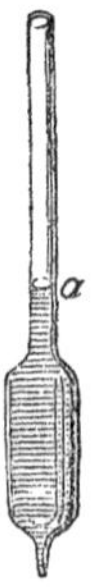

In making an examination of commercial potashes, a fair sample of the mass is first taken, and reduced to powder; of this, 48.16 grammes are accurately weighed out, and dissolved in a quantity of water, so that the volume of the solution is exactly half a liter. If one-tenth of this liquid be taken, that is, fifty cubic centimeters, we shall of course have the quantity which contains 4.816 grammes of the potashes. To draw off this portion conveniently, a pipette is used (fig. 176), which holds fifty cubic centimeters when filled up to a mark *a* on its stem. The pipette is emptied into a plain glass jar, the last drop of liquid being made to flow out by blowing into the pipette. A sufficiently distinct blue tint is given to the liquid in the jar by the addition of a few drops of an infusion of litmus, and the jar placed upon a sheet of white letter-paper, in order to observe the changes of colour afterwards with more facility.

Fig. 177.

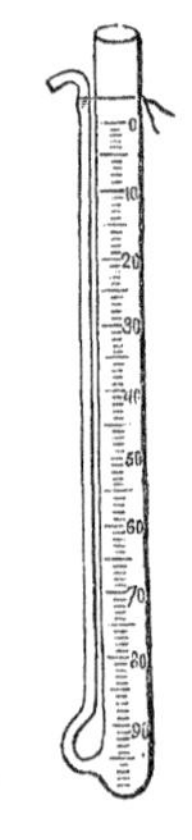

To measure the normal acid fluid, a glass tube of the form fig. 177 is used, 12 or 14 millimeters in internal diameter, which is called a *burette.* It is divided into half cubic centimeters, and the divisions marked on the large tube in an inverse order, as in the former alkalimeter. The beak may be greased below the aperture, to prevent the liquid running

[1] The Gramme is 15.4336 grains; the Cubic Centimeter, 0.06103 English cubic inch; the Liter or 1000 cubic centimeters, 61.03 cubic inches, 0.22017 English imperial gallon, or 1.76133 pint.

down the outside of the glass. The acid is poured from the burette, filled to the division 0, into the jar containing the potassa-solution, the liquid in the latter being constantly stirred. The change to the wine-colour is first observed, and the addition of acid is afterwards continued with the greatest caution, drop by drop, till the liquid assumes at once the onion-skin red. A few drops of acid in excess are inevitably added, owing to the slowness of the action of the last portions of acid upon the colouring matter. The number of these drops in excess is discovered by drawing a line with the liquid upon a slip of blue litmus paper, after the addition of each drop. The lines become red after the lapse of some time, where the acid is in excess, and give the number of drops to be deducted; of these, five are in general equivalent to one measure of the burette. The quantity of potassa is calculated from the measures of normal acid fluid prepared, each measure representing 0.04816 gramme of potassa, as already stated.

The chief objection to the practice of this method is the delicacy, and in some degree uncertainty, of the mode of determining the number of drops of acid always added in excess. This difficulty is best avoided, I believe, by operating upon the alkaline solution while hot and undergoing evaporation, as directed in the preceding method of alkalimetry.[1]

The object of an alkalimetrical process may also be obtained by determining the quantity of carbonic acid in a specimen of soda-ash or potashes. The quantity of carbonic acid is ascertained by decomposing the carbonate by sulphuric acid, and observing the loss of weight occasioned by the escape of the gas. The evolution of hydrosulphuric acid gas at the same time, by the decomposition of sulphide of sodium, is prevented by adding a little bichromate of potassa to the sulphuric acid, so as to oxidize the former acid gas. For every equivalent of carbonic acid, or 22 parts, an equivalent quantity of soda or potassa is allowed to be present; namely, 31 parts of soda or 47 parts of potassa. The process may be conducted by means of the well-devised arrangements of Dr. Will, described in works upon Analytical Chemistry. It would, however, be a subject of regret if this latter method should be allowed to supersede the use of normal fluids and the burette, which are capable of being usefully applied in numerous other investigations besides alkalimetry, and, in fact, form the basis of an interesting department of chemical analysis.

Bicarbonate of soda; $HO.CO_2 + NaO.CO_2$; 84 *or* 1050. — This salt is formed when a stream of carbonic acid gas is transmitted through a saturated solution of the neutral carbonate; it is then deposited as a farinaceous powder, but may be obtained in crystals from a weaker solution, which are rectangular prisms. But it is generally prepared on the large scale by exposing the crystals of neutral carbonate, placed on trays in a wooden case, to an atmosphere of carbonic acid gas: the matter then changes entirely into bicarbonate, which appears in amorphous and opaque masses. One hundred parts of water dissolve of it 10.04 parts at 50° (10° C.) and 16.69 parts at 158° (70° C.), according to M. Poggiale. Although containing two equivalents of acid, this salt is alkaline to test-paper, but its taste is much less unpleasant than the neutral carbonate, and indeed is scarcely perceived when mixed with a little common salt. The crystallized salt is permanent in dry air, but its solution loses carbonic acid, slowly at the temperature of the air, and rapidly above 160°, passing into the state of sesquicarbonate, and ultimately of neutral carbonate. A solution of bicarbonate of soda does not produce a precipitate in salts of magnesia in the cold, nor does it disturb immediately a solution of chloride of mercury; by which properties it is distinguished from the neutral carbonate.

The bicarbonate of soda is obtained otherwise by an interesting reaction. Equal weights are taken of common salt and of the carbonate of ammonia of the shops, which is chiefly bicarbonate; the former is dissolved in three times its weight of

[1] The apparatus and methods of alkalimetry have received much attention from Mr. Griffin. His improved apparatus and test-paper may be procured at the Chemical Museum, 53, Baker Street.

water, and the latter added in the state of fine powder to this solution, the whole stirred well together, and allowed to stand for some hours. The bicarbonate of oxide of ammonium present reacts upon chloride of sodium, producing the more sparingly soluble bicarbonate of soda, which precipitates in crystalline grains and causes the liquid to become thick, with chloride of ammonium (sal-ammoniac), which remains in solution: —

$$HO.CO_2+NH_4O.CO_2 \text{ and } NaCl=$$
$$HO.CO_2+NaO.CO_2 \text{ and } NH_4Cl.$$

The solid bicarbonate of soda is separated from the liquid by pressure in a screw press; but retains a portion of chloride of sodium. Messrs. Hemming and Dyer, who first observed this reaction, proposed it as a process for obtaining carbonate of soda from common salt.

Sesquicarbonate of soda; $2NaO+3CO_2+4HO$; 164 *or* 2050. — This salt presents itself occasionally in small prismatic crystals, but cannot be prepared at pleasure. It is unalterable in the air, but is decomposed in the dry state by a less degree of heat than the bicarbonate, notwithstanding its containing a smaller excess of carbonic acid. The theoretical carbonate of water, supposed to resemble the carbonate of magnesia, will be $HO.CO_2+HO+2HO$; which gives the salt in question, if the last 2HO are replaced by two proportions of protohydrated carbonate of soda. Substitutions of this character appear to be common in the formation of double carbonates and oxalates. The bicarbonate of potassa may be formed by the substitution of carbonate of potassa for the first HO, in the same carbonate of water, while the other 2HO disappear. The sesquicarbonate of soda occurs native in several places, particularly on the banks of the lakes of Soda in the province of Sukena, in Africa, whence it is exported under the name of *Trona;* in Egypt, Hungary, and in Mexico, and has the same proportion of water as the artificial salt.

Double carbonate of potassa and soda. — The carbonates of potassa and soda unite readily by fusion. A compound was also obtained by M. Margueritte, in transparent crystals, by submitting a solution of the two carbonates, in different proportions, to evaporation, of which the formula is $2(NaO.CO_2)+(KO.CO_2)+18HO$. These crystals may be dissolved without injury in a solution of carbonate of potassa, but when dissolved in pure water they are in great part decomposed, and allow crystals of carbonate of soda to be deposited. This double salt may be analysed by evaporating to dryness, after first adding hydrochloric acid, to convert the bases into chlorides of potassium and sodium, and then precipitating the former by means of bichloride of platinum, as described at page 373.

Sulphite of soda; $NaO.SO_2+10HO$; 63 + 90, *or* 787.5 + 1125. — This salt crystallizes in oblique prisms, and is efflorescent like the sulphate of soda, which it much resembles. Its taste is sulphureous, and its reaction feebly alkaline. When heated strongly in a close vessel, it gives sulphate of soda mixed with sulphide of sodium. It is prepared by passing a stream of sulphurous acid through a solution of the carbonate of soda (page 294), or on the large scale by exposing the crystals of carbonate of soda, moistened, to the vapour of burning sulphur. This salt, and also the sulphite of lime, are much employed as an *antichlore,* or to remove the last traces of chlorine from bleached cloth and the pulp of paper. A bisulphite of soda also exists, which appears in irregular and opaque crystals.

Hyposulphite of soda; $NaO.S_2O_2+5HO$; 79 + 45, *or* 987.5 + 562.5. — This salt, of which the preparation and some of the properties have already been described (page 303), is inodorous, persistent in air, very soluble in water, and insoluble in alcohol. It crystallizes in large rhomboidal prisms, terminated by oblique faces, of which the acute angles are replaced by planes. When heated in a covered vessel, it first loses its water, and then undergoes decomposition, and is resolved into sulphate of soda and pentasulphide of sodium. The hyposulphite of soda readily dissolves chloride of silver, forming a double salt of soda and oxide of silver, which has an intensely sweet taste. It also dissolves the red oxide of mercury easily, forming a double salt, which readily decomposes with deposition of sulphide of mercury. With chloride of gold, it gives rise to the formation of chloride of sodium, tetrathionate of soda, and a double hyposulphite of soda and oxide of gold, of which the formula is

$$Au_2O.S_2O_2+3(NaO.S_2O_2)+4HO$$ (Fordos and Gelis).

The use of this last salt is recommended for fixing the daguerréotype image.

Sulphate of soda, Glauber's salt; $NaO.SO_3+10HO$; 71+90, *or* 887.5+1125.—This salt occurs crystallized in nature, and also dissolved in mineral waters, and is formed on neutralizing carbonate of soda by sulphuric acid. But it is more generally prepared by decomposing common salt with sulphuric acid, as in the process for hydrochloric acid (page 335). The sulphate of soda crystallizes readily in long prisms, of which the sides are often channelled, which have a cooling and bitter taste, and contain 55.76 per cent. of water, or 10 equivalents; in which they fuse by a slight elevation of temperature, and which they lose entirely by efflorescence in dry air even at 40°. At 32°, 100 parts of water dissolve 5.02 parts of anhydrous sulphate of soda, 16.73 parts at 64.2° (17.91° C.), 50.65 parts at 91°, which is the temperature of maximum solubility of this salt, and 42.65 parts at the boiling temperature of a saturated solution, which is 217.6° (103.1° C.), as observed by Gay-Lussac. In a supersaturated solution of this salt (page 239), crystals are sometimes slowly deposited, which are different in form and harder than Glauber's salt; they are long prisms with rhombic bases, and contain 8 equivalents of water, or possibly only 7 equivalents (Loewel, Annal. de Ch. et de Phys. 3 sér. xxix. 62; or Chem. Soc. Quart. Journ. iii., 164).

M. Loewel finds these crystals to have a greater solubility than the ten-atom hydrate. The sulphate of soda no doubt exists in the supersaturated solution as eight-atom hydrate, and the salt is induced to crystallize by causes which make it to assume two additional equivalents of water, and form the less soluble hydrate. It is proved that the action of air in causing crystallization is not from its pressure (Gay-Lussac, Annal. de Ch. et de Ph. 2 sér. ii. 296); but, as I have shown, from the solubility of air in the saline solution, carbonic acid exceeding air in activity (Edinb. Trans. xi. 114). Loewel observes, among other curious circumstances, that a rod of glass or metal, which determines the formation of the ten-atom hydrate when plunged into the supersaturated solution, loses this property if it is left in contact with water for twelve hours, or if it has been previously heated to between 40° and 100° C., and continues incapable of inducing crystallization for ten days or a fortnight at the ordinary temperature, if preserved from free contact with the air. I had previously put up clean glass beads into supersaturated solutions contained in jars inverted over mercury, without determining crystallization, and would ascribe the action of the glass surface to adhering soluble matter, rather than the molecular condition of the glass, as supposed by M. Loewel.

A saturated solution of sulphate of soda, kept at a temperature between 91° and 104°, affords octohedral crystals with a rhombic base, which are anhydrous. They are isomorphous with the hypermanganate of baryta. Their density is 2.642. The anhydrous salt fuses at a bright red heat, without loss of acid. Sulphate of soda was at one time the saline aperient in general use, but is now superseded by sulphate of magnesia. It is still, however, occasionally associated with the tartrate of potassa and soda, in Seidlitz powders.

The crystallized sulphate of soda dissolves freely in hydrochloric acid, or in dilute sulphuric acid, and produces a great degree of cold, by which water may easily be frozen in summer. A suitable apparatus for this purpose consists of a hollow cylinder C C (figs. 178 and 179), intended for the reception of the freezing mixture, itself surrounded by a space to contain the water to be frozen, having the external opening *u*, and the whole protected by a double casing, B B, filled with cotton or tow to prevent access of heat. The cylinder A is hollow, and may also have water placed in it to be frozen. This cylinder is turned on a pivot by the handle above,

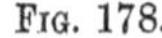
FIG. 178.

FIG. 179.

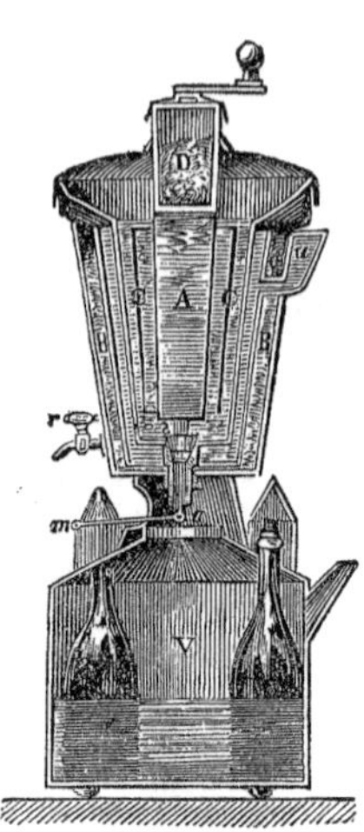

and has projections or vanes, by which the salt and acid are conveniently agitated. The upper part, *D*, of this cylinder is filled with a non-conducting material. The freezing mixture is added in charges of about 3 pounds of pulverized sulphate of soda, and 2 pound measures of hydrochloric acid, at a time; which are repeated after ten minutes, and the stopcock opened to allow the acid solution to flow into the vessel V below, where its low temperature may be further employed to cool wine or other beverages. With 12 pounds of sulphate of soda, and about 10 pounds of acid, from 10 to 12 pounds of ice may be formed in the course of an hour in this manner.

The anhydrous sulphate of soda also forms the mineral *Thenardite,* which was discovered by M. Casasecu in the neighbourhood of Madrid.

PREPARATION OF CARBONATE OF SODA FROM THE SULPHATE.

The sulphate of soda is chiefly formed as a step in the process of preparing soda from common salt. The same manufacture gives rise to the preparation of large quantities of sulphuric acid, of which 80 pounds are required for 100 pounds of salt. From the last, upwards of 50,000 tons of *soda-ash,* and 20,000 tons of crystallized carbonate of soda, were manufactured in 1838; and the production has since greatly increased.

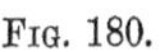
FIG. 180.

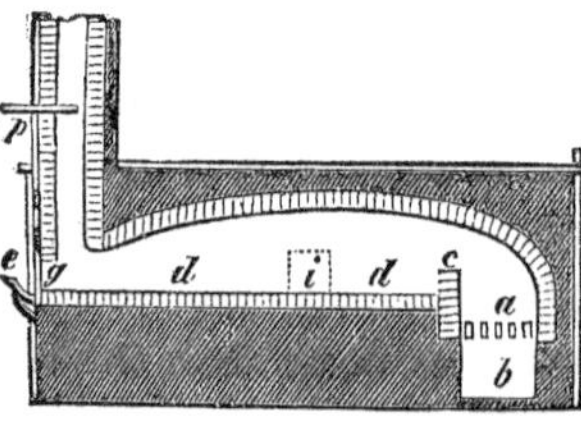

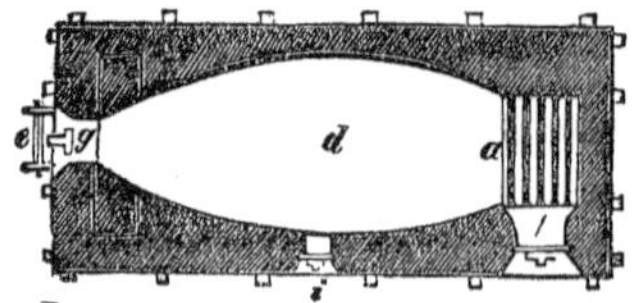

A reverberatory furnace is employed in soda-making and various other chemical manufactures, to afford the means of exposing a considerable quantity of materials to a strong heat, of which a perpendicular and a horizontal section are given in fig. 180. It consists of a fire-place, *a*, in which the fuel is burned, of which *b* is the ash-pit, with a horizontal flue expanding into a small chamber or oven, *d d*, which is raised to a strong red heat by the reverberation on its walls of the flame or heated air from the fire, on its passage to the chimney. The matters to be heated are placed upon the floor of this chamber. It has an opening, *i*, in the side, for the introduction of materials, and another opening, *g*, at the end most distant from the fire. The chimney is provided with a damper, *p*, by which the draught is regulated.

(1.) The sulphate of soda is prepared by throwing 600 pounds of common salt into the chamber of the furnace, already well heated, and running down upon it, from an opening in the roof, an equal weight of sulphuric acid of density 1.600, in a moderate stream. Hydrochloric acid is disengaged and carried up the chimney, and the conversion of the salt into sulphate of soda is completed in four hours. (2.) The sulphate thus prepared is reduced to powder and 100 parts of it mixed with 103 parts of ground chalk, and 62 parts of small coal ground and sifted. This mixture is introduced into a very hot reverberatory furnace, about two hundred weight at a time. It is frequently stirred until it is uniformly heated. In about an hour it fuses; it is then well stirred for about five minutes, and drawn out with a rake into a cast-iron trough, in which is is allowed to cool and solidify. This is called ball soda, or black-ash, and contains about 22 per cent. of alkali. (3.) To separate the salts from insoluble matter, the cake of ball soda, when cold, is broken up, put into vats, and covered by warm water. In six hours the solution is drawn off from below, and the washing repeated about eight times, to extract all the soluble matter. These liquors being mixed together are boiled down to dryness, and afford a salt which is principally carbonate of soda, with a little caustic soda and sulphide of sodium. (4.) For the purpose of getting rid of the sulphur, the salt is mixed with one-fourth of its bulk of sawdust, and exposed to a low red heat in a reverberatory furnace for about 4 hours, which converts the caustic soda into carbonate, while the sulphur also is carried off. This product contains about 50 per cent. of alkali, and forms the soda-salt of best quality. (5.) If the crystallized carbonate is required, the last salt is dissolved in water, allowed to settle, and the clear liquid boiled down until a pellicle appears on its surface. The solution is then run into shallow boxes of cast-iron, to crystallize in a cool place; and after standing for a week the mother liquor is drawn off, the crystals drained, and broken up for the market. (6.) The mother liquor, which contains the foreign salts, is evaporated to dryness, for a soda-salt, which serves for soap or glass making, and contains about 30 per cent. of alkali.

In fig. 181, a soda-furnace is represented, consisting of two compartments: the first, A, in which the sulphate of soda is decomposed, and the second, B, in which

Fig. 181.

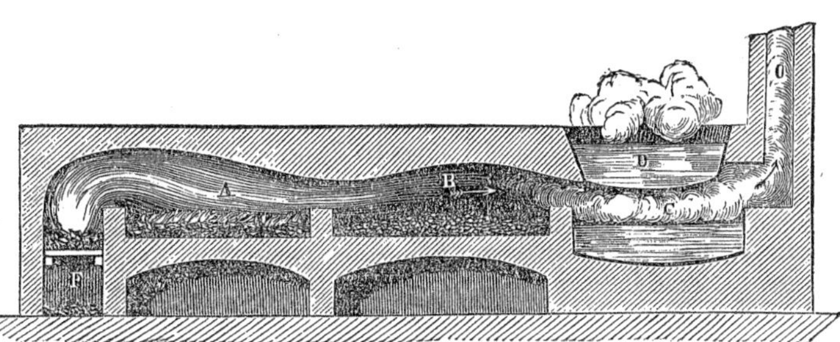

sulphuric acid is applied to the chloride of sodium, and the sulphate of soda formed. The heat from the furnace is further economized by being applied to evaporate solutions of carbonate of soda in C and D.

The most essential part of this process is the fusion of sulphate of soda with coal and carbonate of lime: by the first, the sulphate is converted into sulphide of sodium (page 383); and by the second, the sulphide of sodium is converted into carbonate of soda. These changes may be effected separately to a considerable extent, but not completely, by calcining the sulphate at a higher temperature with coal and carbonate of lime in succession. The lime becomes at the same time sulphide of calcium, or it is more generally supposed to form an oxi-sulphide of calcium, 3CaS.CaO, a compound which would destroy the carbonate of soda, if it was dissolved along with that salt, in the subsequent lixiviation of the ball soda. But the sulphide of calcium being nearly insoluble of itself, or rendered entirely so by its combination with lime, does not dissolve to a sensible extent in the experiment. The application, however, of very hot water to the ball soda is to be avoided. The following diagram is used to represent the chemical changes in this process, supposing for simplicity that charcoal is employed instead of coal, and lime instead of its carbonate; the numbers denoting equivalents:—

REACTION IN THE SODA PROCESS.

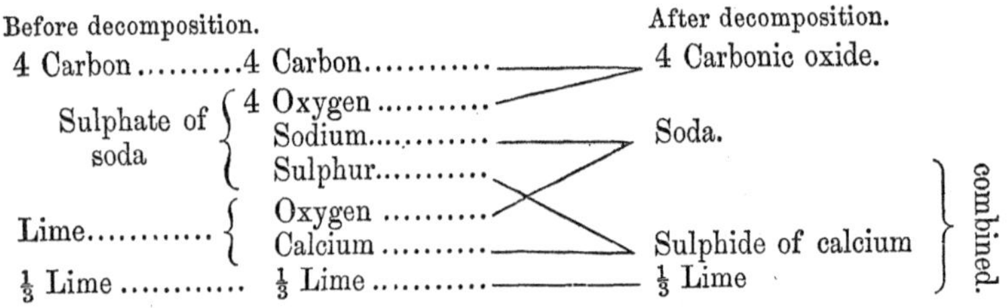

Before decomposition.		After decomposition.	
4 Carbon	4 Carbon	4 Carbonic oxide.	
Sulphate of soda	4 Oxygen		
	Sodium	Soda.	
	Sulphur		combined.
Lime	Oxygen		
	Calcium	Sulphide of calcium	
$\frac{1}{3}$ Lime	$\frac{1}{3}$ Lime	$\frac{1}{3}$ Lime	

Mr. Gossage considers the additional $\frac{1}{3}$ equiv. of lime as superfluous, although not injurious in the process. The soda derives carbonic acid from the carbonate of lime or from the gases of the fire, and is therefore entirely carbonate. No hydrate of soda is dissolved out of the ball soda by alcohol, but a portion of the carbonate appears often to become caustic by the action of the caustic lime, in the subsequent lixiviation.[1]

The insoluble sulphide of calcium of this process is known as soda-waste. It is

[1] The analysis, by Mr. F. Claudet in my laboratory, of a specimen of black-ash from Birmingham, in which a minimum of lime appears to have been used, gave the following results:—

Soluble.	Carbonate of Soda	35.42	
	Sulphide of Sodium	1.45	
	Sulphate of Soda	.78	
	Chloride of Sodium	2.62	
	Silicic acid	.58	
	Oxide of Iron, Alumina	.15	
Insoluble.	Sulphide of Calcium	32.90	= Sulphur 14.6, Calcium 18.9
	Carbonate of Lime	3.73	= Lime 2.09
	Magnesia	.56	
	Oxide of Iron	1.98	
	Alumina	3.59	
	Sand and Silicic acid	4.95	
	Charcoal	10.57	
	Water	.72	
		100.00	

The lime found is not in quantity sufficient to form the oxi-sulphide of calcium, 3CaS.CaO; confirming the view of the process taken by Mr. Gossage. No hydrate of soda, or sulphide of sodium, was dissolved out of this black-ash by alcohol. The portion of the latter salt obtained in the analysis appeared to be the result of over-washing; the sulphide of calcium having a tendency to pass into lime and the soluble hydrosulphate of sulphide of calcium, which decompose a portion of the carbonate of soda. Although this important process has been much studied, its theory is still incomplete. The furnacing of the sulphate of soda is promoted by aqueous vapour, and a loss of sulphur occurs in a way which is not understood. See the papers of Mr. J. Brown (Phil. Mag. xxxiv. 15), of M. B. Unger (Ann. Ch. Pharm., lxi. lxiii. and lxvii.), and the Annual Report on the Progress of Chemistry of Liebig and Kopp, edited by Hoffmann and De la Rue, ii. 292, 1847–48.

not merely valueless, but troublesome to the manufacturer. But the attempt has been made to turn it to account as a source of sulphur. As means are now taken to condense the hydrochloric acid, formerly sent up the chimney, this acid is applied to the soda-waste, from which it disengages hydrochloric and carbonic acids. But hydrochloric acid is not produced, in the soda process, in adequate quantity for this application of it, and the carbonic acid evolved with the hydrosulphuric acid might interfere with the combustion of the latter. These difficulties, however, are in a great degree removed by the discovery of Mr. Gossage, that sulphide of calcium, when moistened with water, is decomposed easily and completely by a single equivalent of carbonic acid. Hence the application of hydrochloric acid to the waste may be made, with the evolution of nothing but hydrosulphuric acid; and the deficiency in the quantity of hydrochloric acid may be made up by a supply of carbonic acid, to be applied to the waste, from any other source. The hydrosulphuric acid would be burned, instead of sulphur, in the leaden chamber, to produce sulphuric acid.

Many changes have been proposed upon the soda process. Sulphate of iron, produced by the oxidation of iron-pyrites, is a cheap salt, and may be applied to convert chloride of sodium into sulphate of soda.—(1.) By igniting a mixture of these salts in a reverberatory furnace, when sulphate of soda, sesquioxide of iron, and volatile sesquichloride of iron are produced. (2.) By dissolving the salts together in water, and allowing the solution to fall to a low temperature, when sulphate of soda crystallizes, and chloride of iron remains in solution (Mr. Phillips); or (3.) By concentrating the last solution at the boiling-point, when the same decomposition occurs, anhydrous sulphate of soda precipitates, and may be raked out of the liquor. The roasting of bisulphide of iron with common salt in a reverberatory furnace may also be made to furnish sulphate of soda. Sulphate of magnesia has been substituted for sulphate of iron, in these three modes of application; but the unavoidable formation of double salts of magnesia and soda makes the separation of the sulphate of soda always imperfect. It has been proposed, instead of furnacing the sulphate of soda, to decompose it by caustic baryta, or by strontia, the last earth being procured by Mr. Tilghmann, for this application of it, by decomposing the native sulphate of strontia from Bristol, by a current of steam at a red heat. Such a process should also furnish the sulphuric acid required to decompose chloride of sodium and form sulphate of soda. Chloride of sodium may also be decomposed by moistening and rubbing it in a mortar with 4 or 6 times its weight of litharge, when an oxichloride of lead is formed, and caustic soda liberated. The decomposition of chloride of sodium by the carbonate of ammonia, with formation of bicarbonate of soda, has already been noticed (page 389). It appears, however, that the soda-process first described, which was invented towards the end of the last century by Leblanc, is still generally preferred to all others.

The old sources of carbonate of soda, namely *barilla*, or the ashes of the salsola soda, which is cultivated on the coasts of the Mediterranean, and *kelp*, the ashes of sea-weeds, have ceased to be of importance, at least in England. Barilla contains about 18, and kelp about 2 per cent. of alkali.

Bisulphate of soda, $HO.SO_3+NaO.SO_3$; 120 *or* 1500. This salt is obtained in large crystals on adding an equivalent of oil of vitriol to sulphate of soda, and evaporating the solution till it attains the degree of concentration necessary for crystallization. If half an equivalent only of oil of vitriol is added, a *sesquisulphate of soda* is obtained in fine crystals, according to Mitscherlich. The ordinary bisulphate of soda contains basic water, but it may be rendered anhydrous by a degree of heat approaching to redness. The salt thus obtained is a true bisulphate of soda, and gives anhydrous sulphuric acid when distilled at a red heat.

Nitrate of soda; $NaO.NO_5$; 85 *or* 1062.5.—This salt crystallizes in the rhomboidal form of calc-spar; density 2.260. It is soluble in twice its weight of water, and has a tendency to deliquesce in damp air. It burns much slower with combustibles than nitrate of potassa, and cannot therefore be substituted for that salt in the manufacture of gunpowder. It is now generally had recourse to, as the source of nitric acid, and is also largely used in agriculture. Nitrate of soda is found abundantly in the soil of some parts of India; and it forms a thin but very extensive bed covered by clay at Atacama in Peru, from which it is exported in great quantity.

Chlorate of soda ($NaO.ClO_5$) is formed by mixing strong solutions of bitartrate of soda and chlorate of potassa, when the bitartrate of potassa precipitates, and the chlorate of soda remains in solution. It crystallizes in fine tetrahedrons, and is considerably more soluble than chlorate of potassa.

Phosphates of soda.—There are three crystallizable phosphates of soda belonging to the tribasic class, which I shall describe under their most usual names.

Phosphate of soda; $HO.2NaO.PO_5+24HO$; 359 *or* 4487.5.—This is the salt known in pharmacy as phosphate of soda, and formed by neutralizing phosphoric acid from burnt bones (page 319) with carbonate of soda. It crystallizes in oblique rhombic prisms, which are efflorescent, and essentially alkaline. M. Malaguti is, I believe, mistaken in ascribing 26 equivs. of water to this salt. The taste of phosphate of soda is cooling and saline, and less disagreeable than sulphate of magnesia, for which it may be substituted as an aperient. It dissolves in 4 times its weight of cold water, and fuses in its water of crystallization, when moderately heated. When evaporated above 90°, this salt crystallizes in another form with 14 instead of 24 atoms of water (Clark). It is deprived of half its alkali by hydrochloric acid in the cold, but not by acetic acid.

Subphosphate of soda; $3NaO.PO_5+24HO$; 381 *or* 4762.5.—Formed when an excess of caustic soda is added to the preceding salt. It crystallizes in slender six-sided prisms, with flat terminations, which are unalterable in air; but the solution of this salt rapidly absorbs carbonic acid, and is deprived of one-third of its alkali by the weakest acids. The crystals dissolve in 5 times their weight of water at 60°, and undergo the watery fusion at 170°. This salt continues tribasic after being exposed to a red heat.

Biphosphate of soda; $2HO.NaO.PO_5+2HO$; 139 *or* 1737.5.—Obtained by adding tribasic phosphate of water to phosphate of soda, till the latter ceases to produce a precipitate with chloride of barium. The solution affords crystals, in cold weather, of which the ordinary form is a right rhombic prism, having its larger angle of 93° 54′. But this salt is dimorphous, occurring in another right rhombic prism, of which the smaller angle is 78° 30′, terminated by pyramidal planes, isomorphous with binarseniate of soda. The biphosphate of soda is very soluble, and has a distinctly acid reaction. Like all the other soluble tribasic phosphates, it gives a yellow precipitate with nitrate of silver, which is tribasic phosphate of silver.

Phosphate of soda and ammonia, Microcosmic salt; $HO.NH_4O.NaO.PO_5+8HO$; 201 *or* 2512.5.—This salt is obtained by heating together 6 or 7 parts of crystallized phosphate of soda, and 2 parts of water, till the whole is liquid, and then adding 1 part of pulverized sal-ammoniac. Chloride of sodium separates, and the solution, filtered and concentrated, affords the phosphate in prismatic crystals. It is purified by a second crystallization. This salt occurs in urine. It is much employed as a flux in blow-pipe experiments. By a slight heat it loses 8HO, by a stronger heat it is deprived of its remaining water and ammonia, and converted into metaphosphate of soda, which is a very fusible salt. It will be observed that the three atoms of base in this phosphate are all different,—namely, water, oxide of ammonium, and soda; of which the two last belong to the same natural family, for bases of the same family may exist together in the salts of bibasic and tribasic acids, forming stable compounds, but not in ordinary double salts. No phosphate exists, corresponding with microcosmic salt, but containing potassa instead of oxide of ammonium; the phosphate of soda, with 14HO, has been mistaken for such a salt.

Pyrophosphate of soda; $2NaO.PO_5+10HO$; 134 + 90, *or* 1675 + 1125.—Procured by heating the phosphate of soda to redness, when it loses its basic water as well as its water of crystallization. The residual mass dissolved in water affords a salt, which is less soluble than the original phosphate, and crystallizes in prismatic crystals, which are permanent in air, and contain ten atoms of water. Its solution is essentially alkaline. This salt is precipitated white, by nitrate of silver. It is to be remarked that insoluble pyrophosphates, including pyrophosphate of silver, are soluble to a considerable degree in the solution of pyrophosphate of soda. The pyrophosphates of potassa and of ammonia can exist in solution, but pass into tribasic salts when they crystallize.

A *bipyrophosphate of soda* ($HO.NaO.PO_5$) exists, obtained by the application of a graduated heat to the biphosphate of soda, but it does not crystallize. Its solution has an acid reaction.

Metaphosphate of soda; $NaO.PO_5$, 103 *or* 1287.5.—The biphosphate of soda, containing only one equivalent of fixed base, affords the metaphosphate of soda, when heated to redness. The metaphosphate of soda fuses at a heat which does not exceed low redness, and on cooling rapidly forms a transparent glass, which is deliquescent in damp air, and very soluble in water, but insoluble in alcohol: its solution has a feebly acid reaction, which can be negatived by the addition of 4 per cent. of carbonate of soda. When evaporated, this solution does not give crystals, but dries into a transparent pellicle, like gum, which retains at the temperature of the air somewhat more than a single equivalent of water. Added to neutral, and not very dilute solutions of earthy and metallic salts, metaphosphate of soda throws down insoluble hydrated metaphosphates, of which the physical condition is remarkable. They are all soft solids, or semifluid bodies; the metaphosphate of lime having the degree of fluidity of Venice turpentine.

The bipyrophosphate of soda appears to undergo several changes under the influence of heat before it becomes metaphosphate. At a temperature of 500°, the salt becomes nearly anhydrous, and affords a solution which is *neutral* to test-paper, but in other respects resembles the bipyrophosphate. But at temperatures which are higher, but insufficient for fusion, the salt being anhydrous, appears to have lost its solubility in water; at least it is not affected at first when thrown in powder into boiling water, but gradually dissolves by continued digestion, and passes into the preceding variety.—(Phil. Trans. 1833, p. 275).

When the fused metaphosphate of soda is slowly cooled, it forms a crystalline mass, as observed by Fleitmann and Henneberg, and gives a crystallizable metaphosphate of soda (page 324).

Borax, Biborate of soda, $NaO.2BO_3+10HO$; $100.8+90$ *or* $1260.+1125.$—This salt is met with in commerce in large hard crystals. It is found in the water of certain lakes in Transylvania, Tartary, China, and Thibet, and is deposited in their beds by spontaneous evaporation. It is imported from India in a crude state, and enveloped in a fatty matter, under the name of *Tinkal,* and afterwards purified. But nearly the whole borax consumed in England is at present formed by neutralizing, with carbonate of soda, the acid from the boracic lagoons of Tuscany. The ordinary crystals of borax are prisms of the oblique system, containing 10 atoms of water, of density 1.692; but it also crystallizes at 133° in regular octohedrons, which contain only 5 atoms of water. This salt has a sweetish, alkaline taste; for, although containing an excess of acid, it has an alkaline reaction, like the bicarbonate of soda, and is soluble in 10 parts of cold, and 2 parts of boiling water.

The anhydrous salt is very fusible by heat, and forms a glass of density 2.367. This glass possesses the property of dissolving most metallic oxides, the smallest portions of which colour it. As the metal may often be discovered by the colour, borax is valuable as a flux in blow-pipe experiments. For this purpose a thin platinum wire is generally used, one end of which is bent into a hook (fig. 182.) The loop being slightly moistened, is dipped into a fine powder of anhydrous borax, and a minute portion of the metallic oxide which we wish to determine is also taken up on the loop. The matter is then fused in the flame of a candle or spirit-lamp directed upon it by means of a mouth blow-pipe (fig. 183.) Often

Fig. 182.

two different colourations are obtained when the metal has more than one oxide, according as the substance is heated in the reducing or white portion of the flame, which, in the blow-pipe flame, is at *b* (fig. 184), or in the oxidating spheres *a a*, and

Fig. 183.

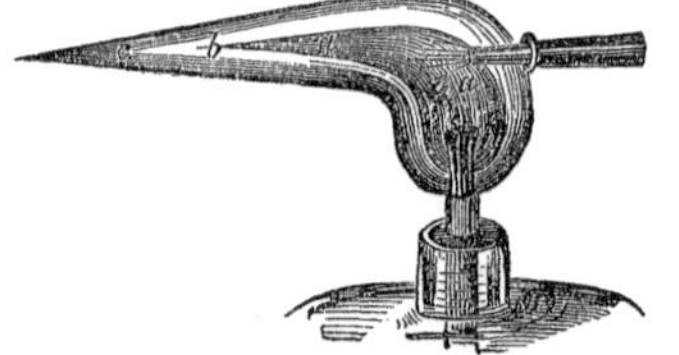

Fig. 184.

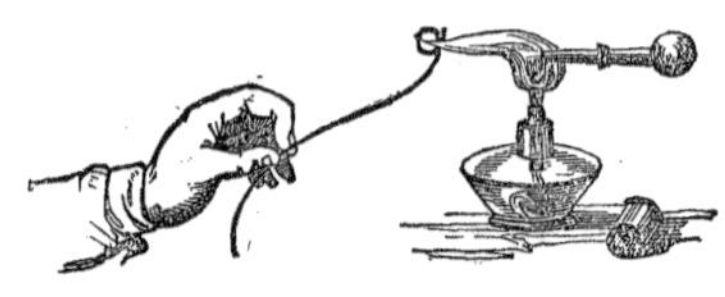

at the point *c*, where there is an excess of atmospheric air. To produce the colour of the protoxide, we expose to the reducing flame; and to produce the colour of the peroxide, we expose to the oxidizing flame.

As pieces of metal could not be soldered together if covered by oxide, borax is fused with the solder upon the surface of the metals to be joined, to remove the oxide. Borax is also a constituent of the soft glass, known as jewellers' paste, which is coloured to imitate precious stones. But the most considerable consumption of this salt is in the potteries, in the formation of a glaze for porcelain.

A neutral borate of soda is formed by calcining strongly 1 eq. of borax with 1 eq. of carbonate of soda, when carbonic acid is expelled. The solution yields a salt belonging to the oblique prismatic system, of which the formula is, $NaO.BO_3+8HO$. When heated, it fuses in its water of crystallization, and is expanded into a vesicular mass of extraordinary magnitude by the vaporization of that water.

When borax is fused with carbonate of soda in excess, the quantity of carbonic acid which escapes indicates the formation of a borate, $3NaO+2BO_3$, but which has not been farther examined. Notwithstanding this, a solution of borax in water is decomposed, and the boracic acid entirely liberated, by a stream of either carbonic or hydrosulphuric acid. Silicic acid, however, in its soluble modification, has no decomposing action upon a solution of borax. Boracic acid, therefore, appears to stand in the scale of acids above silicic, but below carbonic acid. A saturated solution of borax readily dissolves a large amount of arsenious acid, forming a compound remarkable for its great solubility in water. This contains, according to Prof. E. Schweizer, arsenite of soda, borate of soda, and a compound of arsenious and boracic acids, and is probably represented by the formula—

$$NaO.AsO_3+2(NaO.2BO_3)+2(BO_3 2AsO_3)+10HO.$$

A salt is said to exist, formed of $NaO+4BO_3$, but to crystallize with difficulty, produced on combining borax with a quantity of boracic acid equal to what it already contains. M. Laurent has also shown that a sexborate of soda exists in solution, but is not crystallizable. (Ann. de Ch. et de Phys. lxvii., 218.) The borates of potassa have also been examined by Laurent. The sexborate crystallizes well; its formula is $KO.6BO_3+10HO$. A triborate is represented by $KO.3BO_3+8HO$; the biborate corresponds in composition with octohedral borax, but has, notwithstanding, a different and incompatible form.

A simple and very accurate method of analyzing borax is, to add an excess of hydrochloric acid to a solution of the salt, and evaporate to dryness on the water-bath, adding a few more drops of hydrochloric acid towards the end of the operation. The mass, when perfectly dry, is re-dissolved in water, a little nitric acid mixed with the solution, and the chlorine precipitated by nitrate of silver; from the amount of chloride of silver that of the chlorine is deduced, and from the latter the quantity of soda. The alkaline bases of all the other borates may be obtained wholly as chloride by a similar treatment. (Schweitzer, Chem. Gaz. 1850, p. 281.)

Silicates of soda.—The earth silica, or silicic acid, SiO_3 (page 290), is dissolved by caustic soda, and gives, by slow evaporation, a crystallized silicate of soda, $3NaO.2SiO_3$ (Fritzsche). A concentrated solution of caustic soda at a high temperature under pressure dissolves silica freely even in the form of flint or of quartzy sand, and gives a similar silicate, which is used by Mr. Ransome of Ipswich for the induration of plaster and cements, and the formation of artificial stone.

When silicic acid is thrown into carbonate of potassa or soda, in a state of fusion by heat, a fusible silicate is formed, in which, judging from the quantity of carbonic acid expelled, 3 eq. of soda are also combined with 2 eq. of silicic acid, and the oxygen in the soda is to that in the silicic acid as 1 to 2. This silicate dissolves in the clear and liquid carbonate. When, on the other hand, a greater proportion of silicic acid is fused with the carbonate, the whole carbonic acid of the latter is expelled, and the excess of silicic acid then dissolves in the silicate. The silicic acid and silicate of such mixtures do not separate by crystallization, but uniformly solidify together, on cooling, as a homogeneous glass, whatever their proportions may be. It is thus impossible to obtain alkaline silicates, which are certainly definite combinations, in the dry way. A mixture of silicic acid with potassa or soda, in which the oxygen of the former is to that of the latter as 18 to 1, is said still to be fusible by the heat of a forge; but when the proportion is as 30 to 1, the mixture merely agglutinates, or frits. These combinations, even with a large quantity of silicic acid, continue to be soluble in water.

A compound, known as *soluble glass*, is obtained by fusing together 8 parts of carbonate of soda (or 10 of carbonate of potassa) with 15 of fine sand and 1 of charcoal. The object of the charcoal is to facilitate the combination of the silicic acid with the alkali, by destroying the carbonic acid, which it converts into carbonic oxide. This glass, when reduced to powder, is not attacked by cold water, but is dissolved by 4 or 5 parts of boiling water. The solution may be applied to objects of wood, and, when dried by a gentle heat, forms a varnish, which imbibes a little moisture from the air, but is not decomposed by carbonic acid, nor otherwise alterable by exposure. Stuffs impregnated with the solution lose much of their combustibility, and wood is also defended by it, to a certain degree, from combustion.

GLASS.

The alkaline silicates, cooled quickly or slowly, never exhibit a crystalline structure, but are uniformly vitreous (p. 151). They are the bases of the ordinary varieties of glass, which contain earthy silicates besides, but appear to owe the vitreous character to the silicates of potassa and soda. The silicate of lime, and the silicate of the protoxide of iron, crystallize on cooling; so does the silicate of lead, unless it contains a large excess of oxide of lead. The addition of the silicate of potassa or soda deprives them entirely of this property; the silicate of alumina considerably diminishes it. But if silicates of potassa or soda are heated for a long time, the alkali may in part escape in vapour, and if other bases exist in the compound, it then often assumes a crystalline structure on cooling. The alkaline silicates by themselves are soluble in water, and decomposed by acids; the silicate of lime is also dissolved by acids, but the double silicates, on the contrary, resist the action of acids, particularly when they contain an excess of silicic acid, and form an available glass. The following table exhibits the composition of the best known kinds of glass, from the analyses of Dumas and of Faraday:—

COMPOSITION OF VARIETIES OF GLASS.

	Silicic acid.	Potassa.	Lime.	Ox. lead.	Alumina.	Water.
Bohemian glass	69	12	9	0	10	0
Crown-glass	63	22	12	0	3	0
Window-glass	69	11 soda	13	0	7	0
Bottle-glass	54	5	29	6 ox. iron	0	0
Flint-glass	45	12	0	43	0	0
Crystal	61	6	0	33	0	0
Strass	38	8	0	53	1	0
Soluble glass	62	26	0	0	0	12

The analysis, by Mr. T. Rowney, of the superior Bohemian glass, which, on account of its difficult fusibility, is employed for combustion-tubes, gave silicic acid 73.13, potassa 11.49, soda 3.07, lime 10.43, alumina 0.30, sesquioxide of iron 0.13, magnesia 0.26, protoxide of manganese 0.46=99.27. The oxygen of the bases is to that of the silicic acid as 1 to 6. The specimen was decomposed by fusion with carbonate of soda, for the earths, and by fusion with hydrate of baryta for the alkalies (Mem. Chem. Soc. iii. 299).

Silicate of soda and lime.—To form window-glass, 100 parts of quartzy sand are taken, with 35 to 40 parts of chalk, 30 to 35 parts of carbonate of soda, and 180 parts of broken glass. These materials are first fritted, or heated so as to cause the expulsion of water and carbonic acid, and to produce an agglutination of their particles, and afterwards completely fused in a large clay crucible of a peculiar construction; or fused at once, the fritting being now generally discontinued. For the first formation of the glass a higher temperature is required than that at which it is most thick and viscid, and in the proper condition for working it. At the latter temperature the substance possesses an extraordinary degree of ductility, and may be drawn out into threads so fine as to be scarcely visible to the eye. A portion of the plastic mass, on the extremity of an iron tube used as a blow-pipe, may be expanded into a globular flask, and pressed or bent into vessels of any form, which may be pared and fashioned by the scissors. At a lower temperature, glass vessels become rigid, and, when cold, brittle in the extreme, unless they be *annealed*, that is, kept for several hours at a temperature progressively lowered from the highest degree which the glass can bear without softening to the temperature of the atmosphere. The well-known glass tears, or Prince Rupert's drops, as they are called, which are made by allowing drops of melted glass to fall into water, illustrate the peculiar properties of unannealed glass. The surface becoming solid by the sudden cooling, while the interior is still at a high temperature, and consequently dilated, the drop is of greater volume than it would be if cooled slowly and equally throughout its mass. Its particles are thus in a state of extreme tension, and an injury to any part causes the whole mass to fly to pieces. The fracture of unannealed vessels, which is the immediate consequence of scratching their surface, has been compared to the effect upon a sheet of cloth forcibly stretched, of injuring its edge in the smallest degree by a knife or scissors. It then ceases to preserve its integrity by resisting the tension, and is torn across. The relative proportions of the ingredients of this and other species of glass is subject to some variation. But the oxygen in the bases of window-glass is to the oxygen of the silicic acid nearly as 1 to 4; the composition approaching the formula $3NaO.3CaO+8SiO_3$. This glass has a green tint, which is very obvious in a considerable mass of it, occasioned in part, it may be, by the impurities of the materials, but a certain degree of which appears to be essential to a soda-glass. For in all the finer and entirely colourless varieties of glass it is necessary to use potassa.

Silicates of potassa and lime.—Plate-glass used for mirrors, crown-glass, and the beautiful Bohemian glass, are of this composition. In the most remarkable varieties the oxygen of the bases is to that of the acid as 1 to 6, and the oxygen of the lime to that of the potassa in proportions which vary from 1 and $\frac{2}{3}$ to 1 and 1. Its composition approaches the formula $KO.CaO + 4SiO_3$. This is the glass of most difficult fusibility, and therefore most suitable for the combustion-tubes employed in organic analysis. From its purity, and the absence of oxide of lead, it is also made the basis of most coloured glasses, and of stained glass. To produce coloured glasses certain metallic oxides are mixed with the fused glass in the pot; oxide of cobalt, for instance, for a blue colour, oxide of copper for green, binoxide of manganese in small proportion for an amethystine glass, and in large proportion for a black glass, peroxide of uranium for a delicate lemon-yellow tint, and gold for a ruby glass. In stained glass, on the other hand, the metal or metallic oxide is merely applied with a proper flux to the surface of the glass, which is then exposed in an oven to a temperature sufficient to fuse the colouring matter, without distorting the sheet of glass. Different shades of yellow and orange are thus produced by means of silver and antimony, and a superb ruby-red by a proper, but difficult, application of suboxide of copper. The beautiful avanturine glass contains crystals of metallic copper. The green shade of ordinary glass is chiefly due to protoxide of iron, and is corrected by a small addition of binoxide of manganese (hence called *pyrolusite*), which raises the iron to the state of sesquioxide, in which it is not injurious, while, at the same time, the binoxide of manganese, by losing oxygen, passes into the state of the colourless protoxide of that metal.

Silicates of potassa and lead.—These substances enter into the composition of the purer and more brilliant species of glass in use in this country; such as that called crystal, of which most drinking vessels are made, flint-glass for optical purposes, and strass, which is employed in imitations of the precious stones. For crystal, the materials are taken in the following proportions: 120 parts of fine sand, about 40 of purified potashes, 35 of litharge or minium, and 12 of nitre. In this glass the oxygen of the bases is to that of the silicic acid as 1 to a number which may vary from 7 to 9, and the oxygen of the potassa is to that of the oxide of lead as 1 to a number varying from 1 to 2.5. In flint-glass, and in strass, the oxygen of the bases is to that of the silicic acid as 1 to 4, and the oxygen of the potassa is to that of the oxide of lead as 2 to 3 in flint-glass, and as 1 to 3 in strass (Dumas). The more oxide of lead glass contains, the higher its density; the density of this kind of glass exceeding 3.6, while that of the Bohemian glass does not rise higher than 2.4. Glass containing oxide of lead is recommended by its greater fusibility and softness, by which it is more easily fashioned into various forms, and by its great brilliancy, which is remarkable in lustres and other objects of cut glass. The presence of lead in glass is at once discovered by its surface acquiring a metallic lustre when heated to redness in the reducing flame. *Enamel* is a white and very fusible glass, containing a white opaque substance suspended in its mass. It is generally prepared from the stannate of lead, formed by heating and oxidizing together 15 parts of tin and 100 of lead. This is afterwards fused with 50 parts of sand and 40 parts of carbonate of potassa. Besides binoxide of tin, arsenious acid, oxide of antimony, phosphate of lime, and sulphate of potassa, are employed to give opacity to enamel.

Silicates of alumina, of the oxides of iron, magnesia, and potassa or soda.—Green or bottle-glass, of which wine-bottles, carboys, and glass articles of low price consist, is a mixture of these silicates. It is formed of the cheapest materials, such as sand, with soap-makers' waste, lime that has been used to render alkali caustic, &c. In the bottle-glass of this country the small quantity of alkali is chiefly soda. The alkaline sulphates, when fused with silicic and carbonaceous matter, lose their sulphuric acid, and become silicates; even common salt is decomposed by the united action of silicic acid and the aqueous vapour in flame, but much of it is lost from its own volatility. The proportion of silicic acid to the bases is much less in this than in the other kinds of glass, the oxygen of the former being to the latter as 2 to 1; and the oxygen of the alumina and sesquioxide of iron equal to that of the potassa and lime. This glass is, in fact, a mixture of neutral and subsilicates, and, when it contains an excess of lime, is more apt than any of the preceding species to assume a crystalline structure when maintained long in a soft condition by heat.

A bottle of green glass may be devitrified, or converted into what is called Reaumur's porcelain, by enveloping it in sand, and placing it where its temperature is kept high for several weeks, as in a brick-kiln or porcelain-furnace. Glass of all kinds, when strongly and repeatedly heated, loses alkali, from its volatility; the glass then becomes harder and less fusible, and is not so easily wrought,—a circumstance which may sometimes be remarked in blowing a bulb upon a tube which has been too long exposed to the blow-pipe flame. Glass of all kinds, when well manufactured, is supposed to be insoluble in water, but it is eventually acted upon, and soonest when its natural surface is broken; water tending to resolve glass into a soluble alkaline silicate and an insoluble earthy silicate. Glass bottles containing a large proportion of lime may be corroded through by sulphuric acid. An excess of alumina also makes glass very easily attacked by acids, even by the bitartrate of potassa in wines. In common with all natural and artificial silicates, glass is attacked by hydrofluoric acid, with the formation of the volatile fluoride of silicon. (See the Treatise on Glass, in Knapp's Chemical Technology, edited by Ronalds and Richardson, vol. ii.)

Ultramarine.—This beautiful blue pigment is extracted by mechanical operations from the mineral *Lapis lazuli.* The structure of the mineral is granular and slightly laminated: its constituents are, silicic acid 45.40, alumina 31.67, soda 9.09, sulphuric acid 5.89, sulphur 0.95, lime 3.52, iron 0.86, chlorine 0.42, water 0.12=97.92. It was first imitated successfully by M. Guimet in 1827. The process, according to M. Debette, appears to be first the preparation of a polysulphide of sodium, which is afterwards calcined with prepared clay and protosulphate of iron, so as to form sulphide of iron. The last product in fine powder is heated in a muffle with exposure to air for several hours, when it becomes in succession brown, red, green, and blue. The excess of sulphide of sodium and other salts is washed out of the powder, which, dried and washed again at a moderate temperature, gives an ultramarine of a magnificent blue tint. The process is an extremely delicate one, and the nature of the substance which gives the blue colour is very obscure. A sulphide of sodium is supposed to be essential to its composition, as the colour is destroyed by acids, with evolution of the hydrosulphuric acid; while the substitution of carbonate of potassa for carbonate of soda gives a compound corresponding to ultramarine, but which is colourless. (Pelouze et Fremy, Cours de Chim. Génér. ii. 117).

SECTION III.

LITHIUM.

Eq. 6.43 *or* 80.37; Li.

Lithium is the metallic basis of a rare alkaline oxide, lithia, discovered in 1818 by Arfwedson. (Ann. de Ch. et de Ph. x. 82). The name lithia (from λίθειος, stony) was applied to it, from its having been first derived from an earthy mineral. The metal was obtained by Davy by the voltaic decomposition of lithia, and observed to be white, resembling sodium, and to be highly oxidable. The equivalent of lithium is much smaller than that of any other metal, and its oxide has therefore a high saturating power.

Lithia; LiO.—The only known oxide of lithium is a protoxide. It exists in small quantities in the minerals spodumene or triphane, petalite, and lepidolite; but the mineral containing lithia, which is most abundant, is a native phosphate occurring at Rabenstein in Bavaria, and which consists of phosphoric acid 42.64, oxide of iron 49.16, oxide of manganese 4.75, and lithia 3.45. This mineral is dissolved in hydrochloric acid, the iron peroxidized by a little nitric acid, the solution diluted with water, and then ammonia added, which precipitates the insoluble phosphate of sesquioxide of iron. The manganese is afterwards removed by hydrosulphuric acid, the liquid filtered, evaporated to dryness, and the residue calcined to volatilize the ammoniacal salts; the chloride of lithium is then taken up by alcohol.

The hydrate of lithia resembles hydrate of potassa in causticity, but is less soluble in water, and loses its combined water at an elevated temperature. Sulphur acts upon it in the same manner as upon potassa. Its salts are colourless.

The chloride is very soluble in water, as well as in absolute alcohol, and fuses at a high temperature. It crystallizes in cubes containing 4HO.

The carbonate of lithia has a certain degree of solubility, and its solution has an alkaline reaction, properties upon which the claim of lithia to be ranked among the alkalies, instead of the alkaline earths, is chiefly rested. The fluoride of lithium has the sparing solubility of the carbonate.

The sulphate of lithia is soluble, and presents itself in fine crystals, which are persistent in air. It forms a double salt with sulphate of soda, of which the formula is $LiO.SO_3+NaO.SO_3+6HO$. The nitrate and acetate are both very soluble and deliquescent.

The neutral phosphate of lithia is slightly soluble in water, but considerably more so than the double phosphate of lithia and soda, which remains as an insoluble powder when the solution of lithia is evaporated to dryness with that of phosphate of soda. Hence phosphate of soda is used as a test of lithia. The salts of lithia are also recognized, when heated on platinum wire before the blow-pipe, by tinging the flame of a red colour.

ORDER II.

METALLIC BASES OF THE ALKALINE EARTHS.

SECTION I.

BARIUM.

Eq. 68.64 *or* 858; Ba.

Barium, the metallic basis of the earth baryta, was obtained by Davy in 1808, by the voltaic decomposition of moistened carbonate of baryta in contact with mercury: it may likewise be procured by passing potassium in vapour over baryta heated to redness in an iron tube, and afterwards withdrawing the reduced barium, which the residue contains, by means of mercury. The latter metal is separated by distillation in a glass retort, care being taken not to raise the temperature to redness, for the barium then decomposes glass. Barium is a white metal like silver, fusible under a red heat, denser than oil of vitriol, in which it sinks. It oxidates with vivacity in water, disengages hydrogen, and is converted into baryta. It is named barium (from βαρύς, heavy), in allusion to the great density of its compounds.

Baryta; BaO, 76.64 *or* 958.—This earth exists in several minerals, of which the most abundant are sulphate of baryta or heavy-spar, and the carbonate of baryta or witherite. The earth is obtained in the anhydrous condition and pure, by calcinating nitrate of baryta, at a bright-red heat, in a porcelain retort, or in a well-covered crucible of porcelain or silver, but not of platinum. Baryta is a grey powder, of which the density is about 4. When heated to redness in a porcelain tube, and oxygen gas passed over it, it absorbs that gas with avidity, and becomes binoxide of barium, the compound for the preparation of which anhydrous baryta is chiefly required. Baryta slakes and falls to powder when water is thrown upon it, combining with one equivalent of water with the evolution of so much heat as to become incandescent.

Hydrate of baryta is a valuable reagent. Of the different processes for this substance, one of the most convenient is that from the native sulphate. This is a soft mineral, and easily reduced to an impalpable powder, which is intimately mixed with one-eighth of its weight of coal pounded and sifted, or with one-third charcoal-powder and one-fourth resin; the mixture is introduced into a Cornish crucible, and exposed in a furnace to a bright-red heat for an hour. The sulphate is converted by this treatment into sulphide of barium; the last salt is dissolved out of the black residuary mass by boiling water, and the solution, which generally has a yellow tint but is sometimes colourless, is filtered while still hot. The solution, if strong, may crystallize on cooling, in thin plates. As the sulphide absorbs oxygen from the air, and returns to the state of sulphate of baryta, it must not be exposed long in open vessels. To a boiling solution of sulphide of barium in a flask, black oxide of copper from the nitrate is added, in successive small portions, till a drop of the liquid ceases to blacken a solution of lead, and precipitates it entirely white: the liquid then contains only hydrate of baryta in solution. It may immediately be filtered, with little access of air, to prevent absorption of carbonic acid. The decomposition in this process, for which we are indebted to Dr. Mohr of Coblentz, is rather complicated. Six eq. of sulphide of barium and 8 eq. of oxide of copper producing 5 eq. of baryta, 1 eq. of hyposulphite of baryta, and 4 eq. of subsulphide of copper, of which the first only is soluble:

$$6\,BaS \text{ and } 8CuO = 5BaO \text{ and } BaO.S_2O_2 \text{ and } 4Cu_2S.$$

Binoxide of manganese may be substituted in this process for oxide of copper, but generally gives a solution of baryta coloured by some impurity. The reaction is then similar:

$$6BaS \text{ and } 4MnO_2 = 5BaO \text{ and } BaO.S_2O_2 \text{ and } 4MnS.$$

If the solution of sulphide of barium has been concentrated, the greater part of the hydrate of baryta separates on cooling in voluminous and transparent crystals, containing 10HO.

Hydrate of baryta may also be obtained by adding caustic potassa to a saturated solution of chloride of barium; hydrate of baryta precipitates, and must be redissolved in boiling water, and crystallized by cooling, to purify it. It is soluble in 3 parts of boiling water, and in 20 parts of water at 60°. Baryta retains 1 eq. of water with great force like the fixed alkalies. This combination is fusible a little below redness, and runs like an oil; it congeals into a crystalline mass, which attracts carbonic acid very slowly from air, and is therefore the most favourable position in which to preserve hydrate of baryta.

The solution of baryta is strongly caustic, although less so than potassa or soda, and disorganizes organic matter rapidly; it is poisonous, in common with all the soluble preparations of barium. Chlorine decomposes baryta in the same manner as it does the alkalies. Sulphur is dissolved in the solution of baryta with the aid of heat, and, according to the temperature, a sulphate or hyposulphite is formed, with the trisulphide of barium of a green colour. When heated to redness in the vapour of phosphorus, baryta is converted into phosphate of baryta and phosphide of barium. On dropping oil of vitriol upon dry baryta and strontia, the combination is said to produce light with the first, but not with the second. Baryta, whether free or in combination with an acid as a soluble salt, is discovered by means of sulphuric acid, which throws down sulphate of baryta, a compound not decomposed by, nor soluble in, nitric and hydrochloric acids.

Binoxide of barium; BaO_2; 84.64 *or* 1058.—This compound is prepared by exposing anhydrous baryta, from the nitrate, to pure oxygen at a red heat; or by heating pure baryta to low redness in a porcelain-crucible, and then gradually adding chlorate of potassa, in the proportion of about 1 part of the latter to 4 of the former.

The chloride of potassium formed at the same time, is removed, by cold water, from the binoxide of barium, while the latter unites with 6HO. Binoxide of barium, when decomposed by dilute acids with proper precautions, affords binoxide of hydrogen.

Chloride of barium; BaCl+2HO; 104.14+18 *or* 1301.76+225.—A reagent of constant use, which is obtained by dissolving native carbonate of baryta in pure hydrochloric acid diluted with 3 or 4 times its bulk of water, or by neutralizing sulphide of barium by the same acid. It crystallizes from a concentrated solution in flat four-sided tables, bevelled at the edges. The crystals contain 2HO (14.75 per cent. of water), which they lose below 212°. They are said to be soluble in 400 parts of anhydrous alcohol: 100 parts of water dissolve 43.5 parts at 60°, and 78 parts at 222°, which is the boiling-point of the solution.

Carbonate of baryta; $BaO.CO_2$; 98.64 *or* 1233.01.—This salt consists in 100 parts of 22.41 carbonic acid, and 77.59 baryta. The density of the native carbonate is 4.331; it is not attacked by sulphuric acid, and retains its carbonic acid at the highest temperatures. The precipitated carbonate is decomposed by sulphuric acid, and loses its carbonic acid when calcined at a white heat, in contact with carbonaceous matter. It is obtained of greater purity when precipitated by the carbonate of ammonia, than by the carbonate of potassa or soda, portions of which are apt to go down in combination with carbonate of baryta. Although reputed an insoluble salt, carbonate of baryta is soluble in 2300 parts of boiling water, and in 4300 parts of cold water. It is still more soluble in water containing carbonic acid, and is highly poisonous. The precipitated carbonate of baryta, or, better, the hydrate of baryta, is employed in the analysis of silicious minerals, containing an alkali, which are not soluble in an acid. The mineral, in the state of an impalpable powder, is intimately mixed with 4 or 5 times its weight of the hydrate, and exposed in a silver-crucible to a red heat, which occasions a semi-fusion of the mixture and the decomposition of the silicates; the mineral afterwards dissolving entirely in an acid, with the exception of its silica.

Sulphate of baryta; $BaO.SO_3$; 116.64 *or* 1458.01.—This salt consists, in 100 parts, of 34.37 sulphuric acid and 65.63 baryta. The density of heavy-spar, or the native sulphate, varies from 4 to 4.47. It occurs in considerable quantities in trap and other igneous rocks, forming often veins of several feet in thickness, and miles in extent. It is mined for the purpose of being substituted for carbonate of lead, or being mixed with that substance, when used as a pigment. When chloride of barium is added to sulphuric acid, or to a soluble sulphate, at the boiling temperature, sulphate of baryta precipitates readily, in a dense crystalline powder, which may easily be collected and washed on a filter. It is completely insoluble in water and dilute acids, but is soluble in concentrated and boiling sulphuric acid, from which it crystallizes on cooling. Precipitated sulphate of baryta is partially decomposed in a concentrated and boiling solution of carbonate of potassa or soda, and carbonate of baryta formed.

Nitrate of baryta; $BaO.NO_5$; 130.64 *or* 1633.01.—This salt crystallizes in fine transparent octohedrons, which are anhydrous. It is obtained by dissolving carbonate of baryta in nitric acid diluted with 8 or 10 times its weight of water; or by mixing the acid, also in a diluted state, with the solution of sulphide of barium. It requires 12 parts of water at 60°, and 3 or 4 parts of boiling water, for solution; it is insoluble in alcohol. The nitrate of baryta is employed as a reagent, and also in procuring anhydrous baryta.

The chlorate and hyposulphate of baryta are soluble, the iodate, sulphite, hyposulphite and phosphates of baryta, insoluble salts.

SECTION II.

STRONTIUM.

Eq. 43.84 *or* 548.02. Sr.

Strontium is prepared in the same way as barium, which it greatly resembles. It is a white metal, denser than oil of vitriol. It derives its name from Strontian, a mining village in Argyleshire.

Strontia, Strontian, or *Strontites;* SrO; 51.84 *or* 648.02. — The native carbonate of strontia was first distinguished from carbonate of baryta by Dr. Crawford, in 1790, who conceived the idea that the former mineral might contain a new earth. This conjecture was verified in 1793, by Dr. Hope (Edinb. Trans. iv. 14); and much about the same time also by Klaproth. The earth, strontia, is to baryta what soda is to potassa. It occurs in nature as carbonate and more abundantly as sulphate. Strontia may be prepared by a strong calcination of the native carbonate in contact with carbon. It is lighter than baryta, and has a taste which is less acrid and caustic, but stronger than that of lime. It is said not to be poisonous. The hydrate crystallizes with 9HO, but retains only one equivalent at 212° (Mr. Smith). This last hydrate enters into fusion at a very high temperature, without losing its combined water. The anhydrous earth, like baryta, is infusible. The crystallized hydrate requires 52 parts of water to dissolve it at 60°, but only twice its weight at 212°.

The soluble salts of strontia are prepared from the carbonate. They are precipitated by sulphuric acid and by soluble sulphates, but not so completely as the salts of baryta, the sulphate of strontia having a small degree of solubility. Hence, when sulphate of soda is added in excess to a salt of strontia, and the precipitate separated by filtration, so much sulphate of strontia remains in solution, that the liquid yields a white precipitate with carbonate of soda (Dr. Turner). Most of the salts of strontia, when heated on platinum-wire before the blow-pipe, communicate a red colour to the flame. Baryta and strontia in solution may be separated by hydrofluosilicic acid, which precipitates baryta, but forms with strontia a salt very soluble in a slight excess of acid. Hyposulphite of strontia being soluble, while hyposulphite of baryta is insoluble, these earths may also be distinguished by means of hyposulphite of soda.

Binoxide of strontium, obtained by Thénard in brilliant crystalline scales, on adding binoxide of hydrogen to a solution of strontia. It contains two eq. of oxygen.

Chloride of strontium crystallizes in slender prisms, which contain 9HO, and are slightly deliquescent. This salt is soluble in three-fourths of its weight of cold water, and in all proportions in boiling water. At the ordinary temperature it dissolves in 24 parts of anhydrous alcohol, and in 19 parts of alcohol boiling. In this respect it differs from chloride of barium, which is insoluble in alcohol. Chloride of strontium communicates to flame a fine red tint. In the anhydrous condition this chloride absorbs 4 eq. of ammonia, and becomes a white bulky powder.

Carbonate of strontia forms the mineral *strontianite,* which generally has a fibrous texture, and is sometimes transparent and colourless, but generally has a tinge of yellow or green. Its density varies from 3.4 to 3.726. This salt is said to be soluble in 1536 parts of boiling water. It is more soluble in water containing carbonic acid, and occurs in some mineral waters. It retains its carbonic acid when calcined.

Sulphate of strontia is known as *celestine,* and occurs in regular crystals of the same form as sulphate of baryta. Its density is about 3.89. It is soluble in from 3000 to 4000 parts of water, and the solution is sensibly precipitated by chloride of barium. The mineral is found in considerable quantity associated with volcanic sulphur, and in other formations. A large deposit of it exits in the neighbourhood of Bristol, from which it may be obtained in sufficient quantity for any application in the arts. The various compounds of strontium may be prepared from the sulphate of strontia precisely in the same manner as those of barium from the sulphate of baryta.

Hyposulphite of strontia is crystallizable, and soluble in 4 parts of cold, and $1\frac{3}{4}$ parts of boiling water. It loses 31 per cent. of water of crystallization between 122° and 140°, without any other change.

Nitrate of strontia crystallizes at a high temperature in regular octohedrons, of density 2.857, which are anhydrous, but it is generally obtained at a low temperature in crystals, which contain 5HO, of density 2.113 (Filhol). The anhydrous salt dissolves in 5 parts of cold water, and in 1 part of boiling water. A deflagrating mixture, which produces an intensely red illumination, is formed of 40 parts of nitrate of strontia, 13 parts of flowers of sulphur, 5 parts of chlorate of potassa, and 4 parts of sulphide of antimony.

The salts of baryta, strontia, and protoxide of lead, are strictly isomorphous, and greatly resemble each other in solubility and other properties. Hydrofluosilicic acid is employed to separate baryta from strontia, as it precipitates the former but not the latter. Neutral chromate of potassa, which precipitates salts of baryta immediately, precipitates only slowly the salts of strontia. In analysis, strontia is generally estimated as sulphate, but as the latter is not completely insoluble, an addition of alcohol is made to the water employed to wash the precipitate.

SECTION III.

CALCIUM.

Eq. 20, *or* 250; Ca.

Davy obtained evidence of the existence of this metal, and of its analogy to the preceding metals. It is the basis of lime. The name applied to it is derived from calx.

Lime; CaO; 28, *or* 350.—Uncombined lime, or quicklime, as it is termed in the arts, is obtained by heating masses of limestone (carbonate of lime) to redness in an open fire, or lime-kiln. The escape of the carbonic acid is favoured by the presence of aqueous vapour and the gases of the fire, into which that gas can diffuse (page 181). In a covered crucible, carbonate of lime may be fused by heat without decomposition. The lime, properly burnt, remains in porous masses, which may be easily separated from the ashes of the fuel, and are sufficiently hard to be transported from place to place without falling to pieces. Although these masses appear light, the density of lime is not less than 2.3, or even 3.08, according to Royer and Dumas. Water thrown upon them, is first imbibed, and afterwards combines with the lime, which falls to powder in the state of hydrate, and is then said to be slaked. In this combination the temperature may rise to 572°, (300° C.), or sufficiently high to char wood. From its affinity for water, quicklime is applied to deprive certain liquids, such as alcohol, of the water they contain. To obtain pure lime, the crystallized carbonate should be calcined, such as calcareous spar, or Carrara marble. Lime, in common with other infusible earths, phosphoresces strongly when heated to full redness.

The only hydrate of lime known contains 1 eq. of water, which it loses at a low-red heat. It is sparingly soluble in water, but more soluble in cold than in hot water. According to Dalton, lime-water formed at 60°, 130°, and 212°, contains 1 grain of lime in 778, 972, and 1270 grains of water. Hence water saturated in the cold deposits hydrate of lime when boiled. By evaporating the solution in vacuo, Gay-Lussac obtained the same hydrate of lime in small transparent crystals of the hexahedral form. The *milk* or *cream* of lime is merely the hydrate diffused through water. In preparing lime-water, 3 or 4 ounces of slaked lime are agitated several times, during two or three hours, with two quarts of distilled water, and then allowed to settle. The lime-water first drawn off generally contains a little potassa, and should not therefore be considered pure. Lime-water has a harsh acrid taste, is alkaline, and, to a certain extent, caustic. It precipitates carbonic, silicic, boracic, and phosphoric acids from solutions of their alkaline salts. It dissolves oxide of lead. Lime-water absorbs carbonic acid rapidly from the air, and becomes covered by a pellicle of carbonate of lime. Hydrate of lime has the same property, absorbing about half an equivalent of carbonic acid with avidity, but not acquiring quite so much as three-fourths of an equivalent by two or three weeks' exposure to an atmosphere of the gas. Fuchs observed, that when hydrate of lime is exposed to air, it absorbs only half an equivalent of carbonic acid, and that a definite compound of hydrate and carbonate was formed. In the anhydrous condition, lime exhibits no affinity for carbonic acid.

Lime is characterized by affording a bulky precipitate of sulphate of lime, when sulphuric acid is added to its soluble salts. But as the sulphate of lime has a certain degree of solubility, this precipitate does not appear in very dilute solutions of these salts, nor in lime-water, a property by which lime may be distinguished from baryta and strontia. Sulphate of lime may also, when precipitated, be re-dissolved by the addition of nitric acid. Lime is entirely precipitated from neutral solutions by oxalate of ammonia, the oxalate of lime being completely insoluble. In the quantitative estimation of this earth, it is therefore generally thrown down as oxalate, and afterwards obtained as carbonate of lime, by heating the oxalate nearly to redness in a platinum crucible, in which a small fragment of carbonate of ammonia is dissipated at the same time, to prevent any lime becoming caustic by loss of carbonic acid.

Lime is applied to a variety of useful purposes in ordinary life and in the arts, of which the most important are its applications as a manure for land, and as mortar. In the first application, lime appears to be chiefly useful, (1) in promoting the oxidation and decomposition of the insoluble organic matters which the soil contains, and thereby rendering them capable of sustaining vegetable life; (2) in decomposing clay and rendering its potassa soluble, and (3) in restoring to the soil the calcareous element which is annually removed in the crop. In the formation of mortar, the hydrate of lime is mixed with 2 parts of coarse, or 3 parts of fine sand, and made into a paste with water. In building, a stone is laid upon a bed of this paste, which it compresses by its weight, imbibing moisture also from the mortar, which escapes principally through the porous stone. On drying, the mortar binds the stones between which it is interposed, and its own particles cohere so as to form a hard mass, solely by the attraction of aggregation, for no chemical combination takes place between the lime and sand, and the stones are simply united as two pieces of wood are by glue. The sand is useful in rendering insignificant by its mass the contraction of the mortar on drying, and also, from the large size of its grains, in rendering the dry mortar less short and friable. The mortar is subject to an ulterior change, from the slow absorption of carbonic acid, but even in the oldest mortar the conversion of the hydrate of lime into carbonate is never complete. The lime which is called *fat* slakes easily, and with considerable increase of volume; *lean* or poor lime slakes imperfectly, owing frequently to the presence of magnesia in a proportion exceeding 10 or 12 per cent; the latter earth having a comparatively feeble affinity for water. Magnesian lime is also generally considered prejudicial in agriculture, owing, it is supposed, to the magnesia long remaining caustic in the soil.

Some limestones, containing about 20 per cent of clay or silicate of alumina, afford lime which possesses a valuable property, that of forming with water a mass which becomes solid in a few minutes, and therefore hardens in structures covered by water. An excellent hydraulic mortar of this kind is obtained from concretionary masses found in marl, and also as isolated blocks in the bed of the Thames. This lime being burnt, ground, and sifted, when mixed with water to form a paste, sets as quickly as Paris plaster; its solidity increases with the time it has been submerged, and it ends by acquiring the hardness of limestone. Sand is added to it when it is used as common mortar, or in covering buildings to imitate stone. From the minute division of the silicic acid and alumina in this mortar, their combination with lime is more likely to occur than in ordinary mortar. Still the first setting of hydraulic mortar seems to be due simply to the fixation of water, and formation of a solid hydrate like gypsum. Hydraulic mortar is sometimes made by mixing together clay and chalk, and calcining the mixture, or more frequently by adding to hydrate of lime puzzolano ground to fine powder. The latter is a silicious substance of volcanic origin, composed principally of pumice, of which a stratum is excavated in the neighbourhood of Naples. The mortar which it makes with lime has obtained the name of Roman cement.

The hydrate of *binoxide of calcium* precipitates on adding lime-water, drop by drop, to a solution of binoxide of hydrogen. It contains, according to Thénard, 2 eq. of oxygen.

The *protosulphide of calcium* is procured by decomposing sulphate of lime, at a red heat, by hydrogen or charcoal. When newly prepared, it phosphoresces in the dark. It is only very sparingly soluble in water, but it is decomposed by boiling water, according to M. H. Rose, into hydrosulphate of sulphide of calcium, which is soluble, and hydrate of lime. Sulphide of calcium, when moistened with water, is readily decomposed by a stream of carbonic-acid gas, with the evolution of hydrosulphuric acid:

$$CaS.HO \text{ and } CO_2 = CaO.CO_2 + HS.$$

When hydrate of lime is boiled with sulphur and water, and the liquor allowed to cool before it is completely saturated with sulphur, yellow crystals separate from it, which are a *bisulphide of calcium*, combined with 3HO, according to the observations of Herschel. When lime, or protosulphide of calcium, is boiled with excess of sulphur, it dissolves sulphur till a *pentasulphide of calcium* is formed, which resembles in properties the corresponding degree of sulphuration of potassium.

Phosphide of calcium.—Small fragments of quicklime being heated to redness by a spirit-lamp, in a small mattrass with a long neck, and fragments of phosphorus dropped into the same vessel, a mixture is obtained of phosphate of lime and phosphide of calcium. The compound has a chocolate-brown colour. When the temperature is raised too high, the affinities change, and phosphorus escaping in vapour, nothing but lime remains. According to M. P. Thénard, in the reaction which gives phosphide of calcium, 7 eq. of phosphorus act upon 14 eq. of lime:

$$14CaO \text{ and } 7P = 2(2CaO.PO_5) \text{ and } 5Ca_2P.$$

The phosphide, therefore, contains 2 eq. of calcium to 1 eq. of phosphorus, and is analogous to the liquid hydride of phosphorus PH_2. When thrown into water, it is immediately transformed into the hydride of phosphorus referred to, which is spontaneously inflammable, and hypophosphite of lime, which is dissolved.

Chloride of calcium; CaCl; 55.50 *or* 693.75.—Obtained by neutralizing hydrochloric acid with carbonate of lime, or as a residue in several processes; a concentrated solution affords crystals in large striated four-sided prisms, which contain 6 eq. of water. Dried with stirring, above 212°, it affords a crystalline powder, containing 2 eq. of water, which produces an intense degree of cold when mixed with snow (p. 62). The same hydrate was produced on drying the crystals in vacuo over sulphuric acid for ten days. The crystals are soluble in one-fifteenth of their weight of water at 60°, and exceedingly deliquescent. The salt is made anhydrous by heat, and undergoes the igneous fusion at a red heat. The liquid chloride is poured upon a slab, and the transparent cake of solid salt immediately broken into pieces, and preserved in a stoppered bottle. It is much employed, from its great affinity for water, to dry gases and absorb moisture. Chloride of calcium always acquires by fusion a slight but sensibly alkaline reaction from partial decomposition; on which account Liebig prefers the salt strongly dried, but not fused, as the hygrometric agent in organic analysis. Ignited with the sulphates of baryta and strontia, chloride of calcium gives rise to sulphate of lime and the chlorides of barium and strontium. Ten parts of anhydrous alcohol dissolve 7 parts of chloride of calcium, at the boiling-point, and the solution, in cold weather, affords crystals in rectangular scales, which are an alcoholate, containing 2 eq. of alcohol, instead of water of crystallization; $CaCl+2C_4H_6O_2$. Anhydrous chloride of calcium likewise absorbs 4 equivalents of ammoniacal gas, and forms a bulky white powder, $CaCl+4NH_3$, from which the ammonia may be easily expelled again by heat.

A solution of chloride of calcium, when boiled with hydrate of lime, dissolves that substance, and the solution filtered hot, deposits an *oxichloride of calcium*, $3CaO.CaCl+15HO$, in long flat and thin crystals. The salt is decomposed by water and alcohol.

A compound of chloride of calcium with *oxalate of lime* containing water of crystallization, is obtained in good crystals, which are persistent in air, by dissolving oxalate of lime to saturation in hot hydrochloric acid, and allowing the solution to cool. It consists of 1 eq. of each salt, with 7 eq. of water. Oxalate of lime is known to combine with 2 eq. of water, of which 1 eq. appears to remain in this double salt, while the other is replaced by chloride of calcium carrying its 6 eq. of water of crystallization along with it; $CaO.C_2O_3+(HO.CaCl)+6HO$. A similar replacement is observed in the formation of quadroxalate of potassa (p. 164). This salt becomes anhydrous without decomposition at 266° (130° C.) It is decomposed by pure water.

Fluoride of calcium, Fluor-spar; CaF; 38.70 *or* 483.80.—This salt is peculiarly a constituent of mineral veins, and occurs massive, or in transparent crystals which are cubes or octohedrons, and is often of beautiful colours, generally green or purple. It is cut into ornamental forms, and is believed to be the substance of which the *vasa murrina* of the Romans were composed. In minute quantity fluoride of calcium is very generally diffused, being found in the earthy deposit from sea-water when boiled (G. Wilson). It forms a few thousandths of the earth of bones, and a somewhat larger proportion of the enamel of the teeth: in fossil bones the proportion of fluoride of calcium is considerably greater (J. Middleton, Mem. Chem. Soc. ii. 134). It is dissolved to a small extent by water containing carbonic acid, like the other insoluble salts of lime; its density varies from 3.14 to 3.17. When heated gently, on a plate of metal, it becomes luminous in the dark for a short time; the phosphorescent property may be restored by passing electric sparks through the crystal (Griffiths). Fluoride of calcium is obtained in a granular condition, when hydrofluoric acid is neutralized by freshly precipitated carbonate of lime. But when a neutral salt of lime is mixed with a soluble fluoride, the fluoride of calcium appears as a translucent gelatinous mass. This fluoride, whether artificial or natural, is not decomposed by sulphuric acid at a low temperature, but imbibes that acid, and forms a thick ropy liquid. At 104° (40° C.), this mixture begins to decompose, and emits hydrofluoric acid. Fluoride of calcium resists the action of a solution of hydrate of potassa, but is easily decomposed in the dry way by fusion with carbonate of potassa, and fluoride of potassium is formed.

SALTS OF LIME.

Carbonate of lime; $CaO.CO_2$; 50, *or* 625.—This is one of the most abundantly diffused salts in nature, forming the basis of limestones, marbles, marls, coral-reefs, shells, &c. It is anhydrous, and occurs in two incompatible crystalline forms, the rhomboidal crystal of Iceland spar and calc-spar, which, with its numerous modifications, is much the most abundant, and the six-sided prism of arragonite, isomorphous with carbonate of strontia, which last may be readily recognized by falling to powder when heated. The grains of this powder have the form of calc-spar. The density of carbonate of lime in these two forms is sensibly different, that of calc-spar being 2.719, and of arragonite 2.949 (G. Rose). Carbonate of lime consists of 56 lime and 44 carbonic acid in 100 parts.

Carbonate of lime may also be obtained in the state of a hydrate by heating together very slightly 1 part of hydrate of lime, 3 parts of sugar, and 6 parts of water, filtering the solution, and leaving it exposed in a shallow vessel. In twenty-four hours crystals appear upon the surface of the liquid, and in fifteen days the whole lime is generally converted into hydrated carbonate, in the form of sharp transparent rhombs. The carbonic acid is absorbed from the atmosphere. These crystals contain 5 eq. of water; by boiling them in anhydrous alcohol, a second definite hydrate is obtained containing 3 eq. of water, as ascertained by Pelouze. The first of these hydrates has also been found native in a running stream, by Scheerer. The two hydrates of carbonate of lime correspond in composition with two crystalline hydrates of carbonate of magnesia.

Carbonate of lime is considered an insoluble salt, although, according to Fresenius, one part of carbonate of lime dissolves in 8834 parts of boiling water, and in 10601 parts of water at ordinary temperatures: the solution is sensibly alkaline to test-paper. When recently precipitated, carbonate of lime is much more soluble in salts of ammonia: the solution of carbonate of lime in hydrochlorate of ammonia in excess is completely resolved by spontaneous evaporation into chloride of calcium and carbonate of ammonia, which escapes. Sea-water appears to be essentially alkaline from the presence of carbonate of lime, a circumstance calculated, therefore, to prevent the accumulation in the sea of ammonia in the form of fixed salts, and to cause the restoration of that base to the atmosphere. Carbonate of lime is soluble in water containing carbonic acid, and is generally present in the water of wells, and in some mineral waters to a considerable extent. It is deposited from the latter, when exposed to air in a gradual manner and in possession of a crystalline structure, forming stalactites and stalagmites in mountain caverns, and calcareous petrifications, when it flows over wood and other organic and destructible matters, of which it preserves the form. When a current of carbonic-acid gas is passed through lime-water, the greater portion, but not the whole, of the carbonate of lime first precipitated is re-dissolved by the excess of carbonic acid. This solution yields on evaporation the anhydrous carbonate, and no crystalline bicarbonate of lime has been obtained. Carbonate of lime is decomposed with effervescence by acids. At a red heat it parts with carbonic acid, and is converted into quicklime in the manner already described.

A crystalline mineral was discovered by Boussingault at Merida in America, which he ascertained to be a double carbonate of soda and lime, with 5 eq. of water, and named *gaylussite*, in honour of Gay-Lussac. It may be made anhydrous by heat, and its two salts are then separated by water.

The hardness of well and river-water, so far as it is due to carbonate of lime in solution, may be removed by a proper addition of lime-water, the free carbonic acid becoming carbonate of lime, and precipitating together with the portion of carbonate of lime formerly held in solution; colouring and other organic matter is carried down at the same time.[1] This elegant process has been found to act satisfactorily on a large scale. The proportion of carbonate of lime, where it is the only alkaline substance in solution, may be determined with great accuracy by neutralizing 8750 grains of the water (one pint), by means of a normal acid solution containing 0.4562 per cent. of hydrochloric acid (this is 319.37 grs. of HCl in one gallon, or 70000 grs. of water, or as much acid as would neutralize one ounce or 437.5 grs.

[1] Professor Clark: Repertory of Patent Inventions, October 1841; a pamphlet entitled "A New Process for Purifying Waters supplied to the Metropolis," published by R. and J. E. Taylor; and "On the Examination of Water for its Hardness," Pharmaceutical Journal, vi. 526. The instruments and test-liquids required in the examination of waters by Prof. Clark's method may be obtained at Mr. Griffin's, in Baker street, London.

Fig. 185.

of carbonate of lime). This test-acid is prepared by means of pure carbonate of soda, as in the process of alkalimetry (page 386), or from the analysis of the dilute acid by nitrate of silver. The measured quantity of water is placed in an evaporating basin, and being found alkaline by delicate red litmus-paper, the normal acid is added from the small burette (fig. 185) graduated into ten-grain measures, each of which is subdivided into five, till the point of neutralization is reached, the liquid being heated towards the end of the operation. A small portion of 30 or 40 grains of the water is transferred to a small conical wine-glass, and the test-paper left in it for several minutes, to obtain the indication of alkalinity. To save time, a series of six of these wine-glasses is conveniently employed, each containing a sample of the water after successive additions of the test-acid. Each ten-grain measure of the acid required indicates 1 grain of carbonate of lime in 1 gallon of the water, or 0.000014286 per cent. of carbonate of lime. By such means a minutely accurate determination of alkalinity may be obtained; one-hundredth of a grain of carbonate of lime in a pint of water is thus observed. (Prof. Clark).

Sulphate of lime, Gypsum; $CaO.SO_3 + 2HO$; 68 + 18 *or* 850 + 225.—This salt precipitates as a bulky and gritty powder, when sulphuric acid is added to a soluble salt of lime. Sulphate of lime appears to have nearly the same degree of solubility at all temperatures, and requires 460 parts of water for solution, according to Bucholz, or 380 parts of cold, and 388 parts of boiling water, according to Geise. It occurs in nature in well-formed crystals, and also in large crystalline masses, forming beds of gypsum; a mineral which contains 2 eq. of water, and of which the density is 2.322 (Royer and Dumas). Prof. Johnston likewise obtained small prismatic crystals of sulphate of lime, deposited in a steam-boiler, which contain only half an equivalent of water $2(CaO.SO_3) + HO$. Sulphate of lime occurs in a crystalline form, without water, forming the mineral *anhydrite,* of which the density is about 2.96. Sulphate of lime fuses at a strong red heat, without decomposition, and on cooling assumes the crystalline form of the last mineral. To form plaster of Paris, gypsum, in pieces about the size of a pigeon's egg, is heated in an oven till it is nearly anhydrous, and then reduced to a powder. When this is made into a paste with a little water, it forms a hard coherent mass, or sets, in a minute or two, with a slight evolution of heat. This artificial hydrate, or *stucco,* has the same composition as native gypsum. If sulphate of lime has been heated above 300°, in drying, it refuses to set afterwards when mixed with water.

The powder of hydrated gypsum solidifies also when mixed with a solution of potassa, or various salts of potassa, such as the carbonate, bicarbonate (in this case with violent effervescence), sulphate, and silicate, but not with the chlorate or nitrate of potassa, nor with any salt of soda. Double salts are probably formed, as it is the alkaline salts only which are capable of forming double salts, and are considered bibasic by M. Herhardt, that possess the remarkable property in question (Emmet, Am. Journ. of Scien., xxiii. 209).

Hyposulphite of lime is formed by transmitting sulphurous acid through sulphide of calcium, suspended in water, till the solution is neutral and colourless. The solution is decomposed when heated above 140° (60° C.) into sulphur and sulphite of lime. If evaporated below that temperature, it yields large hexagonal prisms of hyposulphite of lime, on cooling, which are colourless. They contain 5 eq. of water, and are persistent in air. The same salt may be obtained very economically by exposing to air the waste-lime of the dry-lime gas purifiers.

Nitrate of lime is a highly deliquescent salt, which crystallizes with 6 eq. of water, like the nitrates of the magnesian class. It is soluble in alcohol.

Phosphates of lime.—On adding chloride of calcium to the tribasic subphosphate of soda, a corresponding phosphate of lime precipitates in bulky gelatinous flakes, of which the formula is $3CaO.PO_5$. This phosphate occurs in nature in combination with fluoride of calcium in the form of hexagonal prisms, in the minerals *apatite* and *moroxite*. The formula of apatite is $CaF + 3(3CaO.PO_5)$. The native phosphates of lead occur in the same form, with chloride of lead in the place of fluoride of calcium. *Hedyphan* is the same mineral, in which a portion of phosphoric acid is replaced by arsenic acid.

Another tribasic phosphate of lime is obtained on adding the solution of common phosphate of soda, drop by drop, to chloride of calcium. This precipitate is slightly crystalline. Its formula, exclusive of its water of crystallization, is $HO.2CaO.PO_5$. Again, when a solution of phosphate of ammonia, supersaturated with ammonia, is treated with a solution of chloride of calcium, till about one-half of the phosphoric acid is precipitated, the precipitate contains 51.263 per cent of lime, and corresponds to the formula $8CaO.3PO_5$ (Berzelius). A biphosphate of lime is also described by Berzelius, obtained on evaporating a solution of any of the preceding salts in nitric acid to the point of crystallization, of which the probable formula is $2HO.CaO.PO_5$. There also exist a pyrophosphate and metaphosphate of lime. The insoluble phosphates of lime are soluble in water containing carbonic acid. It is possibly in this manner that phosphate of lime is dissolved by the alkaline animal fluids.

Hypochlorite of lime; Chloride of lime; Bleaching powder.—This compound, remarkable for its valuable applications in the arts, is generally prepared by exposing hydrate of lime, from the purest lime, to chlorine-gas, the latter being supplied so gradually as to prevent the heat, occasioned by the combination, from rising above 62°. Chlorine is not absorbed by quicklime, nor by the carbonate of lime. When dried at 212°, hydrate of lime, I find, absorbs afterwards little or no chlorine; but dried over sulphuric acid, without heat, it is, on the contrary, in the most favourable condition for becoming chloride of lime. A dry, white, pulverulent compound is obtained by exposing the last hydrate to chlorine, which contains 41.2 to 41.4 chlorine in 100 parts; but of this chlorine about 39 parts only are available for bleaching, owing to 2 parts of that element going to the formation of chloride of calcium and chlorate of lime. A slight addition of moisture to hydrate of lime does not increase the proportion of chlorine absorbed, and renders the compound less stable. The above appears to be the maximum absorption of chlorine by dry hydrate of lime, and is greater than it would be advisable to attempt in the manufacture of bleaching powder, owing to the occurrence of the partial decomposition adverted to. Yet this proportion is considerably short of 1 eq. of chlorine to 1 of hydrate of lime, which are 48.57 chlorine and 51.43 hydrate of lime, in 100 parts. The excess of lime appears to be useful in adding to the stability of the compound. Labarraque mixes the hydrate of lime with $\frac{1}{20}$th of its weight of chloride of sodium, by which means the absorption of chlorine is greatly promoted. The bleaching powder of commerce may contain, when newly prepared, about 30 per cent. of chlorine; but after being kept for several months, the proportion of available chlorine is found more frequently below than above 10 per cent., so much does it deteriorate by keeping.

The reaction which occurs in the formation of hypochlorite of lime is represented as follows:—

$$2CaO \text{ and } 2Cl = Ca.Cl \text{ and } CaO.ClO.$$

Or the product is a mixture of chloride of calcium and hyperchlorite of lime.

The same compound is obtained in solution by transmitting a stream of chlorine-gas through hydrate of lime suspended in water. The lime then absorbs a full equivalent of chlorine, and dissolves entirely.

Ten parts of water take up the bleaching combination from one part of dry chloride of lime, leaving undissolved the hydrate of lime contained in excess. The solution has a slight odour of hypochlorous acid, a rough astringent taste, and alkaline reaction. It destroys most organic matters containing hydrogen, including colouring matters. But its bleaching action is not instantaneous, unless an acid be added to it, which liberates the chlorine. Hence, when Turkey-red cloth, having a pattern printed upon it with tartaric acid thickened by gum, is immersed for about one minute in this solution, it comes out with the colour discharged where the acid was present, but elsewhere uninjured. In this manner white figures are produced upon a coloured ground. The solution of chloride of lime also absorbs and destroys contagious matters in the atmosphere, and is slowly decomposed by carbonic acid, with escape of chlorine. The powder or its solution, when heated, or when kept for a considerable time, undergoes decomposition; 18 eq. of chlorine then leaving 17 eq. of chloride of calcium, and 1 eq. of chlorate of lime, and disengaging 12 eq. of oxygen-gas, according to the observations of M. Morin.

CHLORIMETRY.

The bleaching power of hypochlorite of lime is often estimated by the quantity of a solution of sulphate of indigo, which a constant weight of the substance can deprive of its blue colour, or render yellow. But as the indigo-solution alters by keeping, this method is not unobjectionable. A more exact method is that in which sulphate of iron is used. This method reposes upon the circumstance that the chlorine of hypochlorite of lime converts a salt of the protoxide into a salt or the sesquioxide of iron; half an equivalent, or 222 parts of chlorine, effecting that change upon a whole equivalent, or 1728 parts of cr. protosulphate of iron. Protoxide of iron is convertible into sesquioxide by half an equivalent of oxygen, which the half equivalent of chlorine may be supposed to supply, by decomposing water, in becoming hydrochloric acid. It follows, by proportion, that 10 grains of chlorine are capable of peroxidizing 77.9 grains of cr. protosulphate of iron.

A few ounces of good crystals of protosulphate of iron are reduced to powder, and dried by strong pressure between folds of cloth; the salt may afterwards be preserved in a bottle without change. In a chlorimetric experiment, 78 grains (equivalent to 10 grains of chlorine) of this salt are dissolved in about two ounces of water, which may be acidulated by a few drops of sulphuric or hydrochloric acid. Fifty grains of the chloride of lime to be examined are dissolved in about two ounces of tepid water, by rubbing them together in a mortar, and the whole poured into the alkalimeter (page 386), which is afterwards filled up to 0 on the scale, by the addition of water, and the whole mixed by inverting the alkalimeter upon the palm of the hand. The solution of chloride of lime, being thus made up to 100 measures, is poured gradually into the sulphate of iron, till the latter is completely peroxidized, and the number of measures of chloride required to produce that effect observed. The change in the degree of oxidation of the iron-solution is discovered by means of red prussiate of potassa, which gives a precipitate of Prussian blue with a salt of the protoxide of iron only, and not with a salt of the sesquioxide. By means of a glass-stirrer, a white stoneware plate is spotted over with small drops of the prussiate. A drop of the iron-solution is mixed with one of these, after every addition of chloride of lime, and the additions continued, so long as a deep blue precipitate is produced. The liquid may continue to be coloured green by the iron-salt, but that is of no moment. The richer the specimen of chloride of lime is in chlorine, the fewer measures of its solution are required to peroxidize the iron, the number of measures containing 10 grains of chlorine always producing that effect. The quantity of chlorine in the fifty grains of bleaching powder is now known, being ascertained by the proportion, as m measures (the number poured out by the alkalimeter) is to 10 grains of chlorine, so 100 is to the total grains of chlorine. In a particular experiment the 78 grains of sulphate of iron required 72 measures of the bleaching solution. Hence, as 72 is to 10, so 100 is to 13.89 chlorine in 50 grains of the chloride of lime. The quantity of chlorine in 100 grains of the chloride, or the percentage of chlorine, is obtained by doubling that number; and was therefore, in this instance, 27.78 per cent., or 28 per cent. The arithmetical process may always be reduced to that of dividing 2000 by the number of measures poured from the alkalimeter: thus in the last example —

$$\frac{2000}{72}=27.78.$$

SECTION IV.

MAGNESIUM.

Eq. 12.2, *or* 152.5; Mg.

To obtain magnesium, sodium in a test-tube of hard glass is covered by fragments of anhydrous chloride of magnesium, and heated to redness by a lamp. The alkaline metal unites with chlorine, with strong ignition. After extracting the chloride of sodium by means of water, the magnesium remains in little globules, which may be reunited by fusing them under a stratum of chloride of potassium at a moderate red-heat.

Magnesium has the colour and lustre of silver; it is very ductile, and capable of being beaten into thin leaves, fuses at a gentle heat, and crystallizes in octohedrons. Magnesium is oxidized superficially by moist air, but undergoes no change in dry air or oxygen. Heated to redness, it burns with great brilliancy, forming magnesia. It is evidently more analogous to zinc than to the preceding metals.

Magnesia; MgO; 20.2, *or* 252.5. — This is the only known oxide of magnesium. As usually prepared, by a gentle but long calcination of the artificial carbonate of magnesia, it forms a white soft powder, the *magnesia usta* of pharmacy. Magnesia is of density 3.61 after ignition in a porcelain-furnace (H. Rose), and highly infusible. It combines with water, but with much less avidity than lime does, forming a protohydrate. The native hydrate of magnesia has the same composition, and so has the compound obtained by precipitating magnesia from its soluble salts (by means of hydrate of potassa) and washing well, when dried either without heat or at 212°. These preparations have a silky lustre and a softness to the touch, characteristic of magnesian minerals, such as is observed in asbestos and soapstone.

According to M. Fresenius, magnesia requires for solution 55368 parts of water, either boiling or at ordinary temperatures; the solution is feebly alkaline, and gives a sensible precipitate on the addition of phosphate of soda, followed by ammonia. When this earth and its salts are moistened with nitrate of cobalt, and strongly ignited before the blow-pipe, they assume a fine rose-colour: phosphate of magnesia takes more of a violet tint. Magnesia is precipitated from its soluble salts by lime-water, but is still a strong base capable of neutralizing acids perfectly. Ammonia never throws down more than half of the magnesia from the solution of a salt of magnesia, owing to the formation of a soluble double salt of magnesia and ammonia; and the flaky precipitate produced by ammonia in the solution of a salt of magnesia disappears again completely on the addition of hydrochlorate of ammonia. Magnesia is precipitated from its salts by the carbonates, but not by the bicarbonates, of potassa and soda. It is most correctly estimated by precipitation by the phosphate of soda with caustic ammonia, washing with water containing hydrochlorate of ammonia, and igniting the precipitated phosphate of magnesia and ammonia; the whole magnesia being ultimately obtained in the form of bibasic phosphate of magnesia, $2MgO.PO_5$.

Chloride of magnesium, made by neutralizing carbonate of magnesia with hydrochloric acid, crystallizes in thin needles, which contain 6 eq. of water, and are highly deliquescent. When we attempt to make this salt anhydrous by heat, hydrochloric acid escapes, and magnesia remains. But the pure chloride of magnesium, which is employed in preparing the metal, may be obtained by dividing a quantity of hydrochloric acid into two equal portions, neutralizing one with magnesia and the other with ammonia, mixing and evaporating these two solutions to dryness, when an anhydrous double chloride of magnesium and ammonia is formed. On heating this salt to redness in a covered porcelain-crucible, sal-ammoniac sublimes, and chloride of magnesium remains in a state of fusion, which becomes a translucent, crystalline mass on cooling. This chloride is decomposed by oxygen, which, at a high temperature, displaces its chlorine, and magnesia is formed. According to M. Poggiale, the chloride of magnesium forms with chloride of sodium a double salt, which has the formula $2MgCl.NaCl+2HO$.

Carbonate of magnesia. — This salt occurs native, and then always in the anhydrous condition, as a white, hard, compact mineral of an earthy fracture, which is known as *magnesite*, and sometimes in rhombohedral crystals, similar to those of carbonate of lime. It is prepared artificially by precipitating a soluble salt of magnesia, by means of carbonate of potassa at the boiling-point. The precipitate is diffused in pure water, and a stream of carbonic acid sent through it, by which the carbonate of magnesia is dissolved. On allowing this solution to evaporate spontaneously, the excess of carbonic acid escapes, and carbonate of magnesia is deposited in small hexagonal prisms with right summits. These crystals contain 3 eq. of water. They effloresce in dry air, and then lose 2 eq. of water, according to my own observations. Carbonate of magnesia has also been obtained in crystals, with 5 eq. of water, from the solution in carbonic acid, at a low temperature. There are, consequently, three hydrates of this salt, of which the formulæ are —

$$MgO.CO_2.HO;$$
$$MgO.CO_2.HO+2HO;$$
$$MgO.CO_2.HO+4HO.$$

The fact that the carbonate of magnesia dissolves in carbonic acid-water is not to be held as a proof of the existence of a bicarbonate of magnesia. Various insoluble salts, such as phosphate of lime and fluoride of calcium, dissolve in the same liquid, which appears to possess a specific solvent power. In the analogous solution of carbonate of lime in carbonic acid-water, the proportion of the carbonate was found by Berthollet to have a variable and indefinite relation to the acid. On theoretical grounds, supersalts, of the ordinary constitution, of magnesia, and the magnesian family of oxides, are not to be expected, as they would be double salts of water and another magnesian oxide.

Magnesia alba, or the subcarbonate of magnesia of pharmacy, is prepared by precipitating a boiling solution of sulphate of magnesia or chloride of magnesium, by means of carbonate of potassa. Carbonate of soda is not so suitable as a precipitant of magnesia, as a portion of it is apt to go down in combination with the magnesian carbonate, but it may be used provided the quantity applied be less than is required to decompose the whole magnesian salt in solution. Magnesia alba, when washed with hot water, is very white, light, and bulky. A portion of carbonic acid is lost, the magnesia not being in combination with a full equivalent of that acid. Berzelius found magnesia alba to contain, in 100 parts, 35.77 carbonic acid, 44.75 magnesia, and 19.48 water; or to consist of 3 eq. of carbonic acid, 4 eq. of magnesia, and 4 eq. of water. It is viewed as a combination of 3 eq. of protohydrated carbonate of magnesia with 1 eq. of protohydrate of magnesia; of which the formula is $3(MgO.CO_2.HO)+MgO.HO$. This compound requires 2500 parts of cold, and 9000 of hot water for solution (Dr. Fyfe).

Bicarbonate of potassa and magnesia. — This salt was formed by Berzelius by mixing a solution of nitrate of magnesia or chloride of magnesium (not the sulphate of magnesia) with a saturated solution of bicarbonate of potassa in excess, and allowing the liquor to rest. In the course of a few days, the double salt is deposited in large regular crystals. These crystals are insipid; insoluble in pure water, but slowly decomposed by it. The composition of this salt corresponds with 1 eq. of potassa, 2 of magnesia, 4 of carbonic acid, and 9 of water. It contains the elements of 1 eq. of a hydrated bicarbonate of potassa, and of 2 eq. of hydrated carbonate of magnesia.

$$\begin{cases} MgO.CO_2.HO+2HO \\ HO.CO_2.(KO.CO_2)+2HO \\ MgO.CO_2.HO+2HO. \end{cases}$$

It appears an association, or compound, of three salts of similar constitution. This salt, I find, loses 8HO at 212°, or all its combined water, except the single basic equivalent of the bicarbonate of potassa. A corresponding bicarbonate of soda and magnesia also exists.

Dolomite, a magnesian limestone, very extensively diffused in nature, is a mixture or combination of the carbonates of lime and magnesia, having the crystalline form of calc-spar. The two salts unite in all proportions, but are most frequently found in the proportion of single equivalents. It is remarkable that when this rock is exposed to the solvent action of water containing carbonic acid, the carbonate of lime is dissolved exclusively, and a magnesian limestone remains in the form of a porous and crystalline mass. It is not unusual to find whole mountains of magnesian limestone thus altered.

Sulphate of magnesia; $MgO.SO_3.HO+6HO$; 60.2+63, *or* 752.5+787.5.—This salt exists in many mineral springs, in the waters of Epsom, of Seidlitz in Bohemia, &c., from which it was first procured by evaporation. It is now obtained from the bittern of sea-water, which consists principally of chloride of magnesium and sulphate of magnesia, and is converted wholly into sulphate by the addition of sulphuric acid. Or magnesia is precipitated from sea-water confined in a tank, by means of hydrate of lime, and the earth thus obtained afterwards neutralized by sulphuric acid. Magnesian limestone is also had recourse to for magnesia. It is burned and slaked with water, to obtain it in a divided state, and then neutralized by sulphuric acid. The mixed sulphates are easily separated, that of lime being soluble to a minute extent only, while that of magnesia is highly soluble in water. A solution of sulphate of lime is also decomposed by carbonate of magnesia, with the formation of sulphate of magnesia; and this reaction is often witnessed in beds of magnesian limestone, when water containing sulphate of lime percolates through them.

The crystals of sulphate of magnesia are four-sided rectangular prisms, which, when pure, have a slight disposition to effloresce in dry air. One hundred parts of water at 32° dissolve 25.76 parts of the anhydrous salt, and for every degree above that temperature they take up 0.26564 part additional (see Gay-Lussac's table of the solubility of salts, at page 178). The solution has a bitter disagreeable taste, which is characteristic of all the soluble salts of magnesia. It is not precipitated in the cold by the alkaline bicarbonates, by common carbonate of ammonia, nor by oxalate of ammonia if the solution of sulphate of magnesia be dilute. This salt crystallizes at 32° with 12HO (Fritzsche); it is also generally stated to crystallize about 70°, with 6HO.

Sulphate of magnesia loses 6HO considerably under 300°, but retains 1 eq. of water even at 400.° The last equivalent is replaced by sulphate of potassa, forming the double sulphate of magnesia and potassa, which is considerably less soluble than the sulphate of magnesia, and crystallizes with 6HO. Sulphate of magnesia unites directly with sulphate of ammonia also, at the ordinary temperature, and with sulphate of soda above 100° (Mr. Arrott).

Sulphate of magnesia, when ignited in contact with charcoal, leaves magnesia with very little sulphide of the metal; it is the last of the earths which exhibits any analogy of this kind to the alkalies. The hydrosulphate of sulphide of magnesium is soluble in water, and appears to be formed when sulphate of magnesia is precipitated by sulphide of barium.

Hyposulphate of magnesia forms crystals, which are persistent in air, very soluble, and contain 36.77 per cent. or 6 eq. of water, like the following salt.

Nitrate of magnesia is a very soluble and highly deliquescent salt. It crystallizes with 6HO.

Phosphate of magnesia is formed on mixing cold solutions of common phosphate of soda and sulphate of magnesia, and allowing to stand for 24 hours. The salt appears in tufts of slender prisms, which effloresce in dry air. They are soluble in about 1000 times their weight of water. The composition of this salt, which I carefully examined, may be expressed by the following formula—$HO.2MgO.PO_5 + 2HO + 12HO$. (Phil. Trans. 1837.)

Phosphate of magnesia and ammonia.—This is the well-known granular precipitate which appears when a tribasic phosphate and a salt of ammonia are dissolved together, and any salt of magnesia is added to the mixture. Its formation is had recourse to as a test of the presence of magnesia. Although insoluble in a liquid containing salts, it is so soluble in pure water that it cannot be washed without sensible loss. It is readily dissolved by acids. The same substance forms the basis of the variety of urinary calculus known as the ammoniaco-magnesian phosphate. It is a tribasic phosphate, of which the 3 eq. of base are 1 eq. of oxide of ammonium and 2 eq. of magnesia, with 12 eq. of water of crystallization: ten of the latter may be expelled without any loss of ammonia. The formula of this salt is therefore $NH_4O.2MgO.PO_5 + 2HO + 10HO$. The same salt was found in crystals of considerable magnitude, by Dr. Ulex, in the old soil of the city of Hamburgh, and named *struvite*, as a new mineral species. It has also been found in guano, and hence named *guanite* by Mr. Teschemacher. Dr. Otto has observed a corresponding tribasic phosphate of protoxide of iron and ammonia, which contains only 2 eq. of water; and also an arseniate of manganese and ammonia, of which the water of crystallization appears to be the same as that of the phosphate of magnesia and ammonia. By igniting, without fusing, phosphate of magnesia with a small quantity of carbonate of potassa, an insoluble double salt of similar constitution, $2MgO.KO.PO_5$, was obtained by H. Rose. Corresponding double phosphates, containing 2 eq. of lime, baryta, and strontia, in the place of the 2 eq. of magnesia, were prepared in a similar manner.

Borate of magnesia.—The neutral salt was obtained by M. Wöhler, in the form of crystals, by heating a mixture of the solutions of sulphate of magnesia and borax to the boiling point, so as to form a precipitate, which is re-dissolved on cooling, and leaving the liquid at a temperature only a few degrees above 32° for some months. There were formed on the sides of the vessel thin crystalline needles, transparent, brilliant, hard, and having much of a mineral character, insoluble in hot or cold water, and having the composition $MgO.BO_3 + 8HO$. Boracic acid forms also an insoluble triborate of magnesia, $3MgO.BO_3 + 9HO$; a soluble terborate, $MgO.3BO_3 + 8HO$; and a soluble sexborate, $MgO.6BO_3 + 18HO$.

The mineral *boracite*, which occurs in the cube and its allied forms, is an anhydrous compound of magnesia and boracic acid, in the ratio of 3 eq. of magnesia to 4 eq. of boracic acid, which is represented by $MgO.2BO_3 + 2(MgO.BO_3)$. This mineral becomes electrical by heat. The rare mineral, *hydroboracite*, is, according to Hess, a compound of a borate of lime and borate of magnesia, in both of which the acid and base are in the same ratio as in boracite, with 18 eq. of water.

Silicates of magnesia.—Magnesia is found combined with silicic acid in various proportions, forming several mineral species, of which the formulæ are as follows:—

Steatite	$5(MgO.SiO_3) + 2HO$.
Meerschaum	$MgO.SiO_3 + 2HO$.
Picrosmine and pyrallolite	$6MgO.4SiO_3 + 3HO$.
Peridote (olivine, or chrysolyte)	$3MgO.SiO_3$.
Serpentine (hydrate of magnesia with subsilicate of magnesia)	$2(3MgO + 2SiO_3) + 3(MgO.2HO)$.
Pyroxene or augite (silicate of lime and magnesia)	$3CaO.2SiO_3 + 3MgO.2SiO_3$.
Amphibole, or hornblende (silicate of lime and magnesia)	$CaO.SiO_3 + 3MgO.2SiO_3$.

In these minerals, particularly the two last, the magnesia is often replaced, in whole or in part, by protoxide of iron, which gives them a green, and sometimes black colour. Fine crystals of pyroxene are often found among the scoriæ of blast-furnaces. Serpentine is easily decomposed by acids, and may be employed in the preparation of sulphate of magnesia. A variety of other minerals are formed of silicic acid and magnesia, anhydrous or hydrated; such as talc, metaxite, &c.

ORDER III.

METALLIC BASES OF THE EARTHS.

SECTION I.

ALUMINUM.

Eq. 13.7 *or* 171.2; Al.

This element is named from *alumen*, the Latin term for alum, which is a double salt, consisting of sulphate of alumina and sulphate of potassa.

Like the preceding metal, aluminum is obtained from its chloride by the action of potassium. In order to diminish the violence of the reaction, M. Wöhler recommends that about 20 grains of perfectly dry potassium be introduced into a small platinum-crucible, which is placed within another larger crucible, also of platinum, containing the anhydrous chloride of aluminum. The cover of the larger crucible is then fastened down by an iron-wire, and heat applied with caution. The aluminum is afterwards separated from the chloride of potassium, with which it is mixed, by digesting the crucible and its contents in a considerable quantity of cold water. The metal appears as a grey powder, resembling spongy platinum, but is seen in a strong light, while suspended in water, to consist of small scales or spangles having the metallic lustre. It is not a conductor of electricity when in this divided state, but becomes one when its particles are approximated by fusion. Wöhler finds that iron resembles aluminum in that respect.

Aluminum has no action upon water at the usual temperature, but decomposes it to a small extent at the boiling temperature, with the evolution of hydrogen. It undergoes oxidation more rapidly in solutions of potassa, soda, and ammonia, and the resulting alumina is dissolved by these alkalies. Aluminum requires for fusion a temperature higher than that at which cast-iron melts. Heated in open air, it takes fire and burns with a vivid light, and in oxygen-gas with the production of so much heat as to fuse the alumina, which then has a yellowish colour, and is equal in hardness to the native crystallized aluminous earth, corundum.

Alumina; Al_2O_3: 51.4 *or* 642.5.—This earth is the only degree of oxidation of which aluminum is susceptible, so far as is known at present. In its constitution, alumina is presumed to resemble sesquioxide of iron, because it occurs crystallized in the same form as the native sesquioxide of iron, and the salts into which it enters are strictly isomorphous with the corresponding salts of that oxide. To 3 eq. of oxygen it must, therefore, contain 2 eq. of metal, such being the composition of sesquioxide of iron. Aluminum is not known to enter into combination in any other proportion than that of two equivalents of the metal to three of the halogenous constituent.

Alumina occurs in a state of purity, with the exception of a trace of colouring matter, in two precious stones, the *sapphire* and *ruby;* the first of which is blue, and the other red. They are not inferior in hardness to the diamond. Their density is from 3.9 to 3.97. Alumina may be obtained by calcining the sulphate of alumina and ammonia, or ammoniacal alum, very strongly. But alumina so prepared is insoluble in acids. It is obtained in the state of a hydrate from common alum by dissolving the latter in boiling water, and adding a solution of ammonia (or better, of the carbonate of ammonia), and boiling. This earth is still more perfectly precipitated by the hydrosulphate of ammonia, according to MM. Malaguti and Durocher. The precipitate, which is white, gelatinous, and very bulky, must be carefully washed, by mixing it several times with a large quantity of distilled water, allowing it to settle, and pouring off the clear liquid. By drying in air, alumina is reduced to a few hundredths of the bulk of the humid mass. It is still a hydrate, but, when ignited at a high temperature, it gives anhydrous alumina. One hundred parts of alum furnish 10.3 parts of alumina.

Alumina is white and friable. It has no taste, but adheres to the tongue. Before the oxihydrogen-blow-pipe it melts into a colourless glass. After being ignited, it is dissolved by acids with great difficulty. It is highly hygrometric, condensing about 15 per cent. of moisture from the atmosphere in damp weather. If ignited alumina contains a small portion of magnesia, it becomes warm when moistened with water: this property is very sensible, even when the proportion of magnesia does not exceed half a per cent. It appears to be due, not to chemical combination, but to heat disengaged by humectation,—a phenomenon first observed by Pouillet.

The hydrate of alumina, when moist, is gelatinous and semi-transparent, like starch, but dries up into gummy masses. It is completely insoluble in water, but is readily dissolved by acids, and also by the fixed alkalies; this earth standing in the relation of an acid to the stronger bases. Caustic ammonia dissolves it only in small quantity. The hydrate of alumina is deposited in crystals when the solution of this earth in potassa is allowed to absorb carbonic acid slowly from the air. The crystals are white and transparent at the edges, and contain 3 eq. of water, which they do not lose at 212°. The mineral *gibsite* is a native hydrate of alumina of the same composition, $Al_2O_3 + 3HO$. Another native hydrate exists, containing less water, $Al_2O_3 + 2HO$. It is called *diaspore* by mineralogists, from decrepitating and falling to powder when heated,—a property which the artificial hydrate in gummy masses likewise exhibits.

Hydrated alumina has a peculiar attraction for organic matter, which it withdraws from solution; and hence this earth is apt to be coloured when washed with water not absolutely pure. This affinity is so strong, that, when digested in solutions of vegetable colouring matters, alumina combines with and carries down the colouring matter, which is removed entirely from the liquid, if the alumina is in sufficient quantity. The pigments called *lakes* are such aluminous compounds. The fibre of cotton, when charged with this earth, attracts and retains with force the same colouring matters. Hence the great application of aluminous salts in dyeing, to impregnate cloth or yarn with alumina, and thus enable it to fix the colouring matter, and produce a fast colour. Alumina is then said to be a mordant: binoxide of tin and sesquioxide of iron have an equal attraction for organic colouring matters.

Alumina, it will be observed, is not a protoxide, and is greatly inferior to the preceding earths in basic power. It is dissolved by acids, but never neutralizes them completely. Hence, alum and all the salts of alumina have an acid reaction. Their solutions have an astringent and sweetish taste, which is peculiar to them. Alumina dissolves, to the extent of several equivalents, in some acids, particularly hydrochloric acid, forming feeble compounds, which are even deprived of a portion of their alumina by filtering them through paper. It is usually supposed that alumina does not combine with some of the weaker acids, such as carbonic acid; and that an alkaline carbonate throws down alumina from alum, and not a carbonate of that earth. The carbonate of ammonia, however, according to Mr. Danson, gives a subcarbonate of alumina, which, dried in vacuo at a low temperature, formed a light bulky powder, having the composition $3Al_2O_3.2CO_2 + 16HO$. Alumina dissolves readily in solution of potassa or soda, forming compounds in which it must play the part of an acid. The aluminate of potassa is deposited, on evaporating a solution of alumina in potassa, in white granular crystals, sweet to the taste, and having a strongly alkaline reaction: its formula is $KO.Al_2O_3$, according to M. Fremy. Such combinations occur in nature: *spinell*, a very hard mineral crystallizing in octohedrons, being an aluminate of magnesia, $MgO.Al_2O_3$; and *gahnite*, an aluminate of zinc, $ZnO.Al_2O_3$.

Sulphide of aluminum is formed by burning the metal in the vapour of sulphur. It is a black semi-metallic mass, which is rapidly transformed, by contact with water, into alumina and hydrosulphuric acid. Hydrosulphate of ammonia has the same effect upon the solution of a salt of alumina as ammonia has itself, neutralizing the acid of the salt, and throwing down alumina, while hydrosulphuric acid escapes.

Chloride of aluminum; Al_2Cl_3; 133.9 *or* 1673.75.—When alumina is dissolved in hydrochloric acid, it is to be supposed that water and a chloride of the metal are formed; $3HCl$ and $Al_2O_3 = Al_2Cl_3$ and $3HO$. The solution, when concentrated by spontaneous evaporation in a very dry atmosphere, yields crystals, which Bonsdorff found to contain 12 eq. of water. But it generally forms a saline mass, which deliquesces quickly in the air. When it is attempted to make this salt anhydrous by heat, the chlorine goes off in the form of hydrochloric acid, and pure alumina is left.

The anhydrous chloride was discovered by Oersted, who made known a method of preparing it which has since had numerous applications. Pure alumina, free from potassa, is intimately mixed with oil and lamp-black, made up into pellets, and strongly calcined in a crucible. The alumina is thus made anhydrous, without being otherwise altered. It is then introduced into a porcelain-tube, which is placed across a furnace and exposed to a red heat. Chlorine-gas, carefully dried, is conducted over the materials in the tube, when, under the conjoint influence of carbon and chlorine, the alumina is decomposed; its oxygen is carried off by the carbon as carbonic-oxide gas, and chlorine unites with the aluminum itself. The chloride of aluminum, being volatile, sublimes and condenses in the cool part of the porcelain-tube. A glass-tube, a little smaller than the porcelain-tube, should be introduced into this part of the latter, which may afterwards be drawn out, containing the condensed chloride. The salt is partly in the state of long crystalline needles, and partly in the form of a firm and solid mass, which is easily detached from the glass.

Chloride of aluminum is of a pale greenish-yellow colour, and to a certain degree translucent. In air it fumes slightly, diffuses an odour of hydrochloric acid, and runs into a liquid by the absorption of moisture. It is very soluble in water, but cannot again be recovered in the anhydrous condition. It is equally soluble in alcohol. Chloride of aluminum combines with hydrosulphuric acid, phosphuretted hydrogen, and also with ammonia.

The *fluoride of aluminum* can only be obtained by dissolving pure aluminum in hydrofluoric acid: it does not crystallize. This fluoride unites in two proportions with fluoride of potassium, for which it has a strong affinity. Both the compounds are gelatinous precipitates, which become white and pulverulent after being washed and dried. Berzelius assigned to them the formulæ, $3KF + Al_2F_3$ and $2KF + Al_2F_3$. Fluoride of aluminum exists in two crystalline minerals, one of which, on account of its transparency, hardness, and brilliancy, is reckoned among the precious stones:—

Topaz .... $3(Al_2O_3.SiO_3) + (Al_2O_3 + Al_2F_3)$
Pyknite .... $3(Al_2O_3.SiO_3) + Al_2F_3$.

The *sulphocyanide of aluminum* crystallizes in octohedrons, which are persistent in air.

SALTS OF ALUMINA.

Sulphate of alumina; $Al_2O_3.3SO_3 + 18HO$; 171.4 + 162 *or* 2142.5 + 2025.—Obtained by dissolving alumina in sulphuric acid. It crystallizes in thin flexible plates of a pearly lustre, has a sweet and astringent taste, and is soluble in twice its weight of cold water, but does not dissolve in alcohol. When heated, it fuses in its water of crystallization, swells up, and forms a light porous mass, which appears at first to be insoluble in water, but dissolves completely after a time. Heated to redness, it is entirely decomposed; the residue is pure alumina. This salt has been found, in the crystalline form, in the volcanic Island of Milo in the Archipelago. Sulphuric acid and alumina combine in several proportions, but this is considered the neutral sulphate, as it possesses the same number of equivalents of acid as it contains equivalents of oxygen in the base.

Another sulphate of alumina ($Al_2O_3.3SO_3+Al_2O_3$) was obtained by Maus by saturating sulphuric acid with alumina, which contains twice as much alumina as the neutral sulphate. After evaporation, this subsalt presents itself in a gummy mass, which dissolves in a small quantity of water, but is decomposed when the solution is diluted with a large quantity of water, or boiled; in that case the neutral salt remains in solution, and the following salt precipitates. Subtrisulphate of alumina, $Al_2O_3.3SO_3+2Al_2O_3+9HO$, precipitates, on adding ammonia to the sulphate of alumina, as a white insoluble powder. This subsalt forms the mineral *aluminite,* which is found near Newhaven in England, and at Halle in Germany.

Alum; sulphate of alumina and potassa; $KO.SO_3 + Al_2O_3.3SO_3 + 24HO$; $258.4+216$, *or* $3230+2700$.—Sulphate of alumina has a strong affinity for sulphate of potassa, in consequence of which octohedral crystals of this double salt precipitate when a salt of potassa is added to a strong solution of sulphate of alumina. Alum is a salt of which large quantities are consumed in dyeing. It is prepared by several processes, or derived from different sources. It may be prepared by decomposing clay with sulphuric acid; the decomposition is sometimes effected by igniting pure clay, grinding it afterwards to powder, and mixing it with 0.45 of sulphuric acid, of 1.45 density. This mixture is heated in a reverberatory furnace till the mass becomes very thick; afterwards left to itself for at least a month, and then treated with water to wash out the sulphate of alumina formed. This salt forms, on cooling, a mass of interlaced crystals, being the sulphate of alumina already described, $Al_2O_3.3SO_3+18HO$. Some clays and aluminous schists do not require to be heated before being treated with sulphuric acid. The addition of sulphate of potassa converts the last salt into alum.

The old mode of making alum is still largely practised in England. A series of beds occur low in many of the coal measures, which contain much bisulphide of iron. One of these, known as alum-slate, is a silicious clay, containing a considerable portion of coaly matter, and of the metallic sulphide in a state of minute division. When this mineral is exposed to air and moisture, it soon exfoliates, from the formation of sulphate of iron, the bisulphide of iron absorbing oxygen like a pyrophorus. The excess of sulphuric acid formed attacks the other bases present, of which the most considerable is alumina. Aluminous schists often require to be moderately calcined or roasted before they undergo this change in the atmosphere. The mineral being lixiviated, after a sufficient exposure, affords a solution of sulphate of alumina and protosulphate of iron, from which the latter salt is first separated by crystallization. The subsequent addition of sulphate of potassa to the liquor causes the formation of alum; the chloride of potassium answers the same purpose, and has the advantage over the sulphate that it converts the remaining sulphates of iron into chlorides, which are very soluble, and from which the alum is most easily separated by crystallization. A very pure alum is also obtained in the Roman states from *alum-stone,* which is simply heated till sulphurous acid begins to escape from it, and the residue of this calcination treated with water. This mineral contains an insoluble subsulphate of alumina with sulphate of potassa. The heating has the effect of separating the excess of alumina, so that a neutral sulphate of alumina is formed. Alum-stone appears to be continually produced at the Solfatara, near Naples, and other volcanic districts, by the joint action of sulphurous acid and oxygen upon trachyte, a volcanic rock composed almost entirely of felspar.

The solubility of crystallized alum, according to M. Poggiale, is as follows:—

100 parts of water	at 32° (0° C.)	dissolve	3.29	parts of alum.
—	at 50° (10° C.)	—	9.52	—
—	at 86° (30° C.)	—	22.00	—
—	at 140° (60° C.)	—	31.00	—
—	at 158° (70° C.)	—	90.00	—
—	at 212° (100° C.)	—	357.00	—

It crystallizes very readily in regular octohedrons, of which the apices are always more or less truncated, from the appearance of faces of the cube; their density is 1.71. The taste of alum is sweet and astringent, and its action decidedly acid; it dissolves metals, with evolution of hydrogen, as readily as free sulphuric acid. The crystals effloresce slightly in air, and, when heated, melt in their water of crystallization, which amounts to 45.5 per cent. of their weight, or 24 equivalents. The fused salt, in losing this water, becomes viscid, froths greatly, and forms a light porous mass, known as burnt alum. When submitted to a graduated temperature, alum loses 10 equivalents of water at 212°, and 9 equivalents more at 248° (120° C.); leaving alum combined with 5 eq. of water. This last substance can support a temperature of 320° (160° C.) without losing more water. At 356° (180° C.) it loses 4 equivalents of water; a salt then remains which parts with $\frac{1}{2}$ eq. of water at 392° (200° C.), leaving alum in combination with $\frac{1}{2}$ eq. of water (Hertwig).

A pyrophorus is formed from an intimate mixture of 3 parts of alum and 1 of sugar, which are first evaporated to dryness together, and then introduced into a small stoneware-bottle, and this placed in a crucible and surrounded with sand. The whole is heated to redness till a blue flame appears at the mouth of the bottle, which is allowed to burn for a few minutes, and the mouth then closed by a stopper of chalk. After cooling, the bottle is found to contain a black powder, which becomes red-hot when exposed to air, and catches fire also and burns with peculiar vivacity in oxygen-gas. This property appears to depend upon the highly divided state of sulphide of potassium, which is intermixed with charcoal and sulphate of alumina. A pyrophorus can be produced from sulphate of potassa alone, without the sulphate of alumina; but it does not so certainly succeed.

If the quantity of carbonate of soda necessary to neutralize a portion of alum be divided into three equal portions, and added in a gradual manner to the aluminous solution, it will be found that the alumina at first precipitated is re-dissolved upon stirring, and that no permanent precipitate is produced till nearly two parts of alkaline carbonate are added. It is in the condition of this partially neutralized solution that alum is generally applied as a mordant to cloth. Animal charcoal readily withdraws the excess of alumina from this solution, and so does vegetable fibre, probably from a similar attraction of surface. When this solution is concentrated by evaporation, alum crystallizes from it, generally in the cubic form, and the excess of alumina is precipitated.

Basic alum is a granular crystalline compound, which precipitates when gelatinous alumina is boiled in a solution of alum. The formula of this salt is $HO.SO_3+3(Al_2O_3.SO_3)+9HO$: the alum-stone used in preparing the Roman alum has the same composition.

Sulphate of ammonia may be substituted for sulphate of potassa in alum, giving rise to *ammoniacal alum*,

$$NH_4O.SO_3+Al_2O_3.3SO_3.+24HO.,$$

which agrees very closely in properties with potassa-alum.

Sulphate of alumina also combines with sulphate of soda, forming *soda-alum*, which crystallizes in the same form as common alum, and also contains 24HO, the formula of soda-alum being,

$$NaO.SO_3+Al_2O_3.3SO_3+24HO.$$

Crystals are obtained by mixing the sulphates of soda and alumina, and leaving a concentrated solution to spontaneous evaporation; or by pouring spirits of wine upon the surface of such a solution contained in a bottle, which deposits crystals as the alcohol gradually diffuses through it. This salt effloresces in air as rapidly as sulphate of soda. It is very soluble in water, 10 parts of water at 60° dissolving 11 parts of this salt.

Sulphate of alumina also combines with the sulphate of protoxide of iron, when dissolved with that salt and a considerable admixture of sulphuric acid (Klauer). The double salt was found to contain 1 eq. of protosulphate of iron ($FeO.SO_3$), 1 eq. of sulphate of alumina ($Al_2O_3.3SO_3$), and 24 eq. of water ($24HO$), which indicates a similarity in composition to alum. But it is deposited in long acicular crystals, which do not belong to the octohedral system, and has therefore no claim to be considered an alum. A similar salt with magnesia was obtained in the same way. Another combination of the same class, containing the sulphate of manganese, forms a white fibrous mineral found in a cave upon Bushman's river in South Africa. This native sulphate of alumina and manganese has been carefully examined by Dr. Apjohn and by Sir R. Kane, and found to contain 25HO. It is probable that if the proportion of water in Klauer's salts were accurately determined, it would be found to be the same. These salts may be represented as compounds of a magnesian sulphate, retaining its single equivalent of constitutional water, with sulphate of alumina; the manganese compound thus:—

$$MnO.SO_3.HO + Al_2O_3.3SO_3 + 24HO.$$

Certain salts have been formed, isomorphous with alum, and strictly analogous in composition, in which the alumina is replaced by metallic oxides isomorphous with it, namely, by sesquioxide of iron, sesquioxide of manganese, and sesquioxide of chromium. To these salts the generic term alum is applied, and the species is distinguished by the name of the metallic sesquioxide it contains; as *iron-alum, manganese-alum,* and *chrome-alum.*

Alumina dissolves freely in most acids, but, like metallic peroxides in general, it affords few crystalline salts, except double salts. The oxalate of potassa and alumina is the only other of these that has been fully examined. It is remakable for its composition, containing 3 eq. of oxalate of potassa to 1 eq. of oxalate of alumina, with 6 eq. of water. Its formula is, therefore,

$$3(KO.C_2O_3) + Al_2O_3.3C_2O_3 + 6HO.$$

Like alum it is the type of a genus of double salts. The corresponding oxalates, containing soda, crystallize with 10HO.—(Phil. Trans. 1837, p. 54.)

Nitrate of alumina is said to crystallize with difficulty in prismatic crystals radiating from a centre.

An insoluble *phosphate of alumina* precipitates when phosphate of soda is added to a solution of alum. By fusion it gives a glass, like porcelain: its composition is $2Al_2O_3.3PO_5$ (Berzelius). This salt, dissolved in an acid and precipitated by ammonia in excess, gives a more highly basic phosphate, of which the formula is $4Al_2O_3.3PO_5$ (Berzelius). The last phosphate of alumina occurs in nature, in combination with fluoride of aluminum, in the form of radiating crystals, and is named *wavellite*, of which the formula is $Al_2F_3 + 3(4Al_2O_3.3PO_3) + 36HO$. A phosphate of alumina and lithia, containing the same subphosphate of alumina, forms the rare mineral *amblygonite,* and may be prepared artificially: its formula is $2LiO.PO_5 + 4Al_2O_3.3PO_5$.

SILICATES OF ALUMINA.

The varieties of *clay* are essentially silicates of alumina, but composed as they are of the insoluble matter of various rocks destroyed by the action of water, it is not to be expected that they will be uniform in composition. Mitscherlich considers it probable that the basis of clay is usually a subsilicate of alumina, of which the formula is $2Al_2O_3.3SiO_3$; and which contain 57.42 parts of silicic acid and 42.58 of alumina in 100 parts. But from the analysis of Mosander, the refractory clay of Stourbridge (a fire-clay) is a neutral silicate of alumina, $Al_2O_3.3SiO_3$. *China-clay* or *kaolin*, which is prepared from decaying granite, being the result of the decomposition of the felspar and mica of that mineral, is not uniform in its composition. The clay from a white bed of the Plastic Clay formation, which is worked for the purposes of pottery in the neighbourhood of Farnham, gave Mr. Way the following results:—

White clay dried at 212° contained in 100 parts—

Insoluble in acids, 58.03	Silicic acid	42.28
	Alumina	11.45
	Oxide of iron	3.53
	Lime	0.55
	Magnesia	0.22
Soluble in acids, 41.97	Silicic acid	18.73
	Alumina	12.15
	Oxide of iron	2.11
	Lime	0.27
	Magnesia	0.29
	Potassa	0.86
	Soda	1.41
	Water of combination	6.15
		100.00

Clay, and soils in general from the clay which they contain, possess a remarkable power of separating salts of ammonia and potassa from their solutions, and retaining these bases, first observed with reference to ammonia by Mr. H. O. Thomson, and since ably investigated by Professor Way. A light soil digested with a weak solution of caustic ammonia for two hours, withdrew 0.3438 per cent. of its weight of that base, and 0.3478 per cent. of ammonia from a solution of the hydrochlorate of ammonia, the latter salt being decomposed, and chloride of calcium found in solution. The sulphate of ammonia was decomposed by the same soil and by the clay above described, in a similar manner, sulphate of lime appearing in solution. Hence, when putrid urine and other soluble manures are filtered through clay or soil, the ammonia is entirely retained, while the water drains away containing only earthy salts. This absorptive power of clay is not destroyed by boiling the clay with an acid, nor by drying it between 150° and 200°; but the property is nearly lost in thoroughly burnt clay. The lime present in clay, which appears to be necessary to this action, is not entirely withdrawn by boiling with an acid, as will be observed in the preceding analysis of clay. From the hydrochlorate of ammonia 0.2010 per cent. of ammonia was withdrawn by the white clay, and 0.4366 per cent. of potassa, from the nitrate of potassa, by the same clay. The only solutions of lime which came under the influence of this absorbing power of clay and soils were those of hydrate of lime, and of carbonate of lime in carbonic acid water. Mr. Way does not propose any rationale of this remarkable action of clay, but excludes the supposition of its being due to free alumina and silicic acid (Journal of the Royal Agricultural Society of England, xi. 313, 1850).

A subsilicate of alumina exists, forming a very hard crystallized mineral, *disthene* or *cyanite*, of which the formula is $2Al_2O_3.SiO_3$.

Double silicates of alumina and potassa are extensively diffused in the mineral kingdom, forming a very considerable portion of the solid crust of the globe. The most usual of these double salts are the following:

Potash-Felspar, which is crystallized in oblique rhomboidal prisms, of density 2.5, is composed of single equivalents of the neutral silicates of potassa and alumina.

Its formula is therefore analogous to that of anhydrous alum, silicon being substituted for sulphur; $KO.SiO_3+Al_2O_3.3SiO_3$. It is one of the three principal constituents of granite and gneiss. This species of felspar is named *orthose.* Other varieties of felspar are *albite,* or soda-felspar, containing silicate of soda, $NaO.SiO_3$, in the place of silicate of potassa; lithia-felspar (*petalite, triphane*), $LiO.SiO_3+Al_2O_3.3SiO_3$; and lime-felspar (*labradorite, anorthite*), $CaO.SiO_3+Al_2O_3.3SiO_3$. The alkaline base of felspars is often partially replaced by lime and magnesia, and the most general formula for a felspar would be—

$$\left.\begin{matrix}KO\\NaO\\CaO\\MgO\end{matrix}\right\} SiO_3+Al_2O_3.3SiO_3.$$

Amphigen or *leucite* occurs principally in the lava of Vesuvius in a crystallized state. The relation between the potassa and alumina is the same as in orthose, but it contains one-third less silicic acid. Hence the formula $3KO.2SiO_3+3(Al_2O_3.2SiO_3)$. A similar combination is obtained by precipitating a saturated solution of alumina in potassa, by a solution of silicate of potassa (Berzelius).

When a mixture of silicic acid and alumina is fused with an excess of potassa, and the fused mass washed with water, to withdraw everything soluble, a powder remains in which the potassa and alumina are still in the ratio of single equivalents, but in which the oxygen of the silicic acid is equal to that of the bases. This double salt has consequently the formula, $3KO.SiO_3+3Al_2O_3.3SiO_3$.

Analcime is the soda silicate proportional to amphigen. It is crystallized like amphigen, but contains 6 eq. of water. Its formula is $3NaO.2SiO_3+3(Al_2O_3.2SiO_3)+6HO$.

A third compound may be prepared, corresponding with the artificial potassa-compound above. It occurs also in hexagonal prisms in the lava of Vesuvius, forming the mineral *nephelin.*

Garnet is a double basic silicate of lime and alumina, of which the formula is $3CaO_3.SiO_3+Al_2O_3.SiO_3$.

The *silicates of lime and of alumina* combine in many different proportions, forming a great variety of minerals. Most of them contain water, in consequence of which they froth when heated before the blow-pipe, and hence are called *zeolites.* One of these, named *stilbite,* from its shining lustre, corresponds in composition with felspar, but contains in addition 6 eq. of water: its formula is

$$CaO.SiO_2+Al_2O_3.3SiO_3+6HO.$$

A small portion of one or other of the alkalies is often found in these minerals, besides small quantities of protoxide of iron and other magnesian oxides, replacing, it may be presumed, the lime in part. This extensive class of minerals has been very fully studied by Dr. Thomson, who has added considerably to their number.—(Outlines of Mineralogy and Geology, vol. i.)

EARTHENWARE AND PORCELAIN.

The silicate of alumina is the basis of all the varieties of pottery. When moistened with water, clay possesses a high degree of plasticity, and can be extended into the thinnest plates, fashioned into form by the hand, by pressure in moulds, or, when dried to a certain point, be modelled on the turning lathe. It loses its water also in drying, without cracking, provided it is allowed to contract equally in all directions, and acquires greater solidity. When heated more strongly in the potter's kiln, in which it is not fused nor its particles agglutinated by partial fusion, it becomes a strong solid mass, which adheres to the tongue and absorbs water with avidity. To render it impermeable to that liquid, it is covered with a vitreous matter, which is fused at a high temperature, and forms an insoluble glaze or varnish upon its surface. But the interior mass of ordinary pottery has always an earthy fracture, and presents no visible trace of fusion.

When an addition is made to the clay, of some compound, which softens or fuses at the temperature at which the earthenware is fired, such as felspar in powder, then the clay is agglutinated by the fusible ingredient, and the mass is rendered semi-transparent, in the same manner as paper that has imbibed melted wax remains translucent after the latter has fixed. The accidental presence of lime, potassa, protoxide of iron, or any similar base in the clay, may produce the same effect by forming a fusible silicate diffused through the clay in excess. Such is the constitution of porcelain, and of brown salt-glaze ware of which stoneware bottles are made, which is indeed a sort of porcelain. When these kinds of ware are covered by a fusible material, similar to that which has entered into the composition of their body, and a second time fired, they acquire a vitreous coating. Their fracture is vitreous and not earthy, the broken surface does not adhere to the tongue, and the mass has much greater solidity and strength than the former kinds of earthenware. In combining the ingredients of porcelain, an excess of the fusible material is to be avoided, as it may cause the vessels to soften so much in the kiln as to lose their shape, or even to run down into a glass; while on the other hand if the vitrifiable constituent is in too small a proportion, the heat of the furnace may be inadequate to soften the mass, and to agglutinate it completely.

Felspar mixed with a little clay is used as the glaze for the celebrated porcelain of Levres. Elsewhere a mixture of sulphate of lime, ground porcelain and flint, is sometimes used as a glaze. In painting porcelain, the same metallic oxides are employed as in staining glass. They are combined with a vitrifiable material, generally made thin with oil of turpentine, and applied to the pottery, sometimes under and sometimes above the glaze. To fuse the latter colours, the porcelain must be exposed a third time to heat, in the enamel furnace.

Stoneware. — The principal varieties of clay used here, according to Mr. Brande, are the following: — 1. *Marly clay*, which, with silicic acid and alumina, contains a portion of carbonate of lime: it is much used in making pale bricks, and as a manure, and when highly heated enters into fusion. 2. *Pipe-clay*, which is very plastic and tenacious, and requires a higher temperature than the preceding for fusion: when burned it is of a cream colour, and is used for tobacco-pipes and white pottery. 3. *Potters' clay* is of a reddish or grey colour, and becomes red when heated; it fuses at a bright-red heat; mixed with sand it is manufactured into red bricks and tiles, and is also used for coarse pottery (Manual of Chemistry, p. 1131). The glaze is applied to articles of ordinary pottery after they are fired, and in the condition of biscuit-ware. They are dipped into a mixture of about 60 parts of red lead, 10 of clay, and 20 of ground flint diffused in water to a creamy consistence, and when taken out enough adheres to the piece to give a uniform glazing when again heated. To cover the red colour which iron gives to the common clays when burnt, the body of the ware is sometimes coloured uniformly of a dull green, by an admixture of oxide of chromium, or made black by oxides of manganese and iron; or oxide of tin is added to the materials of the glaze, to render it white and opaque. The patterns on ordinary earthenware are generally first printed upon tissue-paper, in an oily composition, from an engraved plate of copper, and afterwards transferred by applying the paper to the surface of the biscuit ware, to which the colour adheres. The paper is afterwards removed by a wet sponge. The fusion of the colouring matters takes place with that of the glaze, which is subsequently applied, in the second firing. The prevailing colours of these patterns are blue from oxide of cobalt, green from oxide of chromium, and pink from that compound of oxide of tin, lime, and a small quantity of oxide of chromium, known as *pink colour*.

SECTION II.

GLUCINUM, YTTRIUM, THORIUM, ZIRZONIUM.

GLUCINUM.

Eq. 6.97 *or* 87.06; Gl.

Syn. Beryllium.—The compounds of this metal have a considerable analogy to those of aluminum. Glucinum is obtained from its chloride, which is decomposed by potassium in the same manner as the chloride of aluminum. This metal is fusible with great difficulty, not oxidable by air or water at the usual temperature, but it takes fire, in oxygen, at a red-heat, and burns with a vivid light. It derives its name from γλυκὺς, sweet, in allusion to the sweet taste of the salts of its oxide, glucina.

Glucina, Beryllia; Gl_2O_3 is a comparatively rare earth, but contained to the extent of 13⅗ per cent. in the emerald and beryl, of which specimens that are not transparent and well crystallized can be procured in considerable quantity. To decompose this mineral, which is a silicate of glucina and alumina, it must be reduced to an extremely fine powder, the grosser particles which fall first when the powder is suspended in water, being submitted again to pulverization, and the powder calcined with 3 times its weight of hydrate of potassa. The calcined mass is moistened with water, and then treated with hydrochloric acid, added in small portions till it is in excess. The potassa, alumina, and glucina, are thus converted into chlorides, and dissolved. The solution is evaporated to dryness on a water-bath, and the residue acidulated by a few drops of hydrochloric acid: the silicic acid remains undissolved. On adding afterwards carbonate of ammonia in considerable excess to the filtered liquid, the alumina is precipitated together with the lime and oxides of iron and chromium which are usually present, while the glucina alone remains in solution. The liquor is filtered, and the carbonate of ammonia being then expelled from it by ebullition, carbonate of glucina precipitates. The earthy carbonate is ignited, and leaves glucina in the state of a white and light powder, tasteless, infusible by heat, insoluble in water and caustic ammonia, but soluble in caustic potassa and soda. Its density is nearly 3. It is distinguished from alumina, which it greatly resembles, by absorbing carbonic acid from the air, and readily forming a carbonate; and most remarkably by being soluble, when freshly precipitated, in a cold solution of carbonate of ammonia. It is capable of decomposing the salts of ammonia in a hot solution, and replaces that base. The salts of glucina do not form an alum when treated with sulphate of potassa; nor do they become blue, like the salts of alumina, when heated before the blow-pipe with nitrate of cobalt.

Glucina combines with sulphuric acid in several proportions, forming a bisulphate, $Gl_2O_3 . 6SO_3$, which is crystallizable; a neutral sulphate, $Gl_2O_3 . 3SO_3 + 12HO$, which forms fine crystals; a soluble subsalt, $Gl_2O_3 . 2SO_3$, and an insoluble subsalt, $Gl_2O_3 . SO_3$.

Emerald or ***beryl*** is a double silicate of glucina and alumina, of the composition expressed by $Gl_2O_3 . SiO_3 + Al_2O_3 . SiO_3$; but contains besides, lime and some chromium and iron. This mineral crystallizes in six-sided prisms, which are very hard. When coloured green by oxide of chromium it forms the true emerald, and when colourless and transparent *aqua marina,* which are both ranked among the precious stones. The density of the emerald is 2.58 to 2.732.

Euclase is also a silicate of glucina and alumina. It is a very rare mineral, which crystallizes in limpid, greenish prisms.

Chrysoberyl, one of the finest of the gems, consists essentially of 1 equivalent of glucina combined with 6 equivalents of alumina, Gl_2O_3, $6Al_2O_3$.

It is very doubtful whether glucina is a sesquioxide, Gl_2O_3, analogous in composition to alumina. It is indeed quite as probable that glucina is a protoxide, GlO, analogous to magnesia. The equivalent of glucinum would then be reduced to 4.64 on the hydrogen-scale, and 58.04 on the oxygen-scale.

YTTRIUM, ERBIUM, AND TERBIUM.

Eq. 32.20 *or* 402.5; Y.

The earth yttria was discovered in 1794, by Gadolin, in a mineral from Ytterby in Sweden, which is now called gadolinite. It has since been found in several other minerals, but all of which are exceedingly rare. The metal was isolated from its chloride by Wöhler, precisely in the same manner as the two preceding metals. It is of a darker colour than these metals, and in oxidability resembles glucinum.

Yttria is considered a protoxide, YO. Its density is even greater than baryta, being 4.842. It is absolutely insoluble in the caustic alkalies, is precipitated by yellow prussiate of potassa, and its sulphate and some others of its salts have an amethystine tint, properties which distinguish it from the preceding earths. The nitrate of yttria is colourless and crystallizable. The chloride of yttrium is deliquescent, and does not appear to be volatile.

In what has hitherto been distinguished as yttria two new bases have lately been discovered by M. Mosander, which have been named *erbia* and *terbia.* These oxides are less soluble in dilute sulphuric acid than yttria, and are thereby separated from that earth. From a solution in nitric acid of the two new earths, oxide of erbium is precipitated by saturating the liquid with sulphate of potassa, in the form of a sparingly soluble double salt, while the oxide of terbium remains in solution. Each of these bases may then be precipitated singly by means of potassa.

The sulphate and nitrate of terbia readily crystallize; the former salt is efflorescent. The salts of terbia are apt on dessiccation to assume a red amethystine tint.

Erbia assumes a deep-yellow tint when made anhydrous, which appears to be due to oxidation, as the earth becomes colourless in a stream of hydrogen. The sulphate of erbia, which is crystallizable and colourless, does not effloresce in air, like the sulphate of terbia.

THORIUM, OR THORINUM.

Eq. 59.59 *or* 744.9; Th.

This element was discovered by Berzelius, in 1824, in a black mineral, like obsidian, since called *thorite*, from the coast of the North Sea. This mineral contains 57 per cent. of the thorina. This element has been named from the Scandinavian deity Thor. The metal was obtained from the chloride, and exhibited a general resemblance to aluminum. Like yttrium, it burns in oxygen with a degree of brilliancy which is quite extraordinary: the resulting oxide does not exhibit the slightest trace of fusion.

Thorina is considered a protoxide, ThO. Its density is 9.402, and therefore superior to that of all other earths. Thorina forms a hydrate, ThO.HO, which is soluble in alkaline carbonates and in all the acids. It resembles yttria in being insoluble in the caustic alkalies, but differs from that earth in the peculiar property of its sulphate, to be precipitated by ebullition, and to redissolve entirely, although in a slow manner, in cold water. Its sulphate also forms a double salt with sulphate of potassa, which dissolves in water, but is insoluble in a liquid saturated with sulphate of potassa. The solutions of thorina are precipitated white by the ferrocyanide of potassium, a property by which thorina is distinguished from zirconia. Thorina is also precipitated from solutions to which an excess of acid has been added, on afterwards introducing sufficient ammonia, by which it is distinguished from magnesia.

ZIRCONIUM.

Eq. 33.62 *or* 420.2; Zr.

Zirconium is obtained by heating the double fluoride of zirconium and potassium, with potassium, in a glass or iron tube. On throwing the cooled mass into water, the zirconium remains in the form of a black powder, very like charcoal. It contains an admixture of hydrate of zirconia, which may be withdrawn from it by digestion in hydrochloric acid, at 104° (40° C.) The zirconium is afterwards washed with sal-ammoniac to remove completely chloride of zirconium, and then with alcohol to withdraw the sal-ammoniac. If washed with pure water, it is apt to pass through the filter. After being thus treated, the powder assumes, under the burnisher, the lustre of iron, and is compressed into scales which resemble graphite. When heated in air it takes fire below redness. It is very slightly attacked by either alkalies or acids, with the exception of hydrofluoric acid, which dissolves zirconium with evolution of hydrogen.

The constitution of *zirconia* is not certainly known, but it is believed to be analogous to that of alumina, Zr_2O_3. It was first recognized as a peculiar earth by Klaproth in 1789, who discovered it in the zircon of Ceylon, a silicate of zirconia; which is also found in the syenitic mountains of the south-east side of Norway. The *hyacinth* is the same mineral, of a red-colour; it is found in volcanic sand at Expailly in France, in Ceylon, and some other localities. The earth is obtained from this mineral, which is more difficult of decomposition than most others, by processes for which I must refer to Berzelius.[1]

Zirconia is a white earth, like alumina in appearance, of density 4.3. Its hydrate, after being boiled, is soluble with difficulty in acids. When heated, it parts with its water, afterwards glows strongly, from a discharge of heat, becomes denser, and less susceptible of being acted on by reagents. Zirconia forms a carbonate. When once separated from its combinations, it is insoluble in carbonate of potassa or soda, but dissolves in them in the nascent state. The salts of zirconia have a purely astringent taste. It agrees with thorina in being precipitated, when any of its neutral salts are boiled with a solution of sulphate of potassa. The chloride of zirconium is volatile, but less so than the chloride of silicium; a property which has been taken advantage of by M. Wöhler in preparing zirconia.

---

[1] Traité de Chimie, ii. 171. Paris, 1846.

FEBRUARY, 1852.

CATALOGUE

OF

MEDICAL AND SURGICAL WORKS,

PUBLISHED BY

BLANCHARD & LEA,

PHILADELPHIA,

AND

FOR SALE BY ALL BOOKSELLERS.

Druitt's Principles and Practice of Modern Surgery, 1 vol. 8vo., 576 pages, 193 cuts, 4th ed.
Dufton on Deafness and Diseases of the Ear, 1 vol. 12mo., 120 pages.
Durlacher on Corns, Bunions, &c., 12mo., 134 pp
Ear, Diseases of, a new work, (preparing)
Fergusson's Practical Surgery, 1 vol. 8vo., 3d edition, 630 pages, 274 cuts.
Guthrie on the Bladder, 8vo., 150 pages.
Gross on Injuries and Diseases of Urinary Organs, 1 large vol. 8vo., 726 pp., many cuts.
Jones' Ophthalmic Medicine and Surgery, by Hays, 1 vol. 12mo., 529 pp., cuts and plates.
Liston's Lectures on Surgery, by Mütter, 1 vol. 8vo., 566 pages, many cuts.
Lawrence on the Eye, by Hays, new ed. much improved, 863 pp., many cuts and plates.
Lawrence on Ruptures, 1 vol. 8vo., 480 pages.
Miller's Principles of Surgery, 3d ed., much enlarged, 1 vol. 8vo., with beautiful cuts.
Miller's Practice of Surgery, 1 vol. 8vo., 496 pp.
Malgaigne's Operative Surgery, by Brittan, with cuts, one 8vo. vol., 600 pages.
Maury's Dental Surgery, 1 vol. 8vo., 286 pages, many plates and cuts.
Pirie's Principles and Practice of Surgery, edited by Neill, 1 vol. 8vo., 200 cuts, (nearly ready.)
Skey's Operative Surgery, 1 vol. large 8vo., many cuts, 662 pages, a new work.
Sargent's Minor Surgery, 12mo., 380 pp. 128 cuts.
Smith on Fractures, 1 vol. 8vo., 200 cuts, 314 pp.

MATERIA MEDICA AND THERAPEUTICS.

Bird's (Golding) Therapeutics, (preparing.)
Christison's and Griffith's Dispensatory, 1 large vol. 8vo., 216 cuts, over 1000 pages.
Carpenter on Alcoholic Liquors in Health and Disease, 1 vol. 12mo.
Carson's Synopsis of Lectures on Materia Medica and Pharmacy, 1 vol. 8vo., 208 pp. (now ready.)
Dunglison's Materia Medica and Therapeutics, 4th ed., much improved, 182 cuts, 2 vols. 8vo.
Dunglison on New Remedies, 6th ed., much improved, 1 vol. 8vo., 750 pages.
De Jongh on Cod-Liver Oil, 12mo.
Ellis' Medical Formulary, 9th ed., much improved, 1 vol. 8vo., 268 pages.
Griffith's Universal Formulary, 1 vol. 8vo., 560 pages.
Griffith's Medical Botany, a new work, 1 large vol. 8vo., 704 pp., with over 350 illustrations.
Mayne's Dispensatory, 1 vol. 12mo., 330 pages.
Mohr, Redwood, and Procter's Pharmacy, 1 vol 8vo., 550 pages, 506 cuts.
Pereira's Materia Medica, by Carson, 3d ed., 2 vols. 8vo., much improved and enlarged, with 400 wood cuts. Vol. I, ready. Vol. II, in press.
Royle's Materia Medica and Therapeutics, by Carson, 1 vol. 8vo., 689 pages, many cuts.

OBSTETRICS.

Churchill's Theory and Practice of Midwifery, a new and improved ed., by Condie, 1 vol. 8vo., 510 pp., many cuts.
Dewees' Midwifery, 11th ed., 1 vol. 8vo., 660 pp., plates.
Lee's Clinical Midwifery, 12mo., 238 pages.
Meigs' Obstetrics, 2d ed., enlarged, 1 vol. 8vo., (now ready.)
Ramsbotham on Parturition, with many plates, 1 large vol., imperial 8vo., 520 pp. 6th edition.
Rigby's Midwifery, new edition, 1 vol. 8vo., (just issued,) 422 pages.
Smith (Tyler) on Parturition, 1 vol. 12mo., 400 pp.

CHEMISTRY AND HYGIENE.

Bowman's Practical Chemistry, 1 vol. 12mo., 97 cuts, 350 pages.
Brigham on Excitement, &c., 1 vol. 12mo., 204 pp.
Beale on Health of Mind and Body, 1 vol. 12mo.
Bowman's Medical Chemistry, 1 vol. 12mo., many cuts, 288 pages.
Dunglison on Human Health, 2d ed., 8vo., 464 pp.
Fowne's Elementary Chemistry, 3d ed., 1 vol. 12mo., much improved, many cuts.
Graham's Chemistry, by Bridges, new and improved edition. Part 1, (now ready.)
Gardner's Medical Chemistry, 1 vol. 12mo. 400 pp.
Griffith's Chemistry of the Four Seasons, 1 vol. royal 12mo., 451 pages, many cuts.
Knapp's Chemical Technology, by Johnson, 2 vols. 8vo., 936 pp., 460 large cuts.
Simon's Chemistry of Man, 8vo., 730 pp., plates.

MEDICAL JURISPRUDENCE, EDUCATION, &c.

Bartlett's Philosophy of Medicine, 1 vol. 8vo., 312 pages.
Bartlett on Certainty in Medicine, 1 vol. small 8vo., 84 pages.
Dickson's Essays on Life, Pain, Sleep, &c., in 1 vol. royal 12mo., 300 pages.
Dunglison's Medical Student, 2d ed. 12mo., 312 pp.
Taylor's Medical Jurisprudence, by Griffith, 1 vol. 8vo., new edition, 1850, 670 pp.
Taylor on Poisons, by Griffith, 1 vol. 8vo., 688 pp.

NATURAL SCIENCE, &c.

Arnott's Physics, 1 vol 8vo., 484 pp., many cuts.
Ansted's Ancient World, 12mo., cuts, 382 pp.
Bird's Natural Philosophy, 1 vol. royal 12mo., 402 pages and 372 wood-cuts.
Brewster's Optics, 1 vol. 12mo. 423 pp. many cuts.
Broderip's Zoological Recreations, 12mo., 376 p.
Coleridge's Idea of Life, 12mo., 94 pages.
Carpenter's General and Comparative Physiology, 1 large 8vo. vol., many wood-cuts.
Dana on Zoophytes, being vol. 8 of Ex. Expedition, royal 4to., extra cloth.
Atlas to "Dana on Zoophytes," im. fol., col. pl's.
Gregory on Animal Magnetism, 1 vol., royal 12mo.
De la Beche's Geological Observer, 1 large 8vo. vol., many wood-cuts, (just issued.)
Hale's Ethnography and Philology of the U. S. Exploring Expedition, in 1 large imp. 4to. vol.
Herschel's Treatise on Astronomy, 1 vol. 12mo., 417 pages, numerous plates and cuts.
Herschel's Outlines of Astronomy, 1 vol. small 8vo., plates and cuts. (A new work.) 620 pp.
Humboldt's Aspects of Nature, 1 vol. 12mo., new edition.
Johnston's Physical Atlas, 1 vol. imp. 4to., half bound, 25 colored maps.
Kirby and Spence's Entomology, 1 vol. 8vo., 600 large pages; plates plain or colored.
Knox on Races of Men, 1 vol. 12mo.
Lardner's Handbooks of Natural Philosophy, 1 vol. royal 12mo., with 400 cuts, (just issued.)
Müller's Physics and Meteorology, 1 vol. 8vo., 636 pp., with 540 wood-cuts and 2 col'd plates.
Small Books on Great Subjects, 12 parts, done up in 3 handsome 12mo. volumes.
Somerville's Physical Geography, 1 vol. 12mo., cloth, 540 pages, enlarged edition.
Weisbach's Mechanics applied to Machinery and Engineering, Vol. I. 8vo., 486 p. 550 wood-cuts. Vol. II., 8vo., 400 pp., 340 cuts.

VETERINARY MEDICINE.

Clater and Skinner's Farrier, 1 vol. 12mo., 220 pp.
Youatt's Great Work on the Horse, by Skinner, 1 vol. 8vo., 448 pages, many cuts.
Youatt and Clater's Cattle Doctor, 1 vol. 12mo., 282 pages, cuts.
Youatt on the Dog, by Lewis, 1 vol. demy 8vo., 403 pages, beautiful plates.
Youatt on the Pig, 12mo., illustrated.

Other new and important works are in preparation.

GROSS ON URINARY ORGANS—(Just Issued.)

A PRACTICAL TREATISE ON THE

DISEASES AND INJURIES OF THE URINARY ORGANS.

BY S. D. GROSS, M. D., &c.,

Professor of Surgery in the New York University.

In one large and beautifully printed octavo volume, of over seven hundred pages.

With numerous Illustrations.

The author of this work has devoted several years to its preparation, and has endeavored to render it complete and thorough on all points connected with the important subject to which it is devoted. It contains a large number of original illustrations, presenting the natural and pathological anatomy of the parts under consideration, instruments, modes of operation, &c. &c., and in mechanical execution it is one of the handsomest volumes yet issued from the American press.

A very condensed summary of the contents is subjoined.

INTRODUCTION.—CHAPTER I. Anatomy of the Perinæum.—CHAP. II. Anatomy of the Urinary Bladder.—CHAP. III. Anatomy of the Prostate.—CHAP. IV. Anatomy of the Urethra.—CHAP. V. Urine.

PART. I. DISEASES AND INJURIES OF THE BLADDER.

CHAP. I. Malformations and Imperfections.—CHAP. II. Injuries of the Bladder.—CHAP. III. Inflammation of the Bladder.—CHAP. IV. Chronic Lesions of the Bladder.—CHAP. V. Nervous Affections of the Bladder.—CHAP. VI. Heterologous Formations of the Bladder. CHAP. VII. Polypous, Fungous, Erectile, and other Morbid Growths of the Bladder.—CHAP. VIII. Worms in the Bladder.—CHAP. IX. Serous Cysts and Hydatids.—CHAP. X. Fœtal Remains in the Bladder.—CHAP. XI. Hair in the Bladder.—CHAP. XII. Air in the Bladder.—CHAP. XIII. Hemorrhage of the Bladder.—CHAP. XIV. Retention of Urine.—CHAP. XV. Incontinence of Urine.—CHAP. XVI. Hernia of the Bladder.—CHAP. XVII. Urinary Deposits.—CHAP. XVIII. Stone in the Bladder.—CHAP. XIX. Foreign Bodies in the Bladder.

PART II. DISEASES AND INJURIES OF THE PROSTATE GLAND.

CHAP. I. Wounds of the Prostate.—CHAP. II. Acute Prostatis.—CHAP. III. Hypertrophy of the Prostate.—CHAP. IV. Atrophy of the Prostate.—CHAP. V. Heterologous Formations of the Prostate.—CHAP. VI. Cystic Disease of the Prostate.—CHAP. VII. Fibrous Tumors of the Prostate.—CHAP. VIII. Hemorrhage of the Prostate.—CHAP. IX. Calculi of the Prostate.—CHAP. X. Phlebitis of the Prostate.

PART III. DISEASES AND INJURIES OF THE URETHRA.

CHAP. I. Malformations and Imperfections of the Urethra.—CHAP. II. Laceration of the Urethra.—CHAP. III. Stricture of the Urethra.—CHAP. IV. Polypoid and Vascular Tumors of the Urethra.—CHAP. V. Neuralgia of the Urethra.—CHAP. VI. Hemorrhage of the Urethra.—CHAP. VII. Foreign Bodies in the Urethra.—CHAP. VIII. Infiltration of Urine.—CHAP. IX. Urinary Abscess.—CHAP. X. Fistula of the Urethra.—CHAP. XI. False Passages.—CHAP. XII. Lesions of the Gallinaginous Crest.—CHAP. XIII. Inflammation and Abscess of Cowper's Glands.

COOPER ON DISLOCATIONS.—New Edition.—(Just Issued.)

A TREATISE ON

DISLOCATIONS AND FRACTURES OF THE JOINTS.

BY SIR ASTLEY P. COOPER, BART., F. R. S., &c.

EDITED BY BRANSBY B. COOPER, F. R. S., &c.

WITH ADDITIONAL OBSERVATIONS BY PROF. J. C. WARREN.

A NEW AMERICAN EDITION,

In one handsome octavo volume, with numerous illustrations on wood.

After the fiat of the profession, it would be absurd in us to eulogize Sir Astley Cooper's work on Dislocations. It is a national one, and will probably subsist as long as English Surgery.—*Medico-Chirurg. Review.*

WORKS BY THE SAME AUTHOR.

COOPER (SIR ASTLEY) ON THE ANATOMY AND TREATMENT OF ABDOMINAL HERNIA. 1 large vol., imp. 8vo., with over 130 lithographic figures.

COOPER ON THE STRUCTURE AND DISEASES OF THE TESTIS, AND ON THE THYMUS GLAND. 1 vol., imp. 8vo., with 177 figures on 29 plates.

COOPER ON THE ANATOMY AND DISEASES OF THE BREAST, WITH TWENTY-FIVE MISCELLANEOUS AND SURGICAL PAPERS. 1 large vol., imp. 8vo., with 252 figures on 36 plates.

These three volumes complete the surgical writings of Sir Astley Cooper. They are very handsomely printed, with a large number of lithographic plates, executed in the best style, and are presented at exceedingly low prices.

LISTON & MUTTER'S SURGERY.

LECTURES ON THE OPERATIONS OF SURGERY,

AND ON DISEASES AND ACCIDENTS REQUIRING OPERATIONS.

BY ROBERT LISTON, ESQ., F. R. S., &c.

EDITED, WITH NUMEROUS ADDITIONS AND ALTERATIONS,

BY T. D. MÜTTER, M. D.,

Professor of Surgery in the Jefferson Medical College of Philadelphia.

In one large and handsome octavo volume of 566 pages, with 216 wood-cuts.

LIBRARY OF SURGICAL KNOWLEDGE.

A SYSTEM OF SURGERY.

BY J. M. CHELIUS.

TRANSLATED FROM THE GERMAN,

AND ACCOMPANIED WITH ADDITIONAL NOTES AND REFERENCES,

BY JOHN F. SOUTH.

Complete in three very large octavo volumes of nearly 2200 pages, strongly bound, with raised bands and double titles.

We do not hesitate to pronounce it the best and most comprehensive system of modern surgery with which we are acquainted.—*Medico-Chirurgical Review.*
The fullest and ablest digest extant of all that relates to the present advanced state of Surgical Pathology.—*American Medical Journal.*
If we were confined to a single work on Surgery, that work should be Chelius's.—*St. Louis Med. Journal.*
As complete as any system of Surgery can well be.—*Southern Medical and Surgical Journal.*
The most finished system of Surgery in the English language.—*Western Lancet.*
The most learned and complete systematic treatise now extant.—*Edinburgh Medical Journal.*
No work in the English language comprises so large an amount of information relative to operative medicine and surgical pathology.—*Medical Gazette.*
A complete encyclopedia of surgical science—a very complete surgical library—by far the most complete and scientific system of surgery in the English language.—*N. Y. Journal of Medicine.*
One of the most complete treatises on Surgery in the English language.—*Monthly Journal of Med. Science.*
The most extensive and comprehensive account of the art and science of Surgery in our language.—*Lancet.*

A TREATISE ON THE DISEASES OF THE EYE.

BY W. LAWRENCE, F.R.S.

A new Edition. With many Modifications and Additions, and the introduction of nearly 200 Illustrations,

BY ISAAC HAYS, M.D.

In one very large 8vo. vol. of 860 pages, with plates and wood-cuts through the text.

JONES ON THE EYE.

THE PRINCIPLES AND PRACTICE

OF OPHTHALMIC MEDICINE AND SURGERY.

BY T. WHARTON JONES, F. R. S., &c. &c.

EDITED BY ISAAC HAYS, M. D., &c.

In one very neat volume, large royal 12mo. of 529 pages, with four plates, plain or colored, and ninety-eight well executed wood-cuts.

A NEW TEXT-BOOK ON SURGERY—(In Press.)

PLES AND PRACTICE OF SURGERY.

BY WILLIAM PIRIE, F.R.S.E.,

Regius Professor of Surgery in the University of Aberdeen.

EDITED BY JOHN NEILL, M. D.,

Demonstrator of Anatomy in the University of Pennsylvania, Lecturer on Anatomy in the Medical Institute of Philadelphia, &c.

In one very handsome octavo volume, of about 600 pages, with numerous illustrations.

STANLEY ON THE BONES.—A Treatise on Diseases of the Bones. In one vol. 8vo., extra cloth. 286 pp.
BRODIE'S SURGICAL LECTURES.—Clinical Lectures on Surgery. 1 vol. 8vo., cloth. 350 pp.
BRODIE ON THE JOINTS.—Pathological and Surgical Observations on the Diseases of the Joints. 1 vol. 8vo., cloth. 216 pp.
BRODIE ON URINARY ORGANS.—Lectures on the Diseases of the Urinary Organs. 1 vol. 8vo., cloth. 214 pp.

*** These three works may be had neatly bound together, forming a large volume of "Brodie's Surgical Works." 780 pp.

RICORD ON VENEREAL.—A Practical Treatise on Venereal Diseases. With a Therapeutical Summary and Special Formulary. Translated by Sidney Doane, M. D. Fourth edition. 1 vol. 8vo. 340 pp.
DURLACHER ON CORNS, BUNIONS, &c.—A Treatise on Corns, Bunions, the Diseases of Nails, and the General Management of the Feet. In one 12mo. volume, cloth. 134 pp.
GUTHRIE ON THE BLADDER, &c.—The Anatomy of the Bladder and Urethra, and the Treatment of the Obstructions to which those Passages are liable. In one vol. 8vo. 150 pp.
LAWRENCE ON RUPTURES.—A Treatise on Ruptures, from the fifth London Edition. In one 8vo. vol. sheep. 480 pp.
MAURY'S DENTAL SURGERY.—A Treatise on the Dental Art, founded on Actual Experience. Illustrated by 241 lithographic figures and 54 wood-cuts. Translated by J. B. Savier. In 1 8vo. vol., sheep. 286 pp.
DUFTON ON THE EAR.—The Nature and Treatment of Deafness and Diseases of the Ear; and the Treatment of the Deaf and Dumb. One small 12mo. volume. 120 pp.
SMITH ON FRACTURES.—A Treatise on Fractures in the vicinity of Joints, and on Dislocations. One vol. 8vo., with 200 beautiful wood-cuts.

NEW AND IMPORTANT WORK ON PRACTICAL SURGERY.—(JUST ISSUED.)

OPERATIVE SURGERY.

BY FREDERICK C. SKEY, F. R. S., &c.

In one very handsome octavo volume of over 650 pages, with about one hundred wood-cuts.

The object of the author, in the preparation of this work, has been not merely to furnish the student with a guide to the actual processes of operation, embracing the practical rules required to justify an appeal to the knife, but also to present a manual embodying such *principles* as might render it a permanent work of reference to the practitioner of operative surgery, who seeks to uphold the character of his profession as a science as well as an art. In its composition he has relied mainly on his own experience, acquired during many years' service at one of the largest of the London hospitals, and has rarely appealed to other authorities, except so far as personal intercourse and a general acquaintance with the most eminent members of the surgical profession have induced him to quote their opinions.

From Professor C. B. Gibson Richmond, Virginia.

I have examined the work with some care, and am delighted with it. The style is admirable, the matter excellent, and much of it original and deeply interesting, whilst the illustrations are numerous and better executed than those of any similar work I possess.

In conclusion we must express our unqualified praise of the work as a whole. The high moral tone, the liberal views, and the sound information which pervades it throughout, reflect the highest credit upon the talented author. We know of no one who has succeeded, whilst supporting operative surgery in its proper rank, in promulgating at the same time sounder and more enlightened views upon that most important of all subjects, the principle that should guide us in having recourse to the knife.—*Medical Times.*

The treatise is, indeed, one on operative surgery, but it is one in which the author throughout shows that he is most anxious to place operative surgery in its just position. He has acted as a judicious, but not partial friend; and while he shows throughout that he is able and ready to perform any operation which the exigencies and casualties of the human frame may require, he is most cautious in specifying the circumstances which in each case indicate and contraindicate operation. It is indeed gratifying to perceive the sound and correct views which Mr. Skey entertains on the subject of operations in general, and the gentlemanly tone in which he impresses on readers the lessons which he is desirous to inculcate. His work is a perfect model for the operating surgeon, who will learn from it not only when and how to operate, but some more noble and exalted lessons which cannot fail to improve him as a moral and social agent.—*Edinburgh Medical and Surgical Journal.*

THE STUDENT'S TEXT-BOOK.

THE PRINCIPLES AND PRACTICE OF MODERN SURGERY.

BY ROBERT DRUITT, Fellow of the Royal College of Surgeons.

A New American, from the last and improved London Edition.

EDITED BY F. W. SARGENT, M. D., Author of "Minor Surgery," &c.

ILLUSTRATED WITH ONE HUNDRED AND NINETY-THREE WOOD ENGRAVINGS.

In one very handsomely printed octavo volume of 576 large pages.

From Professor Brainard, of Chicago, Illinois.

I think it the best work of its size, on that subject, in the language.

From Professor Rivers of Providence, Rhode Island.

I have been acquainted with it since its first republication in this country, and the universal praise it has received I think well merited.

From Professor May, of Washington, D. C.

Permit me to express my satisfaction at the republication in so improved a form of this most valuable work. I believe it to be one of the very best text-books ever issued.

From Professor McCook, of Baltimore.

I cannot withhold my approval of its merits, or the expression that no work is better suited to the wants of the student. I shall commend it to my class, and make it my chief text book.

FERGUSSON'S OPERATIVE SURGERY. NEW EDITION.

A SYSTEM OF PRACTICAL SURGERY.

BY WILLIAM FERGUSSON, F. R. S. E.,

Professor of Surgery in King's College, London, &c. &c.

THIRD AMERICAN, FROM THE LAST ENGLISH EDITION.

With 274 Illustrations.

In one large and beautifully printed octavo volume of six hundred and thirty pages.

It is with unfeigned satisfaction that we call the attention of the profession in this country to this excellent work. It richly deserves the reputation conceded to it, of being the best practical Surgery extant, at least in the English language.—*Medical Examiner.*

A NEW MINOR SURGERY.

ON BANDAGING AND OTHER POINTS OF MINOR SURGERY.

BY F. W. SARGENT, M. D.

In one handsome royal 12mo. volume of nearly 400 pages, with 128 wood-cuts.

From Professor Gilbert, Philadelphia.

Embracing the smaller details of surgery, which are illustrated by very accurate engravings, the work becomes one of very great importance to the practitioner in the performance of his daily duties, since such information is rarely found in the general works on surgery now in use.

THE GREAT ATLAS OF SURGICAL ANATOMY.

(NOW COMPLETE.)

SURGICAL ANATOMY.

BY JOSEPH MACLISE, SURGEON.

IN ONE VOLUME, VERY LARGE IMPERIAL QUARTO,

With Sixty-eight large and splendid Plates, drawn in the best style, and beautifully colored,

Containing one hundred and ninety Figures, many of them the size of life.

TOGETHER WITH COPIOUS EXPLANATORY LETTER-PRESS.

Strongly and handsomely bound in extra cloth, being one of the best executed and cheapest surgical works ever presented in this country.

This great work being now complete, the publishers confidently present it to the attention of the profession as worthy in every respect of their approbation and patronage. No complete work of the kind has yet been published in the English language, and it therefore will supply a want long felt in this country of an accurate and comprehensive Atlas of Surgical Anatomy to which the student and practitioner can at all times refer, to ascertain the exact relative position of the various portions of the human frame towards each other and to the surface, as well as their abnormal deviations. The importance of such a work to the student in the absence of anatomical material, and to the practitioner when about attempting an operation, is evident, while the price of the book, notwithstanding the large size, beauty, and finish of the very numerous illustrations is so low as to place it within the reach of every member of the profession. The publishers therefore confidently anticipate a very extended circulation for this magnificent work.

To present some idea of the scope of the volume, and of the manner in which its plan has been carried out, the publishers subjoin a very brief summary of the plates.

Plates 1 and 2.—Form of the Thoracic Cavity and Position of the Lungs, Heart, and larger Blood-vessels.
Plates 3 and 4.—Surgical Form of the Superficial Cervical and Facial Regions, and the Relative Positions of the principal Blood-vessels, Nerves, &c.
Plates 5 and 6.—Surgical Form of the Deep Cervical and Facial Regions, and Relative Positions of the principal Blood-vessels, Nerves, &c.
Plates 7 and 8.—Surgical Dissection of the Subclavian and Carotid Regions, and Relative Anatomy of their Contents.
Plates 9 and 10.—Surgical Dissection of the Sterno-Clavicular or Tracheal Region, and Relative Position of its main Blood-vessels, Nerves, &c.
Plates 11 and 12.—Surgical Dissection of the Axillary and Brachial Regions, displaying the Relative Order of their contained parts.
Plates 13 and 14.—Surgical Form of the Male and Female Axillæ compared.
Plates 15 and 16.—Surgical Dissection of the Bend of the Elbow and the Forearm, showing the Relative Position of the Arteries, Veins, Nerves, &c.
Plates 17, 18 and 19.—Surgical Dissections of the Wrist and Hand.
Plates 20 and 21.—Relative Position of the Cranial, Nasal, Oral, and Pharyngeal Cavities, &c.
Plate 22.—Relative Position of the Superficial Organs of the Thorax and Abdomen.
Plate 23.—Relative Position of the Deeper Organs of the Thorax and those of the Abdomen.
Plate 24.—Relations of the Principal Blood-vessels to the Viscera of the Thoracico-Abdominal Cavity.
Plate 25.—Relations of the Principal Blood-vessels of the Thorax and Abdomen to the Osseous Skeleton, &c.
Plate 26.—Relation of the Internal Parts to the External Surface of the Body.
Plate 27.—Surgical Dissection of the Principal Blood-vessels, &c., of the Inguino-Femoral Region.
Plates 28 and 29.—Surgical Dissection of the First, Second, Third, and Fourth Layers of the Inguinal Region, in connection with those of the Thigh.
Plates 30 and 31.—The Surgical Dissection of the Fifth, Sixth, Seventh and Eighth Layers of the Inguinal Region, and their connection with those of the Thigh.
Plates 32, 33 and 34.—The Dissection of the Oblique or External and the Direct or Internal Inguinal Hernia.
Plates 35, 36, 37 and 38.—The Distinctive Diagnosis between External and Internal Inguinal Hernia, the Taxis, the Seat of Stricture, and the Operation.
Plates 39 and 40.—Demonstrations of the Nature of Congenital and Infantile Inguinal Hernia, and of Hydrocele.
Plates 41 and 42.—Demonstrations of the Origin and Progress of Inguinal Hernia in general.
Plates 43 and 44.—The Dissection of Femoral Hernia, and the Seat of Stricture.
Plates 45 and 46.—Demonstrations of the Origin and Progress of Femoral Hernia, its Diagnosis, the Taxis, and the Operation.
Plate 47.—The Surgical Dissection of the principal Blood-vessels and Nerves of the Iliac and Femoral Regions.
Plates 48 and 49.—The Relative Anatomy of the Male Pelvic Organs.
Plates 50 and 51.—The Surgical Dissection of the Superficial Structures of the Male Perineum.
Plates 52 and 53.—The Surgical Dissection of the Deep Structures of the Male Perineum.—The Lateral Operation of Lithotomy.

MACLISE'S SURGICAL ANATOMY—(Continued.)

Plates 54, 55 and 56.—The Surgical Dissection of the Male Bladder and Urethra.—Lateral and Bilateral Lithotomy compared.
Plates 57 and 58.—Congenital and Pathological Deformities of the Prepuce and Urethra.—Structure and Mechanical Obstructions of the Urethra.
Plates 59 and 60.—The various forms and positions of Strictures and other Obstructions of the Urethra.—False Passages.—Enlargements and Deformities of the Prostate.
Plates 61 and 62.—Deformities of the Prostate.—Deformities and Obstructions of the Prostatic Urethra.
Plates 63 and 64.—Deformities of the Urinary Bladder.—The Operations of Sounding for Stone, of Catheterism, and of Puncturing the Bladder above the Pubes.
Plates 65 and 66.—The Surgical Dissection of the Popliteal Space, and the Posterior Crural Region.
Plates 67 and 68.—The Surgical Dissection of the Anterior Crural Region, the Ankles, and the Foot.

Notwithstanding the short time in which this work has been before the profession, it has received the unanimous approbation of all who have examined it. From among a very large number of commendatory notices with which they have been favored, the publishers select the following:—

From Prof. Kimball, Pittsfield, Mass.

I have examined these numbers with the greatest satisfaction, and feel bound to say that they are altogether, the most perfect and satisfactory plates of the kind that I have ever seen.

From Prof. Brainard, Chicago, Ill.

The work is extremely well adapted to the use both of students and practitioners, being sufficiently extensive for practical purposes, without being so expensive as to place it beyond their reach. Such a work was a desideratum in this country, and I shall not fail to recommend it to those within the sphere of my acquaintance.

From Prof. P F. Eve, Augusta, Ga.

I consider this work a great acquisition to my library, and shall take pleasure in recommending it on all suitable occasions.

From Prof. Peaslee, Brunswick, Me.

The second part more than fulfils the promise held out by the first, so far as the beauty of the illustrations is concerned; and. perfecting my opinion of the value of the work, so far as it has advanced, I need add nothing to what I have previously expressed to you.

From Prof. Gunn, Ann Arbor, Mich.

The plates in your edition of Maclise answer, in an eminent degree, the purpose for which they are intended. I shall take pleasure in exhibiting it and recommending it to my class.

From Prof. Rivers, Providence R. I

The plates illustrative of Hernia are the most satisfactory I have ever met with.

From Professor S. D. Gross, Louisville, Ky.

The work, as far as it has progressed, is most admirable, and cannot fail, when completed, to form a most valuable contribution to the literature of our profession. It will afford me great pleasure to recommend it to the pupils of the University of Louisville.

From Professor R. L. Howard, Columbus, Ohio.

In all respects, the first number is the beginning of a most excellent work, filling completely what might be considered hitherto a vacuum in surgical literature. For myself, in behalf of the medical profession, I wish to express to you my thanks for this truly elegant and meritorious work. I am confident that it will meet with a ready and extensive sale. I have spoken of it in the highest terms to my class and my professional brethren.

From Prof. C. B. Gibson, Richmond, Va.

I consider Maclise very far superior, as to the drawings, to any work on Surgical Anatomy with which I am familiar, and I am particularly struck with the exceedingly low price at which it is sold. I cannot doubt that it will be extensively purchased by the profession.

From Prof Granville S Pattison, New York.

The profession, in my opinion, owe you many thanks for the publication of this beautiful work—a work which, in the correctness of its exhibitions of Surgical Anatomy, is not surpassed by any work with which I am acquainted; and the admirable manner in which the lithographic plates have been executed and colored is alike honorable to your house and to the arts in the United States.

From Prof. J. F. May, Washington, D. C.

Having examined the work, I am pleased to add my testimony to its correctness, and to its value as a work of reference by the surgeon.

From Prof. Alden Marsh, Albany, N. Y.

From what I have seen of it, I think the design and execution of the work admirable, and, at the proper time in my course of lectures, I shall exhibit it to the class, and give it a recommendation worthy of its great merit.

From H. H. Smith, M. D., Philadelphia.

Permit me to express my gratification at the execution of Maclise's Surgical Anatomy. The plates are, in my opinion, the best lithographs that I have seen of a medical character, and the coloring of this number cannot, I think, be improved. Estimating highly the contents of this work, I shall continue to recommend it to my class as I have heretofore done.

From Prof. D. Gilbert, Philadelphia.

Allow me to say, gentlemen, that the thanks of the profession at large, in this country, are due to you for the republication of this admirable work of Maclise. The precise relationship of the organs in the regions displayed is so perfect, that even those who have daily access to the dissecting-room may, by consulting this work, enliven and confirm their anatomical knowledge prior to an operation. But it is to the thousands of practitioners of our country who cannot enjoy these advantages that the perusal of those plates, with their concise and accurate descriptions, will prove of infinite value. These have supplied a desideratum, which will enable them to refresh their knowledge of the important structures involved in their surgical cases, thus establishing their self confidence, and enabling them to undertake operative procedures with every assurance of success. And as all the practical departments in medicine rest upon the same basis, and are enriched from the same sources, I need hardly add that this work should be found in the library of every practitioner in the land.

MACLISE'S SURGICAL ANATOMY—(Continued.)

From Professor J. M. Bush. Lexington. Ky.

I am delighted with both the plan and execution of the work, and shall take all occasions to recommend it to my private pupils and public classes

The most accurately engraved and beautifully colored plates we have ever seen in an American book—one of the best and cheapest surgical works ever published.—*Buffalo Medical Journal.*

It is very rare that so elegantly printed, so well illustrated, and so useful a work, is offered at so moderate a price.—*Charleston Medical Journal.*

A work which cannot but please the most fastidious lover of surgical science. In it, by a succession of plates, are brought to view the relative anatomy of the parts included in the important surgical divisions of the human body, with that fidelity and neatness of touch which is scarcely excelled by nature herself. While we believe that nothing but an extensive circulation can compensate the publishers for the outlay in the production of the work—furnished as it is at a very moderate price, within the reach of *all*—we desire to see it have that circulation which the zeal and peculiar skill of the author, the utility of the work, and the neat style with which it is executed should demand for it in a liberal profession.—*N. Y. Jour of Medicine.*

This is an admirable reprint of a deservedly popular London publication. Its plates can boast a superiority that places them almost beyond the reach of competition. And we feel too thankful to the Philadelphia publishers for their very handsome reproduction of the whole work, and at a rate within everybody's reach, not to urge all our medical friends to give it, for their own sakes, the cordial welcome it deserves, in a speedy and extensive circulation.—*The Medical Examiner.*

When the whole has been published it will be a complete and beautiful system of Surgical Anatomy, having an advantage which is important, and not possessed by colored plates generally, viz., its cheapness, which places it within the reach of every one who may feel disposed to possess the work. Every practitioner, we think, should have a work of this kind within reach, as there are many operations requiring immediate performance in which a book of reference will prove most valuable.—*Southern Med. and Surg. Journ.*

The work of Maclise on Surgical Anatomy is of the highest value. In some respects it is the best publication of its kind we have seen, and is worthy of a place in the library of any medical man, while the student could scarcely make a better investment than this.—*The Western Journal of Medicine and Surgery.*

No such lithographic illustrations of surgical regions have hitherto, we think, been given While the operator is shown every vessel and nerve where an operation is contemplated, the exact anatomist is refreshed by those clear and distinct dissections which every one must appreciate who has a particle of enthusiasm. The English medical press has quite exhausted the words of praise in recommending this admirable treatise. Those who have any curiosity to gratify in reference to the perfectibility of the lithographic art in delineating the complex mechanism of the human body, are invited to examine our copy. If anything will induce surgeons and students to patronize a book of such rare value and every-day importance to them, it will be a survey of the artistical skill exhibited in these fac-similes of nature.—*Boston Medical and Surg Journal.*

These plates will form a valuable acquisition to practitioners settled in the country, whether engaged in surgical, medical, or general practice.—*Edinburgh Medical and Surgical Journal.*

We are well assured that there are none of the cheaper, and but few of the more expensive works on anatomy, which will form so complete a guide to the student or practitioner as these plates. To practitioners, in particular, we recommend this work as far better and not at all more expensive, than the heterogeneous compilations most commonly in use, and which, whatever their value to the student preparing for examination, are as likely to mislead as to guide the physician in physical examination, or the surgeon in the performance of an operation.—*Monthly Journal of Medical Sciences.*

We know of no work on surgical anatomy which can compete with it.—*Lancet.*

This is by far the ablest work on Surgical Anatomy that has come under our observation. We know of no other work that would justify a student, in any degree, for neglect of actual dissection. A careful study of these plates, and of the commentaries on them, would almost make an anatomist of a diligent student. And to one who has studied anatomy by dissection, this work is invaluable as a perpetual remembrancer, in matters of knowledge that may slip from the memory. The practitioner can scarcely consider himself equipped for the duties of his profession without such a work as this, and this has no rival, in his library. In those sudden emergencies that so often arise, and which require the instantaneous command of minute anatomical knowledge a work of this kind keeps the details of the dissecting-room perpetually fresh in the memory. We appeal to our readers, whether any one can justifiably undertake the practice of medicine who is not prepared to give all needful assistance, in all matters demanding immediate relief.

We repeat that no medical library, however large, can be complete without Maclise's Surgical Anatomy. The American edition is well entitled to the confidence of the profession, and should command, among them, an extensive sale The investment of the amount of the cost of this work will prove to be a very profitable one, and if practitioners would qualify themselves thoroughly with such important knowledge as is contained in works of this kind, there would be fewer of them sighing for employment. The medical profession should spring towards such an opportunity as is presented in this republication, to encourage frequent repetitions of American enterprise of this kind.—*The Western Journal of Medicine and Surgery.*

MILLER'S PRINCIPLES OF SURGERY.

NEW AND BEAUTIFULLY ILLUSTRATED EDITION—(JUST READY.)

PRINCIPLES OF SURGERY.

BY JAMES MILLER, F. R. S. E., F. R. C. S. E.,

Professor of Surgery in the University of Edinburgh.

THIRD AMERICAN, FROM THE SECOND ENGLISH EDITION.

In one very large and handsome octavo volume, of over six hundred pages,

WITH ABOUT TWO HUNDRED AND FIFTY EXQUISITE WOOD ENGRAVINGS.

The very extensive additions and alterations which the author has introduced into this edition have rendered it essentially a new work. By common consent, it has been pronounced the most complete and thorough exponent of the present state of the science of surgery in the English language, and the American publishers in the preparation of the present edition have endeavored to render it in all respects worthy of its extended reputation. The press has been carefully revised by a competent editor, who has introduced such notes and observations as the rapid progress of surgical investigation and pathology have rendered necessary. The illustrations, which are very numerous, and of a high order of merit, both artistic and practical, have been engraved with great care, and in every point of mechanical execution it is confidently presented as one of the most beautiful volumes as yet published in this country.

BY THE SAME AUTHOR.

THE PRACTICE OF SURGERY.

In one octavo volume, of 496 pages.

HORNER'S ANATOMY.

MUCH IMPROVED AND ENLARGED EDITION.—(*Just Issued.*)

SPECIAL ANATOMY AND HISTOLOGY.

BY WILLIAM E. HORNER, M. D.,

Professor of Anatomy in the University of Pennsylvania, &c.

EIGHTH EDITION.

EXTENSIVELY REVISED AND MODIFIED TO 1851.

In two large octavo volumes, handsomely printed, with several hundred illustrations.

This work has enjoyed a thorough and laborious revision on the part of the author, with the view of bringing it fully up to the existing state of knowledge on the subject of general and special anatomy. To adapt it more perfectly to the wants of the student, he has introduced a large number of additional wood engravings, illustrative of the objects described, while the publishers have endeavored to render the mechanical execution of the work worthy of the extended reputation which it has acquired. The demand which has carried it to an EIGHTH EDITION is a sufficient evidence of the value of the work, and of its adaptation to the wants of the student and professional reader.

NEW AND CHEAPER EDITION OF

SMITH & HORNER'S ANATOMICAL ATLAS.

AN ANATOMICAL ATLAS,

ILLUSTRATIVE OF THE STRUCTURE OF THE HUMAN BODY.

BY HENRY H. SMITH, M. D., &c.

UNDER THE SUPERVISION OF

WILLIAM E. HORNER, M. D.,

Professor of Anatomy in the University of Pennsylvania.

In one volume, large imperial octavo, with about six hundred and fifty beautiful figures.

With the view of extending the sale of this beautifully executed and complete "Anatomical Atlas," the publishers have prepared a new edition, printed on both sides of the page, thus materially reducing its cost, and enabling them to present it at a price about forty per cent. lower than former editions, while, at the same time, the execution of each plate is in no respect deteriorated, and not a single figure is omitted.

These figures are well selected, and present a complete and accurate representation of that wonderful fabric, the human body. The plan of this Atlas which renders it so peculiarly convenient for the student, and its superb artistical execution, have been already pointed out. We must congratulate the student upon the completion of this Atlas, as it is the most convenient work of the kind that has yet appeared; and we must add, the very beautiful manner in which it is "got up" is so creditable to the country as to be flattering to our national pride.—*American Medical Journal.*

HORNER'S DISSECTOR.

THE UNITED STATES DISSECTOR;

Being a new edition, with extensive modifications, and almost re-written, of

"HORNER'S PRACTICAL ANATOMY."

In one very neat volume, royal 12mo., of 440 pages, with many illustrations on wood.

WILSON'S DISSECTOR, New Edition—(Just Issued.)

THE DISSECTOR;

OR, PRACTICAL AND SURGICAL ANATOMY.

BY ERASMUS WILSON.

MODIFIED AND RE-ARRANGED BY

PAUL BECK GODDARD, M. D.

A NEW EDITION, WITH REVISIONS AND ADDITIONS.

In one large and handsome volume, royal 12mo., with one hundred and fifteen illustrations.

In passing this work again through the press, the editor has made such additions and improvements as the advance of anatomical knowledge has rendered necessary to maintain the work in the high reputation which it has acquired in the schools of the United States as a complete and faithful guide to the student of practical anatomy. A number of new illustrations have been added, especially in the portion relating to the complicated anatomy of Hernia. In mechanical execution the work will be found superior to former editions.

WORKS BY W. B. CARPENTER, M. D.

NEW AND IMPROVED EDITION—(Nearly Ready.)

PRINCIPLES OF HUMAN PHYSIOLOGY,

WITH THEIR CHIEF APPLICATIONS TO

PATHOLOGY, HYGIENE, AND FORENSIC MEDICINE.

Fifth American from a New and Revised London Edition.

WITH OVER THREE HUNDRED ILLUSTRATIONS.

In one very large and handsome octavo volume.

This edition, which will be issued simultaneously with that now passing through the author's hands in London, will be found materially altered and improved. The aim of the author has been to bring it thoroughly up with the latest investigations and discoveries, and at the same time to avoid increasing the bulk of the volume, he has omitted some chapters which, treating on comparative anatomy and physiology, have appeared to him unnecessary to the full understanding of the subject of his treatise. A number of new illustrations have been introduced, and the whole is presented as in every way worthy of the very extended favor with which the work has been received.

***COMPARATIVE PHYSIOLOGY*—(*Now Ready.*)**

PRINCIPLES OF PHYSIOLOGY,

GENERAL AND COMPARATIVE.

THIRD EDITION, GREATLY ENLARGED.

In one very handsome octavo volume, of over 1100 pages, with 321 beautiful wood-cuts.

This great work will supply a want long felt by the scientific public of this country, who have had no accessible treatise to refer to, presenting, in an intelligible form, a complete and thorough outline of this important subject. The high reputation of the author, on both sides of the Atlantic, is a sufficient guarantee for the completeness and accuracy of any work to which his name is prefixed; but this volume comes with the additional recommendation that it is the one on which the author has bestowed the greatest care, and on which he is desirous to rest his reputation. Two years have been devoted to the preparation of this edition, which has been thoroughly remoulded and rewritten, so as, in fact, to constitute a new work. The amount of alterations and additions may be understood from the fact that of the ten hundred and eighty pages of the text, but one hundred and fifty belong to the previous edition. Containing, as it does, the results of years devoted to study and observation, it may be regarded as a complete exposition of the most advanced state of knowledge in this rapidly-progressive branch of science, and as a storehouse of facts and principles in all departments of Physiology, such as perhaps no man but its author could have accumulated and classified.

In every point of mechanical execution, and profuseness and beauty of illustration, the Publishers risk nothing in saying that it will be found all that the most fastidious taste could desire.

A truly magnificent work. In itself a perfect physiological study.—*Ranking's Abstract. July* 21, 1851.

This work stands without its fellow. It is one few men in Europe could have undertaken; it is one no man, we believe, could have brought to so successful an issue as Dr. Carpenter. It required for its production a physiologist at once deeply read in the labours of others, capable of taking a general, critical, and unprejudiced view of those labours, and of combining the varied, heterogeneous materials at his disposal, so as to form an harmonious whole. We feel that this abstract can give the reader a very imperfect idea of the fulness of this work, and no idea of its unity, of the admirable manner in which material has been brought, from the most various sources, to conduce to its completeness, of the lucidity of the reasoning it contains, or of the clearness of language in which the whole is clothed. Not the profession only, but the scientific world at large, must feel deeply indebted to Dr. Carpenter for this great work. It must, indeed, add largely even to his high reputation.—*Medical Times.*

CARPENTER'S MANUAL OF PHYSIOLOGY.

NEW AND IMPROVED EDITION—(JUST ISSUED.)

ELEMENTS OF PHYSIOLOGY,

INCLUDING PHYSIOLOGICAL ANATOMY.

SECOND AMERICAN, FROM A NEW AND REVISED LONDON EDITION.

With One Hundred and Ninety Illustrations. In one very handsome octavo volume.

In publishing the first edition of this work, its title was altered from that of the London volume, by the substitution of the word "Elements" for that of "Manual," and with the author's sanction, the title of "Elements" is still retained as being more expressive of the scope of the treatise. A comparison of the present edition with the former one will show a material improvement, the author having revised it thoroughly, with the view of rendering it completely on a level with the most advanced state of the science. By condensing the less important portions, these numerous additions have been introduced without materially increasing the bulk of the volume, and while numerous illustrations have been added, and the general execution of the work improved, it has been kept at its former very moderate price.

To say that it is the best manual of Physiology now before the public, would not do sufficient justice to the author.—*Buffalo Med. Journal.*

In his former works it would seem that he had exhausted the subject of Physiology. In the present, he gives the essence, as it were, of the whole.—*N. Y. Journal of Medicine.*

The best and most complete exposé of modern physiology, in one volume, extant in the English language.—*St. Louis Med. Journal.*

Those who have occasion for an elementary treatise on physiology, cannot do better than to possess themselves of the manual of Dr. Carpenter.—*Medical Examiner.*

PATHOLOGICAL HISTOLOGY—(Preparing.)

PROFESSOR GLUGE'S

ATLAS OF PATHOLOGICAL HISTOLOGY.

TRANSLATED FROM THE GERMAN.

In one very handsome volume, imperial quarto, with several hundred illustrations, plain and colored, on twelve plates, and copious letter-press.

The great and increasing interest with which this important subject is now regarded by the profession, and the rapid advances which it is making by the aid of the microscope, have induced the publishers to present this volume, which contains all the most recent observations and results of European investigations. The text contains a complete exposition of the present state of pathological science, while the plates are considered as the most truthful and accurate representations which have been made of the pathological conditions of the membranes, and the volume as a whole may be regarded as a beautiful specimen of mechanical execution, presented at a very reasonable price.

WILLIAMS' PRINCIPLES—New and Enlarged Edition.

PRINCIPLES OF MEDICINE;

Comprising General Pathology and Therapeutics,

AND A BRIEF GENERAL VIEW OF

ETIOLOGY, NOSOLOGY, SEMEIOLOGY, DIAGNOSIS, PROGNOSIS, AND HYGIENICS,

BY CHARLES J. B. WILLIAMS, M. D., F. R. S.,

Fellow of the Royal College of Physicians, &c.

EDITED, WITH ADDITIONS, BY MEREDITH CLYMER, M. D.,

Consulting Physician to the Philadelphia Hospital, &c. &c.

THIRD AMERICAN, FROM THE SECOND AND ENLARGED LONDON EDITION.

In one octavo volume, of 440 pages.

BILLING'S PRINCIPLES, NEW EDITION—(Just Issued.)

THE PRINCIPLES OF MEDICINE.

BY ARCHIBALD BILLING, M. D., &c.

Second American from the Fifth and Improved London Edition.

In one handsome octavo volume, extra cloth, 250 pages.

We can strongly recommend Dr. Billing's "Principles" as a code of instruction which should be constantly present to the mind of every well-informed and philosophical practitioner of medicine.—*Lancet.*

MANUALS ON THE BLOOD AND URINE,

In one handsome volume royal 12mo., extra cloth, of 460 large pages, with numerous illustrations.

CONTAINING

I. A Practical Manual on the Blood and Secretions of the Human Body. BY JOHN WILLIAM GRIFFITH, M. D., &c.

II. On the Analysis of the Blood and Urine in health and disease, and on the treatment of Urinary diseases. BY G. OWEN REESE, M. D., F. R. S., &c. &c.

III. A Guide to the Examination of the Urine in health and disease. BY ALFRED MARKWICK.

NEW EDITION—(JUST ISSUED.)

URINARY DEPOSITS;

THEIR DIAGNOSIS, PATHOLOGY, AND THERAPEUTICAL INDICATIONS.

BY GOLDING BIRD, A. M., M. D., &c.

A NEW AMERICAN, FROM THE THIRD AND IMPROVED LONDON EDITION.

In one very neat volume, royal 12mo., with over sixty illustrations.

Though the present edition of this well-known work is but little increased in size, it will be found essentially modified throughout, and fully up to the present state of knowledge on its subject. The unanimous testimony of the medical press warrants the publishers in presenting it as a complete and reliable manual for the student of this interesting and important branch of medical science.

ABERCROMBIE ON THE BRAIN.—Pathological and Practical Researches on Diseases of the Brain and Spinal Cord. A new edition, in one small 8vo. volume, pp. 324.

BURROWS ON CEREBRAL CIRCULATION.—On Disorders of the Cerebral Circulation, and on the Connection between Affections of the Brain and Diseases of the Heart. In 1 8vo. vol., with col'd pl's, pp. 216.

BLAKISTON ON THE CHEST.—Practical Observations on certain Diseases of the Chest, and on the Principles of Auscultation. In one volume, 8vo., pp. 384.

HASSE'S PATHOLOGICAL ANATOMY.—An Anatomical Description of the Diseases of Respiration and Circulation. Translated and Edited by Swaine. In one volume, 8vo., pp. 379.

FRICK ON THE URINE.—Renal Affections, their Diagnosis and Pathology. In one handsome volume, royal 12mo., with illustrations.

COPLAND ON PALSY.—Of the Causes, Nature, and Treatment of Palsy and Apoplexy. In one volume, royal 12mo. (Just Issued.)

VOGEL'S PATHOLOGICAL ANATOMY.—Pathological Anatomy of the Human body. Translated by Day. In one octavo volume, with plates, plain and colored.

SIMON'S PATHOLOGY.—Lectures on General Pathology. Publishing in the "Medical News and Library," for 1852.

DUNGLISON'S PRACTICE OF MEDICINE.

ENLARGED AND IMPROVED EDITION.

THE PRACTICE OF MEDICINE.

A TREATISE ON

SPECIAL PATHOLOGY AND THERAPEUTICS.

THIRD EDITION.

BY ROBLEY DUNGLISON, M. D.,

Professor of the Institutes of Medicine in the Jefferson Medical College; Lecturer on Clinical Medicine, &c.

In two large octavo volumes, of fifteen hundred pages.

The student of medicine will find, in these two elegant volumes, a mine of facts, a gathering of precepts and advice from the world of experience, that will nerve him with courage, and faithfully direct him in his efforts to relieve the physical sufferings of the race.—*Boston Medical and Surgical Journal.*

Upon every topic embraced in the work the latest information will be found carefully posted up. *Medical Examiner.*

It is certainly the most complete treatise of which we have any knowledge. There is scarcely a disease which the student will not find noticed.—*Western Journal of Medicine and Surgery.*

One of the most elaborate treatises of the kind we have.—*Southern Medical and Surg. Journal.*

A New Work. Now Ready.

DISEASES OF THE HEART, LUNGS, AND APPENDAGES;

THEIR SYMPTOMS AND TREATMENT.

BY W. H. WALSHE, M.D.,

Professor of the Principles and Practice of Medicine in University College, London, &c.

In one handsome volume, large royal 12mo.

The author's design in this work has been to include within the compass of a moderate volume, all really essential facts bearing upon the symptoms, physical signs and treatment of pulmonary and cardiac diseases. To accomplish this the first part of the work is devoted to the description of the various modes of physical diagnosis, auscultation, percussion, mensuration, &c., which are fully and clearly, but succinctly entered into, both as respects their theory and clinical phenomena. In the second part, the various diseases of the heart, lungs, and great vessels are considered in regard to symptoms, physical signs and treatment, with numerous references to cases. The eminence of the author is a guarantee to the practitioner and student that the work is one of practical utility in facilitating the diagnosis and treatment of a large, obscure and important class of diseases.

THE GREAT MEDICAL LIBRARY.

THE CYCLOPEDIA OF PRACTICAL MEDICINE;

COMPRISING

Treatises on the Nature and Treatment of Diseases, Materia Medica, and Therapeutics, Diseases of Women and Children, Medical Jurisprudence, &c. &c.

EDITED BY

JOHN FORBES, M. D., F. R. S., ALEXANDER TWEEDIE, M. D., F. R. S.
AND JOHN CONNOLLY, M. D.

Revised, with Additions,

BY ROBLEY DUNGLISON, M. D.

THIS WORK IS NOW COMPLETE, AND FORMS FOUR LARGE SUPER-ROYAL OCTAVO VOLUMES,

Containing Thirty-two Hundred and Fifty-four unusually large Pages in Double Columns, Printed on Good Paper, with a new and clear type.

THE WHOLE WELL AND STRONGLY BOUND WITH RAISED BANDS AND DOUBLE TITLES.

This work contains no less than FOUR HUNDRED AND EIGHTEEN DISTINCT TREATISES,

By Sixty-eight distinguished Physicians.

The most complete work on Practical Medicine extant; or, at least, in our language.—*Buffalo Medical and Surgical Journal.*

For reference, it is above all price to every practitioner.—*Western Lancet.*

One of the most valuable medical publications of the day—as a work of reference it is invaluable.—*Western Journal of Medicine and Surgery.*

It has been to us, both as learner and teacher, a work for ready and frequent reference, one in which modern English medicine is exhibited in the most advantageous light.—*Medical Examiner.*

We rejoice that this work is to be placed within the reach of the profession in this country, it being unquestionably one of very great value to the practitioner. This estimate of it has not been formed from a hasty examination, but after an intimate acquaintance derived from frequent consultation of it during the past nine or ten years. The editors are practitioners of established reputation, and the list of contributors embraces many of the most eminent professors and teachers of London, Edinburgh, Dublin, and Glasgow. It is, indeed, the great merit of this work that the principal articles have been furnished by practitioners who have not only devoted especial attention to the diseases about which they have written, but have also enjoyed opportunities for an extensive practical acquaintance with them.—and whose reputation carries the assurance of their competency justly to appreciate the opinions of others, while it stamps their own doctrines with high and just authority.—*American Medical Journal.*

WATSON'S PRACTICE OF MEDICINE—New Edition.

LECTURES ON THE

PRINCIPLES AND PRACTICE OF PHYSIC.

BY THOMAS WATSON, M. D., &c. &c.

Third American, from the last London Edition.

REVISED, WITH ADDITIONS, BY D. FRANCIS CONDIE, M. D.,

Author of "A Treatise on the Diseases of Children," &c.

IN ONE OCTAVO VOLUME,

Of nearly ELEVEN HUNDRED LARGE PAGES, strongly bound with raised bands.

To say that it is the very best work on the subject now extant, is but to echo the sentiment of the medical press throughout the country.—*N. O. Medical Journal.*

Of the text-books recently republished Watson is very justly the principal favorite.—*Holmes' Report to Nat. Med. Assoc.*

By universal consent the work ranks among the very best text-books in our language.—*Ill. and Ind. Med. Journal.*

Regarded on all hands as one of the very best, if not the very best, systematic treatise on practical medicine extant.—*St. Louis Med. Journal.*

Confessedly one of the very best works on the principles and practice of physic in the English or any other language.—*Med. Examiner.*

As a text-book it has no equal; as a compendium of pathology and practice no superior.—*N. Y. Annalist.*

We know of no work better calculated for being placed in the hands of the student, and for a text book, on every important point the author seems to have posted up his knowledge to the day.—*Amer. Med. Journal.*

One of the most practically useful books that ever was presented to the student.—*N. Y. Med. Journal.*

WILSON ON THE SKIN.

ON DISEASES OF THE SKIN.

BY ERASMUS WILSON, F. R. S.,

Author of "Human Anatomy," &c.

SECOND AMERICAN FROM THE SECOND LONDON EDITION.

In one neat octavo volume, extra cloth, 440 pages.

Also, to be had with eight beautifully colored steel plates.
Also, the plates sold separate, in boards.

Much Enlarged Edition of BARTLETT ON FEVERS.

THE HISTORY, DIAGNOSIS, AND TREATMENT OF THE

FEVERS OF THE UNITED STATES.

BY ELISHA BARTLETT, M. D.,

In one octavo volume of 550 pages, beautifully printed and strongly bound.

CLYMER AND OTHERS ON FEVERS.

FEVERS; THEIR DIAGNOSIS, PATHOLOGY, AND TREATMENT.

PREPARED AND EDITED, WITH LARGE ADDITIONS,

FROM THE ESSAYS ON FEVER IN TWEEDIE'S LIBRARY OF PRACTICAL MEDICINE,

BY MEREDITH CLYMER, M. D.

In one octavo volume of six hundred pages.

BENEDICT'S CHAPMAN.—Compendium of Chapman's Lectures on the Practice of Medicine. One neat volume, 8vo., pp. 258.

BUDD ON THE LIVER.—On Diseases of the Liver. In one very neat 8vo. vol., with colored plates and wood-cuts. pp. 392.

CHAPMAN'S LECTURES.—Lectures on Fevers, Dropsy, Gout, Rheumatism, &c. &c. In one neat 8vo. volume, pp. 450.

THOMSON ON THE SICK ROOM.—Domestic management of the sick Room, necessary in aid of Medical Treatment for the cure of Diseases. Edited by R. E. Griffith, M. D. In one large royal 12mo. volume, with wood-cuts, pp. 360.

HOPE ON THE HEART.—A Treatise on the Diseases of the Heart and Great Vessels. Edited by Pennock. In one volume, 8vo., with plates, pp. 572.

LALLEMAND ON SPERMATORRHŒA.—The Causes, Symptoms, and Treatment of Spermatorrhœa. Translated and Edited by Henry J McDougal. In one volume, 8vo., pp. 320.

PHILIPS ON SCROFULA.—Scrofula: its Nature, its Prevalence, its Causes, and the Principles of its Treatment. In one volume, 8vo., with a plate, pp. 350.

WHITEHEAD ON ABORTION, &c.—The Causes and Treatment of Abortion and Sterility; being the Result of an Extended Practical Inquiry into the Physiological and Morbid Conditions of the Uterus. In one volume, 8vo., pp. 368.

WILLIAMS ON RESPIRATORY ORGANS.—A Practical Treatise on Diseases of the Respiratory Organs; including Diseases of the Larynx, Trachea, Lungs, and Pleuræ. With numerous Additions and Notes by M. Clymer, M. D. With wood-cuts. In one octavo volume, pp 508.

DAY ON OLD AGE.—A Practical Treatise on the Domestic Management and more important Diseases of Advanced Life. With an Appendix on a new and successful mode of treating Lumbago and other forms of Chronic Rheumatism. 1 vol. 8vo., pp. 226.

MEIGS ON FEMALES, New and Improved Edition—(Lately Issued.)

WOMAN; HER DISEASES AND THEIR REMEDIES;

A SERIES OF LETTERS TO HIS CLASS.

BY C. D. MEIGS, M. D.,

Professor of Midwifery and Diseases of Women and Children in the Jefferson Medical College of Philadelphia, &c. &c.

In one large and beautifully printed octavo volume, of nearly seven hundred large pages.

"I am happy to offer to my Class an enlarged and amended edition of my Letters on the Diseases of Women; and I avail myself of this occasion to return my heartfelt thanks to them, and to our brethren generally, for the flattering manner in which they have accepted this fruit of my labor."—PREFACE.

The value attached to this work by the profession is sufficiently proved by the rapid exhaustion of the first edition, and consequent demand for a second. In preparing this the author has availed himself of the opportunity thoroughly to revise and greatly to improve it. The work will therefore be found completely brought up to the day, and in every way worthy of the reputation which it has so immediately obtained.

Professor Meigs has enlarged and amended this great work, for such it unquestionably is, having passed the ordeal of criticism at home and abroad, but been improved thereby; for in this new edition the author has introduced real improvements, and increased the value and utility of the book immeasurably. It presents so many novel, bright and sparkling thoughts; such an exuberance of new ideas on almost every page, that we confess ourselves to have become enamored with the book and its author: and cannot withhold our congratulations from our Philadelphia confreres, that such a teacher is in their service. We regret that our limits will not allow of a more extended notice of this work, but must content ourselves with thus commending it as worthy of diligent perusal by physicians as well as students, who are seeking to be thoroughly instructed in the important practical subjects of which it treats.—*N. Y. Med. Gazette.*

It contains a vast amount of practical knowledge, by one who has accurately observed and retained the experience of many years, and who tells the result in a free, familiar, and pleasant manner.—*Dublin Quarterly Journal.*

There is an off-hand fervor, a glow and a warm-heartedness infecting the effort of Dr. Meigs, which is entirely captivating, and which absolutely hurries the reader through from beginning to end. Besides, the book teems with solid instruction, and it shows the very highest evidence of ability, viz., the clearness with which the information is presented. We know of no better test of one's understanding a subject than the evidence of the power of lucidly explaining it. The most elementary as well as the obscurest subjects, under the pencil of Prof. Meigs, are isolated and made to stand out in such bold relief, as to produce distinct impressions upon the mind and memory of the reader.—*The Charleston Medical Journal.*

The merits of the first edition of this work were so generally appreciated, and with such a high degree of favor by the medical profession throughout the Union, that we are not surprised in seeing a second edition of it. It is a standard work on the diseases of females and in many respects is one of the very best of its kind in the English language. Upon the appearance of the first edition, we gave the work a cordial reception, and spoke of it in the warmest terms of commendation. Time has not changed the favorable estimate we placed upon it, but has rather increased our convictions of its superlative merits. But we do not now deem it necessary to say more than to commend this work, on the diseases of women, and the remedies for them, to the attention of those practitioners who have not supplied themselves with it. The most select library would be imperfect without it.—*The Western Journal of Medicine and Surgery.*

He is a bold thinker, and possesses more originality of thought and style than almost any American writer on medical subjects. If he is not an elegant writer, there is at least a freshness—a raciness in his mode of expressing himself—that cannot fail to draw the reader after him even to the close of his work: you cannot nod over his pages; he stimulates rather than narcotises your senses, and the reader cannot lay aside these letters when once he enters into their merits. This, the second edition, is much amended and enlarged, and affords abundant evidence of the author's talents and industry.—*N. O. Medical and Surgical Journal.*

The practical writings of Dr. Meigs are second to none.—*The N. Y. Journal of Medicine.*

The excellent practical directions contained in this volume give it great utility, which we trust will not be lost upon our older colleagues; with some condensation indeed, we should think it well adapted for translation into German.—*Zeitschrift fur die Gesammte Medecin.*

NEW AND IMPROVED EDITION—(Lately Issued.)

A TREATISE ON THE DISEASES OF FEMALES,

AND ON THE SPECIAL HYGIENE OF THEIR SEX.

BY COLOMBAT DE L'ISERE, M. D.

TRANSLATED, WITH MANY NOTES AND ADDITIONS, BY C. D. MEIGS, M. D.

SECOND EDITION, REVISED AND IMPROVED.

In one large volume, octavo, of seven hundred and twenty pages, with numerous wood-cuts.

We are satisfied it is destined to take the front rank in this department of medical science. It is in fact a complete exposition of the opinions and practical methods of all the celebrated practitioners of ancient and modern times.—*New York Journ. of Medicine.*

ASHWELL ON THE DISEASES OF FEMALES.

A PRACTICAL TREATISE ON THE DISEASES PECULIAR TO WOMEN.

ILLUSTRATED BY CASES DERIVED FROM HOSPITAL AND PRIVATE PRACTICE

BY SAMUEL ASHWELL, M. D. WITH ADDITIONS BY PAUL BECK GODDARD, M. D.

Second American edition. In one octavo volume, of 520 pages.

One of the very best works ever issued from the press on the Diseases of Females.—*Western Lancet.*

ON THE CAUSES AND TREATMENT OF ABORTION AND STERILITY. By James Whitehead, M. D., &c. In one volume octavo, of about three hundred and seventy five pages.

NEW AND IMPROVED EDITION—(Lately Issued.)

THE DISEASES OF FEMALES,

INCLUDING THOSE OF PREGNANCY AND CHILDBED.

BY FLEETWOOD CHURCHILL, M. D., M. R. I. A.,

Author of "Theory and Practice of Midwifery," "Diseases of Females," &c.

A New American Edition (The Fifth), Revised by the Author.

WITH THE NOTES OF ROBERT M. HUSTON, M. D.

In one large and handsome octavo volume of 632 pages, with wood-cuts.

To indulge in panegyric, when announcing the fifth edition of any acknowledged medical authority, were to attempt to "gild refined gold." The work announced above, has too long been honored with the term "classical" to leave any doubt as to its true worth, and we content ourselves with remarking, that the author has carefully retained the notes of Dr. Huston, who edited the former American edition, thus really enhancing the value of the work, and paying a well merited compliment. All who wish to be "posted up" on all that relates to the diseases peculiar to the wife, the mother, or the maid, will hasten to secure a copy of this most admirable treatise.—*The Ohio Medical and Surgical Journal.*

We know of no author who deserves that approbation, on "the diseases of females," to the same extent that Dr. Churchill does. His, indeed, is the only thorough treatise we know of on the subject, and it may be commended to practitioners and students as a masterpiece in its particular department. The former editions of this work have been commended strongly in this journal, and they have won their way to an extended, and a well deserved popularity. This fifth edition, before us, is well calculated to maintain Dr. Churchill's high reputation. It was revised and enlarged by the author, for his American publishers, and it seems to us, that there is scarcely any species of desirable information on its subjects, that may not be found in this work. —*The Western Journal of Medicine and Surgery.*

We are gratified to announce a new and revised edition of Dr. Churchill's valuable work on the diseases of females. We have ever regarded it as one of the very best works on the subjects embraced within its scope, in the English language; and the present edition, enlarged and revised by the author, renders it still more entitled to the confidence of the profession. The valuable notes of Prof. Huston have been retained, and contribute, in no small degree, to enhance the value of the work. It is a source of congratulation that the publishers have permitted the author to be, in this instance, his own editor, thus securing all the revision which an author alone is capable of making.—*The Western Lancet.*

As a comprehensive manual for students, or a work of reference for practitioners, we only speak with common justice when we say that it surpasses any other that has ever issued on the same subject from the British press.—*The Dublin Quarterly Journal.*

Churchill's Monographs on Females—(Lately Issued.)

ESSAYS ON THE PUERPERAL FEVER, AND OTHER DISEASES

PECULIAR TO WOMEN.

SELECTED FROM THE WRITINGS OF BRITISH AUTHORS PREVIOUS TO THE CLOSE OF THE EIGHTEENTH CENTURY.

Edited by FLEETWOOD CHURCHILL, M. D., M. R. I. A.,

Author of "Treatise on the Diseases of Females," &c.

In one neat octavo volume, of about four hundred and fifty pages.

To these papers Dr. Churchill has appended notes, embodying whatever information has been laid before the profession since their authors' time. He has also prefixed to the essays on puerperal fever, which occupy the larger portion of the volume, an interesting historical sketch of the principal epidemics of that disease. The whole forms a very valuable collection of papers by professional writers of eminence, on some of the most important accidents to which the puerperal female is liable.—*American Journal of Medical Sciences.*

***MUCH ENLARGED AND IMPROVED EDITION*—(*Lately Issued.*)**

A PRACTICAL TREATISE ON

INFLAMMATION OF THE UTERUS AND ITS APPENDAGES,

AND ON ULCERATION AND INDURATION OF THE NECK OF THE UTERUS.

BY HENRY BENNETT, M. D.,

Obstetric Physician to the Western Dispensary.

Second Edition, much enlarged.

In one neat octavo volume of 350 pages, with wood-cuts.

This edition is so enlarged as to constitute a new work. It embraces the study of inflammation in all the uterine organs, and its influence in the production of displacements and of the reputed functional diseases of the uterus.

Few works issue from the medical press which are at once original and sound in doctrine; but such, we feel assured, is the admirable treatise now before us. The important practical precepts which the author inculcates are all rigidly deduced from facts. . . . Every page of the book is good, and eminently practical. So far as we know and believe, it is the best work on the subject on which it treats.—*Monthly Journal of Medical Science.*

A TREATISE ON THE DISEASES OF FEMALES.

BY W. P. DEWEES, M. D.

NINTH EDITION.

In one volume, octavo. 532 pages, with plates.

MEIGS ON CHILDREN—Just Issued.

OBSERVATIONS ON CERTAIN OF THE DISEASES OF YOUNG CHILDREN.

BY CHARLES D. MEIGS, M. D.,

Professor of Midwifery and of the Diseases of Women and Children in the Jefferson Medical College of Philadelphia, &c. &c.

In one handsome octavo volume of 214 pages.

While this work is not presented to the profession as a systematic and complete treatise on Infantile disorders, the importance of the subjects treated of, and the interest attaching to the views and opinions of the distinguished author must command for it the attention of all who are called upon to treat this interesting class of diseases.

It puts forth no claims as a systematic work, but contains an amount of valuable and useful matter, scarcely to be found in the same space in our home literature. It can not but prove an acceptable offering to the profession at large.—*N. Y. Journal of Medicine.*

The work before us is undoubtedly a valuable addition to the fund of information which has already been treasured up on the subjects in question. It is practical, and therefore eminently adapted to the general practitioner. Dr. Meigs' works have the same fascination which belongs to himself.—*Medical Examiner.*

This is a most excellent work on the obscure diseases of childhood, and will afford the practitioner and student of medicine much aid in their diagnosis and treatment.—*The Boston Medical and Surgical Journal.*

We take much pleasure in recommending this excellent little work to the attention of medical practitioners. It deserves their attention, and after they commence its perusal, they will not willingly abandon it, until they have mastered its contents. We read the work while suffering from a carbuncle, and its fascinating pages often beguiled us into forgetfulness of agonizing pain. May it teach others to relieve the afflictions of the young.—*The Western Journal of Medicine and Surgery.*

All of which topics are treated with Dr Meigs' acknowledged ability and original diction. The work is neither a systematic nor a complete treatise upon the diseases of children, but a fragment which may be consulted with much advantage.—*Southern Medical and Surgical Journal.*

NEW WORK BY DR. CHURCHILL.

ON THE DISEASES OF INFANTS AND CHILDREN.

BY FLEETWOOD CHURCHILL, M. D., M. R. I. A.,

Author of "Theory and Practice of Midwifery," "Diseases of Females," &c.

In one large and handsome octavo volume of over 600 pages.

From Dr. Churchill's known ability and industry, we were led to form high expectations of this work; nor were we deceived. Its learned author seems to have set no bounds to his researches in collecting information which, with his usual systematic address, he has disposed of in the most clear and concise manner, so as to lay before the reader every opinion of importance bearing upon the subject under consideration. We regard this volume as possessing more claims to completeness than any other of the kind with which we are acquainted. Most cordially and earnestly, therefore, do we commend it to our professional brethren, and we feel assured that the stamp of their approbation will in due time be impressed upon it. After an attentive perusal of its contents, we hesitate not to say, that it is one of the most comprehensive ever written upon the diseases of children, and that, for copiousness of reference, extent of research, and perspicuity of detail, it is scarcely to be equalled, and not to be excelled in any language.—*Dublin Quarterly Journal.*

The present volume will sustain the reputation acquired by the author from his previous works. The reader will find in it full and judicious directions for the management of infants at birth, and a compendious, but clear, account of the diseases to which children are liable, and the most successful mode of treating them. We must not close this notice without calling attention to the author's style, which is perspicuous and polished to a degree, we regret to say, not generally characteristic of medical works. We recommend the work of Dr. Churchill most cordially, both to students and practitioners, as a valuable and reliable guide in the treatment of the diseases of children.—*Am. Journ. of the Med. Sciences.*

After this meagre, and we know, very imperfect notice, of Dr. Churchill's work, we shall conclude by saying, that it is one that cannot fail from its copiousness, extensive research, and general accuracy, to exalt still higher the reputation of the author in this country. The American reader will be particularly pleased to find that Dr. Churchill has done full justice throughout his work, to the various American authors on this subject. The names of Dewees, Eberle, Condie, and Stewart, occur on nearly every page, and these authors are constantly referred to by the author in terms of the highest praise, and with the most liberal courtesy.—*The Medical Examiner.*

We know of no work on this department of Practical Medicine which presents so candid and unprejudiced a statement or posting up of our actual knowledge as this.—*N. Y. Journal of Medicine.*

Its claims to merit, both as a scientific and practical work, are of the highest order. Whilst we would not elevate it above every other treatise on the same subject, we certainly believe that very few are equal to it, and none superior.—*Southern Med. and Surg. Journal.*

New and Improved Edition.

A PRACTICAL TREATISE ON THE
DISEASES OF CHILDREN.

BY D. FRANCIS CONDIE, M. D.,

Fellow of the College of Physicians, &c. &c.

Third edition, revised and augmented. In one large volume, 8vo., of over 700 pages.

In the preparation of a third edition of the present treatise, every portion of it has been subjected to a careful revision. A new chapter has been added on Epidemic Meningitis, a disease which, although not confined to children, occurs far more frequently in them, than in adults. In the other chapters of the work, all the more important facts that have been developed since the appearance of the last edition, in reference to the nature, diagnosis, and treatment of the several diseases of which they treat, have been incorporated. The great object of the author has been to present, in each succeeding edition, as full and connected a view as possible of the actual state of the pathology and therapeutics of those affections which most usually occur between birth and puberty.

To the present edition there is appended a list of the several works and essays quoted or referred to in the body of the work, or which have been consulted in its preparation or revision.

Every important fact that has been verified or developed since the publication of the previous edition, either in relation to the nature, diagnosis, or treatment of the diseases of children, have been arranged and incorporated into the body of the work: thus posting up to date, to use a counting-house phrase, all the valuable facts and useful information on the subject. To the American practitioner, Dr. Condie's remarks on the diseases of children will be invaluable, and we accordingly advise those who have failed to read this work to procure a copy, and make themselves familiar with its sound principles.—*The New Orleans Medical and Surgical Journal.*

We feel persuaded that the American Medical profession will soon regard it, not only as a very good, but as the VERY BEST "Practical Treatise on the Diseases of Children."—*American Medical Journal*

We pronounced the first edition to be the best work on the Diseases of Children in the English language, and, notwithstanding all that has been published, we still regard it in that light.—*Medical Examiner.*

From Professor Wm P. Johnston, Washington, D. C.

I make use of it as a text-book, and place it invariably in the hands of my private pupils.

From Professor D. Humphreys Storer, of Boston.

I consider it to be the best work on the Diseases of Children we have access to, and as such recommend it to all who ever refer to the subject.

From Professor M M Pallen, of St Louis.

I consider it the best treatise on the Diseases of Children that we possess, and as such have been in the habit of recommending it to my classes.

Dr Condie's scholarship, acumen, industry, and practical sense are manifested in this, as in all his numerous contributions to science.—*Dr Holmes's Report to the American Medical Association.*

Taken as a whole, in our judgment, Dr. Condie's Treatise is the one from the perusal of which the practitioner in this country will rise with the greatest satisfaction.—*Western Journal of Medicine and Surgery.*

One of the best works upon the Diseases of Children in the English language.—*Western Lancet*

We feel assured from actual experience that no physician's library can be complete without a copy of this work.—*N Y. Journal of Medicine*

Perhaps the most full and complete work now before the profession of the United States: indeed, we may say in the English language It is vastly superior to most of its predecessors.—*Transylvania Med Journal.*

A veritable pædiatric encyclopædia, and an honor to American medical literature.—*Ohio Medical and Surgical Journal.*

WEST ON DISEASES OF CHILDREN.

LECTURES ON THE
DISEASES OF INFANCY AND CHILDHOOD.

BY CHARLES WEST, M. D.,

Senior Physician to the Royal Infirmary for Children, &c. &c.

In one volume, octavo.

Every portion of these lectures is marked by a general accuracy of description, and by the soundness of the views set forth in relation to the pathology and therapeutics of the several maladies treated of. The lectures on the diseases of the respiratory apparatus, about one-third of the whole number, are particularly excellent, forming one of the fullest and most able accounts of these affections, as they present themselves during infancy and childhood, in the English language The history of the several forms of phthisis during these periods of existence, with their management, will be read by all with deep interest.—*The American Journal of the Medical Sciences.*

The Lectures of Dr. West, originally published in the London Medical Gazette, form a most valuable addition to this branch of practical medicine. For many years physician to the Children's Infirmary, his opportunities for observing their diseases have been most extensive, no less than 14,000 children having been brought under his notice during the past nine years. These have evidently been studied with great care, and the result has been the production of the very best work in our language, so far as it goes, on the diseases of this class of our patients. The symptomatology and pathology of their diseases are especially exhibited most clearly; and we are convinced that no one can read with care these lectures without deriving from them instruction of the most important kind.—*Charleston Med. Journal.*

A TREATISE

ON THE PHYSICAL AND MEDICAL TREATMENT OF CHILDREN.

BY W. P. DEWEES, M. D.

Ninth edition. In one volume, octavo. 548 pages.

NEW AND IMPROVED EDITION—(Now Ready.)

OBSTETRICS:

THE SCIENCE AND THE ART.

BY CHARLES D. MEIGS, M. D.,

Professor of Midwifery and the Diseases of Women and Children in the Jefferson Medical College, Philadelphia, &c. &c.

Second Edition, Revised and Improved, with about 150 Illustrations.

In one beautifully printed octavo volume, of seven hundred large pages.

The rapid demand for a second edition of this work is a sufficient evidence that it has supplied a desideratum of the profession, notwithstanding the numerous treatises on the same subject which have appeared within the last few years. Adopting a system of his own, the author has combined the leading principles of his interesting and difficult subject, with a thorough exposition of its rules of practice, presenting the results of long and extensive experience and of familiar acquaintance with all the modern writers on this department of medicine. As an American treatise on Midwifery, which has at once assumed the position of a classic, it possesses peculiar claims to the attention and study of the practitioner and student, while the numerous alterations and revisions which it has undergone in the present edition are sufficient guarantee that it is fully on a level with the most recent improvements of its subject.

As an elementary treatise—concise, but, withal, clear and comprehensive—we know of no one better adapted for the use of the student; while the young practitioner will find in it a body of sound doctrine, and a series of excellent practical directions, adapted to all the conditions of the various forms of labor and their results, which he will be induced, we are persuaded, again and again to consult, and always with profit.

It has seldom been our lot to peruse a work upon the subject, from which we have received greater satisfaction, and which we believe to be better calculated to communicate to the student correct and definite views upon the several topics embraced within the scope of its teachings.—*American Journal of the Medical Sciences.*

We are acquainted with no work on midwifery of greater practical value.—*Boston Medical and Surgical Journal.*

Worthy the reputation of its distinguished author.—*Medical Examiner.*

We most sincerely recommend it, both to the student and practitioner, as a more complete and valuable work on the Science and Art of Midwifery, than any of the numerous reprints and American editions of European works on the same subject.—*N. Y. Annalist.*

We have, therefore, great satisfaction in bringing under our reader's notice the matured views of the highest American authority in the department to which he has devoted his life and talents.—*London Medical Gazette.*

An author of established merit, a professor of Midwifery, and a practitioner of high reputation and immense experience—we may assuredly regard his work now before us as representing the most advanced state of obstetric science in America up to the time at which he writes. We consider Dr. Meigs' book as a valuable acquisition to obstetric literature, and one that will very much assist the practitioner under many circumstances of doubt and perplexity.—*The Dublin Quarterly Journal.*

These various heads are subdivided so well, so lucidly explained, that a good memory is all that is necessary in order to put the reader in possession of a thorough knowledge of this important subject. Dr. Meigs has conferred a great benefit on the profession in publishing this excellent work.—*St. Louis Medical and Surgical Journal.*

TYLER SMITH ON PARTURITION.

ON PARTURITION,

AND THE PRINCIPLES AND PRACTICE OF OBSTETRICS.

BY W. TYLER SMITH, M. D.,

Lecturer on Obstetrics in the Hunterian School of Medicine, &c. &c.

In one large duodecimo volume, of 400 pages.

The work will recommend itself by its intrinsic merit to every member of the profession.—*Lancet.*

We can imagine the pleasure with which William Hunter or Denman would have welcomed the present work; certainly the most valuable contribution to obstetrics that has been made since their own day. For ourselves, we consider its appearance as the dawn of a new era in this department of medicine. We do most cordially recommend the work as one absolutely necessary to be studied by every accoucheur It will, we may add, prove equally interesting and instructive to the student, the general practitioner, and pure obstetrician. It was a bold undertaking to reclaim parturition for Reflex Physiology, and it has been well performed.—*London Journal of Medicine.*

LEE'S CLINICAL MIDWIFERY.

CLINICAL MIDWIFERY,

COMPRISING THE HISTORIES OF FIVE HUNDRED AND FORTY-FIVE CASES OF DIFFICULT, PRETERNATURAL, AND COMPLICATED LABOR, WITH COMMENTARIES.

BY ROBERT LEE, M. D., F. R. S., &c.

From the 2d London Edition.

In one royal 12mo. volume, extra cloth, of 238 pages.

More instructive to the juvenile practitioner than a score of systematic works.—*Lancet.*

An invaluable record for the practitioner.—*N. Y. Annalist.*

A storehouse of valuable facts and precedents.—*American Journal of the Medical Sciences.*

CHURCHILL'S MIDWIFERY, BY CONDIE, NEW AND IMPROVED EDITION—(Just Issued.)

ON THE

THEORY AND PRACTICE OF MIDWIFERY.

BY FLEETWOOD CHURCHILL, M. D., &c.

A NEW AMERICAN FROM THE LAST AND IMPROVED ENGLISH EDITION,

EDITED, WITH NOTES AND ADDITIONS,

BY D. FRANCIS CONDIE, M. D.,

Author of a "Practical Treatise on the Diseases of Children," &c.

WITH ONE HUNDRED AND THIRTY-NINE ILLUSTRATIONS.

In one very handsome octavo volume.

In the preparation of the last English edition, from which this is printed, the author has spared no pains, with the desire of bringing it thoroughly up to the present state of obstetric science. The labors of the editor have thus been light, but he has endeavored to supply whatever he has thought necessary to the work, either as respects obstetrical practice in this country, or its progress in Europe since the appearance of Dr. Churchill's last edition. Most of the notes of the former editor, Dr. Huston, have been retained by him, where they have not been embodied by the author in his text. The present edition of this favorite text-book is therefore presented to the profession in the full confidence of its meriting a continuance of the great reputation which it has acquired as a work equally well fitted for the student and practitioner.

To bestow praise on a book that has received such marked approbation would be superfluous. We need only say, therefore, that if the first edition was thought worthy of a favorable reception by the medical public, we can confidently affirm that this will be found much more so. The lecturer, the practitioner, and the student, may all have recourse to its pages, and derive from their perusal much interest and instruction in everything relating to theoretical and practical midwifery.—*Dublin Quarterly Journal of Medical Science.*

A work of very great merit, and such as we can confidently recommend to the study of every obstetric practitioner.—*London Medical Gazette.*

This is certainly the most perfect system extant. It is the best adapted for the purposes of a text-book, and that which he whose necessities confine him to one book, should select in preference to all others.—*Southern Medical and Surgical Journal.*

The most popular work on Midwifery ever issued from the American press —*Charleston Medical Journal.*

Certainly, in our opinion, the very best work on the subject which exists.—*N. Y. Annalist.*

Were we reduced to the necessity of having but *one* work on Midwifery, and *permitted to choose*, we would unhesitatingly take Churchill.—*Western Medical and Surgical Journal.*

It is impossible to conceive a more useful and elegant Manual than Dr. Churchill's Practice of Midwifery. —*Provincial Medical Journal.*

No work holds a higher position, or is more deserving of being placed in the hands of the tyro, the advanced student, or the practitioner.—*Medical Examiner.*

NEW EDITION OF RAMSBOTHAM ON PARTURITION—(*Now Ready,* **1851.**)

THE PRINCIPLES AND PRACTICE OF

OBSTETRIC MEDICINE AND SURGERY,

In reference to the Process of Parturition.

BY FRANCIS H. RAMSBOTHAM, M.D.,

Physician to the Royal Maternity Charity, &c. &c.

SIXTH AMERICAN, FROM THE LAST LONDON EDITION.

Illustrated with One Hundred and Forty-eight Figures on Fifty-five Lithographic Plates.

In one large and handsomely printed volume, imperial octavo, with 520 pages.

In this edition the plates have all been redrawn, and the text carefully read and corrected. It is therefore presented as in every way worthy the favor with which it has so long been received.

From Professor Hodge, of the University of Pennsylvania.

To the American public, it is most valuable, from its intrinsic undoubted excellence, and as being the best authorized exponent of British Midwifery. Its circulation will, I trust, be extensive throughout our country.

We recommend the student, who desires to master this difficult subject with the least possible trouble, to possess himself at once of a copy of this work.—*American Journal of the Medical Sciences.*

It stands at the head of the long list of excellent obstetric works published in the last few years in Great Britain, Ireland, and the Continent of Europe. We consider this book indispensable to the library of every physician engaged in the practice of Midwifery.—*Southern Medical and Surgical Journal.*

When the whole profession is thus unanimous in placing such a work in the very first rank as regards the extent and correctness of all the details of the theory and practice of so important a branch of learning, our commendation or condemnation would be of little consequence; but, regarding it as the most useful of all works of the kind, we think it but an act of justice to urge its claims upon the profession.—*N. O. Med. Journal.*

DEWEES'S MIDWIFERY.

A COMPREHENSIVE SYSTEM OF MIDWIFERY,

ILLUSTRATED BY OCCASIONAL CASES AND MANY ENGRAVINGS.

BY WILLIAM P. DEWEES, M. D.

Tenth Edition, with the Author's last Improvements and Corrections. In one octavo volume, of 600 pages.

PEREIRA'S MATERIA MEDICA—Vol. I., (Now Ready.)

NEW EDITION, GREATLY IMPROVED AND ENLARGED, to Nov. 1851.

THE ELEMENTS OF
MATERIA MEDICA AND THERAPEUTICS.

COMPREHENDING THE NATURAL HISTORY, PREPARATION, PROPERTIES, COMPOSITION, EFFECTS, AND USES OF MEDICINES.

BY JONATHAN PEREIRA, M. D., F. R. S. AND L. S.,

Third American from the Third and Enlarged London Edition.

WITH ADDITIONAL NOTES AND OBSERVATIONS BY THE AUTHOR.

EDITED BY JOSEPH CARSON, M. D.,

Professor of Materia Medica and Pharmacy in the University of Pennsylvania.

In two very large volumes, on small type, with about four hundred illustrations.

The demand for this new edition of "PEREIRA'S MATERIA MEDICA" has induced the publishers to issue the First Volume separately. The Second Volume, now at press and receiving important corrections and revisions from both author and editor, may be expected for publication in July or August, 1852.

The third London edition of this work received very extensive alterations by the author. Many portions of it were entirely rewritten, some curtailed, others enlarged, and much new matter introduced in every part. The edition, however, now presented to the American profession, in addition to this, not only enjoys the advantages of a thorough and accurate superintendence by the editor, but also embodies the additions and alterations suggested by a further careful revision by the author, expressly for this country, embracing the most recent investigations, and the result of several new Pharmacopœias which have appeared since the publication of the London edition of Volume I. The notes of the American editor have been prepared with reference to the new edition of the U. S. Pharmacopœia, and contain such matter generally as is requisite to adapt it fully to the wants of the profession in this country, as well as such recent discoveries as have escaped the attention of the author. In this manner the size of the work has been materially enlarged, and the number of illustrations much increased, while its mechanical execution has been greatly improved in every respect. The profession may therefore rely on being able to procure a work which, in every point of view, will not only maintain, but greatly advance the very high reputation which it has everywhere acquired.

The work, in its present shape, and so far as can be judged from the portion before the public, forms the most comprehensive and complete treatise on materia medica extant in the English language.

Dr. Pereira has been at great pains to introduce into his work, not only all the information on the natural, chemical, and commercial history of medicines, which might be serviceable to the physician and surgeon, but whatever might enable his readers to understand thoroughly the mode of preparing and manufacturing various articles employed either for preparing medicines, or for certain purposes in the arts connected with materia medica and the practice of medicine.

The accounts of the physiological and therapeutic effects of remedies are given with great clearness and accuracy, and in a manner calculated to interest as well as instruct the reader.—*The Edinburgh Medical and Surgical Journal.*

ROYLE'S MATERIA MEDICA.

MATERIA MEDICA AND THERAPEUTICS;

INCLUDING THE

Preparations of the Pharmacopœias of London, Edinburgh, Dublin, and of the United States,

WITH MANY NEW MEDICINES.

BY J. FORBES ROYLE, M. D., F. R. S.,

Professor of Materia Medica and Therapeutics, King's College, London, &c. &c.

EDITED BY JOSEPH CARSON, M. D.,

Professor of Materia Medica and Pharmacy in the University of Pennsylvania.

WITH NINETY-EIGHT ILLUSTRATIONS.

In one large octavo volume, of about seven hundred pages.

Being one of the most beautiful Medical works published in this country.

This work is, indeed, a most valuable one, and will fill up an important vacancy that existed between Dr. Pereira's most learned and complete system of Materia Medica, and the class of productions on the other extreme, which are necessarily imperfect from their small extent.—*British and Foreign Medical Review.*

POCKET DISPENSATORY AND FORMULARY.

A DISPENSATORY AND THERAPEUTICAL REMEMBRANCER. Comprising the entire lists of Materia Medica, with every Practical Formula contained in the three British Pharmacopœias. With relative Tables subjoined, illustrating by upwards of six hundred and sixty examples, the Extemporaneous Forms and Combinations suitable for the different Medicines. By JOHN MAYNE, M. D., L. R. C. S., EDIN., &c. &c. Edited, with the addition of the formulæ of the United States Pharmacopœia, by R. EGLESFELD GRIFFITH, M. D. In one 12mo. volume, of over three hundred large pages.

The neat typography, convenient size, and low price of this volume, recommend it especially to physicians, apothecaries, and students in want of a pocket manual.

GRIFFITH'S MEDICAL FORMULARY—(Continued.)

From a vast number of commendatory notices, the publishers select a few.

A valuable acquisition to the medical practitioner, and a useful book of reference to the apothecary on numerous occasions.—*American Journal of Pharmacy.*

Dr. Griffith's Formulary is worthy of recommendation, not only on account of the care which has been bestowed on it by its estimable author, but for its general accuracy and the richness of its details.—*Medical Examiner.*

Most cordially we recommend this Universal Formulary, not forgetting its adaptation to druggists and apothecaries, who would find themselves vastly improved by a familiar acquaintance with this every-day book of medicine.—*The Boston Medical and Surgical Journal.*

Pre-eminent among the best and most useful compilations of the present day will be found the work before us, which can have been produced only at a very great cost of thought and labor. A short description will suffice to show that we do not put too high an estimate on this work. We are not cognizant of the existence of a parallel work. Its value will be apparent to our readers from the sketch of its contents above given. We strongly recommend it to all who are engaged either in practical medicine, or more exclusively with its literature.—*London Medical Gazette.*

A very useful work, and a most complete compendium on the subject of materia medica. We know of no work in our language, or any other, so comprehensive in all its details.—*London Lancet.*

The vast collection of formulæ which is offered by the compiler of this volume, contains a large number which will be new to English practitioners, some of them from the novelty of their ingredients, and others from the unaccustomed mode in which they are combined, and we doubt not that several of these might be advantageously brought into use. The authority for every formula is given, and the list includes a very numerous assemblage of Continental, as well as of British and American writers of repute. It is, therefore, a work to which every practitioner may advantageously resort for hints to increase his stock of remedies and of forms of prescription.

The other indices facilitate reference to every article in the "Formulary;" and they appear to have been drawn up with the same care as that which the author has evidently bestowed on every part of the work.—*The British and Foreign Medico Chirurgical Review.*

The work before us is all that it professes to be, viz.: "a compendious collection of formulæ and pharmaceutic processes." It is such a work as was much needed, and should be in the hands of every practitioner who is in the habit of compounding medicines.—*Transylvania Medical Journal.*

This seems to be a very comprehensive work, so far as the range of its articles and combinations is concerned, with a commendable degree of brevity and condensation in their explanation.

It cannot fail to be a useful and convenient book of reference to the two classes of persons to whom it particularly commends itself in the title-page.—*The N. W. Medical and Surgical Journal.*

It contains so much information that we very cheerfully recommend it to the profession.—*Charleston Med. Journal.*

Well adapted to supply the actual wants of a numerous and varied class of persons.—*N. Y. Journal of Medicine.*

CHRISTISON & GRIFFITH'S DISPENSATORY.—(A New Work.)

A DISPENSATORY,

OR, COMMENTARY ON THE PHARMACOPŒIAS OF GREAT BRITAIN AND THE UNITED STATES: COMPRISING THE NATURAL HISTORY, DESCRIPTION, CHEMISTRY, PHARMACY, ACTIONS, USES, AND DOSES OF THE ARTICLES OF THE MATERIA MEDICA.

BY ROBERT CHRISTISON, M. D., V. P. R. S. E.,

President of the Royal College of Physicians of Edinburgh; Professor of Materia Medica in the University of Edinburgh, etc.

Second Edition, Revised and Improved,

WITH A SUPPLEMENT CONTAINING THE MOST IMPORTANT NEW REMEDIES.

WITH COPIOUS ADDITIONS,

AND TWO HUNDRED AND THIRTEEN LARGE WOOD ENGRAVINGS.

BY R. EGLESFELD GRIFFITH, M. D.,

Author of "A Medical Botany," etc

In one very large and handsome octavo volume, of over one thousand closely printed pages, with numerous wood-cuts, beautifully printed on fine white paper, presenting an immense quantity of matter at an unusually low price.

It is enough to say that it appears to us as perfect as a Dispensatory, in the present state of pharmaceutical science, could be made.—*The Western Journal of Medicine and Surgery.*

CARSON'S SYNOPSIS—(*Now Ready.*)

SYNOPSIS OF THE

COURSE OF LECTURES ON MATERIA MEDICA AND PHARMACY,

Delivered in the University of Pennsylvania.

BY JOSEPH CARSON, M. D.,

Professor of Materia Medica and Pharmacy in the University of Pennsylvania.

In one very neat octavo volume of 208 pages.

This work, containing a rapid but thorough outline of the very extensive subjects under consideration, will be found useful, not only for the matriculants and graduates of the institution for whom it is more particularly intended, but also for those members of the profession who may desire to recall their former studies.

THE THREE KINDS OF COD-LIVER OIL,

Comparatively considered, with their Chemical and Therapeutic Properties, by L. J. DE JONGH, M. D. Translated, with an Appendix and Cases, by EDWARD CAREY, M. D. To which is added an article on the subject from "Dunglison on New Remedies." In one small 12mo. volume, extra cloth.

MOHR, REDWOOD, AND PROCTER'S PHARMACY.—Lately Issued.

PRACTICAL PHARMACY.

COMPRISING THE ARRANGEMENTS, APPARATUS, AND MANIPULATIONS OF THE PHARMACEUTICAL SHOP AND LABORATORY.

BY FRANCIS MOHR, PH. D.,
Assessor Pharmaciæ of the Royal Prussian College of Medicine, Coblentz;
AND THEOPHILUS REDWOOD,
Professor of Pharmacy in the Pharmaceutical Society of Great Britain.
EDITED, WITH EXTENSIVE ADDITIONS, BY PROFESSOR WILLIAM PROCTER,
Of the Philadelphia College of Pharmacy.

In one handsomely printed octavo volume, of 570 pages, with over 500 engravings on wood.

To physicians in the country, and those at a distance from competent pharmaceutists, as well as to apothecaries, this work will be found of great value, as embodying much important information which is to be met with in no other American publication.

After a pretty thorough examination, we can recommend it as a highly useful book, which should be in the hands of every apothecary. Although no instruction of this kind will enable the beginner to acquire that practical skill and readiness which experience only can confer, we believe that this work will much facilitate their acquisition, by indicating means for the removal of difficulties as they occur, and suggesting methods of operation in conducting pharmaceutic processes which the experimenter would only hit upon after many unsuccessful trials; while there are few pharmaceutists, of however extensive experience, who will not find in it valuable hints that they can turn to use in conducting the affairs of the shop and laboratory. The mechanical execution of the work is in a style of unusual excellence. It contains about five hundred and seventy large octavo pages, handsomely printed on good paper, and illustrated by over five hundred remarkably well executed wood-cuts of chemical and pharmaceutical apparatus. It comprises the whole of Mohr and Redwood's book, as published in London, rearranged and classified by the American editor, who has added much valuable new matter, which has increased the size of the book more than one-fourth. including about one hundred additional wood-cuts.—*The American Journ. of Pharmacy.*

It is a book, however, which will be in the hands of almost every one who is much interested in pharmaceutical operations, as we know of no other publication so well calculated to fill a void long felt.—*The Medical Examiner.*

The country practitioner who is obliged to dispense his own medicines, will find it a most valuable assistant.—*Monthly Journal and Retrospect.*

The book is strictly practical, and describes only manipulations or methods of performing the numerous processes the pharmaceutist has to go through, in the preparation and manufacture of medicines, together with all the apparatus and fixtures necessary thereto. On these matters, this work is very full and complete, and details, in a style uncommonly clear and lucid, not only the more complicated and difficult processes. but those not less important ones, the most simple and common. The volume is an octavo of five hundred and seventy-six pages. It is elegantly illustrated with a multitude of neat wood engravings, and is unexceptionable in its whole typographical appearance and execution. We take great satisfaction in commending this so much needed treatise. not only to those for whom it is more specially designed, but to the medical profession generally—to every one, who, in his practice, has occasion to prepare, as well as administer medical agents.—*Buffalo Medical Journal.*

NEW AND COMPLETE MEDICAL BOTANY.

MEDICAL BOTANY;

OR, A DESCRIPTION OF ALL THE MORE IMPORTANT PLANTS USED IN MEDICINE, AND OF THEIR PROPERTIES, USES, AND MODES OF ADMINISTRATION.

BY R. EGLESFELD GRIFFITH, M. D., &c. &c.

In one large 8vo. vol. of 704 pages, handsomely printed, with nearly 350 illustrations on wood.

One of the greatest acquisitions to American medical literature. It should by all means be introduced at the very earliest period. into our medical schools, and occupy a place in the library of every physician in the land.—*Southwestern Medical Advocate.*

Admirably calculated for the physician and student—we have seen no work which promises greater advantages to the profession —*N. O. Medical and Surgical Journal.*

One of the few books which supply a positive deficiency in our medical literature.—*Western Lancet.*

We hope the day is not distant when this work will not only be a text-book in every medical school and college in the Union, but find a place in the library of every private practitioner.—*N. Y. Journ. of Medicine.*

ELLIS'S MEDICAL FORMULARY.—Improved Edition.

THE MEDICAL FORMULARY:

BEING A COLLECTION OF PRESCRIPTIONS, DERIVED FROM THE WRITINGS AND PRACTICE OF MANY OF THE MOST EMINENT PHYSICIANS OF AMERICA AND EUROPE.

To which is added an Appendix, containing the usual Dietetic Preparations and Antidotes for Poisons.

THE WHOLE ACCOMPANIED WITH A FEW BRIEF PHARMACEUTIC AND MEDICAL OBSERVATIONS.

BY BENJAMIN ELLIS, M. D.

NINTH EDITION, CORRECTED AND EXTENDED, BY SAMUEL GEORGE MORTON, M. D.

In one neat octavo volume of 268 pages.

CARPENTER ON ALCOHOLIC LIQUORS.—(A New Work.)

A Prize Essay on the Use of Alcoholic Liquors in Health and Disease. By William B. Carpenter, M. D., author of "Principles of Human Physiology," &c. In one 12mo. volume.

TAYLOR'S MEDICAL JURISPRUDENCE.

MEDICAL JURISPRUDENCE.

BY ALFRED S. TAYLOR,

SECOND AMERICAN, FROM THE THIRD AND ENLARGED LONDON EDITION.

With numerous Notes and Additions, and References to American Practice and Law.

BY R. E. GRIFFITH, M. D.

In one large octavo volume.

This work has been much enlarged by the author, and may now be considered as the standard authority on the subject, both in England and this country. It has been thoroughly revised, in this edition, and completely brought up to the day with reference to the most recent investigations and decisions. No further evidence of its popularity is needed than the fact of its having, in the short time that has elapsed since it originally appeared, passed to three editions in England, and two in the United States.

We recommend Mr. Taylor's work as the ablest, most comprehensive, and, above all. the most practically useful book which exists on the subject of legal medicine. Any man of sound judgment, who has mastered the contents of Taylor's "Medical Jurisprudence," may go into a court of law with the most perfect confidence of being able to acquit himself creditably.—*Medico-Chirurgical Review.*

The most elaborate and complete work that has yet appeared. It contains an immense quantity of cases lately tried, which entitle it to be considered what Beck was in its day.—*Dublin Medical Journal.*

TAYLOR ON POISONS.

ON POISONS,

IN RELATION TO MEDICAL JURISPRUDENCE AND MEDICINE.

BY ALFRED S. TAYLOR, F. R. S., &c.

EDITED, WITH NOTES AND ADDITIONS, BY R. E. GRIFFITH, M. D.

In one large octavo volume, of 688 pages.

The most elaborate work on the subject that our literature possesses.—*Brit. and For. Medico-Chirur Review.*

One of the most practical and trustworthy works on Poisons in our language.—*Western Journal of Med.*

It contains a vast body of facts, which embrace all that is important in toxicology. all that is necessary to the guidance of the medical jurist, and all that can be desired by the lawyer.—*Medico-Chirurgical Review.*

It is, so far as our knowledge extends, incomparably the best upon the subject; in the highest degree creditable to the author, entirely trustworthy, and indispensable to the student and practitioner.—*N. Y. Annalist.*

BEALE ON HEALTH—NOW READY.

THE LAWS OF HEALTH IN RELATION TO MIND AND BODY.

A SERIES OF LETTERS FROM AN OLD PRACTITIONER TO A PATIENT.

BY LIONEL JOHN BEALE, M. R. C. S.., &c.

In one handsome volume, royal 12mo., extra cloth.

The "Laws of Health," in relation to mind and body, is a book which will convey much instruction to non-professional readers; they may, from these letters, glean the principles upon which young persons should be educated. and derive much useful information, which will apply to the preservation of health at all ages.—*Med. Times.*

GREGORY ON ANIMAL MAGNETISM—(Now Ready.)

LETTERS TO A CANDID ENQUIRER ON ANIMAL MAGNETISM.

DESCRIPTION AND ANALYSIS OF THE PHENOMENA. DETAILS OF FACTS AND CASES.

BY WILLIAM GREGORY, M. D., F. R. S. E.,

Professor of Chemistry in the University of Edinburgh, &c.

In one neat volume, royal 12mo , extra cloth.

PROFESSOR DICKSON'S ESSAYS—Now Ready.

ESSAYS ON LIFE, SLEEP, PAIN, INTELLECTION, HYGIENE, AND DEATH.

BY SAMUEL HENRY DICKSON, M. D.,

Professor of the Institutes and Practice of Medicine in the Charleston Medical College.

In one very handsome volume, royal 12mo.

DUNGLISON ON HUMAN HEALTH —HUMAN HEALTH, or the Influence of Atmosphere and Locality, Change of Air and Climate, Seasons, Food, Clothing, Bathing, Exercise, Sleep, &c. &c. &c.. on healthy man; constituting Elements of Hygiene. Second edition, with many modifications and additions. By Robley Dunglison, M. D. &c. &c. In one octavo volume of 464 pages.

DUNGLISON'S MEDICAL STUDENT.—The Medical Student, or Aids to the Study of Medicine. Revised and Modified Edition. 1 vol. royal 12mo. extra cloth. 312 pp.

BARTLETT'S PHILOSOPHY OF MEDICINE.—An Essay on the Philosophy of Medical Science. In one handsome 8vo volume. 312 pp.

BARTLETT ON CERTAINTY IN MEDICINE—An Inquiry into the Degree of Certainty in Medicine and into the Nature and Extent of its Power over Disease. In one vol. royal 12mo. 84 pp.

THE GREAT AMERICAN MEDICAL DICTIONARY.

New and Enlarged Edition, brought up to October, 1851.

NOW READY.

MEDICAL LEXICON;

A DICTIONARY OF MEDICAL SCIENCE,

Containing a Concise Explanation of the various Subjects and Terms of

PHYSIOLOGY, PATHOLOGY, HYGIENE, THERAPEUTICS, PHARMACOLOGY, OBSTETRICS, MEDICAL JURISPRUDENCE, &c.

WITH THE FRENCH AND OTHER SYNONYMES.

NOTICES OF CLIMATE AND OF CELEBRATED MINERAL WATERS;

Formulæ for various Officinal, Empirical, and Dietetic Preparations, &c.

BY ROBLEY DUNGLISON, M. D.,

Professor of Institutes of Medicine, &c. in Jefferson Medical College, Philadelphia, &c.

EIGHTH EDITION,

REVISED AND GREATLY ENLARGED.

In one very thick 8vo. vol., of 927 large double-columned pages, strongly bound, with raised bands.

Every successive edition of this work bears the marks of the industry of the author, and of his determination to keep it fully on a level with the most advanced state of medical science. Thus the last two editions contained about NINE THOUSAND SUBJECTS AND TERMS not comprised in the one immediately preceding, and the present has not less than FOUR THOUSAND not in any former edition. As a complete Medical Dictionary; therefore, embracing over FIFTY THOUSAND definitions, in all the branches of the science, it is presented as meriting a continuance of the great favor and popularity which have carried it, within no very long space of time, to an eighth edition.

Every precaution has been taken in the preparation of the present volume, to render its mechanical execution and typographical accuracy worthy of its extended reputation and universal use. The very extensive additions have been accommodated, without materially increasing the bulk of the volume, by the employment of a small but exceedingly clear type, cast for this purpose. The press has been watched with great care, and every effort used to insure the verbal accuracy so necessary to a work of this nature. The whole is printed on fine white paper; and while thus exhibiting in every respect so great an improvement over former issues, it is presented at the original exceedingly low price.

A few notices of former editions are subjoined.

Dr. Dunglison's Lexicon has the rare merit that it certainly has no rival in the English language for accuracy and extent of references. The terms generally include short physiological and pathological descriptions, so that, as the author justly observes, the reader does not possess in this work a mere dictionary, but a book, which, while it instructs him in medical etymology, furnishes him with a large amount of useful information. That we are not over-estimating the merits of this publication, is proved by the fact that we have now before us the seventh edition. This, at any rate, shows that the author's labors have been properly appreciated by his own countrymen; and we can only confirm their judgment, by recommending this most useful volume to the notice of our cisatlantic readers. No medical library will be complete without it. —*The London Med. Gazette.*

It is certainly more complete and comprehensive than any with which we are acquainted in the English language. Few, in fact, could be found better qualified than Dr. Dunglison for the production of such a work. Learned, industrious, persevering, and accurate, he brings to the task all the peculiar talents necessary for its successful performance: while, at the same time, his familiarity with the writings of the ancient and modern "masters of our art," renders him skilful to note the exact usage of the several terms of science, and the various modifications which medical terminology has undergone with the change of theories or the progress of improvement.—*American Journal of the Medical Sciences.*

One of the most complete and copious known to the cultivators of medical science.—*Boston Med. Journal.*

This most complete Medical Lexicon—certainly one of the best works of the kind in the language.—*Charleston Medical Journal.*

The most complete Medical Dictionary in the English language.—*Western Lancet.*

Dr. Dunglison's Dictionary has not its superior, if indeed its equal, in the English language.—*St. Louis Med. and Surg. Journal.*

Familiar with nearly all the medical dictionaries now in print, we consider the one before us the most complete, and an indispensable adjunct to every medical library.—*British American Medical Journal.*

Admitted by all good judges, both in this country and in Europe, to be equal, and in many respects superior to any other work of the kind yet published.—*Northwestern Medical and Surgical Journal.*

We repeat our former declaration that this is the best Medical Dictionary in the English language.—*Western Lancet.*

We have no hesitation to pronounce it the very best Medical Dictionary now extant.—*Southern Medical and Surgical Journal.*

The most comprehensive and best English Dictionary of medical terms extant.—*Buffalo Med. Journal.*

Whence the terms have all been derived we find it rather difficult to imagine. We can only say that, after looking for every new and strange word we could think of, we have not been disappointed in regard to more than a few of most recent introduction, such as the designations given by Professor Owen to the component parts of a Vertebra.—*British and Foreign Medico-Chirurgical Review.*

Dr. Dunglison's masterpiece of literary labor.—*N. Y. Journal of Medicine.*

HOBLYN'S MEDICAL DICTIONARY.

A DICTIONARY OF THE TERMS USED IN MEDICINE

AND THE COLLATERAL SCIENCES.

BY RICHARD D. HOBLYN, A. M., Oxon.

REVISED, WITH NUMEROUS ADDITIONS, FROM THE SECOND LONDON EDITION, BY ISAAC HAYS, M. D., &c. In one large royal 12mo. volume of 402 pages, double columns.

We cannot too strongly recommend this small and cheap volume to the library of every student and practitioner.—*Medico-Chirurgical Review.*

www.ingramcontent.com/pod-product-compliance
Lightning Source LLC
LaVergne TN
LVHW021137110826
845150LV00005B/1053

* 9 7 8 1 4 2 5 5 4 9 7 5 6 *